PLANTS AND PEOPLE

Choices and Diversity through Time

edited by

Alexandre Chevalier, Elena Marinova
and Leonor Peña-Chocarro

Oxbow Books
Oxford & Philadelphia

Other Volumes in this Series

Exploring and Explaining Diversity in Agricultural Technology
(EARTH Volume 2)

Agricultural and Pastoral Landscapes in Preindustrial Society: Choices, Stability and Change
(EARTH Volume 3)

Early Agricultural Remnants and Technical Heritage (EARTH): 8,000 Years of Resilience and Innovation

Volume 1

Series Editors

Patricia C. Anderson and Leonor Peña-Chocarro

Coordinating Editor

Andreas G. Heiss

Published in the United Kingdom in 2014 by
OXBOW BOOKS
10 Hythe Bridge Street, Oxford OX1 2EW

and in the United States by
OXBOW BOOKS
908 Darby Road, Havertown, PA 19083

Hardcover Edition: ISBN 978-184217-514-9
Digital Edition: ISBN 978-1-78297-033-0

A CIP record for this book is available from the British Library

Library of Congress Cataloging-in-Publication Data

Plants and people : choices and diversity through time / edited by Alexandre Chevalier, Elena Marinova and Leonor Peña-Chocarro.

pages cm. -- (Early agricultural remnants and technical heritage (EARTH) : 8,000 years of resilience and innovation ; volume 1)

ISBN 978-1-84217-514-9

1. Plant diversity. 2. Ethnobotany. 3. Human-plant relationships. I. Chevalier, Alexandre, 1966- editor of compilation. II. Marinova, Elena, editor of compilation. III. Peña-Chocarro, Leonor, editor of compilation.

QK46.5.D58P53 2014

581.6'3--dc23

2013049376

Printed in the United Kingdom by Berforts Information Press

For a complete list of Oxbow titles, please contact:

UNITED KINGDOM
Oxbow Books
Telephone (01865) 241249, Fax (01865) 794449
Email: oxbow@oxbowbooks.com
www.oxbowbooks.com

UNITED STATES OF AMERICA
Oxbow Books
Telephone (800) 791-9354, Fax (610) 853-9146
Email: queries@casemateacademic.com
www.casemateacademic.com/oxbow

Oxbow Books is part of the Casemate Group

Cover illustrations

Woman preparing *yufka* (flat bread), in the village of Kavanoluk (Buldan, Turkey). Image: Alexandre Chevalier
Wheat field in Soria (Spain). Image: Leonor Peña-Chocarro
Farmer winnowing einkorn wheat in Zuheros (Córdoba, Spain). Image: Leonor Peña-Chocarro

We would like to dedicate this volume to

Dr Irmeli Vuorela

a much appreciated member of the Steering Committee of the EARTH Programme, who passed away in 2012 at the age of 74 after holding positions at the University of Helsinki, then the Geological Survey of Finland. Irmeli, whose expertise was in using pollen, charcoal and phytolith analysis to reconstruct human-plant relations, in particular for early Finnish agriculture through to the development of urban settlements, advised the programme with her characteristic humour and wisdom, and was supportive to research comprising this book.

ESF MISSION

The European Science Foundation (ESF) is an independent, non-governmental organisation, the members of which are eighty national funding agencies, research-performing agencies, academies and learned societies from thirty countries.
The strength of the ESF lies in the influential membership and in its ability to bring together the different domains of European science in order to meet the challenges of the future.
Since its establishment in 1974, the ESF, which has its headquarters in Strasbourg with offices in Brussels and Ostend, has assembled a host of organisations that span all disciplines of science, to create a common platform for cross-border cooperation in Europe.
The ESF is dedicated to promoting collaboration in scientific research, funding of research and science policy across Europe. Through its activities and instruments the ESF has made major contributions to science in a global context. The ESF covers the following scientific domains:

- Humanities
- Life, Earth and Environmental Sciences
- Medical Sciences
- Physical and Engineering Sciences
- Social Sciences
- Marine Sciences

ESF MEMBER ORGANISATIONS FUNDING THE EARTH NETWORKING PROGRAMME AND PUBLICATIONS

Fonds National de la Recherche Scientifique (FNRS) – National Fund for Scientific Research, Belgium

Fonds voor Wetenschappelijk Onderzoek Vlaanderen (FWO) – Research Foundation Flanders, Belgium

Grantová agentura České republiky (GAČR) – Czech Science Foundation, Czech Republic

Akademie věd České republiky (ASCR) – Academy of Sciences of the Czech Republic, Czech Republic

Det Frie Forskningsråd – Kultur og Kommunikation (FKK) – The Danish Council for Independent Research – Humanities, Demark

Eesti Teadusfond (ETF) – Estonian Science Foundation, Estonia

Suomen Akatemia/Finlands Akademi – Academy of Finland, Finland

Centre National de la Recherche Scientifique (CNRS) – National Centre for Scientific Research, France

Deutsche Forschungsgemeinschaft (DFG) – German Research Foundation, Germany

An Chomhairle um Thaighde sna Dána agus sna hEolaíochtaí Sóisialta (IRCHSS) – Irish Research Council for the Humanities and Social Sciences, Ireland

Nederlandse Organisatie voor Wetenschappelijk Onderzoek (NWO) – Netherlands Organisation for Scientific Research, The Netherlands

Norges Forskningsråd – Research Council of Norway, Norway

Consejo Superior de Investigaciones Científicas (CSIC) – Council for Scientific Research, Spain

Vetenskapsrådet (VR) – Swedish Research Council, Sweden

Schweizerischer Nationalfonds (SNF) – Swiss National Science Foundation, Switzerland

Arts and Humanities Research Council (AHRC), United Kingdom

Contents

SECTION 3: Food and Beyond

Annexes

Foreword

In the beginning, in the late Middle Ages, Botany was a bud on the twig of pharmacy, on the branch of medicine. The tree of science was still small. For centuries, zoological and botanical anatomy remained a division of basic medical studies. With the emergence of more and more perfect and beautiful botanical illustrations, botany became the *scientia amabilis*, a perfect hobby of noble gentlemen and ladies. Archaeology was still a branch of Art History in the 19th century, with three divisions: Egyptology, Mesopotamian Archaeology and Classical Archaeology. Central European Prehistory began with the Alpine lake dwellings and with the Scandinavian scientists. In these early phases, the excavators showed their biological findings, charred crops and animal bones to people interested in Biology, mostly school teachers. Heinrich Schliemann presented his burnt crop stocks from Troy or Tiryns in the sessions of Berlin scientific societies.

The 20th century was the century of Biology. Biology became a tree of science on its own, with dense branching. The first Archaeobotanists and Archaeozoologists were biologists with secondary training in Archaeology, as at that time Archaeobotany was more or less an auxiliary science of Archaeology, especially in central European Prehistory. In the late 20th century, botanical studies at universities became more and more microbiological and physiological. Plant anatomy, taxonomy, vegetation history, ethnobotany, and plant sociology suffered a great decline in importance at the universities. In the course of this, more and more students of Prehistory became interested in Archaeobotany and Archaeozoology and were trained in laboratories in Prehistory departments, for example in Switzerland, the Netherlands, Great Britain and Germany.

Nowadays, Archaeobotany is no longer a botanical science. If you show a charred barley stock to a botanist, he or she will answer: 'It's barley, *Hordeum vulgare*, four-rowed barley'. Archaeobotanical questions are however much more numerous than just species identification: Which soil type was it grown on? Was it autumn- or spring-sown? Was it deposited in an already processed state, or right after harvesting? Were there additional crops? – These questions are archaeological ones, not botanical ones. So it then became time to rename archaeobotany. In the mid 20th century, the archaeobotanists called themselves the 'International Work Group for Palaeoethnobotany (IWGP)'. Especially German archaeobotanists fought for the name Palaeo-Ethnobotany. But *palaeo-* is far too old, somewhere deep in the past, beyond mankind; *archaeo-* is old enough. Now, most scientists prefer 'Archaeobotany'. But to be honest, it's not Botany, it's the Archaeology of plant and animal husbandry. So the word 'archaeology' should be at the end, not 'botany'. Economic and Environmental Archaeology is the right description, and there is already a tradition and a scientific society of this name. It includes, as well as plant remains (*e.g.* macro remains): pollen and spores, charcoal, phytoliths, zoological remains such as bones and shells, and recent disciplines such as isotope analysis.

Archaeobotany is now a branch of the archaeological tree of science. At a handful of universities, there are professorships for Archaeobotany and Archaeozoology. A science of its own needs handbooks. Here is a textbook of Economic and Environmental Archaeology. It is the first volume of the Early Agricultural Remnants and Technical Heritage (EARTH) Programme: 8,000 years of Resilience and Innovation. It is written by the *crème de la crème* of well-known and experienced archaeobotanists. The focus is on the western Old World, with sidelong glances to Africa, Asia, and the Americas.

Methodology, coordinated by A. Chevalier, E. Marinova, and L. Peña-Chocarro, introduces this subject, between nature and culture, botany and archaeology.

The choice of crops cultivated by prehistoric and historic societies is discussed for southeastern Europe by E. Marinova and S. M. Valamoti, for Italy by M. Rottoli, for Spain and adjacent regions by L.Zapata and again L. Peña-Chocarro. – S. Jacomet, C. C. Bakels and F. Sigaut (†) cover western central Europe. L. Scott Cummings and M. C. Bruno contribute articles about the Americas.

The important complex of the fruit trees – wild, selected or domesticated – is unrolled and accurately analysed by some examples: acorns in California by R. Cuthrell; citrus fruits in the Mediterranean by M. Girard and Bui Thi Mai; *Prosopis* on the Peruvian coast by D. J. Goldstein; grapevine in southern France by L. Bouby, Ph. Marinval, and J.-F. Terral; figs in Spain and the Canary Islands by J. Morales and J. Gil; figs and olives in Morocco by Y. Aumeeruddy-Thomas, Y. Hmimsa, M. Ater, and B. Khadari.

The doyenne of the Ethnobotany and Archaeobotany of plants collected from the wild is F. Ertuğ, who covers the East, G. S. Cruz-García covers the West. J. Morales, J. Gil, J. Tardío, and M. Pardo-de-Santayana give reports from Spain and the Canary Islands. C. Selleger presents some impressions of wild food in Mali, in the famous Dogon county. Unusual examples include silverweed, *Argentina* (syn. *Potentilla*) *anserina* in the British Isles by C. Griffin-Kremer, and *Cleome* in southwestern North America by L. Scott Cummings.

The diversity of plant use raises the important question of what is primary, and what is secondary use. There are some answers given here by L. Zapata and L. Peña-Chocarro concerning hulled wheats; by E. Bonnaire about plant remains in *pisé* buildings. C. Griffin-Kremer reports on the different uses of *Ulex europaeus* and *Urtica dioica*, ubiquitous plants in the British Isles. P. C. Anderson introduces perfect artisanal basketry made of the grass *Ampelodesmos mauritanica* in Tunisia. Bui Thi Mai, M. Girard, and F. de Lanfranchi discuss the different functions of *Pistacia lentiscus* in Sardinia. The use of oil/resin/wax/tar, a difficult subject in analytical archaeochemistry, is also present: An ethnological example of resin use in boat caulking in Asia by Bui Thi Mai and M. Girard.

Plants and plant products are important offerings in temples, on altars, in graves, in sacred groves, dug in, laid down, poured out, or burnt: examples are given by A. G. Heiss (Alps) and A.-M. Hansson (Scandinavia). Plants play a crucial role on special days, such as on May Day in the British Isles and elsewhere, shown by C. Griffin-Kremer, or in ceremonies of native tribes in the Americas (North: L. Scott Cummings; Andes: M. Sayre).

The last chapter of this ambitious book is aimed at social status, identity, and context. A. Chevalier and J. Dulanto can distinguish two different social groups in pre-Columbian coastal Peru. Other examples come from Spain's Iberian Iron Age by S. Gonzáles Reyero, from its Middle Ages by J. L. Mingote Calderón; from Roman France by J. Wiethold and F. Durand; from Neolithic northern Germany by W. Kirleis and St. Klooß, and from Mayan Central America by D. J. Goldstein and J. B. Hageman. An ethnological report by G. S. Cruz-García on plant use by poor children in India tops off these considerations.

More than twenty years ago, in 1991, the handbook 'Progress in Old World Palaeoethnobotany' appeared, edited by W. van Zeist, K. Wasylikowa, and K.-E. Behre. The three EARTH volumes will replace it with expansions into Africa, the Far East, and the Americas. It is a welcome comment on and supplement to, the 4th edition (2012) by E. Weiss of D. Zohary and M. Hopf's opus 'Domestication of Plants in the Old World'.

Helmut Kroll
Kiel, autumn 2013

Preface

Patricia C. Anderson and Leonor Peña-Chocarro

Scientific Overview

This book takes an interdisciplinary look at European preindustrial agriculture, including its origins and its diffusion outside Europe. Agriculture and its origins have long been lively and innovative subjects of research, involving people working in a variety of disciplines. Initial impetus to this area came from several quite separate disciplines. **Archaeologists** working in the 1940s and 1950s, studied **flint** blades found on archaeological sites in the Middle East and in Europe, which they hypothesised had been inserts forming the working parts of agricultural tools. Such identifications were based upon intimate knowledge of ethnographic tools (Steensberg 1943), and by examining the microscopic traces of use on their edges (Semenov 1964). **Agronomists**, in turn, interested in the origins of agriculture studied the behaviour of **wild cereals** still growing in their natural habitats and explored the way human activities may have influenced crop evolution (Harlan 1975). This area of study, pioneered by a few researchers like O. Heer (1865), lead agronomists and botanists towards the beginnings of **archaeobotanical research** (*e.g.* Hopf 1954, 1955). Over time, however, the study of agriculture evolved towards an interdisciplinary approach which tried to fully understand its multiple facets: plants, techniques, soil types, environment, agronomic practices, the attendant animal husbandry, impact on landscape, traditions, etc. In other words, the cross-disciplinary character of the subject implied the integration of the various historical, anthropological, archaeological and scientific records in order to fully appreciate the complex and interrelated issues involved.

Thus, **archaeobotanists** have looked at prehistoric **seeds** of ancient crops throughout the world to determine the presence of domesticated plants and explore the possible uses of the species identified. Other studies combined analysis of archaeological plant remains with field and genetic studies of equivalent crops still surviving today (Zohary and Hopf 2012; Harlan 1999), and documented the origin and spread of the domesticated plants from the Near East throughout Europe. Weeds have also been a subject of interest, used to explore crop husbandry practices (Charles *et al.* 1997, Jones *et al.* 2010). In particular, crop processing has been a major topic of interest in the study of agriculture. **Ethnographic observations** of the various steps within the crop processing sequence have allowed researchers to produce models (Hillman 1981, 1984; taken up later by *e.g.* Peña-Chocarro 1999) and statistical methods (*e.g.* Jones 1984, 1987; Pearsall 1988) to identify these from crop remains, which in dry sites are usually preserved by charring. This experimental approach was applied by archaeobotanists using reconstructions of specific archaeological tools preserved in European lake sites, again to see the effect on the crops (Meurers-Balke and Lüning 1999). Waterlogged archaeological sites with exceptional preservation of plant remains have allowed detailed studies of plant use with interesting insights into processing activities, among other things (Herbig 2009, Jacomet 2009) as have rare finds from permafrost (Rollo 2002; Heiss and Oeggl 2009; Dickson 2011).

Pollen analysis has also contributed to treating the question of cereal domestication (Bottema 1999), reconstructing archaeological fields and characterising agricultural practices such as the identification of the slash-and-burn technique in the Nordic areas (Vuorela 1986). And, on a more general level, palynologists have shown the role of both the natural environment and prehistoric human activities in the shaping of landscapes, both based on pollen evidence itself (Kalis *et al* 2003; Carrión *et al.* 2007) and on other microfossils in pollen records, the so-called NPPs ('non-pollen palynomorphs', such as fungal and fern spores, remains of green algae and cyanobacteria, or invertebrate eggs; *e.g.* van Geel *et al.* 2003).

Experimental archaeology has also proven to be a powerful approach to the study of agriculture, with a myriad of examples focusing on specific processes. Agricultural field experiments were carried out in northern Europe by archaeologists, for example using reconstructed ploughs and ards to work fields (Lerche and Steensberg 1980). This led to precise reconstructions of marks in the soil, as well as of archaeological finds of ploughs and ards (Lerche 1994). Other experiments focused on soil, climate and yield of ancient cereal types (Reynolds 1981; Steensberg 1943). By combining soil micromorphology and archaeobotany with an experimental approach (Boissinot and Brochier 1997), it was possible to show field use patterns from as early as the Neolithic in southern Europe (Rösch *et al.* 2002).

Historical records and agronomy were combined to evaluate the cultural influence on concepts of yield (Sigaut 1999) while historians interested in archaeology carried out research into the effect on the soil and the crops of certain farming techniques in large-scale field experiments (Reynolds 1974, 1979, 1981, 1999). Storage pits, for instance, were reproduced and tested using historic and ethnographic references, and were shown to be effective for storing grain similar to ancient varieties (Reynolds 1974). This allowed new interpretations of such archaeological features (Gast and Sigaut 1979–85).

Some **field experiments** were also carried out using reconstructions of harvesting tools based on prehistoric finds which involved measuring the harvesting yield with different kinds of tools, and then studying the experimental tools for traces of use (Korobkova 1981). High-power microscopy, combined with the experiments, allowed finer distinctions of different tool uses and their contact with plant material (Anderson 1999).

New field experiments were designed on a regional basis to test specific tools on ancient crops grown under climatic conditions similar to the past, or on wild cereals (Anderson *et al.* 1991; Anderson 1999; Hillman 1999). These new analyses often disproved or greatly nuanced another archaeological assumption: that all flint tools with gloss were agricultural harvesting tools (Juel Jensen 1988, 1994; van Gijn 1999). Historical and ethnographic studies (*e.g.* Sigaut 1978) came to similar conclusions, that 'sickles' were sometimes used to obtain other materials such as reeds and straw or were not even sickles at all, but rather inserts in threshing tools (Anderson and Inizan 1994). In addition, specialised threshing tools, threshing floors and storage facilities were increasingly being found in the archaeological record in the Near East and southern Europe (Avner 1998; Skakun 1999; Gast and Sigaut 1979–85; Kuijt and Finlayson 2009; Miret i Mestre 2006; Cunningham 2011). These were identified by means of experiments (Anderson *et al.* 2004), ancient descriptions and ethnography (Kardulias and Yerkes 1996; Grégoire in Anderson and Inizan 2004; Ataman 1999). Experiments were also part of the research carried out to investigate the effects of different grinding and pounding tools on the grain and to seek to find a full range of wear and archaeobotanical criteria to identify the function of ancient tools (Procopiou *et al.* 2002).

Examination of historical documents such as the earliest cuneiform tablets, combined with these experiments, gave further insight into grain processing and social organisation during the third millennium in the Near East (Grégoire 1999). The same applies to agricultural techniques and tools from Medieval Europe, thanks to the study of historical texts (Comet 1992) and ethnographic and archaeological investigation (Mingote Calderón 1996).

Diet and the **social context of food** are other rich topics of research, involving historical documents (Carpinschi 2002), ethnoarchaeobotanical studies (Sarpaki 2000), and archaeobotanical analysis including funerary offerings (Marinval 1993; Rottoli and Castiglioni 2011) and of luxury foods (Bakels and Jacomet 2003; van der Veen 2003; Palmer and van der Veen 2002; van der Veen 2008). New methods for identifying microscopic food remains (*e.g.* Winton and Winton 1932, 1935; Gassner *et al.* 1989) used the analysis of plant tissue fragments, phytoliths and starch (Dickson 1987; Hansson and Isaksson 1994; Cummings 1992; Perry *et al.* 2007; Gong *et al.* 2011; Valamoti *et al.* 2008; Henry *et al.* 2009; Revedin *et al.* 2010) or the chemical analyses of pottery residues (Evershed *et al* 2008), all of which opened up new avenues of research.

The complexity of agricultural processes and their organisation within particular communities and societies requires a broad analytical scale on which to investigate them. The archaeology, anthropology and history of the landscape has been popular for some thirty years, linked by a common concern with going 'beyond the site' and situating human activity in its broadest context (Kardulias 1994). During the last twenty years there has been a considerable output of theory and methodology for investigating social and cultural as well as physical landscapes (*e.g.* Anschuetz *et al.* 2001; Ashmore and Knapp 1999; Francovich and Patterson 2000; Behre and Jacomet 1991). This has been complemented by a series of highly intensive, interdisciplinary surveys carrying out empirical analyses of particular landscapes within these theoretical frameworks (*e.g.* Astill and Davies 1999; Barker 1996; Cherry *et al.* 1991; Given and Knapp 2003; Kohler-Schneider 2001; Fischer and Rösch 2010; Stika *et al.* 2008).

This rapid overview shows some of the important scientific results of the above research approaches, and the many new insights they have provided into the human experience of agriculture. These approaches, however, have far greater potential to grow further if allowed to work closely together. Archaeological data and historical records on their own are often not specific enough for the identification of particular agricultural practices. Scientists analysing pollen, phytoliths, seeds and wear traces on tools can now identify particular species and techniques and develop sophisticated taxonomies, but are often forced by practical limitations to work in relative isolation from other interpretative sources. In both cases, disciplines have usually become highly specialised, and paradoxically isolated by their very development and institutionalisation. This has often created a barrier to the investigation of the broader agricultural system with all its actors, strategies and landscapes.

The origin of the programme

During the 1980s and early 1990s, a European working group met which was concerned with 'agro-(for agrarian) archaeology', combining archaeologists, archaeobotanists, ethnographers, historians and agronomists, around experiments in archaeology. The principal impetus was given by Jutta Meurers-Balke in Cologne, the late Sytze Bottema in Groningen, Holland, the late Peter Reynolds of Butser farm in England, Grith Lerche and the late Axel Steensberg in Copenhagen, and the late François Sigaut in Paris, France.

Following on this, in 1998 a new group concerned with preserving the knowledge and cultural heritage of agricultural processes began to meet, its members combining several different strands of research from the earlier group. In addition, its organisers felt it was urgent to record the knowledge and skills of agriculturalists that still use or remember 'traditional', nonindustrial techniques, and to find a means for this knowledge to enter into the mainstream of various disciplines in a dynamic way. At the time, such an ambition was largely outside the mainstream tendencies of the different disciplines.

This group, called 'Early Agricultural Remnants and Technical Heritage' (EARTH) was recognised by the Sub-Committee on Cultural Heritage of the Parliamentary Assembly of the Council of Europe, who formally admitted EARTH as a member of the PACT alliance of networks, and to the Fédération Européenne des Réseaux Européens de Coopération Scientifique et Technique de Coordination. EARTH was also formally accredited to the Parliamentary Assembly of the Council of Europe. Unfortunately, however, these affiliations did not include funding for a project.

Preparation of this ESF (European Science Foundation) programme took several years: The first EARTH committee meetings were held in St. Vallier de Thiey, Southern France in 1998, funded by the French CNRS (Centre National de la Recherche Scientifique), then in Copenhagen in November 1999 and at Butser Ancient Farm in England in January 2001, culminating in an ESF-funded preparatory workshop at St. Martin de Vésubie, Southern France, in November 2001.

In parallel, from 2002 to the present, the CNRS funded a group of twenty French EARTH members working in France in different disciplines and field areas as a 'Groupement de Recherche' (GDR 2517, directed by P. Anderson). Its aims, similar to those of the European-based ESF programme, have been to use various media to record preindustrial agricultural activities from ancient times and the present day, prepare a database for films and images, and to feed this information into the EARTH network as a whole. The funding of new field research for its members has contributed to some of the articles in this book.

The network: working methods leading to this book

The EARTH Monograph Series, entitled 'The dynamics of non-industrial agriculture: 8,000 years of resilience and innovation', was born from a unique opportunity for interdisciplinary and international (especially inter-European) collaboration between 2004 and 2009 as part of the 'Early Agricultural Remnants and Technical Heritage' (EARTH) Scientific 'à la carte' Programme of the European Science Foundation (ESF, Strasbourg, France), financially supported by 15 European organisations, which were represented by Steering Committee Members of the Programme (p. 475).

This ESF-funded EARTH programme was organised and chaired by Patricia C. Anderson, Nice, France and Michael Given, Glasgow (replaced in 2006 by Leonor Peña-Chocarro, Madrid and Rome).

The originality of this programme lay in the means it provided for unusual networking methods and for innovative forms of output. The aim was to find new common ground for integrating different approaches, viewing agriculture from the standpoint of the human actors involved, and fit this together into a form which could be effectively transmitted for research and teaching, as well as for heritage.

In order to achieve the programme's goals to stimulate the creation of new, integrated interdisciplinary approaches, collaborations and networks, researchers and advanced students were chosen from different disciplines but who worked in an interdisciplinary manner. Some were chosen by the programme organisers, whereas others were suggested to them by the contributing European national organisations (refer to page vi with the ESF Member Organisations), meaning that most of the group did not know one another at the outset. It brought together scientists from a variety of different disciplines of the human and natural sciences: archaeology (including archaeobotany and archaeozoology, microwear analysis of tool function, experimental archaeology), ethnography, history, geography and geology, from about twenty European countries, as well as from North America and the Middle East.

The researchers were organised into three teams, each led by two experts from different countries, while respecting the stipulation of the ESF that as many different disciplinary fields and nationalities as possible be combined in each team, as well as researchers of varied age and at different stages of their career. The European Science Foundation emphasised networking and learning through exchange and communication, and funded annual team meetings and three plenary meetings. Such meetings allowed comparison of different points of view and of agricultural systems in different regions.

Each team approached the theme of agriculture from a slightly different standpoint – that of Crop Choice and Diversity (Book 1 in the series), Skills, Processes and Tools (Book 2 in the series), and Agricultural Landscapes (Book 3 in the series). The networking was achieved largely by workshops held in archaeological and ethnographic field areas or local museums that were directly relevant to the research of certain members (see pp. 478–480). They provided an important opportunity to exchange as well as to talk with local farmers and artisans. Although a few grants were provided for members to meet and work on methodological fine-tuning, no funding was provided by this ESF programme for new field research to authors contributing to this book. Nevertheless, the richness and originality of the scope of the articles presented here was achieved in many cases through networking activities among scientists during the programme. The programme also funded activities for advanced students: to attend hands-on summer schools taught by members and run by Leonor Peña Chocarro in a traditional agricultural area of Spain, and grants to attend the Programme's various meetings. All networking activities were efficiently implemented by Marie Russel, Paris, who served as the programme's coordinator from 2004 to 2009.

Some of the papers in this series address field areas and topics emanating from the workshops, and they were developed gradually over the course of the five years of interdisciplinary exchange. Although the authors come originally from various different fields related to the study of agriculture, they shared an interest in developing a common ground where individual research fields could converge into a broader framework, and provide a new knowledge base. Many contributions integrate field and laboratory methodologies into the case studies, in a manner intended to be accessible to students and researchers in different fields, as well as to an interested general public. Ethnoarchaeological studies figure prominently, and help inject the human perspective into the study of agriculture.

Each article in these volumes has received double-blind peer review from two outside experts. Using numerous illustrations, they provide synthetic, interdisciplinary overviews, or detailed accounts of individual and collaborative research, and in some cases relevant experts outside the programme were invited to participate. However, this book series does not seek to – and indeed cannot – provide complete coverage of all disciplines, research themes, time periods, or geographical regions relevant to the study of agriculture, nor could all the relevant experts in each field be included. Each book should instead be seen as a sampling of exciting ways to explore the subject of preindustrial agriculture and its relevance to life today. This book series is an example of the power of academic networking and of the benefits of approaching the theme of agriculture from different angles and perspectives converging into a common space different from that of the researchers' individual fields of expertise. But it would not have seen the light of day without the devotion of the coordinating editor, Andreas G. Heiss, Vienna, and the language editor, Cozette Griffin-Kremer, Paris. Alexandre Chevalier, Brussels, ensured that abundant maps were made to cover the wide range of areas discussed.

The journey has been arduous, sometimes frustrating, but always fascinating and full of good memories. We hope that the final outcome of our programme, this book series, is of value to other scholars and interested people. It is also our hope that it leads to further work by opening new avenues into the study of agriculture.

Patricia C. Anderson
CNRS, Nice, France

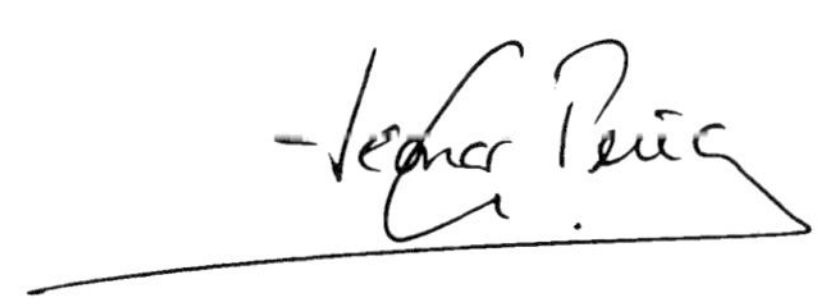

Leonor Peña-Chocarro
CSIC, Madrid, Spain and Rome, Italy

Series Editors of the EARTH Series

Bibliography

Anderson, P. C. (ed.) (1999) *Prehistory of Agriculture. New Experimental and Ethnographic Approaches.* Institute of Archaeology Monograph 40. UCLA, Los Angeles.

Anderson, P. C., Chabot, J. and van Gijn, A. (2004) The Functional Riddle of 'Glossy' Canaanean Blades and the Near Eastern Threshing Sledge. *Journal of Mediterranean Archaeology* 17 (1), 87–130.

Anderson, P. C., Deraprahamian, G. and Willcox, G. (1991) Les premières cultures de céréales sauvages et domestiques au Proche-Orient néolithique: résultats préliminaires d'expériences à Jalès (Ardèche). *Cahiers de l'Euphrate* 5–6, 191–232.

Anderson, P. C. and Inizan, M.-L. (1994) Utilisation du tribulum au début du IIIe millénaire: des lames 'cananéennes' lustrées à Kutan (Nivive V) dans la région de Mossoul, Iraq. *Paléorient* 20 (2), 85–103.

Anschuetz, K., Wilshusen, R. and Scheick, C. (2001) An archaeology of landscapes: perspectives and directions. *Journal of Archaeological Research* 9, 157–211.

Ashmore, W. and Knapp, B. (eds.) (1999) *Archaeologies of Landscape: Contemporary Perspectives.* Blackwell, Oxford.

Astill, G. and Davies, W. (1997) *A Breton Landscape.* UCL Press, London.

Ataman, K. (1999) Threshing sledges and archaeology. In P. Anderson (ed.) *Prehistory of Agriculture.* UCLA, Los Angeles, pp. 211–223.

Avner, U. (1998) Settlement, agriculture, and paleoclimate in Uvda Valley, southern Negev Desert, 6th–3rd millennia BC. In A. Issar and N. Brown (eds.) *Water, Environment and Society in Times of Climate Change. Kluwer,* Dordrecht, pp. 147–202.

Barker, G. (1996) *Farming the Desert: The UNESCO Libyan Valleys Archaeological Survey.* UNESCO, Paris.

Behre, K.-E. and Jacomet, S. (1991) The ecological interpretation of archaeobotanical data. In W. Van Zeist, K. Wasylikowa and K.-E. Behre (eds.) *Progress in Old World Palaeoethnobotany. A retrospective view on the occasion of 20 years of the International Work Group for Palaeoethnobotany* A.A. Balkema, Rotterdam/ Brookfield, pp. 81–108.

Boissinot, P. and Brochier, J.-E. (1997) Pour une archéologie du champ. In G. Chouquer (ed.) *Les Formes du Paysage, Tome II.* Errance, Paris, pp. 35–56.

Bottema, S. (1999) Cereal-type pollen in the Near East as indicators of wild or domestic crops. In P. C. Anderson (ed.) *Prehistory of Agriculture. New Experimental and Ethnographic Approaches.* Institute of Archaeology Monograph 40. UCLA, Los Angeles, pp. 56–66.

Carpinschi, C. (2002) Harvest meals XIXth - XXth centuries. In P. Lysaght (ed.) *Food and Celebration: Proceedings of the 13th International Ethnological Food Research Conference.* Ljubljana, pp. 317–324.

Carrión, J. S., Fuentes, N., González Sampériz, P., Sánchez Quirante, L., Finlayson, J. C., Fernández, S. and Andrade, A. (2007) Holocene environmental change in a montane region of southern Europe with a long history of human settlement. *Quaternary Science Reviews* 26, 1455–1475.

Charles, M., Jones, G. and Hodgson, J. (1997) FIBS in archaeobotany: functional interpretation of weed floras in relation to husbandry practices. *Journal of Archaeological Science* 24, 1151–1161.

Cherry, J., Davis, J., Mantzourani, E. (1991) *Landscape Archaeology as Long-Term History: Northern Keos in the Cycladic Islands from Earliest Settlement Until Modern Times.* UCLA, Los Angeles.

Comet, G. (1992) *Le paysan et son outil: essai d'histoire technique des céréales (France, VIIIe-XVe siècles).* Collection de l'École française de Rome.

Cunningham, P. (2011) Caching your savings: The use of small-scale storage in European prehistory. *Journal of Anthropological Archaeology* 30 (2), 135–144.

Dickson, C. (1987) The identification of cereals from ancient bran fragments. *Circaea* 4 (2), 95–102.

Dickson, J. H. (2011) Why ancient glacier mummies are so special: the ingesta are encoded diaries and maps. In H. Gill-Frerking, W. Rosendahl, A. Zink and D. Piombino-Mascali (eds.) *The Yearbook of Mummy Studies 1.* Verlag Dr. Friedrich Pfeil, München, pp. 45–50.

Evershed, R. P., Payne, S., Sherratt, A. G., Copley, M. S., Coolidge, J., Urem-Kotsu, D., Kotsakis, K., Özdoğan, M., Özdoğan, A. E., Nieuwenhuyse, O., Akkermans, P. M. M. G., Bailey, D., Andeescu, R.-R., Campbell. S., Farid, S., Hodder, I., Yalman, N., Özbaşaran, M., Bıçakcı, E., Garfinkel, Y., Levy, T. and Burton, M. M. (2008) Earliest date for milk use in the Near East and southeastern Europe linked to cattle herding. *Nature* 455, 528–531.

Fischer, E., Rösch, M., Schatz, K. and Stephan, E. (2010) Rinder, Körner, Schweinespeck — Landnutzung und Ernährung in der Eisenzeit. *Archäologie in Deutschland* 2010 (5), 34–37.

Francovich, R. and Patterson, H. (eds.) (2000) *Extracting Meaning from Ploughsoil Assemblages.* The archaeology of the Mediterranean landscape 5. Oxbow Books, Oxford.

Gassner, G., Hohmann, B. and Deutschmann, F. (1989) *Mikroskopische Untersuchung pflanzlicher Lebensmittel.* 5th edition. Verlag Gustav Fischer, Stuttgart.

Gast, M. and Sigaut, F. (1979–1985) *Les Techniques de Conservation de Grains à Long Terme.* 3 vols in 4 parts. CNRS, Paris.

van Geel, B., J. Buurman, O. Brinkkemper, J. Schelvis, A. Aptroot, G. van Reenen and T. Hakbijl (2003) Environmental reconstruction of a Roman Period settlement site in Uitgeest (The Netherlands), with special reference to coprophilous fungi. *Journal of Archaeological Science* 30 (7), 873–883.

van Gijn, A. L. (1999) The interpretation of sickles: a cautionary tale. In P. C. Anderson (ed.) *Prehistory of Agriculture. New Experimental and Ethnographic*

Approaches. Institute of Archaeology Monograph 40. UCLA, Los Angeles, pp. 254–260.

Given, M. and Knapp, B. (2003) *The Sydney Cyprus Survey Project: Social Approaches to Regional Archaeological Survey*. UCLA, Los Angeles.

Gong, Y., Yang, Y., Ferguson, D. K., Tao, D., Li, W., Wang, C., Lü, E. and Jiang, H. (2011) Investigation of ancient noodles, cakes, and millet at the Subeixi Site, Xinjiang, China. *Journal of Archaeological Science* 38 (2), 470–479.

Grégoire, J.-P. (1999) Major units for the transformation of grain: the grain-grinding households of southern Mesopotamia at the end of the third millenium BCE. In P. C. Anderson (ed.) *Prehistory of Agriculture. New Experimental and Ethnographic Approaches*. Institute of Archaeology Monograph 40. UCLA, Los Angeles, pp. 223–238.

Harlan, J. (1975) *Crops and Man*. American Society of Agronomy, Madison.

Harlan, J. (1999) Harvesting of wild-grass seed and implications for domestication. In P. C. Anderson (ed.) *Prehistory of Agriculture. New Experimental and Ethnographic Approaches*. Institute of Archaeology Monograph 40. UCLA, Los Angeles, pp. 1–5.

Hansson, A.-M. and Isaksson, S. (1994) Analyses of charred organic remains. *Laborativ Arkeologi* 7, 21–29.

Heer, O. (1865) Die Pflanzen der Pfahlbauten. *Neujahrsblätter der Naturforschenden Gesellschaft in Zürich* 1866, 1–54.

Heiss, A. G. and Oeggl, K. (2009) The plant macro-remains from the Iceman site (Tisenjoch, Italian-Austrian border, eastern Alps): new results on the glacier mummy's environment. *Vegetation History and Archaeobotany* 18 (1), 23–35.

Henry, A. G., Hudson, H. F. and Piperno, D. R. (2009) Changes in starch grain morphologies from cooking. *Journal of Archaeological Science* 36, 915–922.

Herbig, C. (2009) Recent archaeobotanical investigations into the range and abundance of Neolithic crop plants in settlements around Lake Constance and in Upper Swabia (south-west Germany) in relation to cultural influences. *Journal of Archaeological Science* 36 (6) 1277–1285.

Hillman, G. C. (1981) Reconstructing crop husbandry practices from charred remains of crops. In R. Mercer (ed.) *Farming Practice in British Prehistory*, 123–62. Edinburgh University Press, Edinburgh.

Hillman, G. C. (1984) Interpretation of archaeological plant remains: the application of ethnographic models from Turkey. In W. van Zeist and W. Casparie (eds.) *Plants and Ancient Man. Studies in* Palaeoethnobotany. A. A. Balkema, Rotterdam, pp. 1–41.

Hillman, G. C. and Davies, S. (1999) Domestication rates in wild wheats and barley under primitive cultivation: preliminary results and archaeological implications of field measurements of selection coefficient. In P. C. Anderson (ed.) *Prehistory of Agriculture. New Experimental and Ethnographic Approaches*. Institute of Archaeology Monograph 40. UCLA, Los Angeles, pp. 70–103.

Hopf, M. (1954) Anatomische Untersuchungen an Weizenspelzen und -Körnern verschiedener Polyploidiestufen als Vorarbeit für die Bestimmung prähistorischer Funde. *Der Züchter* 24 (6), 174–180.

Hopf, M. (1955) Formveränderungen von Getreidekörnern beim Verkohlen. *Berichte der Deutschen Botanischen Gesellschaft* 68 (4), 191–193.

Jacomet, S. (2009) Plant economy and village life in Neolithic lake dwellings at the time of the Alpine Iceman. *Vegetation History and Archaeoabotany* 18 (1), 47–59.

Jones, G. (1984) Interpretation of archaeological plant remains: ethnographic models from Greece. In W. van Zeist and W. Casparie (eds.) *Plants and Ancient Man. Studies in Palaeoethnobotany*. A. A. Balkema, Rotterdam, pp. 43–61.

Jones, G. (1987) A statistical approach to the archaeological identification of crop processing. *Journal of Archaeological Science* 14, 311–23.

Jones, G., Charles, M., Bogaard, A. and Hodgson, J. (2010) Crops and weeds: the role of weed functional ecology in the identification of crop husbandry methods. *Journal of Archaeological Science* 37 (1), 70–77.

Juel Jensen, H. (1988) Functional analysis of prehistoric flint tools by high-power microscopy: a review of west European research. *Journal of World Prehistory* 1, 53–88.

Juel-Jensen, H. (1994) *Flint Tools and Plant-Working. Hidden Traces of Stone-Age Technology*. Aarhus University Press, Aarhus.

Kalis, A. J., Merkt, J. and Wunderlich, J. (2003) Environmental changes during the Holocene climatic optimum in central Europe- human impact and natural causes. *Quaternary Science Reviews* 22 (1), 33–79.

Kardulias, P. N. (ed.) (1994) *Beyond the Site: Regional Studies in the Aegean Area*. University Press of America, Lanham.

Kardulias, P. N., and Yerkes, R. (1996) Microwear and metric analysis of threshing sledge flints from Greece and Cyprus. *Journal of Archaeological Science* 23, 657–66.

Kohler-Schneider, M. (2001) *Verkohlte Kultur- und Wildpflanzenreste aus Stillfried an der March als Spiegel spätbronzezeitlicher Landwirtschaft im Weinviertel, Niederösterreich*. Mitteilungen der Prähistorischen Kommission 37. Verlag der Österreichischen Akademie der Wissenschaften, Wien.

Korobkova, G. (1981) Ancient reaping tools and their productivity in the light of experimental tracewear analysis. In P. Kohl (ed.) *The Bronze Age Civilisation of Central Asia*. M. E. Sharpe, New York, pp. 325–349.

Kuijt, I. and Finlayson, B. (2009) Evidence for food storage and predomestication granaries 11,000 years ago in the Jordan Valley. *Proceedings of the National Academy of Sciences* 106 (27), 10966–10970.

Lerche, G. (1994) *Ploughing Implements and Tillage Practices in Denmark from the Viking Period to About 1800, Experimentally Substantiated.* Poul Kristensen, Copenhagen.

Lerche, G. and Steensberg, A. (1980) *Agricultural Field Shapes.* National Museum, Copenhagen.

Meurers-Balke, J., and Lüning, J. (1999) Some aspects and experiments concerning the processing of glume wheats. In P. C. Anderson (ed.) *Prehistory of Agriculture. New Experimental and Ethnographic Approaches.* Institute of Archaeology Monograph 40. UCLA, Los Angeles, pp. 238–254.

Mingote Calderón, J. L. (1996) *Tecnología Agrícola Medieval en España.* Ministerio de Agricultura, Pesca y Alimentación, Madrid.

Miret i Mestre, J. (2006) Sobre les sitges i altres estructures excavades al subsòl. *Cypsela* 16, 213–225.

Palmer, C. and van der Veen, M. (2002) Archaeobotany and the social context of food. *Acta Palaeobotanica* 42 (2), 195–202.

Pearsall, D. M. (1988) Interpreting the Meaning of Macroremain Abundance: The Impact of Source and Context. In C. A. Hastorf and V. S. Popper (eds.) *Current Paleoethnobotany. Analytical Methods and Cultural Interpretations of Archaeological Plant Remains.* Prehistoric Archaeology and Ecology University of Chicago Press, Chicago/London, pp. 97–118.

Peña-Chocarro, L. (1999) *Prehistoric agriculture in Spain: the application of ethnographic models.* BAR International Series 818. Archaeopress, Oxford.

Perry, L., Dickau, R., Zarrillo, S., Holst, I., Pearsall, D. M., Piperno, D. R., Berman, M. J., Cooke, R. G., Rademaker, K., Ranere, A. J., Raymond, J. S., Sandweiss, D. H., Scaramelli, F., Tarble, K. and Zeidler, J. A. (2007) Starch Fossils and the Domestication and Dispersal of Chili Peppers (*Capsicum* spp. L.) in the Americas. *Science* 315, 986–988.

Procopiou, H., Anderson, P. C., Formenti, F. and Juan-Tresseras, J. (2002) Étude des matériaux transformés sur des outils de mouture : identification des résidus par analyse chimique et par observation en microscopie optique et électronique. In H. Procopiou and R. Treuil (eds.) *Moudre et Broyer. L'identification fonctionnelle de l'outillage de mouture dans la Préhistoire et l'Antiquité.* Vol. 1: Editions CTHS, Paris, pp. 111–127.

Revedin, A., Aranguren, B., Becattini, R., Longo, L., Marconi, E., Mariotti Lippi, M., Skakun, N., Sinitsyn, A., Spiridonova, E. and Svoboda, J. (2010) Thirty thousand-year-old evidence of plant food processing. *Proceedings of the National Academy of Sciences* 107 (44), 18815–18819.

Reynolds, P. (1974) Experimental Iron Age Storage Pits. *Proceedings of the Prehistoric Society* 40, 118–131.

Reynolds, P. J. (1979) *Iron Age Farm: the Butser Experiment.* British Museum Publications, London.

Reynolds, P. J. (1981) Deadstock and livestock. In R. Mercer (ed.) *Farming Practice in British Prehistory.* Edinburgh University Press, Edinburgh.

Reynolds, P. J. (1999) Crop yields of the prehistoric cereal types emmer and spelt: the worst option. In P. C. Anderson (ed.) *Prehistory of Agriculture. New Experimental and Ethnographic Approaches.* Institute of Archaeology Monograph 40. UCLA, Los Angeles, pp. 267–275.

Rollo, F., Ubaldi, M., Ermini, L. and Marota, I. (2002) Ötzi's last meals: DNA analysis of the intestinal content of the Neolithic glacier mummy from the Alps. *Proceedings of the National Academy of Sciences of the United States of America* 99 (20), 12594–12599.

Rösch, M., Ehrmann, O., Herrmann, L., Bogenrieder, A., Goldammer, J. P., Hall, M., Page, H. and Schier, W. (2002) An experimental approach to Neolithic shifting cultivation. *Vegetation History and Archaeobotany* 11, 143–54.

Rottoli, M. and Castiglioni, E. (2011) Plant offerings from Roman cremations in northern Italy: a review. *Vegetation History and Archaeobotany* 20 (5), 495–506.

Sarpaki, A. (2000) The study of palaeodiet in the Aegean: food for thought. In S. Vaughan and W. Coulson (eds.) *Palaeodiet in the Aegean.* Oxbow Books, Oxford.

Semenov, S. (1964) *Prehistoric Technology.* Moonraker Press, Bradford-on-Avon.

Sigaut, F. (1999) Yields, sowing, and fertility: analytical significance of yields. In P. C. Anderson (ed.) *Prehistory of Agriculture. New Experimental and Ethnographic Approaches.* Institute of Archaeology Monograph 40. UCLA, Los Angeles, pp. 275–281.

Sigaut, F. (1978) Identification des techniques de récolte des graines alimentaires. *Journal d' agriculture traditionnelle et de la botanique appliquée* 24 (3), 145–161.

Skakun N. N. (1999) Evolution at agricultural techniques in Eneolithic (Chalcolithic) Bulgaria: data from use-wear analysis. In P. C. Anderson (ed.) *Prehistory of Agriculture. New Experimental and Ethnographic Approaches.* Institute of Archaeology Monograph 40. UCLA, Los Angeles, pp. 199–211.

Steensberg, A. 1943. *Ancient harvesting implements. A study in archaeology and human geography.* Copenhagen.

Stika, H.-P., Heiss, A. G. and Zach, B. (2008) Plant remains from the early Iron Age in western Sicily: differences in subsistence strategies of Greek and Elymian sites. *Vegetation History and Archaeobotany* 17 (Supplement 1), 139–148.

Valamoti, S.-M., Samuel, D., Bayram, M. and Marinova, E. (2008) Prehistoric cereal foods from Greece and Bulgaria: investigation of starch microstructure in experimental and archaeological charred remains. *Vegetation History and Archaeobotany* 17 (Supplement 1), 265–276.

van der Veen, M. (2003) When is food a luxury? *World Archaeology* 34 (3), 405–427.

van der Veen, M. (2008) Food as embodied material culture: diversity and change in plant food consumption in Roman Britain. *Journal of Roman Archaeology* 21, 83–109.

Winton, A. L. and Winton, K. B. (1932) *The Structure and Composition of Foods. Volume I: Cereals, Starch, Oil Seeds, Nuts, Oils, Forage Plants. Vol. 1.* John Wiley & Sons, New York.

Winton, A. L. and Winton, K. B. (1935) *The Structure and Composition of Foods. Volume II: Vegetables, Legumes, Fruits. Vol. 2.* John Wiley & Sons, New York.

Sarpaki, A. (2000) The study of palaeodiet in the Aegean: food for thought. In S. J. Vaughan and W. D. E. Coulson (eds.) *Palaeodiet in the Aegean. Papers from a colloquium held at the 1993 meeting of the Archaeological Institute of America.* Oxford, pp. 115–121.

Vuorela, I. (1986) Palynological and historical evidence of slash-and-burn cultivation in south Finland. In K.-E. Behre (ed.) *Anthropogenic indicators in pollen diagrams.* A. A. Balkema, Rotterdam/Boston, pp. 53–64.

Zohary, D., Hopf, M. and Weiss, E. (2012) *The domestication of plants in the Old World.* Oxford University Press, Oxford.

SECTION 1

Methodological Approaches to Plant Use Diversity

1 Factors and Issues in Plant Choice

Alexandre Chevalier, Elena Marinova and Leonor Peña-Chocarro

Because humans are omnivores, they need plants to survive but there is more to it than that: they need different plants in order to fulfil their physiological needs, since no single plant can provide all the elements human beings need. However, among the available edible plants in the environment, humans choose only a small part of those plants as food. Throughout time, humans have shaped, consciously or incidentally, their natural surroundings and an essential part of this is expressed by modifying plant communities and diversity, transforming landscapes into cultivated or managed land. Humans eventually noticed that their influence on some of these wild plants lead to modifications in their physiology, taste specificities and morphology, observations that at some stage lead them to deliberate manipulations of those plants. In this way, people were able to get more from the plant's useful parts, and in a more reliable and systematic way, than from the wild varieties, but in turn they had to work harder and longer to get their food (Sahlins 2003). Plant domestication definitely created plants that were more suitable for human alimentary stability, but this also lead to a reduction in their competitive capacity, and some of these plants eventually became completely dependent upon humans for their reproduction and spread. While limiting ever more their food choices from the wild by focusing on a few very nutritive plants, even running the risk of creating harmful unbalanced diets, at the same time, humans started to promote many domesticated species and varieties of plants different from their wild progenitors in order to adapt them to a diversity of ecological niches and conditions. Parallel to this, people invented a plethora of techniques to get, grow, transform, store and extract the nutritive or useful product components from the plants they needed. Even for the very same plant, strongly varying processes have been applied by different cultural groups across time and space, all of them being as 'efficient' as the others and adapted to the local environmental and social constraints.

Local biodiversity and ecological conditions determined the range of possibilities available to humans, who chose which plant they would use according to a multiplicity of factors (social, cultural, environmental, technical, etc.). According to the importance given to, or imposed by each of these factors, human groups gave different responses even in similar environments, which led to an enormous diversity of practices and plant uses.

What lead to these different choices? A process of adaptation to environmental constraints? Cultural materialism based on the local availability of raw materials and on the group's social structure? Pre-eminence of functional purposes controlled by human biological determinism? Local social categorisation of an objective nature that makes a plant socially acceptable to eat? Adaptive evolution driven by natural selection? Local choices from a specific environment driven by universal human mental categories? Local mental processes that 'think' nature and plants in a specific way which will orientate further plant choices (subjective nature)?

Factors and Issues in Plant Choice

The long-standing relationship between plants and people has always been characterised by a continuous dynamic condition of adaptation, interaction and change. This relationship has not always been mutually beneficial and different degrees of interaction, operating at different scales, have been detected through time. Plant diversity and choice are issues intimately related to the natural, to the technical, as well as to the cultural, social and symbolic realms in which the communities are embedded, and these determine the array of possibilities available to people. Choices are made within these possibilities according to a myriad of different factors, the main one possibly being the social or cultural context.

Environmental Constraints

It would be a truism to state that **environmental constraints** constitute an important limiting factor for plant availability and diversity: crops are and have been throughout time geographically limited, due to various natural factors such as altitude, humidity, temperature, soil type, and humans often had a range of dietary and technical choices limited by the natural conditions and resources they were living in. There are endless studies of human and cultural ecology showing the enormous variability of the practices and systems developed by indigenous communities across the world, not only to adapt to specific environmental conditions but also to overcome them (Amisah *et al.* 2009; Berkes *et al.* 2000; Rex Immanuel *et al.* 2010; Wilken 1987). The choice of specific plants, or the selection of different varieties from one species that will be better adapted to local ecological conditions, constitutes a widespread solution to overcome specific environmental limits, such as the hulled wheats in Mediterranean mountain areas (Peña-Chocarro 1996, Peña-Chocarro and Zapata 1998). Another example is the promotion of certain species through different strategies (pruning, burning, sowing in fallow fields, etc.) in order to increase their population density, as is the case of many species in areas of Mesoamerica, such as for palms or grasses (Casas *et al.* 2007; González-Insuasti and Caballero 2007; González-Insuasti *et al.* 2008) or the huge number of varieties of maize – about 350 different native varieties (the exact number varies according to authors and date of publication) or of potatoes, that may total as many 3,500 different genotypes (both wild and domesticated *Solanum* tubers, divided into eight species according to Ochoa 1999).

Throughout their history, humans have generally had to cope with environmental fluctuations and unpredictability. The success and failure of the various subsistence systems through time, whether based on agriculture or on the intensive use of wild plants, were greatly dependent on buffering mechanisms to cope with dangerous fluctuations due to both natural and cultural reasons. Studies addressing buffering tactics are numerous both in archaeological and historical literature (Allen 2004; Galant 1989; Halstead and O'Shea 1989; Kellhofer 2002; Kohler and van West 1996; Roberts and Rosen 2009; Stopp 2002; van der Veen 2008) as well as in human ecology (Balée 1994; Barlett 1980; Berkes *et al.* 2000; González-Insuasti *et al.* 2008) and provide interesting insights into the strategies aimed at subsistence and social stability.

Ethnobotanical and anthropological research across the world has highlighted the importance of these types of studies as an efficient approach (Altieri 2004; Bellon 1991; Eakin 2005; Wale and Virchov 2003). However, some authors (Winterhalder *et al.* 1999) have stressed the qualitative character of most archaeological and anthropological literature on subsistence risk, criticising the use of risk-averse behaviours without taking into account formal models.

One of the buffering mechanisms against environmental changes is plant diversity, both wild and domesticated, and the exploitation of different ecological niches, if not ecosystems. Indeed, diversity and diversification are key in addressing issues related not only to biodiversity and genetic resource conservation, on which research has largely focused (Orlove and Brush 1996; Brush and Meng 1998; Smith and Wishnie 2000), but also in dealing with subsistence systems in the past. Recent research on diversity, focusing mostly on crops grown by indigenous communities across large parts of the planet, has shown that diversity levels are greatly influenced by farmers' decisions within household contexts (Brush and Meng 1998), although these also have effects at a larger scale. However, this has to be understood as a liberty given within the constraints of one's cultural identity. In fact, Brush and Meng (1998) suggested that the decisions taken at household level determine the choice of the selected species, varieties or landraces to be grown depending on their likelihood of meeting the household's physiological needs, social status membership and identity requirements. Indeed, yield stability and crop responsiveness are fundamental elements in domestic agricultural production and when these become unstable, farmers develop a wide range of adaptive strategies aiming to minimise shortfalls and lessen variability. Diversification appears, thus, as a common strategy to manage risk and uncertainty better and avoid crop failures in communities characterised by subsistence economies. In the past, it was also a strategy for exploiting the diversity of available ecosystems.

Semi-domesticates and wild plants also played an important role in the everyday life of agricultural management and food intake, and this could be a crucial role when dealing with food uncertainty, shortages and/or agricultural regression, either because of ecological issues or social problems (Balée 1994; Ertuğ 2009). Two chapters of the current book, namely Chapter 4 on trees (Bouby and Ruas) and Chapter 5 on wild plants (Cruz-García and Ertuğ), show that the contribution of cultivated or gathered fruits, wild or semi-wild rhizomes and greens to the everyday diet is far from being insignificant and that they were not only used as famine foods.

Technological Developments and Solutions

The development of **technological solutions** is another answer to ecological constraints. The creation and maintenance of ponds, drainage systems, levees, floating gardens, raised fields, sunken fields, irrigation canals carved in rock or simple wooden planks hanging along cliffs, aqueducts, or terraces (Chakravarty *et al.* 2006; Denevan *et al.* 1987; Lentz 2000; Marcus and Stanish 2006) are only a few among the many examples of different landscape engineering strategies that have allowed humans to overcome some of their environmental constraints. Volume 3 of the EARTH Series, *Agricultural and pastoral landscapes in pre-industrial society: choices, stability and change* (Retamero *et al.* 2014) takes up these issues, as well as social strategies developed to maximise the output from nature, such as seasonal occupation of specific ecologies, or the exploitation of several different ecologies year-round if it is not possible to develop technical solutions, or if the input required proves to be too high or socially inadequate for a group. Among these social strategies, trade is probably the most effective way to overcome environmental constraints, thus making available in a particular location plants from completely different ecologies.

Of course trade was limited until very recently to non-perishable food plants or to processed plants, such as dried, cured, or smoked products, which could be transported for weeks, at times for months. These preservation practices are part of the plant-processing techniques and field maintenance mentioned in Volume 2 of the EARTH Series, *Exploring and explaining diversity in agricultural technology* (van Gijn *et al.* 2014).

Objective technical constraints, such as the availability of raw materials to elaborate the tools necessary to deal with agricultural methods and plant processing are probably less related to plant choices than environmental constraints and social factors. Indeed, the diversity of techniques developed and applied to plants by humans is quite impressive and enabled people to make the most of almost any plant, as is illustrated in Chapter 6 (Griffin-Kremer).

However, the diversity of technical choices made in growing, storing and processing plants explains the diversity of food plant preparations across time and geographic location, as well as the diversity of plant remains in archaeological contexts. Technical choices are therefore highly relevant in archaeobotany in correctly interpreting plant diversity, as is highlighted in Chapter 3 (Marinova) of this volume, but this applies less in ethnography in the efforts to explain choices made.

Cultural Factors

An important key to understanding human choices regarding plants is also represented by the **cultural motivations** for selecting a specific plant for food (even toxic ones), for creating a new cultivar, or maintaining a particular crop over centuries. Choice is triggered first by what has been called the 'omnivore dilemma' (Fischler 1993); precisely because they are omnivores, humans are autonomous and free to choose whatever they want from nature. They are therefore highly adaptable to any ecology in securing their food supply. However, they cannot get all the necessary dietary elements from a limited range of food, and they specifically have to look for this alimentary diversity. Innovation, exploration and changes are crucial for their survival, but this is precisely a source of anxiety and rejection, since some foods may be poisonous and kill them. Interestingly, humans also include plants that are toxic or some of whose parts may be toxic in their food supplies, through elaborate processes of transformation, such as parching acorns, squeezing out the juice of the bitter manioc, or taking out the *testa* of beans and chenopods, not to mention the well-known nixtamalisation – a process of dehulling of maize grains through an alkali product – that allows a maize-based diet which would otherwise cause pellagra. Humans also modified initially toxic wild plants into domesticated edible ones, such as the almond or the potato. Thus, humans decide which plants have to be accepted and included in their symbolic and dietary realms beyond any potential neophobia vs. neophilia attitude as Fischler (1993) would describe it, and beyond known or potential toxicities.

In fact, choices are intimately related to how people perceive their environment and project themselves within it, and how people perceive themselves within a cultural and social group and interact with their social milieu (Descola 2005, 2011). Cultural representations will transform a natural element into a social one that otherwise would remain outside of the social realm and would not be considered by humans. From the vast array of plants present in a given environment, humans elaborate mentally and socially which plants are good to eat and which ones are taboo (Lahlou 1998; Lévi-Strauss 1964, 1967, 1968), as well as which are appropriate for humans and which are reserved for the gods, as is illustrated in Chapter 7 (Hansson and Heiss).

Within a given cultural group, social norms will tell people what to eat, when, how, where and with whom, according to their social position (Goody 1982; Douglas 1984; Fischler 1988; Mennell 1992; Counihan 1998). At the same time, these foods will define people and indicate their social identity (Tajfel and Turner 1986; Turner 1989). Together, social norms and identity will define how to secure and manage the household needs. They will also define what and when specific plants will be cultivated or gathered. For instance, wild plants are usually stigmatised by most of the social groups within a society and neglected, but they may be prized by low social status people for providing a complementary ingredient to domestic economies, and widely used by all social segments of a cultural group during famine times. These wild plants have now made a comeback among high social status people, because of their current representation of nature and health.

Many examples across the world provide useful insights into the socio-economic and cultural factors that influence plant choices, and therefore diversity in traditional farming systems (Dercon 1996; Rana *et al.* 2007). A particular landrace, a local variety of a plant, may be cultivated because of its biological adaptation to local environmental conditions and productivity, but also because of its gustative qualities praised by the whole cultural group or only by a specific social sub-segment, or because of its social and cultural values. Motivations are all eventually linked with identity issues, whether cultural or social in nature (Tajfel 1982; Lyons 2007). Particular species may be connected to local 'traditions' that are rooted in cultural identities or symbolisms (see examples in Papa 1996, Palmer 2002, Chevalier 2013) and their cultivation strongly encouraged according to one's social membership. Additionally, the need to maintain and back up beliefs and cultural continuity – so-called cultural inertia – seem to have influenced certain choices, such as the many examples of wild plants collected due to long-standing traditions showing resistance to change (Begossi 1998). Chapter 8 (Chevalier) illustrates some of these issues.

In traditional systems, diversity or its opposite, uniformity, is the outcome of social identity and context. In other words, within a given environment with its own limits and ecological constraints, individuals (whether farmers or foragers) choose what species to collect, encourage, plant, maintain or simply use according to multiple social factors and cultural perceptions of nature from which some may become driving forces, such as group membership and identity, while others may have a more or less significant opportunistic role depending on circumstances and historical contexts.

Historical Dimensions

Although plant 'choice' has been a subject of interest and addressed by various disciplines (ecology, anthropology, genetic resource conservation, etc.), little attention has been paid to the topic's historical dimension. Most studies on plant choice have focused on present-day examples of indigenous communities still maintaining traditional landraces and/or exploiting wild plants or on ethnographic accounts of human environmental interaction during the recent past. Archaeological or historical data has seldom been used to provide the necessary time-depth to these studies, thus hindering the possibility of gaining new insights into patterns of diversity through time. This has happened in spite of the fact that intentionality and choice are central to archaeological interpretation and present in most archaeological enquiries (David 2004; Tilley 2004).

Debates on plant use in the past, particularly those involving archaeobotanical studies, usually involve discussions of plant diversity and – less commonly – of choice. The variety of species represented in an archaeological site is often seen as a mere reflection of the biodiversity existing in the past. However, their presence results from an initial choice made by humans that is often overlooked for two main reasons. First, the difficulties in identifying plant remains to detailed taxonomic level due to preservation factors often do not allow identifying species or varieties. As a consequence, quantifying diversity becomes a complicated task, if not a hypothetical one. Second, despite the uncontested importance of scale issues in archaeology (Lock and Molyneaux 2006; Escalona and Reynolds 2011), it is still a complex issue to deal with. When discussing diversity and choice, we move between different time scales: from the individual decision of a particular farmer in a particular place and time or single events to processes developing over periods of time longer than a lifetime's experience. Most especially, when addressing the issue of choice and its motivation in the past, and in particular for prehistoric cultures and cultures without history, beyond the mere ecological possibilities and human physiological needs, we are usually allowed only broad social categorisation inferences, such as 'high social class', 'exotic', or 'feast'. If it is very difficult to reconstruct the transformation processes of a product, and to master, even to some extent, the taphonomic issue of why and how certain remains were preserved, it is impossible to reconstruct the social motivations that led to a specific plant choice, unless we resort to cultural materialism, ecological determinism or functionalism. When texts are available, we are able to establish this link between social necessity and choice, as can be seen in some examples given in Chapters 6 (Griffin-Kremer), 7 (Hansson and Heiss) and 8 (Chevalier).

The EARTH Team 1 Book and its Structure

This book is a result of the work of one team involved in the Collaborative Research Program funded by the European Science Foundation (ESF) and entitled 'Early Agricultural Remnants and Technical Heritage' (EARTH). Within this framework, senior and younger scientists from Europe and a few guest contributors from beyond met from 2005 to 2009 in different venues around Europe to exchange views, data and experiences on choices made by humans in selecting plants for multiple uses such as alimentation, fodder or raw materials for crafts and construction. These scientific exchanges led to this book, the goal of which is to portray the diversity of plant uses and choices, as well as address specific issues related to this diversity.

The use of the word 'plants' instead of 'crops' in many places of this monograph is a deliberate choice. First, this is because many contributions in the present volume do not make any clear-cut separation between the 'traditional' classification of 'domesticated', 'tended', 'semi-wild', 'wild', etc. plants: most of the authors deal with a continuum of physiological states of plants used by humans, from wild plants to GMOs (genetically modified organisms), with as many possible intermediate states of adaptation to human needs, and introgressions (infiltration of the genes of one species into the gene pool of another). In addition, there are as many definitions of what a 'crop' is as there are authors using this word. And finally, every human group classifies and uses plants according to its own cultural taxonomies that can make our current Western academic classifications useless in most cases.

Dye plants as well as spices are only marginally mentioned – which some readers may regret – and not all the European countries are included, while some contributions deal with countries in Asia, Africa, North America and South America. In the same way, plants used as medicine are not directly addressed. This is due mainly to the area of expertise of this EARTH team and does not result from a deliberate choice to leave out these topics or consider them as less important or less worthy of interest.

The present publication is divided into seven chapters, each focusing on plant use-related issues.

Each chapter consists of an introductory section, in which the chapter leader summarises important aspects and issues related to the chapter topic, and then includes individual contributions to illustrate the main issues of the topic.

In lines with this plan, L. Zapata coordinates an introductory methodological chapter with different contributions focusing on the various analytical tools used to address the study of plant use, both in the past and in the present. Ranging from archaeobotanical techniques (L. Scott Cummings' contribution) to current ethnobotanical methods (G. S. Cruz-García), on to the analyses of historical sources (J. L. Mingote Calderón, M. Russel and the late F. Sigaut) and those of ancient images (S. González Reyero), the chapter explores the ways in which each discipline has tackled the topics under discussion.

Covering a large area, Europe and America, although focusing mostly on the former, the third chapter, coordinated by E. Marinova, explores crop diversity through time by looking at the archaeobotanical record in different regions. The European southeast is addressed by E. Marinova and T. Valamoti for a period spanning from the early Neolithic to the late Bronze Age. Sites in Bulgaria and Greece are examined showing the diversity of prehistoric agriculture. The patterns observed are explained by taking into account physical factors such as climate or geographical barriers as well as cultural ones. M. Rottoli makes a valuable and welcome contribution to the crop history of Italy from the beginnings of agriculture to the Roman period. By looking at the main crops involved in agriculture through time, this contribution explores the influence of population movements in the spread of crops to new areas. A survey with broader chronological scope (Neolithic to Middle Ages) is presented by S. Jacomet for central Europe, examining various crop introductions through time. Hexaploid and tetraploid wheats, spelt, millet, hulled barley or rye are some of the crop species considered, each introduced in the region at a certain point in the past. Crop diversity during the Neolithic in the Iberian Peninsula is addressed in the contribution by L. Peña-Chocarro and L. Zapata. The different combinations of species found in the varied territory of Iberia are explored and both ecological and socio-cultural motivations are discussed in order to explain the underlying reasons for such diversity. A time-span of 1400 years (from 500 BCE to 900 CE) is surveyed in C. Bakels' paper on western Europe, which provides an overview of the various factors (physical, socio-economic and cultural) influencing diversity in the region. With the same focus in western Europe, the contribution by the late F. Sigaut analyses the various crop and non-crop introductions which occurred in the region, and contributed to enlarging crop diversity. In addition, technical innovations such as the scythe are also analysed, providing a better framework to explore diversity. The American section includes two contributions, the first by L. Scott Cummings focusing on the American southwest and the second by M. Bruno with emphasis on the Andean area of the Titicaca Lake basin. In the first case, an overview of plant diversity through the analyses of coprolites from several sites is presented, showing the variability of species identified through pollen, macroremains and phytolith analysis. The work emphasises the role wild plants played in the Puebloan communities. Through archaeobotanical data, M. Bruno's paper looks at the various processes involved in producing diversity: crop domestication, risk reduction and agricultural intensification, as well as political economic strategies, thus exploring the dynamic interactions between production, environment, culture and politics.

In Chapter 4, L. Bouby and M.-P. Ruas examine the role of fruit-tree domestication and cultivation throughout time by looking at different aspects related to unnoticed uses and practices resulting from choices and intentional decisions. The other contributors lead the readers through various fruit tree cultures. First, R. Cuthrell reviews the uses of acorns by indigenous groups in California by looking at different types of data: ecological, ethnographical and archaeological. Second, D. J. Goldstein illustrates the role of some *Prosopis* tree species in the development of ancient agricultural systems in the Pre-Hispanic period in Peru. At the same time, the author guides the reader to the importance of these species in sustaining modern production in coastal Peru. M. Rottoli's paper examines the changing role of fruits in human subsistence in Italy, from the gathering of wild fruits to their cultivation proper at the end of the Bronze Age or beginning of the Iron Age. The contribution of Bui Thi Mai and M. Girard focuses on an exotic species (*Citrus*) in Antiquity in the Mediterranean area. After an exhaustive introduction to the species' botany, the authors describe and discuss critically the various sites where this plant has been attested in the pollen record. Based on this evidence, they propose the hypothesis that *Citrus* was present in southern Italy, probably before the first century of the Christian era, and that it has indeed been cultivated in the western part of the Ancient World since this period. The origins of viniculture during the prehistoric period in southern France are presented by L. Bouby, P. Terral and J.-F. Terral. The authors show that in addition to the specialised viticulture directly or indirectly promoted by Massaliotes and especially Romans in Mediterranean France, the indigenous Iron Age inhabitants seem to have grown grape more as a secondary crop. The growing of figs as a staple food in the pre-Hispanic period of the Canary Islands is discussed by J. Morales and J. Gil, based on archaeobotanical remains, ethnohistorical documents and palaeodiet studies, which indicate that this fruit played an important role in the diet of the first inhabitants of Gran Canaria. Fig and olive domestication and cultivation are the topic of the paper by Y. Aumeeruddy-Thomas, Y. Hmimsa, M. Ater and B. Khadari. The authors explore the processes of incorporating forest, wild or spontaneous trees into agricultural systems in the Mediterranean. Based on their observations made in the Rif Mountain agroecosystems, they conclude that tree domestication is not as distinct as that of cereal crops, as wild, spontaneous and domesticated tree forms are still very closely linked. This situation suggests a possible continuum in tree domestication processes over a long period of time.

In Chapter 5, G. S. Cruz-García and F. Ertuğ give an overview of several topics related to the use of wild plant resources in the past and present. An example of exploring wild plant resources by colonisation of new areas is given in J. Morales' and J. Gil's contribution, focused on the Canary Islands. Then, J. Tardío and M. Pardo-de-Santayana present the regional patterns of traditional human consumption of wild plants in Spain. Their research is based on a database with more than 4600 records. Wild food plants in the Dogon country are explored by C. Selleger in a study which shows that these plants could be considered as a mainstay of agricultural society in sub-Saharan Africa, enabling continuity of the agricultural system. C. Griffin-Kremer gives the example of silverweed as a plant that may have been on its way from a 'wild' plant gathered knowledgeably to a 'tended' one in a garden plot. The paper draws attention to the potential of wild plant resources used in the past for food and abandoned in recent times. The chapter ends with the contribution of L. Scott Cummings, who presents palynological evidence on the use of *Cleome* (beeweed) in North America, not only as a famine food resource when cultivated plants become too scarce to sustain human needs, but also as a complement to maize consumption.

In Chapter 6, C. Griffin-Kremer recalls the multiplicity of non-human food uses of plants, whether as fodder, for their medicinal or intoxicating properties, for crafting everyday life objects, as well as for clothing (flax, hemp), dyes, for industrial goals, such as fuel, ropes, fishing nets or buildings, or as sentinels to prevent plagues or intoxications. In the first contribution illustrating the versatility of plant uses, C. Griffin-Kremer emphasises the diversity of uses of furze (*Ulex* spp.) in both rural and urban contexts as fuel, for livestock production or for crafting domestic artefacts. In the same way nettles (*Urtica* spp.) were valued as much as a medicinal, as fodder, and as a textile fibre. Like furze, Nettles had associations with ritual and, beyond that, with saintliness. L. Peña-Chocarro and L. Zapata survey the uses of hulled wheats (*Triticum monococcum* L., *T. dicoccum* Schübl. and *T. spelta* L.) in Mediterranean mountain areas. These species, highly resistant to disease, provide both food and fodder, as well as a rich variety of domestic uses. E. Bonnaire takes the example of by-products produced during threshing of various cereals that have been used as temper in ceramics, mud walls, or mud bricks, thus helping to elucidate the entire sequence of cereal processing. P. Anderson's contribution deals with diss (*Ampelodesmos mauritanica* (Poir.) T. Durand and Schinz), a grass used in Tunisia in basketry, as fodder for animals, roof thatching and even as a snack for children when the stems are young. Bui Thi Mai *et al.* present the possible ancient uses of the mastic tree (*Pistacia lentiscus L.*), and compare it with recent ethnological inquiries carried out in a Sardinian community. In particular, this contribution presents the uses of the oil from its berries, its action as a pesticide and medicine from its leaves, and most especially, the *mastic* or sap from its trunk used in medicines, cosmetics and foods as preservatives. In the final contribution, Bui Thi Mai and M. Girard launch onto the trail of the oleoresins produced by two *Dipterocarpaceae* trees in Vietnam, most particularly for boat caulking. They observed present-day practices in their ethnographic inquiries and discuss the pertinent literature about modern round-boats, as well as the remains and techniques implied by the palynological analysis of the late fifteenth-century Brunei wreck.

In Chapter 7, A.-M. Hansson and A. G. Heiss do not attempt to embrace all the uses of plants in ritual and festive contexts, but focus mainly on plant offerings as an expression of rituals performed with regard to supernatural forces, aimed at winning their support for agricultural yields (fertility rites), protection for the living and the dead (propitiatory rites) or thanking them for all they provide (thanksgiving rites). Indeed, the material components of plant offerings allow researchers to trace these practices with a broad range of methods, and across a large time-scale. A.-M. Hansson explores a stone bun offering from an early medieval cremation grave at Lovö in Sweden, and concludes it may imply a protection rite based on traditional oral sources from northern Europe and from other similar findings elsewhere in Europe. A. G. Heiss investigates plant offerings as agricultural fertility rites in the Alpine region, ranging from the Alpes-de-Haute-Provence in France to Styria in Austria, that are found from the Copper Age to the Roman Iron Age. Initially focused on meat from domesticated animal offerings, studies now show that food plants are as important as the zoological remains. Botanical remains show that burnt offerings reflect an important part of daily nutrition, with a common and constant representation of cereals and a clear dominance of cultivated crops. C. Griffin-Kremer presents plant use in propitiatory rituals performed up to the 20th century at various times, including the eleven days eliminated when Britain adopted the Gregorian calendar in 1752, and in particular May Day and Maying celebrations. Centred around the British Isles, but with quick incursions into other European countries that hold such May festivals, Griffin-Kremer provides a panorama of plants used and their prophylactic, medicinal, decorative or devotional role for this one 'holiday'. L. Scott Cummings provides examples of documented ceremonial and ritual plant use by a native group in the American southwest, the Hopi, for fertility, propitiatory and thanksgiving rites. She presents the different plants used for specific festivals and rituals that enliven the Hopi calendar year. M. Sayre focuses on the use of ceremonial plants in the Andes at the site of Chavín de Huántar, located in central Perú, dated 1000–200 BCE. In particular, he uses both iconographic sources from the site itself and botanical remains to address the variety of psychoactive plants in South American ceremonial sites.

Finally, Chapter 8 addresses the issues of social status, identity, and the social as well as the physical contexts that shape and drive the food plant choices made by humans. The chapter points out that no one is really free to choose, but that food plant selection is shaped by a whole set of cognitive and symbolic associations rooted in cultural and social settings that create an entire taxonomy of plants that people are allowed to eat. Thus, the contributions in this chapter illustrate the tight relationship between social norms, beliefs and values, linked to social status, identity and context, on the one hand, with plant choices, on the other hand. A. Chevalier shows that the Peruvian site of Pampa Chica, dated between 800/400 and 150 BCE presents duplicated spaces that may have been used by two groups with different social status within an egalitarian pre-Columbian society. The different plant remains point toward the affirmation of each group's identity based on their respective mythical origins. In this case, social status does not refer to social or economic power, in other words, to social hierarchy, but is related to social and cultural identity through mythical origin and kinship. In the opposite case, S. González, focusing on plant representation in the Iron Age Iberian world between the 6th and the 2nd century BCE, shows that Iberian elites used plant images to justify their political and economical power: wild plants and an exuberant fantasised nature are attributes of the elites as, in other contexts and periods, would be the case of wild animals such as the lion and the mythical unicorn. Exploring social status in the oppidum of Bibracte, France, between the end of the first century BCE to 14 CE, F. Durand and J. Wiethold demonstrate that the diet of the elites of one of the richest and most powerful Gaulish tribes, the Aedui, does not seem to be based only on luxury food plants, even if it does include Roman plants such as olives and coriander. Instead, diversity and the presence of huge quantities of gathered wild fruits seem more indicative of high social status markers among the Aedui. W. Kirleis and S. Klooß show the same, but for northern Germany in the Neolithic Funnel Beaker Culture, ranging from 4100 to 2800

BCE: wild plants in association with inhumations would indicate high social status. Moreover, social and physical contexts would define what kind of plants are allowed or prohibited: for the living world of farmers, the presence of wild plants is discrete, whereas in the world of the after-life, wild plants seem to be of particular importance. In contrast, G. S. Cruz-García shows that wild plant consumption is related to low social status in the Wayanad district of the Western Ghats of India. Wild plants, as opposed to the case of the Iberians and the Gauls, are often seen as symbols of poverty and 'tribalness'. Their gathering and consumption tend to be hidden to outsiders of the group, due to the social stigma that is currently associated with using wild plants for food. J. L. Mingote Calderón demonstrates that the choice of plants is highly codified and restricted according to one's social status in Spain, from the Middle Ages through the 19th century: local ordinances obliged farmers to cultivate only certain plants, and punishments or fines were imposed on them if they did not comply with rules set by the elites.

Finally, D. J. Goldstein and J. B. Hageman, in their study of a Maya site, link plant choices not only to social class, but also to the context of use: their work focuses on the social codes used to actively establish and maintain social hierarchy, as expressed by two residential units within a Late Classic (CE 600–900) Mayan lineage at the site of Guijarral, Belize. Archaeobotanical remains associated with periodic feasting near ancestor shrines were distinct from daily domestic activities near house mounds. They come to the conclusion that the consumption of specific foods in specific places created and reinforced in-group social inequality during the Late Classic.

After four years of meeting and intense discussion in the framework of the EARTH scientific program, this book attempts to explore plant diversity and choice through time by looking at a variety of related topics involved in the study of plant use. According to the authors' fieldwork, initial training and experience, the individual contributions are based on very different theoretical backgrounds. The diversity of the viewpoints, as well as of the human cultures, regions and periods considered, constitutes the strength and the interest of the present publication. The book does not claim to provide a state-of-the-art of the topics addressed, nor to be all-comprehensive. It is in no way an encyclopaedia on plants and their uses through time and across geographic location, but it does attempt to provide new insights and a useful framework for approaching past and modern plant choice and diversity of use.

Chapter Bibliography

Allen, M. S. (2004) Bet-hedging strategies, agricultural change, and unpredictable environments: historical development of dryland agriculture in Kona, Hawaii. *Journal of Anthropological Archaeology* 23, 196–224.

Altieri, M. (2004) Linking ecologists and traditional farmers in the search for sustainable agriculture. *Frontiers in the Ecology and the Environment* 2 (1), 35–42.

Amisah, S., Gyampoh, A. B., Sarfo-Mensah, P. and Quagrainie, K. K. (2009) Livelihood trends in response to climate change in forest fringe communities of the Offin Basin in Ghana. *Journal of Applied Sciences and Environmental Management* 13 (2), 5–15.

Balée, W. (1994) *Footprints of the forest: Ka'apor ethnobotany - the historical ecology of plant utilization by an Amazonian people.* Columbia University Press, New York.

Barlett, P. F. (1980) Adaptive strategies in peasant agricultural production. *Annual Review of Anthropology* 9, 545–573.

Begossi, A. (1998) Resilience and neo-traditional populations: the caicaras (Atlantic Forest) and Caboclos (Amazon, Brazil). In F. Berkes and C. Folke (eds.) *Linking Ecological and Social Systems for Resilience and Sustainability.* Cambridge University Press, Cambridge, pp. 129–157.

Bellon, M. (1991) The ethnoecology of maize variety management. A case study from Mexico. *Human Ecology* 19 (3), 389–418.

Berkes, F., Colding, J. and Folke, C. (2000) Rediscovery of traditional ecological knowledge as adaptive management. *Ecological Applications* 10 (5), 1251–1262.

Brush, S. B. and Meng, E. (1998) Farmers' valuation and conservation of crop genetic resources. *Genetic resources and Crop Evolution* 45, 139–150.

Casas, A., Otero-Arnáiz, A., Pérez-Negro, E. and Valiente-Banuet, A. (2007) In situ management and domestication of plants in Mesoamerica. *Annals of Botany* 100, 1101–1115.

Chakravarty, K. K., Badam, G. L. and Paranjpye, V. (eds.) (2006) *Traditional water management systems of India.* Aryan Books International, New Delhi.

Chevalier, A. (2013) From Ecological Constraints To Cultural Identities: Pre-Columbian attitudes toward food. In M. I. J. Davies and F. Nkirote (eds.) *Humans & the Environment: New Archaeological Perspectives for the Twenty-First Century.* Oxford University Press, Oxford, pp. 97–116.

Counihan, C. and Kaplan, S. (eds.) (1998) *Food and gender: identity and power.* Harwood Academic Publishers, Amsterdam.

David, B. (2004). Intentionality, agency and an archaeology of choice. *Cambridge Archaeological Journal* 14 (1), 67–71.

Denevan, W., Mathewson, K. and Knapp, G. (eds.) (1987) *Pre-Hispanic agricultural fields in the Andean region.* BAR International Series 359, ii. Archaeopress, Oxford.

Dercon, S. (1996). Risk, crop choice and savings: evidence from Tanzania. *Economic development and cultural change* 44 (3), 485–513.

Descola, P. (2005) *Par-delà nature et culture.* Gallimard, Paris.

Descola, P. (2011) *L'écologie des autres. L'anthropologie et la question de nature.* Quae, Versailles.

Douglas, M. (1984) *Food in the social order: studies of food and festivities in three American communities.* Russell Sage Foundation, New York.

Eakin, H. (2005) Institutional change, climate risk and rural vulnerability: Cases from central Mexico. *World Development* 33 (11), 1923–1938.

Ertuğ, F. (2009). Wild plant foods: routine dietary supplement or famine foods? In A. Fairbairn and E. Weiss (eds.) *From foragers to farmers: Gordon C. Hillman Festschrift.* Oxbow Books, Oxford, pp. 64–70.

Escalona, J. and Reynolds, A. (eds.) (2011). *Scale and scale change in the early Middle Ages. Exploring landscape, local society, and the world beyond.* The Medieval Countryside 6. Brepols, Turnhout.

Fischler, C. (1988) Food, self and identity. *Social Science Information* 27 (2), 275–292.

Fischler, C. (1993) *L'homnivore.* Odile Jacob, Paris.

Gallant, T. (1989) Crisis and response: risk-buffering behaviour in Hellenistic Greek Communities. *Journal of Interdisciplinary History* 19 (3), 393–413.

González-Insuasti, M. and Caballero, J. (2007) Managing plant resources: How intensive can it be? *Human Ecology* 35, 303–314.

González-Insuasti, M., Martorell, C. and Caballero, J. (2008) Factors that influence the intensity of non-agricultural management of plant resources. *Agroforestry Systems* 74, 1–15.

Goody, J. (1982) *Cooking, cuisine and class: a study in comparative sociology.* Cambridge University Press, Cambridge.

Halstead, P. and O'Shea, J. (eds.) (1989) *Bad year economics. Cultural responses to risk and uncertainty.* Cambridge University Press, Cambridge.

Kelhofer, L. (2002) Changing Perceptions of Risk: The Development of Agro-Ecosystems in Southeast Asia. *American Anthropologist* 104 (1), 178–194.

Kholer, T. A. and van West, C. R. (1996) The calculus of self-interest in the development of cooperation: sociopolitical development and risk among the northern Anasazi. In J. A. Tainter, and M. B. Tainter (eds.) *Evolving complexity and environmental risk in the prehistoric southwest.* Addison-Wesley, Reading, MA, pp. 169–196.

Lahlou, S. (1998) *Penser manger.* PUF, Paris.

Lentz, D. (ed.) (2000) *Imperfect balance: landscape transformations in the Precolumbian Americas.* Columbia University Press, New York.

Lévi-Strauss, C. (1964) *Le cru et le cuit.* Mythologiques 1. Plon, Paris.

Lévi-Strauss, C. (1967) *Du miel au cendres.* Mythologiques 2. Plon, Paris.

Lévi-Strauss, C. (1968) *L'origine des manières de table.* Mythologiques 3. Plon, Paris.

Lock, G. and Molyneaux, B. (eds.) (2006) *Confronting scale in archaeology: Issues of theory and practice.* Springer, New York.

Lyons, D. (2007) Integrating African cuisines. Rural cuisine and identity in Tigray, highland Ethiopia. *Journal of Social Archaeology* 7 (3), 346–371.

Marcus, J. and Stanish C. (eds.) (2006) *Agricultural strategies.* Cotsen Institute of Archaeology, UCLA, Los Angeles.

Mennell, S., Murcott, A. and Van Otterloo, A. (1992) *The sociology of food: eating, diet, and culture.* Sage, London/Newbury Park.

Ochoa, C. (1999) *Las papas de Sudamérica: Perú.* CIP, Lima.

Orlove, B. and Brush, B. (1996) Anthropology and the conservation of biodiversity. *Annual Review of Anthropology* 25, 329–352.

Palmer, C. (2002) Milk and cereals: identifying food and food identity among Fallahuin and Bedouin in Jordan. *Levant* 34, 173–195.

Papa, C. (1996) The 'farre de Montelione': landrace and representation. In S. Padulosi, K. Hammer and J. Heller (eds.) *Hulled Wheats. Promoting the conservation and use of underutilized and neglected crops.* IPGRI, Rome, pp. 154–171.

Peña-Chocarro, L. (1996) In-situ conservation of hulled-wheat species: the case of Spain. In S. Padulosi, K. Hammer and J. Heller (eds.) *Hulled Wheats. Promoting the conservation and use of underutilized and neglected crops.* IPGRI, Rome, pp. 129–146.

Peña-Chocarro, L. and Zapata, L. (1998) Hulled wheats in Spain: history of minor cereals. In A. A. Jaradat (ed.) *Triticeae III.* Oxford & NBH, New Delhi; Science Publishers, New Hampshire, pp. 45–52.

Rana, R. B., Garforth, C., Sthapit, B. and Jarvis, D. (2007) Influence of socio-economic and cultural factors in rice varietal diversity management on-farm in Nepal. *Agriculture and Human Values* 24, 461–472.

Retamero, F., Schjellerup, I. and Davies, A. (eds.) (2014) *Agricultural and pastoral landscapes in pre-industrial society: choices, stability and change.* EARTH Series vol. 3. Oxbow Books, Oxford.

Rex Immanuel, R., Imayavaramban, V., Lyla Elizabeth, L., Kennan, T. and Murugan, G. (2010) Traditional farming knowledge on agroecosystem conservation in northeast coastal Tamil Nadu. *Indian Journal of Traditional Knowledge* 9 (2), 366–374.

Roberts, C. N. and Rosen, A. M. (2009) Diversity and complexity in early farming communities of southwest Asia: New insights into the economic and environmental basis of Neolithic Çatalhöyük. *Current Anthropology* 50 (3), 393–402.

Sahlins, M. (2003) *Stone Age economics.* 2nd edition. Routledge, London.

Smith, E. A. and Wishnie, M. (2000) Conservation and subsistence in small-scale societies. *Annual Review of Anthropology* 29, 493–524.

Stopp, M. P. (2002) Ethnohistoric analogues for storage as an adaptive strategy in northeastern subartic prehistory. *Journal of Anthropological Archaeology* 21, 301–328.

Tajfel, H. (ed.) (1982) *Social identity and interhroup relations.* Cambridge University Press, Cambridge.

Tajfel, H. and Turner, J. (1986) The social identity theory of intergroup behaviour. In S. Worchel and W. Austin (eds.) *Psychology of intergroup relations.* Nelson-Hall, Chicago, pp. 7–24.

Tilley, C. (2004) Mind and body in landscape research. *Cambridge Archaeological Journal* 14 (1), 77–80.

Turner, J. (1989) *Rediscovering the social group: A self-categorization theory.* Blackwell, Oxford.

van der Veen, M. (2008) Food as embodied material culture: diversity and change in plant food consumption in Roman Britain. *Journal of Roman Archaeology* 21, 83–109.

van Gijn, A., Whittaker, J. and Anderson, P. (eds.) (2014) *Exploring and explaining diversity in agricultural technology.* EARTH Series vol. 2. Oxbow Books, Oxford.

Wale, E. and Virchow, D. (2003) *Crop diversity as the derived outcome of farmers' 'survival first' motives in Ethiopia: what role for on-farm conservation of sorghum genetic resources?* Contributed paper at the 25th International Conference of Agricultural Economists (August 16–22, 2003, Durban, South Africa). URI: http://purl.umn.edu/25882.

Wilken, G. C. (1987) *Good farmers. Traditional agricultural resource management in Mexico and Central America.* University of California Press, Berkeley.

Winterhalder, B., Lu, F. and Tucker, B. (1999) Risk-sensitive adaptive tactics: Models and evidence from subsistence studies in biology and anthropology. *Journal of Archaeological Research* 7 (4), 301–348.

2 Exploring Diversity in the Past and in the Present

2.1. EXPLORING DIVERSITY IN THE PAST: AN INTRODUCTION

Lydia Zapata

This chapter will deal with the methods and different ways in which the use of plants, crops and diversity can be explored and studied in the past and in the present. The origin of human plant use is a poorly known subject which has relied on data from different sources: assumptions on hominin dietary evolutionary trends, palaeodietary analyses, the identification of archaeobotanical remains and the study of lithic artefacts. Some of the most impressive information on Palaeolithic plant use comes from the Near East and Egypt, thanks to the conditions of preservation in those areas and the systematic work of different authors and research groups (Hillman 1989; Kislev *et al.* 1992; Lev *et al.* 2005; Nadel *et al.* 2004; Nadel and Werker, 1999; Weiss *et al.* 2004). A very early case is that of the Acheulian site of Gesher Benot Ya'aqov, where a unique association of edible nuts with pitted hammers and anvils was retrieved for the Early-Middle Pleistocene in OIS19.[1] This probably represents the first example of palaeobotanical and lithic evidence together, that is, the plant foods eaten by some early hominins (seven terrestrial and aquatic nuts) and the technologies used for processing these foods (Goren-Inbar *et al.* 2002). The Upper Palaeolithic site of Ohalo II, near the Sea of Galilee, dated *ca.* 23,000 BP offers good evidence for the use of wild grasses both for food and for a grass bedding on the floor in a dwelling hut which consists of bunches of *Puccinellia* (alkali grass) stems and leaves, covered by a thin layer of clay arranged in a repeated pattern around a central hearth (Nadel *et al.* 2004, 6821). The staple foods represented in this assemblage, thousands of years before domestication took place, are small-grained wild grasses, wild wheat and barley (Weiss *et al.* 2004, 9551).

For the Mesolithic, from an initial period of methodology when plant foods were virtually ignored and interpretation was completely meat-biased, we have moved on to a second where – under the influence of Clarke's work (1978) – the importance of plants has been reassessed. However, we still lack sound archaeological data which explores what plants were being used for and how. The exceptions come again from areas where expert research teams work and where preservation allows for a better recovery of all types of remains. Some of these results stress the role that foods derived from roots and rhizomes may have had for hunter-gatherers, but the correct recovery and identification procedures for the parenchymatous tissues[2] that derive from these foods is not yet a standard protocol in our analyses (with some exceptions, see for example Kubiak-Martens 1999 and Mason *et al.* 1994). Chapter 5 in this book provides a good insight into the diversity of the use of plants from the wild.

Regarding the study of diversity in relation to agriculture, 'crops' are the result of a very long relationship between humans and the vegetal world, the product of a series of events that occurred at different places over long periods of time (Tanno and Willcox 2006, 1886). Nowadays, on a global scale, it is suggested that there were not single events of domestication and that different trials and agrarian practices carried out over a long time led to the morphological and genetic changes that we define as domestication. In the different homelands of agriculture, it might have taken centuries and even millennia for societies to evolve from gathering, through pre-domestic cultivation to the exclusive use of morphologically domestic plants (Bellwood 2009, 623). In the case of the Near East, systematic sampling and the increasing amount of archaeobotanical data from Natufian and Neolithic sites now allows us to understand how, when and where this happened (*e.g.* Colledge and Conolly 2007; Fuller 2007; Willcox 2005). However, why it happened is another complex issue, and different proposals have been put forward depending on research traditions and backgrounds, and changes in theoretical agendas in archaeology (among many, see Hillman 2000; Cohen 2009; Belfer-Cohen and Goring-Morris 2009).

G. Hillman, for example, makes a good case for the origin of crops in the Euphrates, based on sound archaeobotanical remains from Tell Abu Hureyra and on ethno-ecological modelling for the region. We can see how a huge diversity of wild plants was being used by Epipalaeolithic inhabitants of the site. Seed foods appear to have included over 120 types and the total number of plant-food species consumed probably exceeded 250 among the last hunter-gatherer populations (Hillman 2000, 397). This extremely diverse diet in terms of plant foods is something that we are very rarely able to attest archaeologically, even if we may presume this was often the case. The cultivation of plants – or the fact of humans planting in primary or secondary habitats – derived according to Hillman from changing climatic conditions that forced people to take up this new strategy for obtaining food. In turn, this resulted in a narrowing of plant diversity in human diet. After thousands of years during which humans probably exploited an enormous diversity of plant foods, agriculture resulted in a dramatic contraction of the number of plants that were being used. In recent times at least the diet enjoyed by most hunter-gatherer societies has been much more diverse than that of most farming groups. For prehistoric hunter-gatherers, cereals were seasonal plant foods, but agriculture made them permanent staples. Cereals are very demanding in terms of input and work and most probably people had less free time for gathering. It is likely that traditional knowledge about plant foods started being lost or at least that it was transformed. This process is something that we can easily observe nowadays with globalisation and with the introduction of new crops in rural areas. Knowledge about the use of wild plants and crops – something that took many generations to build up – is quickly lost in a single generation when people stop gathering plants or when they start focusing on more productive or industrial crops, as is the case of some areas of the Maghreb with the present extension of hemp (*Cannabis sativa*). Traditional knowledge is lost rapidly by young people and turning back to ways of the past is almost impossible. Archaeobotanical methods, with the help of ethno-ecological models, have been able to show that the transition from foraging to farming in southwestern Asia resulted in a first collapse of dietary diversity by *ca.* 8000 BP when agrarian systems were fully developed.

This process is not only attested in the Near East. According to archaeobotanical information, cereals – and domestic animals as well – soon became staples in the human diet (*e.g.* Bogaard and Jones 2007 and Zapata *et al.* 2004 for western Europe and many other regions in the volume by Colledge and Conolly 2007). Notwithstanding this general idea, we must not forget that there are good arguments for wild plant foods having played an important role in some areas at least during the Neolithic – see for example the wonderful examples of Arbon Bleiche (Jacomet *et al.* 2004) and in general in the Alpine region where plant remains are preserved by waterlogging (Jacomet 2009), or wild plant seed storages at Çatal-Höyük (Fairbairn *et al.* 2007).

Among the huge diversity of plants that humans were using in Eurasia when they decided to cultivate, why did they focus on cereals? The narrowing in diversity in favour of cereals is most probably determined by different factors – that of course are also present on other continents where they became predominant plant foods. They are annually cultivated in dense fields, easily storable, and may produce important grain crops rich in carbohydrates. Carbohydrates are very important, as they provide a significant proportion of energy in most human diets, fundamental to keeping the body alive (Hardy 2007). It is very likely that cereals became rapidly accepted foods in places where the diversity of plants rich in carbohydrates was not that great, as in western and northern Europe. Of course, there were plant foods such as acorns, roots and tubers, which are rich in starch, but cereals probably lent diversity to this array as well as adding the possibility of easily increasing production by growing plants in secondary habitats and new plots. Cereals also produce very important by-products such as chaff and straw that can be used for animal food, fuel, building purposes or crafts. Also, they do not require heavy investment in processing and detoxifying – in order to get rid of substances such as the tannins we find in important wild foods such as acorns.

Hence, archaeobotany and ethno-ecological modelling show that the very fact of plant cultivation resulted in the first narrowing in plant-use diversity. This process is starting to be understood for the Near East (Hillman 2000) but can only be suggested for Europe, due to the lack of archaeobotanical research from Palaeolithic and Mesolithic contexts (Mason *et al.* 1994).

Once agriculture expanded from the Near East, and thanks to archaeobotanical methods, we can see different situations regarding crop diversity, as we can see in several contributions in this volume. There are regions or sites where agriculture focuses on hulled wheats, others where free-threshing cereals predominate, sites where they are absent, regions where legumes are not present and others where they are very abundant and varied. If we consider taphonomical[3] problems that affect plant preservation and the low number of sites that have been properly sampled in some regions, these situations may be explained. Legumes are traditionally poorly represented in archaeobotanical records for reasons which may run from their real lack of importance in human diet during prehistory to taphonomical problems related to processing not done close to any fire. Other situations, like the predominance of free-threshing or hulled wheats in some regions and sites, may reflect real regional or site patterns related to different reasons. Ecology, risk management, culture and function can be suggested as the most obvious ones.

Starting with the ecological factors that may be involved in plant diversity, crops have different growing requirements and tolerances that may limit or support their thriving in specific regions. This is a key issue in marginal environments, under wet or cold conditions, very different from the ones where they originated. A good case is the growing of barley in Iceland soon after its settlement in the ninth century. The cultivation of domestic crops ceases in the fourteenth century apparently due to climatic deterioration and the inhabitants of the island start to rely on wild plant foods (Guđmundsson 1996). In an environment where the cultivation of cereals is barely feasible, climatic change may completely impede its viability. It may be an extreme case of virtually no human choice that has been traced back thanks to a combination of written records, ethnobotany and archaeobotany.

Risk management may be another factor to explain the diversity of crops we can see in the Neolithic of places like the Iberian Peninsula thanks to archaeobotanical sources. Due to their different requirements, the sowing of mixed crops of cereals or legumes in the same or in different fields is a common practice in order to reduce the risks of crop failure (Jones and Halstead 1995). Experiments or trials carried out by the first farmers have also been suggested in order to explain a high diversity of crops during this period. Thus, the first people who took agriculture into different regions would experiment with all available crops and assess the results and farming practices in a new ecological setting. For Iberia, and thanks to archaeobotanical methods, the proposal has been put forward that after a first trial phase, the cereals which responded best were more widely grown while others disappeared – such as naked barley – or took on a secondary role – as did hulled wheats (Buxó *et al.* 1997, 22).

Since different crops have different properties – whether the flour is good for bread or if the straw is good for thatching – crop use is another important factor in selecting which crops are going to be grown, something that can be traced back and assessed thanks to a great extent to ethnobotanical examples (see G. S. Cruz-García in this chapter).

Diversity and crop choices on the other hand may also be guided by cultural decisions. The consumption of specific crops may derive from their being socially valued foods, even if yields are not high or if they are very labour-intensive – as is the case of emmer in Spain (Peña-Chocarro 1999) and Ethiopia (d'Andrea 2003) – or for symbolic reasons (Hayden 1996; 2003). When it is available, iconography (see S. González Reyero, Chapter 2.4) is an excellent source of information to throw light on certain aspects of ideology and the cultural meaning of specific plants.

Diversity is therefore an important issue when dealing with the history of human plant use, crops and people's choices. In this chapter, we are going to summarise some of the main sources that help us trace the history of crop and plant use, resorting sometimes to case studies that may be particularly significant. For prehistory and for regions of the world where written and iconographical records are missing, **archaeology** is indeed the only source of information for the relationship between crops and people. For some other times and places, archaeology can be combined with other methods of information, including ethnobotany and the study of the present use of plants. From the beginning of archaeological practice in the nineteenth century, archaeological plant remains have been recognised and identified. However, it was from the second half of the twentieth century on – in part due to the contribution of processualist and science-based archaeology – that many techniques have been developed which enable us to recover and identify different parts of plant remains, not only the most common ones such as seeds, wood charcoal and pollen, but also others such as diatoms or phytoliths, the study of which is quite recent and very promising. L. S. Cummings discusses in Chapter 2.2 these techniques and the possibilities and limits when using archaeology as a tool to approach this problem. As we can see in Chapter 2.3 by J. L. Mingote Calderón, M. Russel and the late F. Sigaut, **written sources** are another way of approaching this issue with the question of how diversity was dealt with in the past. Historians working on agrarian history, particularly in Europe, have been tackling this issue using different sources that give us information for the last two thousand years. Classical authors and their agronomic and encyclopaedic writings have been examined and are still the object of much reinterpreting and contrasting with other sources such as the archaeological ones. Authors like Hippocrates for the classical period, Theophrastus and Caton for the Hellenistic one and Pliny, Varro and Columella for the Imperial period give us good insights on what agriculture and plant use were like in their periods or on the ideals of how they should be used. Written sources for Andalusian agronomy are particularly interesting as a source of information for the role they play in transmitting classical culture and the Byzantine oriental tradition. During **the medieval and modern periods**, historians have also used a wide array of written documents from those institutions where they have been preserved: monasteries, municipal archives and notary registers, among others.

S. González-Reyero in Chapter 2.4 deals with **iconographic sources** and the way people have represented nature in the past. She points out that images do not reflect reality but materialise certain aspects of ideology and the society of the people who produced them. In order to explore these ideas further in another chapter of this book, she uses as a case study the analysis of plant images in the social strategies and the construction of the political territory of the Iberian Iron Age culture. G. S. Cruz-García in Chapter 2.5 approaches diversity in the present by explaining the methods and aims of **ethnobotanical studies**, which refer not only to the use of plants but also to their socio-cultural and economic context as well as people's perception and values.

We must always keep in mind that for different reasons all sources we may use to explore diversity of crops and crop choices are necessarily incomplete and that the contexts in which they were created must be taken into account. For example, archaeology routinely faces the major problem of plant preservation, while written sources may not be objective and can be biased by the author's own perception and aims when writing. Iconographic sources carry a cultural interpretation of the environment and materialise specific aspects of the ideology and codes of the people who produced the images. Ethnobotany studies are very often restricted to present-day people and may have problems observing long-term trends or changes in plant-people interaction through time. This is why the study of plant diversity, particularly in the past, cannot be restricted to one single source of information. However obvious this may be, the overspecialisation that researchers tend to have nowadays pushes us very often to work and go into depth using a single source of information and interdisciplinarity is rarely achieved. There are periods and places such as Medieval or Modern Europe where the use of different sources has turned out to be particularly appropriate, but there is no tradition of archaeologists, historians and ethnographers collaborating, discussing or contrasting results and ideas. This volume tries to overcome this research deficiency by bringing together people with different disciplinary backgrounds and skills.

2.2. EXPLORING DIVERSITY THROUGH ARCHAEOBOTANY

Linda Scott Cummings

Archaeobotany and Past Crop Diversity

Exploring diversity in the past through examination of the archaeobotanical record is imbued with advantages and disadvantages. One of the most striking advantages is the fact that archaeobotany gives direct evidence on past crops. Some of the research methods applied use the most advanced tools in biotechnology and chemistry, such as the analysis of biochemical markers from organic residues, DNA extraction and analysis of crop remains, and many more. Disadvantages lie within the variable quality of archaeobotanical evidence, which in many cases is rather reduced due to varying preservability of certain crop plants in the archaeological deposits. However, the methodological advances in archaeobotany continuously expand the capacity of acquiring information on crops used by people in the past. Early investigations began with the identification of visible pieces of botanical remains from archaeological sites and first have demonstrated that plants – and especially cultivated crops – always played a significant role in people's lives throughout prehistory. From the very beginnings of archaeobotanical research in the nineteenth century, scholars have been describing the diversity and variability of the discovered remains and trying to compare them with contemporary plants (Heer 1865). These attempts are still going on today, but in addition to the classical methods of morphological description and measurements also utilise the above-mentioned biomolecular approaches (Schlumbaum *et al.* 2008).

Preservation of Plant Remains in Archaeological Deposits (see also Chapter 7.1)

Various plant remains get incorporated in the archaeological layers during the existence of a site. Plants which were used by a settlement's inhabitants and brought to the site have a higher chance to become part of the archaeological deposits, so that the archaeobotanical finds represent the human choices and purposes. Furthermore, it depends on the deposition conditions of the site which types of organic matter will preserve through the centuries and eventually will be recovered during archaeobotanical research. Even seeds, belonging to the most durable plant parts, hardly live longer than a century in the soil, and most live for a much shorter period of time (Harrington 1972; Justice and Bass 1978; Minnis 1981, 147; Quick 1961). Once seeds have died, decomposing organisms act to break them down. But in cases the sediments are waterlogged, there is a much higher chance that the plant remains get preserved (Fig. 2.1) as the microorganisms are hindered from growing. This means excellent preservation of the organic matter and provides a great diversity of archaeobotanical information from such environments. Another kind of preservation which offers rich and diverse archaeobotanical information is desiccation, which occurs in very dry environments such as deserts. In all other kinds of environments where the organic matter is exposed to changing wet and dry spells, the plant materials most commonly decay unless if before their deposition they became charred. Fire sufficiently hot to char seeds and other botanical remains, but without or low oxygen access, is required to preserve remains in charred state.

Fig. 2.1. Charred (A) and waterlogged (B–F) macrofloral remains from two sites in central Europe (Images: A. G. Heiss). A) charred grain of emmer wheat (*Triticum dicoccum* Schübl.) from a Late Iron Age settlement in Lower Austria. Three views (from left to right): ventral, lateral, dorsal (from: Kohler-Schneider and Heiss 2010). Waterlogged remains from a Late Bronze Age well in Styria, Austria: B) cupule of common agrimony (*Agrimonia eupatoria* L.), C) mericarp of wild carrot (*Daucus carota* L.), D) seed of St. John's wort (*Hypericum perforatum* L.), E) nutlet of bugleweed (*Lycopus europaeus* L.), F) nutlet of hairy buttercup (*Ranunculus sardous* Crantz) (from: Heiss and Drescher-Schneider 2012).

The study of plant macrofossils is still one of the main sources of information on past crop diversity, and has provided extremely important information for understanding the history and prehistory of agriculture across time and different regions (Zohary *et al.* 2012).

Main Archaeobotanical Approaches and their Contribution to the Knowledge on Past Crop Diversity

Depending of the preservation conditions at an archaeological site different kind of plant remains are preserved, and serve as sources of archaeobotanical interpretations and models. According to their particle sizes, these remains are usually divided into two main groups: 1) plant macrofossils (seeds/fruits, wood, leaf parts, etc. – remains larger than 0.1 mm, and visible with the naked eye); and 2) plant microfossils (pollen, phytoliths and starch granules – remains which usually only rarely exceed particle sizes of 0.1 mm). This relates also to the methodologies for extraction and study of these plant remains, and for their interpretation. In the following, the main approaches for the analysis of these different groups of plant remains will be described in the light of their potential to explore the past crop diversity.

Plant Macrofossil Analysis

Analysis of plant macroremains (also called macrobotanical, or macrofloral analysis) focuses on plant remains such as seeds/fruits, wood and charcoal remains observable under low magnification, and has been providing extremely important information for understanding crop diversity for more than a century. Recovery of charred, desiccated, or waterlogged remains from soil sediments allows species determination and sometimes variety-level identification for many crops, but often is hindered by matters of preservation specific to the respective plant organs and tissues.

Pollen Analysis

Pollen (Fig. 2.2) is the male generation of seed plants and together with spores produced by ferns and mosses are the main subject of pollen analysis (palynology). Good preservation of pollen is expected in conditions that include desiccation and waterlogging. Changing wet and dry conditions together with good oxygen supply (as found in most dry soils) oxidises and destroys pollen. Unlike macrofossil remains, pollen is completely destroyed by fire.

Cereal pollen is usually distinct from most native grasses, but is nearly impossible to separate into

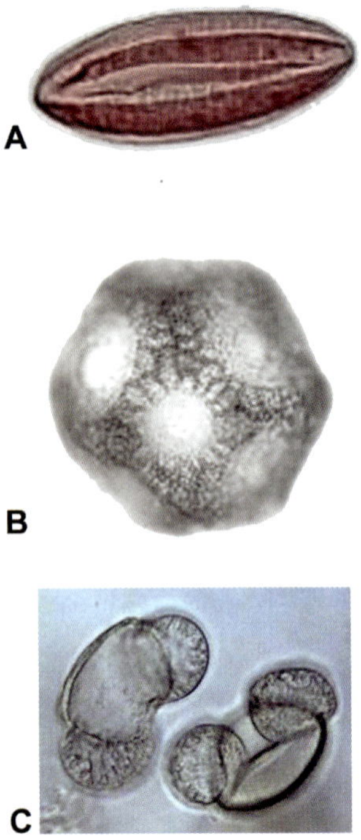

Fig. 2.2. Three examples of pollen structures. A) joint-fir (*Ephedra*), B) prickly pear (*Opuntia*), C) pine (*Pinus*). Images: L. S. Cummings.

genera and species of certain cultigens. Therefore, for some crops, pollen analysis offers little other than establishing presence of crops of a particular category or genus. For others, pollen may be an excellent way of tracking the introduction and presence of crops. *Ipomoea batatas* (sweet potato) pollen, for example, has distinctive bacula (column-like structures) between the pores and is similar only to a few weedy *Ipomoea* species. If those weedy species are not native (nor introduced) to the area of study, then recovery of even 10% of an individual *Ipomoea batatas* pollen grain is sufficient to establish presence of this crop.

One advantage of examining pollen from sediment profiles immediately at or near the archaeological site is that the pollen record contains evidence of plants growing at the site, and available to the former occupants. This is often not true for pollen records from bogs and lakes, which may be located many kilometres from the archaeological sites, but provide information on the general trends of the vegetation and environmental change in the study area. Pollen analysis of cultural layers (Cummings 1998) and features contributes valuable information concerning selection and use of plants by occupants of the site. Sampling for pollen analysis of such structures also offers the opportunity to recover starch granules which are valuable for the interpretation of the subsistence record. However, profound knowledge of the mechanisms of pollen deposition, relocation and preservation is vital for the interpretation of pollen assemblages from anthropogenic deposits.

Starch Analysis

Starch granules are basically intracellular structures of a plant for storing energy. The most common places for starch to occur are in seeds and in roots/tubers. As a general rule, starches produced in seeds tend to have centric hila,[4] while those produced in roots/tubers have eccentric (off-centre) hila. It is most important to note that starches with eccentric hila can present as having centric hila in some rotations. Therefore, examining the starches in a fluid mounting medium in which they may be rolled is extremely important. A review of starches and methods of their extraction may be found in *Ancient Starch Research* edited by Torrence and Barton (2006), numerous articles such as Perry *et al.* (2007) and at http://www.paleoresearch.com.

Starch analysis has developed into a tool for identifying crops in the prehistoric record rather recently. Many of the crops produce distinctive starches. For instance, among cereal grains, barley (*Hordeum*), wheat (*Triticum*), and rye (*Secale*) all produce starches of similar shapes and sizes and with similar extinction crosses[5] (Fig. 2.3). However, some of the barley starches exhibit concentric rings that distinguish them from starches produced by wheat and the other cultigens.

Even though many starches survive pollen extraction methods, minimal processing of samples is suggested for the recovery of starches, particularly when examining grinding stones. For a simple wash of the surface with pure reverse osmosis de-ionised (RODI) water is recommended. A toothbrush using high-frequency sound waves is most efficient in removing remains from grinding surfaces. Certainly, removal of any calcium carbonates on the surface of the grinding stone precedes washing the grinding surface, as they have served to 'seal in' the archaeobotanical evidence of grinding.

Phytolith Analysis

Phytoliths (Fig. 2.4; also see below) have become an important resource for examining the presence of crops in both the Old and New Worlds. Morphometric studies of phytoliths produced in the glumes

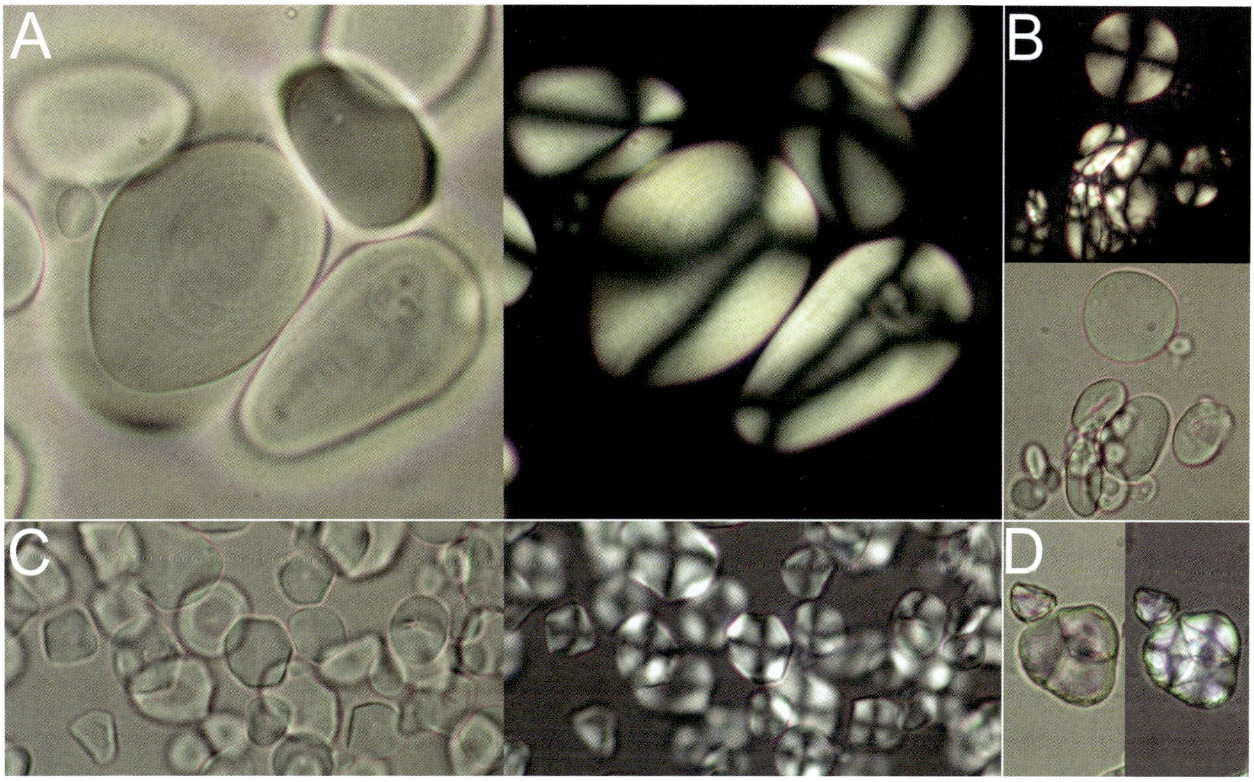

Fig. 2.3. Starch grains of various plant species, each specimen in bright field illumination, and in polarised light: A) potato (*Solanum tuberosum* L.) stem tuber; B) bread wheat (*Triticum aestivum* L.); C) maize (*Zea mays* L.) grain; D) biscuit root (*Lomatium* sp.) root tuber. Images: C. Yost.

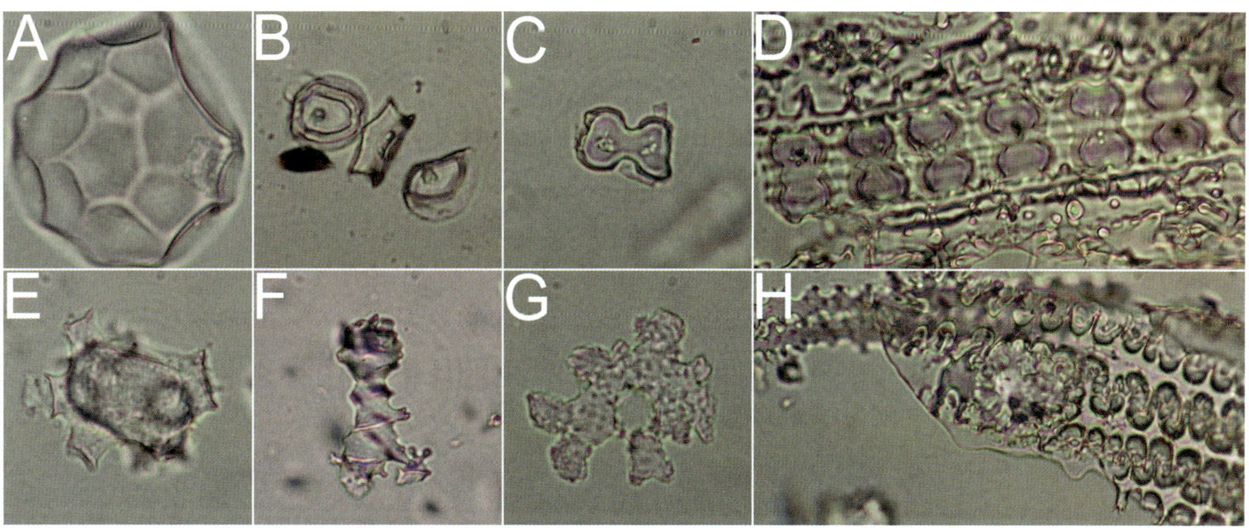

Fig. 2.4. Phytoliths of various plant species: A) buffalo gourd (*Cucurbita foetidissima* Kunth) fruit rind; B) rondels from maize (*Zea mays* L.) glumes; C) bilobate phytolith from jungle rice (*Echinochloa colona* (L.) Link) leaf; D) bilobate phytoliths *in situ* in rice (*Oryza sativa* L.) leaf; E) phytolith from bulrush (*Scirpus* sp.) rhizome; F) phytolith from Ponderosa pine (*Pinus ponderosa* Douglas ex P. et C. Laws.) needle; G) phytolith from heliconia (*Heliconia stricta* Huber) seed coat; and H) dendriform phytoliths from bread wheat (*Triticum aestivum* L.) glume. All phytoliths are part of the phytolith reference collection at PaleoResearch Institute, Golden, Colorado. Images: C. Yost.

of cereals have resulted in establishment of the expectation that for pure populations it is possible to distinguish between the major cereals (Ball *et al.* 1996).

Phytoliths contribute to both the economic and environmental records at archaeological sites. The role of phytolith analysis at a site is dependent, at least in part, on the local vegetation and plants that might have been exploited economically. Plants growing in the tropics appear to need more protection or armour against predation. For at least some plants, this resulted in the accumulation of silica within certain parts of the plants. Accumulation of opal silica inside cells yields a cast of the interior of the cells, or, in some cases, the space between cells. The shapes of phytoliths are particularly important in identifying the plants that they represent. Recent examination for both pollen and phytoliths of agricultural field terraces from the Mayan Mountains in Belize resulted in the conclusion that many of the fields had been burned, suggesting the practice of swidden agriculture. Pollen analysis recovered large quantities of microscopic pieces of charcoal and very little pollen. One midden yielded sufficient pollen to state that it was not simply poor preservation conditions in the sediments that resulted in the absence of pollen. Phytolith recovery, however, was excellent. Phytoliths identified probable crops including *Maranta* (in the arrowroot family) roots (Cummings and Yost 2008).

In Central and South America, Piperno (2006) and Pearsall (1989) have employed measurement of *Zea mays* leaf phytoliths to determine the presence of maize. Cummings *et al.* (2009a), Pearsall *et al.* (2003), Piperno (1993) and Thompson and Stallern (2001), have employed morphometry or typology to characterise *Zea mays* cupule phytoliths to identify the presence of maize within an archaeological record. Cummings (2007) and Cummings *et al.* (2009a) also used morphometry to characterise the race of maize present and begin a search of the distribution of individual races of maize in North America.

Organic Residues

Chemical residues are becoming more common for identifying the presence of crops. Ceramics absorb residues of the foods cooked in them that may be released through use of a chemical (often a combination of chloroform and methanol) extracting solution. The resulting residue may be identified through use of various techniques such as high-pressure liquid chromatography (HPLC), Fourier Transform Infrared Spectroscopy (FTIR) (Isaksson 1999), gas chromatography mass spectrometry (GCMS), and sometimes paper chromatography. Chemical analyses are relative newcomers to the field of identifying the presence of food residues, having been employed only for the past few decades. Cummings (2007) has shown that agave plants have unique signatures that allow tracing use of this desert resource in archaeological sites. Agave is a relative of the aloe, which is known to have unique medicinal properties, both when examined through historic literature (*e.g.* Moldenke and Moldenke 1952) and also through modern chemical analysis (Murray *et al.* 2000; Stephen 2006). It is likely that the agave also will prove to be a resource that provided important nutritional components to the diet. Organic residues have been shown to mirror other lines of evidence indicating the presence of maize as well (Cummings *et al.* 2009b; Cummings and Yost 2008).

Sampling

One of the crucial prerequisites in order to guarantee reliable and representative results of archaeobotanical analysis is the provenience of the samples. During the early, exploratory phases of archaeobotany, samples were collected and examined from areas of a site with obvious concentrations of larger plant remains, as observed during screening by the archaeologists. Archaeobotanical practice has however shown that samples should be collected from many areas and features of a site.

Often the full story is available only through analysis of multiple data sets. A brief discussion of recovery of evidence of foods eaten by examining coprolites from the American southwest (L. S. Cummings, Chapter 3 in this volume) shows evidence of the recovery of pollen, phytoliths, and macrofloral remains of individual cultigens and native plants. No single data set succeeded in identifying all of the types of plants consumed. Nor did any single data set excel in identifying the most evidence of consumption of plants. Examination of multiple data sets is necessary to understand any economic or subsistence record because different foods will leave different traces. Roots, for instance, will be more regularly represented by starches unless they are collected or harvested while the plant is in flower, in which case pollen might be the most abundant evidence. Seeds are usually best represented in the macrofloral record, unless they are not subject to processing such as parching or cooking that results in seed loss and accidental charring in hearths. Some seeds ripen while the plant is still partly in flower, meaning that pollen is introduced with the seeds that are collected and thus transported into any processing or storage setting. Greens are particularly difficult to detect in the archaeobotanical record because most greens are collected before the plant flowers, so pollen is often not introduced with the greens. Greens do not include seeds, so no seeds are expected. Most greens do not produce starches, so no starch record is expected. Finally, even when greens produce calcium oxalates they do not survive in sediments, and so are rarely, if ever, recovered during phytolith analysis. It is primarily when greens are collected late in the season and the plant has begun to pollinate that evidence of the use of greens is recovered in the archaeobotanical record. Preservation conditions usually are not sufficient for recovery of plant cuticle,[6] meaning that even this analysis will not identify the use of greens. One exception is the presence of evidence for consumption of greens in coprolites. Calcium oxalates survive well in coprolites and plant cuticles from some plants also survive.

Using multiple data sets to examine subsistence records gains importance with the increasing integration of the different archaeobotanical and archaeometric approaches. The archaeobotanical analyses that we have been relying upon for decades will continue to be important in understanding human selection, utilisation and processing of foods. However, together with new detection methods, our understanding of food use and processing technology will expand.

Hearths and Living Surfaces

Extracting plant macrofossils by means of flotation methods have led to a large and ever-increasing macrobotanical data set from many areas within archaeological sites. In general, if the site is in the open, areas that were burned, whether they represent hearths, roasting pits or structures that burned accidentally (catastrophically) or intentionally (upon abandonment or perhaps as an act of war or aggression) are the best locations to sample for macrofloral remains. Because direct flame destroys pollen, some of these same areas are not the best locations to collect pollen samples, although pollen samples collected from areas that have been burned still may be valuable. For instance, sampling near the rim of hearths, which is away from direct flames and also experiences less heat, may provide excellent information concerning foods that were processed (and dropped) in the hearth. The floor, when it is visible, within approximately 0.5 metre of the edge of the hearth is another excellent area to sample, since this area would have been used when preparing foods for cooking. Medicinal plants and/or foods might have been piled on floors or contained within vessels within approximately one metre of hearths. A particularly good example of this is methodical sampling of a floor of a catastrophically burned pit house in the American southwest that yielded a large quantity of *Ephedra* shrub pollen on the floor approximately 1 m west of the hearth (Cummings 1998; Scott 1983). Since broken pottery was recovered from this area, it is probable that *Ephedra* stems and flowers were stored in the vessel awaiting further processing, probably to make tea, at the hearth. Methodical sampling of living surfaces also may provide evidence of food processing areas, such as grinding grains, storing foods in ceramic vessels that either spilled or were broken and even hanging plants from the ceiling for storage (Cummings 1998; Scott 1983). Systematic sampling of the floor in squares thus provided much more complex information on plant use than simple sampling across the living surface.

Food Consumption – Coprolites

Evidence for food consumption may be derived with most confidence through examination of coprolites or palaeofaeces, since they represent digested portions of foods expelled from the body. Although intuition has lead several scholars to postulate that people do not eat silica-rich diets or foods heavy in silica, anyone who has examined cereal grains knows that glumes or 'bran' from cereals, which provide fibre in the diet, are silica-rich. Therefore, phytoliths are an important element of the diet, if for no other reason that they are resident in many of the fibres that people consume. An example is provided by L. S. Cummings in the discussion of palaeofaeces (coprolites) from Step House in Mesa Verde National Park (Chapter 3, this volume).

Vessels

Ceramics or other containers that held food are excellent for sampling in order to recover evidence of food preparation or storage. For some analyses, isolation of the samples from food containers means that the pollen, phytoliths and starch recovered most likely represent foods and include further evidence on the past crops. In this case, identification may be more specific, particularly in the case of phytoliths. Morphometry of phytoliths from maize cobs or glumes, for instance, may identify not only the fact that maize was present, but also the cultivar of maize. Use of a vessel to prepare maize beer, for instance, may be established through phytolith analysis. FTIR analysis of organic residues contained in the ceramics also would be very helpful in identifying brewing beer as it detects alcohol, among other compounds. In the case of examining records from ceramic vessels, little of value is readily visible or apparent. Valuable data comes from examining the 'charred' residue adhering to ceramics and from recovering the organic residues that had soaked into the fabric of the ceramic shard during use (Evershed 2008).

Grinding Stones

The study of grinding stones often has focused primarily on traces on the surface of the stones to determine what was ground. More direct evidence of grinding may be obtained from examining the surfaces of grinding stones using pollen, phytolith and starch analysis, since these microscopic remains are present as a result of the grains being ground. Traces left on the stone through the grinding process are secondary indicators of the grains ground. In the event that grinding stones were used to grind more than one type of grain, pollen, phytolith and starch analysis (combined) would be the most likely way of determining this use.

Conclusions

Understanding use and processing of foods, as well as domestication and selection of wild foods has rested on the discoveries of existing archaeobotanical and palaeoethnobotanical studies. These studies will continue to contribute to our understanding of past peoples. New technologies will increase our understanding of people and their relationship to their environment, including their manipulation of that environment through the domestication of plants and control of valuable resources. As people today seek to better understand change in their own environment, they will probably be more successful in identifying change and stability in the past, which will result in increasing technology for examining the past. Use of the commonly recognised areas of study including pollen analysis (palynology), phytolith analysis, macrofloral analysis and the newer analyses such as starch analysis, organic residue analysis and protein residue analysis, has contributed to our current understanding of how people domesticated, processed and utilised many plants.

2.3. EXPLORING DIVERSITY THROUGH WRITTEN SOURCES

José Luis Mingote Calderón, Marie Russel and François Sigaut (†)

Problems Involved in Working with Historical Sources

Written sources include a huge array of documents that can be used to approach subjects such as the ones we deal with in this volume: crops, diversity, agricultural choices and plant use in the past. In any of the periods under consideration, the sources are of two types and the approaches to be taken to them differ: the texts actually concerning plants and the references to them in otherwise scattered sources. Historical sources can also be primary, in the case of documents which give us first-hand information on a subject, and secondary, when such documents were not written contemporaneously to the subjects they deal with. A primary source is not necessarily more reliable than a secondary one, since every type of source poses its own problems and implies a certain amount of subjectivity.

In many areas of research, the main difficulty is that written sources are few and far between (although, as we shall see, the exact opposite can be the case!). As far as plants are concerned, one complexity may lie in the fact that their number is overwhelming. Forty years ago, Uphof's *Dictionary of Economic Plants* (1968) provided a bibliography of about 1400 references. A few years later, S. Rehm and G. Espig listed exactly 1577 references in *Die Kulturpflanzen der Tropen and Subtropen* (1976). However, the number of references given in a book is necessarily limited by factors such as the point of view of the author, the languages he or she can read, the time he or she could spare for it, the constraints of the publisher and so on. The real number may be much larger, although nobody really knows. On useful plants in modern times since, say, the seventeenth century, the number of written sources must be on the order of several tens (if not hundreds) of thousands, in a score of different languages. No one, even after having spent a whole life devoted to the matter, which would be impossible anyway, can be expected to have more than a very partial view of the whole.

One cause of this state of affairs is that the number of 'useful' species is itself enormous. In *Les fondements biologiques de la géographie humaine*, published in 1943, Max Sorre quotes figures on the order of 2,700 to 2,900 edible species. The number of species mentioned in *Die Kulturpflanzen* is still in the order of 2,500. The total number of 'economic' plants that have an entry in Uphof's *Dictionary* is about 10,000. The most recent evaluation we were provided with (Chauvet 2009) is between 7,000 and 8,000 cultivated plants and between 20,000 and 30,000 useful plants. But this is only a 'reasonable' estimate. Depending on the number of species each author identifies within the same genus, and on what is thought to make a species 'useful', real estimates may be much higher; there is one approaching 90,000 species.

Going back in time, in a European context, the earliest texts we have at our disposal come from Greco-Roman Antiquity, with texts from the Archaic Greek to the Roman Imperial periods (see a recent compilation in Segura and Torres 2009). The first attestations to plants in a European source are to be found in the *Iliad* and the *Odyssey*. It is worthwhile to make a survey of the occurrences of such information, on the one hand, in order to analyse the continuity of sources and, on the other, to have particular points of comparison.

For scattered sources in Antiquity, we need to examine all literary genres involved, including tales of war, medical or culinary writings, theatre, philosophical or legal texts, and refer to dictionaries or concordances to find the first occurrences of the terms. Moreover, the entire period of Antiquity provides us with texts that are fundamental for the study of plants, such as agronomic or encyclopaedic writings that must be examined exhaustively: Hippocrates for the Classical period, Theophrastus and Cato for the Hellenistic, Pliny and Columella for the Imperial period.

The Example of Roman Agronomic Treatises

The search for information in agronomic treatises and some sources related to them in the Classical world obviously has its limits, although we can often find important details within them. R. Martin's synthesis (1971) generally highlighted the intentions and contexts underlying these various authors' work. Hence we must emphasise the fundamental premise that the majority of these writings were meant to speak of rural economy within the scope of how the overall economy functioned, not only in regard to its agricultural aspects. This involved a series of ideological aspects that we must always be aware of in order not to fall into the error of an overly 'modern' reading of these texts, one example among others concerning production associated with divine will (Mingote Calderón 2007). The clichéd repetition of topics or rhetoric exercises is another aspect usually to be taken into account in such texts.

The result of this is that along with the temporal and local peculiarities to be found in the vision of each author – as R. Martin stressed – such texts are laced with information that is repeated almost automatically. Likewise, the frequent omission of citing the author a writer may be referring to lends a certain tint of personal experience – even if this is an unconscious impression – and gives the impression that only one person is conveying the information instead of several, as we can see clearly in the later *Geoponica* (Meana *et al.* 1998).

It is sufficient to consult R. Martin's list (1971, 21–22) to have some idea of the number of authors who took up the topic of agrarian practices in the Roman world, which makes it difficult to provide even a minimal description of the contents of all their works. It is quite enough to mention the list he cites for the Republican period alone: Cato (1979), the Sasernae and Tremelius Scroffa (whose work has not come down to us, but is cited by other authors), as mentioned by Varro. For the Augustan period, we have Virgil (1986) and Sabinus of Tyre; for the Julio-Claudian period, Celsus, Julius Atticus, Julius Graecinus, Columella (1968; 1979a; 1979b; 1988) and Pliny the Elder (1949; 1958; 1964; 1972), as well as the later authors such as Julius Frontinus, Apuleius, Sextus Julius Africanus, Florentinus, Gordianus the Great, Quintilinus, Gargilius Martialis and Palladius. We should also not forget that some of them refer to numerous Greek authors – 40 or 50 are quoted, some of them very well known such as Hesiod or Xenophon – who preceded them or the Punic ones such as Mago, the first author to have written a synthesis of agricultural treatises.

In making such a survey, whatever the ancient source, we must always remember their incomplete character and take the context in which they were written into account. They have come down to us only through the whims of time and human beings, such texts carry the bias of their own time and probably are not representative of all that was written about plants. In addition, it can often be difficult to grasp to what extent the author carried out an objective examination of reality. When translators are sufficiently specialised in these matters – which is at times wishful thinking – they can contribute greatly to contextualising and understanding the texts, as well as to the botanical identification of the plants involved. In spite of these reservations, the number of attestations enables us to see the conditions in which plants were consumed and even the importance of some of them as foodstuffs.

Medieval and Modern Sources

For the medieval and modern periods, historians have used a wide array of written documents from those institutions through which they were preserved: monasteries, municipal archives and notary registries among others. Monastery archives, for example, provide insight into everyday aspects such as crops and plant foods and woodland exploitation, always to be understood within the broader context of information on the origins and meanings of this type of property (see a good example in García de Cortázar 1969, with information on crops, horticulture, plant foods and woodland use in the dominions of the monastery of San Millán de la Cogolla in northern Iberia during 930–1250 CE). We must take into account that the farming activities of monasteries tend to focus on certain crops, such as grapevines, which may be far from the realities of a family holding.

Cadastral surveys include written records with comprehensive registers of properties which may record the ownership, location and dimensions of land and cultivations. Fiscal records in general are very important since they enable us to study agricultural spaces, stockbreeding activities, demographic variables, territorial and legal organisation, transport, communications, industrial and commercial activities, land use in general and particularly of agricultural lands, the areas taken up by pasture, meadows, fields and forested lands, soil qualities, the location of plots and rotation systems (Erdozain 1993). Documents such as the *Domesday Book*, completed in 1086, are an essential source of information, recording as it does an extensive survey of much of England and parts of Wales, to find out what or how much each landholder had in terms of land and livestock, and exactly what it was worth.

Notarial sources are another extremely valuable source of information, since they give us a quick glimpse of everyday life including data on crops and their techniques, animal husbandry, land holdings and their transfer, transport tolls and so on. These sources may include deeds of sale or obligations relating to particular plots such as vineyards or olive groves and products bought or sold by deferred payment. Rental or leasing agreements, censuses, transactions on credit and wills enumerate the movable goods and furnishings left to heirs, as do inventories carried out in order to avoid litigation over inheritance. All these may include land, livestock, technical implements and tools, foods stocks, the domestic trousseau, dowry guarantees and matrimonial charters that set up family or personal pacts between the parents of a couple or between the two marriage partners (Acosta 1993; Pereira and Rodríguez 1982).

Among the diverse sources cited above, we can single out the death inventories that have been extensively utilised by historians in analysing what crops were being cultivated at a given time and their importance. These documents may cover quite long periods and show changes in patterns of consumption. Handling such sources ably implies taking into account the time of death of the person in the agricultural year and the impact this has on how to interpret the presence or absence of particular plants or animals that 'should' be there. As an example, the gradual substitution of one cereal crop for another or others has been analysed, as in the case of the introduction of maize in Galicia for human consumption, or of the varying importance of different cereal crops and their geographic distribution (Pérez García 1981; 1982).

The Case of the Andalusi Agronomic Treatises and their Implications

For the medieval and modern periods in a country such as Spain, Andalusian agronomic written sources are particularly interesting for the role they play in transmitting classical culture and most especially in the transmission of Byzantine oriental tradition more than the occidental one. The texts span a period going well beyond the Middle Ages so that they run from the tenth to the fifteenth century. Various Andalusian agronomical treatises were partially translated into Romance languages. Hence, they were written in Castilian from the Alfonsian period in the thirteenth century on, all of which were copied during the fourteenth and fifteenth centuries. Likewise, there are some translations into Catalan from the end of the fifteenth and beginning of the sixteenth century (Millás Vallicrosa 1943; 1948; Carabaza Bravo 1994; Ibn al-Wafid 1997), and there is the case of some translations of Ibn al-Wafid being used by an agronomist as well known as the Toledan Gabriel Alonso de Herrera (1470–1539, see Herrera 1981), who did not know Arabic and whose influence in Spain extended into the nineteenth century.

Fortunately, the vast majority of these medieval texts were published and translated into Castillian. Simply enumerating the authors provides a clear idea of their importance: from the tenth to the eleventh centuries there is the Kitab fi Tartib, an anonymous Andalusian agricultural treatise (López and López 1990). From the eleventh century, we have the works of the Toledan Ibn al-Wafid and those of the Sevillian authors Abu l-Jayr (1991) and Ibn-Hayyay (1988); Ibn Bassal (1955) wrote his treatise on the classification of soils from 1095 on and the work of the Granadan al-Tignari was written at the turn of the eleventh to the twelfth centuries (Malik al-Tignari 2007). The Sevillian Ibn al-Awwam (1988) was active at the turn of the twelfth to the thirteenth centuries and, finally, the Almeriensian Ibn Luyun composed his text between the thirteenth and fourteenth centuries (Eguaras 1976).

It is essential to mention the existence of calendars which, along with information about starlore or the solar rising and setting times so important to the practices linked to Muslim prayer, may also contain references to agricultural practices. Among all of these, the famous 961 CE Hispano-Mozarabic *Calendar of Cordoba* stands out (Dozy 1961). It was written by Bishop Rabi ben Zaid and dedicated to his Caliph. This is not the only example and another calendar, possibly of Cordoban origin, has been published and may be dated approximately to the thirteenth century (Navarro 1990).

As is true of agricultural treatises in other periods, the composition of the Andalusi texts is linked to the political power of the time. We know that some of the authors worked in the gardens or orchards of the magnates of the period. This was the case of Ibn Bassal for the Taifa king of Seville, al-Mutamid, or of Ibn al-Wafid, a pharmacologist and physician who was also the vizier of the king of Toledo, al-Mamum. The latter was known as Abenguefit and his *Kitab al-adwiya al-mufrada* was translated into Latin as *De medicamentis simplicibus,* a work that was printed more than fifty times in Latin under the title of *De Medicinis universalibus et particularibus.* These relations between authors and their patrons place such treatises within the framework of an elite production, which is generally the case for this kind of writing.

The interrelations among these various authors are complex and they highlight the existence of a thematic tradition that led each of the authors to read what was written previously (Bolens 1981). This often implicit familiarity with earlier and contemporary writings is what accounts for the many transtextual quotations and copies. This means that researchers investigating this type of treatise must devote a part of their attention to unravelling this textual tradition and raises the issue of how to localise and chronologically locate this information, something that must be analysed case by case. In spite of the fact that these treatises belong to a previous written tradition, there are many references to personal experimentation, so it is relatively easy to find references to trials carried out with a certain sort of experimental intent. Also, we find critiques of, or discrepancies with, opinions from earlier authors, who had experimented with various alternatives. We might add that irrigated crops occupy a fundamental place in these texts.

Translations of many of these treatises into Castillian, such as the works of Ibn al-Wafid or Ibn Bassal, were motivated by their prestige and 'authority'. The early sixteenth-century Christian author de Herrera lavished praise on his fellow native of Iberia Ibn Wafid, just as he did on the Hispano-Roman Columella. Here we see that devotion to this agronomic tradition repeatedly crosses over the borders of Islamic culture itself. Pietro de Crescenzi is another author who wrote an agricultural treatise based largely on classical and medieval sources, as well as on his own experience as a landowner. His work *Liber ruralium commodorum* (*The Book of Rural Benefits*) was completed some time between 1304 and 1309. It includes chapters on the botanical properties of plants and horticultural techniques, cereal agriculture, arboriculture, vines and winemaking, plants useful for food and medicine and a monthly calendar of tasks.

Using these Texts as Sources for (Historical) Inquiry

In addition to this interest in identifying the sources used by each author, often due to the philological preoccupations of the editors of the treatises, a specific interest has recently arisen in emphasising some particular subjects dealt with in these works. It will suffice for our purposes here to mention a few examples taken from the eight volumes of the *Ciencias de la Naturaleza en al-Andalus*, published by the School of Arabic Studies in Granada CSIC (for detailed analyses, see García Sánchez and Álvarez de Morales 1990–2004, then Guardiola 1990; 1992;

Álvarez 1994; Carabaza Bravo 1994). The list of agricultural tools seems to have been copied from one treatise to another so that this topic must be approached with considerable caution. There are also various entries concerning grafting, use of the harrow in Andalusia, irrigation, and numerous references to various types of food or medicinal crops, including citrus fruits, olives and hemp, as well as diverse animal products.

We must also note that, along with discussion of standard practice in cultivation, we also encounter references to practices that we today would not qualify as technical. Thus, C. Álvarez de Morales (1994) called attention to the fact that some magical procedures are to be found in these treatises. As was pointed out years ago, it is possible to find practices that fall between technical and 'magic', although we must resist any temptation to apply to the Middle Ages the values which these terms hold in our own culture. A good example might be the case of threatening a non-fruiting tree, which is widely attested both within and beyond the European sphere, even in recent ethnographic inquiries. This is a practice which perfectly illustrates how, in certain mental conceptions, ritualised beliefs and strictly technical behaviours are combined, as attested by Abu l-Jayr from his personal experience (Mingote Calderón 1995, 163–174).

Here, cautious comparison of ethnographic literature with archaeological remains and historical documents can be particularly fruitful. When analysing specific cases, such as the use of perforated ceramic vases after firing for agricultural purposes, we see the wealth of the information they attest to, related both to their various particular uses and to the typologies of objects. Furthermore, when this information is compared with that found in the Roman agronomists, we can observe the existence of continued attitudes which simultaneously highlights the importance of the discontinuities (Mingote Calderón 1993).

Modern Written Sources and their Complexity: the Example of a Dye Plant and Oils

In contrast to the dearth of information we may lament in the case of practices and plants in Antiquity and the Medieval period, the opposite is often the case of modern written records. Whatever the plant we are interested in, there may often be more written sources about it than we can imagine. A good example of this is the tinctorial plant called *maurelle* or *tournesol* (*Croton tinctorius* L., *Chrozophora tinctoria* A. Juss., nothing to do with the sunflower) in French and dyer's crotone in English. This was a wild plant, gathered by the inhabitants of Grand-Gallargues (halfway between Nîmes and Montpellier) within an area extending two to three hundred kilometres east and west of their village. Once brought home, the plants were crushed and pressed; cloth (locally termed *drapeaux*) was soaked in their juice which, after drying, was finally laid out on fresh manure heaps. When they had acquired the desired blue colour, the *drapeaux* were sent to the Netherlands where the blue dye was extracted and used, mainly in confectionery. Some authors assert that the business was already attested in the fifteenth century. It went on without important changes (or so it seems) until the 1830s, when, because the wild grounds where the *maurelle* grew naturally became scarce, some plant growers began to put it into cultivation – which was of little use anyway since it was soon to be made obsolete by the development of synthetic dyes.

This is only a summary of a short note written recently (Sigaut 2007), after the author found a few nineteenth century written sources on the subject entirely by chance. As is so often the case, if a complete inquiry could be undertaken on this rather unimportant plant, the number of written sources would be found to be in the order of several hundred, in at least three languages (Latin, French, Dutch). This is another way, perhaps more concrete than databanks, to pose the problem of written sources.

We have just said that the *maurelle* was an unimportant plant. From an economic point of view this is undeniable. There were never more than one or two thousand inhabitants in Grand-Gallargues, they did not all make their living from the plant, and even if we add the merchants who carried the *drapeaux* to the Netherlands and the workers who extracted and used the dye there, the grand total of people concerned must never have exceeded, say, four to five thousand. On a European scale this is negligible, but it is not without interest. Here we have a plant that was always gathered wild for centuries, although it was the object of a regular commerce and of a sophisticated industry. The *maurelle* does not fit into our nice but artificial categories. As we see it, the true question is: how many such 'unimportant' plants were there which are practically never heard of nowadays? And if they look unimportant taken one by one, are they still unimportant taken all together?

If we turn to oil-plants, the question is about the same, only on a still larger scale, since most plants produce seeds and most seeds contain a certain percentage of oil, so that most plants can be oil-producing, at least potentially. In fact, what makes a plant oil-producing or not is, firstly, the ease or difficulty with which its oil can be extracted, which is a matter of implements and techniques and, secondly, the uses to which this oil can be put. Both are rather complex matters, on which there is an extensive literature. But what we know of this literature has been found by chance (exactly as in the *maurelle* case), not by consulting bibliographies or databanks. A book such as the one by J. Fritsch (1905), although by all accounts totally obsolete, is the best introduction into the matter we have been able to find. It deals mainly with the techniques for extracting and refining oils, but it also records more than 120 species of oil-producing plants of economic importance all over the world. This is only one example among many and books like this one also exist in most European languages (some of them being quoted by Fritsch himself).

The Fritsch handbook was written at a time when trade in such products was flourishing on a world-wide scale, with the consequence that the local production of vegetable oils in 'developed' countries was at a minimum. But there were times when the situation was reversed. During World War II, for instance, France was cut off from foreign countries, so that the government made efforts to revive the production of oil from old-fashioned sources such as beechnuts, horse-chestnuts, pumpkin seeds, grape seeds, etc. In fact, oil had been extracted from some of these plants for ages but only in particular areas and it took the constraints of scarcity to make them an object of serious attention again. A booklet like *Les plantes à l'huile*, by M. Jouven (1942) is a pretty good example of this 'scarcity literature'. Similarly, a number of pamphlets about enhancing the production of oil from indigenous plants were published during the French Revolution in the 1790s, when political events caused exceptionally long shortages. Since all European countries have had their own times of scarcity, it is to be guessed that they have also had the attendant literature.

In the last decades, most written sources have come from ethnologists or museum keepers. One of the best examples we know of is *Tradycyjne olejarstwo w Polsce* by Henryk Olszański (1989). It is in this book that we can find a very important detail, namely that the seeds of weeds, resulting from the winnowing and sieving of the main cereals, were usually carried to the oil mill. It would be quite interesting to check whether this practice was in use in countries other than Poland.

Conclusion: a Plea for Serendipity

Many other observations could be made. As a conclusion, a final remark we want to stress is that as far as the uses of plants are concerned, the main problem is that the number of written sources is huge. To cope with the problem one immediately thinks of databanks. But even databanks might not be a miracle solution. More than thirty of them have been recorded by Chauvet (2009), some of general use, some restricted to one continent, one country or one category of plants, etc. Sifting through them could quickly become a full-time assignment, so this casts some doubts on the usefulness of extant bibliographies and databanks. For the time being at least, it may still be more expedient to rely on flair and chance to find one's way into the mass of relevant documents. We must also stress the geographical and historical particularities of written sources. A Roman agricultural treatise, the medieval census of the *Domesday Book* or a French scientific dictionary of the eighteenth century are documents of quite divergent natures. They must each be read and understood in a different way, even if in some cases – agricultural treatises for example – they may have an internal continuity, since it is normal that they quote each other continuously, even if explicit references to authors are omitted.

2.4. REPRESENTING NATURE: IMAGES AND SOCIAL DYNAMICS IN ANCIENT SOCIETIES

Susana González Reyero

Approaches to the Analysis of Ancient Images

Plants were omnipresent in all spheres of life in societies of the past, from the most instrumental and utilitarian of practices to tales of the extraordinary, from narratives of heroic deeds of ancestors to tales of the afterlife. The world of plant life was a part of the way various societies constructed their space, be it real or imaginary.

Written documents from some of these societies have come down to us, while from others we have only images as the sole message-bearers. If we consider this iconography in the light of the theme that concerns us here, the key question is: to what extent can these images help us to examine the relation between plants and societies of the past? The first response to this question was based on how ancient images were commonly analysed in themselves. According to this perspective, their testimony enabled us to recognise concrete uses of plants for some of these past societies. This perspective was based on a direct and immediate reading: iconography makes it possible to recognise some types of crops, some forms of processing, the production of foods and so on. This is to say that, according to this tradition of studies, images attest to how particular peoples in the past utilised plants.

There are many examples of this approach. Forms of work and uses of plants are illustrated in many areas in the ancient Mediterranean such as Egypt, Greece and Rome. For example, the black-figure amphora by the Antimenes painter, made about 520 BCE and found in Vulci (Italy), shows a group of young people knocking off and collecting the olive crop, a typical task undertaken in the Greek countryside. From ancient Tanagra (Greece) of the early fifth century BCE, we have some terracotta figurines that show various tasks, such as kneading dough to make bread. There are many examples from ancient Egypt as well, which illustrate tasks associated with processing plants and making food, and these sometimes include the model granaries that were deposited in tombs to ensure abundance in the afterlife. The Metropolitan Museum of Art holds some of these models, such as the one from the Meketre tomb of the twelfth dynasty (1981–1975 BCE), in which a sequence of food production tasks is depicted. There are also Roman mosaics which portray various agrarian techniques, such as the Zliten (Libya) mosaic of the late first century CE, in which there is a scene representing grain threshing (Dunbabin 1978, 17, fig. 96). Other mosaics show, for example, particular foods such as the vegetables represented in a second-century CE work in the Tor Marancia villa, near the Catacombs of Domitilla on the outskirts of Rome.

The conservation of these images is, in many cases, extraordinary and represents a documentation of great value, a veritable window onto the uses of plants in the past. Nonetheless, as we shall put forward here, this window can also represent a trap, even an illusion. In fact, utilisation of images within a broader archaeological and palaeobotanical context has raised important issues but also some problems because we know that, as happens with literary sources, images can transform reality to bend it to the objectives of some individuals. For instance, images can manipulate practices and customs in order to show an idealised picture of what life was like in a Greek *chorá* or in a Roman *territorium*. Images can be a far too easy form of testimony, much too visible and only apparently direct: for example, when an ancient image shows the task of weaving was it really done in this way at the time?

This is the argument that recent research directions have attempted to clarify or even refute: the fact that an image may show an activity does not mean this was the way it was usually done at the time. Indeed, it may be an exaggerated or exceptional practice. The dangers involved in extrapolating such information for a specific time are clear. Also if we attach ourselves solely to what is exceptional we will be distorting history. We could be writing only the history of those sections of society who usually controlled the making of images, the elites, when our own objective is broader. It is at the heart of archaeological investigation to concentrate on the more general and recurrent, rather than on the exceptional and unique.

This tradition that supported an immediate reading of images has had various consequences and led to the progressive discrediting of these iconographic sources. When comparing the iconographic testimony with the progressively improving knowledge of archaeological contexts we could frequently see that their testimonies did not coincide. So what happened then? Could it be that images lie to us?

The problem highlighted here has to be considered along with another fundamental question: how can we analyse a picture that may be twenty-five centuries old? And, when we consider that images are quite often the sole voluntarily-provided messages of societies that did not use writing, we must emphasise that this issue is central to our initial question.

What later studies have demonstrated is that this 'direct' interpretation of images is dangerous and frequently erroneous. The dazzling evidence that a particular image carries may have influenced us to take as normal something that could have been exceptional at the time. And what if such an image is interpreted as representing a regular working process – specific ways of weaving, of handling bread and so on – while in reality what is represented is a specific ritual, a celebration regulated by more particular norms, etc? That is to say, this vision of images might be generalising and describing a work process on the basis of a visual testimony that may have been exceptional.

An example drawn from weaving is paradigmatic of the problem, in the two-fold testimony from a *pelike* or Attic red-figure jar and a piece of Iberian

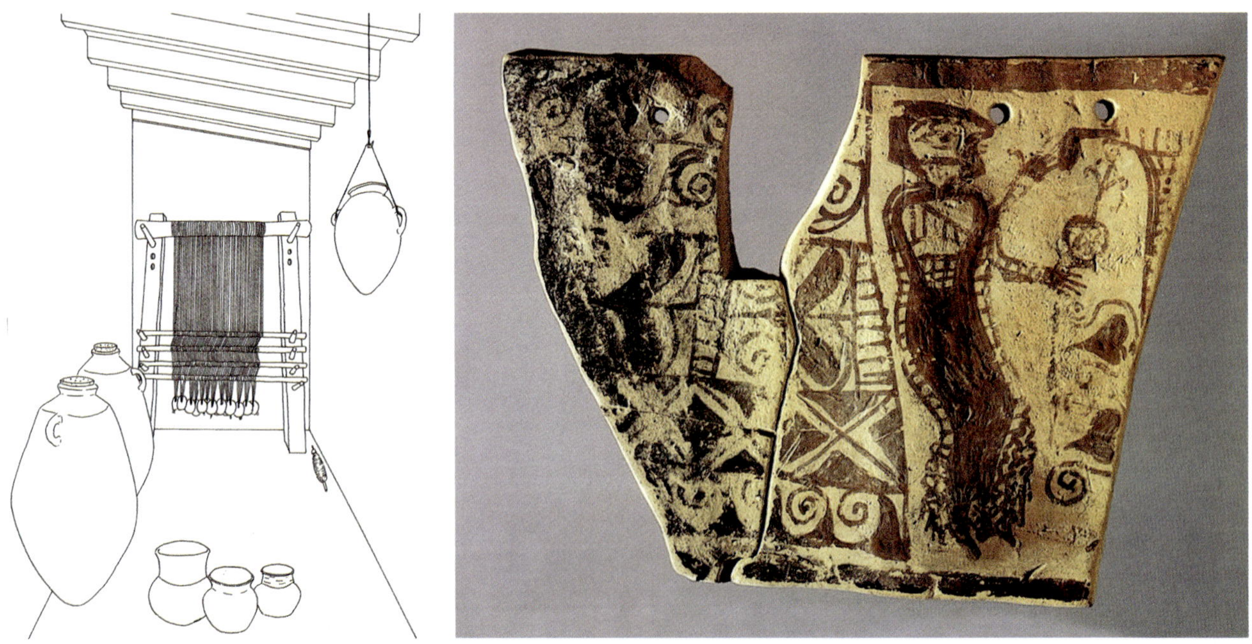

Fig. 2.5. Weaving in Iberian times: image from La Serreta de Alcoi (Alicante, third century BCE), and proposed restitution of a loom from space 3 in Cancho Roano (Badajoz), after Celestino and Jiménez Ávila (1996, 135).

pottery. The Greek *pelike* was made between 480 and 470 BCE and is attributed to the Argos painter. It shows a woman with two small objects: she is looking in a mirror and has a basket of wool, possibly an allusion to weaving being an important task for Athenian women. The Iberian example comes to us from the Serreta de Alcoi, in Alicante (Spain), where a ceramic lid was found with a picture showing a woman weaving on a footloom (Fig. 2.5). However, if we observe some of the elements of this scene, we see that they remove us from any scene of everyday work. The ankle-length robe which the woman is wearing is exceptional, as is the exuberant vegetation surrounding her. Likewise, in the Athenian case, this representation of a woman with a basket of wool is not really related to a daily activity, but to the importance held by the weaving of the *peplos* dedicated to Athena every four years. These factors lead us to the hypothesis that these scenes were not reflections of daily lives. The Iberian pottery and the Greek *pelike* represent weaving as partaking of a celebration important in defining a woman's role in society. In this case, the image bears witness to an exceptional event.

However, this 'direct' approach to analysing images raises yet another problem and a major one. If we were to think that this fragmentary and selective testimony is all that images have to provide us we would also be losing out on an important part of their real value, because of the question: what can images truly tell us? Just as Mitchell (2005) asked in his well-known work – 'what do pictures want?'

Towards a Contextual and Social Reading of Ancient Images

In societies of the past images were means through which people acted socially. They represent a significant investment of resources, capacities and time and their presence in the local landscape is far from accidental. We are dealing here with societies that did not experience the constant presence of images which is typical in our own present-day culture and that has made us accustomed to them. In the ancient world the use of images in the social sphere is a powerful and active agent and responds to highly concrete rules. What is actually reflected is how the social relationships which define every society are structured. In this sense, by examining these images, the rest of material culture and the forms through which a community expresses itself and interacts with the environment, enables us to approach how the social relationships are constructed within the community.

Therefore, regarding pictures only as testimony to the use of a loom, for example, or of how a task was carried out in the past, would be reductionist. Certainly what an image does express is the intentionality of certain groups or individuals to act within a social context. Acting, but what for? Acting both to maintain an inherited social order through the telling of stories or beliefs which, in fact, conceal the ideology that underwrites the social order and, in certain historical circumstances, acting to make the social order and social relationships change by expressing new beliefs, new heroic actions, legends or myths. If we take into account who might commission or have rights over these images, what the iconographic record provides us with is discourses frequently associated with dominant groups such as the social elites or those who are struggling to become dominant.

Such images have their field of action within changing social relationships and are the keys to the forms of domination within a community. Their use, therefore, was key to the ways in which social continuity and change were produced. That is to say, images are central to the ideological discourses necessary to enable certain families and groups to maintain their specific social roles. Images are most particularly useful in these social manoeuvres because of certain characteristics inherent in them. For instance, an image is polysemic and is capable of being arranged into various discourses. It is also versatile and polyvalent, and has, finally, an interesting transformative capacity in relation to the physical place where it is made to appear. Images contain within them a host of multi-layered and multi-faceted meanings. They can influence society through the diffusion of their message. In turn, society can transform images, preserve them as relics, destroy or mutilate them.

Nonetheless, what has changed since the first direct readings are the ways in which we approach the historical analysis of images. And this change has provided us with new possibilities and perspectives to integrate pictures into historical discourse and, in our case, to analyse which uses and meanings were given to various plants by peoples of the past.

Archaeological research on the images produced by different cultures has been carried out by formalists, processualists or structuralists. Structuralism has been fundamental because of its understanding of symbols as mental structures that shape cultural reality. The contributions of French iconology, as exemplified by the *Cité des Images* (Bérard *et al.* 1984; Shanks 2004, 2), has been especially valuable. Critics of this school have attacked its tendency to homogenise territorial or contextual differences and not see images as products of historical processes (Buxton 1994, 17). In this sense, post-structuralism has focused on both abstract symbolic representation and on social practice as a continuous process of making and remaking social reality (Ingold 1992; 1993). The meaning of the symbols does not exist beyond the moment in which people learn about them and compile them in meaningful ways (Thomas 1996, 97). According to Robb's (1998, 160) metaphor, they might then resemble the *tesserae* of mosaics, fragments that are put together temporally for communities who give them meanings, while at the same time experimenting with them.

Since the late 1980s, post-processual critics have insisted on some key aspects. A fundamental one is the essential contextualisation of the images, which discards interpretation solely centred on their autopsic analysis. This has further given rise to post-processual archaeology's most important methodological contribution, the contextual approach (Hodder 1987; Trigger 1991, 66). One of its fundamental tenets is that human expression through images is a dynamic component of ideology and social practice, rather than just aesthetic. Thus, material culture is an active agent in the social process.

Fig. 2.6. The representation of ploughing in two *kalathoi* found in Cabezo de Alcalá (Azaila, Teruel) below and Alcorisa (Teruel) on the top. Second century BCE. Image: Léxico de Iconografía Ibérica, CSIC.

An example from the Iberian Peninsula for the Roman period will illustrate how images have action within a social context as their objective. A *kalathos*, a characteristic Iberian pottery from Alcorisa (Teruel), represents an ard and a team of plough oxen (Fig. 2.6). It was an exceptional illustration, in a context in which Iberian societies had become subject to the power of Rome. This unusual vase also recalls other more common scenes among the Iberians, such as the hunt carried out by mounted warriors in a wood full of wild boar (Olmos 2008, 261). A first reading might tempt us to emphasise which ard and ox team were utilised by people in the late second century BCE. However, the find of an almost identical vase, in both form and iconographic representation, in the neighbouring city of Azaila, leads us to think that this image expresses something of an important territorial and historical significance. If we also grant that we encounter here a social context of reformulation established by the recent Roman conquest, we may agree upon the consequent interpretations of these vases (Olmos 1996; Aranegui 1999). These necessitate highlighting the territorial and historical context of the on-going social transformation and redefinition established by Rome. Thus, we can understand the exceptional nature of this image of an ox team as well as the considerable space devoted to the scene on these vases. These unusual vases may be showing us how the re-foundation of the indigenous cities, on which Rome imposed its new political rules, was ritualised. The presence of the two vases in privileged places in the two cities may recall the agreement made between these communities and Rome.

More recently, the analysis of images has been considerably enriched by the contributions of post-processual methodology. Working on the basis of this post-processual perspective, both archaeologists and anthropologists have thoroughly engaged with the symbolic aspects of material culture. Objects are conceived as metaphors that link different cultural spheres and construct meanings. Art would thus be a meta-language that formulates identities and creates referents (Tilley 1999, 263–265). Structural and symbolic approaches such as this have been reinforced by Hodder's (1982) ethnoarchaeological demonstration that material culture is no more than the epiphenomenal reflection of social organisation and, at times, can be used to invert or conceal it. From this point of view, material culture is not a mere reflection but is profoundly involved in making and transforming society: 'material culture transforms, rather than reflects, social organisation according to the strategies of groups, their beliefs, concepts and ideologies' (Hodder 1982, 212). It is not material culture which transforms social organisation, but the individuals using material culture who do so. It is in this way that they can maintain or change social structures. From this standpoint of the constructed nature of the social, it is paramount to place archaeological finds within the context of their production, exchange and consumption, which is also corroborated in the fact that every object can contain, along with its social life, many and diverse uses.

In the Greek world, interest was increasingly devoted to a global and symbolic interpretation of landscape and territory as a human construction in which animals and plants participate (Alcock and Osborne 1994; Osborne 1987; Buxton 1994). What stands out in these approaches is the semantic and metaphorical value of animals and plants and their relationship with the social imaginary (Darmon and Schnapp-Gourbeillon 1992; Giebel 2003). Much progress in this field has been made thanks to indispensable iconographic collections such as the *Lexicon Iconographicum Mythologiae Classicae* (LIMC: *www.rzuser.uni-heidelberg.de/~m99/; www.mae.u-paris10.fr/limc-france*) and the *Thesaurus Cultus et Rituum Antiquorum* (ThesCRA).

At the same time, the closer dialogue between anthropology and archaeology has enhanced the contributions of transcultural comparisons or Alfred Gell's (1992; 2009) theory of the anthropology of art. Beyond considering the work of art as an aesthetic result, Gell proposes to situate it within a network of relations between 'agents', which manifest agency through the intentionality of each image used, and 'patients', or receivers. These phenomena should, therefore, take into account the concurrence of agencies in each object, whether real or perceived. Gell replaces the always difficult notion of aesthetics with the manifold intentionality of social agents (Bloch 2009, VII). This ample spectrum of intentionalities which converge in the object means that it can be perceived as having a will or a power of its own, as, for example, sculptures that represent and commemorate the ancestors. This is how images come to be an active link in the chain that connects the members of a society.

An example of this process of commemorating the ancestors is provided by the so-called heads of Edetania (Fig. 2.7), the figures found in various settlements in this Mediterranean region of Iberia. In Tossal de Sant Miquel de Llíria and Puntal del Llops, both in the present-day province of Valencia, we find these terracotta heads in which the slip and the painting underline specific traits and colours. They occur within the domestic sphere and, in spite of their relative modesty, they abound in details about clothing, such as some anatomical details and the crossed strips over the neck, which is a sign associated with male prestige (Aranegui 1996; Olmos 1999). They emphasise certain features as a possible allusion to deceased members of the family group. Making images of the ancestors indicates to us the importance of commemorating them, thus seeking to perpetuate the social role of their descendants.

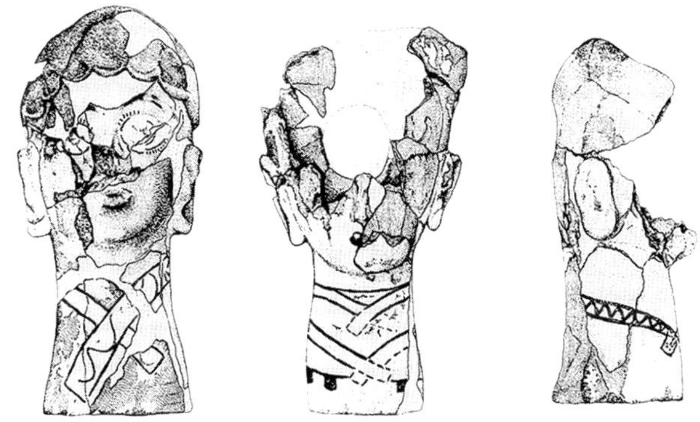

Fig. 2.7. Votive head from El Puntal dels Llops (Valencia) (Bonet Rosado and Mata Parreño, 1990, 195, fig. 2). Drawing after Francisco Chiner.

It is fundamental to consider that images materialise certain aspects of ideology, the set of strategies, ideas, tactics and symbols used to promote, perpetuate or change a social or cultural order. As elements participating in the social sphere, their visibility could be varied, depending on the framework used – architecture, artefacts made of precious metal, ceramics, and so on. They could be displayed in various ways: publicly, privately or directed at particular segments of society. Some of them were exhibited, but others were kept private and used exclusively for social differentiation, restricting their access and their use to the private circle of a few.

The fabrication of an image, therefore, is tied to the strategic manipulation that operates in the social and ritual dimensions of the human experience. It has a fundamentally social purpose. In this sense, it is a social construction that acquires its full meaning within a specific context. It implies a will to manipulate the ways in which the ideas and beliefs of the group are materialised (Aldhouse-Green 2004, XVI).

The themes represented through images in ancient societies cover a broad range and always depend on the way a particular society in the past perceived and interpreted the world. Images, therefore, play an active role in how ancient landscapes, political processes, agreements and coercions were constituted, in how each community constructed a symbolic geography or its own space in relation to 'the Other', etc. It is essential that we take into account the choice of a particular support, which is frequently motivated by the technological and social processes which underlie the creation of images. Among these, the transformation of metals is a field whose development enables us to appreciate the fundamental social factor which motivates the making of images. Thus, from a vision of artisans as a group subject to power, we have moved on to, in various contexts, appreciating more fully the social role of specialised artisans. Nonetheless, and in parallel to the processes of social differentiation and the consolidation of aristocratic groups, the figure of the goldsmith has come to be portrayed as a prestige craftsman who is, if not the owner of the raw material, at least the proprietor of the tools that can transform it, as well as probably of the workshop itself (Perea *et al.* 2007).

From the foregoing, we can argue that the image is, in fact, a social product that must be methodologically approached through its context. Its material character gives the image a fundamental characteristic which is its historicity: images, once created, enjoy a certain autonomy and can be interpreted in various ways (Fig. 2.8). They have a great capacity to change meanings through their potential use by cultures and peoples who may be far away temporally and spatially. This is a fundamental point and implies that we must analyse the making of particular symbols in specific spatial and temporal contexts. In order to do so, it is essential to integrate this analysis of images with that of the archaeological record, from the immediate context where the image was found to the settlement pattern that constitutes the territory which we are analysing. Symbols have a whole spectrum of meanings depending on the context. In the same way, the presence of the same image in other places and periods must be dealt with very carefully, insofar as temporality and physical space are essential to the historical construction of the image. For these reasons, we cannot approach images as a finished product: they are objects with a biography or a social life and they belong to the historical process of specific societies.

Fig. 2.8. The fight between a man and a griffin in a die found in Tomb 100 of the Iberian necropolis at Cabezo Lucero, Alicante, fourth century BCE. Image: Léxico de Iconografía Ibérica, CSIC.

Fig. 2.9. Alabaster sculpture of Galera, plan of its burial mound and found objects. Image: Léxico de Iconografía Ibérica, CSIC. Plan after Rodríguez-Ariza (Rodríguez-Ariza *et al.* 2008, 172, Fig. 2).

The alabaster sculpture known as the Lady of Galera (Granada, Spain; Fig. 2.9) is a clear example of an image with a very long social life. Its originality lies in the fact that this statue sums up many features that we also find separetely elsewhere in the ancient Mediterranean. We have argued that it was manufactured in a Phoenician context in Lower Andalusia during the seventh century BCE (González Reyero 2010a). This alabaster divinity is a sculpture into which liquids of animal or vegetal origin were poured. It was designed to enable control of an action: the precise moment in which a liquid was poured, surely in the course of a ritual. Only one person could control this libation. The sculpture was probably associated with a lineage and a cult that was aristocratic, minoritary and restricted. Not just anyone would have access to and be able to possess this kind of image. The figure is characterised by meticulous and expert workmanship and careful design: the lady appears seated and flanked by two sphinxes, holding in her hands a bowl into which the liquid which covers her body and flows from her breasts has been poured. The importance of this image would have lent it a long social life and meant it was used over several generations. It was eventually deposited in a tomb located in the necropolis of Tútugi (Granada, Spain) in the fifth century BCE (Rodríguez-Ariza *et al.* 2008). It is possible that the specific meaning of such a statue was different over its almost two hundred years of use, but what interests us today is how it points to the long duration of the practice of liquid libations in Phoenician and Iberian communities in the Iberian peninsula.

This perspective that I defend involves integrating the archaeological record and the analysis of the image. The significance and social role of images are constructed by a plurality of contextual factors: those present in their creation (type of support, technological process), those affecting their use, and their position in the landscape and the social space. Certain characteristics, such as polysemy, enable their insertion into changing social discourses and make them subjectively meaningful in the eyes of the beholder (Descola 2010). Symbolisation provides a social definition to objects and concepts on which the identity of the group is sustained.

We have pointed out previously how an ancient image can provide us more often with a window on exceptional practices carried out by a privileged social class. However, and without analysing the context, we can set aside the idea that what the image represents was a typical way of processing or emphasising plants.

Let us return now to the question asked at the beginning: what can pictures contribute to a better understanding of the relationships established between human communities and plants? Or better still: was this solely a utilitarian and instrumental relationship? What the analysis of iconography enables us to say is an emphatic 'no'. Plants appear frequently in the images of many ancient societies (Fig. 2.10), even when they do not figure in any process or production, even when no growing, harvesting and so on are being illustrated. For example, in the Iberian *oppidum* of Tossal de Sant Miquel de Llíria (Valencia), plants and writing are the protagonists and practically the only things represented on some large vases that have been documented in special prestige areas within the Iberian city. This is already an example of how plants are represented not only for illustrating a task, but for transmitting important messages.

Plants appear because they help define the real or imaginary spaces in which actions take place: they define both the cultivated fields as well as the distant and imagined forest where the ancestor-hero fought (Fig. 2.11).

We believe this is a fundamental contribution of ancient images. It enables us to show how plants play a role in a particular worldview and how various societies in the past developed ideological models that helped them make sense of the environment that surrounded them.

In addition to the valuable information about everyday work, what we can and must analyse in ancient images is how plants participated in discourse, be it the discourse of power or of opposition, which appear in every culture. When images represent domestic or everyday tasks, we carefully examine them to see how they also fit into the chain of actions of, for example, a ritual or an exceptional celebration. Placing its analysis into a broader context is essential.

Nonetheless, the presence of plants in images of hunting, combat, collective celebrations and other important events in the life of a community enables us to argue that the plant world was fundamental to the relationships which made up community. That is to say, because they were included in images, we know that plants were part of the cultural construction that made it possible to answer fundamental questions: where did we come from, what was the land of our ancestors, how did we reach the land we live in today, what explanation do sacred objects offer to us, why is it that certain people have the right to accumulate the fruits of the community's work, and so on. Images enable us to observe how plants make up a

Fig. 2.10. Plants and writing as protagonists in singular vases from Tossal de Sant Miquel de Llíria (Valencia), second century BCE. After Mata Parreño *et al.* (2010, 137).

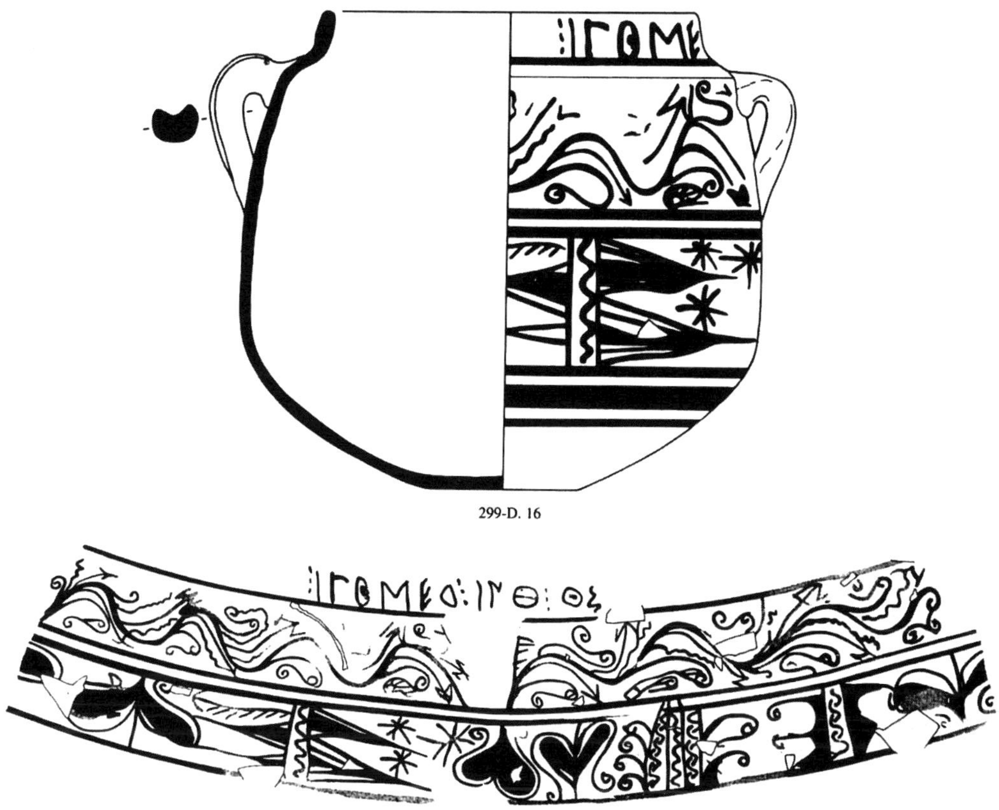

Fig. 2.11. Plants and writing as protagonists in a singular vase from Department 111 at Tossal de Sant Miquel de Llíria (Valencia), second century BCE. After Bonet Rosado (1995, 87).

part of the social imaginary of communities. They can thus represent the idea of the *physis* or fruitful nature of the Greek world, in which qualities were attributed to certain vegetables or fruits, such as the Mediterranean association between pomegranates and immortality, etc.

These are some of the main questions which we can address, thanks to the integration, *a priori* non-hierarchical, of images in the rest of the archaeological record. The purpose is to analyse the image within historical structures and processes, from the perspective that images were means utilised by groups or individuals to control, manipulate or subvert the social order.

Examining the universe represented by images leads us to perceive the great diversity of forms in which peoples of the past experienced their world and constructed their social identities. Therefore, this approach allows us to examine some of the key elements in the construction of social hierarchies in the ancient Mediterranean and it also places us at the frontiers of knowledge, bridging various scientific disciplines. From this standpoint, we are in a privileged position from which we can attain something fundamental: understanding the specific forms that various groups utilise to confer materiality to the perception of the natural environment in which their lives unfolded. Hence, through images, their experience of the world was made material and present, serving to influence the spaces they lived in, as well as the continuous construction of their history.

2.5. EXPLORING DIVERSITY IN THE PRESENT: ETHNOBOTANY STUDIES

Gisella S. Cruz-García

What is Ethnobotany?

Ethnobotany is the scientific discipline that studies the dynamic relationships between people and plants (Fig. 2.12). The term ethnobotany comes from Harshberger who defined it in the late 1890s as 'the use of plants by aboriginal peoples' (Cotton 1996, 3). Today it is considered that ethnobotany not only refers to the use and management of plants, but also incorporates the socio-cultural and economic context, as well as people's perception, conceptualisation, values and views (Alcorn 1995; Balick and Cox 1996). Ethnobotany, which is part of the discipline of ethnobiology, incorporates two main inter-related approaches: anthropology and botany. Nevertheless, due to its interdisciplinary nature (Prance 1995), it constitutes an interface of several other disciplines such as ecology, economics, linguistics, geography, agriculture and pharmacology (Cunningham 2000; Martin 2004), in addition to archaeology, nutrition, bioinformatics and mathematical biology (Ethnobiology Working Group 2003).

Fig. 2.12. Ethnobotany studies the dynamic relationships between people and plants: collection of fruits and herbs in Limburg, The Netherlands. Image: G. S. Cruz-Garcia.

Nowadays, a major focus of the ethnosciences is the integration of scientific and traditional (or indigenous) knowledge (Rist and Dahdouh-Guebas 2006). Ethnobotany explores products and processes of knowledge systems, such as their creation, acquisition, transformation and transmission (Ellen 2006; Ethnobiology Working Group 2003). Ethnobotany studies how a group of people (ethnic, cultural or linguistic group) classify and organise their knowledge about the environment (Price 2001), given that the way humans classify the world surrounding them influences the way they interact with it. This aspect of ethnobotany has often been overlooked in the past, when this discipline used to have a more utilitarian approach to research.

History of Ethnobotany

The roots of human culture are entwined with plants. The study of people-plant interactions has been very important throughout human history, since all human cultures are dependent on plant resources for their subsistence. Plants have profoundly influenced the course of civilisation, starting with the use of plants in pre-history by hunter-gatherers and the development of agriculture later on (Balick and Cox 1996). Cross-cultural interest in plant use is clearly seen in the spread of several agricultural crops throughout the world. For instance, maize, cassava and potato from the New World are nowadays staples in the Old World, whereas wheat and rice are basic foods in the New World (Minnis 2000).

At an initial stage, ethnobotany was generally focused on the utilitarian study of useful plants by botanists (Ford 1978). For instance, Greek, Roman and Islamic societies were interested in the study of plant use. In the year 77 CE, one of the first studies, called *De Materia Medica* by the Greek Pedanius Dioscorides of Anazarbus, was published, consisting of a compilation of useful flora from the Mediterranean (Pardo de Santayana *et al.* 2010). Ethnobotanical explorations were carried out not only by ancient Egyptians in Syria and Somalia, but also had a long history in China (Minnis 2000). During the Medieval and Renaissance periods, numerous European naturalists, such as the Swedish botanist Linnaeus, conducted botanical explorations around the world (Pardo de Santayana *et al.* 2010). Botanical gardens were established in Renaissance Europe, as well as by the Incas in South America and the Aztecs in Mesoamerica. Ethnobotanical research was intensified after European colonisation (Minnis 2000).

The pre-Classical period, according to Clément (1998), began in the nineteenth century when the foundations and branches of ethnobiology were established. A botanist called J. M. Harshberger, who studied ancient plant remains in southwest North America, coined the term ethnobotany in 1896. Other important scholars of this period were Fewkes, Palmer, Powers, Barrow and Hough (Minnis 2000). The diversity of scholars and ethnobotanical reports increased in the following decades. At this time, ethnobotany was mainly focused on making inventories of flora, with their respective uses and preparations, which could be used by Western civilisation (Ellen 2006; Ethnobiology Working Group 2003; Prance 1995). Indigenous peoples were thought not to have any form of scientific knowledge (Clément 1998).

In the 1950s, the classical period of ethnobiology began with the shift from an *etic* to an *emic* approach (*etic* and *emic* are explained later in this section). Marked by the work of Conklin, importance was given to local perceptions, vernacular nomenclature and systematic classification of plants (Clément 1998; Ellen 2006). From the 1980s onwards, the post-classical period was characterised by the emergence of cooperation between ethnobotanists and indigenous communities (Clément 1998). The acknowledgement of the importance of traditional knowledge by Western scientists increased, along with much of the expansion and diversification of ethnobotany (Cotton 1996). The increasing attention to indigenous communities by scholars occurred once scientists recognised the increasing loss of traditional knowledge of indigenous and folk cultures, as well as the destruction of natural ecosystems and diversity, due to the encroachment of development (Prance 1995). Nowadays, ethnobotany is becoming more technological, experimental and participative, while indigenous peoples are being more empowered to collaborate in defining research, conservation and development priorities (Ethnobiology Working Group 2003). Furthermore, ethnobotany goes beyond the study of indigenous or traditional societies to the investigation of people from any cultural tradition in the world (Ellen 2006; Ford 2000). Fig. 2.13 represents a summary of the most important events in the development of ethnobotany.

Main Areas of Research

Ethnobotany has an ample scope, characterised by both academically-driven and practice-driven research, providing a platform for the convergence of diverse scientific disciplines (Ellen 2006). The most popular areas of ethnobotanical research are ethnoecology, traditional agriculture, cognitive ethnobotany, material culture, palaeoethnobotany and archaeobotany, environmental history, ethnomedicine, and ethnopharmacology.

Ethnoecology is focused on understanding local people's constructions according to their own ethnoscience categories (Frake 1962) recognising human-environment interactions as ecological (Minnis 2000). The study of traditional ecological knowledge (or traditional environmental knowledge) is a main component of ethnoecological

Year	Events in the Development of Ethnobotany
1492	The discovery of the New World initiates the identification of several plants of considerable economic value based on observation of native peoples.
1663	John Josselyn studies the natural history of New England, later publishing his text on native herbal medicine, 'New England realities discovered'.
1785	Withering's publications on ethnopharmacology trigger the development of this field in Europe.
1800s	**The 'pre-classical period' of ethnobotany**
1803	Ehrenberg reports fossil pollen preserved in sedimentary rocks, initiating a rapid development in the field of palynology in Europe, a powerful tool used by palaeoethnobotanists.
1871–78	Seminal works by botanists Palmer and Powers are published. This period is dominated by economic botanists.
1893	Anthropological interest in aboriginal botany leads to increasing emphasis on the cultural significance of plants.
1895	Harshberger introduces the term "Ethnobotany"
1896	Fewkes introduces the term 'Ethnobotany' in anthropological literature
1898	The Ethnology Department of the US National Museum (today: Department of Anthropology, National Museum of Natural History, Smithsonian Institution) endeavors to document all useful plants of North American Indians
1900	The first PhD in ethnobotany is awarded to David Barrow.
1919	Traditional peoples' resource management is pioneered by Gilmore.
1930	Castetter establishes a masters programme in ethnobotany at the University of New Mexico
1947	Foundation of the Society for Economic Botany
1950s	**Beginning of the 'classical period' of ethnobotany**
1950–1970	Increasing interest in linguistic concepts and classifications, while Conklin highlights the practical significance of understanding folk classification systems. Palaeoethnobotany emerges and archaeobotanical techniques improve.
1981	The Society of Ethnobiology publishes the first issue of the *Journal of Ethnobiology*
1988	The *Declaration of Belem* was signed in the First International Congress of Ethnobiology in Brazil. The *International Society of Ethnobiology* was founded with the objective of understanding the complex relationships between human societies and their environments, recognising indigenous peoples as critical actors in the conservation of biological, cultural and linguistic diversity.
1990s	Both post-graduate and undergraduate programs in Ethnobotany become increasingly available, while many research projects focus on practical applications of plant knowledge. Establishment of *People and Plants* initiative of WWF, UNESCO and the Kew Royal Botanical Gardens, aiming to increase community-based plant conservation programs worldwide.
2000s	Launch of the journal *Ethnobotany Research and Applications* and the *Journal of Ethnobiology and Ethnomedicine.*
2006	The *Code of Ethics* for Ethnobiological research was adopted by all the members of the *International Society of Ethnobiology.*

Fig. 2.13. Important events in the development of ethnobotany. Adapted from Cotton (1996), with references also from Clément (1998); Cunningham (2000), the International Society of Ethnobiology (2010), the Society for Economic Botany (2011), and the Society of Ethnobiology (2011).

research, including issues such as traditional vegetation management, ethnopedology and ethnoclimatology. Another important area is landscape ethnoecology, comprising the study of cultural views of the landscape, local classifications of its components in ethnoecological systems, and their significance for local people (Johnson and Hunn 2010).

Traditional agriculture, also called ethnoagronomy, refers to the study of resource management and subsistence economies (Pieroni *et al.* 2005), as well as understanding the cultural, economic and genetic reasons underlying agriculture (Ford 2000). Traditional agriculture is aimed at investigating traditional knowledge about crop varieties and agricultural resources, as well as the environmental impact of variety selection (Cotton 1996). In parallel, ethnobotany also contributes to the understanding of the ecological relationships and human management involved in plant domestication (Alcorn 1995; Casas *et al.* 1997; Minnis 2000). For example, Casas *et al.* (1996) studied the management and domestication of plants among the Nahua and the Mixtec in Mexico from an ethnobotanical perspective, including wild, weedy and domesticated plants (see Chapter 5.1 for a discussion on domestication and management of wild food plants).

Cognitive ethnobotany is focused on researching the organisation of knowledge systems or folk classifications, which is also called ethnotaxonomy (Cotton 1996), considering how people view their own environment (Minnis 2000). Brent Berlin (1992) proposed a set of principles to describe the traditional systems of ethnobiological classification (modified from his previous results from 1972). Cognitive ethnobotany also deals with traditional perceptions of the natural world (symbolism, ritual and myth).

Material culture deals with traditional knowledge of plants in art and technology. Material culture refers to the *artefacts* or objects made by existing traditional societies. These artefacts include tools, shelter, clothing, boats, containers, as well as decorative arts and craft, such as toys and ornaments. In many societies, not only is timber essential to the construction of houses, shelters, furniture and fences, but also non-timber plant products derived from leaves, pigments, fibres and resins are remarkably important (Cotton 1996).

Palaeoethnobotany and archaeobotany investigate past interactions of peoples and plants, based on the interpretation of archaeobotanical remains. This field is very important for documenting human-plant interactions that occurred in pre-literate societies, until five thousand years ago (Minnis 2000).

Environmental history is focused on the understanding of prehistoric human action on the environment (Stahal 1996), such as the study of ancient crops and respective agricultural techniques (Minnis 2000).

Ethnomedicine is the study of the cultural interpretation of disease, illness and health, healing systems and traditional health care (Pieroni *et al.* 2005).

Traditional phytochemistry studies traditional knowledge about plant chemicals, for instance, for medicine or pest control (Cotton 1996). *Ethnopharmacology* is aimed at understanding the pharmacological and cultural scopes of the uses of medicinal plants (International Society for Ethnopharmacology 2011). *Ethnopharmacy* and *ethnopharmacognosy* are also important related fields.

Methods in Ethnobotanical Research

The first step in ethnobotanical research is to define the *domain* (mental category or subject of interest). For example, a *domain* could consist of the 'wild food plants' consumed by a certain indigenous group). Borgatti (1999, 115) calls it *cultural domain* and defines it as 'a set of items or things that are all of the same type or category'. The study of people's interpretation of *domains* is called *cultural domain analysis* (Bernard 2002). *Cultural domain analysis* involves not only investigating the structure and arrangement rules of the *domain*, but also the relations among its components, their associated values and variability. *Cultural domain analysis* aims to understand knowledge systems in addition to their similarities and differences among or within groups of people (Puri and Vogl 2004). Knowledge systems could certainly be affected by inter-cultural factors (livelihood strategies, natural resources, level of external contact or acculturation, ethnicity, religion) and intra-cultural factors (age, gender, class, education, literacy, occupation, migration, kinship, language ability).

The study of the components of a *domain*, also called *categorisation*, involves two approaches: (a) *etic* approach related to the researcher's perception and classification of the study object, (b) *emic* approach referred to the classifications of local people based on the way they perceive the world in their own language (Martin 2004; Cotton 1996). In both cases, it is important to start with an *emic* perspective, for example, referring to the local names of plants and understanding their local classification systems (Martin 2004).

Given that ethnobotany is at the interface of several disciplines, there is a wide variety of research methods (Ethnobiology Working Group 2003). The most common ethnobotanical methods, drawn from anthropology, botany, ecology, linguistics, palaeoethnobotany and archaeobotany, are described in this section. Nevertheless, given the interdisciplinary nature of ethnobotany, research studies might also require the use of more specialised methods such as art history, molecular biology, economic anthropology, development studies, environmental economics, ethics and law, communication and education, among others (Cotton 1996). Ethnobotanists usually combine different methods according to their specific research needs.

Anthropological Methods

Different methods used in anthropological research might also be applied to ethnobotany. These methods could be qualitative or quantitative. Qualitative methods are used to acquire an in-depth understanding of human behaviour through general ethnographic accounts, whereas quantitative methods consist of a systematic empirical investigation of measurable verbatim answers allowing the use of statistical methods (Martin 2004; Verschuren *et al.*

1999). Both qualitative and quantitative methods are important and complement each other permitting, to a certain degree, triangulation of sources as well as greater depth (Verschuren *et al.* 1999). The most common anthropological methods for ethnobotanical data collection are basically grouped in interviews and observation. Additionally, special cognitive and linguistic analytical tools, which are specific semi-structured interview techniques (Borgatti 1999), are widely used to facilitate quantification, cross-verification and choosing participants for specific projects (Cotton 1996).

According to the level of control of the researcher on the interview, interviews are classified in structured, semi-structured, unstructured and informal. Structured interviews are based on a set of instructions and questions (or questionnaire) allowing to conduct quantitative analysis. Semi-structured interviews are more flexible, as the researcher follows an interview guide but allows respondents to express their opinions and ideas in their own way. In unstructured interviews the researcher has a clear plan in mind, but both researcher and respondent can follow new leads. Informal interviews are characterised by a total lack of structure and control, producing qualitative data. In general, questions could be closed-ended, when respondents have to choose from a list of options provided in the interview; or open-ended, when respondents are free to give the answer in their own words. Specific kinds of interviews are key informant interviews and focus group discussions. Key informant interviews are carried out with experts on the research topic, whereas focus groups are discussions where a group of people has a joint interview session (Bernard 2002).

Observation could be participant or non-participant, depending on the researcher's involvement in the daily life activities of the people in the study area. Participant observation is usually carried out over a long period of time, taking part not only in community life, but also in local events and processes (Bernard 2002; Cunningham 2000; Kottak 2008; Martin 2004).

Analytical tools, mostly developed by linguistic and cognitive anthropologists in the 1960s, are important for assessing the empirical knowledge and cultural preferences of the informants. Many of these tools were incorporated in response to the need for systematic data collection methods to understand people's perceptions and classification of the world (Martin 2004). The most commonly used analytical tools in *cultural domain analysis* are freelistings, triadic comparisons, paired comparisons, pile sorting, ratings and rankings.

- Freelistings: informants are asked to list plants that fulfil a particular criterion (such as 'wild', 'food' or 'medicine'). Freelistings are used to identify the components of a *cultural domain.*
- Triadic comparisons: informants are presented the items of the *domain* in groups of three and asked to pick out of each group the item they perceive as the most different. Picking an item means that they consider the other two similar. Triads are used to discover the arrangements of a *cultural domain.*
- Pile sorting: informants are asked to sort items of a *domain* into groups according to how similar they are. This method is also used to discover the arrangements of a *cultural domain.*
- Paired comparisons: informants are presented pairs of items and asked to indicate which of the two is more related to a specific criterion. This method is important for identifying the arrangement rules of the groups of items in a *domain.*
- Ratings: informants are asked to place each plant or item of the *domain* along an abstract scale.
- Rankings: informants are asked to compare plants or items to each other and to list them according to a specific criterion. Both ratings and rankings are used for identifying the arrangement rules of the groups of items in a *domain.*

(Bernard 2002; Borgatti 1999; Cotton 1996; Martin 2004; Puri and Vogl 2004; Ryan and Bernard 2000).

The data produced by any of these tools consist of numeric values assigned to each component of the *domain* (plant or item), allowing researchers not only to compare them and understand the structure of the *domain*, but also to compare knowledge, values and practices of different groups of informants. For example, from freelistings it is possible to obtain a salience index and from rankings a preference ranking index, whereas indexes of agreement can be produced by any tool applied.

- The salience index quantifies the importance of a plant in relation to the number of uses mentioned by informants (Cotton 1996). The most common indexes are Smith's salience index (Smith 1993) and Sutrop's cognitive salience index (Sutrop 2001).
- The preference ranking index estimates the preference or importance of a plant in relation to a particular criterion (Cotton 1996), such as taste or availability (Price 1997).
- Indexes of agreement compare the level of agreement between informants with respect to the components and/or structure of a *domain* (Boster 1985).

In addition, ethnobotanists have also been interested in gaining a deeper understanding of the comparative cultural value of plant species. This is reflected in the development of use-value and cultural significance indexes. These indexes have been modified in the last years: researchers add or change variables, edit formulas and adapt them to their particular research question and study area.

- The use-value index estimates the overall usefulness of a plant in terms of number of uses mentioned by each informant and total number of informants (Cotton 1996; Cunningham 2001; Phillips and Gentry 1993). Data could be obtained not only through interviews, but also by examining utilisation surveys, which might involve walking with informants in vegetation transects (Anderson 1991).
- The cultural significance index is defined as 'the importance of the role that [a plant] plays within a particular culture' (Turner 1988, 275). Initially, it included the quality, intensity and exclusivity of plant use (Turner 1988), and afterwards the contemporary plant use was added (Stoffle *et al.* 1990). Later on, Pieroni (2001) presented the cultural food significance index, including other variables such as quotation index (frequency of citation of a species in a freelisting) and perceived availability. On the other hand, Reyes-García *et al.* (2006) introduced the cultural, economic and practical value indexes, which were summed up to equal the total value of a species. Tardío and Pardo de Santayana (2008) conducted a comparative study of different indexes of cultural importace and use-value of plants.

Botanical Survey

The botanical survey is essential for ethnobotanical research. It consists of collecting voucher specimens of plants and conducting their taxonomical identification (Cotton 1996; Cunningham 2000; Martin 2004). Basic plant characteristics should be registered at the time of specimen collection, such as: life-form, height, diameter, colour of flower and fruit and local name (for a detailed explanation about specimen collection, pressing, drying, labelling and preservation, see Martin 2004). Depending on the objective of the research, other botanical data should be collected, such as stem diameter, height, bark thickness, biomass, volume, in addition to leaf measurements (culm length, petiole width, foliage mass, specific leaf area), canopy measurements (biomass, volume, area, density, crown position), dry weight, production of flowers, fruits and seeds, age (of trees, bulbs, corms, stem tubers), phenology,[7] as well as characteristics related to the plant's reproductive biology and yields (Cunningham 2000).

Ecological Methods

Ecological surveys are mainly devoted to understanding the distribution, diversity, occurrence, abundance and structure of plant populations and communities (Fig. 2.14), as well as landscape and ecosystem processes. The purposes of carrying out ecological surveys could vary from assessing harvesting quantities, pressures and impact on the vegetation, to commercialisation of plant resources, conservation and management. The first step in ecological research is to conduct a general

Fig. 2.14. Ecological surveys are important for understanding the distribution, diversity, abundance and structure of plant populations and communities. Image: G. S. Cruz-García.

description of the ecosystem(s) in terms of soil, vegetation types, climate, land form, stages of ecological succession and land use zones. The next step is to carry out quantitative research of plant resources (Martin 2004). Most common ecological methods include vegetation surveys, systematic participatory mapping, transect walks, time lines and seasonal calendars.

Vegetation surveys could be conducted at species, population or community level. Systematic surveys of vegetation start with the selection of ecological sampling units (Cotton 1996) that are sub-divided into quadrants or transects (Martin 2004). Sampling units should, preferably, be randomly established and carefully defined regarding their distribution, number, environmental gradients, size and type (see Cunningham 2000; Krebs 1999). Seasonality is important; hence sampling units should be surveyed in different seasons. Most common analysis consists in calculating species richness, spatial patterns and dispersion, diversity indexes, rank abundance curves, as well as their comparison across space and through time (Krebs 1999; Magurran 2004).

Participatory mapping assesses the local knowledge of resources, not only in relation to perceived abundance and harvesting patterns, but also to access to resources, tenure and territoriality. The complementary use of aerial photographs and satellite images will help to understand the distribution of vegetation, historical change, degree of threat and disturbances in the landscape (Cunningham 2000). They can be compared to the knowledge of the landscape and values attributed to it by local people, as shown in the research carried out by Fagerholm and Käyhkö (2009) in Tanzania and Barrera-Bassols *et al.* (2006) in Mesoamerica.

Transect walks (or walks in the woods) are conducted in the research area with local informants, in combination with interviews. Time lines are usually aimed at identifying important historical events, while seasonal calendars illustrate the change of uses, availability and preferences throughout the year (Cunningham 2000).

Linguistic and other Symbolic Analysis

In ethnobotany, it is necessary to understand how information is communicated within a particular group of people, therefore linguistic and symbolic analyses are important. On the one hand, language is the 'gateway to knowledge and perceptions' (Price 2001, 159) and on the other hand, much information is transmitted using symbolic representations such as myths, rituals and art (Cotton 1996). Particularly, linguistic analysis is essential for ethnotaxonomical studies that investigate traditional systems of classification. This involves the use of analytical tools and linguistic evidence (see Berlin 1992; Cotton 1996; Martin 2004).

Palaeoethnobotanical and Archaeobotanical Methods

Palaeoethnobotany involves the study of botanical evidence found in archaeological sites, in addition to analysis of historical texts and interpretation of plants in ancient art. For the study of botanical evidence, it is necessary to apply archaeobotanical techniques. Archaeobotanical evidence includes plant fossils, grain impressions in baked clay and microwear polishes found on tools. In addition to these, other archaeological samples – such as coprolites, phytoliths in grinding tools, residues in pottery – could give information on the uses of the plants, as well as on how they were harvested and processed in the past (Cotton 1996). Palaeoethnobotanical research involves sampling and collection in archaeological sites, plant material identification and plant material dating. Dating can be done using pollen profiles, isotope analysis and thermoluminescence (Cotton 1996; Pearsall 1989).

Ethnobotanical Issues and Imperatives

Intellectual and Theoretical Imperatives

Given that ethnobiology is a field that is rapidly growing and creatively expanding, the NSF funded the Biocomplexity Workshop aiming to explore recent scientific developments and critical issues of the discipline. As an outcome of this workshop, the report 'Intellectual imperatives in ethnobiology' (2003) listed the main contemporary intellectual and theoretical imperatives of ethnobiology, which can certainly be applied to ethnobotany, in relation to the areas of knowledge systems; medicine, health and nutrition; ecology, evolution and systematics; landscapes and global trends; and the role of social-cultural-political systems in biocomplexity research. Moreover, Cunningham (2000) emphasised that it is also imperative to apply efficiently quantitative methods and predictive models to conservation strategies.

Clearly, important concerns of study are evolution, systematics and ecology, where it is necessary to understand more than purely 'artificial selection'. Ethnobotany incorporates the complexity of human interactions with nature, allowing more thorough research on the management of flora and domestication processes. This is certainly necessary for addressing major intellectual concerns such as:

> 'How does human use and management of biodiversity affect ecological processes and patterns? How have human interactions with taxa – from gathering to domestication influenced evolution and systematics, and what trends or differences are there within and among taxa? In the evolutionary process, how are "natural" and "artificial" selection similar and different?' (Ethnobiology Working Group 2003, 3).

The study of domestication can particularly profit from an ethnobotanical perspective, especially in relation to the 'ability to work cross-culturally' thus uncovering the 'indigenous insight on complex questions' (Ethnobiology Working Group 2003, 4).

Ethnobotanical research can be very useful for revealing 'positive examples of human mediated biodiversity creation and management' (Ethnobiology Working Group 2003, 3), which is also an important intellectual imperative of the discipline. In this regard, Alcorn (1995, 29) emphasises the importance of studying plant management systems, so called ethnobotanical dynamics, and how in traditional agro-ecosystems 'human activities influence both the crops and the natural vegetation occupying the region'.

Ethnobotany, Conservation, and Sustainable Development

The inter-relations between biodiversity and human cultures, or the study of bio-cultural diversity, is an important imperative of ethnobotany. It was emphasised by the Ethnobiology Working Group (2003, 3) that 'landscape transformations are dependent on distributions of culture, biota, and environments' concluding that 'biodiversity is correlated with human cultural diversity'. Given the on-going problem of loss of biodiversity, as well as the associated loss of traditional knowledge and cultures, it is essential to understand the complex dynamics of humans with the environment (Prance 1995). The impacts of ethnobotanical research are important for cultural survival and biodiversity conservation (Ethnobiology Working Group 2003; Prance 1995) and should contribute to the formulation of practices and policies for sustainable development (Prance 1995).

It has been demonstrated in the last decades that many top-down, centralised conservation actions planned by outsiders (policy-makers or urban-based planners) have failed, because they do not include local perceptions and views (Cunningham 2000). Nevertheless, in the last decades, many ethnobotanists have been keen on applying their research results to conservation and development strategies (Cotton 1996; Martin 2004; Rist and Dahdouh-Guebas 2006; Sillitoe 2006) and have much to offer to policy-makers (Alcorn 1995; Sillitoe 2006). Moreover, the incorporation of indigenous insights and indigenous peoples in decision-making and problem-solving has also been demonstrated to be important (Balick and Cox 1996; Cunningham 2000; Ethnobiology Working Group 2003; Rist and Dahdouh-Guebas 2006). There are many examples of such initiatives, such as community projects, establishment of markets for non-timber forest products (Cotton 1996; Martin 2004), conservation of wild crop relatives and endangered plants, healthcare, social forestry, educational programs for the young, popular workshops and publications, ecotourism, and so on (Martin 2004). Sillitoe (2006) also explains the importance of ethnobiology for development, for instance, through the promotion of participatory initiatives, the facilitation of local solutions to development and the support to alternative development strategies, as a response to the critiques of development as an exported capitalist concept.

Results from palaeoethnobotanical research could also be applied for effective resource management. Palaeoethnobotanical evidence gives us lessons to be learned from the past, such as those drawn from the ancient Mayan agricultural fields or *chinampas* in Mexico, which proved to be economically and ecologically sustainable (Cotton 1996).

Ethical Considerations

Bio-prospecting – or searching for new products

from nature with commercial objectives – led in the mid-1990s and early 2000s to the exploitation of these resources (and their associated traditional knowledge) without any compensation to indigenous or local peoples. Furthermore, third parties accessed ethnobotanical information published by researchers without the consent of the local knowledge-holders (Bannister 2007; Cotton 1996; Ethnobiology Working Group 2003; Posey 1990). In this perspective, Bannister (2007, 16) emphasises that 'indigenous communities across the world, consequently, have been put in the position of contesting patent applications related to their traditional plant uses, copyright over associated stories, and trademarks over use of indigenous names and designs'. Today, unintended consequences of scientific research can no longer be ignored by ethnobotanists. There are global debates going on regarding indigenous rights and cultural misappropriation (Bannister 2007; Martin 2004), as well as over-exploitation of plant resources and loss of biodiversity (Balick and Cox 1996; Cotton 1996).

Nowadays there is much emphasis in the international arena on benefit-sharing, prior informed consent, intellectual property rights (Bannister 2007; Ethnobiology Working Group 2003) and the importance of returning the research results to the community (Ethnobiology Working Group 2003; Martin 2004). Professional societies have developed codes of ethics, professional standards and research guidelines (Ethnobiology Working Group 2003). For example, the International Society of Ethnobiology promotes a Code of Ethics which provides a framework of conduct for ethnobiological (including ethnobotanical) research and a framework for decision-making (2006).

2.6. CONCLUSIONS

Lydia Zapata

Choices and diversity of human plant use can be explored through different materials and methods. Some of them, such as archaeobotany, are available for any chronological framework, whereas some others are linked to specific time periods and materials. Written sources and iconography are only available in some parts of the world for the last few millennia of human history, whereas ethnobotany focuses on the study of present populations, thus providing insight into the dynamic relationships between people and plants today. Some sources are unlikely to increase significantly, for example, written documents of early historical periods. On the other hand, new methods and techniques in archaeology are continuously increasing and it is becoming difficult even for specialists to keep up with current new developments. The analyses of hardly visible remains such as phytoliths, starch or organic residues are now becoming common in the bibliography, and this helps us approach complex subjects related to human subsistence and plant economy through the most modern technology.

Most of us will agree that the change to farming gave rise to a real revolution in human history, something that most probably resulted in many places in a narrowing of plant diversity in favour of cereals. However, the diversity of crops also varied through time. A wide spectrum of cultivated plants was used in some places even during the early Neolithic, but new crops continued to be added progressively to agrarian systems (millets from the Far East, tree crops, American cultigens...). How far these changes in plant use depended on real human choices and culture or on environmental triggers is a subject of debate which often is rooted in theoretical orientations and traditions. But in order to discuss these issues, robust sets of raw data are needed. In this chapter dealing with the methodological approach, we have tried to show that the necessary over-specialisation of the researchers as well as the limitations and incomplete nature of many of our sources makes interdisciplinarity obligatory.

Chapter Acknowledgements

The work by Lydia Zapata was carried out within the projects AGRIWESTMED (ERC-AdG-230561), HAR2008-03976/HIST, HAR2008-01920/HIST and HAR2011-23716 *Nuevos cultivos, nuevos paisajes* and Programa Consolider TCP-CSD2007-00058 from the Spanish Government. Lydia Zapata is part of the Research Group of the Basque Government IT622-13 and UPV/EHU UFI 11-09.

Chapter Notes

1 Oxygen Isotope Stages (also: MIS, Marine Isotope Stages) are alternating warm and cool periods in the Earth's palaeoclimate, deduced from oxygen isotope data deriving from the analysis of deep sea core samples. OIS19/MIS19 refers to the period from *ca.* 798,000–755,000 years before today, in which a 'short' warmer interglacial (MIS19c) of *ca.* 10,000 years duration took place.

2 Parenchyma is the ground tissue in a plant, forming the greater part of leaves, roots, fruit pulp and the pith of stems.

3 Taphonomy is the study of decaying or fossilising organisms, involving the processes prior to and after their deposition.

4 In starch granules, the hilum (plural: hila) marks the area around which the layers of starch are deposited. Hila can be centric or excentric.

5 Extinction crosses are cross-like structures visible under a microscope when starch grains stained with iodine are illuminated from behind with polarised light.

6 The cuticle consists of the outermost cell Wall layers of the plant epidermis. The cellulose fibres in the cuticle are incrusted with water-repelling wax-like substances.

7 Phenology deals with the study of periodic life cycle events of plants such as flowering, and their relation to environmental factors, such as day length and climate.

Chapter Bibliography

Abu L-Jayr (1991) *Kitab al-Filaha. Tratado de agricultura.* Carabaza Bravo, J. M. (ed.). Ministerio de Asuntos Exteriores. Agencia Española de Cooperación Internacional. Instituto para la Cooperación con el Mundo Árabe, Madrid.

Acosta, F. (1993) Fuentes y métodos para la historia rural (siglos XVIII–XX). *Noticiario de Historia Agraria* 6, 151–163.

Alcock, S. and Osborne, R. (eds.) (1994) *Placing the Gods. Sanctuaries and Sacred Space in Ancient Greece,* Oxford.

Alcorn, J. B. (1995) The scope and aims of ethnobotany in a developing world. In R. E. Schultes and S. v. Reis (ed.) *Ethnobotany: evolution of a discipline* Dioscorides Press/Timber Press, Portland, pp. 23–39.

Aldhouse-Green, M. (2004) *An archaeology of images. Iconology and cosmogony in Iron Age and Roman Europe.* Routledge, London.

Álvarez de Morales, C. (1994) Magia y superstición en la literatura agrícola andalusí, *Ciencias de la Naturaleza en al-Andalus. Textos y Estudios.* III, editados por Expiración García Sánchez. Consejo Superior de Investigaciones Científicas, Granada, pp. 391–402.

Anderson, A. B. (1991) Forest management strategies by rural inhabitants in the Amazon estuary. In A. Gómez-Pompa, T. C. Whitmore and M. Hadley (ed.) *Rainforest regeneration and management.* UNESCO, Paris, pp. 351–360.

Aranegui, C. (1999) Personaje con arado en la cerámica ibérica (ss. II–I a.C.). Del mito al rito, *Pallas* 50, 109–120.

Balick, M. J. and Cox, P. A. (1996) *Plants, people and culture:*

the science of ethnobotany. Scientific American Library, New York.

Ball, T., Gardner, J. S. and Brotherson, J. D. (1996) Identifying phytoliths produced by the inflorescence bracts of three species of wheat (*Triticum monococcum* L., *T. dicoccon* Schrank., and *T. aestivum* L.) using computer-assisted image and statistical analysis. *Journal of Archaeological Science* 23, 619–632.

Bannister, K. (2007) The ethics of engagement: an ethnobiologist's perspective. *Academic Matters* October 2007. Ontario Confederation of University Faculty Associations, Ontario, pp. 16–17.

Barrera-Bassols, N., Zinck, J. A. and Van Ranst, E. (2006) Local soil classification and comparison of indigenous and technical soil maps in a Mesoamerican community using spatial analysis. *Geoderma* 135, 140–162.

Belfer-Cohen, A. and Goring-Morris, N. (2009) For the First Time. *Current Anthropology* 50 (5), 669–672.

Bellwood, P. (2009) The Dispersals of Established Food-Producing Populations. *Current Anthropology* 50 (5), 621–626.

Bérard, C., Durand, J.-L. and Lissarrangue, F. (1984) La Cité des Images: religion et société en Grèce antique. F. Nathan, Paris.

Berlin, B. (1992) *Ethnobiological classification: principles of categorization of plants and animals in traditional societies.* Princeton University Press, Princeton (NJ).

Bernard, H. R. (2002) *Research methods in anthropology. Qualitative and quantitative approaches.* Altamira Press, Walnut Creek/London/New Delhi.

Bloch, M. (2009) Une nouvelle théorie de l'art. In A. Gell (ed.) *L'art et ses agents, une théorie anthropologique.* Collection Fabula. Les presses du réel, París, pp. 7–17.

Bogaard, A. and Jones, G. (2007) Neolithic farming in Britain and Central Europe: contrast or continuity? *Proceedings of the British Academy* 144, 357–375.

Bolens, L. (1981) *Agronomes andalous du Moyen Age.* Droz, Genève.

Bonet Rosado, H. (1995) El Tossal de Sant Miquel de Llíria: la antigua Edeta y su territorio. Servicio de Investigación Prehistórica, Valencia.

Bonet Rosado, H. and Mata Parreño, C. (1990) Cabezas votivas y lugares de culto edetanos. *Verdolay* 2, 185–199.

Borgatti, S. (1999) Elicitation techniques for cultural domain analysis. In J. J. Schensul, M. D. L. Compte, B. K. Nastasi and S. P. Borgatti (eds.) *Enhanced ethnographic methods: Audiovisual techniques, focussed group interviews and elicitation techniques* Altamira Press, Walnut Creek, (CA), pp. 115–151.

Boster, J. (1985) Requiem for the omniscient informant: There is life in the old girl yet. In J. Dougherty (ed.) *Directions in cognitive anthropology.* University of Illinois Press, Illinois, pp. 177–198.

Buxó, R., Alonso, N., Canal, D. Echave, C. and González, I. (1997) Archaeobotanical remains of hulled and naked cereals in the Iberian Peninsula. *Vegetation History and Archaeobotany* 6, 15–23.

Buxton, R. (1994) *Imaginary Greece. The Contexts of Mythology*, Cambridge University Press, Cambridge.

Carabaza Bravo, J. M. (1994) Una versión catalana de un tratado agrícola andalusí, In E. García Sánchez (ed.) *Ciencias de la naturaleza en al-Andalus. Textos y Estudios III.* Consejo Superior de Investigaciones Científicas, Escuela de Estudios Árabes, Granada, pp. 169–192.

Casas, A., Caballero, J., Mapes, C. and Zárate, S. (1997) Manejo de la vegetación, domesticación de plantas y origen de la agricultura en Mesoamérica. *Boletín de la Sociedad Botánica de México* 61, 31–47.

Casas, A., Del Carmen Vázquez, M., Viveros, J. L. and Caballero, J. (1996) Plant management among the Nahua and the Mixtec in the Balsas River Basin, Mexico: an ethnobotanical approach to the study of plant domestication. *Human Ecology* 24 (4), 455–478.

Cato, M. P. and Varro, M. T. (1979) *On agriculture.* W. D. Hooper and H. B. (eds.) Harvard University Press, Cambridge/London.

Chauvet, M. (2009) *Note sur les systèmes d'information sur les plantes utilesI.* (Unpublished paper, 9 pp).

Celestino, S. and Jiménez Ávila, F. J. (1996) *El palacio-santuario de Cancho Roano V. El sector oeste.* Diputación Provincial de Badajoz.

Clarke, D. L. (1978) *Mesolithic Europe: the economic basis.* Duckworth, London.

Clément, D. (1998) The historical foundations of ethnobiology. *Journal of Ethnobiology* 18 (2), 161–187.

Cohen, M. N. (2009) Introduction: Rethinking the Origins of Agriculture. *Current Anthropology* 50 (5), 591–595.

Colledge, S., Conolly, J. (eds.) (2007) *The Origins and Spread of Domestic Plants in Southwest Asia and Europe.* University College London, Institute of Archaeology Publications. Left Coast Press, Inc., Walnut Creek, (CA).

Columella, L. J. M. (1979) *On Agriculture.* With a recension of the text and an English translation by E. S. Forster and Edward H. Heffner, Volume III: Res Rustica X–XII. De arboribus. Harvard University Press, London/Cambridge, VII + 435 pp.

Columella, L. J. M. (1979) *On Agriculture.* With a recension of the text and an English translation by Harrison Boyd Ash. Volume I: Res Rustica I–IV. Harvard University Press, London/Cambridge, XXXI + 461 pp.

Columella, L. J. M. (1988) *De los trabajos del campo.* Edición preparada por Antonio Holgado Redondo. Ministerio de Agricultura, Pesca y Alimentación, Madrid, 339 pp.

Columella, L. J. M. (1968) *On agriculture.* With a recension of the text and an English translation by E. S. Forster and E. H. Heffner, Volume II: Res rustica V–IX. Harvard University Press, London/Cambridge, XI + 505 pp.

Cotton, C. M. (1996) *Ethnobotany.* John Wiley & Sons, London.

Cummings, L. S. (1998) Sampling Prehistoric Structures for Pollen and Starch Granules. In *New Developments in*

Palynomorph Sampling, Extraction, and Analysis. Edited by V. M. Bryant and J. W. Wrenn. American Association of Stratigraphic Palynologists Foundation, Contributions Series Number 33, pp. 35–51.

Cummings, L. S. (2007) Phytoliths as Artifacts: Evidence of Threshing on Silica Bodies. *Plants, People and Places: Recent Studies in Phytolith Analysis*. Oxbow Books, Oxford.

Cummings, L. S. and Yost, C. (2008) *Pollen, Phytolith, and Organic Residue Analysis (FTIR) of Samples from Sites FB17153, FB17155, FB5042, and FB5044, Fort Bliss, New Mexico*. Ms. on file with Lone Mountain Archaeological Services, El Paso, Texas. PRI Technical Report 08–48.

Cummings, L. S., Puseman, K. and Dexter, J. (2009a) Pollen, Macrofloral, and Coprolite Analysis of Samples from the Bluff Great House Site, 42SA22674, Southeast Utah. Appendix 4. In C. M. Cameron (ed.) *Chaco and After in the Northern San Juan: Excavations at the Bluff Great House*. University of Arizona Press.

Cummings, L. S., Chad Yost, Puseman, K. and Logan, M. K. (2009b) *Pollen, Starch, Phytolith, Organic Residue (FTIR), and Macrofloral Analyses; Charcoal Identification; and AMS Radiocarbon Dating of Samples from Sites 5DA1656 and 5DA2703, Colorado*. Ms. on file with URS Corporation, Denver, Colorado. PRI Technical Report 09–155.

Cunningham, A. B. (2000) *Applied Ethnobotany: People, wild plant use and conservation*. Earthscan, London.

Cunningham, A. B. (2001) Leaves, lifestyles and landscapes: people and ecological factors in leafy vegetable use. In V. Machakaire (ed.) *Proceedings of the Workshop 'Comparing and contrasting different research approaches on semi and uncultivated food plants'*. Harare.

d'Andrea, A. C. (2003). Social and technological aspects of non-mechanised emmer processing. In Anderson, P. C., Cummings, L. S., Schippers, T. K., and Simonel, B. (eds.) *Le traitement des récoltes: un regard sur la diversité, du Néolithique au présent. Actes des XXIIIe rencontres internationales d'archéologie et d'histoire d'Antibes*, 17–19 Octobre 2002, Éditions APDCA, Antibes, pp. 47–60.

Darmon, J. P. and Schnapp-Gourbeillon, A. (1992) The semantic value of animals in Greek Mythology. In Y. Bonnefoy and W. Doniger (eds.) *Greek and Egyptian Mythologies*. Chicago, pp. 128–133.

Descola, P. (2010) *La Fabrique des images. Visions du monde et formes de la representation*. Somogy, Paris.

Dozy, R. (1961) *Le calendrier de Cordoue*. Nouvelle édition accompagnée d'une traduction française annotée par Ch. Pellat. E. J. Brill, Leiden.

Dunbabin, K. M. D. (1978) The Mosaics of Roman North Africa: Studies in Iconography and Patronage. Clarendon Press, Oxford.

Eguaras, J. (1975) *Ibn Luyun: Tratado de agricultura*. Patronato de la Alhambra, Granada, 276 pp.

Ellen, R. (2006) Ethnobiology and the science of humankind: introduction. *Journal of the Royal Anthropological Institute (N.S.)* (Special issue), S1–S22.

Erdozain, P. (1993) Los espacios rurales y urbanos a través de las fuentes catastrales. Aspectos teóricos y metodológicos. *Noticiario de Historia Agraria* 5, 205–212

Ethnobiology Working Group (2003) *Intellectual imperatives in Ethnobiology*. NSF Biocomplexity workshop report. Missouri Botanical Garden, (MO).

Evershed (2008) Organic residue analysis in archaeology: the archaeological biomarker revolution. *Archaeometry* 50, 895–924.

Fagerholm, N. and Käyhkö, N. (2009) Participatory mapping and geographical patterns of the social landscape values of rural communities in Zanzibar, Tanzania. *Fennia* 187 (1), 43–60.

Fairbairn, A., Martinoli, D, Butler, A. and Hillman, G. C. (2007) Wild plant seed storage at Neolithic Çatal Höyük East, Turkey. *Vegetation History and Archaeobotany* 16, 467–479.

Ford, R. I. (1978) Ethnobotany: historical diversity and synthesis. In R. I. Ford (ed.) *The nature and status of Ethnobotany*. Anthropological papers 67. Museum of Anthropology, University of Michigan, Ann Arbor, pp. 33–50.

Ford, R. I. (2000) Agriculture: an introduction. In P. E. Minnis (ed.) *Ethnobotany: a reader* University of Oklahoma Press, Oklahoma, pp. 243–246

Frake, C. O. (1962) Cultural Ecology and Ethnography. *American Anthropologist* 64 (1), 53–59.

Fritsch, J. (1905) *Fabrication et raffinage des huiles végétales: Manuel à l'usage des Fabricants, Raffineurs, Courtiers et Négociants en Huiles*. H. Desforges, Paris.

Fuller, D. Q. (2007) Constracting patterns in crop domestication and domestication rates: Recent archaeobotanical insights from the Old World. *Annals of Botany* 100, 903–924.

García de Cortázar, J. A. (1969) *El dominio del monasterio de San Millán de la Cogolla (siglos X a XIII): introducción a la historia rural de Castilla altomedieval*, Universidad de Salamanca, Salamanca.

García Sánchez, E. and Álvarez de Morales, C. (eds.) (1990–2004) *Ciencias de la naturaleza en al-Andalus: textos y estudios*. 7 vols. Consejo Superior de Investigaciones Científicas / Escuela de Estudios Árabes, Madrid/ Granada.

Gell, A. (1992) The Technology of Enchantment and the Enchantment of Technology. In J. Coote and A. Shelton (eds.) *Anthropology, Art and Aesthetics*. Clarendon, Oxford pp. 40–66.

Gell, A. (2009) *L´art et ses agents, une théorie anthropologique*, Collection Fabula, Les presses du réel, Paris.

Giebel, M. (2003) *Tiere in der Antike: von Fabelwesen, Opfertieren und treuen Begleitern*. Theiss, Stuttgart.

González Reyero, S. (2010a) La Dama de Galera. Creación, transformación iconográfica e incidencia en las dinámicas sociales, *Rivista di Studi Fenici* 35, 141–160.

González Reyero, S. (2010b) Nature and the symbolization of landscape in the Iberian iconography (III–II BC).

Images and social dynamics. In L. Levêque, L., M. Ruiz del Árbol and L. Pop (eds.) *Heritage, Images Memory of European Landscapes.* L'Harmattan, Paris, pp. 41–57.

Goren-Inbar, N., Sharon, G., Melamed, Y. and Kislev, M. (2002) Nuts, nut cracking, and pitted stones at Gesher Benot Ya'aqov, Israel. *Proceedings Natural Academy of Sciences USA* 99 (4), 2455–2460.

Guardiola, M. D. (1990) Instrumental agrícola en los tratados andalusíes, *Ciencias de la naturaleza en al-Andalus. Textos y estudios I.* Editados por E. García Sánchez. Consejo Superior de Investigaciones Científicas. Escuela de Estudios Árabes, Granada, pp. 107–149.

Guardiola, M. D. (1992) Utillaje de uso agrícola en los tratados andalusíes, *Ciencias de la naturaleza en al-Andalus II. Textos y estudios.* Editados por E. García Sánchez. Consejo Superior de Investigaciones Científicas. ICMA, Madrid, pp. 171–220.

Guđmundsson, G. (1996) Gathering and processing of lyme-grass (*Elymus arenarius* L.) in Iceland: an ethnohistorical account. *Vegetation History and Archaeobotany* 5 (1–2), 13–23.

Hardy, K. (2007) Food for thought: starch in Mesolithic diet. *Mesolithic Miscellany* 18 (2), 2–11.

Harrington, H. D. (1972) *Western Edible Wild Plants.* The University of New Mexico Press, Albuquerque.

Hayden, B. (1996) Feasting in Prehistoric and Traditional Societies. In P. Wiessner and W. Schietenhovel (eds.), *Food and the Status Quest: an interdisciplinary perspective,* Berghahn Books, Oxford, pp. 127–147.

Hayden, B. (2003) Were luxury foods the first domesticates? Ethnoarchaeological perspectives from Southeast Asia. *World Archaeology* 34 (3), 458–469.

Heer, O. (1865) Die Pflanzen der Pfahlbauten. *Neujahrsblätter der Naturforschenden Gesellschaft in Zürich* 1866, 1–54.

Heiss, A. G. and Drescher-Schneider, R. (2012) Umwelt- und Landwirtschaftsgeschichte im Laßnitztal: Erste Ergebnisse von Pollen- und Großrestanalysen aus Brunnen- und Moorablagerungen. *Forum Archaeologiae* 63 (6).

Herrera, G. A. de (1981) *Agricultura general.* Ministerio de Agricultura y Pesca, Madrid.

Hillman, G. C. (1989) Late Palaeolithic plant foods from Wadi Kubbaniya in Upper Egypt: dietary diversity, infant weaning, and seasonality in a riverine environment. In D. R. Harris and G. C. Hillman (eds.) *Foraging and Farming. The Evolution of Plant Exploitation.* Unwin Hyman Ltd, London, pp. 207–239.

Hillman, G. C. (2000) Abu Hureyra 1: the Epipalaeolithic. In A. M. T. Moore, G. C. Hillman and A. J. Legge *Village on the Euphrates. From Foraging to Farming at Abu Hureyra.* Oxford University Press, Oxford, pp. 327–399.

Hodder, I. (1982) *Symbols in action. Ethnoarchaeological studies of material culture.* Cambridge University Press, Cambridge.

Hodder, I. (1987) *The archaeology of contextual meanings.* Cambridge University Press, Cambridge.

Ibn al-Awwam (Ebn El Awwam, Abu Zacaria Iahia Aben Mohamed Ben Ahmed) (1988) *Libro de agricultura. Su autor El doctor excelente... Traducido al castellano y anotado por Josef Antonio Banqueri...* Facsimile of the 1802 edition. Ministerio de Agricultura, Pesca y Alimentación, Madrid.

Ibn Bassal (1955) *Libro de agricultura.* J. M. Millás and M. Aziman (eds.) Instituto Muley el-Hasan, Tetuán.

Ibn Hayyay (1988) *Ahmad b. Muhammad b. Hayyay al-Isbili: al-Muqni fi-filaha.* Universidad de Granada, Granada.

Ibn Wafid (1997) *Tratado de agricultura. Traducción castellana (Ms. s. XIV).* Universidad de Málaga, Málaga.

Ingold, T. (1992) Culture and the perception of environment. In E. Croll and D. Parkin (eds.) *Bush Base: Forest Farm: Culture, Environment and Development,* Routledge, London, pp. 39–56.

Ingold, T. (1993) The temporality of landscape. *World Archaeology* 25, 152–174.

International Society for Ethnopharmacology (2011) *International Society for Ethnopharmacology,* http://www.etnobotanica.de/index.htm

International Society of Ethnobiology (2006) *ISE Code of Ethics. International Society of Ethnobiology,* http://ethnobiology.net/code-of-ethics/

International Society of Ethnobiology (2010) *International Society of Ethnobiology,* http://ethnobiology.net/

Isaksson, S. (1999) Guided By Light: The Swift Characterization of Ancient Organic Matter by FTIR, IR-Fingerprinting and Hierarchical Cluster Analysis. *Laborativ Arkeologi* 12, 35–43.

Jacomet, S. (2009) Plant economy and village life in Neolithic lake dwellings at the end of the time of the Alpine Iceman. *Vegetation History and Archaeobotany* 18, 47–59.

Jacomet, S., Leuzinger, U. and Schibler, J. (eds.) (2004) *Die neolithische Seeufersiedlung Arbon Bleiche 3: Umwelt und Wirtschaft* vol. 12, Archäologie im Thurgau, Frauenfeld.

Johnson, L. M. and Hunn, E. (2010) Introduction. Landscape ethnoecology. Concepts of biotic and physical space. In L. M. Johnson and E. Hunn (ed.) *Landscape ethnoecology. Concepts of biotic and physical space* Berghahn Books, New York/Oxford, pp. 1–26.

Jones, G. and Halstead, P. (1995) Maslins, Mixtures and Monocrops: on the Interpretation of Archaeobotanical Crop Samples of Heterogeneous Composition. *Journal of Archaeobotanical Science* 22, 103–114.

Jouven, M. (1942) *Les plantes à huile.* Éd. de Montsouris, Paris.

Justice, O. L. and Bass, L. N. (1978) *Principles and Practices of Seed Storage.* Agriculture Handbook ed 506. U.S. Department of Agriculture, Washington D.C.

Kislev, M. E., Nadel, D. and Carmi, I. (1992) Epipalaeolithic (19,000 BP) cereal and fruit diet at Ohalo II, Sea of Galilee, Israel. *Review of Palaeobotany and Palynology* 73, 161–166.

Kohler-Schneider, M. and A. G. Heiss (2010) Archäobotanische Untersuchung der latènezeitlichen Siedlung von Michelstetten, Niederösterreich. In E. Lauermann (ed.) *Die latènezeitliche Siedlung von Michelstetten. Die Ausgrabungen des Niederösterreichischen Museums für Urgeschichte in den Jahren 1994-1999* Archäologische Forschungen in Niederösterreich 7. NÖ Institut für Landeskunde, St. Pölten, pp. 116-147.

Kottak, C. P. (2008) *Anthropology. The exploration of human diversity.* 12th edition. Mc Graw Hill International, New York.

Krebs, C. J. (1999) *Ecological methodology.* Benjamin/Cummings Series in the Life Sciences Benjamin/Cummings, California.

Kubiak-Martens, L. (1999) The plant food component of the diet at the late Mesolithic (Ertebølle) settlement at Tybrind Vig, Denmark. *Vegetation History and Archaeobotany* 8 (1–2), 117–127.

Lev, E., Kislev, M. E. and Bar-Yosef, O. (2005) Mousterian vegetal food in Kebara Cave, Mt. Carmel. *Journal of Archaeological Science* 32 (3), 475–484.

López y López, Á. C. (1990) *Kitab fi Tartib awqat al-girasa wa-l-magrusat. Un tratado agrícola andalusí anónimo.* Consejo Superior de Investigaciones Científicas. Escuela de Estudios Árabes, Granada.

Magurran, A. E. (2004) *Measuring biological diversity.* Blackwell, Oxford/Victoria.

Malik Al-Tignari, M. (2007) *Kitab zuhrat al-bustan wa-nuzhat al-adhan. (Esplendor del jardín y recreo de las mentes).* Edición e introducción de Expiración García Sánchez (Ed). Consejo Superior de Investigaciones Científicas, Madrid.

Martin, G. J. (2004) *Ethnobotany: A Methods Manual.* Chapman and Hall, London.

Martin, R. (1971) *Recherches sur les agronomes latins et leurs conceptions économiques et sociales.* Les Belles Lettres, Paris.

Mason, S., Hather, J. G. and Hillman, G. (1994) Preliminary investigation of the plant macro-remains from Dolní Vestonice II, and its implications for the role of plant foods in Palaeolithic and Mesolithic Europe. *Antiquity* 68, 48–57.

Mata Parreño, C., Badal García, E., Collado Mataix, E. and Ripollès i Alegre, P. P. (eds.) (2010) Flora Ibérica: De lo real a lo imaginario. Diputación de València, València.

Meana, M. J., Cubero, J. I. and Sáez, P. (1998) *Geopónica o extractos de agricultura de Casiano Baso.* Ministerio de Agricultura, Pesca y Alimentación, Madrid.

Millas Vallicrosa, J. M. (1943) Tratado de Agricultura by Ibn Wafid, *Al-Andalus* VIII (2), 281–332.

Millas Vallicrosa, J. M. (1948) Tratado de Agricultura by Ibn Bassal, *Al-Andalus* XIII (2), 347–430.

Mingote Calderón, J. L. (1993) *La necesidad de una visión etnológica en los estudios arqueológicos. El mundo agrícola, IV Congreso de Arqueología Medieval Española. Sociedades en transición. Tomo I. Ponencias.* Diputació Provincial d'Alacant. Asociación Española de Arqueología Medieval, Alicante, pp. 57–84.

Mingote Calderón, J. L. (2007) Una aproximación etnológica a aspectos inmateriales asociados a los aperos agrícolas romanos. *Sautuola XIII, Metalistería de la Hispania Romana* (C. Fernández coord.), 589–603.

Mingote Calderón, J. L. (1995) *No todo es trabajo. Técnicas agrícolas tradicionales.* Diputación de Salamanca, Salamanca.

Minnis, P. E. (1981) Seeds in Archaeological Sites: Sources and Some Interpretive Problems. *American Antiquity* 46 (1), 143–152.

Minnis, P. E. (2000) Introduction. In P. E. Minnis (ed.) *Ethnobotany: a reader.* University of Oklahoma Press, Oklahoma, pp. 3–10

Mitchell, W. J. T. (2005) *What do Pictures Want? The Lives and Loves of Images.* University of Chicago Press, Chicago, (IL).

Moldenke, H. N. and Moldenke, A. L. (1952) *Plants of the Bible.* Dover Publications, Inc., New York.

Murray, R. K., Granner, D. K., Mayes, P. A. and Rodwell, V. W. (2000) *Harper's Biochemistry.* 25th Edition ed. McGraw-Hill, New York.

Nadel, D. and Werker, E. (1999) The oldest ever brush hut plant remains from Ohalo II, Jordan Valley, Israel (19000 BP). *Antiquity* 73, 755–764.

Nadel, D., Weiss, E., Simchoni, O., Tsatskin, A., Danin, A. and Kislev, M. (2004) Stone Age hut in Israel yields world's oldest evidence of bedding. *PNAS (Proceedings of the National Academy of Sciencs of the United States of America)* 101 (17), 6821–6826.

Navarro, M. Á. (ed.) (1990) *Risala fi awqat al-sana: un calendario anónimo andalusí.* Escuela de Estudios Árabes, Consejo Superior de Investigaciones Científicas, Granada.

Olmos, R. (1996) Metáforas de la eclosión del cultivo. Imaginarios de la agricultura en época ibérica. *Archivo Español de Arqueología* 69, 3–16.

Olmos, R. (ed.) (1999) *Los Iberos y sus imágenes.* CSIC-Micronet, D.L Madrid.

Olmos, R. (2007–2008) Ex Ilice dictum. La fundación mitológica de la Colonia Iulia Ilici Augusta. *Rendiconti della Pontifica Accademia Romana di Archeologia*, vol. LXXX, 193–215.

Olmos, R. (2008) La simbolización del espacio sagrado en la iconografía ibérica. In Dupré, X.; Ribichini, S.; Verger, S. (eds.) *Saturnia Tellus: definizioni dello spazio consacrato in ambiente etrusco, italico, fenicio-punico, iberico e celtico.* Rome, pp. 251–266.

Olszánski, H. (1989) *Tradycyjne olejarstwo w Polsce.* Muzeum Budownictwa Ludowego, Sanok.

Osborne, R. (1987) *Classical Lanscape with Figures.* London.

Palladius, Rutilius Taurus Aemilianus (1975) *Opus agriculturae. De veterinaria medicina. De insitione.* R. H. Rodgers (ed.). BSB B. G. Teubner Verlagsgesellschaft, Leipzig.

Palladius, Rutilius Taurus Aemilianus (1990) *Tratado de agricultura. Medicina Veterinaria. Poema de los injertos.* A. Moure (ed.). Gredos, Madrid.

Pardo de Santayana, M., Pieroni, A. and Puri, R. K. (2010) The ethnobotany of Europe, past and present. In M. Pardo de Santayana, A. Pieroni and R. K. Puri (eds.) *Ethnobotany in the new Europe. People, health and wild plant resources.* Studies in environmental anthropology and ethnobiology – vol. 14. Berghahn Books, New York/Oxford, pp. 1–15.

Pearsall, D. M. (1989) *Paleoethnobotany: A handbook of procedures.* Academic Press, London.

Pearsall, D. M., Chandler-Ezell, K. and Chandler-Ezell, A. (2003) Identifying maize in neotropical sediments and soils using cob phytoliths. *Journal of Archaeological Science* 30, 611–627.

Peña-Chocarro, L. (1999) *Prehistoric Agriculture in Southern Spain during the Neolithic and the Bronze Age. The application of ethnographic models.* BAR International Series 818. Archaeopress, Oxford.

Perea, A., Williams, D. and Olmos, R. (2007) *El Héroe y el Monstruo. Catálogo de la exposición: El héroe y el monstruo.* Museo Arqueológico Nacional, Ministerio de Cultura 27 de abril–29 de julio de 2007, Madrid.

Pereira, J. L. and Rodríguez, M. (1982) Estructura y tipología de las fuentes notariales en Cáceres y su tierra durante los tiempos modernos. *Norba. Revista de Arte, Geografía e Historia* 3, 191–204.

Pérez García, J. M. (1981) Aproximación al estudio de la penetración del maíz en Galicia. In A. Eiras Roel (ed.) *La historia social de Galicia en sus fuentes y protocolos.* Universidad de Santiago de Compostela, Santiago de Compostela.

Pérez García, X. M. (1982) O millo en Galicia: un estado de la cuestión. *Revista Galega de Estudios Agrarios* 7–8, 87–104.

Perry, L., Dickau, R., Zarillo, S., Holst, I., Pearsall, D., Piperno, D., Berman, M., Cooke, R., Rademaker, K., Ranere, A., Raymond, J., Sandweiss, D., Scaramelli, F., Tarble, K. and Zeidler, J. (2007) Starch Fossils and the Domestication and Dispersal of Chilli Peppers (*Capsicum* spp. L.) in the Americas. *Science* 315, 986–988.

Phillips, O. and A. H. Gentry (1993) The useful plants of Tambopata, Peru. I. Statistical hypotheses tests with a new quantitative technique. *Economic Botany* 47 (1), 15–32.

Pieroni, A. (2001) Evaluation of the cultural significance of wild food botanicals traditionally consumed in northwestern Tuscany, Italy. *Joumal of Ethnobiology* 21 (1), 89–104.

Pieroni, A., Price, L. and Vandebroek, I. (2005) Welcome to Journal of Ethnobiology and Ethnomedicine. *Journal of Ethnobiology and Ethnomedicine* 1 (1), 1.

Piperno, D. R. (2006) *Phytoliths: A Comprehensive Guide for Archaeologists and Paleoecologists.* Alta Mira Press, Lanham, (MD).

Piperno, D. R. and Pearsall, D. M. (1993) Phytoliths in the Reproductive Structures of Maize and Teosinte: Implications for the Study of Maize Evolution. *Journal of Archaeological Science* 20, 337–362.

Pliny the Elder (1949) *Histoire naturelle. Livre XII.* A. Ernout (ed.). Société d'Édition Les Belles Lettres, Paris.

Pliny the Elder (1958) *Histoire naturelle. Livre XV.* J. André (ed.). Société d'Édition Les Belles Lettres, Paris.

Pliny the Elder (1964) *Histoire naturelle. Livre XVII.* J. André (ed.). Société d'Édition Les Belles Lettres, Paris.

Pliny the Elder (1972) *Histoire naturelle. Livre XVIII.* H. Le Bonniec and A. Le Boeuffle (eds.). Société d'Édition Les Belles Lettres, Paris.

Posey, D. (1990) Intellectual Property Rights: And Just Compensation for Indigenous Knowledge *Anthropology Today* 6 (4), 13–16.

Prance, G. (1995) Ethnobotany today and in the future. In R. E. Schultes and S. V. Reis (ed.) *Ethnobotany: evolution of a discipline.* Dioscorides Press/Timber Press, Portland, pp. 60–80.

Price, L. L. (1997) Wild plant food in agricultural environments: A study of occurrence, management, and gathering rights in northeast Thailand. *Human Organization* 56 (2), 209–221.

Price, L. L. (2001) Demystifying farmers' entomological and pest management knowledge: A methodology for assessing the impacts on knowledge from IPM-FFS and NES interventions. *Agriculture and Human Values* 18, 153–176.

Puri, R. K. and Vogl, C. (2004) *A methods manual for ethno-biological research and cultural domain analysis with analysis using ANTHROPAC.* Course Manual. University of Kent, Canterbury.

Quick, C. R. (1961) How Long Can a Seed Remain Alive? In A. Stefferud (ed.) *Seeds, the Yearbook of Agriculture.* U.S. Government Printing Office, Washington D.C., pp. 94–99.

Reyes-García, V., Huanca, T., Vadez, V., Leonard, W. and Wilkie, D. (2006) Cultural, practical, and economic value of wild plants: A quantitative study in the Bolivian Amazon. *Economic Botany* 60 (1), 62–74.

Rist, S. and Dahdouh-Guebas, F. (2006) Ethnosciences – a step towards the integration of scientific and indigenous forms of knowledge in the management of natural resources for the future. *Environment, Development and Sustainability* 8, 467–493.

Robb, J. G. (1998) The ritual landscape concept in archaeology: a Heritage construction. *Landscape research* 23, 159–175.

Rodríguez Ariza, M. O., Gómez Cabeza, F. and Montes Moya, E. (2008) El Túmulo 20 de la necrópolis Ibérica de Tútugi (Galera, Granada). *Trabajos de Prehistoria* 65 (1), 169–180.

Ryan, G. W. and H. R. Bernard (2000) Data Management and Analysis Methods. In N. K. Denzin and Y. S. Lincoln (eds.) *Handbook of Qualitative Methods* Sage

Publications, Thousand Oaks, (CA), pp. 769–802.

Schlumbaum, A., Tensen, M. and Jaenicke-Despre, V. (2008) Ancient plant DNA in archaeobotany. *Vegetation History and Archaeobotany* 17, 233–244.

Scott, L. J. (1983) *A Model for the Interpretation of Pitstructure Activity Areas at Anasazi Sites (Basketmaker III-Pueblo I) Through Pollen Analysis.* Master's Thesis, Department of Anthropology, University of Colorado, Boulder.

Segura, S. and Torres, J. (2009) *Historia de las Plantas en el mundo antiguo.* Universidad de Deusto/Consejo Superior de Investigaciones Científicas, Bilbao/Madrid.

Shanks, M. (2004) *Art and the Early Greek State. An interpretative archaeology.* Cambridge University Press, Cambridge.

Sigaut, F. (2007) Quelques mots sur le tournesol. In A. Durand (ed.) *Plantes exploitées Plantes cultivées.* Publications de l'Université de Provence, Aix-en-Provence, pp. 171–177.

Sillitoe, P. (2006) Ethnobiology and applied anthropology: rapprochement of the academic with the practical. *Journal of the Royal Anthropological Institute* 12 (S1), 119–142.

Smith, J. J. (1993) Using ANTHROPAC 3.5 and a spreadsheet to compute a free-listing salience index. *Cultural Anthropology Methods* 5 (3), 1–3.

Society for Economic Botany (2011) *Society for Economic Botany:* http://www.econbot.org/_welcome_/to_seb.php

Society of Ethnobiology (2011) *Society of Ethnobiology:* http://www.ethnobiology.org/

Stahal, P. W. (1996) Holocene biodiversity: and archaeological perspective from the Americas. *Annual Review of Anthropology* 25, 105.

Stephens, A. M., Phillips, G. O. and Williams, P. A. (2006) Food *Polysaccharides and Their Applications.* CRC Taylor & Francis, Boca Raton, (FL).

Stoffle, R. W., Halmo, D. B., Evans, M. J. and Olmsted, J. E. (1990) Calculating the cultural significance of American Indian plants: Paiute and Shoshone ethnobotany of Yucca Mountain Nevada. *American Anthropologist* 2, 416–432.

Sutrop, U. (2001) List task and cognitive salience index. *Field Methods* 13, 263–276.

Tanno, K. and Willcox, G. (2006) How Fast Was Wild Wheat Domesticated? *Science* 311, 1886.

Tardío, J. and Pardo de Santayana, M. (2008) Cultural importance indices: A comparative analysis based on the useful wild plants of southern Cantabria (northern Spain). *Economic Botany* 62 (1), 24–39.

Teofrasto (1988) *Historia de las plantas.* Díaz-Regañón, J.M. (ed.) Gredos, Madrid.

Thomas, J. (1996) *Time, Culture and Identity.* Routledge, London.

Thompson, R. G. and Staller, J. E. (2001) An Analysis of Opal Phytoliths from Food Residues of Selected Sherds and Dental Calculus from Excavations at the Site of La Emerenciana, El Oro Province, Ecuador. *The Phytolitharien* 13 (2–3), 8–14.

Tilley, C. (1999) *Metaphor and material culture.* Blackwell, Oxford.

Torrence, R. and Barton, H. (eds.) (2006) *Ancient Starch Research.* Left Coast Press, Inc., Walnut Creek, (CA).

Trigger, B. G. (1991) Post-processual development in Anglo-American Archaeology. *Norwegian Archaeological Review* 24 (2), 65–76.

Turner, N. J. (1988) 'The Importance of a Rose': Evaluating the Cultural Significance of Plants in Thompson and Lillooet Interior Salish. *American Anthropologist* 90 (2), 272–290.

Verschuren, P., H. Doorewaard and R. Poper (1999) *Designing a research project* LEMMA, Utrecht.

Virgil (1986) *Eclogues. Georgics. Aeneid, 1-6.* With an English translation by H. Rushton Fairclough. Harvard University Press. William Heinemann Ltd., Cambridge, (MA)/London.

Weiss, E., Wetterstrom, W., Nadel, D. and Bar-Yosef, O. (2004) The broad spectrum revisited: Evidence from plant remains. *Proceedings of the National Academy of Sciences of the United States of America* 101 (26), 9551–9555.

Willcox, G. (2005). The distribution, natural habitats and availability of wild cereals in relation to their domestication in the Near East: multiple events, multiple centres. *Vegetation History and Archaeobotany* 14, 534–541.

Zapata, L., Peña-Chocarro, L., Pérez Jordà, G. and Stika, H.- P. (2004) Early Neolithic agriculture in the Iberian Peninsula. *Journal of World Prehistory* 18 (4), 283–325.

Zohary, D., Hopf, M. and Weiss, E. (2012) *Domestication of Plants in the Old World.* Oxford University Press, Oxford.

SECTION 2

Food Plants

3 Crop Diversity Through Time

3.1. INTRODUCTION

Elena Marinova

Crop diversity is a key element for maintaining sustainable agriculture and reducing risk and uncertainty in crop production. It is also an important factor for economic stability, as it provides the genetic source for better adaptation to changing environments, resistance to pathogens and crop improvement in general. Moreover, crop diversity has a fundamental role in providing a higher quality of life through variation in tastes, flavours, colours, and the size of certain plant products, all of which feed back into food diversity. Studies of the genetic diversity of crops point to the existence of a wide variety of factors affecting farmers' strategies which, in turn, determine their choices. Factors influencing crop choice and diversity fall into the realm of different worlds: natural, social, cultural, economic and political, in which communities are embedded.

Many studies have emphasised the role of environmental factors to explain crop diversity. Temperature, rainfall, altitude, wind, soil types, hydrology, etc., are some of the basic constraints affecting land use, and therefore crop diversity. On the other hand, anthropological and sociological research focuses increasingly on human societies and their behaviour, elucidating a strong link between crop diversity and culture and society. Farmers' decisions on what to grow have an impact on the variability of species and varieties grown at household level, and this was probably also the case in the past. However, households are part of much larger complex systems which in turn influence decisions. In this complex scenario, crop diversity appears as the outcome of the interaction and co-evolution between domestic species and human cultures over millennia.

In the following chapter, the changes in crop diversity over the past 8,000 years will be illustrated with examples from Europe and America. Evidence for crop diversity in the past has been primarily collected from the archaeobotanical record in various regions. Seeds, fruits and other plant remains represent one of the main proofs of the crops used by past communities. The archaeobotanical material considered in this chapter consists mainly of plant macro-remains preserved in charred, mineralised or desiccated states. Of course, these remains do not represent the full range of crops used by past populations; instead, they only correspond to a small part of what was actually consumed. To a large extent, preservation and taphonomy – the ways in which plant and animal remains are deposited in the soil and the different ways in which they are preserved – play a key role in their conservation in the archaeological record (for more on these issues see Chapter 2). In any case, the most common and important useful plants appear to be represented in the archaeological layers and their presence is traceable, thus archaeobotanical data can be a valuable source for exploring crop diversity in the past. Yet, it should be stressed that the archaeobotanical methods dealing with charred plant material have strong limitations for estimating crop diversity at landrace level.

Only broader tendencies at higher taxonomic level (species, genus) can be observed on different spatial scales. In addition, the temporal scale at which we operate is large, ranging from several hundreds of years to millennia.

This chapter focuses on crop diversity through time. The contributions show the different crop associations and combinations and highlight factors affecting such variability from the very beginning of agriculture on and focus on different parts of Europe and America. The chapter begins with an overview of past crop diversity in southeastern Europe (Bulgaria and northern Greece) for the period 6500–1200 cal. BCE (see Chapter 3.2) by E. Marinova and S.M. Valamoti. The area under consideration is characterised by different environmental conditions (climate, geology, flora, fauna etc.) including examples from true Mediterranean settings to temperate and continental climates. But also the geographical position of the region, at a crossroads of cultural exchange of elements between Europe and the East, provides an excellent opportunity for tracing changes in crop diversity through time. Several common trends have been identified or re-confirmed, such as the predominance of the glume wheats, the important role of pulses, the clear visibility of flax and the introduction of various crops during the Bronze Age, probably through cultural contacts. At the same time, interesting patterns of crop diversity start to emerge which in the future will allow a broadening of the range of issues to be explored using archaeobotanical data. Some examples are the predominance of einkorn in certain parts of Bulgaria and northern Greece, the appearance of free-threshing wheats on the Black Sea coast of Bulgaria during the Chalcolithic, the presence of chickpea in the later phases of the Early Neolithic of southern Bulgaria, the absence of Celtic bean and millet in the Neolithic of both regions, and the total absence in the Bulgarian archaeobotanical record of a wide range of crops encountered in Greece during the Bronze Age, in particular oil, medicinal and hallucinogenic plants.

Crop diversity in northern Italy, presented in Chapter 3.3, is discussed within the context of the Mediterranean and European cultural influences that shaped agriculture from the Neolithic to the Roman age. According to M. Rottoli, a key factor determining crop diversity is population movement, either as a result of invasions or of peaceful migrations like those related to the establishment of trading outposts or foundation of colonies. By exploring different contexts, studied through archaeobotanical analysis, such as the Neolithisation process, the interaction between colonies or military camps and the local population, or the spread of rye in northern Italy, the author shows the influence which people's movements had on crop diversity at different times. Other scenarios in which crop diversity may have been affected are also considered, such as those related to institutional impositions or even those involving single individuals spreading crop varieties across particular territories.

For southwestern central Europe (Chapter 3.4), S. Jacomet summarises the results of archaeobotanical research covering a period of more than of 7,500 years. The author uses data from *ca.* 140 sites studied over thirty years illustrating the range of different plant species cultivated for food and other purposes since the Neolithic period and exploring the reasons for changes through time. Cultural factors appear as a driving force for changes in crop diversity. The existence of trade networks, in some cases over long distances, is one of the factors put forward by the author to explain the arrival of some crops and, in turn, the increase in crop diversity. Naked wheats, both tetraploids and hexaploids, as well as some weedy taxa and other species such as parsley, dill, celery, etc., seem to enter the Alpine zone between 4600 and 4000 BCE through networks already existing with other European regions. This would be the case of the millets during the Bronze Age, as well. However, environmental constraints are also considered for the spread of some species such as spelt, which according to Jacomet, may be related to soil impoverishment and climate.

A less known region, southwestern Europe, is introduced in Chapter 3.5 by L. Peña-Chocarro and L. Zapata. The authors explore early agriculture in the Iberian Peninsula for the period between 5700–4000 cal. BCE. The region under consideration is a large territory characterised by a multiplicity of environmental settings to which farming had to adapt. Early Neolithic agriculture in the Iberian Peninsula was one of the most varied on the whole of the continent. L. Peña-Chocarro and

L. Zapata emphasise the multiple factors that could have influenced farmers' decisions to grow specific crops giving rise to the various situations encountered in the archaeological record, which reflects the complexity of Neolithic farming communities. Amongst the factors considered for explaining patterns of crop diversity, crop ecological requirements could partially explain specific crop combinations. In addition, diversity may also aim at reducing risks of crop failure. However, the authors point out other factors guided mostly by human decisions related to social and cultural traditions, which could also have influenced diversity.

From a later period and from another European region come the examples presented by C. Bakels in (Chapter 3.6).Crop choice and the underlying reasons for it in western central Europe from 500 BCE to CE 900 are the focus of this contribution. Bakels looks at different factors determining crop diversity in the period considered. Soil types, such as sandy as opposed to loess, account for variability in crop species. In addition, the profound economic and social transformation which occurred during Roman times resulted in changes in production as specialisation was probably imposed by a new market economy. Availability of labour and cultural preferences are issues also investigated in this chapter.

Moving forward in time, the contribution by the late F. Sigaut (Chapter 3.7) focuses on the major crops and agricultural developments in western Europe, highlighting the various changes that occurred in the Middle Ages and Early Modern times, both in crops and in technical aspects. The paper draws attention to crucial elements, like the appearance of iron, for the development of new agricultural tools and innovations, but also to those elements that persisted through millennia. Two main aspects are at the core of the paper: innovation in techniques and the appearance of new crops, which can help to better understand crop diversity and crop choice in the past.

From Europe, discussion moves to America with a contribution by L. Scott-Cummings (see Chapter 3.8) on crop diversity and choices in the prehistoric American southwest. The paper introduces a new data set to understand crop diversity: the analyses of coprolites which included pollen, plant macrofossils and phytoliths. Coprolites provide evidence of the rich and nutritive diet of Puebloan people, which included both cultivated and wild plants. The analyses also demonstrate some improvements in food preparation which enhanced food taste. The chapter brings in a new region and new techniques which enable us to make progress in the study of crop diversity.

M. Bruno presents an excellent example of crop diversity in her contribution (Chapter 3.9). The region under consideration is the basin of Lake Titicaca during the period between 1500 BCE–1000 CE. Through the archaeobotanical study of three different sites, M. Bruno deals with the various factors affecting crop diversity which came to light after a process of interaction between environment, culture, politics and production that started during the Formative Period.

As these papers all show, crop diversity should be understood as part of the dynamic processes that characterise the interaction between humans and their natural, social and cultural environments. The aim of the chapter, whilst contributing to the better understanding of the mechanisms involved in crop diversity, has a specific focus in looking at the processes through time. The various contributions illustrate the way different past cultures have dealt with crop diversity according to the changes in their natural and social environments.

3.2. CROP DIVERSITY AND CHOICE IN PREHISTORIC SOUTHEASTERN EUROPE: CULTURAL AND ENVIRONMENTAL FACTORS SHAPING THE ARCHAEOBOTANICAL RECORD OF NORTHERN GREECE AND BULGARIA

Elena Marinova and Soultana-Maria Valamoti

Introduction

Southeastern Europe, due to its geographical location, is at the crossroads where cultural elements from Europe and the East interact. The area is also characterised by quite diverse environmental conditions (climate, geology, flora, fauna, etc.) encompassing climate regimes as diverse as the true Mediterranean to temperate and continental climatic conditions and transitional ones between those three. Moreover, various geographical barriers such as numerous mountain ranges add further to the environmental diversity of the area (Fig. 3.1), and also influence communication routes between different cultural groups. In this paper, we will discuss crop diversity in a time period ranging from the earliest Neolithic in the area to the Final Bronze Age, *i.e.* approximately between 7000–1000 BCE in this region, and attempt to investigate the factors shaping it.

Chronological Frameworks

The approximate chronological correlation between the regions considered is given in Fig. 3.2. The terminology used to refer to time periods is the one used for each country, not a unified one, thus dates and immediate comparison are given. The beginning of the Neolithic in Greece is placed in the middle of the seventh millennium BCE (though earlier dates are available for some sites, Perlès 2001) and ends in the middle of the fourth millennium BCE (*ca.* 3500–3200 BCE). The period between 4500–3500 BCE is also known as the Final Neolithic (Andreou *et al.* 1996; Papathanasopoulos 2001). The Bronze Age covers the time span between 3500–1100 BCE.

In Bulgaria, the Neolithic Period starts at the end of the seventh millennium BCE and continues until 4900 BCE, followed by the Chalcolithic period (4900–3800 BCE). The latter corresponds roughly to the last phases of the Late and the Final Neolithic in Greece (Fig. 3.2). The Bronze Age follows, covering the time span from approximately 3000–1200 BCE.

Natural Conditions

The modern climate in northern Greece has a gradient from a Mediterranean climate along the coast of the Aegean to a more continental climatic regime further inland. In the territory of southern Bulgaria today, the climate is sub-Mediterranean, sub-continental to continental from south to north. The mean annual temperatures run from 16°C to 10°C from the south to the north of the study area. The precipitation regimes in the south have two maxima (April and November) and average annual precipitation of 550–700 mm. In the northern part of the region, they have one maximum in June and an annual precipitation of 450–580 mm. This affects vegetation which ranges from Mediterranean *maquis* to a mixed deciduous oak woodland (see Bohn *et al.* 2003).

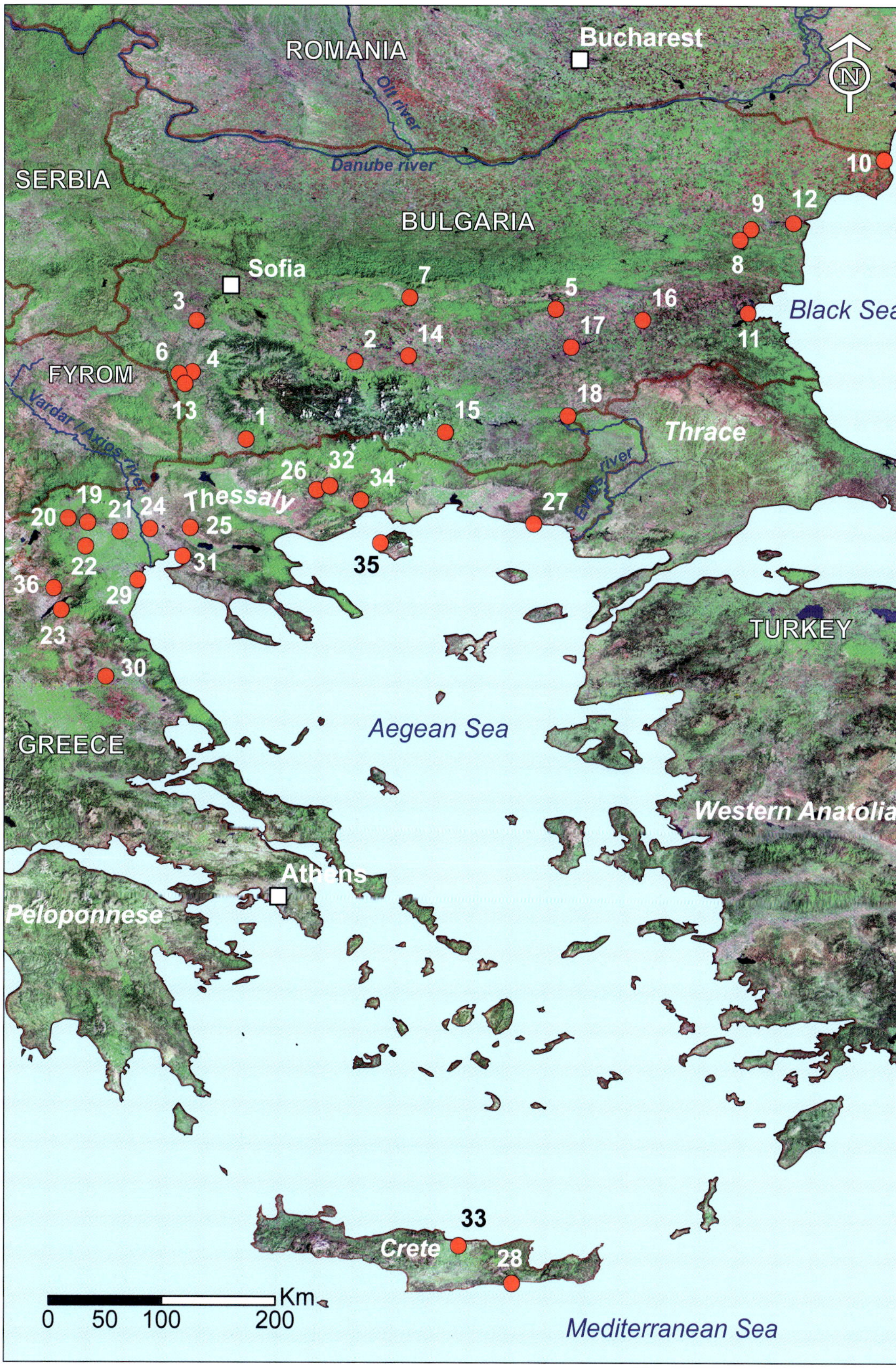

Fig. 3.1. Map of the study area. Sites mentioned in the text: 1) Kovacevo; 2) Tell Kapitan Dimitrievo; 3) Galabnik; 4) Balgarchevo; 5) Tell Karanovo; 6) Drenkovo; 7) Dabene Sarovka; 8) Provadia; 9) Sava; 10) Durankulak; 11) Burgaska Mogila; 12) Varna Chalcolithic necropolis; 13) Kamenska Chuka; 14) Nebet Tepe; 15) Adata; 16) Okrazhna Bolnitsa; 17) Goljama Detelina; 18) Kush Kaja; 19) Mandalon; 20) Apsalos; 21) Archontikon; 22) Angelochorion; 23) Servia; 24) Kastanas; 25) Assiros Toumba; 26) Sitagroi; 27) Makri; 28) Mirtos; 29) Makrygialos; 30) Otzaki; 31) Toumba Thessalonikis; 32) Arkadikos; 33) Knossos; 34) Dikili Tash; 35) Skala Sotiros; 36) Kremasti Koilada. Map: R. Lugon, J.-C. Loubier and A. Chevalier.

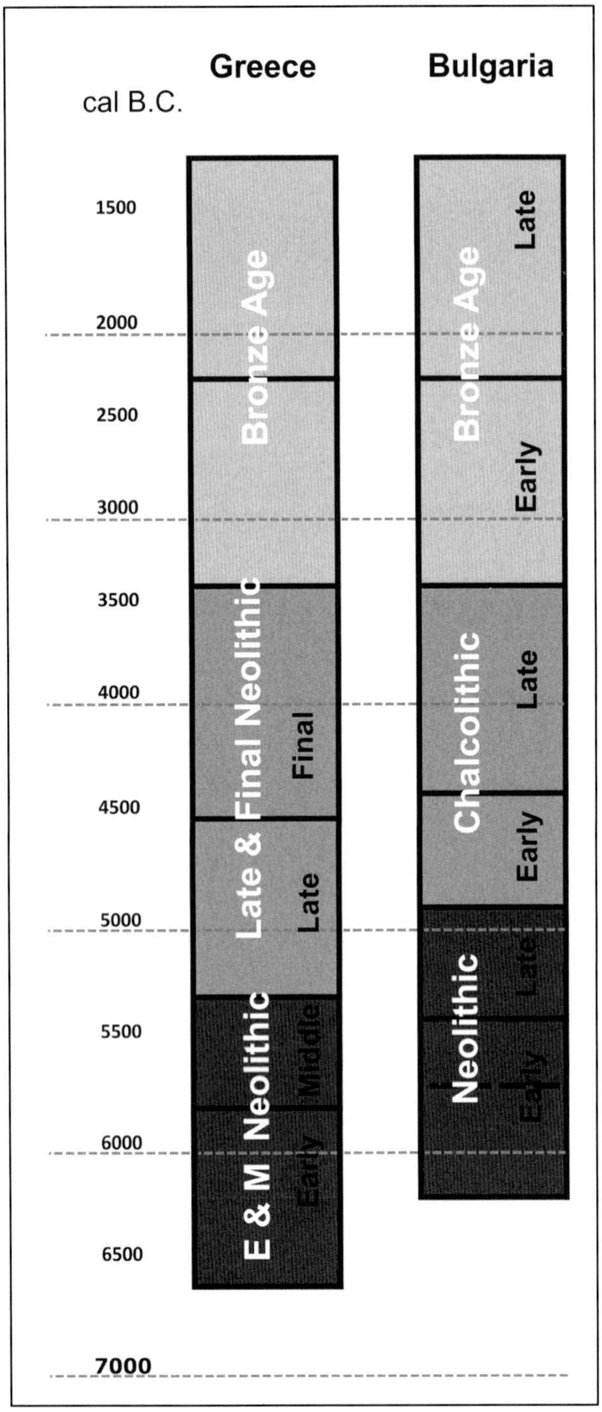

Fig. 3.2. Chronological framework of the considered regions - the main periods and their duration.

In order to avoid repetition, the evidence for past environmental conditions, climate and vegetation change will be given in the last section of the paper dealing with the interaction between human societies and their natural environment. The study area with the regions under consideration is represented in Fig. 3.1.

Crop Diversity and its Diachronic Development

In order to discuss crop diversity, it is necessary to set out the criteria used to identify a species present at a site as a crop used by its inhabitants. Obviously species present in pure, dense rich concentrations provide the main criterion for identifying these crops, yet, in certain cases, the frequent occurrence of certain species, albeit in small numbers, is also discussed. As we will be comparing different regions of southeastern Europe, it is also important to mention that not all periods or areas are equally well represented by archaeobotanical remains, thus some of our observations may be altered in the future. For example, although approximately 79 sites from Greece are represented by archaeobotanical remains (*e.g.* Hansen 2000; Megaloudi 2006), those with many samples, retrieved by flotation and fully published, come mainly from northern Greece (Valamoti 2009). For Bulgaria, the main amount of systematically studied sites (many samples retrieved with flotation) belongs to its southern part (to the south from the Stara Planina Mountains) (Marinova 2006; Popova 2009) – thus far, over 30 sites out of a total of 80 sites with archaeobotanical information are available, covering the period under consideration. For the northern part, systematic studies are available from nine sites out of a total of 21 with archaeobotanical evidence published, all of them situated in the north-central and northeastern part of Bulgaria, with the exception of one Neolithic site (Marinova 2009) from the northwest. For the adjacent areas of southeastern Europe, we mainly rely on the data published by Borojević (2006), Monah and Monah (2008) and Fischer and Rösch (2004).

Cereals and Pulses

From the beginning of the Neolithic, cereals and pulses characterise the archaeobotanical assemblages in Greece and Bulgaria. The so-called Neolithic Near Eastern 'crop package' (according to Zohary 1996) consists of emmer (*Triticum dicoccum* Schübl.), einkorn (*Triticum monococcum* L.), hulled barley (*Hordeum vulgare* L.), lentil (*Lens culinaris* Medik.), pea (*Pisum sativum* L.), bitter vetch (*Vicia ervilia* (L.) Willd.), chick pea (*Cicer arietinum* L.) and flax (*Linum usitatissimum* L.). In Greece, the Early Neolithic crop species encountered are einkorn, emmer, free-threshing wheat, two-row hulled and naked barley, and hulled six-row barley, lentils, bitter vetch, pea, grass pea, and chickpea (represented by a few seeds only) (Valamoti and Kotsakis 2007). Grass pea (*Lathyrus sativus* L.), does not belong to the package, but according to Zohary *et al.* (2012) probably was included relatively early in it, as charred concentrations of it occurred in the pre-pottery Neolithic of southeastern Anatolia (Gritille) – pointing at least to its use if not explicitly domestication (Zohary *et al.* 2012). The main finds of pure dense concentrations of seeds of grass pea occur in several Neolithic sites in the archaeobotanical record of southeastern Europe. Thus, it is possible that the geographical area under consideration, due to its rich floral diversity, might have contributed plant species in the Near Eastern 'crop package'. In the southern part of Greece, some of the wild progenitors of the first cultivated plants (barley and lentils) were available during Mesolithic times, so the inhabitants were already familiar with harvesting wild plants and perhaps cultivating them (Valamoti and Kostakis 2007). Although one could question the concept of a clearly formed package for some early Neolithic sites in Greece, further to the north, at the Bulgarian Neolithic sites it seems that for the Neolithic, archaeobotanical assemblages correspond to the Near Eastern crop assemblage (with the addition of grass pea), especially in view of the increasingly up-to-date archaeobotanical information (Marinova 2006). In neighbouring regions like Serbia, Romania and eastern Hungary, belonging to related Neolithic cultures, this is visible in the archaeobotanical assemblages from the Early Neolithic, as well (Colledge and Conolly 2007a). There is also some reduction of the crops used, such as the absence of barley, free-threshing wheat and several leguminous crops at the earliest Neolithic sites in western Hungary, Austria and Germany (see Kreuz *et al.* 2005; Coward *et al.* 2008; Bogaard *et al.* 2008). This reduction in crop diversity observed for the Neolithic, with a gradient from the south to the north, is intriguing indeed and an interaction of cultural and environmental factors may have led to this pattern.

The cereal crops in the study area were dominated by the hulled wheats. The evidence from northern Greece from sites dated to the sixth and fifth millennia BCE (Late and Final Neolithic levels) indicates that einkorn is dominant at most sites in the area. A similar predominance of einkorn is observed at certain sites at least during the Bronze Age (*e.g.* Assiros, Kastanas, Archondiko: Jones *et al.* 1986; Kroll 1983; Valamoti *et al.* 2008; Valamoti 2009). Concerning the contemporary sites in Bulgaria, a similar picture is observed. During the Bulgarian Early Chalcolithic (4900–4500 cal. BCE), dominance of emmer is visible in the archaeobotanical record from a few sites in southern Bulgaria (Marinova *et al.* 2002; Marinova 2006), but at the majority of sites from this period, einkorn is the most important cereal crop. This is also true for the Late Chalcolithic (4500–3900 cal. BCE). For the Early Bronze Age (*ca.* 3500 cal. BCE) of Bulgaria, the tendency of domination of einkorn is visible in many sites, together with quite high quantities of hulled barley, as well (Hajnalová 1980; Popova 1995; Marinova 2003; Marinova 2004). In the earliest stages of the Neolithic of Bulgaria, the cereal crops were also dominated by einkorn and in certain areas this continues throughout the whole prehistoric period. In the Thracian plain, it seems that at least during the second part of the Early Neolithic (5700–5400 BCE) and, during the Bulgarian Late Neolithic (5400–4900 cal. BCE), the dominant crop in most sites was emmer.

Until recently, einkorn and emmer were considered to be the only wheats used in the prehistoric period, but it gradually became evident that another wheat type was also present: the 'new glume wheat-type' (Fig. 3.3), resembling *Triticum timopheevi* Zhuk., as defined by Jones *et al.* (2000). The finds of this wheat first described in several sites from the Neolithic and Bronze Age of Greece provide an example of early crop diversity, unrecognised before this. Nearly all Neolithic sites investigated archaeobotanically in northern Greece have yielded chaff concentrations of the new glume wheat type such as Kremasti Koiladas, Mandalo, Apsalos, Makriyalos, Makri, Arkadikos (Jones *et al.* 2000; Valamoti 2004; Valamoti 2006; Karathanou and Valamoti, 2011). In the Bulgarian Early Neolithic, the recent studies of storages from Tell Kapitan Dimitrievo, *ca.* 5700 cal. BCE (Marinova 2011), revealed that about one third of all (over 120 items) of the glume bases of hulled wheat found have the morphological features of the 'new wheat-type'. Another example from the early Chalcolithic Tell Provadia (4700 cal. BCE) showed that in a storage site dominated by einkorn about 14% of the glume bases found have the morphological characteristics of the 'new wheat type'. There is also strong evidence that the

Fig. 3.3. New glume wheat type, spikelet fork, from Apsalos, Greece; after Valamoti 2006.

materials, once published as *Triticum spelta* L., from the Early Bronze Age of Dabene Sarovka (Marinova 2003) could be of the new wheat type.

Further examples for this *T. timopheevi*-like hulled wheat in the adjacent areas represent the finds from the Chalcolithic (*ca.* 4700–4500 cal. BCE) in southwestern Romania (Fischer and Rösch 2004) or Neolithic of eastern Hungary (Bogaard *et al.* 2008). This evidence indicates that probably with increasing 'up-to-date' studies in the region, the picture of the importance of the 'new' wheat-type (also 'new glume wheat') will become clearer and points toward wider spread of this crop in the study area.

Another hulled wheat which is of certain importance in the study area is spelt (*Triticum spelta*). Evidence for its presence as a crop comes from Archondiko and is dated to the end of the third millennium BCE, yet it is a grain concentration (Valamoti *et al.* 2008) and doubts on the reliability of identification of spelt based on grain have been raised. Sporadic spelt glume bases come from Bronze Age Kastanas (Kroll 1983), while it is certainly identified as a stored crop at Assiros (Jones *et al.* 1986). Sporadic finds of spelt have also been published from Bronze Age sites (2500–1200 cal. BCE) in the Thracian plain (Popova 1995). It should be stressed that several identifications, published as *Triticum* cf. *spelta*, from earlier Bulgarian prehistory (Late Neolithic to Chalcolithic) could probably belong to the 'new' wheat-type. Such identification issues make any discussion of spelt in the context of prehistoric southeastern Europe premature and problematic at present. The identification issues will be not considered further as they are not the focus of the current paper.

Evidence for the presence of free-threshing wheat (tetra- or hexaploid) in the study region is available from the earliest stages of the prehistoric period on (Hopf 1978; Marinova 2007; Evans 1964, Valamoti 2009).

The relatively scarce evidence for naked wheat from the Neolithic to the Bronze Age in the area under consideration is based in many cases on cereal grains with the corresponding morphological features. In very few cases the more diagnostic rachis fragments of the ears were also found. Free-threshing wheats in Neolithic Greece, although occasionally present, mainly as a few rachis internodes (*e.g.* at Late Neolithic Makri), are very scarce indeed with the exception of the Early Neolithic find from Knossos consisting of a rich grain concentration (Evans 1964). It is during the Bronze Age that they are more prominent in the archaeobotanical record of Greece at sites such as Archondiko (end of third millennium BCE) and Assiros (end of second millennium BCE) (Valamoti *et al.* 2008; Jones *et al.* 1986). In contrast, these wheats are encountered in rich charred concentrations at sites earlier than those of northern Greece: from the Bulgarian Middle/Late Chalcolithic storage finds are known, so the period between 4500–3900 cal. BCE is one of the peaks of its importance, especially for the Black Sea coastal area and eastern Thracian plain. A similar tendency is also observed in the Chalcolithic period of western and eastern Romania and well-illustrated at several sites (Fischer and Rösch 2004; Monah 2007; Monah and Monah 2008). During the Early Bronze Age, the free-threshing wheats seem to disappear from the archaeobotanical record of Bulgaria, to reappear again only from the first millennium BCE onwards at several sites in Bulgaria. Thus, the situation between Greece and Bulgaria is different concerning the use and importance of free-threshing wheats. Their predominance in the eastern part of Bulgaria among sites belonging to the Varna culture, well known for rather early gold finds, might suggest that this is related to very culture-specific influences, connections and exchange of technological knowledge affecting these crops.

Besides wheat, another cereal crop of importance that played a role in crop diversity during the period considered was barley (*Hordeum vulgare*). In Greece, both naked and hulled, two-row and six-row barley have been identified among Neolithic contexts (Hansen 2000; Megaloudi 2006; Valamoti 2004). Several varieties of it were also recorded for Bulgaria: it seems that in the Early Neolithic the hulled forms are more important and later, during the late Neolithic and Chalcolithic (5500–3500 cal. BCE), the naked form gains importance. During the Bronze Age in the study area, barley – especially its hulled form – is more frequently encountered.

Millet (*Panicum miliaceum* L.) is a cereal crop which played a certain and probably quite important role in the study area, but only from the Late Bronze Age onwards both in Greece and Bulgaria. Rich finds of millet have been recovered at LBA Assiros (Jones *et al.* 1986), Kastanas, (Kroll 1983), Angelochori (Valamoti 2010) and five LBA sites in southern Bulgaria: Kamenska Chuka, Nebet Tepe, Adata, Okrazna Bolnitza, Goljama Detelina (see Popova 2009; 2010). Early Neolithic finds of millet from Greece are most likely intrusive with the exception of the only millet concentration from Olynthus, possibly dated to the end of the Neolithic period (Mylonas 1929, Valamoti 2007). Single finds for the Neolithic are known from two sites (Drenkovo and Kapitan Dimitrievo) in Bulgaria (Kreuz *et al.* 2005, Marinova, *forthcoming*), but we cannot exclude the possibility that some of them may be of intrusive character, considering both that they consist of single grains and the evidence for the plant's origin (Hunt *et al.* 2008). Late Chalcolithic finds of millet from the Black Sea coast of Bulgaria – Durankulak (Marinova 2006) and Provadia (Marinova 2008) – could represent some of the routes of migration of this crop from the steppe area situated north of the Black Sea.

The leguminous crops or pulses are a very important group of cultivated plants grown in the area, most abundant and numerous of which were lentils (*Lens culinaris*), pea (*Pisum sativum*), grass pea (*Lathyrus sativus/cicera*) and bitter vetch (*Vicia ervilia*) throughout the Neolithic and the Bronze Age. Celtic bean, however, is only found at Greek Bronze Age sites (*e.g.* Mandalo, Skala Sotiros, Angelochori) and not in Bulgaria. During the prehistoric period, they played an important role throughout southeastern Europe (*e.g.* Sarpaki 1992; Halstead 1981), but also in Bulgaria and at least Romania and Serbia (see data summary in Kreuz *et al.* 2005; Fischer and Rösch 2004; Borojević 2006).

The grass pea – already mentioned in the introduction and considered to be a southeastern European addition to the founder crop assemblage – is of special interest (Kislev 1989). Furthermore, for the whole period under consideration (6100–1200 cal. BCE) in the study area, this crop is present in continuous evidence of storage finds indicating its cultivation not only during the Neolithisation processes, but throughout the prehistoric period. This evidence also shows its importance for ancient subsistence in the region. A certain increase in pulse crop diversity in the northern part of the area under consideration is represented in the finds of chick pea (*Cicer arietinum*). They appear during the final stages of the Bulgarian Early Neolithic (5700–5500 cal. BCE) at sites in southern Bulgaria (Marinova and Popova 2008). The finds of chick pea come from four sites with a ubiquity of *ca.* 25% of the studied samples from three of the sites and a find of a concentration of chick pea in a vessel containing over 300 seeds. From Greece and particularly Thessaly, only one early Neolithic find is known from Otzaki, approximately 6000 cal. BCE (Kroll 1981), otherwise it is absent (Valamoti 2009). These finds cover a period slightly preceding the Bulgarian early Neolithic finds of *C. arietinum*. Until now, there has been no evidence for *C. arietinum* from the Neolithic of neighbouring areas further to the northwest, in Serbia and Romania (around 5800–5600 cal. BCE). This could be connected with the state of knowledge and may not necessarily mean that chick pea was absent there. This hypothesis can also be supported by the finds of chick pea in the layers of two of the sites (Galabnik and Balgarchevo), which are strongly related to the Neolithic cultures of Serbia and adjacent territories to the west (Marinova and Popova 2008). Yet, its absence from northern Greece is unlikely to be due to lack of adequate data. Therefore, a probable option is that the finds from the Bulgarian early Neolithic correspond to direct contacts with Anatolia, which took place through the eastern part of the study area and not necessarily through Thessaly. Given the scarce evidence available from eastern Thrace, further studies are needed to confirm or reject this suggestion.

Crops other than Cereals and Pulses

Flax (*Linum usitatissimum*) is found in rich concentrations from the first half of the fifth millennium BCE in Greece onwards at sites such as Makriyalos (Fig. 3.4), Arkadikos (Valamoti 2004) and Dikili Tash (Valamoti 2011). For Bulgaria, several finds of flax seed concentrations were published for two Early Neolithic, one Chalcolithic and one Bronze Age site. Textile remains, consisting of flax fibres conserved in the copper patina of a grave inventory from the Varna Chalcolithic necropolis (*ca.* 4200 cal. BCE), indicate the use of a fibre-producing variety of flax (Fig. 3.5). From the Late Bronze Age of the site Kush Kaja in southwestern Bulgaria, charred food remains from a vessel also provide evidence for the use of flax as food (Marinova in Popov and Ilev 2007).

For the Bronze Age in the third millennium, the oil-rich seeds of *Lallemantia* sp. Fisch. and C.A. Mey have been identified in rich pure concentrations at Mandalo and Archondiko and later from a fourteenth-century BC destruction layer at Assiros (Jones and Valamoti 2005). As this plant is not native to Europe, it is suggested that it arrived either as imported seed or as an introduced crop from the east through cultural contacts (Jones and Valamoti 2005; Valamoti and Jones 2010)

The occurrence of the crop in the territory of Bulgaria is still less explored due to the rather recent identification of the plant in the archaeobotanical record. Revisions of earlier finds and future studies will give a more detailed picture and most probably extend the oil crop diversity with this plant. Fig. 3.6 provides an example from an earlier, not recognised (Marinova 2003) Bulgarian find of *Lallemantia* at the site Dabene-Sarovka EBA II (2700–2500 cal. BCE). The seed shown in Fig. 3.6 comes from a concentration of over 80 charred seeds from the storage area of a house. Other crops yielding seeds rich in oil have been identified from Bronze Age contexts in Greece, but not in Bulgaria. These crops are *Brassica* sp. L., *Papaver somniferum* L. ssp. *somniferum*, and *Camelina* sp. Crantz., suggesting a rather wide variety of

Fig. 3.4. Seeds from flax from Makriyalos, Greece Late Neolithic.

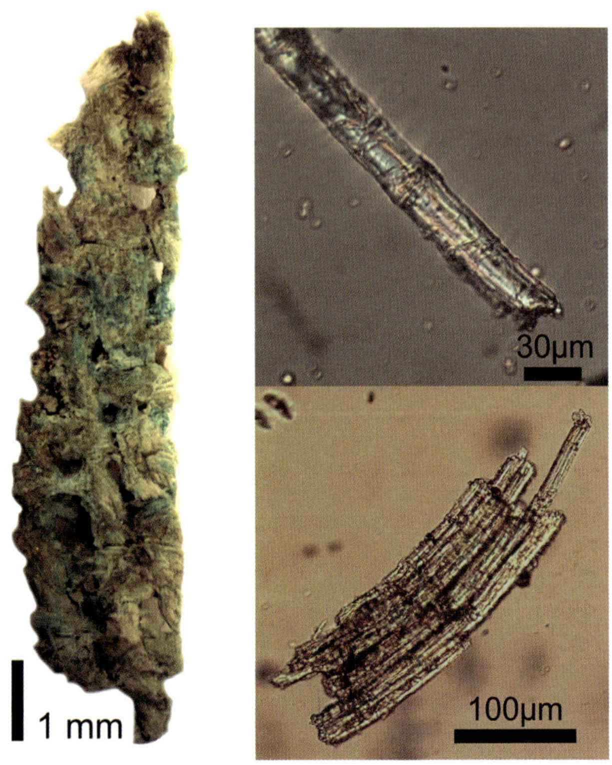

Fig. 3.5. Textile and fibre of flax from patina on copper object from the Chalcolithic necropole, Varna, Bulgaria.

Fig. 3.6. *Lallemantia* sp. from Early Bronze Age II storage at the site Dabene-Sarovka, Bulgaria.

crops potentially used for oil in the Bronze Age of northern Greece (Jones and Valamoti 2005; Andreou *et al.* 2013).

The Bronze Age occurrence of safflower (*Carthamus tinctorius* L.) in several parts of the study area suggests far-reaching contacts and exchange routes between the region and the Near East, and that this plant was introduced into areas such as Hungary (Gyulai 1993), Serbia (Kroll 1990), Greece (Becker and Kroll 2008) and Bulgaria (Marinova and Riehl 2009). The earliest safflower finds (comprising a concentration of fragmented and *ca.* 28 whole charred fruits) originate from the last third of the early Bronze Age in the eastern part of the Thracian plain, at Tell Karanovo at about 2800–2600 cal. BCE (Görsdorf and Bojadziev 1996). More convincing evidence for its cultivation is furnished by the finds from Hungary representing a concentration of *ca.* 1245 safflower fruits. In the agronomic literature, the Aegean region and eastern Europe are usually not included in the area of its ancient cultivation (Hannelt 1961). *Carthamus tinctorius* is considered among the plants mentioned in Linear B texts from Crete (Sarpaki 2001), yet the attribution of specific plants to different Linear B words can be problematic, while archaeobotanical finds (lacking at the moment from the area) would be a much more reliable line of evidence to use for the identification of the areas to which ancient cultivation of safflower extended.

During the period considered after 5500 cal. BCE in the study area, some hints for the use of coriander (*Coriandrum sativum*) are available. It was recorded with just a few fruits for the Late Neolithic of Bulgaria (*ca.* 5500-4900 cal. BCE) in one site in the Thracian Plain (Tell Kapitan Dimitrievo, Marinova 2006) and the Late Neolithic of Greece (*ca.* 5200 cal. BCE) in northern Greece at Servia (Hubbard and Housley 2002), yet the Greek finds could be intrusive from later Bronze Age levels. More reliable evidence of coriander is identified as a crop at Sitagroi and is dated to the Early Bronze Age where it is present in significant numbers (Renfrew 2003). Coriander is available in Chalcolithic layers from eastern Romania. There, the oldest finds are from the Early Chalcolithic (4800–4500 cal. BCE) consisting of a few seeds, but the most spectacular is the evidence of coriander storage from the Late Chalcolithic period (*ca.* 4200–4000 cal. BCE) found in a vessel (Monah and Monah 2008), which clearly suggests cultivation of the plant. Coriander is a plant which has its natural distribution in the eastern Mediterranean area. In the northern parts of southeastern Europe except the most southerly area of modern-day Bulgaria, coriander is spread only in secondary plant societies developed under anthropogenic influence. Soon after this influence had stopped, its wild form disappears from nature. This means that this plant thrives in those parts of the study area located at some distance from the Mediterranean only through human impact. Coriander was therefore introduced in these areas during the Late Neolithic and Chalcolithic.

As regards fruit species, the grape and the olive have been closely linked with prehistoric communities of Greece. The grapevine is present throughout the Neolithic in Greece and there are clear indications that it was used for juice extraction as early as the end of the fifth millennium at Dikili Tash (Valamoti *et al.* 2007) and later in other parts of Greece, northern and southern (Valamoti 2009). The issue of whether it was cultivated or not remains unresolved, although measurements from Bronze Age sites suggest that morphologically domesticated pips were present, thus cultivation seems a strong possibility. The olive is closely associated with the south of Greece and from the fourth millennium onwards, there are indications for its use, yet problems related to the archaeobotanical distinction between the wild and the domesticated forms render the identification of cultivation through the archaeobotanical record problematic (Valamoti 2009). The climate of northern Greece is marginal for the cultivation of the olive, thus it is no surprise that no olive remains have been encountered among Neolithic and Bronze Age assemblages from the region (Valamoti 2009). Likewise, further to the north outside the Mediterranean in the territory of Bulgaria, no olives can thrive. Concerning the grape, the finds of single pips indicate most probably that no cultivation took place during the period considered. For more information on arboriculture see Chapter 4.

Our preceding discussion has shown that during the Neolithic and the Bronze Age in southeastern Europe, the glume wheats were the staple crops consumed by its ancient population. In Greece, einkorn, emmer and the new glume wheat-type were grown as separate crops, though the possibility of their having been cultivated as a maslin (as mixed crops) remains plausible (Valamoti 2004; 2009). In Bulgaria during the Early Neolithic, predominantly mixed crops of emmer and einkorn (in storages) occur. The finds of separated crops are rather rare, but this changes during the following periods (Chalcolithic and Bronze). The general picture reflecting the Early Neolithic crops used in Bulgaria are rather similar to that already outlined for Greece. In both countries, a wide range of pulses was consumed and they seem to have constituted important crops. This crop variety is widened further by other cultivated plants, like flax, and plants harvested from the wild (not discussed here). During the Bronze Age, many more crops were introduced into Greece (Valamoti 2007; 2009), in particular during the Late Bronze Age, such as Celtic bean, millet, opium poppy, *Brassica* sp., *Lallemantia* sp. The Bronze Age is also a period when new crops were introduced to Bulgaria, but with a slightly different spectrum than in Greece: also including *Lallemantia* sp., and millet, as well as safflower. Yet, despite similarities in the general trends, it is becoming apparent that we are not dealing with a uniform crop pattern throughout southeastern Europe, Greece and Bulgaria in particular, but with certain regional differences such as the absence of chickpea from Greece, the absence of Celtic bean and of a wide range of plants such as opium poppy, *Brassica* and *Camelina* from Bulgaria during the Bronze Age, as well as the very limited presence of the grapevine and a total absence of the olive. On the other hand, a much more prominent presence of free-threshing wheat is observed in Bulgaria. Overall, crop diversity appears to be greater in the southern parts of southeastern Europe, as the archaeobotanical evidence from Greece indicates.

Temporal Changes of Crop Diversity: Interaction Between Social Changes and Natural Environment

As already stated in the introduction, geographical linkages and barriers are generally of importance for crop diversity. The archaeobotanical methods dealing with charred plant material still have strong limitations for estimating crop diversity on the level of landraces and local varieties, so what can be observed are mainly the broader tendencies connected with the changes in crop diversity over larger time scales such as several hundred years to a millennium. Considering the environmental constraints in the study area, it is especially the geographical barriers and natural connection between the regions which could have played an important role for variations in crop diversity. For example, for the Bulgarian Neolithic, a clear east-west division can be observed in the archaeobotanical record, which fits well with the different relief of the two parts – the west rather mountainous and the east dominated by lowlands. Another important geographical barrier is the Stara Planina mountain range, responsible for a north-south division, but due to the lack of evidence from the areas on the north of the mountain range up to the Danube, no reliable conclusions can be drawn at present. For Greece a north–south differentiation in the animal and plant economy of the past has been observed (see Halstead 1994).

Climate can of course have a significant impact on crop choice and certain crops are more tolerant of certain growth conditions, *e.g.* drought, heavy rainfall, etc. Besides this, climatic conditions, although similar in a broad area, may affect smaller regional units differently due to their relief, aspect, soil conditions etc. Recent studies have linked climatic influences on the plant economy in the past through a combined examination of the archaeobotanical record and climatic indicators such as ratios of stable carbon isotopes ($\delta^{13}C$) indicating drought stress (Riehl 2008; 2009; Araus *et al.* 2007). It is believed that during periods characterised by aridity, the archaeobotanical record would show reduced crop variability and the appearance of drought-resistant crops (*e.g.* safflower, *Carthamus tinctorius*; Marinova and Riehl 2009). For the area under consideration, where no clearly pronounced drought events are recorded yet for the period (6000–1000 BP) and annual precipitation is never under 400 mm, most probably other factors such as seasonality changes, early frost and heavy rains during the growing season, would have influenced crop choice. The diversity observed in our study area would have acted as a buffer against such seasonal climatic fluctuations (*cf.* Halstead 1989; 1990; 1994). A pronounced influence of climatic change would be difficult to trace, as the reconstructed Holocene precipitation and temperature for the region never reached extreme variability (*e.g.* Davis *et al.* 2003). However, the same study indicated changes in seasonality in the period around 8000 BP expressed in cooler, wetter summers and milder winters compared to today's situation in southeastern Europe and this could have influenced crop diversity during the beginnings of the Neolithic there.

For northern Greece, there are indications for climatic fluctuations in precipitation during the Neolithic (Bottema 1990) and for changes in precipitation regimes during the Bronze Age; yet no systematic integration of various lines of evidence of environmental conditions has been attempted for the Neolithic and the Bronze Age for the whole of northern Greece. For Bulgaria, the palaeoecological records coming mainly from the mountain areas (see Tonkov and Marinova 2005; Stancheva and Temniskova 2004; Lazarova 2003; Filipovitch *et al.* 1998) show a certain climatic deterioration (cooling or reduced water availability) during the period 4300–3600 cal. BCE roughly contemporary with the transition between the Chalcolithic and Bronze Age. Furthermore, during the Late Bronze Age at several sites in Bulgaria, increasing humidity or water availability was recorded (for discussions, see Marinova and Atanassova 2006). Our examination of the crop repertoire of southeastern Europe does not demonstrate significant changes in the staples used in the Neolithic and Bronze Age in the region, *i.e.* the glume wheats dominate throughout, together with pulses. At present, our evidence is of rather unequal quality and therefore any correlation between climatic changes and crop diversity changes is still likely to be problematic.

Apart from environmental factors, increasing evidence is available for the role of cultural factors in shaping crop diversity. This could be illustrated by different crops, for example the chickpea mentioned above, *Lallemantia*, safflower or millet. An interesting example in this connection is chick pea (*Cicer arietinum*). This crop occurs in the archaeobotanical record in the area of consideration during two rather short phases. Its appearance is most probably connected with cultural processes that took place during the period of *ca.* 5700–5500 BCE, in which repeated contacts with Anatolia have been observed in the archaeological record. Therefore, the most probable option is that the finds from the Bulgarian early Neolithic correspond to direct contacts with Anatolia. One hint for this could be that the earliest radiocarbon date for the Bulgarian chick pea finds comes from the site of Kapitan Dimitrievo. Also, the evidence based on pottery of contacts with Anatolia and Thrace established at Kovacevo Ic and Id (Lichardus-Itten *et al.* 2006) could argue for this hypothesis.

Lallemantia, as well as a whole range of Bronze Age crops, is indicative of cultural contacts leading to the introduction of new crops to a region. Long-distance contacts with communities to the east have been suggested, for example for *Lallemantia* (Jones and Valamoti 2005), safflower (Marinova and Riehl 2009) and to the north for opium poppy, spelt wheat and millet (Valamoti 2007; Valamoti 2009). The nature of these contacts, however, is rather difficult to grasp and the presence of these introduced species does not necessarily imply direct acquisition from the area of origin.

It is also interesting to mention the period of the Bulgarian Middle/Late Chalcolithic period (*ca.* 4500-3900 cal. BCE), when the palaeoeconomical significance of free-threshing wheats increases on the Black Sea coast, as reflected by storage finds. Roughly contemporary to these finds (*ca.* 4800–3700 cal. BCE), we also perceive in other regions of Europe an increasing importance and cultivation of free-threshing wheat, as in the lake shore sites in the Alps and adjacent areas (see Jacomet 2007a; Maier 1996), the Mediterranean Neolithic at the Iberian Peninsula (Buxo 2007) and Middle/Late Neolithic of Italy (see Rottoli and Castiglioni 2009 and literature cited there). Concerning the far-reaching trade routes of spondylus and obsidian during the period, between at least the Alpine area, north Italy and the Black Sea coast (Sériadès 1995; Kotsakis 1996; Müller 1997), the occurrence of free-threshing wheat could also be interpreted as triggered by cultural changes and interactions.

Conclusions

Our paper has outlined a general pattern regarding crops and their diversity in northern Greece and Bulgaria, and set out our observations in relation, where possible, to other areas of the Balkans. Although at present our body of data is not uniformly rich and reliable for comparisons between different periods within the Neolithic and the Bronze Age to be made, it has been possible to draw a rough picture of crops for these two broad periods and to compare the archaeobotanical evidence between certain parts of southeastern Europe. Thus, several common trends have been identified or re-confirmed, such as the predominance of the glume wheats, the important role of pulses, the clear visibility of flax and the introduction of various crops during the Bronze Age, probably through cultural contacts. At the same time, several interesting differences are beginning to emerge and in the future would be worth exploring further on the basis of more archaeobotanical data: the predominance of einkorn in certain parts of Bulgaria and in many sites of northern Greece, the appearance of free-threshing wheat as a crop on the Black Sea coast of Bulgaria during the Chalcolithic, the presence of chickpea in the later phases of the Early Neolithic of southern Bulgaria, the absence of Celtic bean and millet in the Neolithic in both regions, and the total absence of a wide range of crops encountered in Greece during the Bronze Age, in particular oil, medicinal and hallucinogenic plants. It is evident that many crops are common in this part of southeastern Europe, shared among the different cultures for a considerable length of time. At the same time, certain areas are slightly different and in the future we need to consider the broader cultural context and environments of these areas to understand divergences in the crop pattern. Such observations would also greatly benefit from a consideration of a wide range of environmental factors, particular to individual sites or regions, which may have interacted with human choices as regards the crops on which they relied. Moreover, independent information from palaeoclimate proxies (as, for example, the $\delta^{13}C$ measurements) and models could help us to clarify climate-driven change in crop diversity.

3.3. CROP DIVERSITY BETWEEN CENTRAL EUROPE AND THE MEDITERRANEAN: ASPECTS OF NORTHERN ITALIAN AGRICULTURE

Mauro Rottoli

'*Quell'anno, sulla via del ritorno, si fermò in Austria presso la famiglia di contadini dove aveva lavorato per la semina delle patate e, visto il raccolto eccezionalmente abbondante e la qualità, ne chiese una decina di chili da portare a casa come semenza. [...] Alla vigilia di quel natale arrivò a casa con pochi gulden d'argento ma con una razza di patate che poi per tanti e tanti anni diede buoni raccolti e si diffuse tra le nostre montagne.*' Mario Rigoni Stern, *Storia di Tönle*, Einaudi Ed., Torino, 1978, p. 27

Introduction

The geographical position and environmental characteristics of northern Italy (Fig. 3.7), which varies progressively from a Mediterranean to a continental climate, make it well-adapted to the cultivation of species of markedly differing ecological characteristics. From the Neolithic onwards, southern influences (overwhelmingly prevalent) and northern components have alternated and been combined in this part of Italy, giving rise to an agriculture which has in time developed its own characteristics, thanks to the fertility of its soils and the wide variety of environments it contains.

An excellent example for this is provided by the Romanisation of the Po Plain. Rimini, the first Roman *colonia* in north Italy, was founded in 268 BCE. During the course of the next two centuries, all of northern Italy was progressively Romanised by means of the foundation of *coloniae*, the assignment of land confiscated from enemies to veterans, or the granting of rights to local populations. Settlement by 'colonists' from the south and the desire of the newly-Romanised to take up the ways of the capital in agricultural production (as in other matters) had the effect that typically Mediterranean crops were introduced into the Po Plain. The Romans' remarkable technological capacity allowed them to grow southern Italian species in the north: not just in the zones with greater climatic affinity (the Adriatic coast, the region of the Insubri, Liguria and southern Piedmont), but also in hilly areas and the central part of the plain. The olive (*Olea europaea* L.), for example, was grown during late Roman and medieval times not only in the region of the pre-Alpine lakes (see Castiglioni *et al.* 2007), but also in the hills around Parma, those of the Veneto and even in Cremona, at the centre of the Po Plain (Ferrari 1988); the common myrtle (*Myrtus communis* L.) was almost certainly grown at Aquileia in the northeast (Castiglioni and Rottoli 2010).

This is not the only occasion on which aspects of the southern economy and southern agriculture were imported into northern Italy. In this article we will consider the history of introduction and spread of some crops from Neolithic to recent times. Archaeobotanical data are examined to understand the way of introducing crops into a new geographic area in relation to social and economic aspects.

The Neolithic Tradition

The distinguishing features of agriculture and gathering in northern Italy since the Neolithic are a varied cultivation of cereals, an abundance of legumes and the collection of a plentiful variety of

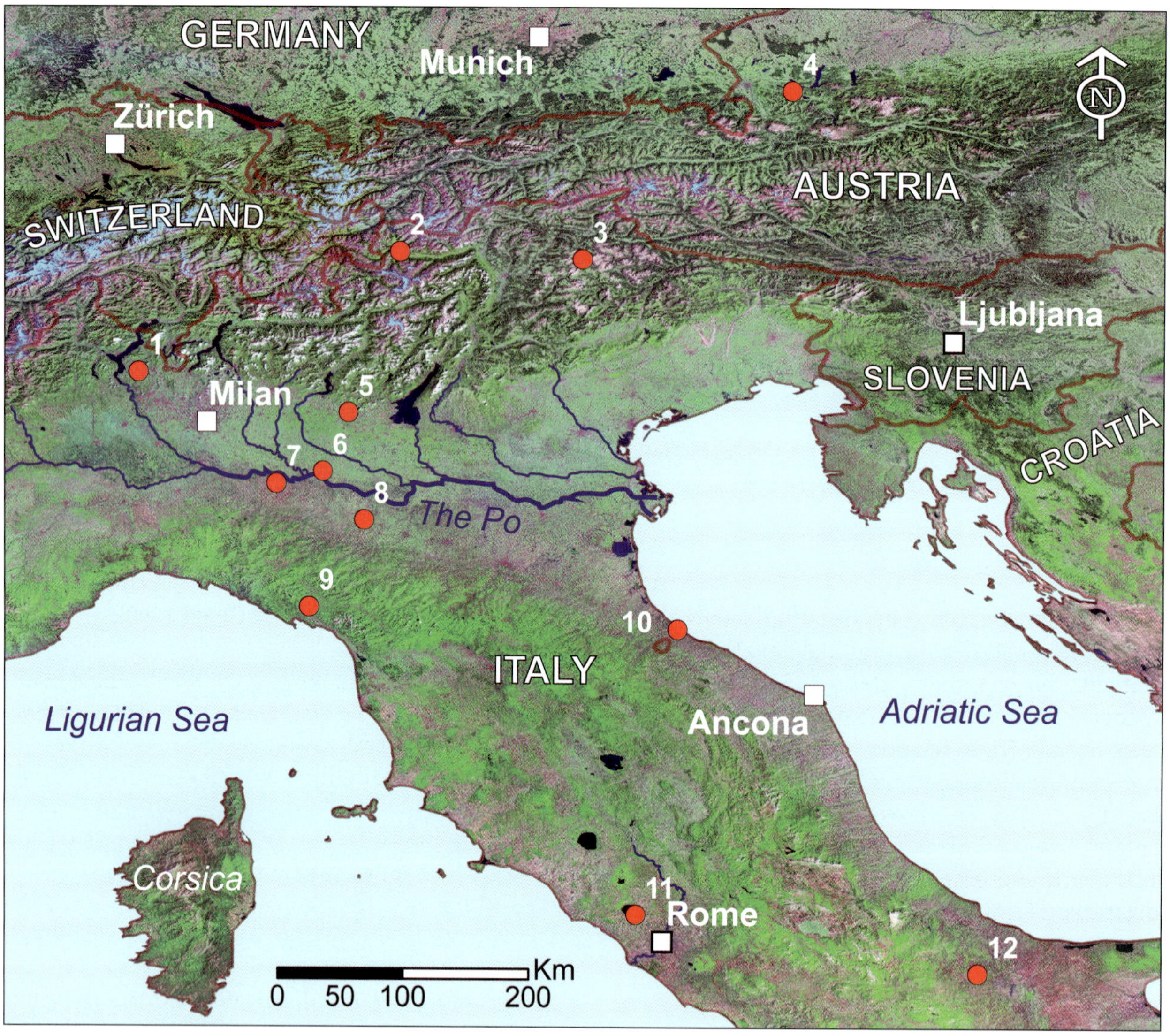

Fig. 3.7. Sites mentioned in the text: 1) Isolino di Varese; 2) Ganglegg; 3) Sotćiastel; 4) Aquileia; 5) Brescia; 6) Cremona; 7) Piacenza; 8) Parma; 9) Filattiera; 10) Rimini; 11) La Marmotta; 12) Masseria Mammarella. Map: R. Lugon, J.-C. Loubier and A. Chevalier.

fruit. Data concerning the Neolithic (Rottoli and Castiglioni 2009) confirm that agriculture began with species from the Near East: barley (*Hordeum distichum/vulgare* L.), emmer wheat (*Triticum dicoccum* Schübl.) and einkorn wheat (*Triticum monococcum* L.); bitter vetch (*Vicia ervilia* (L.) Wild.), lentil (*Lens culinaris* Medik.) and pea (*Pisum sativum* L.), and flax (*Linum usitatissimum* L.). Only the chickpea (*Cicer arietinum* L.) is absent from both northern and southern Italy until the Bronze Age. In the oldest northern Italian settlements, this first group of cultivated plants is accompanied by other more recently domesticated species of which the origin is often not known with certainty (Zohary *et al.* 2012): the free-threshing wheats (*Triticum aestivum* L./*durum* Desf./*turgidum* L.), the 'new glume wheat' (Jones *et al.* 2000), grass pea/red pea (*Lathyrus sativus/cicera* L.), common vetch (*Vicia sativa* agg.) and perhaps the broad bean (*Vicia faba* var. *minor*).

With respect to the cultivation of cereals, the most recent data suggest the existence of a difference between the northeast and west-central portions of northern Italy: in the northeast, free-threshing wheats are rare and 'new glume wheat' abundant. In the centre and west, the free-threshing wheats were decidedly more plentiful, whilst the 'new glume wheat' has not yet been recorded (Rottoli 2012). The old distinction between northern and southern Italy, featuring the contrast between a slowly evolving agriculture dominated by barley, emmer and einkorn, and a more developed panorama in which a major role was played by free-threshing wheats, appears in the light of recent work to

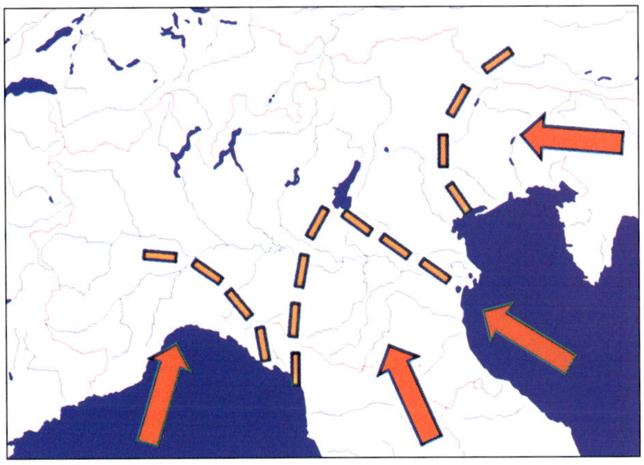

Fig. 3.8. The introduction of the first agriculture into northern Italy: possible routes and areas of influences.

be only partially true. The situation seems to be more complicated at a regional level than had been previously thought (Mottes *et al.* 2010). The precocious onset of various agricultural economies seems to have depended upon contacts or wholesale immigration along routes which cannot yet be traced with precision. Penetration into the Po Plain probably occurred from several routes to the west, south and east (Liguria, the Apennines, the Adriatic and Friuli), which themselves depended on contacts via the sea – along the eastern Ionic and Adriatic coasts, from the south and west (Ionian, Tyrrhenian Ligurian Seas) – and overland, via the Balkans (Fig. 3.8).

Neolithic innovations soon reached Alpine regions, although it is not clear how the introduced southern crops became integrated with northern cultivations or any Mesolithic inheritance, even if we note that Mediterranean influences arriving especially from the southwest have been emphasised by archaeological and botanical studies in the Swiss Alps (Jacomet 2007a).

Spelt

Recent research suggests that this species (*Triticum spelta* L.) may have developed in various parts of Europe by means of the introgression of a tetraploid wheat (emmer) into free-threshing hexaploid wheats (bread wheat); the oldest finds date to the end of the third millennium BCE and a well-documented case of its cultivation in Switzerland is placed at 2300 cal. BCE (Akeret 2005; Kohler-Schneider and Caneppele 2009). Older examples are known from Italy, but these specimens are incomplete and all somewhat doubtful (Rottoli and Castiglioni 2009); there is, therefore, no solid evidence for an earlier introduction or an independent origin.

In the history of Italian agriculture, spelt occupies a decidedly secondary role in both the north and south, as confirmed by archaeobotanical data and historical agricultural accounts. In the Po Plain, it turns up fortuitously in the Bronze Age Terramare (Rottoli 2001; Mercuri *et al.* 2006) and occasionally on Iron Age sites (*e.g.* Aquileia, on the Adriatic coast: Maselli Scotti and Rottoli 2007), whereas in mountainous regions, above all in the South Tyrol, its presence is markedly more consistent during both the Bronze and Iron Ages (*e.g.* Ganglegg/Schluderns: Schmidl and Oeggl 2005; Schmidl *et al.* 2007; Sotćiastel: Swidrak and Oeggl 1998).

The Romans were familiar with spelt wheat, but did not appreciate it particularly; the Latin term *far*, which is generally used by classical authors in reference to ritual usage, is employed to indicate emmer, but seems to have been applied indifferently to all hulled wheats. Among the possessions listed in the tenth century CE inventory of the monastery of Santa Giulia in Brescia, spelt constitutes only 1% of the food reserves; cultivation was concentrated on the hills near Piacenza and the Ligurian Apennines towards France (Montanari 1979). On the basis of current information, the species seems to have emerged (or been widespread) in central Europe and been introduced into Italy from the north, across the Alps; while in northern provinces of the Roman Empire spelt regionally was the dominant cereal, it was important only in mountain areas in Italy, probably due to its rustic qualities.

Rye

Rye (*Secale cereale* L.) was for a long time present as a weed in fields in which other cereals were grown. The few grains of rye and pollen records that have been found show a limited presence in various parts of Europe before it was cultivated intentionally (Behre 1992). In northern Italy, finds before the Roman period are scarce and uncertain (Mercuri *et al.* 2006) and in the first centuries of the Empire its use was limited (Castelletti 1975). The success of the species is evident, though, in Early Medieval times, first in lowland areas and then in

Fig. 3.9. A medieval assemblage of cereals from northern Italy. From left: *Avena sativa*, *Secale cereale*, *Triticum aestivum/durum*, *Setaria italica* (top), *Panicum miliaceum* (right) and *Sorghum bicolor* (bottom).

the Alps (*cf.* Castiglioni *et al.* 1999; Castiglioni and Rottoli 2010), while little was cultivated in the south. There has been much discussion about whether the fortune of rye in northern Italy was associated with a deterioration of the climate, or if other economic and social factors (such as a collapse in technical capabilities caused by the fall of the Roman Empire) were important (Montanari 1979). The hypothesis that new dietary habits imposed by invaders (Goths, Lombards, Franks) from north of the Alps took root seems worthy of consideration, bearing in mind that the success of rye perhaps caused a reduction in the cultivation of emmer, but not of the more demanding free-threshing wheats (Castiglioni *et al.* 1999; Castelletti *et al.* 2001, Fig. 3.9).

Millet and Foxtail Millet

Classical writers, perhaps with a certain disdain, described northern Italy as a zone characterised by the production of millet (*Panicum miliaceum* L.) and foxtail millet (*Setaria italica* (L.) Pal. Beauv.). References to these crops are found in Polybius (*ca.* 203–120 BCE), Strabo (*ca.* 64 BCE–CE 21) and Pliny (first century CE). This obviously does not imply that these were the only crops grown in northern Italy, but indicates their rarity in the south. The cultivation of millet is in fact recorded from the Bronze Age onwards also in southern Italy, whereas foxtail millet, which was rarer in the Po Plain, appears to have been absent from central and south Italy in both the Bronze (Fiorentino *et al.* 2004) and Iron Ages (Costantini 2002), and is known from only a few localities during the Roman period (*e.g.* Filattiera, Tuscany: Rottoli and Negri 1998).

The most recent re-evaluation of information concerning the spread of millet and foxtail millet across Asia and Europe (Hunt *et al.* 2008) did not propose definite conclusions with respect to the details of its introduction into Europe.

Millet has recently been found in Austria in layers dated to 2930–2880 cal. BCE (Kohler-Schneider and Caneppele 2009) and Italian Copper Age finds have become ever more numerous (*ca.* 3500–2100 cal. BCE, Castiglioni and Rottoli 2009; Barfield *et al.* 2002). The hypothesis that millet (and perhaps foxtail millet, too) arrived during this period from north of the Alps would at present appear the most likely.

Opium Poppy

The distribution of the wild ancestor of the opium poppy (the wild *setigerum* poppy, or *Papaver somniferum* L. ssp. *setigerum*) indicates that its domestication occurred in the coastal areas of the western Mediterranean (Zohary *et al.* 2012). The La Marmotta settlement (*ca.* 5690–5260 cal. BCE, near Rome, (Fugazzola Delpino 1998), which is situated in the area of origin, has yielded the oldest seeds and seedpods with characteristics intermediate between the wild and cultivated forms (Rottoli 2002; Fig. 3.10). It is not known whether cultivation took place only at this location, or also happened elsewhere (Spain? *cf.* Stika 2005; Peña Chocarro 1999; Peña-Chocarro and Zapata, Chapter 3.5). The spread of the species northwards is of great antiquity; introduction into the Netherlands does not seem to have occurred through contact with France, but rather by means of eastern connections with the Linearbandkeramik culture (Bakels 2000).

Fig. 3.10. Poppy (*Papaver somniferum* subsp. *setigerum*) stigmatic disc from an early Neolithic site (La Marmotta, Rome)

Nor is it possible to follow with precision the plant's arrival in northern Italy, or to establish any direct relationship with La Marmotta. The species has recently been identified at Isolino di Varese (near Varese) in layers dated to 4800–4600 cal. BCE (Banchieri and Rottoli 2009), almost a millennium older than other northern Italian sites at which it had previously been found (Rottoli and Castiglioni 2009). It is not possible to determine whether or not these sites may have functioned as centres from which spreading northwards – in particular towards Austria and Switzerland – took place; the opium poppy has, in any case, occasionally been recovered from older contexts north of the Alps (Ulm-Eggingen, *ca.* 5200 cal. BCE, Upper Swabia, including Federsee: Jacomet 2007a; Kujawy, central Poland, LBK feature: Bieniek 2007).

Legumes

There is today a marked difference between northern and southern-central Italy in the production and consumption of legumes. Northern Italy produces traditional crops of peas, beans (*Phaseolus* spp. L. of American origin) and cowpeas (*Vigna unguiculata* (L.) Walp., of African origin, introduced in the Roman period), whereas in the south the established crops are chickpeas, lentils, grass peas and red peas (*Lathyrus sativus* and *Lathyrus cicera*), white lupins (*Lupinus albus* L.) and broad beans (*Vicia faba* L.). Other species of ancient dietary importance have almost disappeared or been reduced to the role of livestock fodder; their distribution is largely southern (Hammer *et al.* 1999). It is unclear when this sharp distinction came into being, although it seems likely that it is quite, or perhaps even very, recent (after the introduction of the American bean or still later than this?). It seems likely that the distribution of legumes was more homogeneous in Antiquity, although several species (such as the lupin) grow only with difficulty in northern Italy and the cultivation of these was restricted to Liguria until modern times. Archaeobotanical records, which are often poor for leguminous plants, present further difficulties of interpretation. It is undeniable that the oldest data indicate a broad similarity between northern and peninsular Italy with respect to the consumption of legumes (in the south, though, the broad bean is known with certainty from the Middle Neolithic, Costantini 2002). This basic uniformity, with the odd exception, seems to continue in the periods that followed: the chickpea (absent in the Neolithic, as mentioned above) is first found in the south in the Late Bronze Age (Masseria Mammarella, Campobasso: Costantini 2002) and appears in northern Italy only in the Roman period (Castiglioni *et al.* 2007). Data regarding the Bronze Age, whilst not abundant, demonstrate the presence of broad beans, peas and lentils. In the Iron Age, when one might have expected a greater dissimilarity between the north and the south due to Greek colonisation, significant differences are not recorded; in northern Italy traditional species (pea, lentil, bitter vetch, grass pea, broad bean) continue to be widely used and the pattern seems similar in the south. Lentil cultivation appears to have been widespread in the plains and hilly zones of northern Italy, but more restricted in the Alpine area (Schmidl *et al.* 2007). The most recent studies on the Roman and medieval periods seem to confirm a similar diet of legumes in both the north and the south, but in Roman times it is often not possible to distinguish between local production and transport from other regions.

Other Crops

The routes by which some species arrived in northern Italy (as also in other parts of the world) are difficult to evaluate; in many cases, introduction did not occur directly from the areas of origin but through zones of secondary diffusion. This is the case with regard to many species of American origin, which were introduced into Italy – together with other parts of western Europe – not directly from America, but from European colonies in Asia, where cultivation had been established and cultivars better suited to European climatic conditions had been selected (see, for example, Hancock 2004).

Prior to the discovery of America, the periods in which greater numbers of new species were introduced seem to have been those when invasions or particular increases in trading occurred. Such developments did not always lead to notable consequences, however; the effects of Arab expansion in the Mediterranean, which resulted in the introduction of new species into Sicily (citrus fruits), seem in general to have been relatively limited. Earlier, the Roman Empire had acted as a powerful vehicle for the transmission of knowledge, with the introduction of various species of fruit and vegetables, sorghum (*Sorghum bicolor* (L.) Moench) and rice (*Oryza sativa* L.), through contacts and exchange with Asiatic and African peoples. These did not always have immediate success. Rice, for example, although known since the time of the Greeks and found in central Europe in the Roman epoch (Knörzer 1970), was cultivated in northern Italy only from the fifteenth to sixteenth centuries onwards in accordance with the wishes of the Sforza family (Castelletti 1978). Sorghum appears to have attracted some interest during the Early Medieval period in northeastern Italy, but remained completely unknown in the south (Castiglioni and Rottoli 2010).

Hemp cultivation has been demonstrated in Lazio, not far from Rome, from the first century CE. Despite the fact that the wild plant grew in Italy, the species does not seem to have been cultivated before this date (for a possible exception in the Middle Bronze Age, *cf.* Ravazzi *et al.* 2004). It is still unknown whether the onset of cultivation was due to contacts with the Near East or whether it arrived from central Europe via northern Italy (Mercuri *et al.* 2002).

Temporal Changes of Crop Diversity: Human Societies and Historical Background

The study of these events, especially the more recent and better known examples, suggests that the introduction of crops into new territories depends upon various factors, perhaps acting together, which have had diverse results in different cases. For many crops, the role of climate does not seem to have had a particular importance.

The most influential factors depend upon population movement, either as a result of invasions or in the form of peaceful emigrations, such as for the opening of trading stations or foundation of colonies. In the first case, the plants and agricultural practices brought by new arrivals, when compatible with the adaptability of the species concerned and the novel environmental conditions, may be imposed on the pre-existing inhabitants, leading to additions and substitutions. In the second case, the transformation is gradual and different systems of production and consumption may survive, perhaps for centuries, in neighbouring territories. Such situations are also susceptible to archaeobotanical analysis: the process of becoming Neolithic is the classic example of an 'invasion', although it is well known that the mode of diffusion is not all that linear (Guilaine 2003). Relations between colonies or military settlements and local populations have been analysed in various regions (see, *e.g.* for Sicily: Stika *et al.* 2008; for the Roman border in Germany: Kreuz 1999; for the Iberian Peninsula: Buxó 2008; for the Romanisation of Gaul: Wiethold 2003a). The spread of rye in northern Italy could, at least partially, be due to a situation of this sort (as mentioned previously), inasmuch as the species was already known and appreciated by central European barbarian peoples who moved into Italy.

A prince or central administration, perhaps influenced by scholars or philanthropists, could also introduce or impose new crops in a particular area (*cf. e.g.* Charlemagne and the *Capitulare de Villis*). In more recent times, this phenomenon has occurred a number of times; it seems, for example, to have been the cause of the introduction of maize, rice and potatoes in northern Italy, but is less easy to investigate archaeologically, especially when precise written evidence is lacking. Research into the introduction of European species into the Americas is of both historical and botanical interest; often abuse of power and philanthropy (real or presumed) are inextricably mixed.

Another phenomenon which seems even less well-defined (and perhaps impossible to evaluate archaeologically) is that involving an individual, who may have been rich or powerful, a merchant or a humble farmer. Historical examples of this type, recent and ancient, are known. For example, Pliny (*N.H.* XV, 14, 47; Bostock 1855) attributes the introduction of the jujube (*Ziziphus jujuba*) into Italy from Africa to the consul Sextus Papinius (early first century CE). Many cases are known from the last two centuries of 'migrants', single or small groups which move and live in distant lands and bring with them (or have sent) seeds from their country of origin; this mechanism may also have acted in the distant past. A variant, also well documented, is the return of the migrant who brings home with him products or agricultural procedures which he has got to know in distant parts (for a literary version of the introduction of a new type of potato into the Veneto Alps, Italy, see Rigoni Stern 1978, see quotation above).

Conclusion

In northern Italy, crops were introduced progressively by contacts with Mediterranean and northern Europe. Actually, no crop seems to have an independent and autonomous origin in this area, but thanks to the great fertility of the soils and the good climate conditions, agriculture in northern Italy was able to develop characteristics that set it apart from other areas. Archaeobotanical and genetic data enable us to understand the origin of plants, overland and sea route of crop diffusion and the occasions when a single crop was introduced; thanks to literary and historical documents, it is sometimes possible to understand the mechanism and motivation of introductions.

The recognition of such processes in the ancient past by means of archaeological and archaeobotanical study is clearly not straightforward; frequently it is only possible to record the results without being able to fully understand the relative causes.

3.4. CROP DIVERSITY IN SOUTHWESTERN CENTRAL EUROPE FROM THE NEOLITHIC ONWARDS

Stefanie Jacomet

Introduction

In the following, an overview of the changes in crop diversity over a time span of 7,500 years is given. The region considered in this chapter comprises a larger area north of the Alps (Fig. 3.11). It was chosen on the one hand because, since before the Neolithic, this area has been in a zone where influences from the southwest/west and the southeast/east met. This becomes clear *e.g.* during Neolithisation in the sixth and fifth millennia cal. BCE (Stöckli 2009), and therefore, the history of crop diversity may reflect cultural influences. On the other hand, there are many very well preserved wetland settlements, especially during the prehistoric periods, which allow extremely detailed insights into subsistence. Thirdly, this was the region where in the last thirty years much archaeobotanical research has been done by several archaeobotany laboratories, including the Basel University Archaeobotany team.

The results presented here are based on on-site archaeobotanical data and summarise publications recently compiled by the author and colleagues (Jacomet 2006; 2007a; 2008; 2009; Jacomet and Brombacher 2009). In addition, compilations for the series 'SPM' (= *die Schweiz vom Paläolithikum bis zum frühen Mittelalter*/Switzerland from the Palaeolithic to the High Middle Ages) have been used (Brombacher and Kühn 2005; Jacomet *et al.* 1998; 1999b; 2002; Meylan-Krause *et al.* 2002; Ebnöther *et al.* 2002).

The database used consists of several hundred investigated settlements for which plant macro-remains were analysed (for basics in archaeobotany, we refer to the literature: Jacomet 2007b; Jacomet and Kreuz 1999; Van der Veen 2007 and literature cited there). In archaeological layers, plants used by humans are well represented and therefore give a reliable overview about what was eaten and exploited in different ways.

The region considered here contains mostly rather flat to hilly areas (Fig. 3.11, Maps A and B). Most of them were glaciated during the last ice age and the geological underground is therefore mostly glacial till. However, there are also some areas with loess, *e.g.* in the Upper Rhine valley. Altogether, these were regions suitable for agriculture. In addition, we have included results from mountainous areas, the Jura and the Alps. The latter – the Alpine sites – are not shown on the maps (for this, see Jacomet *et al.* 1999a).

In this contribution, the focus is on cultivated (in the sense of domesticated) crops (cultivars) which were planted for food or other (such as technical) purposes from the Neolithic period on. Besides describing the spectra of cultivars in the different archaeological periods, the main aim of this study is to discuss possible reasons for the observed changes in the cultivar spectra over the course of time.

The nomenclature of scientific plant names for cereals follows van Zeist (1984) and for most other cultivars Zohary *et al.* (2012), for wild plants and cultivars not mentioned in Zohary *et al.* (2012), we refer to Oberdorfer (2001) and Mabberley (1990).

Results

Generalities

Before the appearance of Neolithic agricultural societies around 5500 cal. BCE – visible in the archaeological materials in the form of ceramics or polished stone axes – the area was settled by different groups of late Mesolithic populations (see for an overview *e.g.* Mazurié de Keroualin 2003). These had wide-ranging trade networks (*e.g.* Mauvilly *et al.* 2008). The earliest Neolithic influences reached the area from two directions which can be traced back to the two different main 'Neolithisation' streams known today (Lüning 2000; Gronenborn 2006). The eastern way is that through the Balkans until the Carpathian basin where LBK (Linearbandkeramik) developed. LBK and the succeeding groups are the so called 'Danubian cultures'. The western way is that from the western part of the Mediterranean basin. Westerly influenced Neolithic cultures have their roots in the Impressa/Cardial-Cultures which developed mainly during the sixth millennium cal. BCE in the western Mediterranean (overviews *e.g.* Guilaine 2003). In addition, there were also connections to the south via passes in the Alps, as finds like the Iceman (Jacomet and Oeggl 2009) or newly detected artefacts on a high Swiss pass (Schnidejoch) show (Suter *et al.* 2005). These influences from the east on the one hand ('Danubian') and from the (south-) west on the other hand ('Mediterranean'), are an important characteristic of the region discussed here during at least the Neolithic period.

As regards establishing the Neolithic 'way of life' in the area concerned, several different theories exist, mainly based on weak evidence (see the discussions in Mauvilly *et al.* 2008; Behre 2007; Tinner *et al.* 2007). Whether late Mesolithic populations introduced some cereal growing or not will remain unclear until good on-site evidence comes to light. In Fig. 3.12, an overview of the importance of several crops in the region discussed here is given.

Neolithic (ca. *5500–2200 cal. BCE*)

With the onset of the Neolithic, a restricted set of cultivars reached the area considered here. Most of them belong to the so called Near Eastern 'founder crops' (see *e.g.* Zohary 1996): These are several wheat taxa (*Triticum* div. spec.), barley (*Hordeum vulgare* L.), pea (*Pisum sativum* L.), lentil (*Lens culinaris* Medik.) and flax (*Linum usitatissimum* L.). As far as we can judge, only remains of multi-rowed forms of barley are present. An exception might be poppy (*Papaver somniferum* L.) which seems to have originated in the western Mediterranean (Bakels 1982, but see a PPNC find from the Israeli coast: Kislev *et al.* 2004; for an up-to-date overview see Salavert 2010). Examples of their archaeobotanical finds are given in Fig. 3.13.

Early Neolithic (LBK, *ca.* 5500–5000 cal. BCE): most of the few investigated settlements of this period included in this study lie north of the Rhine valley ('Hochrhein') or in the Upper Rhine area as well as the Neckar region in southwestern Germany. They are situated on fertile soils, mostly above loess (for details see Jacomet 2007a). Although Alpine areas like the dry inner Alpine valley of Valais were also settled during this period, there are no archaeobotanical investigations available from there.

In all the settlements, einkorn (*Triticum monococcum* L.) and emmer (*Triticum dicoccum* Schübl.) are by far the dominant cereals (see more in the articles compiled in Colledge and Conolly 2007b). Other cereals like naked wheat (*Triticum aestivum* L./*durum* Desf./*turgidum* L.) have only rarely been found. Since the investigations of the LBK well of Kückhoven in the Rhineland (significantly farther to the north), we know that it was at least partly of the hexaploid type (*Triticum aestivum* L.) (Knörzer 1998). In our region, however, the ploidy-level is not specifiable, because only a few grains were detected.

Barley (*Hordeum vulgare* L.) is present in only a few places, too, and mostly in very low amounts. In most cases, it was identified as the six-rowed, naked form. All in all, the cereal spectrum is very uniform during the early Neolithic in the whole area; it is questionable whether, beside einkorn and emmer, other cereals were grown in early Neolithic times (for more details see *e.g.* Kreuz 2007; 2012; Kreuz *et al.* 2005). It is worth mentioning one find of millet (*Panicum miliaceum* L.) in one of the oldest LBK sites of Bavaria (Kreuz 1990) and two grains of rye (*Secale cereale* L.) in another early site in Baden-Württemberg. At least, the latter most probably represents weeds.

Beside einkorn and emmer, pulses obviously played an important role as food. Pea (*Pisum sativum* L.) is found in most of the settlements, partly even in large amounts. Lentil (*Lens culinaris* Medik.) is much

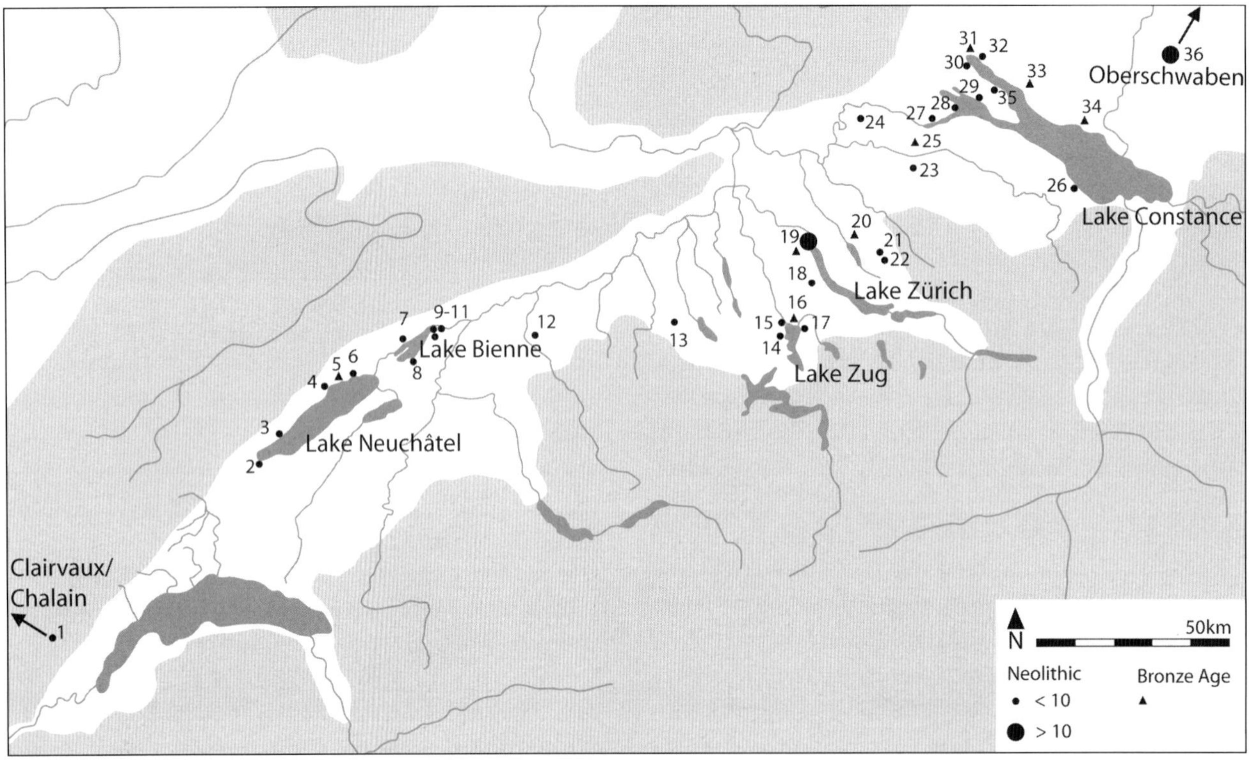

Fig. 3.11. Maps of the region with the archaeological sites included in this study.

A (above): Map of the lake dwellings of the Neolithic and Bronze Age included in this study (from Jacomet 2004). Dot sizes: number of samples investigated (<10 or >10). 1) Clairvaux et Chalain; 2) Yverdon-Avenue des Sports; 3) Concise-Sous Colachoz; 4) Auvernier; 5) Hauterive-Champréveyres; 6) St. Blaise; 7) Twann; 8) Lüscherz; 9) Lattrigen (inkl. Sutz); 10) Port; 11) Nidau BKW; 12) Burgäschisee-Süd; 13) Egolzwil; 14) Risch-Oberrisch; 15) Cham (Eslen, St. Andreas); 16) Zug-Sumpf; 17) Zug-Vorstadt; 18) Horgen; 19) Zürich; 20) Greifensee-Böschen; 21) Pfäffikon-Burg; 22) Robenhausen; 23) Gachnang-Niederwil; 24) Thayngen-Weier; 25) Uerschahusen-Horn; 26) Arbon Bleiche 3; 27) Wangen; 28) Hornstaad; 29) Allensbach; 30) Bodman; 31) Bodman-Schachen; 32) Sipplingen; 33) Unteruhldingen; 34) Hagnau-Burg; 35) Wallhausen; 36) Several sites in the Federsee/Oberschwaben region (Alleshausen, Oedenahlen, Reute, Stockwiesen, Torwiesen, Aichbühl, Riedschachen, Wasserburg Buchau, among others). B: Map of the Basel Region and surroundings with all the settlements investigated archaeobotanically from the Neolithic to the Modern era (from Jacomet and Brombacher 2009). ▲: sites of prehistoric (Neolithic, Bronze Age, Iron Age) periods, approx. 5500 BCE cal to 50 BCE cal).

B (opposite): Map of the Basel Region and surroundings with all the settlements investigated archaeobotanically from the Neolithic to the Modern era (from Jacomet and Brombacher 2009). ▲: sites of prehistoric (Neolithic, Bronze Age, Iron Age) periods, approx. 5500 BCE cal to 50 BCE cal): 1) Rekingen, Rekingen-Bierkeller ; 2) Wittnau, Hüttenweg; 12) Bennwil, Ötschberg; 13) Binningen, Friedhofstr.; 14) Therwil, Baslerstr./Fichtenrain; 56) Chevenez, Combe Ronde; 57) Chevenez, Combe en Vaillard; 58) Courtételle, Tivila; 59) Delémont, En la Pran; 67) Schallstadt-Wolfenweiler, Mengen-Abtsbreite; 73) Vogtsburg-Burkheim, Burgberg; 94) Matzenheim; 95) Schaeffersheim. ■: sites of the Roman period (approx. 50 cal. BCE until 400 CE): 3–8 numerous excavations in Kaiseraugst (AG); 9) Rheinfelden, Augarten-West; 10) Zurzach; 15–17 and 19–28) numerous excavations in Augst (BL); 29) Reinach, Reinacherhof; 30) Reinach, Langrüttiweg; 31) Reinach, Mausackerweg; 60) Alle, Noir Bois; 61) Alle, Les Aiges; 70) Baden-Baden, Gernsbacher Str. 13; 71) Baden-Baden, Gernsbacher Str. 30; 75) Stettfeld, Marcellusplatz; 76) Stettfeld, Mühlberg; 77) Stettfeld, Talstr. 4; 78) Riegel, Feldgasse; 79) Riegel, Am Fronhofbuck; 80) Riegel, Spitalstr.; 92) Brombach; 93) Lahr, Dinglingen; 96) Strasbourg, Grenier d'Abondance; 97) Stephansfeld-Brumath; 99) Biesheim-Kunheim, Oedenburg; 100) Horbourg-Wihr, Kreuzfeld; 101) Horbourg-Wihr, Nouvelle Mairie. ●: sites of the Middle Ages and the Modern era (400 until 1800 cal. CE): 18) Arlesheim, Mühle; 32) Böckten; 33) Eptingen, Riedfluh; 34) Füllinsdorf, Altenberg; 35) Laufen, Rathausplatz; 36) Lausen-Bettenach, Gartenweg; 63) Courtedoux, Creugenat; 66) Develier-Courtételle, La Pran/Tivila; 68) Bruchsal, Altstadt, Bischofsburg; 69) Schallstadt-Wolfenweiler, Mengen am Tuniberg; 82) Freiburg i.Br., Augustinerplatz; 83) Freiburg i.Br., Gauchstr. 21; 84) Freiburg i.Br., Gauchstr. 23a; 85) Freiburg i.Br., Gauchstr. 23b; 86) Freiburg i.Br., Gauchstr. 25; 87) Freiburg i.Br., Grünwälderstr. 16; 88) Freiburg i.Br., Grünwälderstr. 18b; 89) Freiburg i.Br., Grünwälderstr. 21; 90) Freiburg i.Br., Oberlinden 19; 91) Freiburg i.Br., Salzstr. 20; 103) Mulhouse, Eglise St. Etienne. ★: sites of prehistoric and of the Roman period: 11) Allschwil, Neuweilerstr.; 98) Lutter, St. Joseph. ✱ sites of the Roman period and of the Middle Ages and the Modern era: 64) Delémont, La Communance.

Sites within the city of Basel (not indicated on the map): Sites with occupation of prehistoric (Neolithic, Bronze Age, Iron Age) periods, approx. 5500 until 50 cal. BCE): 37–40) Gasfabrik. Sites with occupation of the Roman period (approx. 50 cal. BCE until 476 cal. CE): 47) Münsterplatz Trafostation. Sites with occupation from the Middle Ages and the Modern era (476 until 1800 cal. CE): 43) Deutschritterkapelle Rittergasse; 44) Blumenrain; 45) Gerbergässlein; 46) Kleinhüningen Fischerhaus; 49) Augustinergasse 2; 50) Reischacherhof; 52) Rosshof; 53) Spalenberg 40; 54) Stadthausgasse 28; 55) Teufelhof. Sites with occupation of prehistoric and of the Roman period: 41) Rittergasse 4. Sites with occupation from the prehistoric to the Modern era: 42) Martinsgasse 6 and 8. Sites with occupation from the Roman period to the Modern era: 51) Rittergasse/Bäumleingasse. Map: R. Lugon, J.-C. Loubier and A. Chevalier.

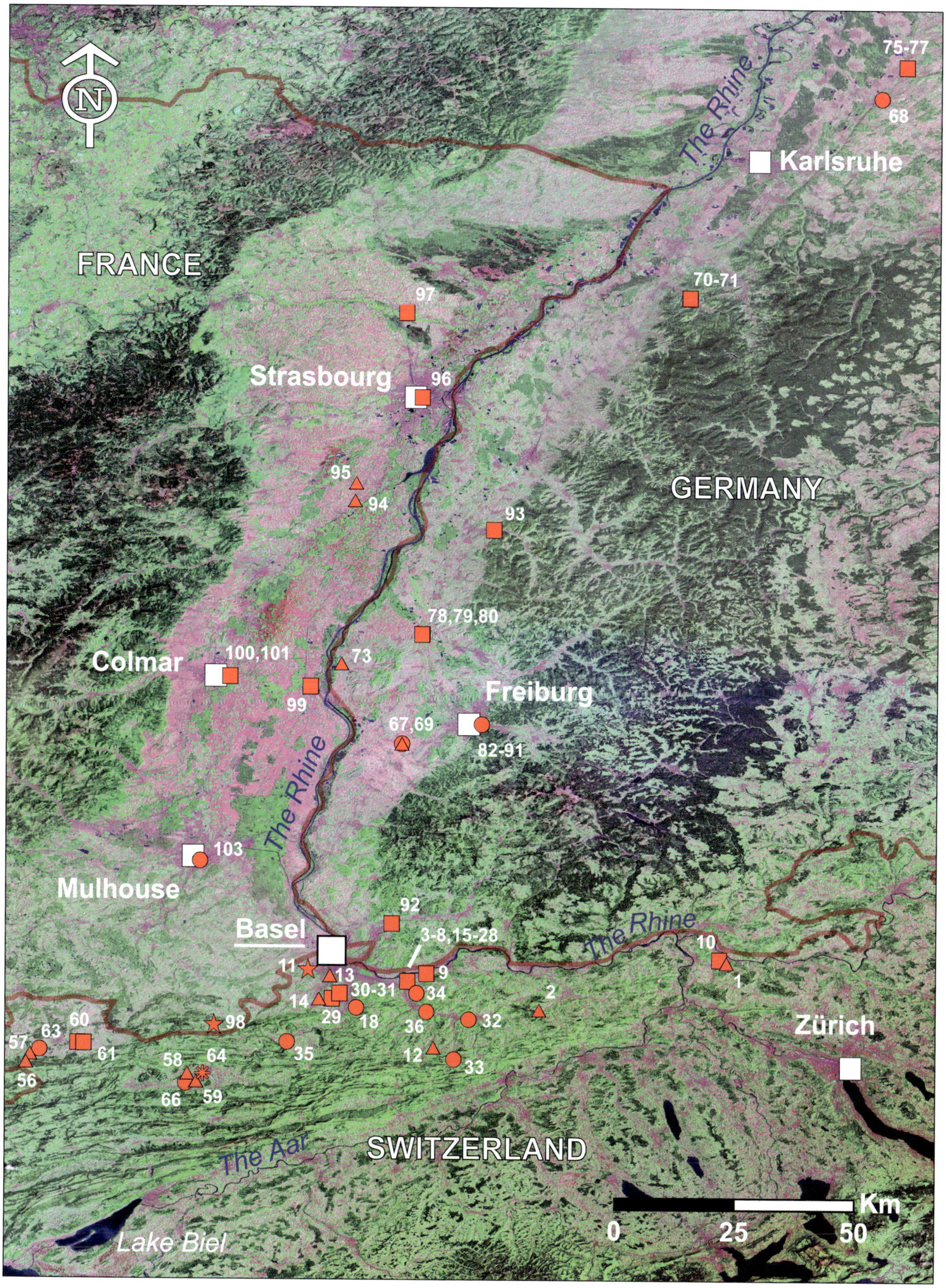

rarer. In some places, flax (*Linum usitatissimum* L.) seeds were found, too. Because all of the LBK-sites concerned are situated on mineral soils where only charred remains are preserved, flax is very probably clearly underrepresented. All in all, one gets the impression that flax-growing was important during early Neolithic times, in the whole of the area. In the LBK-well of Kückhoven in the Rhineland (Knörzer 1998) flax is present in considerable amounts (in total almost 2000 remains).

Opium poppy (*Papaver setigerum* DC./*somniferum* L.) has been found very rarely, from the middle LBK onwards. Distinguishing between the wild form, *Papaver setigerum* and the cultivar, *Papaver somniferum* based on seeds is not possible. Usually, it is assumed that outside of the range of the wild form (which is Mediterranean) the cultivar is present. However, its role cannot really be judged because of the above-mentioned, unfavourable preservation conditions in dry mineral soils. For example, in the Kückhoven well already cited, over 6000 seeds were found (Knörzer 1998).

Middle Neolithic (ca. 4900–4300 cal. BCE)

In the area considered here, quantifiable archaeobotanical investigations from only a very few settlements of the middle Neolithic exist. Favourable areas inside the Alps like the Alpine Rhine valley were also settled (for an overview, see Martin 2010).

In the Middle Neolithic settlements, einkorn (*Triticum monococcum* L.) and emmer (*Triticum dicoccum* Schübl.) continue to be the most important cereals. In some of the sites however, naked wheat reaches much higher numbers than in all the early Neolithic sites of the area considered here. The type of the naked wheat (*Triticum aestivum* L./*durum* Desf./*turgidum* L.) is known from at least two places, where rachis remains were found (Piening 1998; Dieckmann *et al.* 1998). It is clearly a hexaploid type (*Triticum aestivum* L.). Also in other middle Neolithic settlements in Europe, hexaploid naked wheat was found for the first time in larger amounts (for an overview, see Maier 1998). In almost all of the middle Neolithic settlements, barley (*Hordeum vulgare* L.) also has a clearly higher importance compared with the early Neolithic. It was mostly identified as

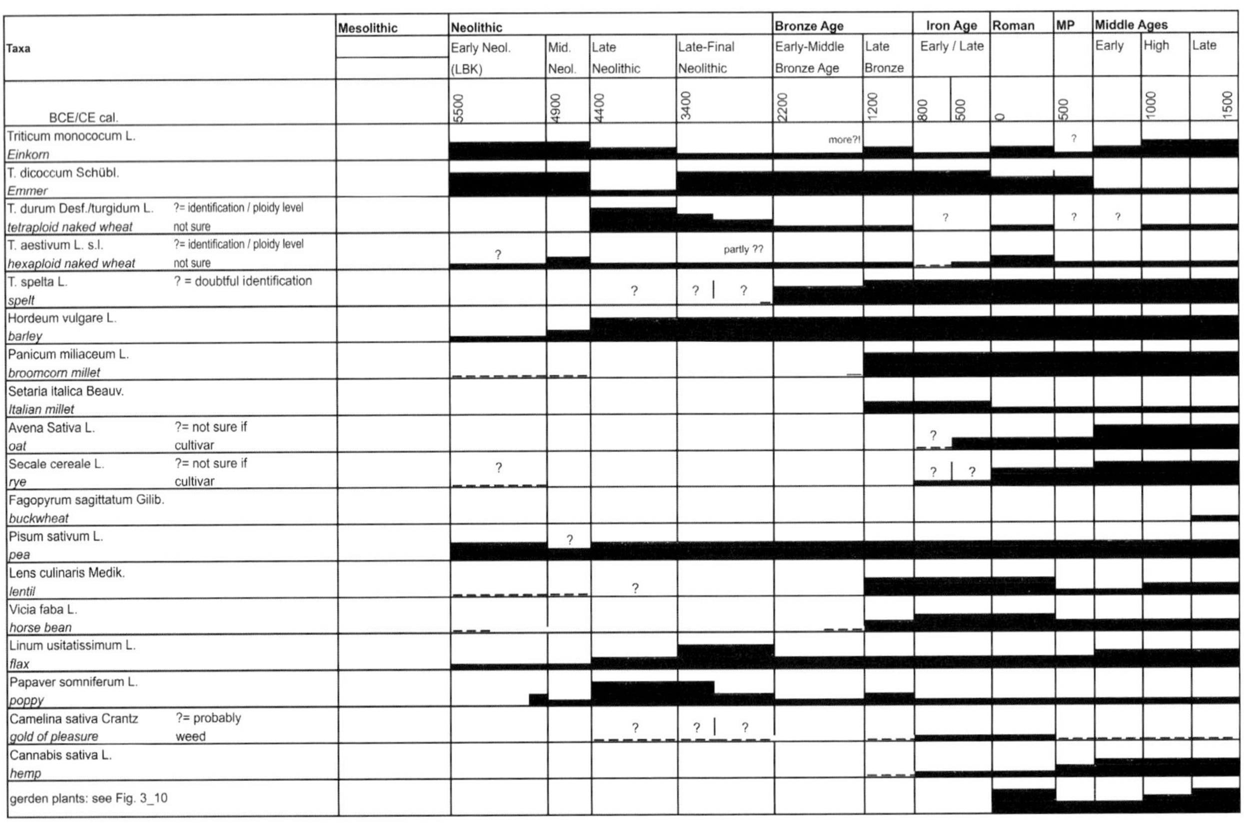

Fig. 3.12. The importance of cultivars beginning with the Neolithic in the region discussed in this paper. Taxa already present in prehistoric periods. The thickness of the bars shows the approximate importance. Idea from Willerding 1980, modified.

Fig. 3.13. Archaeobotanical finds of crops from the Late Neolithic lakeshore settlement of Arbon Bleiche 3 (canton of Thurgau, lake Constance): A. *Triticum* (*durum/turgidum* or *dicoccum*) – ear (impression from the bottom of a pot); Photo: AATG D. Steiner, www.archaeologie.tg.ch © Amt für Archäologie Thurgau); B. *Linum usitatissimum* – seeds; C. *Silene cretica* – seed; D. *Papaver somniferum* – seed. Photos B–D: Georges Haldimann, la Chaux-de-Fonds, © IPNA, Basel University.

a multi-rowed, naked form. To sum up, there seem to be some basic changes in the importance of the cereal spectra compared with the early Neolithic.

Other cultivars are very rare. There are some hints of a continuation of the growing of lentil, poppy and flax. Pea is lacking, which most probably is mere chance.

Late to Final Neolithic (ca. 4300–2200 cal. BCE)

From the late to final Neolithic (4300–2200 cal. BCE), there are many more settlements (>100) for which plant remains were investigated. Most of them are located on lake shores or in peat bogs (the famous 'Lake Dwellings'; see *e.g.* Menotti 2004; Fig. 3.11). North of the Alps, they are spread from the French Jura over the whole Swiss Plateau until Bavaria and

Austria, and belong to different cultural groups (for details about this period, see Jacomet 2006; 2007a; 2008; 2009). Therefore, we know much more about the plant economy of this time period than for the previous periods, because preservation is much better. Due to waterlogging, the spectra are extremely diverse, and above all, gathered plants are much better represented (Tolar *et al.* 2010). In addition, it has become more and more clear that inner parts of the Alps, including higher altitudes, were also intensively used during the Late Neolithic (Martin 2010). There are several cultural groups and regional particularities – including climatic differences – within the region concerned here (for details see the papers cited above).

In the late Neolithic, at a few places within the area considered here, einkorn and emmer continue to be important to dominant; in most of the sites, however, tetraploid naked wheat (*Triticum durum* Desf./*turgidum* L.) appears as the main wheat species (for details, see Jacomet 2007a; 2006). In most of the settlements from Upper Swabia westwards, this type of naked wheat is by far the most important wheat species until around 3000 cal. BCE (Jacomet 2006). Later, above all with the appearance of the Corded Ware culture from around 2800 cal. BCE onwards in the whole of the area, tetraploid naked wheat declines and emmer again becomes of higher importance (Jacomet 2009).

In a few of the older settlements (before 3500 cal. BCE) in the eastern part of the region, there are some mostly doubtful identifications of spelt (*Triticum spelta* L.). Unequivocal identifications from this time span are lacking from our region. The first true spelt appears during the final phases of the Neolithic, in a site of the Bell Beaker culture (Akeret 2005). In eastern central Europe, spelt had already appeared earlier (Kohler-Schneider 2007). There is an ongoing discussion about the origin of spelt (for some details, see Jacomet 2008 or Stika and Heiss 2013).

Barley – still mainly naked, six-rowed barley – is present and found almost always in large amounts.

Flax (*Linum usitatissimum* L.) and opium poppy (*Papaver somniferum* L.) are very widespread and mostly present in immense amounts in the waterlogged layers. However, flax is much rarer in sites before 4000 cal. BCE. The highest numbers are reached in sites after around 3800 cal. BCE (often in those of the Pfyn culture), for which stocks have also been detected. In this period, for the first time, the typical Mediterranean flax weed *Silene cretica* L. appears (Brombacher and Jacomet 1997, Fig. 3.12). This weed becomes more common with the intensification of flax-growing around the middle of the fourth millennium cal. BCE. Extremely high numbers of flax finds are reached during the second half of the fourth millennium cal. BCE, in sites of the so-called Horgen culture. Flax continues to be important until the end of the Neolithic, whereas opium poppy becomes more rare after around 2800 cal. BCE.

It is also worth noting the fairly regular finds of some garden plants like celery (*Apium* spec.) and dill (*Anethum graveolens* L.) after 4000 cal. BCE. Also, lemon balm (*Melissa officinalis* L.) was found in some sites, as well as parsley (*Petroselinum crispum* (Mill.) A. Hill) in one site. The seeds of these taxa are (mostly) found only in very small amounts, although they have thus far been found in several places (see tables in Jacomet 2006; 2007a). Millets and lentil have been found extremely rarely.

Bronze Age (ca. *2200–800 cal. BCE)*

During the Bronze Age, more unhospitable areas like the inner Alps – at even higher altitudes – also become settled (for more details, see Zoller and Erny-Rodman 1994; Jacomet *et al.* 1998; 1999a). The settlements included here are partly dryland sites, partly lake dwellings. The latter existed until the latest phases of the Late Bronze Age – then, they disappeared from the Alpine foreland.

The earlier phases of the Bronze Age mostly do not show large differences in comparison with the latest Neolithic phases. An important wheat species in many places is emmer (*Triticum dicoccum* Schübl.). However, spelt (*Triticum spelta* L.) now becomes the most important wheat in a larger part of the area considered. It is also grown in higher altitudes above 1000 m above sea level. There is a rather regular presence of einkorn and naked wheat, the latter of both ploidy levels. Naked wheat, however, is rather a rare taxon during the Bronze Age (there might be regional differences). Barley (*Hordeum vulgare* L.) is very important. Here, for the first time, hulled barleys of the multi-rowed type seem to prevail (see *e.g.* Jacomet and Behre 2009).

Like spelt, barley is also grown inside the Alps at higher altitudes.

There are fundamental changes in the cereal spectrum from the onset of the Late Bronze Age onwards (around 1250 cal. BCE): new cultivars now appear in the region considered here. There is a wide range of cereals grown (for pictures, see *e.g.* Jacomet and Karg 1996). Beside the taxa that are already important during the early Bronze Age (Fig. 3.11), above all spelt, millets now appear massively. Both broomcorn millet (*Panicum miliaceum* L.) and Italian millet (*Setaria italica* (L.) Beauv.) are grown regularly, and therefore found in large amounts (with stocks known from several burnt layers of lakeshore settlements like Zug-Sumpf). Millets were first domesticated in China (see *e.g.* Nasu *et al.* 2007; Lu *et al.* 2009) and are the first cereal cultivars to appear in central Europe which do not belong to the Near Eastern founder crop package. There is almost nothing known about spectra of the Middle Bronze Age – it might be possible that some of these changes are already present slightly earlier.

There is also a significant increase in the importance of pulses: from the Bronze Age (at least the LBA) onwards, pea (*Pisum sativum* L.), lentil (*Lens culinaris* Medik.) and horse bean (*Vicia faba*) are found in large amounts, and all of them were grown regularly, at least pea and horse bean also at higher altitudes in the Alps. In addition, there are some scattered finds of bitter vetch (*Vicia ervilia* (L.) Willd.), as in other parts of central Europe, where bitter vetch may even have played a larger role in some places (Becker and Kroll 2008).

Besides the newly grown cultivars, flax and poppy continue to be of certain importance. For flax, there are hints it was grown at higher altitudes, as well (Jacomet *et al.* 1999a).

Iron Age (ca. 800 until ca. 50 cal. BCE)

Most of the Iron Age sites considered here are dryland sites, and waterlogged preservation is rare. This might influence the representation of the taxa somewhat. Most of the sites considered are located in flat or hilly areas, some also inside the Alps (for the latter see *e.g.* Schmidl *et al.* 2007).

The differences with the Late Bronze Age are not too large. Concerning wheat and barley, there is no difference – spelt continues to be the most important wheat species. Naked wheats are astonishingly rare – however, there are large regional differences, as we could show for the Basel region (Jacomet and Brombacher 2009). There, on good soils and under favourable climatic conditions of the southern Upper Rhine area, there are finds of stocks of hexaploid naked wheat (*Triticum aestivum* L.). Tetraploid naked wheat is also present, although in small amounts (Kühn and Iseli 2008). A new cereal which appears during the Late Iron Age is oat (*Avena sativa* L.; see *e.g.* Jacomet *et al.* 1999b), though in very tiny amounts. Rye is present only as single grains, most probably representing weeds.

The pulses of the Late Bronze Age continue to be important, as do flax and poppy. Also during the Iron Age there are some finds of bitter vetch (*Vicia ervilia* (L.) Willd.). Gold-of-pleasure, *Camelina sativa* (L.) Crantz, is now of some importance as an oil plant, however, not necessarily in the region considered here (more to the north, see *e.g.* Knörzer 1978; Kreuz 2004). For the first time, sure traces of hemp (*Cannabis sativa* L.) are found, too, mainly during the Late Iron Age. This agrees with other finds in Europe (see *e.g.* Dörfler 1989; Bouby 2002; Rösch 1999). The origin of *Cannabis* is in eastern to central Asia (Zohary *et al.* 2012). It reached western central Europe during the Early Iron Age, perhaps only as a product, as the finds of hemp-textiles in the rich 'prince's grave' near Hochdorf/Stuttgart show (sixth century BCE) (Körber-Grohne 1985). In addition, flax and poppy continue to be of some importance.

During the Late Iron Age, some garden plants appear in the spectra, too. In this paper, we do not consider fruit remains for which we cannot tell, based on the morphology of their seeds or fruits, whether they are the wild form or cultivars like apple, pear or cherry. These had mostly been present since Neolithic times, probably as plants gathered in the wild. For further information on the topic, also see Chapter 4. There are, however, single finds of cultivars like celery (*Apium graveolens* L.), plum (*Prunus domestica* L. and its different subspecies), fig (*Ficus carica* L.), dill (*Anethum graveolens* L.), fennel (*Foeniculum vulgare* Mill.) or coriander (*Coriandrum sativum* L.). These come mostly from the Upper Rhine area (Kreuz 2004; Kreuz 2003; Jacomet and Brombacher 2009). Also, there are some finds of grape (*Vitis* sp.), although it is very hard to judge whether of cultivars or not. New palynological data show that in the dry inner Alpine Valais, wine-growing may have already started during the Iron Age (Curdy *et al.* 2009).

Roman Period (ca. *50 cal. BCE–400 CE)*

Astonishingly, there are no fundamental changes in the cereal spectrum between the Late Iron Age and the Roman period. The most important cereals continue to be a multi-rowed, hulled barley (*Hordeum vulgare* L.), as well as spelt and millets (above all broomcorn millet, *Panicum miliaceum* L.). Naked wheat is common in certain regions, hexa- and tetraploid varieties are present (see Rösch *et al.* 1992; Jacomet and Brombacher 2009). Emmer must have had a certain significance, too, whereas einkorn and Italian millet are rarer than before. Oat is found regularly, although not in large amounts.

The only new introduction of a cereal during the Roman period is astonishingly rye (*Secale cereale* L.), because this cereal is considered to be of minor quality by Roman written sources (André 1998). For the first time, rye is found in larger amounts, with stocks even appearing in the third century CE – since then it must have been cultivated, at least in certain regions. The same results show up in regions slightly more to the north (Kreuz 2004).

During the Roman period, the same pulses continued to be important as during the Iron Age, mainly horsebean and lentil. Flax and poppy are present too and must have had a certain importance. The frequency of hemp rises somewhat, and ubiquities are higher. The cultivation of hemp was obviously spread by the Romans (Dörfler 1989).

The most fundamental innovation of the Roman period was the widespread introduction of gardening. We see an enormous rise in the importance of garden plants, in terms of numbers of taxa but also in terms of numbers of finds (Fig. 3.11); this is not only the case in the region considered here, but in the whole of the Roman Empire north of the Alps. Therefore, together with other colleagues like Kreuz (2004), we can conclude that widespread gardening began soon after the onset of the Roman era. We have to emphasise that most of the finds of garden plants are only well represented when cultural layers are preserved in a waterlogged state. However, in the meantime, there are many regions where locations with such preservation have been investigated, not only of Roman layers but also for the Iron Age (for more details, see Jacomet and Brombacher 2009).

Figure 3.14 shows a list of the most important garden plants of the Roman period (as well as earlier and later periods). As an example, we have chosen the Basel region, which is very well investigated and might be representative for other places in the Roman provinces, as well (also see Bakels and Jacomet 2003). Important vegetables included celery (*Apium graveolens* L.), die beet (*Beta vulgaris* L.), bottle gourd (*Lagenaria vulgaris Ser.* (= *L. siceraria* (Molina) Standley) and parsnip (*Pastinaca sativa* L.). There may have been many more taxa, but at least for some, the identifiability of their seeds is difficult (like of *Brassica*-species or garden orache, *Atriplex hortensis* L.). Also warmth-loving taxa like bottle gourd were most probably grown in some climatically favoured regions like the Upper Rhine valley, because not only seeds, but also whole fruits were found (Reddé *et al.* 2005; Vandorpe and Jacomet 2011). Our own experiments have shown that the production of bottle gourd is quite possible in the Upper Rhine area. Besides vegetables, many spices were also grown in gardens. Here, we may mention dill (*Anethum graveolens* L.), summer savory (*Satureja hortensis* L.), coriander (*Coriandrum sativum* L.), garlic (*Allium sativum* L.) and fennel (*Foeniculum vulgare* Mill.). There were also spices imported from far away like pepper (*Piper nigrum* L.) from India (for more details, see Jacomet and Brombacher 2009).

From the Roman period onwards, the widespread growing of fruit trees and shrubs had also begun (Willerding 2002). From this time on, over ten different woody fruit species are known from the region considered here (Fig. 3.14). We have to reckon with the onset of the growing of apple, pear and cherry cultivars from Roman times onwards, even when the morphology of the finds still does not enable us to clearly identify wild forms and cultivars. In other cases, such as walnut (*Juglans regia* L.), plums (*Prunus domestica* L. and its different subspecies), almond (*Amygdalus communis* L. (= *Prunus dulcis* (Miller) D. A. Webb), peach (*Persica vulgaris* Miller (=*Prunus persica* (L.) Barsch), mulberry (*Morus nigra* L.) and also wine (*Vitis vinifera* L.), it is clear that the finds are of cultivars. Medlar (*Mespilus germanica* L.) was also already known, as finds from

Scientific name	Kinds of preservation	Total of remains	all Tx.	carb. Tx.	wlTx.	min. Tx.	Neolithic n=5, all ME ubiquity	Neolithic number of remains	Bronze Age n=9, 7MB, 2 FB ubiquity	Bronze Age number of remains	Iron Age n=19, 16MB, 3 FB ubiquity	Iron Age number of remains	Roman p. n=53, 39 MB,14FB ubiquity	Roman p. number of remains	Early Med. n=8, 5MB, 3FB ubiquity	Early Med. number of remains	High-/Late Med. n=29, 17MB, 12FB ubiquity	High-/Late Med. number of remains	Modern n=8, 4MB, 2FB, 2TR ubiquity	Modern number of remains
Apium graveolens	various	10821	1	1	1	1					5	6	25	10630	13	1	13	57	25	7
Atriplex hortensis	various	5	1	1	1												5	5		
Beta vulgaris	various	33224	1	1	1	1							11	212			13	32974	13	38
Brassica oleracea	various	2	1	1	1						5	1					3	1		
Cichorium endivia/intybus	waterlogged	9	1		1								2	1			3	1	25	7
Cucumis sativus	various	57	1		1	1							2	5			10	44	25	8
Lagenaria siceraria	various	78	1		1	1							13	78						
Pastinaca sativa	various	151	1	1	1						5	1	4	143	13	1	5	6		
Petroselinum crispum	various	6	1		1	1											3	1	13	5
vegetables (cultivars) number of remains		**44353**										**8**		**11069**		**2**		**33089**		**65**
vegetables (cultivars) number of taxa		**9**	**9**	**5**	**9**	**5**	**0**	**0**	**0**	**0**	**3**	**3**	**6**	**6**	**2**	**2**	**8**	**8**	**5**	**5**
Allium sativum	charred	53	1	1									2	2			3	51		
Anethum graveolens	various	1401	1	1	1	1							23	1028	38	11	21	361	13	1
Brassica nigra	various	86	1	1	1	1					5	2			13	1	13	76	25	7
Coriandrum sativum	various	4519	1	1	1	1							28	4439	13	5	8	16	50	56
Foeniculum vulgare	various	315	1	1	1	1							11	105			13	75	38	134
Lepidium sativum	waterlogged	3	1			1							2	3						
Melissa officinalis	waterlogged	1	1		1								2	1						
Piper nigrum	waterlogged	28	1		1								4	28						
Satureja hortensis	various	1128	1	1	1	1							17	1101			10	19	13	8
Satureja montana	waterlogged	1	1			1							2	1						
spices, number of remains		**7535**										**2**		**6708**		**17**		**598**		**206**
spices, number of taxa		**10**	**10**	**6**	**7**	**7**	**0**	**0**	**0**	**0**	**1**	**1**	**9**	**9**	**3**	**3**	**6**	**6**	**5**	**5**
Castanea sativa	waterlogged	1	1		1								2	1						
Juglans regia	various	1210	1	1	1								45	790	25	69	33	311	50	41
Ficus carica	various	38956	1	1	1	1							30	36398			13	2266	38	269
Malus sylvestris/domestica	various	962	1	1	1	1					5	3	15	869	25	2	23	51	25	87
Mespilus germanica	waterlogged	16	1		1												3	4	25	12
Morus nigra	various	1512	1		1	1							9	586			13	139	38	787
Phoenix dactylifera	charred	18	1	1									6	18						
Pinus pinea	various	1038	1	1	1								6	1038						
Prunus armeniaca	mineralisiert	1	1			1											3	1		
Prunus avium/cerasus	various	7026	1	1	1	1					5	1	15	318			23	5213	63	1462
Prunus domestica/insititia	various	5522	1	1	1	1							17	664			18	4713	50	141
Prunus dulcis	waterlogged	1	1		1								2	1						
Prunus persica	various	170	1	1	1	1							17	141			10	18	25	11
Pyrus communis/pyraster	various	1373	1	1	1								4	3			23	860	25	510
Sorbus domestica	waterlogged	13	1		1														25	13
Vitis vinifera	various	34716	1	1	1	1					5	3	36	16566	13	8	49	12116	63	6006
nuts and fruit trees (cutivars) number of remains		**92535**										**7**		**56355**		**79**		**25692**		**9339**
nuts and fruit trees (cutivars) number of taxa		**16**	**16**	**10**	**14**	**8**	**0**	**0**	**0**	**0**	**3**	**3**	**13**	**13**	**3**	**3**	**11**	**11**	**11**	**11**
Carthamus tinctorius	waterlogged	179	1		1								2	178					13	1
Claviceps purpurea	charred	13	1	1											13	8	8	5		
Ruta graveolens	waterlogged	2	1		1														13	2
Cannabis sativa	various	311	1	1	1	1					5	7	17	149	13	44	10	104	13	7
Linum usitatissimum	various	3887	1	1	1	1	40	2	22	2	16	92	23	266	50	531	26	2983	25	6
Olea europaea	waterlogged	73	1		1								8	73						
Papaver somniferum	various	1521	1	1	1	1			22	5	11	8	19	1103	25	7	23	374	25	24
Consolida ajacis	charred	6	1	1											25	6				
Prunus cerasifera	waterlogged	5	1		1												3	5		
Various, number of remains		**5997**						**2**		**7**		**107**		**1769**		**596**		**3471**		**40**
Various, number of taxa		**9**	**9**	**5**	**7**	**3**	**1**	**1**	**2**	**2**	**3**	**3**	**5**	**5**	**5**	**5**	**5**	**5**	**5**	**5**

Fig. 3.14. The importance of cultivars (and some other taxa) which were mainly newly introduced in the Roman period. Numbers of remains and ubiquities (proportion (%) of sites in which a taxon is appearing), sites from the Basel region (from Jacomet and Brombacher 2009). N = number of sites; Tx = Taxa; MB = Mineralbodenerhaltung (preservation by charring); FB = waterlogged preservation; TR = desiccated preservation; min = mineralised.

Eschenz at Lake Constance show (Pollmann and Jacomet 2012). It is a matter of debate whether for instance fig (*Ficus carica* L.) was grown locally. In good years, single fig trees at advantageous locations may produce ripe fruits. Fig, however, is so widespread in Roman layers that most of the figs must have been imported from Mediterranean regions as dried fruit.

Besides locally grown taxa, there is a whole range of species which were surely imported from Mediterranean regions, like date (*Phoenix dactylifera* L.), pomegranate (*Punica granatum* L.), pine (*Pinus pinea* L.) or olives (*Olea europaea* L.; Bakels and Jacomet 2003; Livarda and Van der Veen 2008). The latter might have been mostly used as spices or snacks. Finally, there is an interesting attestation of an exotic dye plant, the safflower, *Carthamus tinctorius* L., the only find of this species north of the Alps (P. Vandorpe, *pers. comm.*).

Migration period and Middle Ages (ca. *400–1500 CE)*

The early medieval periods (until around 800 CE) show a high diversity of cereal species being grown. There are slight changes visible compared with the Roman period. Whereas spelt, and in some regions also naked wheat, continue to be significant, the importance of rye and in some regions also einkorn increases. This also becomes visible, for example, in the first appearance of ergot (*Claviceps purpurea*), a fungus typically (but not only!) growing on rye (Willerding and Teegen 2002). Oat, emmer and broomcorn millet were also of some importance.

During the later medieval periods (Late Middle Ages, after around 800 CE), there is a clear shift towards growing fewer cereal species (see *e.g.* Rösch *et al.* 1992). This reflects the introduction of the so-called three-field-rotation system, with growing winter cereals in the first year, summer cereals in the second and a short fallow in the third year (Willerding 1996; Ernst and Jacomet 2006; Karg 1996; Rösch *et al.* 1992). The main winter cereals were now rye and spelt; the main summer crops barley and oat, and perhaps also broomcorn millet. Beside these main cereals, in some regions einkorn was still important; this was the case until the nineteenth century, for example, in northwestern Switzerland (Ernst and Jacomet 2006). Because of its starch-containing fruits, buckwheat (*Fagopyrum esculentum* Moench) is also considered as a cereal. It appears very late in the region considered here, during the Late Middle Ages (for example, fifteenth century in Basel; Kühn 1996). This late appearance is in agreement with many other localities over Europe (see *e.g.* Wiethold 2003b). A cereal which was never grown in the region considered here is rice (*Oryza sativa* L.). Rice was imported from southern regions like Italy. In some medieval town centres, remains of rice were found in latrines, *e.g.* in Freiburg im Breisgau in the southwesternmost area of Germany (Sillmann 2002). The earliest traces date to 1300–1500 CE, and they are related to wealthy households or abbeys.

Concerning pulses, there is no difference compared with the Iron Age and Roman period. There are many finds of pea, lentil and horse bean. In some regions, however, from high-/late medieval times onwards, a new species appears to become important, common vetch (*Vicia sativa* L.). Poppy is still present, as is flax and hemp. There are hints of an intense production of both fibre plants, in the form of remains of retting and breaking, respectively. Hints of retting include mass finds of flax and hemp pollen in small lakes (Dörfler 1989; Rösch 1999). There are also remains of breaking flax, *e.g.* from Early Medieval settlements in the Swiss Jura (Brombacher 2008).

Most of the vegetables, spices and fruit trees known from the Roman period continue to exist in the Middle Ages (Fig. 3.14). However, there is a significant decline during the migration period and the following sixth and seventh centuries, although not all cultivars of these groups introduced by the Romans disappeared during the migration period, as Fig. 3.14 shows (see also Rösch 2008). There was some continuity, above all in fruit-tree-growing, as finds of walnut and wine from northern Switzerland indicate (for more details, see Jacomet and Brombacher 2009). Other taxa, like bottle gourd, disappear with the end of the Roman period. Also 'exotic' imported taxa like olives do not seem to be present anymore. However, some new taxa, such as apricot (*Armeniaca vulgaris* Lam. (=*Prunus armeniaca* L.) appear during the Middle Ages.

Discussion

Archaeobotany enables us to identify diaspores of most of the taxa of cultivars to species level. We understand the term 'species' here in a traditional, morphological sense. However, for judging the role of cultivars, it should be possible to identify varieties. Although this may be possible when preservation conditions are favourable, it is almost never possible to assign an archaeological variety to a modern one (see example for prunes, Pollmann *et al.* 2005). This would only be possible with aDNA (ancient DNA) investigations, which in most cases turn out to be very difficult, although more and more results are being obtained (for an overview, see Schlumbaum *et al.* 2008). For the moment, however, we have to rely on the relatively rough identifications we have. Therefore, for instance, it is extremely hard to judge the influence of climate on the spectrum of cultivars, because some varieties of a cultivar may be rather resistant to cold, while others are not. There is, nonetheless, some general knowledge which shows for instance that naked wheat in general is more susceptible to cold-wet climate conditions than some of the glume wheats (see, for example, the discussion in Jacomet *et al.* 1989 and literature cited there). This is to some degree shown in our spectra by the fact that a higher importance of hexaploid naked wheat appears only in low-lying regions, with favourable climate and good soils, *e.g.* during the Iron Age and Roman Period. However, within the framework of this paper, we cannot discuss such differences in detail. In order to do that, one would have to thoroughly evaluate the spectra of smaller regions in detail. Here, we concentrate on chronological differences and discuss possible reasons for those. It has to be emphasised, however, that finding reasons for the observed developments is not easy – mainly because of the present lack of the existence of a Eurasian archaeological plant database. Such a database would be the precondition for reconstructing the origin of cultivars in detail. It should be targeted for the future by linking the existing regional databases (*e.g.* Kreuz and Schäfer 2002).

One important change during the Middle Neolithic, mainly with the onset of the Rössen culture around 4600 cal. BCE, is the massive appearance of hexaploid naked wheat (*Triticum aestivum* L.). This is not only the case in our area, but also in other places in Europe (*e.g.* Bakels 1991; Bakels *et al.* 1993; Maier 1998). The direction from which this type of naked wheat penetrated the western part of central Europe is a matter of debate and basically not known. Maier (1998) thinks that hexaploid wheat is perhaps part of the cultivar set of the 'Danubian' cultures, and derives from somewhere in the East.

A striking phenomenon is also the appearance of tetraploid naked wheat in the northern Alpine foothills from the Late Neolithic (*ca.* 4300 cal. BCE) onwards. The direction from which this taxon reached the Alpine foreland has become somewhat clearer only very recently. For over twenty years, there have been theories that it must have come from the western Mediterranean, from the influence sphere of the Early Neolithic Cardial/Impressa and the succeeding cultural groups (Jacomet and Schlichtherle 1984; Schlichtherle 1997; Maier 1998). However, until a few years ago, no rachis remains had been found, although recent excavations show such remains also in the Mediterranean lake shore settlements from the Cardial period. At least two sites have been investigated: the La Marmotta site in central Italy (Rottoli and Pessina 2007) and La Draga in northwest Spain (Bosch i Lloret *et al.* 2000; Buxó *et al.* 2000). There, tetraploid naked wheat is present, as remains of ears, spikelets and rachis remains show (pers. comm. R. Buxo, F. Antolin and M. Rottoli). So, *Triticum turgidum* Desf./*durum* L. must have penetrated our area from the south and southwest, together with many other influences (see below), also visible in the archaeological artefacts. The most recent results from Slovenian lake dwellings of the fourth millennium cal. BCE show that indeed farther to the east, naked wheat is lacking in the spectra (Jeraj *et al.* 2009; Tolar *et al.*, 2010). However, astonishingly, tetraploid naked wheat is not restricted to the influence sphere of westerly-based cultural groups (like, for example, the Egolzwil and Cortaillod-Cultures). It spread very rapidly until the Federsee region in Upper Swabia and penetrated the influence spheres of more easterly influenced cultural groups (such as, above all, the Pfyn culture). There, signs of strong influence from the east are not only the presence of copper but also of smelters. On the other hand, there are also westerly characteristics. Farther to the east, tetraploid naked wheat seems to disappear (Kohler-Schneider and Caneppele 2009).

The case of tetraploid naked wheat (and most probably also of the hexaploid naked wheat in the Middle Neolithic) shows that the spread of cultivars is mainly due to cultural influences, probably connected with long-distance trade. This is also shown by the appearance of some other plants of

Mediterranean origin around 4000 cal. BCE, such as parsley, dill, celery or lemon balm. These taxa are basically restricted to the regions where tetraploid naked wheat is the dominant wheat form (see Jacomet *et al.* 1989). Seeds must have reached the region discussed here regularly or these species were grown, although this is hard to judge. Celery (*Apium*) is originally a plant of salty places, near the sea coast or near salt springs in the interior. It is also possible that it reached our area with the salt trade. Introduction of new weedy taxa like the flax weed *Silene cretica* around 4000 cal. BCE (Brombacher and Jacomet 1997) points in the same direction.

From as early as the middle of the fourth millennium BCE onwards, a decline of tetraploid naked wheat begins, and at the same time, the importance of emmer increases. This is accentuated towards the final phases of the Neolithic – after 2800 cal. BCE. Again, we see the reason for this in cultural influences: it is connected with the onset of the Horgen culture and later the Corded Ware, which reached the study area from the north and east (for more details, see Jacomet 2006; 2008). Regarding the Corded Ware culture (*e.g.* 'mitteldeutsche Schnurkeramik', Hopf 1982), emmer (in addition to einkorn and barley) was the most important cereal species in northeastern areas. It is therefore not surprising that emmer became the most important wheat species in the study area during the Corded Ware culture. Parallel to the tetraploid naked wheat, opium poppy and 'exotic' species such as dill and celery also became rarer or disappeared. This can be seen perhaps in connection with decreasing (south)western cultural influences. At Lake Biel, for instance, the import of flint from (south)western sources was interrupted for some centuries after 3200 BCE (Hafner and Suter 2004).

There is another, very important innovation towards the end of the Neolithic – the appearance of spelt. This crop is extremely resistant to disease or wetness, and grows on a wide range of soils. The origins of spelt still lie in the dark, even though recent DNA-investigations have shown that European spelt must have originated somewhere in Europe and that it goes back to a hybridisation between hexaploid naked wheat and emmer (for a description of the state of the art, see Jacomet 2008; Dvorak *et al.* 2012; Stika and Heiss 2013). Once spelt existed in many regions, it became the most important cereal rather quickly. The spread of spelt, therefore, may not be connected with cultural drift, but rather with soil deterioration, the spread of disease, bird damage and maybe also climatic shifts.

From the onset of and during the Bronze Age, remarkable shifts in the spectrum of cultivars appear. This has to be seen in connection with the emergence of a series of long-distance trade networks between the central Asian steppes, the Carpathians, the Aegean, Anatolia and western Europe, mainly during the third millennium BCE. This is the appearance of the so-called 'steppe-corridor', which finally connected China with Europe (see *e.g.* Kristiansen and Larsson 2005). Therefore, it is not at all astonishing that during the Bronze Age for the first time cultivars of Eastern Asian origin – millets – become part of the set of cultivated plants in central Europe. Concerning millet, there are some scattered earlier finds, although there are no sure signs of cultivation. In our region, this is the case around 1300–1200 cal. BCE – mainly with the beginning of the Late Bronze Age. Somewhat earlier, pulses also become a very important part of subsistence, whereas before they were rather rare and only pea is found in larger amounts and regularly. The Bronze Age pulse spectra, with large amounts of horse bean, lentil and pea, and partly also the appearance of bitter vetch, remind us of Bronze Age spectra of the Aegean (see Becker and Kroll 2008). Both millets and pulses are summer crops.

These changes in the spectrum of cultivars are also visible in the spectrum of weeds. A whole range of weeds, typical for winter and summer crops, reached the region considered here during this period (for details, see Jacomet and Brombacher 2009). This is to be seen in connection with the diversification of agriculture during the Bronze Age, mainly from the Late Bronze Age onwards. The weed spectra of winter cereals such as spelt now point – in contrast to the Neolithic (this is highly debated, see the 'extremes' described in Bogaard 2004) – to an extensive cultivation on larger fields, the so-called 'outfield', which were tilled with ards (Bogaard 2002; 2012). These ards, pulled by oxen, are henceforth often displayed in rock art in the Alpine region. In contrast, millets and pulses must have been cultivated in an intense way on small plots, the so-called 'infield'.

Another introduction of the Bronze Age seems to be a new type of barley, hulled barley. It is hard to judge what might be the reasons for this. Perhaps the new barley types, as well as crops like spelt, were more resistant to environmental influences of various types, and – another important fact – they grew in higher altitudes. The latter became very important during the Bronze Age when the Alps were settled intensively. The same holds for summer crops with a short growing season like millets and also pulses (above all pea and horse bean). The latter appear in large amounts in southern and inner Alpine sites during the Early Bronze Age, and in the northern Alpine foothills, mainly during the Late Bronze Age. Therefore, they must have spread from south to north as our investigations show for the moment (ongoing analyses of Early Bronze Age sites in northern Italy, Perego, oral comm.). More research is needed to understand the spread of new cultivars during the Bronze Age in greater detail. All in all, we can conclude that through the introduction of new cultivars during the Bronze Age, it may have become possible to produce surpluses.

The introduction of widespread gardening surely goes back to the Roman conquest of land north of the Alps. However, some rare garden plants appear as early as the (mainly Late) Iron Age. Some of these finds may go back to imports from the Mediterranean world, others may have been locally grown – this cannot be decided on the basis of archaeobotanical data. It is clear from other archaeological finds that the Celts – inhabiting most of the regions considered here – had many contacts with the Mediterranean and imported wine, for instance (*e.g.* Kimmig 2000).

Another striking innovation which appears during the Roman period is the introduction of rye – another crop extremely well suited to poor soils and growing in higher altitudes (Behre 1992). The direction from which rye came into our region is not clear at all. One possibility would be regions around the North Sea and Baltic Sea coasts, perhaps with soldiers of the Roman army who came from there. This should be investigated in future projects.

Finally, there are some crops which are already present in the earliest Neolithic and played a certain role until the nineteenth century, for example, einkorn. Of the latter, it is known that in the Middle Ages and the Modern Era, mainly the straw was used, for binding grapes or for producing decoration.

Conclusion

The history of cultivars in the region discussed here seems primarily driven by cultural drift. On the one hand, humans must have moved and transported their set of cultivars. On the other hand, there are signs of widespread trade networks which also included trade in seeds. However, there are some signs that environmental constraints (including climate, disease...) had some influence on the spread of cultivars, as the appearance of spelt may indicate. On the other hand, the development of varieties (landraces...) is probably mainly due to environmental constraints. Unfortunately, we cannot reconstruct this with the help of traditional archaeobotany. Perhaps in the future, ancient DNA investigations will highlight some more aspects.

3.5. CROP DIVERSITY IN THE NEOLITHIC OF THE IBERIAN PENINSULA

Leonor Peña-Chocarro and Lydia Zapata

Introduction

Archaeobotanical work carried out over the past 20 years has demonstrated that by *ca.* 5600–5500 BCE, domestic plants were already present in the Iberian Peninsula (Fig. 3.15) showing an enormous diversity when compared with other European regions, particularly central Europe. First farmers from Iberia grew a large variety of crops which included hulled and free-threshing wheats, barley and a wide variety of legumes such as peas, lentils, broad beans, vetches and grass peas. Flax and poppy were also part of the broad range of crops cultivated in the region. Although most of the available data focuses on domestic crops, there is growing evidence of wild foodstuffs being consumed by early farmers (Buxó 1997; Zapata 2000), increasing the range of plant resources used for subsistence. However, their importance in the overall diet is difficult to evaluate.

Located in the most southwestern corner of Europe, occupying a marginal position within the European continent, Iberia was the last region to adopt agriculture in the Mediterranean world. The territory occupied by the Iberian Peninsula shows a complex physiography with high plateaus and mountain chains. Besides, as a consequence of both Mediterranean and Atlantic influences, there is an enormous variety of climates and landscapes which offer different ecological settings to which farming had to adapt.

Research into early agriculture in Iberia has considerably increased over the past years (Buxó 2007; Peña-Chocarro 2007; Peña-Chocarro *et al.* 2005b; Stika 2005; Zapata 2007; Zapata and Peña-Chocarro 2005; Zapata *et al.* 2004). However, this has followed different research traditions which vary from region to region, so areas like the Mediterranean and Portuguese coasts have been the focus of interest for a long period, even if plant remains have not always been recovered, whereas other areas like inner Iberia still remain unknown or have only recently started to be studied from an archaeobotanical point of view (Peña-Chocarro *et al.* 2005a; Stika 2005). On the other hand, strategies for sampling and recovering plant remains have been very varied, so data is not homogeneous. In any case, even if much more data is still needed to get a full picture of early farming in Iberia, the growing interest in the subject and the production of new data allow some progress in the study of crop diversity.

Crop Diversity in the Iberian Neolithic

Archaeological research in Iberia, with its most recent emphasis on archaeobotany, is generating an important dataset on plant remains. Archaeobotanical evidence suggests that early Neolithic agriculture in this region was one of the most varied in the whole of the continent (Zapata *et al.* 2004). A wide variety of crops was cultivated, including cereals and legumes as well as some fibre and oil plants. Hulled wheats (*Triticum monococcum* and *T. dicoccum*), free-threshing wheats (*T. durum* and *T. aestivum*), naked barley (*Hordeum vulgare* var. *nudum*), and hulled barley (*H. vulgare*) were the main cereal species, while peas (*Pisum sativum*), lentils (*Lens culinaris*), broad bean (*Vicia faba*), vetch (*Vicia sativa*) and grass pea (*Lathyrus sativus/*

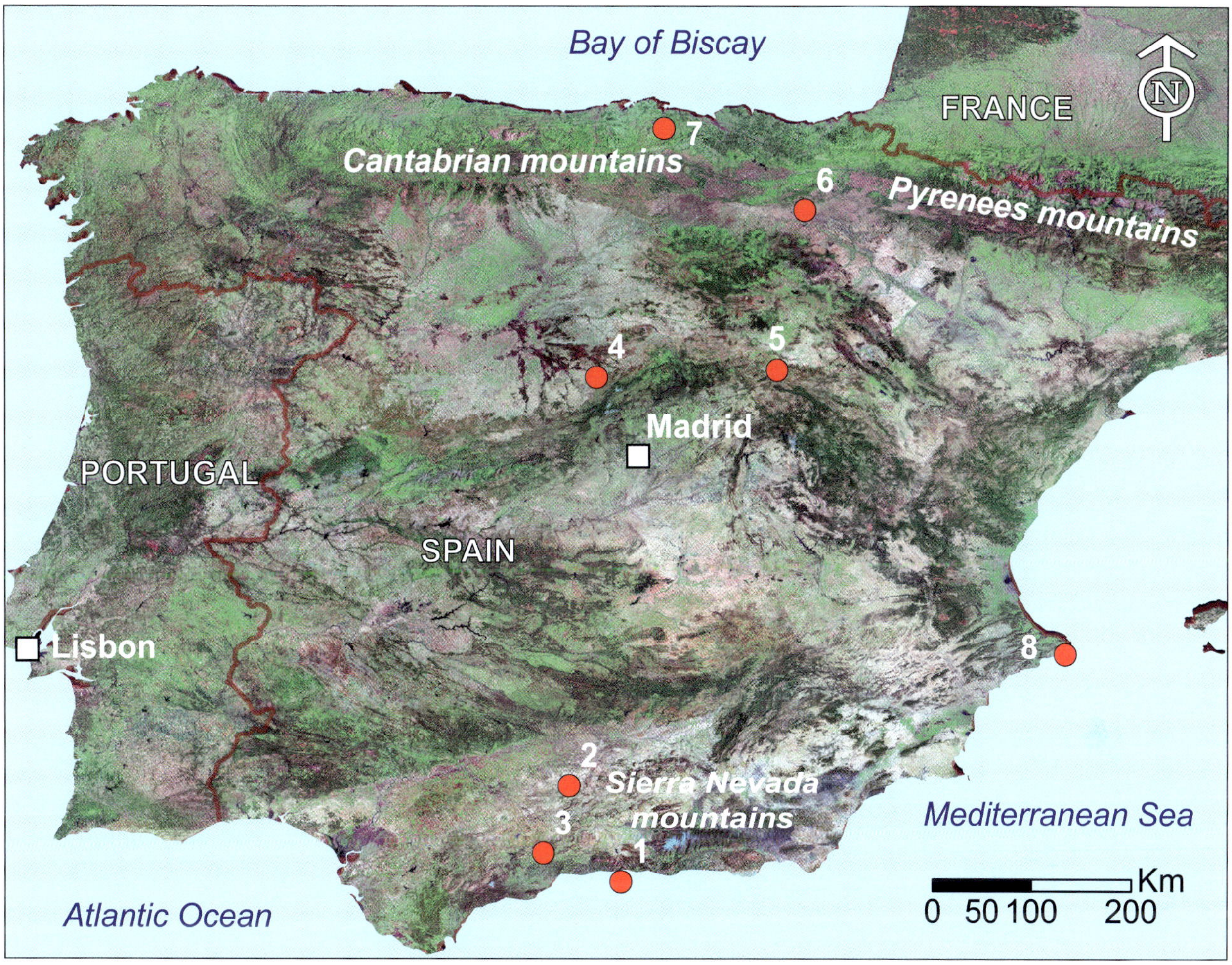

Figure 3.15. Map of the Iberian Peninsula with Neolithic sites (5600–4000 BCE). 1) Cueva de Nerja; 2) Cueva de Los Murciélagos; 3) Cueva del Toro; 4) La Vaquera; 5) La Lámpara; 6) Los Cascajos; 7) El Mirón; 8) Cova de Cendres. R. Lugon, Map: R. Lugon, J.-C. Loubier and A. Chevalier.

cicera) were the most common legumes. Flax (*Linum usitatissimum*) and perhaps opium poppy (*Papaver somniferum/setigerum*) were also cultivated. On arrival in Iberia, all these species (except poppy, see below) encountered different climates, soils, diseases, cultural traditions etc., so the newcomers had to deal with many different environments, at times succeeding or failing, depending on their ability to cope with the new situations.

Cereals

Hulled Wheats: Einkorn and Emmer

The term hulled wheats describes a group of species: *Triticum monococcum* (einkorn), *T. dicoccum* (emmer) and *T. spelta* (spelt), whose tough glumes continue to adhere to the grain after the wheat has been threshed. The final product is, therefore, whole spikelets as opposed to the free-threshing species, whose grains easily separated from the chaff. Their tough glumes give excellent protection to the crop in the field and in storage. Hulled wheats still survive today as relic crops in some mountain areas in Europe, the Near East and Africa (D'Andrea and Haile 2002; Hillman 1984; papers in Padulosi *et al.* 1996; Peña-Chocarro 1996; 1999; Peña-Chocarro and Zapata 1997; 1998; Peña-Chocarro *et al.* 2009), where they appear well adapted to poor soils and harsh conditions. Flour or whole or fragmented grain is still consumed by humans and animals. In addition, einkorn straw provides an excellent raw material for crafts and thatching, while emmer straw is mainly used for animal bedding (Peña-Chocarro 1999).

Einkorn and emmer were the main crops in large areas of Europe during the early Neolithic. In central Europe, the LBK (*Linearbandkeramik*, linear pottery) culture groups cultivated these two species as the main cereals (Bakels 2000; 2007; Jacomet 2007a; Kreuz 2007; Kreuz *et al.* 2005; Marinova 2007). In other Mediterranean areas, such as Italy, they were also an important element of the plant diet. Free-threshing wheats appear frequently in the northern and western parts of northern Italy, but also in some regions (Rottoli and Castiglioni 2009) of the southern part of the Apennine peninsula. However, in Iberia, hulled wheats played only a secondary role (Buxó *et al.* 1997), emmer being dominant in areas of central Spain (Stika 2005; Estremera 2003; Peña-Chocarro 2007) and the Basque Country (Zapata 2007). Along the eastern coast, however, both species were seldom major crops.

FREE-THRESHING WHEATS:DURUM WHEAT AND BREAD WHEAT

These two species, grouped together by their behaviour during threshing, are different at ploidy level (*T. durum* is tetraploid[1] and *T. aestivum* is a hexaploid[2]). They show a significant overlapping in grain shape and also overlap considerably in some other morphological features, so most identifications refer to *T. aestivum/T. durum* unless chaff is recovered. Free-threshing wheats are the dominant wheats in Iberia during the Neolithic and they are documented in southern and central Spain as well as in Portugal. In addition, rachis[3] remains have allowed identification of both species in several contexts across the region. Even in areas where archaeobotanical research had not previously evidenced free-threshing cultivation during the Neolithic, like the Cantabrian fringe, these species have been identified (Peña-Chocarro *et al.* 2005a and b)

BARLEYS

Different varieties of barley, naked and hulled, as well as different forms, two and six row, have been documented. Naked barley seems to be the preferred species in large parts of southern Iberia. Current research shows that key sites such as Nerja (Málaga) and Cueva de Los Murciélagos (Códoba) are overwhelmingly dominated by naked forms of barley.

Legumes

The advantages of combining cereal and legumes for both diet and soil fertility have been widely acknowledged (Zohary *et al.* 2012). Their rich content in proteins matches well with the starch contained in grains. Besides, legumes are able to increase soil fertility by fixing the atmospheric nitrogen. It is remarkable that from the inception of agriculture legumes were already present. By the mid-sixth millennium BCE, a total of six legumes have been identified in Iberia: pea (*Pisum sativum*), lentil (*Lens culinaris*), broad bean (*Vicia faba*), bitter vetch (*Vicia ervilia*), common vetch (*Vicia sativa*) and grass peas (*Lathyrus sativus* and *L. cicera*). Most of the remains come from the eastern and southern coast from sites such as Cova de Cendres (Alicante), Cueva del Toro (Málaga), Cueva de Los Murciélagos (Córdoba) (Buxó 1997; Peña-Chocarro 1999). Peas and broad beans appear as the most common species, whereas lentils and bitter vetch are found only occasionally. Some of the species (peas, lentils, broad beans) are immediately associated with human food. Others like the grass pea (*L. sativus*) have been widely consumed as human food (Peña-Chocarro and Zapata 1999), but the chickling vetch (*L. cicera),* bitter vetch (*Vicia ervilia*) and common vetch (*V. sativa*) are generally linked to animal fodder.

Other possible crops

Early farmers from Iberia did cultivate other plants as well. Poppy (*Papaver somniferum*) and flax (*Linum usitatissimum*) are two additional crops of the period. Flax is a rather uncommon species in the Iberian Neolithic. It has been documented in La Revilla del Campo (Soria), in an early Neolithic context (Stika 2005), in Los Castillejos (Granada) during the 5th millennium BCE (Rovira i Buendía 2007) and in La Vaquera (Segovia) in a late Neolithic level (Estremera 2003). On the contrary, poppy is present in the early Neolithic levels of Cueva de Los Murciélagos (Córdoba) (Peña-Chocarro 1999) (Fig. 3.16). Other early poppy finds include the capsules

Figure 3.16. Seeds of *Papaver setigerum/somniferum* from Cueva de Los Murciélagos (Córdoba). Photo: Jacob Morales.

belonging to the wild type recorded by Neuweiler (1935) from Cueva de Los Murciélagos (Granada) and the seed retrieved from La Lámpara (Soria) (Stika 2005). In later Neolithic phases, poppy seeds were also present at other sites. The overlapping between the wild *P. setigerum* and the domesticated *P. somniferum* makes identification of one or the other very difficult. In addition, the possible domestication of *Papaver somniferum* in the western Mediterranean region has been suggested by several authors (Schultze-Motel 1979; Bakels 1982; Bakels *et al.* 1992; Zohary *et al.* 2012; Salavert 2010) and we have discussed elsewhere (Peña-Chocarro 1999; 2007; Zapata *et al.* 2004) the role Iberia may have had, as well as the problems related to the correct identification of poppy seeds.

Mapping and Exploring Crop Diversity

Iberia was characterised by substantial crop diversity at the beginning of the Neolithic, when compared to other European regions. The number of crops identified indicates that early farmers in this region had at their disposal a considerable array of domestic plants whose presence is unevenly documented across the territory. There are regions like the northern Meseta, the Ebro valley and perhaps the Basque Country (further data is still needed for this region) where hulled wheats (particularly emmer) seem to have played an important role, or at least a predominant position when archaeobotanical assemblages are studied; other regions show an important presence of legumes (southern part), and others seem to be dominated by free-threshing cereals (Mediterranean). This multiplicity of situations and combination of possibilities (for details see Zapata *et al.* 2004) largely reflects the unevenness of research. Neolithic studies have focused for a long time on the eastern coast, and therefore the number of sites investigated is larger and data is more abundant in that region. However, as research advances, new scenarios become visible providing new information and helping to clarify the factors involved in crop diversity. For example, recent research in central Iberia suggests the co-existence of sites where the fully Neolithic crop assemblage is present, as in Cueva de La Vaquera (Segovia), with other sites where free-threshing wheats and legumes are absent and hulled wheats are the dominant species, as in Los Cascajos (Navarra), La Lámpara and La Revilla (Soria). In addition to hulled wheats, barley-rich samples associated with ritual and funerary contexts were retrieved in Los Cascajos (Peña-Chocarro *et al.* 2005a).

Although we are aware of issues of data quality (conservation, taphonomy, sampling strategies, recovery techniques, dating, etc.) which may distort or enhance special features or conceal patterns, the available information allows a good start on exploring the uncharted territory of crop diversity.

The absence of certain species or the predominance of others may be explained, taking different factors into account. Tolerance of certain ecological conditions and crop-growing requirements (water, nutrients, temperature, solar energy, etc.) of the different species was certainly a key element but probably not the only one. The spread of farming through Europe was undoubtedly a process of selection of the stocks more suitable for the specific environmental situations.

The crops found in Iberia show different growing requirements and were adapted to many different ecological settings, suggesting a certain varietal diversity. Still, there is not enough resolution on our morphological identifications to pin down the exact varieties cultivated. At a more general level, it is possible to extrapolate some ecological requirements of present-day species to the past. For example, hulled wheats are tough species, well adapted to poor soil conditions, and highly resistant to fungal diseases (Nesbitt and Samuel 1996). Nowadays, they grow in marginal mountain areas, where other species fail to develop due to the loss of fertility. Barley is also a resistant species able to survive in dry and poor soils. It is likely, therefore, that the combination of hulled wheats and barley at sites like La Lámpara and La Revilla (Soria) is not a coincidence. Both sites are situated at *ca.* 1000 m asl, in an area dominated by harsh conditions which may have led human groups to select for specific crop species which could better withstand the region's particular environmental conditions.

Moving to the north, the coastal fringe of Iberia, characterised by its mountainous character and oceanic climate, shows some peculiarities. Most of the crop species identified in the region are hulled wheats and barley (Zapata 2007). This led us to suggest that Atlantic agriculture in Iberia could show specific features due to the ecological setting where it developed: mountain areas in wet environments, so that hulled wheats and barley were preferred (Zapata and Peña-Chocarro 2005). The study of the plant remains from El Mirón (Cantabria) (Peña-Chocarro *et al.* 2005a and b), however, has shown that free-threshing wheats were also amongst the species identified in the region, opening an interesting debate about the spread of free-threshing wheats to humid mountain areas. New finds and continuity in research will enable us to obtain new data for exploring crop diversity in this region in further detail.

Since the specific crop-growing requirements may have had an effect on the diversity of the species selected for that particular area, it is likely that farmers chose those better adapted for their plots. Apart from the specific crop demands, Neolithic farmers had to cope with short-term environmental variability which may have operated at different scales, causing food shortages and scarcity. Resource diversification then may have become a possible buffering strategy against environmental uncertainty and risk. Crop diversity, in particular, appears as a useful measure to cope with seasonal variability, as this is generally predictable, avoiding crop failures and food shortages and, therefore mitigating their effects.

However, although it is true that ecological requirements of crops may account for diversity and that environmental conditions are important to understand crop combinations, there are other factors guided mostly by human decisions that may also influence diversity. Crop use, for instance, may be an important factor accounting for diversity. A clear example is that of einkorn in present-day traditional agriculture; its straw is highly appreciated for thatching purposes, and this explains that in areas of North Africa and Europe it has been cultivated until the present day only for this purpose (Peña-Chocarro *et al.* 2009). Closely related to crop use are the social and cultural traditions of people which may influence the species or particular varieties farmers choose to grow. Some species/varieties may have particular qualities or traits that are absent in others, or they may play a specific role in the local traditions of the community, or might be linked to certain festivities. Cultural preferences thus become an important element for explaining crop diversity. Examples of crops cultivated for their particular role in societies are, for example, the case of emmer in Spain and Ethiopia (D'Andrea 2003), cultivated because it has a role within society, although it may be extremely labour-intensive. There are also examples of symbolic or festive reasons for growing a particular species (Hayden 1996; 2003; Hansson and Heiss this volume Chapter 7).

Most of our discussion has focused on the wide range of factors affecting crop diversity, emphasising the role of cultural and social factors. However, it should not be forgotten that in particular periods or areas, diversity may have been a question of necessity more than an option.

Conclusions

Crop diversity in the Iberian Neolithic was high; farmers grew a wide variety of crops which probably fulfilled their basic food needs and satisfied specific requirements in the sphere of human beliefs and practices. Crop diversity implied a good knowledge of the growth needs, processing and uses of the particular species cultivated, as well as considerable skills in coping with environmental forces. Apart from environmental constraints and plant growth requirements, there are a myriad of other culturally induced practices that may account for such diversity and that need to be highlighted when discussing crop diversity in the past. In fact, these are aspects frequently included in current debates of sustainability and conservation, and in the agendas of policy-makers. Archaeobotanical data in Iberia is still too scarce to provide full understanding of plant use during the Neolithic; there are many aspects which have been little explored and that may help to understand crop diversity. Issues such as the interaction between agriculture and animal husbandry are underdeveloped and offer good possibilities of revealing hidden patterns that may contribute to better understanding the causalities of crop diversity. In spite of these reservations, archaeobotanical remains are starting to offer interesting insights into geographical patterns and possible causes of diversity.

3.6. CHOICE OF A CROP AND ITS UNDERLYING REASONS: EXAMPLES FROM WESTERN CENTRAL EUROPE 500 BCE–CE 900

Corrie Bakels

Introduction

Anyone wanting to grow something is confronted with the question of what to sow or plant. Situations in which there are hardly any plants to choose from are almost non-existent, but the actual choice is not entirely free. It is guided by several factors:

1 The range of crops available to the person who has to make the choice. It goes without saying that European farmers living before the voyages of Columbus could not consider growing maize or tomatoes, as these crop plants have their origin in Mesoamerica.

2 The physical environment, for instance soil and climate.

3 The socio-economic environment. In regard to this factor, the model introduced by the nineteenth-century agricultural economist Johann Heinrich von Thünen should be referred to (von Thünen 1826). The basic idea behind his model is that the nature of agricultural production in a specific area is determined by its distance to the market (or the place where the products will be consumed). In his model, the market is a central place such as a town. The central place is surrounded by ring-shaped areas, each of which provides a specific kind of product (Fig. 3.17), the kind of product being determined by its keeping qualities or the cost of transport. Starting from this model, it may be expected that farming units in a region with a centralised organisation operate in one of the rings surrounding a central place, and therefore specialise in growing crops which are the most profitable in their specific setting. If a central organisation is weak or lacking, farmers produce for a more restricted community or even for their own families only. In such a setting, the range of products will be large, because all ring-shaped areas, in the sense of von Thünen, will be centred on one single farm (Fig. 3.17). Distance to the place where the products are in demand is one thing. Another is the factor of labour. Some crops require more hands than others and also the nature of the available labour may differ. Sowing and harvesting crops on large fields demand a different kind of care than tending crops in gardens. The kind of person making the crop choice is a factor not to be disregarded.

4 The last factor to be mentioned here is cultural preference. Some plants may be more valued than others and the influence of preferences on crop choice must not be underestimated.

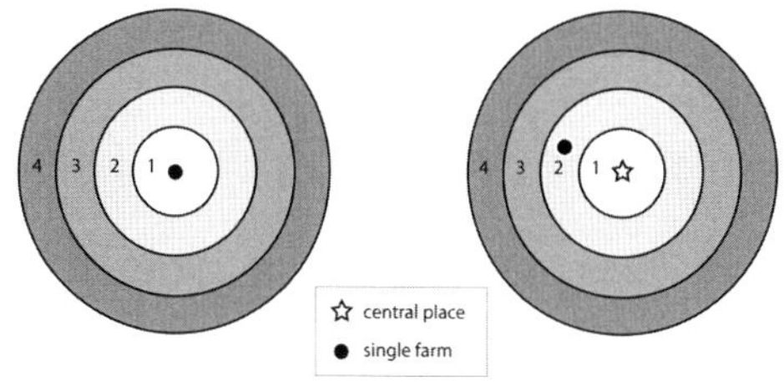

Fig. 3.17. The model by von Thünen. Left: four ring-shaped areas around a single farm; right: a single farm in a location with a central place.

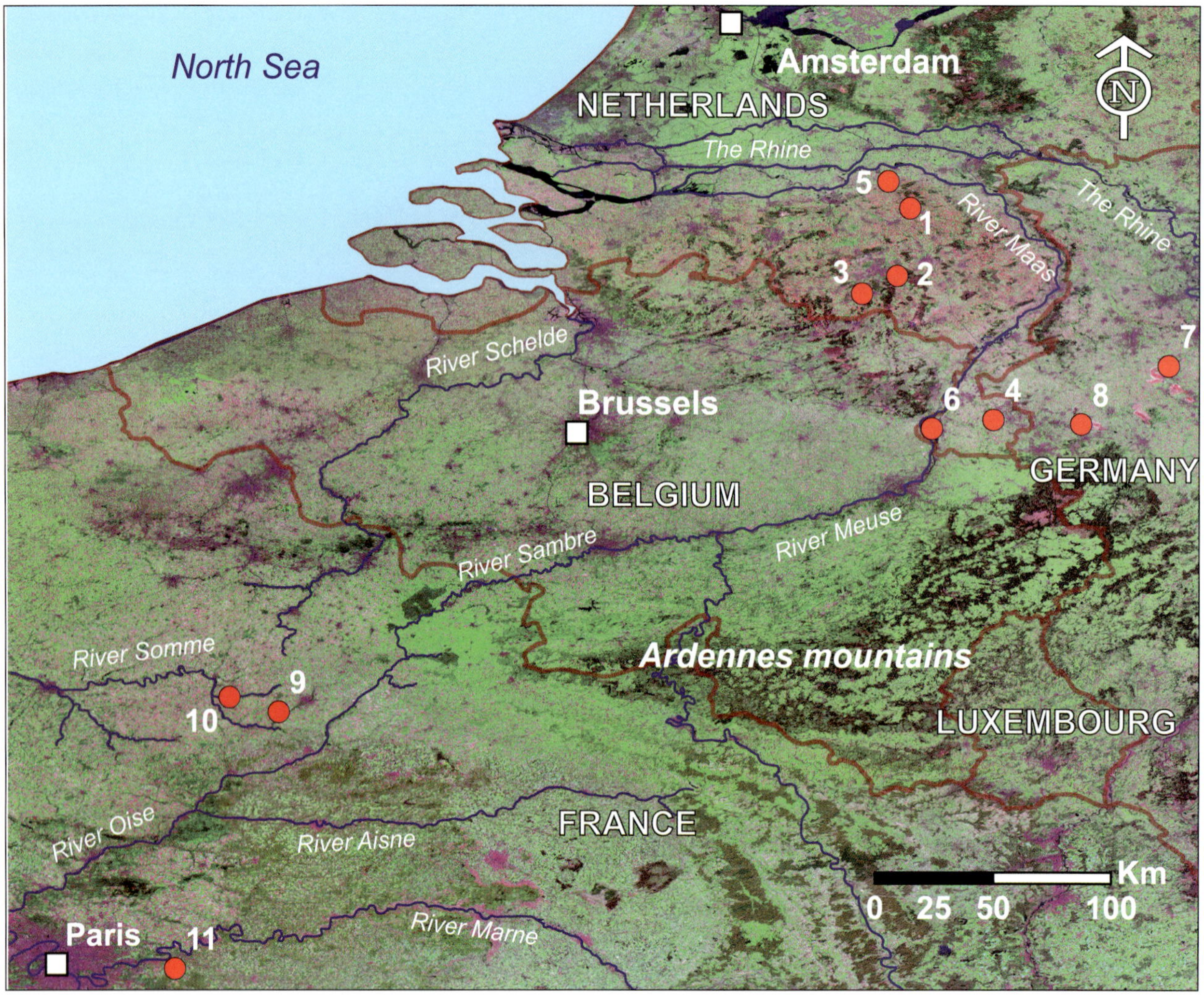

Fig. 3.18. Map with the location of the settlements mentioned in the text. 1) Uden; 2) Geldrop; 3) Dommelen; 4) Voerendaal; 5) Oss; 6) Maastricht; 7) Grevenbroich; 8) Laurenzberg; 9) Savy; 10) Athies; 11) Serris. Map: R. Lugon, J.-C. Loubier and A. Chevalier.

In the following, I shall try to give examples of the influence of factors 2, 3, and 4 on past crop choice in western central Europe, to be more specific, in northern France, the German Rhineland and the southern parts of the Netherlands (Fig. 3.18). In order to avoid the difficulty presented in factor 1, I will stay within a specific slice of time, namely 500 BCE–CE 900. According to the archaeobotanical finds, all crops mentioned below were available, even if they were not always grown. Although the Romans introduced new plants during the first centuries CE, these were not the kind of crops providing the examples presented below (Bakels 2009).

Physical Constraints

The example dealing with the influence of the physical environment concerns the soil factor. The oceanic climate offers some variations in the region under review, but these are not strong. What is compared here are charred remnants of crops retrieved during excavations in rural settlements. For this case study the comparison is restricted to charred remains, because comparison (any comparison) should be restricted to one single mode of preservation in order to prevent biases due to different means of preservation. Fig. 3.19 presents the results obtained in three early Medieval (sixth to ninth century CE) settlements on sand and three of such settlements on loess. The graphs show the frequency with which a different crop plant appears in the archaeological record, *i.e.* the percentage of the samples revealing this plant calculated on the basis of all samples. The samples were taken in places where plant material was dumped, for instance in pits. The idea is that the more common the plant, the more often it turns up in the waste.

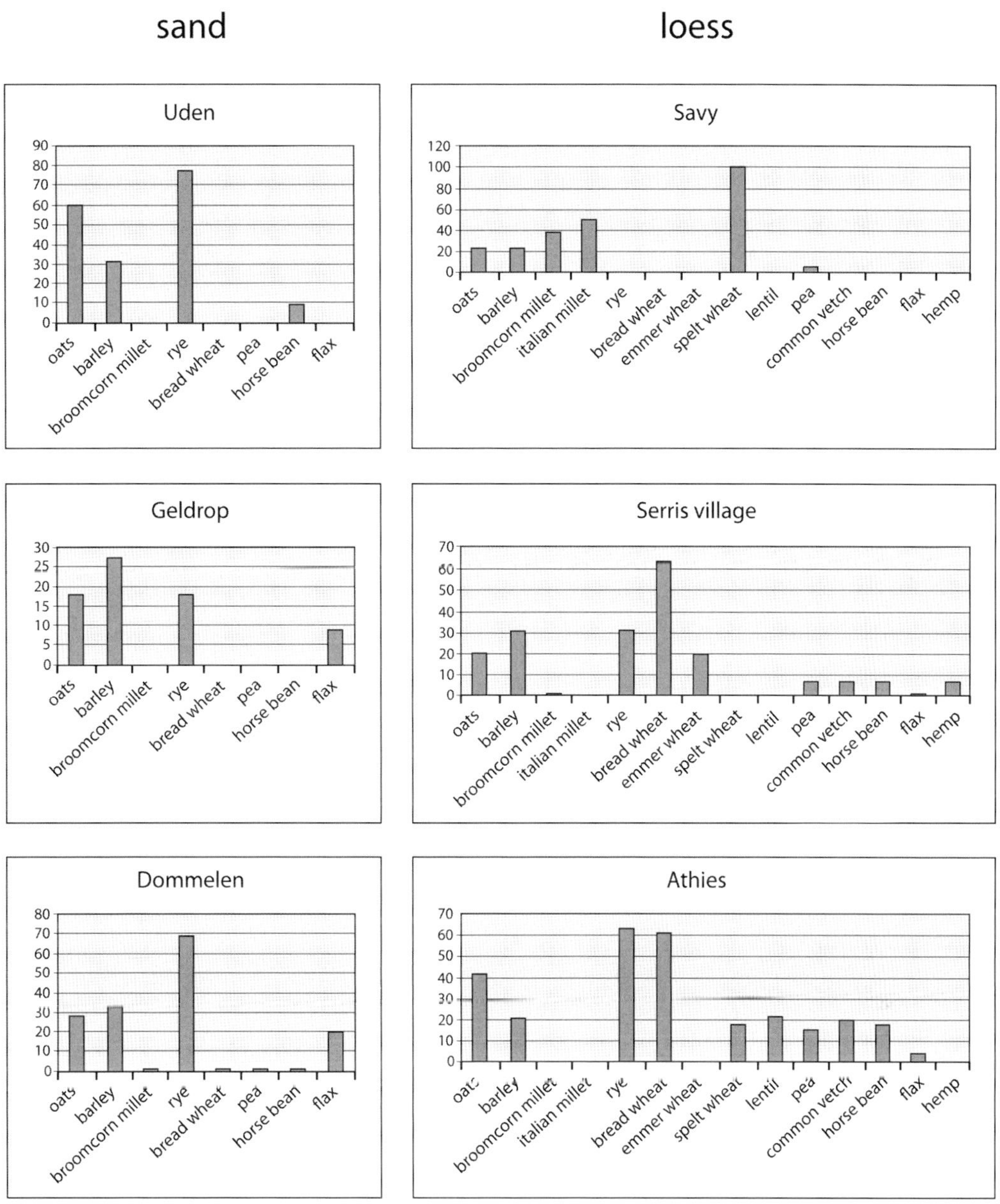

Fig. 3.19. The frequencies of charred remains of crop plants. Left column: early Medieval farming communities on sand; right column: early Medieval farming communities on loess. Data from research done by the author.

It is obvious that the three sites on sand contained fewer species than the sites on loess. It is also obvious that the farmers on sand did not grow wheat. There are no reasons to attribute the difference to an economic or cultural background (Bakels 2005). This must be a physical constraint on crop choice, most probably soil.

Socio-Economic Factors

Obviously, loess offers more possibilities for growing a large range of crops than sand. Fig. 3.19 also shows that the farmers on loess did not make quite the same choices. The time range 500 BCE–CE 900 offers an opportunity to study whether their choice was guided by the socio-economic environment.

Until the Roman occupation in 50 BCE, the inhabitants of the region functioned in a tribal system. During the first centuries, the territories of the tribes were rather restricted in size. Although exchange of products took place, it is assumed that people produced food for a restricted area, which implies that everybody grew a wide range of crops. In the last century BCE, tribal territories seem to have become larger and more centralised organisations. The prediction is that local farmers began to specialise. Indeed, such a trend has been observed by Véronique Matterne in northern France (Matterne 2001).

When the Romans invaded the region, they introduced a central authority and a market economy. This should result in still more specialisation. After CE 400, Roman rule collapsed, whilst Germanic tribes invaded the region. For some time, central authority disappeared altogether, but at the start of the sixth century one of the local leaders, Chlodovec (also known as Clovis) succeeded in uniting many small territories. In CE 750, the still more powerful leaders of the Carolingian dynasty, with Charlemagne as its most renowned member, took over. In the same period, organisations of the Christian Church, especially monasteries, grew to be important landowners. More and more central rule was imposed again. We may therefore expect that the range of crop plants grown by local communities rose after the collapse of the Roman Empire, only to shrink again when centralised power regained importance.

Figure 3.20 presents the trend as seen in charred waste retrieved from rural settlements on loess in the German Rhineland and the adjacent Dutch loess area. The farms at Grevenbroich and Laurenzberg were of the restricted size common in protohistoric times. After the introduction of the Roman economy, farms became larger and Voerendaal is an example of such a large Romanised farm. Such farms go under the name of villa rustica.

In the fourth century, the Voerendaal farm fell into disrepair and Germanic people settled in the surviving structures. The predicted trend in the diversity of crops is indeed visible. A wide range of crops makes way for a crop choice mainly directed toward cereal growing. Roman people valued spelt wheat (*Triticum spelta* L.) over emmer wheat (*Triticum dicoccum* Schübl.) and this fact is seen as well. In earlier periods, emmer wheat was the most important wheat. After the heyday of Roman rule, the diversity in crops tends to widen again.

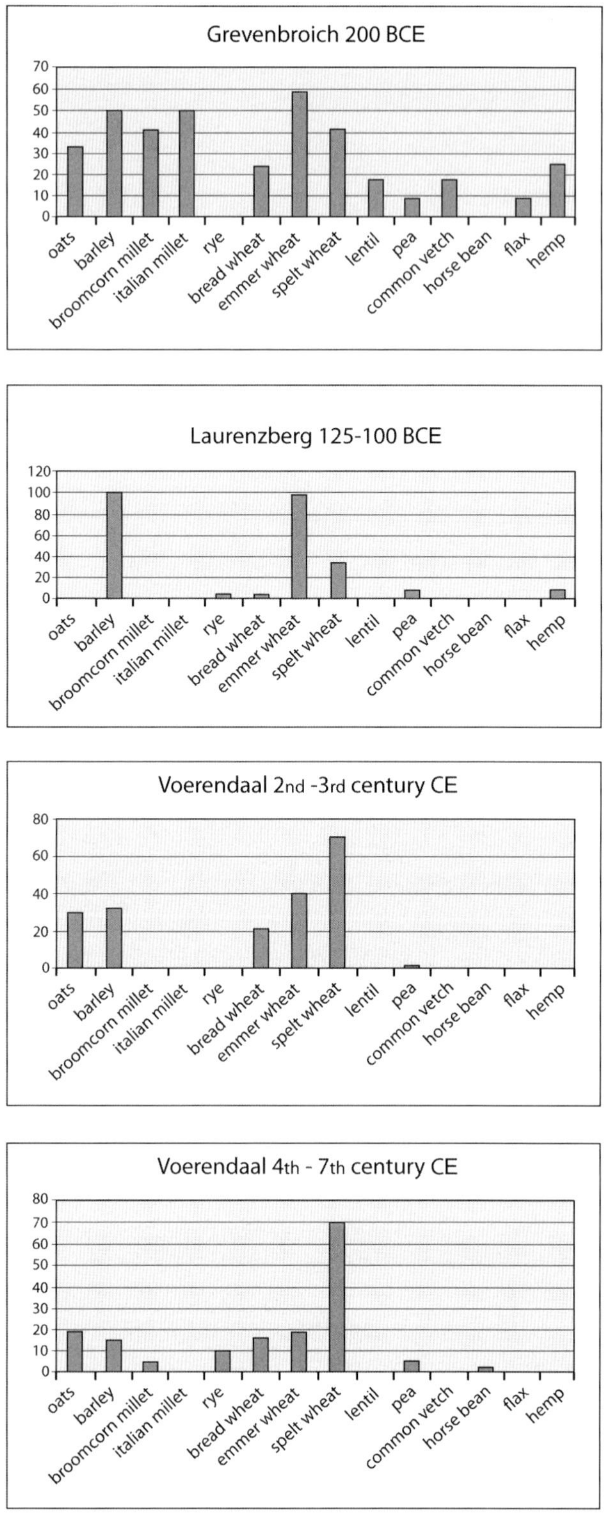

Fig. 3.20. The frequencies of charred remains of crop plants on loess in the German Rhineland and adjacent southern Netherlands. Data from Knörzer 1979; 1980; Kooistra 1996.

Figure 3.21 offers the situation at one single site in northern France, namely Savy. The series starts in the first half of the first century when Roman rule and the Roman market economy were not yet well established. The second graph shows the frequencies during the heyday of the Roman Empire. After its collapse, there is a gap in the record, but the return to more diversity is clearly present in the samples dated to the end fifth to the middle sixth century. The last graph depicts the slow return to a more centralised society.

In my opinion, the data recovered from the loess region in Germany-the Netherlands and northern France do indeed offer grounds for stating that a socio-economic factor such as the degree of organisation plays a role in the choice of a crop.

The Labour Factor

So far I have considered only the grade of organisation of society, but the availability of labour is important as well. A special case of problems with labour was recognised by Carol van Driel-Murray, who wrote about the effect of recruitment for the Roman army on local communities: 'for a self-sufficient agricultural society the permanent removal of able-bodied men will have created grave problems (...); loss of male labour shifts agricultural tasks onto women, assisted by children and the elderly, causing changes in practice which should be archaeologically traceable' (van Driel-Murray 2008). She supposes that cultivation shifted to horticultural regimes with more emphasis on vegetables and fruit, since small plots are better manageable with the labour available. She stretches the argument even further to suggest that fruits and nuts which are in general listed as 'gathered', were actually grown in gardens at that time.

A case from the sandy soils in the southern part of the Netherlands is presented in Fig. 3.22, where water-logged plant remains retrieved from Pre-Roman and Roman Iron Age farms (indigenous farms within the Roman Empire) in a small-sized area could be compared. Water-logged remains found in the bottom of wells were chosen for this kind of analysis instead of charred remains, because fruits are better preserved in this way.

The difference in fruit content before and after the arrival of the Romans is striking. Hazelnuts, apples and four kinds of berries appear in the lists, whilst in the earlier periods only sloe plums could be recorded (Fig. 3.22). All of them belong to the native flora, and whether such fruits must be interpreted as cultivated in gardens as suggested by Carol van Driel-Murray, cannot as yet be proven. They may

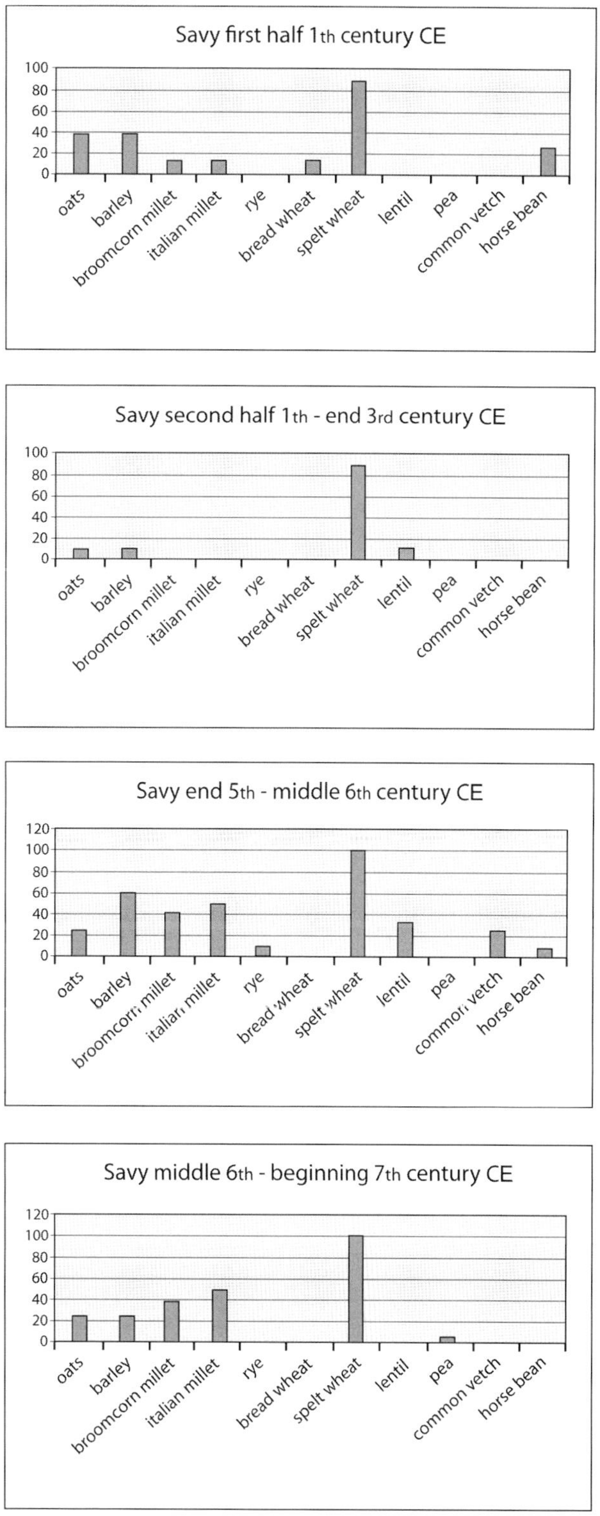

Fig. 3.21. The frequencies of charred remains of crop plants retrieved from Savy, northern France. Data from research done by the author.

have been gathered in the natural environment. But even then, it is a shift in crop diversity, if wild crops are considered crops as well.

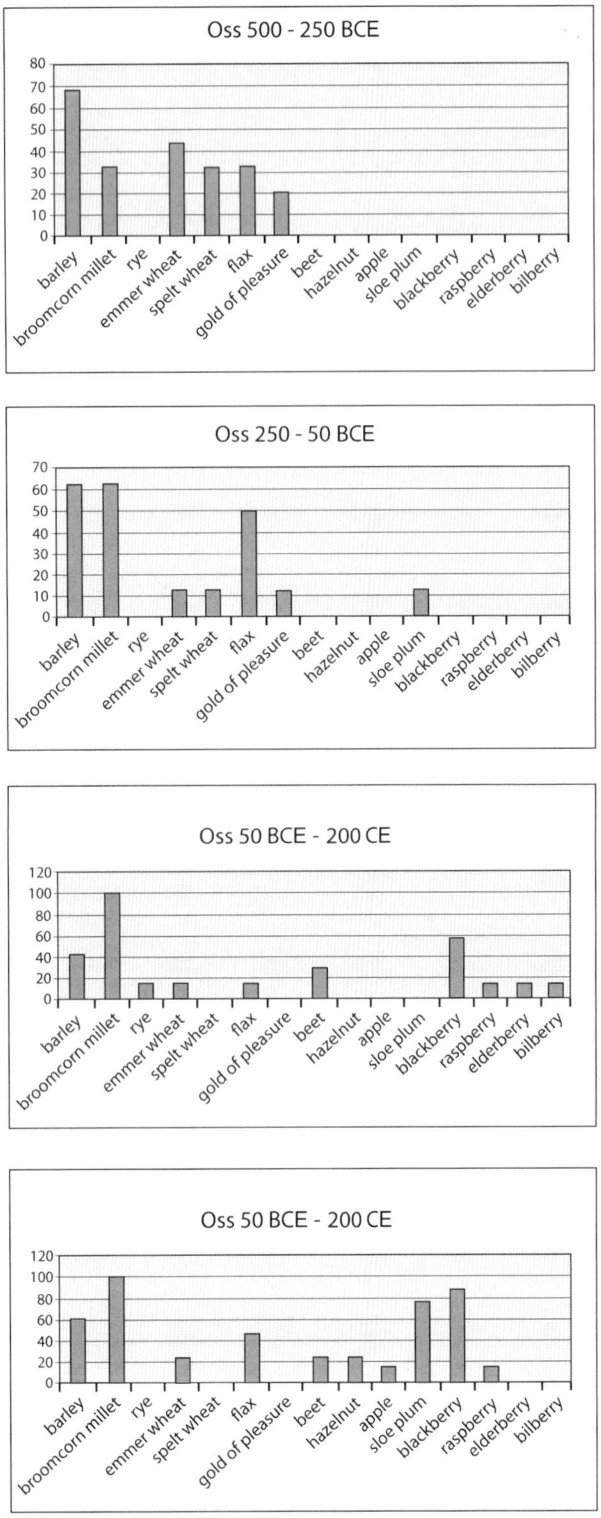

Fig. 3.22. The frequencies of waterlogged plant remains found in four different rural settlements near Oss, the Netherlands, two of which are contemporaneous. Data after Bakels *et al.* (1997), supplemented by new data obtained by the author.

Fig. 3.23. Charred rye dated to the fourth century CE, found in Maastricht, the Netherlands. Photo J. Pauptit/C. C. Bakels.

Cultural Preference

One specific choice of a crop was already referred to: spelt wheat grown for the Roman market. But the case of rye (*Secale cereale* L.) is more striking. As can be observed in Fig. 3.20 and 3.22, rye was already known before the Early Middle Ages. Fig. 3.19 shows, however, that rye became important only during this period. Growing rye seems to be a cultural choice. The earliest important finds in the region are connected with a Germanic tribe (Fig. 3.23). Romans considered rye as poor food and only fit to avert starvation, if the Roman author Pliny is to be believed. Also, earlier populations did not cultivate rye. As soon as the Germanic people arrive, rye cultivation starts almost everywhere. Rye must have been part of the Germanic heritage.

Conclusion

The diversity of crops does not depend on one single factor. Even in a rather restricted part of Europe, the part studied here, at least four factors did play a role. Environmental constraint is one of them, but the three others are of human origin. In many cases, people made the actual choice, not the environment. Economic and social transformation resulted in changes in production. Crops and the land-use strategies changed, also reflecting changes in available labour and cultural preferences. What was true for the past may also apply to the future.

3.7. CROPS AND AGRICULTURAL DEVELOPMENTS IN WESTERN EUROPE

François Sigaut (†)

Introduction

One of the problems we meet when we are discussing 'agriculture' today is that in the last 150 years, our idea of what the word means has changed beyond recognition. The term 'agricultural revolution' has been put to far too many uses (and misuses). But what resulted from the development of machines, fertilisers, genetically selected crops and pesticides in Europe and North America from the middle of the nineteenth century on can properly be called a revolution. Yields have increased about ten times, and labour productivity by a factor of 100 to 1000, according to the task concerned. Nothing like that is recorded from any other period in the past. True, assessing yields for older times is a rather tricky business, but the orders of magnitude are fairly well known. Figures given by Roman agricultural writers in the last two centuries BC are not very different from those known from eighteenth century sources, so it would be possible to assume that nothing really important did actually happen in between. Of course, this is not quite true. There were many changes. But there were also many things that did not change. The main difficulty is to draw a fair picture of both, never forgetting that non-changes may be as significant and important as changes.

Changes and Non-Changes

One of the things that did not change much was the reliance of people on cereals for food. A large number of famine plants were known and consumed in times of scarcity, but in normal times, cereals provided the main food for nearly everybody. Of course, there were some differences among regions, especially according to the greater or lesser importance of animal husbandry in the area. But as a rule, only the very rich could afford a diet that was based on something other than cereals – on meat mainly, with an emphasis on poultry and game. The main exception is that of chestnuts. In some areas of central and southern France and Italy, chestnuts became the staple food of peasants from the fifteenth to the nineteenth century (Pitte 1986). It is difficult to tell whether this reliance on chestnuts was a quite new thing or only the development of earlier, less specialised practices. Chestnuts, like acorns (especially the so-called *glands doux* or 'soft acorns', without tannins, well known in Spain and in some areas of southern France and Italy), may have been formerly used either as famine foods or as snacks for shepherds, travellers, outlaws, etc. What is not fully understood is how and why chestnuts became a staple as late as they did, and only in some areas.

Granted, people never lived by bread alone, but bread (in its broadest sense, including all other cereal foods that are not technically 'bread') was for them a matter of life or death. The failure of other crops could be a more or less serious nuisance. The failure of 'corn' (the word everywhere means the most important cereal, be it wheat, rye, oats, etc.) meant famine. And famines were an irregular, but frequent occurrence; they came back at least once every 5 to 10 years. Again, there were exceptions. Maritime areas, like the Netherlands or England, were up to a point insured against famines, inasmuch as they occupied strategic positions in the international trade of grain, which went on along seacoasts and rivers. But the maritime trade of grain only became important during the Middle Ages. In most landlocked areas famines, or at least severe food scarcities, remained a fact of life until as late as the years 1860 to 1880. This situation, which could be called Malthusian, was one of the things that probably did not change much between the Neolithic and the nineteenth century.

Among the things that did not change much, at least since the coming of iron, is also the tool-kit. We know very little about Neolithic agricultural tools, anyway, and nothing at all, so to speak, on tillage tools (if any). By the Final Neolithic and the Bronze Age, our knowledge improves somewhat, but not enough to provide us with any clear idea of how people really tilled their fields, planted, weeded and harvested their crops, etc. Only with the coming of iron does this state of affairs begin to improve, for two main reasons. First, there are more and better data. And second, these data are more likely to be rightly interpreted, because ancient iron tools are (with proper caution) comparable to modern ones.

Of course, 'comparable' does not mean 'similar'. But iron tools, ancient and recent, have much in common. This is not to say that once iron is there changes become insignificant. Quite the contrary. The seemingly infinite diversity of agricultural tools and implements to be seen in any local museum today is obviously the result of a similarly infinite number of local innovations, the importance of which should not be underestimated. For they are our only means to understand how cultivation techniques were more and more finely tuned to local conditions. But on the other hand, the existence and performance of most tools depends on one main factor: the availability of good quality metal – iron and steel. Archaeologists date the Iron Age from the twelfth or eleventh centuries BCE on. But what is relevant for us would be to know when iron became cheap and abundant enough to be used in the manufacture of tools, not only of weapons or luxury objects. The exact date may be assumed to differ somewhat from one area to the other. On the whole, it can be assumed that this stage was everywhere reached by the fifth century BCE. From that time on, European peasants were in possession, not of a complete tool-kit, but of the means to develop it according to local needs and circumstances.

One of the most important of these developments was that of the scythe. Iron blades more or less like scythe blades have been found in late La Tène sites (second century BCE) in Switzerland, but it is not known with any certainty what they were used for or how they were handled. Some of them may have been used for harvesting hay already, but which ones and what were the uses of the others? We just do not know. The language of the Roman writers provides a pretty good image of the situation. The word *falx* meant any incurved cutting implement or weapon. There were about a dozen different *falces*, some of which were sickles (*falces mesorias*), others billhooks (*falces arborarias*), etc. There were also *falces fenarias*, used to harvest hay, which can reasonably be called scythes, but they were not similar to the modern ones. Pliny wrote that they were handled with one hand only in Italy and with two hands in Gaul (e.g. Bostock 1855), and this is the only evidence we have of the way they were wielded. Neither the tools themselves, nor the iconography give us better clues. 'True' scythes, looking so much like modern ones that their interpretation cannot reasonably be disputed, have been found in the lower Rhine valley and dated to the seventh or eighth centuries CE (Henning 1991). It took of course some time for them to diffuse outwards. For example, scythes only reach Ireland in the thirteenth century, introduced by the first Anglo-Norman conquerors (Kelly 1998). But as concerns the tool itself, there are only three major dates: the second century BCE when it makes its first appearance; the seventh century CE when it reaches its definite, modern form, and the twentieth century when it is replaced by machines. These three dates provide a good idea of the kind of time scale that is needed for understanding agricultural history.

Scythes are important in at least three respects: craftsmanship, animal husbandry, and landscape. Craftsmanship: scythes are a masterpiece of pre-industrial metallurgy – a point that cannot be further discussed here. Animal husbandry: scythes mean hay and hay storage, which is a solution for one of the most acute problems of keeping animals in countries with harsh and long winters. And landscape: scythes mean meadows, that is a part of the landscape put aside for grass to be mown. Most (if not all) European languages have a special word for it (English meadow, German Wiese, French pré and prairie, etc.), not to be confused with grazing areas (pasture, *Weide*, *pâture* or *pâtis*, etc.). Meadows are also grazed after the hay harvest, but so are all other fields after their main crop has been taken off. It could be said indeed that meadows are fields too, with the only difference that their crops are not sown. It is not usual to consider meadow-grasses as cultivated plants, but in a sense, they should be. For it is obvious that after a few years of regular mowing, the set of plants growing in the meadow is not 'wild' any more. This is only one more case where drawing a line between wild and cultivated plants does not make much sense.

The scythe appears to be the last great innovation of what could be called the European Iron Age agriculture. With it, the tool-kit of European peasants is virtually complete. As already said, it does not mean that there would not be any further innovations, quite the contrary. But these were to be adaptations to local conditions and needs. In France for example, vineyards were usually tilled by hand, with hoes, with the result that local museums show a bewildering diversity of hoes – an implement nearly unknown in Britain, where vineyards were rare or unimportant. Local innovations are to be found everywhere. Each country, each province, each village perhaps developed its own tool-kit according to its needs and possibilities. And the same is certainly true of ploughs, carts and wagons and all other farm implements. But by late Roman times or the early Middle Ages, the basic models are there. They are the result of the availability of iron which was obtained in the preceding centuries. And notwithstanding their many further developments, they would be there until the advent of machines in the nineteenth century. That makes a grand total of about two millennia during which permanencies (non-changes) were at least as important as changes.

Oat and Rye: Iron Age Cereals?

Let's go back to cereals. In prehistoric Europe, the only cereals grown belonged to the genera *Hordeum* and *Triticum* (leaving aside the millets for the moment). By the tenth century CE, two more species had been added to the list: rye (*Secale cereale*) and oats (*Avena sativa*), and they had reached about the same level of importance as wheat and barley. The paradox is that we know nearly nothing about when, where, how and why this happened. For the Roman agricultural writers, rye and oats (if mentioned) were little more than weeds. And the idea that they were indeed weeds which had been slowly put into cultivation is an old one. The trouble is that this idea explains nothing. People had ignored rye and oats for millennia, why should they have suddenly put them into cultivation as late as the first centuries BCE or CE? The sad fact is that we not only lack the answers, but we also lack the questions, inasmuch as nobody seems to have really bothered to ask. In comparison with the enormous mass of literature on the domestication of cereals in the early Neolithic, the literature on the domestication of rye and oats seems nearly inexistent.

If we consider the chronology, we can hypothesise that the domestication of rye and oats must have some relationship with the advent of the Iron-Age tool-kit discussed above. But what could those relations have been?

Let's have a look at the case of oats. Until the nineteenth century and with few local exceptions, cereals were reaped (that is, cut with sickles, handful by handful), never mown (with scythes). But spring oats – not winter oats, which are another case – were mown, not reaped, and this peculiarity is attested from Carolingian times on. So it is not unreasonable to suppose that the growing of oats may have something to do with the diffusion of the scythe.

But why should spring oats have been mown and not the other cereals? A part of the answer could be that in many parts of Europe, oats were grown for feeding horses rather than humans. So they were less valued than the main cereals and their harvest had to be dispatched as quickly as possible so as not to interfere with the 'true' harvest of the main cereals. But another part of the answer may be that spring oats were sown in March, after only one ploughing which was done flat, the seeds being buried by harrowing, whereas other cereals were sown after several ploughings, usually ridge-and-furrow, the last ploughing being intended both to finish the ridges and to bury the seeds. Now, a prerequisite for the regular use of scythes is that the ground be as flat as possible. Here, it is the plough which comes into the picture. Flat ploughing for oats required a plough (a 'true' plough, with a wide and flat share, a coulter and a mouldboard), not an ard. So it is quite possible that the development of the plough and that of the scythe both had something to do with the cultivation of oats – of spring oats for horses to be precise. For it must be stressed again that this hypothesis is not relevant for winter oats (grown in northern France, for instance) nor for spring oats in the northern parts of Europe, where only spring cereals were grown anyway.

The case of rye is much less clear. The ploughing and harvesting techniques for rye do not seem to have differed significantly from those used for wheat – the less so since both were often sown together (*maslin, méteil, Mengkorn...*). Rye can produce crops in soils that are unfit for wheat, but this remark is of little use to explain when and why it was put into cultivation. The only possible hypothesis for the moment comes from a rarely observed fact, which I would propose to call 'the rye exception'. As far as is known to me, rye was never transformed into anything other than bread proper (and beer, meaning by 'beer' any fermented liquid product), whereas all other cereals can be accommodated into a large number of products other than bread (porridge being one among many). If, to be made edible, rye had to be made into bread, and bread only, it leads to the hypothesis that its domestication was made possible by the diffusion into northern Europe of Mediterranean bread-making techniques, and especially of the closed bread oven which seems to have been a Roman innovation.

Once again, it should be stressed that our purpose is not to give answers but to ask questions. The rapid expansion of rye and oats in the first centuries BCE has to be explained. The only chance we have of finding explanations is to ask as many questions as possible. This is what we have tried to do here.

Rice, Buckwheat, Maize and Millets

After rye and oats, the only cereals that were introduced into European agricultures were rice, buckwheat, and maize, in that order. Rice was already known in Europe in Roman times, but it never became important, except in some very limited areas of Italy and Spain. Buckwheat was of first-rank importance in Asia and in eastern Europe (Russia, Poland, etc.). It only reached western Europe toward the end of the Middle Ages, via the Baltic countries. In northern Germany, Denmark and the Netherlands, buckwheat was found to thrive on the paring and burning of moors and peat-bogs, which were very extensive and until then completely unproductive. (By the way, let us not forget that paring and burning is a technique that requires hoes or spades with very strong and sharp iron blades.) In France, where moors and peat-bogs were much less extensive, buckwheat was grown in other conditions, either in very poor soils like the chalky wastes of the 'Champagne pouilleuse', or as a substitution for a fallow (that is, the several ploughings before a crop of rye) in Brittany.

The history of maize is fairly well known, since it was brought back from America by none other that Cristobal Colon himself. But it seems to have taken some time, between one and two centuries, for maize to become really important anywhere. For the most part, maize found its place in areas with hot and wet summers, where it superseded millet. Unfortunately, we do not really know how the replacement process went on, one of the reasons being that millet itself is poorly documented in written sources.

There are a lot of difficulties with millet. The word itself is misleading because, in current English, it refers to any one of nearly twenty cereal species that have little in common, except that they are more extensively grown in hot climates than in Europe. In ancient times, the name millet was restricted to the 'true' millet, *Panicum miliaceum* (Latin *milium*, Spanish. *mijo*, French *mil, millet*, Italian *miglio*, German *Hirse*, etc.), whereas the other species of European importance, *Setaria italica*, had a quite different name (Latin *panicum*, French *panis*, Italian *panico*, German *Pfenig...*). One cause of the modern confusion is obvious: both species were practically unknown in Britain, which is also a plausible explanation for why most economic historians, following the British lead, have always considered millets as negligible. There is another plausible explanation: in many cases, peasants grew millet by hand, as a snatch-crop and for their own consumption only, so that it was not recorded in written documents. But this second explanation is not quite convincing, for there are many other cases where millets (either *Panicum* or *Setaria* or both) were really important and remained so even after the coming of maize, well into the nineteenth century (Hörandner 1995). The problem of millets may be said to be the reverse of the problem of rye or oats. We know that rye and oats became of first-rank importance in the first centuries of our era, but we do not know why. As for millets, we know they were always there, but we do not know whether they were important or not.

The case of millets is probably not unique. A similar story could perhaps be told about tuber plants, with the triumph of the potato in the nineteenth century pushing the traditional uses of roots and tubers (beets, carrots, parsnips, turnips, etc.) into oblivion. Some of them found new uses as fodder or industrial plants (turnips, sugar-beets...). But their significance and importance in older times has often been lost and accordingly remains to be rediscovered.

Domestication is Not Always the Answer

Some plants pose a question of another kind: what is the difference between the regular use of a plant and its cultivation (or domestication)? This is the question we have already asked about meadow grasses. Another example is that of furze or gorse (*Ulex europaeus*). In many regions of Atlantic Europe, from northwestern Spain to the British Isles, furze was a first-rank fodder plant. The problem was to get rid of its spines, which prevent animals eating it in its natural state. The solution was to crush the plant in a mortar or a handmill before feeding it to animals. This developed into a regular practice in Brittany since at least the seventeenth century, for the feeding of young horses that were one of the main riches of the country as an export product. Was furze cultivated, *i.e.* sown in fields regularly ploughed, etc.? It was sometimes, but not always, all depending on local circumstances and opportunities. And as far as is known, there were no significant biological differences between 'wild' and 'cultivated' furzes; both remained spiny enough to require crushing before being fed to animals.

The Spanish broom (*Spartium junceum*), is often confused with furze. However, the plant has a different geographical area, and more uses: for fodder too, but also for basketry, for textile and even perfumes. A recent study (Olivier 2005) has shown how important Spanish broom was in a small region of southern France (around Lodève, Languedoc); but it also reminded us that Spanish broom was important in other areas of Spain and Italy. What furze and Spanish broom have in common is that both plants were sown or not, according to local circumstances, and that there are no visible differences between the sown and the 'wild' plants – with the consequence that their geographical area of use more or less coincides with their natural area of reproduction. Contrary to plants of world-wide distribution like wheat, barley, rice, maize, etc., furze and Spanish broom have not been carried by men beyond their natural habitat.

Spanish broom is a plant with many uses: it should always be kept in mind that this is the rule, not the exception. Flax, one of the oldest known domesticated plants, is a good example: it produces edible grains, oil, and fibres, one of its peculiarities being that linseed oil is a regular food item in some countries (Germany for instance), whereas it is not regarded as edible in others (France). Most cereals too had uses other than food, and these non-food uses, although they may look quite secondary, may have been decisive in the choice of some cereals instead of others. The main cereals (wheat, rice, barley, rye...) do produce useful kinds of straw (for fodder; thatching, mats, baskets, etc.), whereas the 'millets' (including maize, sorghum, etc.) do not. This may have had an incidence on harvesting techniques: 'straw cereals' (*céréales à paille* in French) are ordinarily harvested with a sickle, whereas millets are not. My hypothesis is that the sickle was not primarily an implement designed to harvest just the grain, for there are a lot of other techniques for doing that (Sigaut 1991), but to harvest something that could be called 'grain-with-the-straw'.

The number of non-food plants, or of food plants with non-food uses, is enormous. My last remark will be that in many cases, the distinction edible/not edible applies, not to the parts of the plant, but to one of its products. This is typically the case of oil. As we have just seen, linseed oil is edible in some countries, not in others. But olive oil, now one of the most fashionable ingredients of modern cuisine, seems to have first been exclusively used for body ointments, long before being found to be edible: when and how did the discovery take place? In fact, oils (and animal fats) have always had a large number of different purposes, such as quenching hot steel, greasing axles, treating cloth, lighting lamps, etc. If we are to understand why one particular plant was important (or not) in this or that time/place, we have to ask questions about all the possible uses of all its possible products.

Conclusions

Western European agriculture provides many examples of increasing crop diversity through time. From the beginnings of agriculture when wheat and barley were the dominant cereal species, Europe has witnessed the arrival of several new cereals: rye, oats, millets, rice, maize, as well as the contribution of other non-cereal species such as the buckwheat which enlarged the human diet. Some of them played a role in animal fodder, too. At the same time, innovations like the scythe allowed technical developments which helped in improving aspects of animal husbandry, for instance. In any case, introductions and innovations were also accompanied by static periods during which no fundamental changes occurred. Long periods of continuity with no major turning points contributed similarly to the development of agriculture in the region by encouraging permanencies of values and traditions.

Despite the the great contribution of written sources to these issues, there are still aspects such as the concrete introduction of certain species like rye or oats, which greatly depend on interdisciplinary approaches that allow integration of different types of data and enable us to reach more advanced levels of knowledge.

3.8. CROP DIVERSITY AND CHOICE IN THE PREHISTORIC AMERICAN SOUTHWEST

Linda Scott Cummings

Crop diversity may be assessed through multiple data sets. Identifying crops and identifying the plants that people ate are sometimes two different issues. The best evidence of what people ate comes from examination of coprolites or palaeofaeces, since they represent the digested remains of foods that were eaten. That is most often done by examining the pollen, phytolith, and macrofloral evidence for foods eaten. Other interpretation of crops often rests on recovering pollen, phytoliths, and/or macrofloral remains from archaeological sites or agricultural fields and assuming that they represent elements of the diet. Even farther removed, in terms of direct evidence, are the interpretations of crops based on tools recovered at sites.

Two sets of coprolites from southwestern Colorado provide a good look at the diet of people living in the northern portion of the American southwest during the period of prehistory when agriculture was practiced. An introduction to the area and relevant time period are important to understanding this discussion. Both archaeological sites that contained the coprolites examined are located in southwestern Colorado – one inside and one just outside Mesa Verde National Park. Both sites are best described as 'cliff dwellings', meaning that the residences were constructed in rock shelters on the steep sides of drainages. The overhanging rock provided protection from weather. Hoy House, a 56-room complex, is located near the southern boundary of the park, just outside the park, while Step House is located in the southwestern portion of the park. Hoy House was occupied between CE 1140 and the early 1200s. Step House was occupied at a similar time. Structures at both sites were made from shaped sandstone blocks set in an earthen mortar.

At Step House, the coprolites were recovered from Room 21, probably an unroofed work area located in the upper part of the cave. This room is associated with Puebloan occupation of the site, but could not be narrowed further. There are doorways into Room 21 from two adjacent rooms and a window into a third room. The wall near the window also displayed holes that were probably to accommodate wall pegs, presumably for hanging items. Coprolites and trash were deposited in the southern two-thirds of the room, constituting fill above the upper (latest) floor. The northern portion of this layer of fill contained sterile sand and rock. Many artefacts including a stone knife blade fragment, hammerstones, a notched hammer, unifacial *manos* and another grinding stone, a wooden arrow point, fire drill 'points' made from juniper, an oak axe or hammer handle, fire pokers, and three twill-plaited sandals were found on the floor of this room. Twelve coprolites from Step House in Mesa Verde National Park were examined for pollen, phytoliths, macrofloral remains, and parasites. Food remains from maize (*Zea mays*) cobs and kernels, squash (*Cucurbita*) rinds, seeds, and peduncles (stems), bean (*Phaseolus*) pods and one bean seed, an amaranth (*Amaranthus*) root, and prickly pear cactus (*Opuntia*) remains were noted in the trash, as were bones from turkey, cottontail rabbit, mule deer, bighorn sheep, marmot and packrat.

Fifty-nine coprolites from Hoy House and Lion House were examined for pollen and macrofloral remains in the late 1970s when phytolith analysis of human coprolites was not done. Therefore, recovery of evidence of plants that people ate relies on the pollen and macrofloral data bases. Cultigens in the diet consisted of *Zea mays* (maize), *Cucurbita* (squash/pumpkin), and *Phaseolus* (common bean). This triad of cultivated plants has been known to be the staple of prehistoric peoples in the American southwest who practiced agriculture.

The coprolites examined from Hoy House and Lion House are dated only by their association with the sites from which they were recovered. These sites were occupied between the mid-CE 1100s and the early 1200s. Examination of a stratigraphic pollen record through the trash midden at Hoy House (Scott 1979) indicates that a nearly climaxed forest existed on the mesa top near Johnson Canyon at the time of initial occupation of Hoy House around mid-CE 1100. Extensive clearing of arable lands on the mesa top began shortly thereafter, continuing with greater efficiency into the early 1200s.

Cultigens

Pollen, seeds, and phytoliths recovered from the coprolites indicate that the diet included maize on a regular basis (95% of the coprolites at Hoy House and 100% of the coprolites examined from Step House). In contrast, *Cucurbita* (squash) pollen was observed in 41% of the coprolites (Hoy House) and up to 58% of the coprolites at Step House. Finally, *Phaseolus* (bean) pollen was recorded in a mere 7% of coprolites at Hoy House, while the macrofloral remains were noted in 17%. In contrast, by including phytolith analysis, *Phaseolus* evidence was recovered in 92% of the coprolites at Step House, even though macrofloral remains were noted in only 42% (Cummings 1994, Scott 1979). It is likely that beans were under-represented in the pollen and macrofloral records. Recovery of *Phaseolus* hook-shaped hairs in the phytolith record for the Step House coprolites indicates that people were eating not only the dried beans, but also the bean pods, since the hairs are part of the pods. Bean pods exhibit hook-shaped hairs on their exterior. If the beans are removed from the pods and eaten, these hairs are not expected to show up in coprolites. However, if the bean pods are eaten along with the beans, the hairs will be introduced into the human digestive tract. This is the only evidence that will distinguish between eating just the 'beans' and eating the 'beans and pods'. Hook-shaped hairs were recovered from 92% of the coprolites examined from Step House, indicating that beans, including the pods, were a common food. This recovery also provides evidence that beans were stored whole, in their pods, since it is not reasonable to expect that this room was used primarily during the summer when bean pods were available fresh. Instead, it is far more likely that this room was used as a latrine during inclement weather, which in this area is usually during the winter.

Cultigens usually provide a more consistent food source than relying upon gathered, native foods. For this reason, many cultures often gravitate towards extensive reliance upon cultigens at the expense of native, wild foods that often are more nutritious. Cultigens provide a regular calorie base for the diet. In addition, in the American southwest, maize is the staple that provides the highest carbohydrate and calorie input to the diet. The combination of maize and beans provides all the amino acids in appropriate ratios, since each individual food does not contain a full complement of amino acids in desirable concentrations, making this combination an excellent source of protein. In addition, native saltbush leaves and fruit cases were burned to produce a salty ash that could be added to maize. Addition of this type of alkali, termed *nixtamalisation* after the Aztec words *nixtli* (ashes) and *tamalli* (uncooked dough made from corn), softens the outer coating of the kernel, releasing enzymes that act on starches and proteins. The action of denaturing proteins by the alkali makes them more available to human digestion systems. This includes release of niacin that is otherwise bound to proteins in the corn. One more benefit is the increase in minerals that are absorbed into the corn while being boiled in the alkali water. Quantities of calcium, iron, zinc, potassium, and copper are increased. Finally, the interaction between the fat in the maize kernel and the alkali can produce a soapy taste, which is easily removed by discarding the liquid used to boil the maize (http://kitchenscience.sci-toys.com/acids, accessed August 15, 2012.) Whether or not prehistoric people understood the improvement in nutrition, use of ash made cooking dried maize kernels easier and more complete and added valuable salt to the diet. The change in taste alone might account for much of the preference to add ash. Squash was a good source of Vitamin A, although it does not contribute much as a calorie or carbohydrate source. The choice to eat squash flowers also appears to reflect preference, as it does not add much volume or calories to the diet. Beans (pulses) provide slightly more calories and fewer carbohydrates, when compared with squash. Maize, then, probably was the major staple for providing calories and carbohydrates to the diet.

Encouraged or Tolerated

Some native foods grow well in disturbed areas, so are constantly part of the weedy plant population. These plants were easily encouraged or tolerated around the fringes of cultivated plots and even villages. In many areas, including the American southwest, these plants constituted a large part of the diet. Plants that probably were encouraged, but not domesticated, include goosefoot, a member of the beet/spinach family that was valued for both its leaves and seeds. Chenopodiaceae-Amaranthaceae (Cheno-ams) pollen (a term used in North America to denote pollen from the Chenopodiaceae family and the genus *Amaranthus*) from this and related plants was present in 100% of the coprolites at both Hoy House and Step House. Recovery of calcium oxalate crystals from 100% of the Step House coprolites indicates that people were eating greens of plants in the Chenopodiaceae family on a consistent basis. The absence of phytolith analysis makes it difficult to confirm consumption of greens at Hoy House, highlighting the importance of phytolith analysis when interpreting the importance of greens, such as goosefoot greens, or saltbush fruits, both of which produce calcium oxalate crystals that often survive the digestive process.

Cleome (beeweed, see Fig. 3.24) was noted in 95% of the samples examined for pollen at Hoy House and 83% at Step House. This abundance indicates that beeweed greens were a common element of the local diet. According to ethnographic literature, the greens and flowers were collected together when gathering material that would be boiled up and eaten or boiled further to make viscous pottery paint. The mass could be dried and stored, to be reconstituted later and used either as pottery paint or food. The reconstituted material was usually fried in grease when it was to be eaten (Whiting 1939). Understanding this type of food storage is important when interpreting the food resources available to people of the past. Plants eaten for their greens usually provide relatively few calories and were probably viewed as a supplement, rather than a staple, by most people. Greens are usually available seasonally, as they tend not to dry well. Cooking *Cleome*, then storing the cooked 'guano' appears to be an exception to this.

Members of the mustard family, such as *Lepidium* (pepper grass) or *Descurainia* (tansy mustard), also might have been encouraged to grow in the vicinity of agricultural fields. Both pepper grass/tansy mustard and goosefoot (*Chenopodium*) are weedy plants that would need little encouragement other than failure to uproot or remove them.

Another weedy plant that was valued for both its leaves and its seeds, which were ground into flour,

Fig. 3.24. *Cleome* sp. (beeweed) a) plant and b) cooked greens

is *Portulaca* (purslane). The succulent leaves have a slightly lemony taste and are good eaten raw. It is likely that members of the nightshade family, such as *Physalis* (ground cherry) also were either encouraged or tolerated, since these plants are not particularly abundant growing wild on the landscape. Recovery of pollen representing this family and seeds identified as ground cherry from 25% and 75% of the coprolites, respectively, attests to the importance of plants in this group in the diet. Ground cherry fruits have an adstringent quality that is unique. They are particularly good when included in meals or dishes that also contain maize, beans, and/or light meats such as turkey or rabbit.

Native

Native *Opuntia* (prickly pear cactus) produces edible, sweet fruits, as well as edible, succulent pads (Fig. 3.25). Recovery of calcium oxalate druses[4] in both the phytolith and macrofloral records, as well as pollen, indicates consumption of various parts of this cactus. *Opuntia* pollen was noted in 24% of the coprolites at Step House, while macrofloral evidence in the form of epidermal tissue, glochids, and calcium oxalate crystals were observed in 12%, 8%, and 5% of the coprolites (total of 33%), respectively.

Grass seeds from native grasses also were eaten. Evidence of their consumption may be found in any of the data bases. However, since grass seeds often were ground into meal, recovery of seeds is expected to be relatively rare. Therefore, recovery of grass seeds in 33% of the coprolites at Step House probably under-represents the number of people eating grass seeds.

Fig. 3.25. *Opuntia* sp. (prickly pear cactus) with fruit (tunas)

Another important discovery from the phytolith analysis was the importance of yucca leaves as something that was chewed. Although yucca pollen was recovered occasionally at Step House (8%), calcium oxalate raphids[5] were present in 92% of the coprolites, indicating that prehistoric people living in this area probably chewed yucca leaves much as modern people chew gum. Recovery of *Yucca* raphids is probably the result of chewing yucca quids and/or processing yucca leaves with the teeth to remove the fibres for basketry or other uses. The raphids present in *Yucca* leaves are not present in the edible fruit and flowers, so consumption of these parts is not expected to leave the same type of evidence.

Conclusions

Although the diet of Puebloans living in the American southwest has long been thought to have been dominated by the cultivated plants maize, beans, and squash, there is also considerable evidence for consumption of native plants. Certainly the cultigens played an important part in the diet, providing important sources of calories and nutrients. It is likely that they also added greatly to stability for the Puebloans living in hamlets and villages.

Richness of diet was noted primarily in the variety of non-cultivated plant remains recovered. Evidence for Cheno-ams was noted in 75% of the pollen and 100% of the phytolith samples at Step House. Recovery of druses representing Cheno-ams, such as goosefoot or saltbush, indicates that people ate not only ground seeds, but also ate leaves of these plants, probably much like people eat spinach today, and possibly also fruits of saltbush. The seeds of these plants were ground into meal, providing a protein-rich meal with fewer carbohydrates than are present in grass seeds or cereals such as maize. Although small, they are very easy to collect, making them a valuable resource when considering the number of calories expended to harvest and prepare this resource compared to the calories ingested.

The importance of native plants, whether encouraged to grow on disturbed land or not, is also evidenced by recovery of remains from several other plants. *Cleome* pollen was noted with regularity, indicating that people living in Step House also ate beeweed greens, probably on a regular basis. Prickly pear (*Opuntia*) cactus also was part of the diet. This might have included the succulent pads, after the spines had been removed by singeing, or the fruits, many of which are sweet. Eating the sweet fruits is noted to provide a lot of sugar in the diet and also result in significant tooth decay (Watson 2008). Native grass seeds also were part of the diet and, in fact, were probably eaten more commonly than appears from this recovery. Evidence for *Portulaca* was not as abundant in samples from Step House as it was at Hoy House, suggesting that the people living in Step House probably did not consume as much purslane or that they ate purslane when it was young and had not yet flowered.

Although evidence for diet is more direct when obtained from coprolites, microscopic (pollen, phytolith and starch) and macroscopic (macrofloral) remains from dwellings, features, vessels, and other locations also provide valuable information on diet. For instance, pollen recovered from a pit-house (dwelling) floor near Mesa Verde yielded a similar complex of plants to those observed in these coprolites that appear to have been processed. Examination of samples collected every quarter-metre from the floor surface provides valuable information concerning economic activities within the structure. Two types of pollen that were widely distributed over the floor (*Cleome* and *Apiaceae*) suggest hanging beeweed and plants from a member of the *Apiaceae* family from the ceiling either to dry or for storage. The cultigens maize, beans (pulses), and squash were processed in the vicinity of the cooking hearth and also the 'kitchen' area along the south wall and behind small walls separating this area from the larger room used for living and sleeping. Other native foods noted to have been processed or stored in this structure included wild onion (*Allium*, Fig. 3.26), which probably was hung from the wall or ceiling in a corner; prickly pear cactus (*Opuntia*); *Ephedra* that was processed as a medicine and probably contained within a ceramic vessel to the west of the hearth; cattails (*Typha*); and others (Cummings 1998).

Fig. 3.26. *Allium* sp. (wild onion) a) harvested bulb, b) plant

Analysis of microscopic and macroscopic botanic remains from coprolites and dwellings provides evidence that the Puebloan people of southwestern Colorado were successful agriculturalists who also gathered or collected native plants that were important parts of their diet. Whether these native plants were weedy plants that grew in disturbed areas around their hamlets or villages or they were actively encouraged to grow in disturbed areas or even along the edges of cultivated gardens or larger agricultural fields, they appear to have had an important role in the diet of the local people. Even plants expected to have been less abundant on the landscape, growing in special habitats, such as cattails in wetlands or wild onion or cactus on the mesa tops, thus requiring more effort to collect, were sufficiently valued to be retained in the diet. Analysis of diet of these people provides an example of continuity of diet with the past, reaching back to pre-agricultural times, and retention of importance and use of native plants that grew locally.

3.9. PROCESSES OF PREHISTORIC CROP DIVERSIFICATION IN THE LAKE TITICACA BASIN OF THE SOUTH AMERICAN ANDES

Maria C. Bruno

Introduction

The Lake Titicaca basin has long been recognised as an area of crop diversity in the Andes of South America (Beck and García 1991; Cardenas 1989; La Barre 1947; Weberbauer 1945). People began to farm this region as early as 1500 BCE and communities of indigenous Aymara and Quechua subsistence farmers continue today. The lake is located in the *altiplano*, a high (3500–4000 m asl) plain that extends between 15° and 22° S between the eastern and western Andean mountain ranges (Allmendinger *et al.* 1997; Clapperton 1993). The *altiplano* is one of the driest and coldest inhabited zones of the Andes with an annual rainfall of approximately 200–800 mm per year (Vuille *et al.* 2000) and a mean annual temperature of between 7º C–10° C (Montes de Oca 1995). The Lake Titicaca basin, however, is an oasis within the *altiplano*. The immense body of water (approximately 8562 km^2) absorbs solar radiation making the water temperature warmer (10–14° C) than the surrounding land and air (Wirrmann 1992). This radiation of warmth generates 'thermal effects' (Boulange and Aquize 1981) that create warmer yearly temperatures and more rainfall. This temperate micro-climate supports very productive agricultural systems around the lake (Vacher *et al.* 1992).

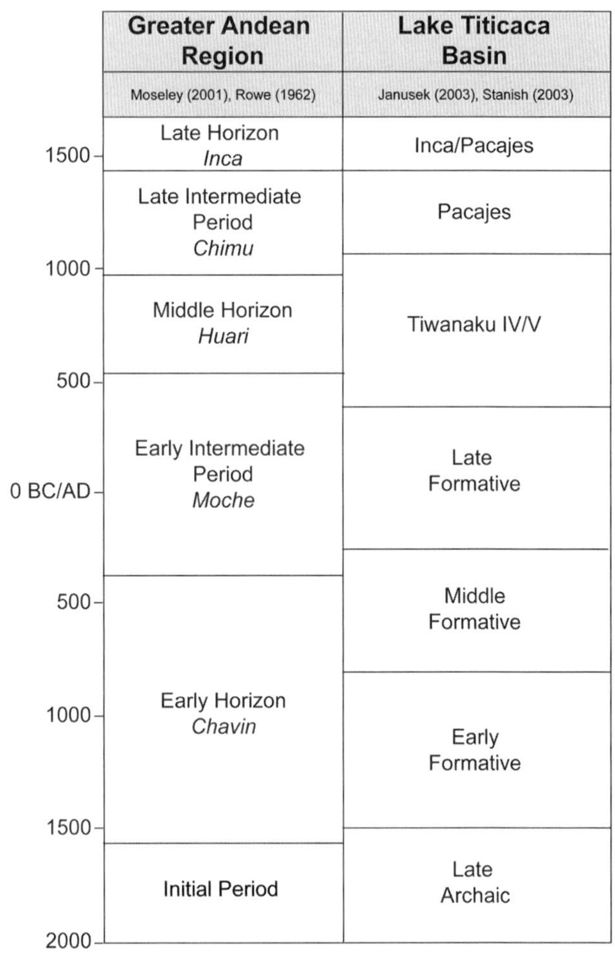

Fig. 3.27. Lake Titicaca Basin chronology. (assembled after Janusek 2003; Moseley 2001; Rowe 1962; Stanish 2003).

Although its cool, dry climate prevents cultivation of Andean species such as chili peppers (*Capsicum baccatum* Jacq.), squash (*Cucurbita maxima* Duchesne, *C. moschata* Duchesne ex Poir.) and coca (*Erythroxylon coca* Lam.), the region has a great varietal diversity of

Fig. 3.28. South America showing Andes mountain range and Lake Titicaca.

crop species that can be grown: *quinoa* (*Chenopodium quinoa* Willd.), *kañawa* (*Chenopodium pallidicaule* Aellen), potatoes (*Solanum tuberosum* L., *Solanum stenotomum* Juz.), and the Andean tubers *oca* (*Oxalis tuberosa* Molina), *isañu* (*Tropaeolum tuberosum* Ruíz and Pavón), and *ullucu* (*Ullucus tuberosus* Caldas). Farmers here also cultivate unique varieties of maize (*Zea mays* L.), *tarwi* (*Lupinus mutabilis* Sweet), and several Eurasian crops, principally fava beans (*Vicia faba* L.) and barley (*Hordeum vulgare* L.).

The goal of this paper is to explore some of the processes by which this diversity arose. I examine archaeobotanical data from the southern Lake Titicaca Basin to evaluate three models: 1) diversification due to the domestication of a crop in a particular area (a Vavilovonian model); 2) diversification of crop species as a strategy of risk reduction and agricultural intensification; and 3) diversification as the result of a political-economic strategy.

The Andean region as a whole was home to several pre-Columbian civilisations (Fig. 3.27, Fig. 3.28).

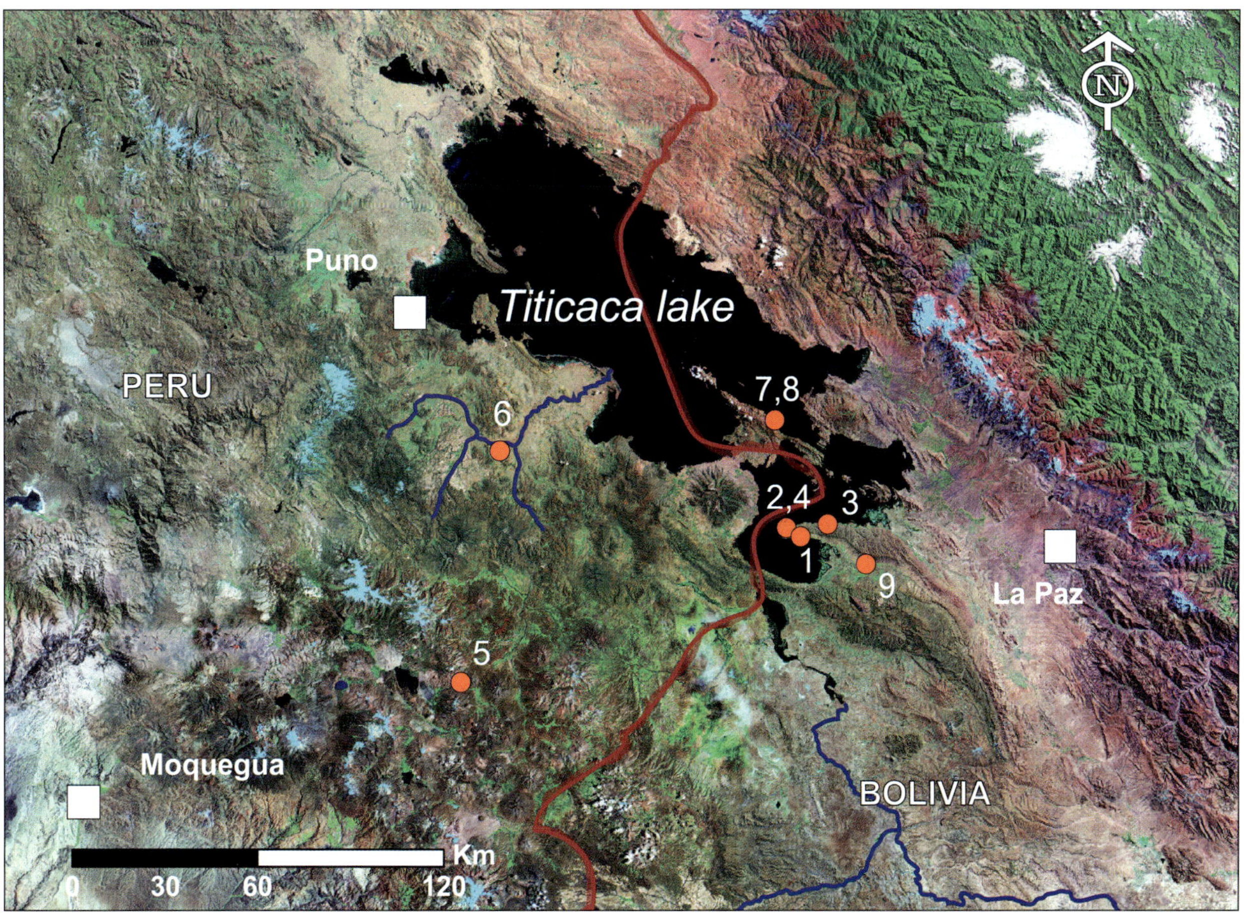

Fig. 3.29. The Lake Titicaca Basin with the sites mentioned in text. 1) Kala Uyuni; 2) Kumi Kipa; 3) Chiripa; 4) Sonaji; 5) Quelcatani Cave; 6) Jiskairumoko; 7) Ch'isi; 8) Q'ota Pata; 9) Tiwanaku. Map: R. Lugon, J.-C. Loubier and A. Chevalier.

In the Lake Titicaca Basin, complex societies first developed during the Formative Period, approximately 1500 BCE to CE 300 (Hastorf 2008; Janusek 2004a). This gave rise to one of the first Andean states, Tiwanaku, between CE 300 and 1100 (Fig. 3.27, Fig. 3.29) (Kolata 1993; Ponce Sangines 1977). Its monumental, ceremonial, and urban centre was located approximately 20 km from the southern lakeshore in modern-day Bolivia. Its influence, however, spread to southern Peru and northern Chile. After the fall of the Tiwanaku state, the population split into small, independent groups, often called *señorios* or kingdoms (CE 1150–1450). By 1475, the Inca Empire conquered these groups and dominated the region until the arrival of the Spanish in 1532. In this discussion, I focus primarily on the first two periods of Titicaca Basin prehistory, the Formative and Tiwanaku periods, and their roles in producing the diversity of crop varieties found in the region today.

Based on recent genetic and archaeobotanical studies, I argue that the lake basin was likely not the centre of domestication for several of the crops, but that their diversity emerged out of dynamic productive, environmental, and cultural processes throughout its prehistory. First, I examine patterns in *quinoa* and tuber remains during the Formative period (1500 BCE–CE 300) and discuss how early farmers may have increased crop diversity as a strategy of risk reduction and agricultural intensification. Second, I examine how political factors during the Tiwanaku period (CE 300–1100) may have lead to the development of new maize varieties specifically adapted to the lake basin.

Domestication as an Explanation of Diversity

Following concepts developed by evolutionary biologist and botanist de Candolle (1884), the Russian agronomist N. I. Vavilov (1992) proposed that the place where a crop was domesticated could be located by identifying the area with greatest diversity of crop varieties. Vavilov (1992), following Cook (1925), proposed that the Andean region of South America was one of the world's primary centres of crop domestication. Subsequent researchers documented great diversity in varieties of chenopods, potatoes, and other Andean tubers in the Lake Titicaca basin, suggesting this may be their place of origin (Gandarillas 1979; Hawkes 1990; León 1967; Ugent 1970). Thus, a Vavilovian model might explain the diversity of crop varieties found in the Lake Titicaca Basin.

Unlike some of the other primary centres of domestication (Near East, Mexico, eastern North America, tropical South America), there has been relatively little archaeological and archaeobotanical research focused on the origin and spread of Andean highland crops. Although limited, available archaeobotanical data do provide some insight into the early history of the domesticated pseudocereal *quinoa*, potatoes, and the Andean tuber *oca* in the Lake Titicaca Basin. Current data show that fully agricultural societies were established by the Formative period (Browman 1989; Bruno and Whitehead 2003; Eisentraut 1998; Whitehead 2006), thus many researchers suggest that the process of plant domestication must have begun in the Late Archaic period (approximately 4000–1500 BCE) (Aldenderfer 1989; Bruno 2006; Pearsall 1992) (Fig. 3.27).

Chenopodium quinoa grains have been found in two Archaic sites in the western Lake Titicaca Basin, Quelcatani Cave and Jiskairumoko, Peru. Although Phyllisa Eisentraut (1998) identified domesticated *quinoa* grains in Archaic levels at Quelcatani Cave, direct AMS[6] dating of these seeds showed they derived from the Formative period approximately 900 cal. BCE (2740±50 bp). Radiocarbon dates were calibrated using OxCal 4.1, IntCal 04 (Reimer *et al.* 2004). Andrea Murray (2005) identified *quinoa* grains from Jiskairumoko and the site directors (Dr. Nathan Craig and Mark Aldenderfer) are currently waiting for the results of direct AMS dates in order to determine their age. *Quinoa* seeds occur in sites in the central Peruvian highlands and Chilean coast but have not been directly dated. They derive from contexts ranging from 5000 to 3000 BCE (Aldenderfer 1999; Iriarte 2007; Kuznar 1993; Nordstrom 1990; Pearsall 1992; 2008; Planella *et al.* 2005). Currently, the earliest direct date for domesticated *quinoa* in the lake basin is from the Formative period site of Chiripa, Bolivia, and is approximately 1500 cal. BCE (3200±60 bp) (Bruno and Whitehead 2003; Whitehead 1999) (Fig. 3.29).

Tuber remains are generally scarce in the archaeological record because they do not preserve well due to their high water content and thin-celled walls (Wright *et al.* 2003). In the Lake Titicaca basin, it is rare to find large diagnostic specimens, but fragments of parenchyma tissue derived from tubers are common (Browman 1989; Erickson 1976; Whitehead 2006; Wright *et al.* 2003). To date, no tuber or parenchyma remains have been identified in Archaic period sites in the Lake Titicaca Basin; however, Claudia Rumold (2010) has identified *Solanum* spp. starch grains from Late Archaic (approximately 3200–2300 BCE) ground stone at the site of Jiskairumoko. She also identified *Solanum* spp. starch grains from Early Formative ground stone (Rumold 2010).

Like *quinoa*, the earliest tuber macroremains appear in Formative period sites, although none have been directly dated. At Chiripa, archaeologists have found whole charred potatoes, perhaps even *chuño* (freeze-dried potatoes), *oca*, and *ullucu* (Browman 1989; Towle 1961; Towle, *pers. comm.*). Midori Lee (1997) identified remains of *chuño* and *oca* at Ch'isi on the Copacabana Peninsula, Bolivia. Both of these sites pertain to the Middle Formative period (800–250 BCE), and are much later than the earliest identified potato or Andean tuber remains encountered in the central highlands or coast (earliest approximately 2000 BCE) (Pearsall 2008).

Given the fragmentary nature of the archaeological record, molecular genetic studies of crop populations can provide a better manner with which to identify where plants were domesticated (Pickersgill 2007; Zeder *et al.* 2006). To date, molecular studies have placed the domestication of two of these aforementioned crops outside of the Lake Titicaca Basin: *quinoa* in the eastern Andean slopes or plains of Argentina, Uruguay, and Paraguay (Wilson 1990) and *oca* on the eastern slopes of the Andes (Emshwiller 2006). Molecular studies of the potato suggest a single origin in southern Peru, which could include the Titicaca Basin (Spooner 2005).

Future excavations, archaeobotanical analyses, and molecular studies will improve our understanding of early plant use in the Lake Titicaca Basin. Current data, however, suggests that initial domestication may have occurred outside of the basin, requiring us to look beyond a Vavilovian explanation for diversity here (Brush *et al.* 1995; Harris 1990; Hawkes 1983). These data do show us, however, that farmers have been cultivating these crops in the lake basin for over 2000 years. I now explore how factors such as risk reduction, agricultural intensification, environmental characteristics, and socio-political processes may have contributed to the high crop diversity in the region.

Crop Diversification as a Strategy of Risk Reduction and Agricultural Intensification

Unlike modern, industrial monocrop agriculture, traditional farming systems strategically incorporate a diversity of crops and varieties. Increasing crop diversity not only decreases the risk of crop failure, but may also contribute to greater overall productivity (Hawkes 1983; Netting 1993). For example, many of the crop varieties cultivated in the lake basin today reflect the different risks present in various micro-environments within the region. There are sturdy frost- and drought-resistant varieties of potatoes, *ullucu, quinoa*, and *kañawa* that are grown at higher elevations and in the open plains. They also have varieties of *oca, isañu*, and potatoes that thrive in more temperate settings on the lakeshore and in protected valleys. Farmers plant each of these varieties across the landscape each year. This diversity usually guarantees that some of the crops will be successful despite poor conditions (frosts, drought, flood), and, in the best-case scenario, result in a successful harvest (Bruno 2011; Carter and Mamani 1982; Ochoa 1990). Thus, strategies of risk reduction and agricultural intensification provide another possible explanation for the diversity of crop varieties seen in the Lake Titicaca Basin.

I evaluate this scenario using results from my study of Formative period macrobotanical remains from the Taraco Peninsula, Bolivia. I analysed charred plant remains from three sites: Kala Uyuni, Sonaji, and Kumi Kipa (Fig. 3.30). The Taraco Archaeological Project, directed by Drs. Christine Hastorf and Matthew Bandy, excavated these sites between 2003 and 2005. Here, I consider 198 samples in total, 61 samples from the Middle Formative (MF) period and 137 samples from the Late Formative (LF) period.

Quinoa is one of the most common and abundant seeds recovered from macrobotanical samples in the Lake Titicaca Basin (Browman 1989, 147; Eisentraut 1998, 171, 187; Whitehead 2006, 268; Wright *et al.*

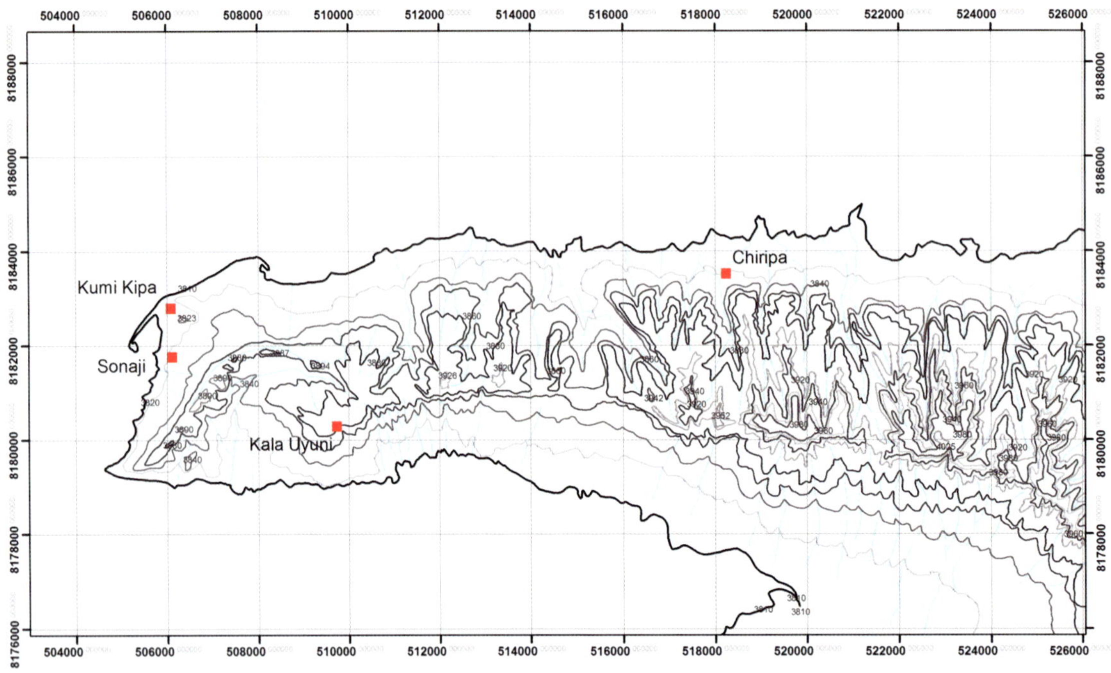

Fig. 3.30. The Taraco Peninsula, Bolivia, with sites mentioned in text.

2003, 387). In my study, *quinoa* seeds were present in 90% of the MF samples and 97% of the LF samples (see Bruno 2008 for full description of data) (Fig. 3.31). Its prevalence is due in part to the plant's production of thousands of small, sturdy seeds that preserve very well when charred. It was also likely a very important crop early in farming history because it does not require much effort to cultivate. The soil does not need to be heavily tilled before planting, and the seed can be broadcast sown. To harvest *quinoa*, the stalk is simply pulled out of the ground by hand.

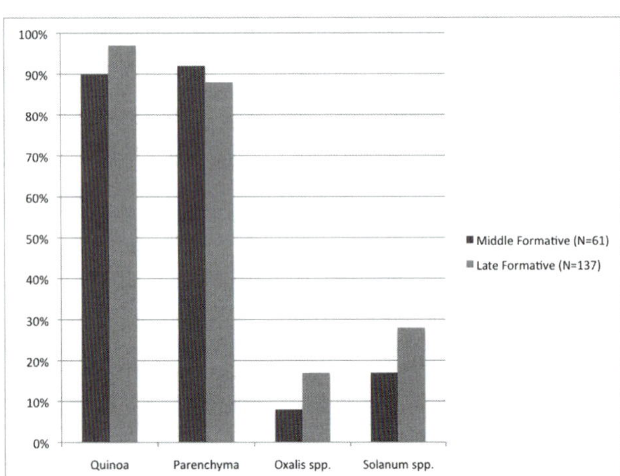

Fig. 3.31. Ubiquity of crops through time.

Assessing the role of tubers is more difficult given poor preservation, but tentative trends are emerging based on patterns in tuber/parenchyma fragments, as well as seeds derived from these species. Tuber and parenchyma tissue fragments are common in the samples I studied, occurring in 92% of the MF samples and 88% of the LF samples (Fig. 3.31). Although there were no diagnostic fragments, there are seeds of potentially two domesticated species, *Oxalis* spp. and *Solanum* spp., in the samples. Humans do not consume the seeds of these crops, but the plant is often fed to livestock before the tubers are harvested. Grazing herds also feed on wild populations (Bruno 2008). Thus, these seeds likely entered the archaeological record through the burning of domesticated camelid dung, a practice still common today (Hastorf and Wright 1998; Miller and Smart 1984; Winterhalder *et al.* 1974). Wild and domesticated species of *Oxalis* spp. and *Solanum* spp. have similar seeds; therefore, it is not possible to determine if the archaeological specimens are from crops or natural populations. Despite these difficulties, there is a notable change in their presence through time: the ubiquity of *Oxalis* spp. seeds increases from 8% in the MF to 17% in the LF and *Solanum* spp. increases from 17% to 28% (Fig. 3.31). The increase in their ubiquity suggests that both *Oxalis* and *Solanum* became more common on the landscape through time. Wild and domesticated species of these crops do have a close relationship (Brush *et al.* 1981; Emshwiller 2002; Johns and Keen 1986). It is common to find wild species growing in cultivated fields and studies show that the great diversity of potatoes, in particular, is due in part, to high introgression between wild and domesticated populations (Brush *et al.* 1995; Ochoa 1990).

These patterns suggest that during the Late Formative period tuber crops increased in importance on the Taraco Peninsula. The slower adoption of tuber crops may be due to the labour needed to cultivate them. Unlike *quinoa*, the soil must be excavated to plant and harvest tubers. Interestingly, we do find an increase in the density of stone agricultural tools during the Late Formative period (Miller *et al.* 2008), supporting the hypothesis that tuber production increased at this time. Increasing the cultivation of tubers would have increased the overall diversity of crop species produced on the Taraco Peninsula. Although these additional species would provide greater protection against crop failure from frost, pests, droughts or flooding, they would also increase overall food production. Based on survey (Bandy 2006) and excavations (Bandy 2007), we know that Late Formative populations were larger and more complex than previous Taraco inhabitants. It is, therefore, possible that crop diversification as a strategy of both risk reduction and agricultural intensification in this region has its roots in the Formative period (Bruno 2008, Chapter 12).

Political Economy and the Diversification of Altiplano Maize

While farmers must meet the needs of their household and village, there are certain circumstances when they must also meet demands of a larger political entity (Brookfield 1972; Sahlins 1972, 101–102). For example, the importance of maize (*Zea mays* L.) production for the political economy of the Inca Empire has been well-documented for the central Andes (Hastorf 1990; Hastorf and Johannessen 1993). Thus, the demands of an influential leader or a state apparatus may also motivate farmers to develop new varieties or adopt new crops.

This may have been the impetus for the development of maize varieties that can grow in the Lake Titicaca basin. Genetic data verify that maize was domesticated in Mexico and was introduced into the Andes and the Lake Titicaca basin (Matsuoka *et al.* 2002). Andean farmers adopted this tropical crop and developed dozens of new varieties, several of which are adapted to the more temperate areas of the Lake Titicaca basin (Confite Puñeno, Altiplano, Patillo and Kulli) (Cutler 1946; Ramírez *et al.* 1960). A growing body of data on maize from archaeological sites in the Titicaca basin permits us to track when it was introduced into the basin and began to be cultivated (Logan *et al.* 2012).

The earliest evidence of maize in the Titicaca Basin is from microbotanical remains. Amanda Logan (2006) identified maize phytoliths and starch grains from the sites Chiripa, Kala Uyuni, and Kumi Kipa on the Taraco Peninsula. She encountered them in a variety of contexts, including public, special-use buildings (sunken courts), domestic middens, and ground stone. Robert Thompson identified maize phytoliths from sites at Ch'isi on the Copacabana Peninsula, Bolivia (Chávez and Thompson 2006) (Fig. 3.28). None of these remains have been directly dated but derive from contexts dating to the Middle and Late Formative periods. There is little macrobotanical evidence for maize in the Formative period. Midori Lee (1997) identified one maize kernel and one maize glume from the Middle Formative period site of Qhot'a-Pata on the Copacabana Peninsula. I found two fragmented maize kernels at the Late Formative period site Kala Uyuni (Bruno 2008).

The scant evidence of maize throughout the Formative period suggests that it was not cultivated in the basin during this time. There is excellent preservation of charred seeds and even parenchyma tissue in these sites, thus the lack of maize macrobotanical remains should not be attributed to poor preservation. Instead, it may have been traded or brought in from temperate zones to the east and west. While maize may have been consumed as a food, the kernels boiled and eaten, as is common today, it may have been fermented to produce an alcoholic beverage known as *chicha* (Cutler and Cárdenas 1947; La Barre 1938). Recent studies of carbon isotopes from human remains suggests that maize was only consumed in significant quantities by individuals associated with Late Formative ceremonial architecture at sites such as Tiwanaku, Khonkho Wankane, and Kala Uyuni (Berryman 2010; Miller *et al.* 2008). Given the available data, maize was probably obtained from other regions as a special-use plant, not a food crop, during the Formative period.

Maize took on a much greater role in the Tiwanaku period. Analysis of macrobotanical remains from the site of Tiwanaku show that maize (cupules, grains and cobs) occurs in approximately 20% of all samples (Hastorf *et al.* 2006; Wright *et al.* 2003). An increase in maize consumption during the Tiwanaku period is also evident in human stable isotopes. Tiwanaku period individuals from sites in the lake basin have higher C4 signatures than those from the Formative period, particularly among people associated with monumental architectural spaces in the Tiwanaku urban core (Berryman 2010).

The increase in maize during the Tiwanaku period could be the result of it becoming a common food or that the consumption of *chicha* increased. Evidence for the latter can be found in the Tiwanaku ceramic assemblage, which includes specialised drinking mugs, known as *kerus*, and large jars possibly used in fermenting great quantities of alcohol (Alconini Mujica 1995; Janusek 2003). Scholars believe that an important part of Tiwanaku's political strategy was to host large feasts and celebrations for people who travelled to and lived in the ceremonial centre (Janusek 2002; 2004b). There is also evidence for many of these practices in the Tiwanaku colonies in Moquegua and Cochabamba (Goldstein 2003).

A study of morphological variability in grain and cupules demonstrates that the Tiwanaku had access to a wide range of maize varieties (Hastorf *et al.* 2006). Morphological comparisons of maize from the Moquegua Valley in Peru and Cochabamba Valley in Bolivia suggests that these were two source regions, but there are other morphological variants whose geographical origins have not been identified (Hastorf *et al.* 2006). It is possible that some of these unidentified varieties could have been cultivated in the lake basin. Although more research is needed, I suggest that the cultural and political importance of maize for the Tiwanaku state may have instigated the development of varieties that could be grown in the basin. After the state fell, this variety would still have been available to local farmers, and may have been integrated into the household farming system alongside varieties of *quinoa* and tubers, as it is today.

Conclusions

The Lake Titicaca Basin is a locus of great crop diversity within the Andean region of South America. While some of this diversity could be attributed to it being a centre of crop domestication, the reasons why farmers integrated and developed new crops and varieties lie in the dynamic interactions of production, environment, culture, and politics. While it is possible that potatoes may have been domesticated in the Lake Titicaca Basin, other crops such as *quinoa*, *oca*, and maize were not, thus their presence and varietal diversity might be explained using other models, such as those addressing risk reduction, agricultural intensification, and political economy.

Archaeobotanical data from the southern Lake Titicaca Basin, particularly the Taraco Peninsula, Bolivia, suggests that crop diversification began as early as the Formative period with increasing integration of tuber crops, such as potatoes and *oca*, to systems primarily based on *quinoa* cultivation. The data also indicate that maize was not part of the Titicaca Basin farming systems until the Tiwanaku period. Given its important role in the politics of the Tiwanaku state, it is possible that farmers were encouraged to develop varieties that could also be grown locally. These are presented as a hypothesis to be tested, for much more research is needed to truly understand the character of crop domestication and diversification in the region. In addition to tracking the introduction of new crops, another important avenue of research for the future is the study of morphological changes within particular crops such as *quinoa* and *kañawa*. With these data, we might be able to document the appearance of new crop varieties. We have begun to do this with maize, but more work is needed. An advantage to conducting such research in this region is that indigenous farmers continue to cultivate many of these crops and are conserving this great diversity in their fields. Collaborations between farmers, agronomists, and archaeologists promise to reveal more information regarding the origins of the diversity in the Lake Titicaca Basin of the Andes and preserving it for the future.

3.10. CONCLUSIONS

Elena Marinova

The current chapter took up a specific focus on looking at crop diversity through time and to seeking out a better understanding of the processes involved in this diachronic perspective. The various contributions illustrate the way different communities have maintained, enlarged or decreased diversity according to the changing natural and social environments they were part of. The studies presented have shown that the link between crop diversity and culture is unmistakable and that their interaction is often an extremely dynamic process. Crop diversity appears, therefore, as one of the multiple results of the dynamic nature of human decisions and natural resources.

Chapter Acknowledgements

The work by Leonor Peña-Chocarro and Lydia Zapata (Chapter 3.5) was carried out within the projects AGRIWESTMED (ERC-AdG-230561), HAR2008-03976/HIST, HAR2008-01920/HIST and HAR2011-23716 *Nuevos cultivos, nuevos paisajes* and Programa Consolider TCP-CSD2007-00058 from the Spanish Government. Lydia Zapata is part of the Research Group of the Basque Government IT622-13 and UPV/EHU UFI 11-09.

Research of the Taraco Peninsula materials by Maria C. Bruno (Chapter 3.9) was funded by grants from the United States National Science Foundation (Dissertation Improvement Grant #0321720), the Wenner Gren Foundation (#7073), and IIE Fulbright. The author's special thanks go to Christine Hastorf, Matthew Bandy, the Taraco Archaeological Project members, José Capriles, and Eduardo Machicado.

Chapter Notes

1 With four sets of chromosomes.
2 With six sets of chromosomes.
3 The central stalk of the cereal ear.
4 Spherical crystalline aggregations of calcium oxalate.
5 Bundles of needle-shaped calcium oxalate crystals.
6 Accelerator Mass Spectrometry, a high-resolution variant of radiocarbon dating.

Chapter Bibliography

Akeret, Ö. (2005) Plant remains from a Bell Baker site in Switzerland, and the beginnings of *Triticum spelta* (spelt) cultivation in Europe. *Vegetation History and Archaeobotany* 14, 279–286.

Alconini Mujica, S. (1995) *Ríto, Simbolo e Historia en la Pirámide de Akapana, Tiwanaku*. Editorial Accion, La Paz.

Aldenderfer, M. S. (1989) Archaic Period in the South Central Andes. *Journal of World Prehistory* 3 (2), 117–158.

Aldenderfer, M. S. (1999) The Late Preceramic-Early Formative Transition on the South-Central Andean Littoral. In M. Blake (ed.) *Pacific Latin America in prehistory: the evolution of archaic and formative cultures*. Washington State University Press, Pullman, pp. 213–222.

Allmendinger, R. W., Jordan, T. E., Kay, S. M. and Isacks, B. L. (1997) The Evolution of the Altiplano-Puna Plateau of the Central Andes. *Annual Review of Earth and Planetary Sciences* 25, 139–174.

André, J. (1998) *Essen und Trinken im alten Rom*. Reclam, Stuttgart.

Andreou, S., Fotiadis, M. and Kotsakis, K. (1996) Review of Aegean Prehistory V: The Neolithic and Bronze Age of Northern Greece. *American Journal of Archaeology* 100, 537–597.

Andreou, S., Heron, C., Jones, G. Kiriatzi, V., Psaraki, K., Roumbou, M. and Valamoti, S. (2013) Smelly barbarians or perfumed natives? An investigation of oil and ointment use in Late Bronze Age northern Greece. In S. Voutsaki and S.-M. Valamoti (eds.) *Diet, Economy and Society in the Anciet Greek World: Towards a Better Integration of Archaeology and Science*. Pharos Supplement. Peeters, Leuven, pp. 173–185.

Araus, J. L., Ferrio, J. P., Buxo, R. and Voltas, J. (2007) The historical perspective of dryland agriculture: lessons learned from 10000 years of wheat cultivation. *Journal of Experimental Botany* 58, 131–145.

Bakels, C. C. (1982) Der Mohn, die Linearbandkeramik und das Westliche Mittelmeergebiet. *Archäologisches Korrespondenzblatt* 12, 11–13.

Bakels, C. C. (1991) The Crops of the Rössen Culture.

In E. Hajnalová (ed.) *Palaeoethnobotany and Archaeology. International Work-Group for Palaeoethnobotany 8th Symposium Nitra-Nové Vozokany 1989.* Acta Interdisciplinaria Archaeologica 7, 23–36.

Bakels, C. C. (2000) The Neolithization of the Netherlands: Two ways, one result. In A. Fairbairn (ed.) *Plants in Neolithic Britain and beyond.* Oxbow Books, Oxford, pp. 101–106.

Bakels, C. C. (2005) Crops produced in the southern Netherlands and northern France during the early medieval period: a comparison. *Vegetation History and Archaeobotany* 14, 394–399.

Bakels, C. C. (2007) Nature or culture? Cereal crops raised by Neolithic farmers on Dutch loess soils. In S. Colledge and J. Conolly (eds.) *The origins and spread of domestic plants in Southwest Asia and Europe.* Left Coast Press, Walnut Creek, pp. 343–348.

Bakels, C. C. (2009) *The Western European Loess belt, Agrarian History 5300 BC–AD 1000.* Springer Verlag, Berlin.

Bakels, C. C. and Jacomet, S. (2003) Access to luxury foods in Central Europe during the Roman period: the archaeobotanical evidence. *World Archaeology* 34, 542–557.

Bakels, C. C., Alkemade, M. J. and Vermeeren, C. E. (1993) Botanische Untersuchungen in der Rössener Siedlung Maastricht-Randwijck. *Archaeo-Physika* 13, 49–62.

Bakels, C., Constantin, C. and Hauzeur, A. (1992) Utilisation de graines de pavot comme degraissant dans un vase du groupe de Blicquy. *Archäologisches Korrespondenzblatt* 22, 473–479.

Bakels, C. C., Wesselingh, D. and van Amen, I. (1997) Acquiring a taste: the menu of Iron Age and Roman period farmers in Oss-Ussen, the Netherlands. *Analecta Praehistorica Leidensia* 39, 193–211.

Banchieri, D. and Rottoli, M. (2009) Isolino Virginia: una nuova data per la storia della coltivazione del papavero da oppio (*Papaver somniferum* subsp. *somniferum*). *Sibrium* 25 (2004–2009), 31–49.

Bandy, M. S. (2006) Early Village Society in the Formative Period in the Southern Lake Titicaca Basin. In W. Isbell and H. Silverman (ed.) *Andean Archaeology III: North and South.* Springer, New York, pp. 210–236.

Bandy, M. S. (2007) Kala Uyuni and the Titicaca Basin Formative. In M. S. Bandy and C. A. Hastorf (ed.) *Kala Uyuni: An Early Political Center in the Southern Lake Titicaca Basin: 2003 Excavations of the Taraco Archaeological Project.* University of California, Berkeley, pp. 135–143.

Barfield, L., Borrello, M., Buteux, S. and Ciaraldi, M. (2002) Scavi preistorici sulla Rocca di Manerba, Brescia. In A. Ferrari and P. Visentini (eds.) *Il declino del mondo neolitico. Ricerche in Italia centro-settentrionale fra aspetti peninsulari, occidentali e nord-alpini. Atti del Convegno, Pordenone 5–7 aprile 2001.* Quaderni del Museo Archeologico del Friuli Occidentale 2, pp. 291–310.

Beck, S. G. and García, E. (1991) Flora y Vegetación en los Diferentes Pisos Altitudinales. In E. Forno and M. Baudoin (ed.) *Historia Natural de un Valle en Los Andes: La Paz.* Servicio Grafico Quipus, La Paz, pp. 65–208.

Becker, C. and Kroll, H. (2008) *Das Prähistorische Olynth. Ausgrabungen in der Toumba Agios Mamas 1994–1996. Ernährung und Rohstoffnutzung im Wandel.* Prähistorische Archäologie in Südosteuropa 22. Verlag Marie Leidorf, Rahden/Westfalen.

Behre, K.-E. (1992) The history of rye cultivation in Europe. *Vegetation History and Archaeobotany* 1, 141–156.

Behre, K.-E. (2007) Evidence for Mesolithic agriculture in and around central Europe. *Vegetation History and Archaeobotany* 16, 203–219.

Berryman, C. A. (2010) *Food, Feasts, and the Construction of Identity and Power in Ancient Tiwanaku: A Bioarchaeological Perspective.* Ph.D. thesis. Vanderbilt University, Nashville.

Bieniek, A. (2007) Neolithic plant husbandry in the Kujawi region of central Poland. In S. Colledge and J. Connelly (eds.) *The Origin and Spread of Domestic Plants in Southwest Asia and Europe.* Left Coast Press, Walnut Creek, pp. 327–342.

Bogaard, A. (2002) Questioning the relevance of shifting cultivation to Neolithic farming in the loess belt of Europe: evidence from the Hambach Forest experiment. *Vegetation History and Archaeobotany* 11, 155–168.

Bogaard, A. (2004) *Neolithic Farming in Central Europe.* Routledge, London.

Bogaard, A. (2011) Farming practice and society in the Central European Neolithic and Bronze Age: an archaeobotanical response to the secondary products revolution model. In A. Hadjikoumis, E. Robinson and S. Viner (eds.) *The Dynamics of Neolithisation in Europe. Studies in honour of Andrew Sherratt.* Oxbow Books, Oxford, pp. 266–283.

Bogaard, A., Bending, J. and Jones, G. (2008) Crop husbandry and its social significance in the Körös and LBK cultures. In D. Bailey, A. Whittle and D. Hofmann (eds.) *Living Well Together? Settlemet and Materiality in the Neolithic of South-East and Central Europe.* Oxbow Books, Oxford, pp. 131–138.

Bohn, U., Gollub, G., Hettwer, C., Neuhäuslova, Z., Schlüter, H. and Weber H. (2003) *Karte der natürlichen Vegetation Europas 1:2.500.000.* Landwirtschaftsverlag, Bonn.

Borojević, K. (2006) *Terra and silva in the Pannonian Plain: Opovo Agro-Gathering in the Late Neolithic.* BAR International Series 1563. Archaeopress, Oxford.

Bosch i Lloret, À., Chinchilla Sánchez, J. and Tarrús i Galter, J. (2000) *El Poblat Lacustre Neolitic de la Draga. Excavaciones de 1990 a 1998.* Monografies del CASC 2. Girona.

Bostock J (transl.) (1855) *Pliny the Elder. The Natural History.* In G. R. Crane (ed.) Perseus Digital Library Project. Tufts University website. http://www.perseus.tufts.edu/cgi-bin/ ptext?lookup=Plin.+Nat.+toc.

Bottema, S. (1990) Holocene environment of the southern

Argolid: a pollen core from Kiladha Bay. In T. J. Wilkinson and S. T. Suhon (eds.) *Excavations at Franchthi Cave, Greece* 6. Indiana University Press, Bloomington/ Indianapolis, pp. 117–138.

Bouby, L. (2002) Le chanvre (*Cannabis sativa* L.): Une plante cultivée à la fin de l'âge du Fer en France du Sud-Ouest? *Comptes rendus palevol* 1, 89–95.

Boulange, B. and Aquize, J. E. (1981) La Sédimentation Actuelle Dans Le Lac Titicaca et de Son Bassin Versant. *Rev. Hydrobiologic Tropical* 14 (4), 269–287.

Brombacher, C. (2008) Les macrorestes végétaux, reflets des pratiques agricoles, de l'alimentation et de l'environnement. In M. Guélat, C. Brombacher, C. Olive and L. Wick (eds.) *Develier-Courtételle - un habitat rural mérovingien. Environnement et exploitation du terroir au début du Moyen Age.* Cahiers d'Archéologie Jurassienne 16. Porrentruy, pp. 103–141.

Brombacher, C. and Jacomet, S. (1997) Ackerbau, Sammelwirtschaft und Umwelt: Ergebnisse archäobotanischer Untersuchungen. In J. Schibler, H. Hüster-Plogmann, S. Jacomet, C. Brombacher, E. Gross-Klee and A. Rast-Eicher (eds.) *Ökonomie und Ökologie neolithischer und bronzezeitlicher Ufersiedlungen am Zürichsee. Ergebnisse der Ausgrabungen Mozartstrasse, Kanalisationssanierungen Seefeld, AKAD/Pressehaus und Mythenschloss in Zürich.* Monographien der Kantonsarchäologie Zürich 20. Zürich/Egg, pp. 220–279.

Brombacher, C. and Kühn, M. (2005) Archäobotanische Funde. In R. Windler, R. Marti, U. Niffeler and L. Steiner (eds.) *SPM VI. Frühmittelalter. Die Schweiz vom Paläolithikum bis zum frühen Mittelalter. Vom Neandertaler bis zu Karl dem Grossen.* Basel, pp. 229–231.

Brookfield, H. (1972) Intensification and Disintensification in Pacific Agriculture. *Pacific Viewpoint* 13, 30–48.

Browman, D. L. (1989) Chenopod Cultivation, Lacustrine Resources, and Fuel Use at Chiripa, Bolivia. In E. E. Voigt and D. M. Pearsall (ed.) *New World Paleoethnobotany: Collected Papers in Honor of Leonard W. Blake* Special Issue *Missouri Archaeologist* 47. Missouri Archaeological Society, Columbia, pp. 137–172.

Bruno, M. C. (2006) A Morphological Approach to Documenting the Domestication of *Chenopodium* in the Andes. In D. B. Melinda, A. Zeder, E. Emshwiller, and B. D. Smith (eds.) *Documenting Domestication: New Genetic and Archaeological Paradigms.* University of California Press, Berkeley, pp. 32–45.

Bruno, M. C. (2008) *Waranq Waranqa: Ethnobotanial Perspectives on Agricultural Intensification in the Lake Titicaca Basin (Taraco Peninsula, Bolivia).* Ph.D. thesis. Washington University, St. Louis.

Bruno, M. C. (2011) Farmers' Experience and Knowledge: Utilizing Soil Diversity to Overcome Rainfall Variability on the Taraco Peninsula, Bolivia. In N. Miller, K. Moore and K. Ryan (eds.) *Sustainable Lifeways: Cultural Persistence in an Ever-changing Environment.* University of Pennsylvania Press, Philadelphia. pp. 210–243.

Bruno, M. C. and Whitehead, W. T. (2003) *Chenopodium* Cultivation and Formative Period Agriculture at Chiripa, Bolivia. *Latin American Antiquity* 14, 339–355.

Brush, S. B., Carney, H. J. and Huaman, Z. (1981) Dynamics of Andean Potato Agriculture. *Economic Botany* 35, 70–88.

Brush, S. B., Kesseli, R., Ortega, R., Cisneros, P., Zimmerer, K. and Quirós, C. F. (1995) Potato Diversity in the Andean Center of Crop Domestication. *Conservation Biology* 9, 1189–1198.

Buxó, R. (2008) The agricultural consequences of colonial contacts on Iberian peninsula in the first millennium BC. *Vegetation History and Archaeobotany* 17, 145–154.

Buxó, R. (2007) Crop evolution: new evidence from the Neolithic of west Mediterranean Europe. In S. Colledge and J. Conolly, (eds.) *The origins and spread of domestic plants in Southwest Asia and Europe.* Left Coast Press, Walnut Creek, pp. 155–171.

Buxó, R., Alonso, N., Canal, D., Echave, C. and González, I. (1997) Archaeobotanical remains of hulled and naked cereals in the Iberian Peninsula. *Vegetation History and Archaeobotany* 6, 15–23.

Buxó, R., Rovira i Buendia, N. and Saüch, C. (2000) Les Restes Vegetals de Llavors i Fruits. In A. Bosch i Lloret, J. Chinchilla Sánchez and J. Tarrús i Galter (eds.) *El Poblat Lacustre Neolític de la Draga. Excavaciones de 1990 a 1998.* Monografies del CASC 2, 129–140.

Buxó, R. (1997) *Arqueología de las plantas.* Crítica, Barcelona.

Cardenas, M. (1989) *Manual de Plantas Economicas de Bolivia.* Editorial Los Amigos del Libro, La Paz.

Carter, W. and Mamani, M. (1982) *Irpa Chico: Individuo y Comunidad en la Cultura Aymara.* Libreria Editorial Juventud, La Paz.

Castelletti, L. (1975) Segale (*Secale cereale* L.) subfossile a Lomello. Atti Ce.S.D.I.R. VI, 1974–75. Como, pp. 55–71.

Castelletti, L. (1978) Resti vegetali macroscopici del XII secolo nella torre civica di Pavia. *Archeologia Medievale* 5, 239–248.

Castelletti, L., Castiglioni E. and Rottoli, M. (2001) L'agricoltura dell'Italia settentrionale dal Neolitico al Medioevo. In O. Failla and G. Forni (eds.) *Le piante coltivate e la loro storia. Dalle origini al transgenico in Lombardia nel centenario della riscoperta della genetica di Mendel.* Franco Angeli, Milano, pp. 33–84.

Castiglioni, E. and M. Rottoli (2010) Il sorgo (*Sorghum bicolor*) nel medioevo in Italia settentrionale. *Archeologia Medievale* 37, 485–496.

Castiglioni, E., Cottini, M. and Rottoli, M. (1999) I resti botanici di Santa Giulia a Brescia. In G. P. Brogiolo (ed.) *S. Giulia di Brescia, gli scavi dal 1980 al 1992. Reperti preromani, romani e altomedievali.* All'Insegna del Giglio, Firenze, pp 401–424.

Castiglioni, E., Cottini, M. and Rottoli, M. (2007) Ambiente e alimentazione dal I millennio a.C. al Medioevo, In M. Fortunati and R. Poggiani Keller (eds.) *I primi millenni. Dalla Preistoria al Medioevo, I. Storia economica e sociale di*

Bergamo. Fondazione per la Storia Economica e Sociale di Bergamo-Istituto di Studi e Ricerche. Cenate Sotto, Bergamo, pp. 277–285.

Chávez, S. J. and Thompson, R. G. (2006) Early Maize on the Copacabana Peninsula: Implications for Archaeology of the Lake Titicaca Basin. In J. Staller, J. Tykot and B. Benz (eds.) *Histories of Maize: Multidisciplinary Approaches to the Prehistory, Linguistics, Biogeography, Domestication, and Evolution of Maize*. Elsevier Press, Oxford, pp. 415–428.

Clapperton, C. M. (1993) Quaternary Geology and Geomorphology of South America. Elsevier, Amsterdam.

Colledge, S. and Conolly, J. (2007a) The Neolithisation of the Balkans: a review of the archaeobotanical evidence. In M. Spataro and P. Biagi (eds.) *A Short Walk through the Balkans: the First Farmers of the Carpathian Basin and Adjacent Regions*. Società Preistoria Protostoria Friuli-V.G., Trieste, 12, 25–38.

Colledge, S. and Conolly, J. (eds.) (2007b) The Origins and Spread of Domestic Plants in Southwest Asia and Europe. Left Coast press, Walnut Creek.

Cook, O. F. (1925) Peru as a Center of Domestication: Tracing the Origin of Civilization Through the Domesticated Plants. *The Journal of Heredity* 16 (2–3), 33–46 and 95–110.

Costantini, L. (2002) Italia centro-meridionale. In G. Forni and A. Marcone (eds.) *Storia dell'agricoltura italiana. I: L'età antica. 1: Prestoria*, Accademia dei Georgofili, Edizioni Polistampa, Firenze, pp. 221–234.

Coward, F., Shennan, S., Colledge, S., Conolly, J. and Collard, M. (2008) The spread of Neolithic plant economies from the Near East to northwest Europe: a phylogenetic analysis. *Journal of Archaeological Science* 35, 42–56.

Cummings, L. S. (1994) Chapter 9. Anasazi Diet: Variety in the Hoy House and Lion House Coprolite Record and Nutritional Analysis. In K. D. Sobolik (ed.) *Paleonutrition. The Diet and Health of Prehistoric Americans*. Occasional Paper Series 22. Southern Illinois University Press, Carbondale, pp. 134–150.

Cummings, L. S. (1998) Sampling Prehistoric Structures for Pollen and Starch Granules. In V. M. Bryant and J. W. Wrenn. (eds.) *New Developments in Palynomorph Sampling, Extraction, and Analysis*. American Association of Stratigraphic Palynologists Foundation, Contributions Series Number 33. Milne, Los Angeles, pp. 35–51.

Curdy, P., Paccolat, O. and Wick, L. (2009) Die ersten Weinbauern im Wallis. *Archäologie Schweiz* 32, 2–19.

Cutler, H. C. (1946) Races of maize in South America. *Botanical Museum Leaflets, Harvard University* 12, 257–291.

Cutler, H. C. and Cárdenas, M. (1947) *Chicha: A Native South American Beer.* Botanical Museum of Harvard University, Cambridge.

D'Andrea, C. (2003) Social and Technological Aspects of Non-Mechanised Emmer Processing. In P. C. Anderson, L. S. Cummings, T. S. Schippers and B. Simonel (eds.) *Le traitement des récoltes: un regard sur la diversité, du Néolithique au présent*. Actes des XXIIIe rencontres Internationales d'archéologie et d'histoire d'Antibes, 17–19 octobre 2002. Éditions APDCA. Antibes,pp. 47–60.

D'Andrea, A. C. and Haile, M. (2002) Traditional emmer processing in highland Ethiopia. *Journal of Ethnobiology* 22, 179–217.

Davis, B., Brewer, S., Stevenson, A. and Guiot, J. (2003) The temperature of Europe during the Holocene reconstructed from pollen data. *Quaternary Science Reviews* 22, 1701–1716.

De Candolle, A. (1884) *Origin of Cultivated Plants*. Kegan Paul, Trench and Co, London.

Dieckmann, B., Hoffstadt, J., Maier, U. and Spatz, H. (1998) Zum Stand der Ausgrabungen auf den 'Offwiesen' in Singen, Kr. Konstanz. *Archäologische Ausgrabungen in Baden-Württemberg 1997*, 43–47.

Dörfler, W. (1989) Die Geschichte des Hanfanbaus in Mitteleuropa aufgrund palynologischer Untersuchungen und von Grossrestnachweisen. *Praehistorische Zeitschrift* 64, 218–244.

Dvorak, J., Deal, K. R., Ming-Cheng, L., You, F. M., von Borstel, K. and Denghani, H. (2012) The origin of Spelt and Free-Threshing Hexaploid Wheat. Journal of Heredity 103, 426–441.

Ebnöther, C., Monnier, J., Jacomet, S., Deschler-Erb, S., Schibler, J., Hüster-Plogmann, H. and Luginbühl, T. (2002) Ländliche Besiedlung und Landwirtschaft. In L. Flutsch, U. Niffeler, and F. Rossi (eds.) *SPM V: Römische Zeit. Die Schweiz vom Paläolithikum bis zum frühen Mittelalter* 5, Basel, pp. 136–177.

Eisentraut, P. J. (1998) *Macrobotanical Remains from Southern Peru: A Comparison of Late Archaic-Early Formative Period Sites from the Puna and Suni Zones of the Western Titicaca Basin*. Ph.D. thesis. University of California, Santa Barbara.

Emshwiller, E. (2002) Biogeography of the *Oxalis tuberosa* Alliance. *The Botanical Review* 68, 128–152.

Emshwiller, E. (2006) Origins of Polyploid Crops: The Example of the Octoploid Tuber Crop *Oxalis tuberosa*. In M. A. Zeder, D. G. Bradley, E. Emshwiller and B. D. Smith (eds.) *Documenting Domestication: New Genetic and Archaeological Paradigms* University of California Press, Berkeley, pp. 153–168.

Erickson, C. L. (1976) *Chiripa Ethnobotanical Report: Flotation Recovered Archaeological Remains from an Early Settled Village on the Altiplano of Bolivia*. Ph.D. thesis. Washington University, St. Louis.

Ernst, M. and Jacomet, S. (2006) The value of the archaeobotanical analysis of desiccated plant remains from old buildings: methodological aspects and interpretation of crop weed assemblages. *Vegetation History and Archaeobotany* 15, 45–56.

Estremera, M. S. (2003) *Primeros agricultores y ganaderos*

en la Meseta Norte. El neolítico de la Cueva de La Vaquera (Torreiglesias, Segovia). Memorias. Arqueología en Castilla y León 11, Junta de Castilla y León, Zamora.

Evans, J. D. (1964) Excavations at the neolithic settlement of Knossos 1957–60, Part 1. *The Annual of the British School at Athens* 59, 132–240.

Ferrari, V. (1988) Vegetazione e flora nell'ecosistema medievale (secoli VIII–XV). In B. R. Ferrari and R. Groppali (eds.) *Natura e ambiente nella provincia di Cremona dall'VIII al XIX secolo*, Cremona, pp 9–55.

Filipovitch, L., Lazarova, M., Stefanova, I. and Petrova, M. (1998) Development of vegetation in Mt. Sredna Gora during the Holocene. *Phytologica Balcanica* 4, 13–29.

Fiorentino, G., Castiglioni, E., Rottoli, M. and Nisbet, R. (2004) Le colture agricole in Italia nel corso dell'età del Bronzo: sintesi dei dati e linee di tendenza. In D. Cocchi Genick (ed.) *L'età del Bronzo Recente in Italia*. Atti del Congresso Nazionale, Lido di Camaiore, 26–29 ottobre 2000. Baroni editore, Viareggio (Lucca), pp. 219–226.

Fischer, E. and Rösch, M. (2004) 8. Archäobotanische Untersuchungen. In W. Schier and F. Draşovean (eds.) Vorbericht über die rumänisch-deutschen Prospektionen und Ausgrabungen in der befestigten Tellsiedlung von Uivar, jud. Timiş, Rumänien (1998–2002). *Praehistorische Zeitschrift* 79 (2), 209–220.

Fugazzola Delpino, M. A. (1998) La vita quotidiana del neolitico. Il sito della Marmotta sul Lago di Bracciano. In A. Pessina and G. Muscio (eds.) *Settemila anni fa il primo pane, ambienti e culture delle società neolitiche*. Catalogo della Mostra, Udine dicembre 1998–maggio 1999. Udine, pp. 185–191.

Gandarillas, H. (1979) Genética y origen. In M. Tapia, H. Gandarillas, S. Alandia, A. Cardozo, A. Mujica, R. Ortiz, V. Otazu, J. Rea and B. Salas (eds.) *Quinua y Kañihua, Cultivos Andinos* IICA, Bogota, pp. 45–64.

Goldstein, P. S. (2003) From Stew-Eaters to Maize-Drinkers. The Chicha Economy and the Tiwanaku Expansion. In T. L. Bray (ed.) *The Archaeology and Politics of Food and Feasting in Early States and Empires*. Kluwer Academic/Plenum Publishers, New York, pp. 143–172.

Görsdorf, J. and Bojadziev, J. (1996) Zur absoluten Chronologie der bulgarischen Urgeschichte. Berliner ^{14}C-Datierungen von bulgarischen archäologischen Fundplätzen. *Eurasia Antiqua* 2, 105–173.

Gronenborn, D. (2006) Letzte Jäger – erste Bauern. *Archäologie in Deutschland* 3, 18–23.

Guilaine, J. (2003) De la vague à la tombe, la conquète néolithique de la méditerranée. Paris.

Gyulai, F. (1993) Environment and agriculture in Bronze Age Hungary. *Archaeolingua Series Minor 4*. Hungarian Academy of Sciences, Budapest, pp. 7–59.

Hafner, A. and Suter, P. J. (2004) *Abgetaucht – Aufgetaucht. 1984–2004*. Archäologischer Dienst des Kantons Bern, Bern.

Hajnalová, E. (1980) Paleoethnobotanical findings from the multi-layer Nova Zagora settlement. *Studia Praehistorica* 4, 91–98. [in Russian with English summary]

Halstead, P. (1981) Counting sheep in Neolithic and Bronze Age Greece. In I. Hodder, G. Isaac and N. Hammond (eds.), *Patterns of the past: studies in honour of David Clarke*. Cambridge University Press, Cambrige, pp. 307–339.

Halstead, P. (1989) Like Rising Damp? An Ecological Approach to the Spread of Farming in South East and Central Europe. In A. Milles, D. Williams and N. Gardner (eds.), *The Beginnings of Agriculture*. BAR International Series 496. Archaeopress, Oxford, pp. 23–53.

Halstead, P. (1990) Waste not, Want not: Traditional Responses to Crop Failure in Greece. *Rural History* 1, 147–164.

Halstead, P. (1994) The north-south divide: Regional paths to complexity in prehistoric Greece. In C. Mathers and S. Stoddart (eds.), *Development and Decline in the Mediterranean Bronze Age*. Sheffield Archaeological Monographs 8. J. R. Collis Publications, Sheffield, pp. 195–219.

Hammer, K., Knüppfer, H., Laghetti, G. and Perrino, P. (1999) *Seeds from the Past. A Catalogue of Crop Germplasm in Central and North Italy*. CNR, Bari.

Hancock, J. F. (2004) *Plant evolution and the origin of crop species*. CABI Publishing, Cambridge (MA).

Hanelt, P. (1961) Zur Kenntnis von *Carthamus tinctorius* L. *Die Kulturpflanze* 9, 114–145.

Hansen, J. (2000) Palaeoethnobotany and palaeodiet in the Aegean region: notes on legume toxicity and related pathologies. In S. J. Vaughn and W. D. E. Coulson (eds.) *Palaeodiet in the Aegean*. Oxbow Books, Oxford, pp. 13–27.

Harris, D. R. (1990) Vavilov's Concept of Centres of Origin of Cultivated Plants: Its genesis and its Influence on the Study of Agricultural Origins. *Biological Journal of the Linnean Society* 39, 7–16.

Hastorf, C. A. (2008) The Formative Period in the Titicaca Basin. In H. Silverman and W. H. Isbell (eds.) *The Handbook of South American Archaeology*. Springer, New York, pp. 545–561.

Hastorf, C. A. (1990) The effect of the Inka state on Sausa agricultural production and crop consumption. *American Antiquity* 55 (2), 262–290.

Hastorf, C. A. and S. Johannessen (1993) Pre-Hispanic Political Change and the Role of Maize in the Central Andes of Peru. *American Anthropologist* 95, 115–138.

Hastorf, C. A. and Wright, M. F. (1998) Interpreting Wild Seeds From Archaeological Sites: A Dung Charring Experiment From the Andes. *Journal of Ethnobiology* 18 (2), 211–227.

Hastorf, C. A., Whitehead, T. W., Bruno, M. C. and Wright, M. (2006) The Movements of Maize into Middle Horizon Tiwanaku, Bolivia. In J. Staller, J. Tykot and B. Benz (ed.) *Histories of Maize: Multidisciplinary*

Approaches to the Prehistory, Linguistics, Biogeography, Domestication, and Evolution of Maize. Elsevier Press, Oxford, pp. 429–448.

Hawkes, J. G. (1983) *The Diversity of Crop Plants*. Harvard University Press, Cambridge (MA).

Hawkes, J. G. (1990) *The Potato: Evolution, Biodiversity, and Genetic Resources*. Belhaven Press, London.

Hayden, B. (1996) Feasting in Prehistoric and Traditional Societies. In P. Wiessner and W. Schietenhovel, (eds.) *Food and the Status Quest: An interdisciplinary perspective*. Berghahn Books, Oxford, pp. 127–147.

Hayden, B. (2003) Were luxury foods the first domesticates? Ethnoarchaeological perspectives from Southeast Asia. *World Archaeology* 34 (3), 458–469

Henning, J. (1991) Fortleben und Weiterentwicklung Spätrömanischen Agrargerätetraditionen in Nordgallien. Eine Mähsense der Merowingerzeit aus Kerkhoven (Belgien). *Acta Archaeologica Lovaniensa* 30, 49–59.

Hillman, G. C. (1984) Traditional husbandry and processing of archaic cereals in recent times: the operations, products and equipment which might feature in Sumerian texts. Part I: the glume wheats. *Bulletin of Sumerian Agriculture* 1, 114–152.

Hopf, M. (1978) Frühe Kulturpflanzen aus Südeuropa. *Berichte der Deutschen Botanischen Gesellschaft*, 91, 31–38.

Hopf, M. (1982) *Vor- und frühgeschichtliche Kulturpflanzen aus dem nördlichen Deutschland*. Kataloge vor- und frühgeschichtlicher Altertümer 22. Mainz.

Hörandner, E. (ed.) (1995) *Millet-Hirse-Millet*. Peter Lang, Bern.

Hubbard, R. and Housley, R. (2000) The agriculture in prehistoric Servia. In C. Ridley, K. Wardle and C. A. Mould (eds.) *Servia I: Anglo-Hellenic rescue excavations 1971–73*. British School at Athens, Supplementary Volume 32. Oxbow Books, Oxford, pp. 330–336.

Hunt, H., Vander Linden, M., Liu, X., Motuzaite-Matuzeviciute, G., Colledge, S. and Jones, M. K. (2008) Millets across Eurasia: chronology and context of early records of the genera *Panicum* and *Setaria* from archaeological sites in the Old World. *Vegetation History and Archaeobotany* 17 (Supplement 1), 5–18.

Iriarte, J. (2007) New Perspectives on Plant Domestication and the Development of Agriculture in the New World. In T. Denham, J. Iriarte and L. Vrydaghs (eds.) *Rethinking Agriculture: Archaeological and Ethnoarchaeological Perspective*. Left Coast Press, Walnut Creek, pp. 167–188.

Jacomet, S. (2004) Archaeobotany. A vital tool in the investigation of lake-dwellings. In F. Menotti (ed.) *Living on the lake in prehistoric Europe. 150 years of lake-dwelling research*. London, pp. 162–177.

Jacomet, S. (2006) Plant Economy of the Northern Alpine Lake Dwelling area – 3500–2400 cal. BC. *Environmental Archaeology* 11, 64–83.

Jacomet, S. (2007a) Neolithic plant economies in the northern Alpine Foreland from 5500–3500 cal. BC. In S. Colledge and J. Conolly (eds.) *The Origins and Spread of Domestic Plants in Southwest Asia and Europe*. Left Coast Press, Walnut Creek, pp. 221–258.

Jacomet, S. (2007b) Plant Macrofossil Methods and Studies. Use in Environmental Archaeology. In S. A. Elias (ed.) *Encyclopedia of Quaternary Science 3*. Oxford, pp. 2384–2412.

Jacomet, S. (2008) Subsistenz und Landnutzung während des 3. Jahrtausends v. Chr. aufgrund von archäobotanischen Daten aus dem südwestlichen Mitteleuropa. In W. Dörfler and J. Müller (eds.) *Umwelt - Wirtschaft - Siedlungen im dritten vorchristlichen Jahrtausend Mitteleuropas und Südskandinaviens*. Offa-Bücher 84. Neumünster, pp. 355–377.

Jacomet, S. (2009) Plant economies and village life in Neolithic lake dwellings at the time of the Alpine Iceman. *Vegetation History and Archaeobotany* 18, 47–59.

Jacomet, S. and Behre, K.-E. (2009) Der mittelbronzezeitliche Getreidefund. In A. N. Lanzrein (ed.) *Die befestigte Höhensiedlung Toos-Waldi von der Frühbronzezeit bis in die Spätantike*. Archäologie im Thurgau 15. Frauenfeld, pp. 71–79.

Jacomet, S. and Brombacher, C. (2009) Geschichte der Flora in der Regio Basiliensis seit 7500 Jahren: Ergebnisse von Untersuchungen pflanzlicher Makroreste aus archäologischen Ausgrabungen. *Mitteilungen der Naturforschenden Gesellschaften beider Basel* 11, 27–106.

Jacomet, S. and Karg, S. (1996) Ackerbau und Umwelt der Seeufersiedlungen von Zug-Sumpf im Rahmen der mitteleuropäischen Spätbronzezeit. Ergebnisse archäobotanischer Untersuchungen. In Regierungsrat des Kantons Zug (ed.) *Die spätbronzezeitlichen Ufersiedlungen von Zug-Sumpf, Band 1: Die Dorfgeschichte*. Zug, pp. 198–303 and 365–368.

Jacomet, S. and Kreuz, A. (1999) Archäobotanik. Aufgaben, Methoden und Ergebnisse vegetations- und agrargeschichtlicher Forschungen. Ulmer Verlag, Stuttgart.

Jacomet, S. and Oeggl, K. (eds.) (2009) Special Issue: Palaeoethnobotany at the time of the Tyrolean Iceman. *Vegetation History and Archaeoabotany* 18, 1–103

Jacomet, S. and Schlichtherle, H. (1984) Der kleine Pfahlbauweizen Oswald Heer's – Neue Untersuchungen zur Morphologie neolithischer Nacktweizen-Ähren. In W. A. van Zeist and W. A. Casparie (eds.) *Plants and Ancient Man. Proceedings of the 6th symposium of the international work group for palaeoethnobotany 1983 in Groningen*. Balkema, Rotterdam, pp. 153–176.

Jacomet, S., Brombacher, C. and Dick, M. (1989) *Archäobotanik am Zürichsee. Ackerbau, Sammelwirtschaft und Umwelt von neolithischen und bronzezeitlichen Seeufersiedlungen im Raum Zürich. Ergebnisse von Untersuchungen pflanzlicher Makroreste der Jahre 1979–1988*. Zürcher Denkmalpflege Monographien 7. Zürich.

Jacomet, S., Brombacher, C. and Schraner, E. (1999a) Ackerbau und Sammelwirtschaft während der Bronze- und Eisenzeit in den östlichen Schweizer

Alpen – vorläufige Ergebnisse. In P. Della Casa (ed.) *Prehistoric alpine environment, society, and economy. Papers of the International Colloquium PAESE '97 in Zürich.* Universitätsforschungen zur prähistorischen Archäologie 55. Zürich, pp. 231–244.

Jacomet, S., Jaquat, C., Winter, M. and Wick, L. (1999b) Umwelt, Ackerbau und Sammelwirtschaft. In F. Müller, G. Kaenel and G. Lüscher (eds.) *SPM IV, Eisenzeit. Die Schweiz vom Paläolithikum bis zum frühen Mittelalter 4.* Basel, pp. 98–115.

Jacomet, S., Rachoud-Schneider, A.-M., Zoller, H. and Burga, C. A. (1998) Vegetationsentwicklung, Vegetationsveränderung durch menschlichen Einfluss, Ackerbau und Sammelwirtschaft. In S. Hochuli, U. Niffeler and V. Rychner (eds.) *SPM III: Bronzezeit. Die Schweiz vom Paläolithikum bis zum frühen Mittelalter 3.* Basel, pp. 141–170.

Jacomet, S., Schibler, J., Maise, C., Wick, L. and Deschler-Erb, S. (2002) Mensch und Umwelt. In L. Flutsch, U. Niffeler and F. Rossi (eds.) *SPM V: Römische Zeit. Die Schweiz vom Paläolithikum bis zum frühen Mittelalter 5.* Basel, pp. 21–40.

Janusek, J. W. (2002) Out of Many, One: Style and Social Boundaries in Tiwanaku. *Latin American Antiquity* 13 (35–61):

Janusek, J. W. (2003) Vessels, Time, and Society: Toward a Ceramic Chronology in the Tiwanaku Heartland. In A. L. Kolata (ed.) *Tiwanaku and its Hinterland: Archaeological and Paleoecological Investigations of an Andean Civilization 2, Urban and Rural Archaeology.* Smithsonian Institution Press, Washington D.C., pp. 30–89.

Janusek, J. W. (2004a) Tiwanaku and Its Precursors: Recent Research and Emerging Perspectives. *Journal of Archaeological Research* 12, 121–183.

Janusek, J. W. (2004b) *Identity and Power in the Ancient Andes.* Routledge, London.

Jeraj, M., Velušček, A. and Jacomet, S. (2009) The diet of Eneolithic (Copper Age, fourth millennium cal B.C.) pile dwellers and the early formation of the cultural landscape south of the Alps: A case study from Slovenia. *Vegetation History and Archaeobotany* 18, 75–89.

Johns, T. and Keen, S. L. (1986) Ongoing Evolution of the Potato on the Altiplano of Western Bolivia. *Economic Botany* 40, 409–424.

Jones, G. and Valamoti, S. M. (2005) *Lallemantia*, an imported or introduced oil plant in Bronze Age northern Greece. *Vegetation History and Archaeobotany* 14, 571–577.

Jones, G., Valamoti, S. M. and Charles, M. (2000) Early crop diversity: a 'new' glume wheat from northern Greece. *Vegetation History and Archaeobotany* 9, 133–146.

Jones, G., Wardle, K., Halstead, P. and Wardle, D. (1986) Crop Storage at Assiros. *Scientific American* 254, 96–103.

Karathanou, A. and Valamoti, S. M. (2011) Αρχαιοβοτανικά κατάλοιπα από την Κρεμαστή Κοζάνης (Plant remains from Kremasti Kozanis). AEMTH 2008.

Karathanou, A. and Valamoti, S. M. (2011) Η αρχαιοβοτανική έρευνα στη νεολιθική θέση Κρεμαστή Κοιλάδας (Archaeobotanical investigations at the Neolithic settlement Kremasti Koiladas). AEMTH 22, 57–70.

Karg, S. (1996) *Ernährung und Agrarwirtschaft in der spätmittelalterlichen Stadt Laufen (Schweiz). Paläoethnobotanische Funde aus der Holzhäuserzeile am Rathausplatz.* Dissertationes Botanicae 262. Gebrüder Borntraeger, Stuttgart.

Kelly, F. (1998) *Early Irish Farming.* Dublin Institute for Advanced Studies, Dublin.

Kimmig, W. (ed.) (2000) *Importe und mediterrane Einflüsse auf der Heuneburg.* Römisch-Germanische Forschungen 59. Philipp von Zabern, Mainz.

Kislev, M. E. (1989) Origins of the cultivation of *Lathyrus sativus* and *L. cicera* (Fabaceae) *Economic Botany* 43, 262–270.

Kislev, M. E., Hartmann, A. and Galili, E. (2004) Archaeobotanical and archaeoentomological evidence from a well at Atlit-Yam indicates colder, more humid climate on the Israeli coast during the PPNC period. *Journal of Archaeological Science* 31, 1301–1310.

Knörzer, K. H. (1970) *Novaseum IV. Römerzeitliche Pflanzenfunde aus Neuss.* Limesforschungen X. Gebrüder Mann Verlag, Berlin.

Knörzer, K. H. (1978) Entwicklung und Ausbreitung des Leindotters (*Camelina sativa* s.l.). *Berichte der Deutschen Botanischen Gesellschaft* 91, 187–195.

Knörzer, K. H. (1979) Pflanzliche Großreste des latènezeitlichen Siedlungsplatzes Grevenbroich-Gustorf, Kreis Neuss. *Rheinische Ausgrabungen* 19, 601–610.

Knörzer, K. H. (1980) Subfossile Pflanzenreste aus der jüngerlatènezeitlichen Siedlung bei Laurenzberg, Gem. Eschweiler, Kr. Aachen. *Bonner Jahrbücher* 180, 443–457.

Knörzer, K. H. (1998) Botanische Untersuchungen am bandkeramischen Brunnen von Erkelenz-Kückhoven. In Koschik, H. (ed.) *Brunnen der Jungsteinzeit. Internationales Symposium Erkelenz 27. bis 29. Oktober 1997.* Materialien zur Bodendenkmalpflege im Rheinland 11. Rheinland-Verlag, Köln/Bonn, pp. 229–246.

Kohler-Schneider, M. (2007) Early agriculture and subsistence in Austria. In S. Colledge and J. Conolly (eds.) *The Origins and Spread of Domestic Plants in Southwest Asia and Europe.* Left Coast Press, Walnut Creek, pp. 209–220.

Kohler-Schneider, M. and Caneppele, A. (2009) Late Neolithic Agrculture in Eastern Austria: archaeobotanical results from sites of the Baden and Jevišovice cultures (3600–2800 BC). *Vegetation History and Archaeoabotany* 18, 61–74.

Kolata, A. L. (1993) The Tiwanaku: Portrait of an Andean Civilization. Blackwell, Oxford.

Kooistra, L. I. (1996) *Borderland Farming.* Van Gorcum, Assen.

Körber-Grohne, U. (1985) Die biologischen Reste aus dem hallstattzeitlichen Fürstengrab von Hochdorf, Gemeinde Eberdingen (Kreis Ludwigsburg). In U. Körber-Grohne and H. Küster (eds.) *Hochdorf I.* Forschungen und Berichte zur Vor- und Frühgeschichte in Baden-Württemberg 19. Theiss, Stuttgart, pp. 87–164.

Kostakis, K. (1996) Ανταλλαγές και σχέσεις (Exchange and relationships). In C. Papathanasopoulos (ed.) *Νεολιθικός Πολιτισμός στην Ελλάδα (Neolithic Civilisation in Greece).* Goulandris Foundation, Athens, pp. 168–170.

Kreuz, A. (1990) Die ersten Bauern Mitteleuropas – eine archäobotanische Untersuchung zu Umwelt und Wirtschaft der ältesten Bandkeramik. *Analecta Praehistorica Leidensia* 23, 1–256.

Kreuz A. (1999) Becoming a Roman farmer: preliminary report on the environmental evidence from the Romanization project. *Journal of Roman Archaeology* Supplementary series number 32, 71–98.

Kreuz, A. (2003) Unerwartete Pflanzenfunde aus der keltischen Saline in Bad Nauheim. *Hessen Archäologie 2002,* 66–68.

Kreuz, A. (2004) Landwirtschaft im Umbruch? Archäobotanische Untersuchungen zu den Jahrhunderten um Christi Geburt in Hessen und Mainfranken. *Bericht der Römisch-Germanischen Kommission* 85, 97–292 and 9 tables.

Kreuz, A. (2007) Archaeobotanical perspectives on the beginning of agriculture north of the Alps. In S. Colledge and J. Conolly (eds.) *The Origins and Spread of Domestic Plants in Southwest Asia and Europe.* Left Coast Press, Walnut Creek, pp. 259–294.

Kreuz, A. (2012) Die Vertreibung aus dem Paradies? Archäobiologische Ergebnisse zum Frühneolithikum im westlichen Mitteleuropa. *Berichte der Römisch-Germanischen Kommission* 91 (2010), 23–196.

Kreuz, A. and Schäfer, E. (2002) A new archaeobotanical database program. *Vegetation History and Archaeobotany* 11, 177–179.

Kreuz, A., Marinova, E., Schäfen, E. and Wiethold, J. (2005) A comparison of early Neolithic crop and weed assemblages from the Linearbandkeramik and the Bulgarian Neolithic cultures: differences and similarities. *Vegetation History Archaeobotany* 14, 237–258.

Kristiansen, K. and Larsson, T. (2005). The rise of Bronze Age society: travels, transmissions and transformations. Cambridge University Press, New York.

Kroll, H. (1981) Thessalische Kulturpflanzen. *Zeitschrift für Archäologie* 15, 97–103

Kroll, H. (1983) *Kastanas. Ausgrabungen in einem Siedlungshuegel der Broze- und Eisenzeit Makedoniens 1975-1979. Bd. 2. Die Pflanzenfunde.* Praehistorische Archaeologie in Suedosteuropa 2. Volker Speiss, Berlin

Kroll, H. (1990) Ein Fruchtfund von *Carthamus tinctorius* belegt diese Färbepflanze für die Bronzezeit Jugoslaviens. *Archaeologischer Korrespomdenzblatt* 20, 41–46.

Kühn, M. (1996) *Spätmittelalterliche Getreidefunde aus einer Brandschicht des Basler Rosshof-Areales (15. Jahrhundert A.D.).* Materialhefte zur Archäologie in Basel 11. Archäolog. Bodenforschung Basel-Stadt, Basel.

Kühn, M. and Iseli, M. (2008) Botanische Makroreste aus der spätlatènezeitlichen Siedlung Basel-Gasfabrik, Grabung 1989/5. In P. Jud (ed.) *Die Töpferin und der Schmied. Basel-Gasfabrik, Grabung 1989/5.* Materialhefte zur Archäologie in Basel 20. Basel, pp. 293–324.

Kuznar, L. A. (1993) Mutualism between *Chenopodium,* Herd Animals, and Herders in the South Central Andes. *Mountain Research and Development* 13, 257–265.

La Barre, W. (1938) Native American Beers. *American Anthropologist* 40, 224–234.

La Barre, W. (1947) Potato Taxonomy Among the Aymara Indians of Bolivia. *Acta Americana* 5: 83–103.

Lazarova, M. (2003) The Holocene vegetation history of the Central Rhodopes Mountains, Southern Bulgaria. In S. Tonkov (ed.) *Aspects of Palynology and Palaeoecology: Festschrift in Honour of Elissaveta Bozilova.* Pensoft Publishers, Sofia/Moscow, pp. 245–256

Lee, M. (1997) Paleoethnobotanical Report of Five Yaya-Mama Sites from the Lake Titicaca Basin. Ph.D. thesis. University of Missouri.

León, J. (1967) Andean Tuber and Root Crops: Origin and Variability. In E. A. Tai, W. B. Charles, E. F. Iton, P. H. Haynes and K. A. Leslie (ed.) *Proceedings of the International Symposium on Tropical Root Crops held at University of the West Indies.* University of the West Indies, St. Augustine, Trinidad, pp. 76–88

Lichardus-Itten, M., Demoule, J.-P., Perničeva, L. and Grebska-Kulova, M. (2006) Kovačevo, an early Neolithic site in South-West Bulgaria and its importance for the European Neolithization. In I. Gatsov and H. Schwarzberg (eds.) *Aegean-Marmara-Black Sea: the present state of research on the early Neolithic.* Schriften des Zentrums für Archäologie und Kulturgeschichte des Schwarzmeerraumes 5. Beier and Beran, Weißbach, pp. 83–94.

Livarda, A. and Van der Veen, M. (2008) Social access and dispersal of condiments in North-West Europe from the Roman to the Medieval period. *Vegetation History and Archaeoabotany* 17, 201–209.

Logan, A. L. (2006) *The Application of Phytolith and Starch Grain Analysis to Understanding Formative Period Subsistence, Ritual, and Trade on the Taraco Peninsula, Highland Bolivia.* Master's thesis. University of Missouri.

Logan, A. L., Hastorf, C. A. and Pearsall, D. M. (2012) 'Let's Drink Together': Early Ceremonial Use of Maize in the Titicaca Basin. Latin American Antiquity 23 (3), 235–258.

Lu, H., Zhang, J., Liu, K., Wu, N., Li, Y., Zhou, K., Ye, M., Zhang, T., Zhang, H., Yang, X., Shen, L. and Li, Q. (2009) Earliest domestication of common millet (*Panicum miliaceum*) in East Asia extended to 10'000 years ago. *PNAS* 106, 7367–7372.

Lüning, J. (2000) *Steinzeitliche Bauern in Deutschland. Die Landwirtschaft im Neolithikum.* Universitätsforschungen zur Prähistorischen Archäologie 58. Habelt, Bonn.

Mabberley, D. J. (1990) *The plant-book. A portable dictionary of higher plants.* Cambridge University Press, Cambridge.

Maier, U. (1996) Morphological studies of free-threshing wheat ears from a Neolithic site in southwest Germany, and the history of naked wheats. *Vegetation History and Archaeobotany* 5, 39–55.

Maier, U. (1998) Der Nacktweizen aus den neolithischen Ufersiedlungen des nördlichen Alpenvorlandes und seine Bedeutung für unser Bild von der Neolithisierung Mitteleuropas. *Archäologisches Korrespondenzblatt* 2, 205–218.

Marinova, E. (forthcoming) Archaeobotanical Studies. In Nikolov, V. (ed.) *Kapitan Dimitrievo - the excavations of 2001-2004.* BAN, Sofia (in Bulgarian with English summary)

Marinova, E. (2003) Paleoethnobotanical study of Early Bronze Age II in the Upper Styama Valley (Dubene-Sarovka) IIB. In L. Nikolova (ed.) *Early Communication Systems in South Eastern Europe.* BAR International Series 1139 (2), 499–504.

Marinova, E. (2004) Archäobotanische Ergebnisse aus der Bronzezeit von Tell Karanovo und ihr regionaler Kontext. *Diomedes* 3, 53–58.

Marinova, E. (2006) Vergleichende paläoethnobotanische Untersuchung zur Vegetationsgeschichte und zur Entwicklung der prähistorischen Landnutzung in Bulgarien. Dissertationes Botanicae 401. Gebrüder Borntraeger, Stuttgart.

Marinova, E. (2007) Archaeobotanical data from the early Neolithic of Bulgaria. In S. Colledge and J. Conolly (eds.) *The origins and spread of domestic plants in Southwest Asia and Europe.* Left Coast Press, Walnut Creek, pp. 93–109.

Marinova, E. (2008) Archaeobotanical studies (contributions to Chapter 5, 6, 7, 8 and 9). In V. Nikolov (ed.) *Provadia - Solnitsata Praehistoric Salt Producing Center. The 2005-2007 Excavation Seasons.* National Archaeological Institute with Museum, BAS, Sofia (in Bulgarian with English summary) pp. 79–81, 173–174, 203, 225, 273–274

Marinova, E. (2009) Archaeobotanical materials from the Neolithic site Ohoden-Valoga: Structure 1 and Grave 1. In G. Ganetsovski (ed.) *Ohoden: a site from the Early Neolithic, excavations 2002-2006*, Craft Hous Publisher, Sofia, pp. 76–80.

Marinova, E. and Atanassova, J. (2006) Anthropogenic impact on vegetation and environment during the Bronze Age in the area of Lake Durankulak, NE Bulgaria: pollen, microscopic charcoal, non-pollen palynomorphs and plant macrofossils. *Review of Paleobotany and Palynology* 141, 165–178.

Marinova, E. in Popov, H. and Iliev, S. (2007) Settlement from the Late Bronze Age and Early Iron Age in the area of Kush Kaja, village Valche Pole, municipality Ljubimetz. *Archarological findings and excavations*, 2006: 199–201 [in Bulgarian]

Marinova, E. and Popova, Tz. (2008) *Cicer arietinum* (chick pea) in the Neolithic and Chalcolithic of Bulgaria: implications for cultural contacts with the neighbouring regions? *Vegetation History and Archaeobotany* 17 (Supplement 1), 73–80

Marinova, E. and Riehl, S. (2009) *Carthamus* species in the ancient Near East and southeastern Europe: archaeobotanical evidence for their distribution and use as a source of oil. *Vegetation History and Archaeobotany* 18, 341–349.

Marinova, E., Tchakalova, E., Stoyanova, D., Grozeva, S. and Dočeva, E. (2002) Ergebnisse archäobotanischer Untersuchungen aus dem Neolithikum und Chalkolithikum in Südwestbulgarien. *Archaeologia Bulgarica* 6 (3): 1–11.

Martin, L. (2010) *Agriculture et alimentation végétale en milieu montagnard durant le Néolithique: Nouvelles donnés carpologiques dans les Alpes françaises du nord.* Ph.D. thesis. Université Paris I (Panthéon-Sorbonne) and Basel University.

Maselli Scotti, F. and Rottoli M. (2007) Indagini archeobotaniche all'ex Essiccatoio Nord di Aquileia: i resti vegetali protostorici e romani. In G.Cuscito and C. Zaccaria (a cura di) *Aquileia dalle origini alla costituzione del ducato longobardo. Territorio - Economia - Società.* Antichità Altoadriatiche LXV. Centro di Antichità Altoadriatiche, Casa Bertoli Aquileia. Editreg, Trieste, pp 783–816.

Matsuoka, Y., Vigouroux, Y., Goodman, M. M., Sanchez G., J., Buckler, E. and Doebly, J. (2002) A single domestication for maize shown by multilocus microsatellite genotyping. *Proceedings of the National Academy of Science of the United States of America (PNAS)* 99 (9), 6080–6084.

Matterne, V. (2001) *Agriculture et alimentation végétale durant l'âge du Fer et l'époque gallo-romaine en France septentrionale.* Éditions Monique Mergoil, Montagnac.

Mauvilly, M., Jeunesse, C. and Doppler, T. (2008) Ein Tonstempel aus der spätmesolithischen Fundstelle von Arconciel/La Souche (Kanton Freiburg, Schweiz). *Quartär* 55, 151–157.

Mazurié de Keroualin, K. (2003) *Genèse et diffusion de l'agriculture en Europe.* Agriculteurs, Chasseurs, Pasteurs. Errance, Paris.

Megaloudi, F. (2006) *Plants and Diet in Greece from Neolithic to Classic Periods: the archaeobotanical remains.* Archaeopress, Oxford.

Menotti, F. (ed.) (2004) *Living on the lake in prehistoric Europe. 150 years of lake-dwelling research.* Routledge, London.

Mercuri, A. M., Accorsi, C. A. and Bandini Mazzanti, M.

(2002) The long history of *Cannabis* and its cultivation by the Romans in central Italy, shown by pollen records from Lago Albano and Lago di Nemi. *Vegetation History and Archaeobotany* 11, 263–276.

Mercuri, A. M., Accorsi, C. A., Bandini Mazzanti, M, Bosi, G., Cardarelli, A., Labate, D., Marchesini, M. and Trevisan Grandi, G. (2006) Economy and environment of Bronze Age settlements - Terramaras - on the Po Plain (Northern Italy): first results from the archaeobotanical research at the Terramara di Montale. *Vegetation History and Archaeobotany* 16, 43–60.

Meylan Krause, M.-F., Jacomet, S. and Schibler, J. (2002) Essen und Trinken. In L. Flutsch, U. Niffeler and F. Rossi (eds.) *SPM V: Römische Zeit. Die Schweiz vom Paläolithikum bis zum frühen Mittelalter* 5. Basel, pp. 231–242.

Miller, M. J., Bruno, M. C., Moore, K. M., Hastorf, C. A., Logan, A. L., Capriles Flores, J., Steadman, L., Roddick, A. P. and Bandy, M. S. (2008) *Food Traditions and Cooking Practices in the Formative Settlements.* Meeting of the Society for American Archaeology, Vancouver, Canada.

Miller, N. F. and Smart, T. L. (1984) Intentional Burning of Dung as Fuel: A Mechanism For the Incorporation of Charred Seeds into the Archaeological Record. *Journal of Ethnobiology* 4 (1), 15–28.

Monah, F. (2007) The spread of cultivated plants in the region between the Carpatians and Dniester, 6th–4th millenia cal. BC. In S. Colledge and J. Conolly (eds.) *The Origins and Spread of Domestic Plants in Southwest Asia and Europe.* Left Coast Press, Walnut Creek, pp. 111–123.

Monah, F. and Monah, D. (2008) *Cercetari archaeobotanice in tell-ul chalcolithic Poduri-Dealul Ghindaru* [Archaeobotanical Research at the Chalcolithic Tell Poduri-Dealul Ghindaru]. Constantin Matasa, Piatra-Neamt (in Romanian with English summary)

Montanari, M. (1979) *L'alimentazione contadina nell'alto Medioevo.* Liguori editore, Napoli.

Montes de Oca, I. (1995) Geografía y Clima de Bolivia. *Bulletin de l'Institut Français d'Études Andines* 24, 357–368.

Moseley, M. E. (2001) The Incas and Their Ancestors: The Archaeology of Peru. Thames & Hudson, London.

Mottes, E., Petrucci, G., Rottoli, M. and Visentini, P. (2010) Evolution of the Square Mouthed Pottery Culture in Trentino-Alto Adige, Veneto and Friuli: cultural, chronological, palaeoeconomic and environmental aspect. *Gortania* 31, 97–124.

Müller, J. (1997) Neolithische und chalkolithische Spondylus-Artefakte. Anmerkungen zu Verbreitung, Tauschgebiet und sozialer Funktion. In C. Becker, M. L. Dunkelmann, C. Metzner-Nebelsick, H. Peter-Röcher, M. Roeder and B. Teran (eds.) *Xrovos. Beiträge zur prähistorischen Archäologie zwischen Nord- und Südosteuropa. Festschrift für Bernhard Hänsel.* Verlag Marie Leidorf, Espelkamp, pp. 91–106.

Murray, A. P. (2005) *Chenopodium* Domestication in the South-Central Andes: Confirming the Presence of Domesticates at Jiskairumoko (Late Archaic-Formative), Peru. Thesis. California State University, Fullerton.

Mylonas, G. E. (1929) *Excavations at Olynthus I: The Neolithic settlement.* John Hopkins Press, Baltimore.

Nasu, H., Momohara, A., Yasuda, Y. and He, J. (2007) The occurence and identification of *Setaria italica* (L.) P. Beauv. (foxtail millet) grains from the Chengtoushan site (ca. 5800 cal B.P.) in central China, with reference to the domestication centre in Asia. *Vegetation History and Archaeobotany* 16, 481–494.

Nesbitt, M. and Samuel, D. (1996) From staple crop to extinction? The archaeology and history of the hulled wheats. In S. Padulosi, K. Hammer and J. Heller (eds.) *Hulled Wheats. Promoting the conservation and use of underutilized and neglected crops* 4. Proceedings of the First International Workshop on Hulled Wheats. IPGRI, Rome, pp. 4–41.

Netting, R. M. (1993) *Smallholder, Householders: Farm Families and the Ecology of Intensive, Sustainable Agriculture.* Stanford University Press, Stanford.

Neuweiler, E. (1935) Nachträge urgeschichtlicher Pflanzen. *Vierteljahrsschrift der Naturforschenden Gesellschaft Zürich* 80, 98–122.

Nordstrom, C. (1990) *Evidence for domestication of Chenopodium in the Andes.* Paleoethnobotany Laboratory Reports 19. University of California, Berkeley.

Oberdorfer, E. (2001) *Pflanzensoziologische Exkursionsflora für Deutschland und angrenzende Gebiete.* Ulmer Verlag, Stuttgart.

Ochoa, C. M. (1990) *The Potatoes of South America: Bolivia.* Cambridge University Press, Cambridge.

Olivier, S. (2005) Le genêt textile en Lodévois. *Histoire et Sociétés rurales* 23: 137–168.

Padulosi, S., Hammer, K. and Heller, J. (eds.) (1996) *Hulled Wheats. Promoting the conservation and use of underutilized and neglected crops* 4. Proceedings of the First International Workshop on Hulled Wheats. IPGRI, Rome

Papathanasopoulos, C. (ed.) (1996) *Νεολιθικός Πολιτισμός στην Ελλάδα (Neolithic Civilisation in Greece).* Goulandris Foundation, Athens.

Pearsall, D. M. (1992) The Origins of Plant Cultivation in South America. In W. Cowan and P. J. Watson (eds.) *The Origins of Agriculture: An Interntaional Perspective.* Smithsonian Institution Press, Washington D.C., pp. 173–205.

Pearsall, D. M. (2008) Plant Domestication and the Shift to Agriculture in the Andes. In H. Silverman and W. H. Isbell (ed.) *The Handbook of South American Archaeology.* Springer, New York, pp. 105–120.

Peña-Chocarro, L. (1996) In-situ conservation of hulled-wheat species: the case of Spain. In S. Padulosi, K. Hammer and J. Heller (eds.) *Hulled Wheats. Promoting the conservation and use of underutilized and neglected crops 4.* Proceedings of the First International Workshop on Hulled Wheats. IPGRI, Rome, pp. 129–146.

Peña-Chocarro, L. (1999) *Prehistoric agriculture in Southern Spain during the Neolithic and the Bronze Age: the application of ethnographic models*. BAR International Series 818. Archaeopress, Oxford.

Peña-Chocarro, L. (2007) Early Agriculture in southern and central Spain. In S. Colledge and J. Conolly (eds.) *The origins and spread of domestic plants in Southwest Asia and Europe*. Left Coast Press, Walnut Creek, pp. 173–187.

Peña-Chocarro, L. and Zapata, L. (1997) El *Triticum dicoccum* (ezkandia) en Navarra: de la agricultura prehistórica a la extinción de un trigo arcaico). *Zainak* 14, 249–262.

Peña-Chocarro, L. and Zapata, L. (1998) Hulled wheats in Spain: history of minor cereals. In J. Jaradat (ed.). *Triticeae III. Proceedings of the Third International Triticeae Symposium, Aleppo, Syria, 4–8 May 1997*. Science Publishers, New Hampshire, pp. 45–52.

Peña-Chocarro, L. and Zapata, L. (1999) History and traditional cultivation of *Lathyrus sativus* L. and *Lathyrus cicera* L. in the Iberian peninsula. *Vegetation History and Archaeobotany* 8, 49–52.

Peña-Chocarro, L., Zapata, L., García Gazólaz, J. González Morales, M., Sesma Sesma, J. and Straus, L. (2005a) The spread of agriculture in northern Iberia: new archaeobotanical data from el Mirón Cave (Cantabria) and the open-air site of Los Cascajos (Navarra). *Vegetation History and Archaeobotany* 14, 268–278.

Peña-Chocarro, L., Zapata, L., Iriarte, M. J., González Morales, M. and Straus, L. (2005b) The oldest agriculture in Northern Atlantic Spain. *Journal of Archaeological Science* 32, 579–587.

Peña-Chocarro, L., Zapata, L., González Urquijo, J. E. and Ibáñez Estévez, J. (2009) Einkorn (*Triticum monococcum* L.) cultivation in mountain communities of the Western Rif (Morocco): An ethnoarchaeological project. In E. Weiss and A. Fairbairn (eds.) *Ethnobotanist of distant pasts: Papers in honour of Gordon Hillman*. Oxbow Books, Oxford, pp. 103–111.

Perlès C. (2001) *The Early Neolithic in Greece. The first farming communities in Europe*. Cambridge World Archaeology. Cambridge University press, Cambridge

Pickersgill, B. (2007) Domestication of Plants in the Americas: Insights from Mendelian and Molecular Genetics. *Annals of Botany* 100, 925–940.

Piening, U. (1998) Die Pflanzenreste aus Gruben der Linearbandkeramik und der Rössener Kultur von Ditzingen, Kr. Ludwigsburg. *Fundberichte aus Baden-Württemberg* 22 (1), 125–160.

Pitte, J.-R. (1986) *Terres de castanide, Hommes et paysages du châtaignier de l'Antiquité à nos jours*. Fayard, Paris.

Planella O., Cornejo B. and Tagle B. A. (2005) Alero Las Morrenas 1: Evidencias de Cultígenos entre Cazadores Recolectores de Finales del Período Arcaico en Chile Central. *Chungará (Arica)* 37, 59–74.

Pollmann, B. and Jacomet, S. (2012) First evidence of *Mespilus germanica* L. (medlar) in Roman Switzerland. *Vegetation History and Archaeobotany* 21 (1), 61–68.

Pollmann, B., Jacomet, S. and Schlumbaum, A. (2005) Morphological and genetic studies of waterlogged *Prunus* species from the Roman vicus Tasgetium (Eschenz, Switzerland). *Journal of Archaeological Science* 32, 1471–1480.

Ponce Sangines, C. (1977) *Tiwanaku, Tiempo, y Cultura Ensayo de Sintesis Arqueologia*. 2nd edition. Editorial Los Amigos del Libro: La Paz.

Popova, T. (1995) Plant remains from Bulgarian prehistory. In D. W. Bailey and I. Panayotov (eds.) *Prehistory of Bulgaria*. Monographs in World Archaeology 19. Prehistory Press, Madison, Wisconsin, pp 193–207.

Popova, T. (2009) Paleobotanic catalogue of the studied sites and studied remains (debries) in the territory of Bulgaria (1980–2008). *Interdisciplinary Studies* 20/21, 71–165.

Popova, T. (2010) *Plant Environment of Man between 6000 and 2000 BC in Bulgaria*. BAR International Series 2064. Archaeopress, Oxford.

Ramírez, E. R., Timothy, D. H., Díaz B., É., Grant, U. J., Nicholson G. E., Anderson, E. and Brown, W. L. (1960) *Races of maize in Bolivia*. National Academy of Sciences-National research Council Publication 474. Washington D.C.

Ravazzi, C., Cremaschi, M. and Forlani, L. (2004) Studio archeopalinologico della terramara S. Rosa. Nuovi dati, analisi floristica e sintassonomica della vegetazione nell'età del Bronzo. In M. Bernabò Brea amd M. Cremaschi (eds.) *Il Villaggio Piccolo della terramara di S. Rosa di Poviglio (Scavi 1987/1992)*. I.I.P.P., Firenze, pp. 703–736.

Reddé, M., Nuber, H. U., Jacomet, S., Schibler, J., Schucany, C., Schwarz, P.-A. and Seitz, G. (2005) Oedenburg. Une agglomération d'époque romaine sur le Rhin Supérieur. Fouilles Françaises, Allemandes et Suisses sur les communes de Biesheim et Kunheim (Haut-Rhin). *Gallia* 62, 215–277.

Reimer, P. J, Baillie, M. G. L., Bard, E., Bayliss, A., Beck, J. W., Bertrand, C. J. H., Blackwell, P. G., Buck, C. E., Burr, G. S., Cutler, K. B., Damon, P. E., Edwards, R. L., Fairbanks, R. G., Friedrich, M., Guilderson, T. P., Hogg, A. G., Hughen, K. A., Kromer, B., McCormac, G., Manning, S., Ramsey, C. B., Reimer, R. W., Remmele, S., Southon, J. R., Stuiver, M., Talamo, S., Taylor, F. W., van der Plicht, J. and Weyhenmeyer, C. E. (2004) IntCal04 Terrestrial Radiocarbon Age Calibration, 0–26 cal kyr BP. *Radiocarbon* 46, 1029–1059.

Renfrew, J. M. (2003) Grains, seeds, and fruits from prehistoric Sitagroi. In E. S. Elsner and C. Renfrew (eds.) *Prehistoric Sitagroi: Excavations in Northeast Greece, 1968–1970. vol. 2: The Final Report*. Monumenta Archaeologica 20. UCLA Press, Los Angeles, pp. 1–29.

Riehl, S. (2008) Climate and agriculture in the ancient Near East: a synthesis of the archaeobotanical and stable carbon isotope evidence. *Vegetation History and Archaeobotany*. 17 (Supplement 1), 43–52

Riehl, S. (2009) Archaeobotanical evidence for the interrelationship of agricultural decision-making and climate change in the ancient Near East. *Quaternary International* 197: 93–114

Rigoni Stern, M. (1978) *Storia di Tönle*. Einaudi, Torino.

Rösch, M. (1999) Hanf. In H. Beck, D. Geuenich, H. Steuer and D. Timpe (eds.) *Reallexikon der Germanischen Altertumskunde* 13. Berlin and New York, pp. 630–631.

Rösch, M. (2008) New aspects of agriculture and diet of the early Medieval period in central Europe: waterlogged plant material from sites in south-western Germany. *Vegetation History and Archaeoabotany* 17 (Supplement 1), 225–238.

Rösch, M., Jacomet, S. and Karg, S. (1992) The history of cereals in the region of the former Duchy of Swabia (Herzogtum Schwaben) from the Roman to the Post-Medieval period: results of archaeobotanical research. *Vegetation History and Archaeobotany* 1 (4), 193–231.

Rottoli, M. (2001) Analisi archeobotaniche: i macroresti vegetali. In P. Frontini (ed.) *Castellaro del Vhò. Campagne di scavo 1996–1999*. Comune di Milano, Settore Cultura Musei e Mostre, Raccolte Archeologiche e Numismatiche, Milano, pp. 175–195.

Rottoli, M. (2002) Zafferanone selvatico (*Carthamus lanatus*) e cardo della Madonna (*Silybum marianum*), piante raccolte o coltivate nel Neolitico antico a 'La Marmotta'? *Bullettino di Paletnologia Italiana* 91–92, 47–61.

Rottoli, M. (2012) I materiali botanici del Canale Anfora. In F. Maselli Scotti (ed.) Gli scavi del 2004 al Canale Anfora di Aquileia.

Rottoli, M. and Negri, S. (1998) I resti vegetali carbonizzati. In Giannichedda E. (ed.) *Filattiera-Sorano: l'insediamento di età romana e tardo antica - Scavi 1986–1995*. All'Insegna del Giglio: Firenze, pp 198–212.

Rottoli, M. and Pessina, A. (2007) Neolithic agriculture in Italy: an update of archaeobotanical data with particular emphasis on northern settlements In Colledge, S. and Conolly, J. (eds.) *The Origins and Spread of Domestic Plants in Southwest Asia and Europe*. Walnut Creek, pp. 140–154.

Rottoli, M., Castiglioni, E. (2009) Prehistory of plant growing and collecting in Northern Italy, based on seed remains from the Early Neolithic to the Chalcolithic (*c*. 5600–2100 cal B.C.). *Vegetation History and Archaeobotany* 18, 91–103.

Rovira I Buendía, N. (2007) Agricultura y gestión de los recursos vegetales en el sureste de la Península ibérica durante la Prehistoria Reciente. Unpublished PhD Thesis, Universitat Pompeu Fabra, Barcelona. (http://www.tdx.cat/handle/10803/7468).

Rowe, J. H. (1962) Stages and Periods in Archaeological Interpretation. Southwestern *Journal of Anthropology* 18, 40–54.

Rumold, C. U. (2010) *Illuminating Women's Work and the Advent of Plant Cultivation in the Highland Titicaca Basin of South America: New Evidence from Grinding Tool and Starch Grain Analyses*. Ph.D. thesis. University of California, Santa Barbara.

Sahlins, M. (1972) *Stone Age Economics*. Aldine, Chicago.

Salavert, A. (2010) *Apport de l'archéobotanique à la compréhension des sociétés néolithiques. Analyses anthracologiques et carpologiques de neuf sites rubanés de Moyenne-Belgique (5200–5000 av. J.-C.)*. Ph.D. thesis. Université de Paris I (Panthéon-Sorbonne).

Salavert, A. (2010). Le pavot (*Papaver somniferum*) à la fin du 6e millénaire av. J.-C. en Europe occidentale'. *Anthropobotanica* 1, 3–16

Sarpaki, A. (1992) The palaeoethnobotanical approach: the Mediterranean triad or is it a quartet? In B. Wells (ed.) *Agriculture in Ancient Greece*. Swedish Institute at Athens, Stockholm, pp. 62–76.

Sarpaki, A. (2001) Condiments, perfume and dye plants in Linear B: a look at the textual and archaeobotanical evidence. In A. Michailidou (ed.) *Manufacture and measurement: counting, measuring and recording craft items in early Aegean societies*. Research Center for Greek and Roman Antiquity, National Hellenic Research Foundation, Athens, pp 195–265.

Schlichtherle, H. (1997) Pfahlbauten rund um die Alpen. In H. Schlichtherle (ed.) *Pfahlbauten rund um die Alpen*. Archäologie in Deutschland Special Issue. Theiss, Stuttgart, pp. 7–14.

Schlumbaum, A., Tensen, M. and Jaenicke-Després, V. (2008) Ancient plant DNA in Archaeobotany. *Vegetation History and Archaeobotany* 17, 233–234.

Schmidl, A., Jacomet, S. and Oeggl, K. (2007) Distribution patterns of cultivated plants in the Eastern Alps (Central Europe) during Iron Age. *Journal of Archaeological Science* 34, 243–254.

Schmidl, A. and Oeggl, K. (2005) Subsistence strategies of two Bronze Age hill-top settlements in the eastern Alps-Friaga/Bartholomäberg (Voralberg, Austria) and Ganglegg/Schluderns (South Tyrol, Italy). *Vegetation History and Archaeobotany* 14, 303–312.

Schultze-Motel, J. (1979) Die urgeschichtlichen Reste des Schlafmohns (*Papaver somniferum* L.) und die Entstehung der Art. *Die Kulturpflanze* 27, 207–215.

Scott, Linda J. (1979) Dietary Inferences from Hoy House Coprolites: A Palynological Interpretation. *The Kiva* 44, 257–281.

Sériadès, M. (1995) *Spondylus gaederopus*: The Earliest European Long Distance Exchange System; A Symbolic and Structural Archaeological Approach to Neolithic Societies. *Porocilo o raziskovanju paleolitika, neolitika in eneolitika v Slovenia 22*, 233–56.

Sigaut, F. (1991) Les techniques de récolte des grains. Identification, localisation, problèmes d'interprétation. In M.-C. Cauvin (ed.) *Rites et rythmes agraires*. Travaux de la Maison de l'Orient, 20. Lyon, pp. 31–43.

Sillmann, M. (2002) Botanische Grossreste mittelalterlicher und frühneuzeitlicher Latrinen und Gruben aus Freiburg im Breisgau. In L. Galiato, F. Löbbecke and M. Untermann (eds.) *Das Haus 'zum Roten Basler Stab' (Salzstrasse 20) in Freiburg in Breisgau.* Forschungen und Berichte der Archälogie des Mittelalters in Baden-Württemberg 25. Theiss, Stuttgart, pp. 623–770.

Spooner, D. M. (2005) A single domestication for Potato based on Multilocus Amplified Fragment Length Polymorphism Genotyping. *Proceedings of the National Academy of Sciences* 102, 14694–14699.

Stancheva, R. and Temniskova, D. (2004) Paleoecology of Holocene diatoms from *Sphagnum* peat bogs in central Sredna Gora Mountains (Bulgaria). *Geologica Carpathica* 55, 65–76.

Stanish, C. (2003) Ancient Titicaca: The Evolution of Complex Society in Southern Peru and Northern Bolivia. University of California Press, Los Angeles.

Stika H.-P. (2005) Early Neolithic agriculture in Ambrona, Provincia Soria, central Spain. *Vegetation History and Archaeobotany* 4, 189–197.

Stika, H.-P. and Heiss, A. G. (2013) Plant cultivation in the Bronze Age. In H. Fokkens and A. Harding (eds.) *The Oxford Handbook of the European Bronze Age.* Oxford University Press, Oxford, pp. 340–361.

Stika H.-P., Heiss A. G. and Zach B. (2008) Plant remains from the early Iron Age in western Sicily: differences in subsistence strategies of Greek and Elymian sites. *Vegetation History and Archaeobotany* 17 (Supplement 1), 139–148.

Stöckli, W. E. (2009) *Chronologie und Regionalität des jüngeren Neolithikums (4300-2400 v. Chr.) im Schweizer Mittelland, in Süddeutschland und in Ostfrankreich.* Antiqua 45. Reihardt, Basel.

Suter, P. J., Hafner, A. and Glauser, K. (2005) Lenk – Schnidejoch. Funde aus dem Eis – ein vor- und frühgeschichtliche Passübergang. In P. J Suter. and M. Ramstein (eds.) *Fundberichte und Aufsätze. Chronique archéologique et textes.* Archäologie im Kanton Bern 6B, Bern, pp. 499–522.

Swidrak I. and Oeggl K. (1998) Paläoethnobotanische Untersuchungen von Bodenproben aus der bronzezeitlichen Siedlung von Sotciastel. In U. Tecchiati (ed.) *Sotciastel – Un abitato fortificato dell'età del Bronzo in Val Badia. Istitut Cultural Ladin 'Micurà de Rü'.* Soprintendenza Provinciale ai Beni Culturali, Bolzano, pp. 334–346.

von Thünen J. H. (1826) *Der Isolierte Staat in Beziehung auf Landwirtschaft und Nationalökonomie.* Hamburg.

Tinner, W., Nielsen, E. H. and Lotter, A. F. (2007) Mesolithic agriculture in Switzerland? A critical review of the evidence. *Quaternary Science Reviews* 26, 1416–1431.

Tolar, T., Jacomet, S., Velušček, A. and Čufar, K. (2010) Recovery techniques of waterlogged archaeological sediments – a comparison of different treatment methods of samples from Neolithic Lake shore settlements. *Vegetation History and Archaeoabotany* 19, 53–67

Tonkov, S. and Marinova, E. (2005) Pollen and plant macrofossil analyses of radiocarbon dated mid-Holocene profiles from two subalpine lakes in the Rila Mountains, Bulgaria. *The Holocene* 15, 663–671.

Towle, M. A. (1961) *The Ethnobotany of Pre-Columbian Peru. vol. 30.* Viking Publications in Anthropology. Aldine, Chicago.

Ugent, D. (1970) The potato. *Science* 170 (3963), 1161–1166.

Vacher, J., Brasier de Thuy, E. and Liberman, M. (1992) Influence of the Lake on Littoral Agriculture. In C. D. Joux and A. Iltis (ed.) *Lake Titicaca: A Synthesis of Limnological Knowledge.* Kluwer Academic Press, Dordrecht, pp. 511–522.

Valamoti, S. (2004) Plants and People in Late Neolithic and Early Bronze Age Northern Greece. An archaeobotanical investigation. BAR International Series 1258. Archaeopress, Oxford.

Valamoti, S. M. (2006) Η συμβολή των αρχαιοβοτανικών καταλοίπων στη διερεύνηση των εκτεταμένων οικισμών: τα δεδομένα από το νεολιθικό οικισμό Άψαλος-Γραμμή [The contribution of archaeobotanical evidence to the investigation of flat-extended settlements]. *EGNATIA (Annals of the Dept. of History and Archaeology, Aristotle University of Thessaloniki)* 10, 9–130.

Valamoti, S. M. (2007) Food across borders: a consideration of the Neolithic and Bronze Age archaeobotanical evidence from northern Greece. *Aegeum* 30, 281–293.

Valamoti, S. M. (2009) *An Archaeobotanical Investigation of Prehistoric Diet in Greece* [in Greek]. University Studio Press, Thessaloniki.

Valamoti, S. M. (2010) Η ανθρώπινη δραστηριότητα στο Αγγελοχώρι μέσα από τα φυτά και τις χρήσεις τους: η συμβολή των αρχαιοβοτανικών δεδομένων [Human activity at Angelochori through plant remains and their uses: the archaeobotanical evidence]. In E. Stefani (ed.) Αγγελοχώρι Ημαθίας: Οικισμός της Ύστερης Εποχής τους Χαλκού, Τόμος 1 [Angelochori Hemathias: a Late Bronze Age settlement, Volume 1]. Kyriakides publication, Thessaloniki, pp. 171–197, 269–288, 327–330.

Valamoti, S. M. (2011) Flax in Neolithic and Bronze Age Greece: archaeobotanical evidence. *Vegetation History and Archaeobotany* 20 (6), 549–560.

Valamoti, S. M. and Jones, G. (2010) Bronze and oil: a possible link between the introduction of tin and *Lallemantia* to northern Greece. *Annual of the British School at Athens* 105, 83–96.

Valamoti, S. M., Mangafa, M., Koukouli-Chrysanthaki, Ch. and Malamidou, D. (2007) Grape-pressings from northern Greece: the earliest wine in the Aegean? *Antiquity* 81, 54–61.

Valamoti, S. M., Papanthimou, A. and Pilali, A. (2008) Cooking ingredients from Bronze Age Archondiko: the archaeobotanical evidence. In Y. Facorelli, N.

Zacharias, K. Polikreti (eds.) *Proceedings of the 4th International Symposium of the Hellenic Society for Archaeometry, 28-31 May 2003*. National Hellenic Research Foundation, Athens, 187–194.

Valamoti, S.-M., Kostakis, K. (2007) Transitions to agriculture in the Aegean: the archaeobotanical evidence. In S. Colledge and J. Conolly (eds.) *The Origins and Spread of Domestic Plants in Southwest Asia and Europe*. Left Coast Press, Walnut Creek, pp. 75–92.

Van der Veen, M. (2007) Formation processes of desiccated and carbonized plant remains – the identification of routine practice. *Journal of Archaeological Science* 34, 968–990.

van Driel-Murray, C. (2008) Those who wait at home: the effect of recruitment on women in the Lower Rhine area. In U. Brandl (ed.) *Frauen und Römisches Militär Beiträge eines Runden Tisches in Xanten vom 7. bis 9. Juli 2005*. BAR International Series 1759. Archaeopress, Oxford, pp. 82–91.

van Zeist, W. A. (1984) List of names of wild and cultivated cereals. *Bulletin on Sumerian Agriculture* 1, 8–16.

Vandorpe, P. and Jacomet, S. (2011) Plant Economy and Environment. In M. Reddé (ed.) *Oedenburg. Fouilles Françaises, Allemandes et Suisses à Biesheim et Kunheim, Haut-Rhin, France. Vol. 2. L'agglomération civile et les sanctuaires. 2: Matériel et études*. Monographien des RGZM Mainz 79 (2). RGZM, Mainz, pp. 3–72.

Vavilov, N. I. (1992) *Origin and Geography of Cultivated Plants*. Cambridge University Press, New York.

Vuille, M., Bradley, R. S. and Keimig, F. (2000) Interannual Climate Variability in the Central Andes and its Relation to Tropical Pacific and Atlantic Forcing. *Journal of Geophysical Research* 105 (D10) 12, 447–412, 460.

Watson, J. T. (2008) Prehistoric dental disease and the dietary shift from cactus to cultigens in northwest Mexico. *International Journal of Osteoarchaeology* 18, 202–212.

Weberbauer, A. (1945) *El Mundo Vegetal de los Andes Peruanos*. Estación Experimental Agrícola de la Molina, Lima.

Whitehead, W. T. (1999) Radiocarbon Dating. In C. A. Hastorf (ed.) *Early Settlement at Chiripa, Bolivia: Research of the Taraco Archaeological Project No. 57*. Contributions of the University of California Archaeological Research Facility. Berkeley, pp. 17–22.

Whitehead, W. T. (2006) Redefining Plant Use at the Formative Site of Chiripa in the Southern Titicaca Basin. In W. Isbell and H. Silverman (ed.) *Andean Archaeology III: North and South*. Springer, New York, pp. 258–278.

Whiting, Alfred F. (1939) *Ethnobotany of the Hopi. Museum of Northern Arizona, Flagstaff.* http://kitchenscience.sci-toys.com/acids. Accessed November 13, 2010.

Wiethold, J. (2003a) How to trace the 'Romanisation' of central Gaul by archaeobotanical analysis? Some considerations on new archaeobotanical results from France centre-est. In F. Favori and A. Vignot (eds.) *Actualité de la Recherche en Histoire et Archéologie agraires, Actes du colloque international AGER V, septembre 2000*. Annnales Littéraires 764, Série 'Environnement, sociétiés et archéologie' 5. Presses Universitaires Franc-Comtoises, Besançon, pp. 269–282.

Wiethold, J. (2003b) Archäobotanische Untersuchungen zur Ernährungs- und Wirtschaftsgeschichte des Mittelalters und der frühen Neuzeit. In R. Noel, I. Paquay, and J.-P. Sosson (eds.) *Au-delà de l'écrit. Les hommes et leurs vécus matériels au Moyen Age à la lumière des sciences et des techniques. Nouvells perspectives.* Typologie des Sources du Moyen Age occidental, hors série. Louvain-la-Neuve, pp. 461–499.

Willerding, U. (1980) Anbaufrüchte der Eisenzeit und des frühen Mittelalters, ihre Anbauformen, Standortverhältnisse und Erntemethoden. In H. Beck, D. Denecke and H. Jankuhn (eds.) *Untersuchungen zur eisenzeitlichen und frühmittelalterlichen Flur in Mitteleuropa und ihrer Nutzung*. Abhandlungen der Akademie der Wissenschaften in Göttingen 116. Göttingen, pp. 126–196.

Willerding, U. (1996) Fruchtfolge. In H. Beck, H. Steuer and D. Timpe (eds.) *Reallexikon der Germanischen Altertumskunde* 10. Walter de Gruyter, Berlin, pp. 133–138.

Willerding, U. (2002) Obst und Obstbau. In H. Beck, D. Geuenich and H. Steuer (eds.) *Reallexikon der Germanischen Altertumskunde 21*. Walter de Gruyter, Berlin, pp. 517–523.

Willerding, U. and Teegen, W. R. (2002) Mutterkorn. In H. Beck, D. Geuenich and H. Steuer (eds.) *Reallexikon der Germanischen Altertumskunde 20*. Walter de Gruyter, Berlin, pp. 449–453.

Wilson, H. D. (1990) Quinua and relatives (*Chenopodium* sect. *Chenopodium* subsect. *Cellulata*). *Economic Botany* 44 (Supplement), 92–110.

Winterhalder, B., Larsen, R. and Thomas, R. B. (1974) Dung as an Essential Resource in a Highland Peruvian Community. *Human Ecology* 2, 89–104.

Wirrmann, D. (1992) Morphology and Bathymetry. In C. Dejoux and A. Iltis (eds.) *Lake Titicaca: A Synthesis of Limnological Knowledge*. Kluwer Academic Publishers, Dordrecht, pp. 16–22.

Wright, M. F., Hastorf, C. A. and Lennstrom, H. A. (2003) Pre-Hispanic Agriculture and Plant Use at Tiwanaku: Social and Political Implications. In A. L. Kolata (ed.) *Tiwanaku and its Hinterland: Archaeological and Paleoecological Investigations of an Andean Civilization 2, Urban and Rural Archaeology*. Smithsonian Institution Press, Washington D.C., pp. 384–403.

Zapata, L. (2000) La recolección de plantas silvestres en la subsistencia mesolítica y neolítica. Datos arqueobotánicos. *Complutum* 11, 157–169.

Zapata, L. (2007) First farmers along the coast of the

Bay of Biscay. In S. Colledge and J. Conolly (eds.). *The origins and spread of domestic plants in Southwest Asia and Europe*. Left Coast Press, Walnut Creek, pp. 189–208.

Zapata, L. and Peña-Chocarro, L. (2005) L'agriculture néolithique de la Façade Atlantique Européenne. Unité et diversité des processus de néolithisation sur la façade atlantique de l'Europe (6–4 millénaires avant J.-C.). *Bulletin de la Société Préhistorique Française Memoire* 36, 189–199.

Zapata, L., Peña-Chocarro, L., Pérez Jordá, G. and Stika, H. P. (2004) Early Neolithic agriculture in the Iberian Peninsula. *Journal of World Prehistory* 18, 285–326.

Zeder, M. A., Emshwiller, E., Smith, B. D. and Bradley, D. G. (2006) Documenting Domestication: The Intersection of Genetics and Archaeology. *Trends in Genetics* 22, 139–155.

Zohary, D. (1996) The mode of domestication of the founder crops of Southwest Asian agriculture. In D. R. Harris (ed.) *The origins and spread of agriculture and pastoralism in Eurasia*. University College London Press, London, pp. 142–158.

Zohary, D., Hopf, M. and Weiss, E. (2012) *Domestication of Plants in the Old World*. Oxford University Press, Oxford.

Zoller, H. and Erny-Rodman, C. (1994) Epochen der Landschaftsentwicklung im Unterengadin. In A. F. Lotter and B. Ammann (eds.) *Festschrift Gerhard Lang. Beiträge zur Systematik und Evolution, Floristik und Geobotanik, Vegetationsgeschichte und Paläoökologie*. Dissertationes Botanicae 234. Gebrüder Borntraeger, Berlin/Stuttgart, pp. 565–581.

4 Adding Diversity. Between Occasional Food and Speculative Productions: Diversity of Fruit Uses, Diversity of Practices Regarding Fruit Tree Cultivation

4.1. INTRODUCTION

Laurent Bouby and Marie-Pierre Ruas

Work on the origins and history of agriculture in Eurasia long paid little attention to the case of fruit trees. Particularly in archaeobotany, literature has concentrated on cereals and legumes, which make up the staple vegetal foods in Europe and played a major role in the beginnings of agriculture in the Near East. Today, the origins, the processes of domestication and diffusion, and the agrarian practices connected with these plants are fairly well known. On the contrary, fruits have all too often been considered simply as gathered or cultivated in archaeobotanical work.

From a historiographical perspective, fruit trees have been especially regarded as taking on a significant place in the history of European agriculture with the emergence of complex societies, which took place in the eastern Mediterranean basin during the Bronze Age in the third and second millennia BCE (*e.g.* Childe 1958; Renfrew 1972; Gilman 1981; Fall *et al.* 1998; Janick 2005). These authors emphasise the capacity of fruit trees, principally those typical of Mediterranean arboriculture (*i.e.* the grapevine, *Vitis vinifera* L., the olive, *Olea europaea* L., and the fig-tree, *Ficus carica* L.), to provide food products (wine, oil and dried fruit) that are easy to conserve, transport and which have a high commercial value. The aspect of capital intensification supposedly represented by adoption of fruit arboriculture has also been highlighted. Trees do not bear fruit until they are several years old, when they begin bearing for dozens of years, even centuries, but they require continual care from the time they are first planted. Thus, they represent an investment to guarantee production in the future (*e.g.* Gilman 1981). Such an investment is seen as implying settled attachment to a particular place and long-term care and utilisation. This development is connected with the hierarchisation of society, since the ruling classes possess the capacity to carry out such investments, to ensure the social stability and security necessary for their viability, the redistribution of food indispensable to agrarian specialisation and the ability to set up commercial networks to handle the products involved. Furthermore, urbanisation creates the markets for consumption of these products. This can be seen, for example, in northern France where fruit-growing prevailed in the vicinity of urban centres during Roman times (Zech-Matterne 2010).

The other major implication in adopting fruit tree cultivation that has been put forward in the literature is the use of vegetative propagation as opposed to the cultivation of annual plants domesticated in the beginnings of Near-Eastern agriculture, which relies on sowing. Zohary and Spiegel-Roy (1975; also see Zohary 2004; Zohary *et al.* 2012) most especially see the shift from reproduction by seed to vegetative propagation as the keystone in the domestication of fruit trees. They consider species easily propagated vegetatively as pre-adapted to domestication and emphasise the fact that the classic fruit trees of the Mediterranean (date palm, *Phoenix dactylifera* L., fig, olive and grape) all fit into this category.

Objectives

It is not our aim in this chapter to question these overall characteristics of the history of arboriculture, nor even to emphasise their importance. What we wish to do here is to illustrate the great diversity of uses, of situations and of alternative practices, which go unnoticed but may have been crucial in some cases and must be taken into account when we examine the history of arboriculture and the domestication of fruit trees. Although archaeobotanical data do argue for the development of arboriculture, from the fourth to third millennia on, in the eastern Mediterranean basin (Van Zeist 1991), recent work suggests the existence of fruit tree cultivation far earlier than previously thought; for example, as early as the beginnings of the Neolithic in the Near East for the fig-tree (Kislev *et al.* 2006), or during the Neolithic for the olive in Spain (Terral and Arnold-Simard 1996; Terral 2000).

These examples raise the question of a probable form of arboriculture in the Neolithic in the Old World, a phenomenon that was long overlooked. It is true that such agroecological practices are very hard to assess with archaeobotany, it often being difficult to distinguish gathered, cultivated and domesticated fruit, insofar as the domestication syndromes are not very conspicuous for most fruit trees and the morphological differences are not very significant or visible in archaeological remains. In fact, to deal with the question of arboriculture, archaeobotanists often have to develop specific methods, but also – and perhaps especially – to appeal to inter-disciplinary work. Although archaeobotanists initiated most of the case studies proposed in the chapter, cross-disciplinary cooperation has been vital to this examination. Ethnographic, historical or ecological sources have been highlighted, as have other archaeological approaches. We felt it was interesting to present examples taken from diverse chronological, cultural, ecological and geographic contexts. The nomenclature utilised in this chapter is based on that of Zohary *et al.* (2012) for plants cultivated in the Old World and on the *Flora Europaea* (Tutin *et al.* 1964–1983) for spontaneous plants in Europe.

Fruit trees have multiple functions and the operations undertaken in the technical chains involved in the cultivation, treatment, transformation and consumption of fruits can consequently be highly diverse. Considering the limitations of space to devote to the subject here, this chapter will concentrate only on two main aspects: on the one hand, the economic role of fruits as food and, on the other, the practices connected with the cultivation and management of fruit trees, especially their propagation, the way they are set out in a given area and to the relationships between wild and cultivated. It is, nonetheless, important not to lose sight of the fact that cultivation of a fruit tree in an agrosystem can be explained by the diversity and complementarity of products involved (woodworking, firewood, fodder, medicinals, resins...) and the services provided (shade, fencing, boundary-marking, prevention of erosion...), rather than solely by the nutritional properties of fruits, which can be of secondary interest in some cases. D. Goldstein's presentation (Chapter 4.3) of the role of the woody genus *Prosopis* (algarrobo) for pre-Hispanic and modern societies of the Peruvian coast felicitously shows the multiplicity and complexity of uses and function, both economic and ecological, that a tree may have in an agrosystem. In the same vein, harvesting, transformation, storage and consumption practices, as well as preparation of fruits as food, are only dealt with in passing in this chapter. R. Cuthrell's article (Chapter 4.2), dealing with the use of acorns by indigenous societies in California, nonetheless develops these latter aspects, clearly illustrating the diversity of technical choices proposed even by hunter-gatherer populations for whom acorns also represented mercantile goods and objects of exchange between groups.

Fruits and Arboriculture: Some General Considerations

It is not possible to propose a uniformly valid definition of what a fruit is (see Ruas 1996; Alexandre-Bidon, *forthcoming*). For a botanist, a fruit is the result of the transformation of the ovary of a flower after pollination (Robert *et al.* 1998), but this definition is not necessarily appropriate to the category of fruits when they are considered as food. As an example, cereals, the caryopses of which are – botanically speaking – fruits, represent a food category distinct from that of fruits. When defined in relation to food use, fruits as a group vary according to the socio-anthropological context that determines culinary practices and consumption habits. For example, in the medieval Occident, the artichoke (*Cynara scolymus* L.) was regarded as a fruit and was still eaten as a dessert in the seventeenth century (Alexandre-Bidon, *forthcoming*). Thus, for the purposes of this chapter, we will adopt a flexible definition of fruit that agrees both with botanical criteria and the ethno-historical context of the examples under consideration. We shall potentially take into account the ensemble of fruits and false fruit produced by trees, shrubs or bushes, be they fleshy fruits or nuts, dry or fresh, on the condition that they are edible by humans.

Fruit tree cultivation includes highly varied practices and techniques (Janick 2005). We consider any action favouring the production of fruit, the maintenance and propagation of a species or selected forms as entering into this category. Hence, it is necessary to refer not only to quite elaborate and specialised technical systems, but also to take into account practices utilised in systems considered more unsophisticated, such as forms of agroforestry or even hunting-gathering.

Certain practices are peculiar to a particular tree or to rare species, such as the traditional manual pollination of date palms. Nonetheless, a certain number of features, gestures and techniques are common in arboriculture and can be cited as characteristic of this form of agriculture. Since they are perennials, fruit tree cultivation above all supposes deferred production. Pruning is one of the most characteristic practices of arboriculture and consists of eliminating parts of the tree, mainly branches but also buds, leaves, flowers or fruit, and is meant, on the one hand, to control and orient the growth of the tree and, on the other, to invigorate it and promote both the quantity and quality of fruit production. Recourse to vegetative reproduction (cuttings, marcotting, grafting) is also common and often indispensable to the maintenance and propagation of cultivars that are none other than selected clones.

Fruit in Food and Economic Systems

Fruits provide important sources of sugars, vitamins, minerals and fibre, but also and often, of lipids and proteins (as in the case of hard-shelled fruit such as walnuts or hazelnuts). They can be eaten fresh, kept by simply drying or by various preparation and transformation procedures (cooking, fermentation...), eaten by themselves or as part of culinary preparations. In some cases, fruits are staple foods. In addition to the example of acorns utilised in California, figs may have played a similar role in the Canary Islands in pre-Hispanic times. This is what J. Morales and J. Gil (Chapter 4.7) propose in an article that also furnishes an example of the contribution of interdisciplinarity in resolving questions pertaining to the use of fruits in the past.

Other situations of this sort could be mentioned, although in our world today, fruits are usually a complementary element to the major caloric contribution mainly composed of herbaceous plants (cereals, roots and tubers, pulses). They can then be seen as an interesting source of diversity. Most of the case studies in this chapter focus on a particular species and hence may not necessarily highlight this diversity. The assessment provided by M. Rottoli (Chapter 4.4) for northern Italy adopts a different perspective and reminds us of all the variety of fruits consumed in a particular area over time.

This diversity is attested from the Neolithic through the Roman Empire, but underwent major changes with the development of classic Mediterranean arboriculture as early as the Iron Age.

Diversity is expressed both in the panoply of fruit consumed but also in the multitude of varietal types to be found in a single species; a phenomenon that is especially difficult to perceive through the methods of archaeobotany, but the antiquity of which is clearly to be seen in texts. In this context, the grapevine provides a particularly fine example. Theophrastus, a Greek author of the fourth century BCE notes that there are as many types of grape as there are varieties of ground they are grown on. The Latin agronomists claim that it is not even possible to compile the complete list (André 1952). In the ways they consume, transform or cook fruits, people add cultural diversity to this biodiversity. Continuing with the example of the grape in Antiquity, the Romans made not only a multitude of different wines – dry, sweet or aromatic, heated-must wines (*vin cuit*) that can be used as a honey substitute, low-alcohol domestic wines made from grape skins (*piquette*) and vinegar – but also ate the grapes as fresh or dried fruit, alone or in cooked dishes with meats, vegetables or pastries (André 1981; Bouvier 2001).

Fruit provides a source of diversity for both food and cooking; a diversity that can be used to affirm cultural or social distinctions through some marked symbolic value or status as a luxury food possessed by some fruit. In the hierarchical conception of the natural and social world and its relationship to God, typical of the Middle Ages and the modern era in Europe, fruits were considered the most noble products of the vegetal world and, hence, as the most appropriate food for the upper classes (Grieco 1996). The luxury status of foods depends directly on the socio-cultural context, but among the features noted in the literature to define such products in many European and Asiatic societies, what is frequently flagged up is the unessential nutritional aspect, the difficulty involved in obtaining them, especially because they are of exotic origin or labour-intensive to produce or prepare. Sweetened products and alcoholic beverages are often to be found among luxury foods (van der Veen 2003). Many fruits fall into these categories. The majority of plants considered by Bakels and Jacomet (2003) as luxury foods in the early Roman period in central Europe are fruits and the main criterion lending them this status is the fact of being imported products.

In traditional Andean societies, alcoholic beverages – the *chichas* – play a fundamental role in social processes, notably the maintenance of power structures, due to being consumed during feasts and to ostentatory gift exchanges between elites (see various contributions in Jennings and Bowser 2008). These fermented drinks were most often made from maize (*Zea mays*) but also from other grains, roots and various fruit (Goldstein *et al.* 2008). M. Dietler (1990) emphasised the role of wine imported from the Mediterranean by Celtic societies of central Europe to be used in feasts and for the affirmation of elite status. This social dimension appears not only in consumption but also, at times, in the very fact of cultivating certain fruit trees or in the ways they are cultivated (Quellier 2006). What are we to think of a tree that comes from southeast Asia such as the *Citrus* (probably *Citrus medica*), the presence of which is revealed by palynology as early as the seventh century BCE in the port cities of the western Mediterranean, the Greek Cumae or the Punic Carthage, as explained in Bui Thi Mai and M. Girard's article (Chapter 4.5)? Was cultivating this tree in an urban context not some kind of ostentatory sign? The citron was mainly used in medicine and perfumery in Antiquity. The Greeks were familiar with it, but an examination of classical sources (texts and iconography) does not enable us to point out a sure attestation of cultivation of the species before the fourth century CE in Italy. It was considered by the Latins to be an exotic and still brought high prices at that time (André 1981; Amigues 2007).

The symbolic character of a fruit is most usually manifest to the archaeobotanist by its presence in a funerary or cult context. For example, fruits represent a large portion of the taxons found as food offerings in Roman funerary ensembles in western Europe and certain fruits are very frequent (for example hazel, *Corylus avellana* L., fig-tree, *F. carica*, walnut, *Juglans regia* L., date palm, *Phoenix dactylifera*, stone pine, *Pinus pinea* L. and gravepine, *Vitis vinifera*; *e.g.* Marinval 1993; Petrucci-Bavaud and Jacomet 1997; Bouby and Marinval 2004). The highly symbolic status in Roman cult practices of several of these species is well known from texts. For a single example, we can take dates, which are fairly often found in funerary contexts and extremely rare in contemporary domestic quarters: this exotic fruit must have had a certain luxury character (Bakels and Jacomet 2003), in spite of the testimony in texts (André 1981), and must have possessed both a symbolic function and been a social marker.

Fruit Tree Cultivation: Propagation and Domestication Methods

The fundamental character of vegetative propagation techniques has already been mentioned, among which grafting is certainly the most emblematic in the European sphere. This truly fascinated people in the Middle Ages and modern times, if we are to believe the agronomic and horticultural treatises, because of both the real competence and techniques – at least those applied to work for the aristocracy and the bourgeoisie – and for the impact on a magical and marvelous world of the 'phantasmogoric graft' (Mane 2006; Quellier 2006).

Domestication of most fruit trees is regarded as implying vegetative propagation so that, in this case, domestication is largely equivalent to selecting individuals with particular qualities and replicating them by cloning over long periods of time, which would involve a very slow renewal of cultivars (Zohary and Spiegel-Roy 1975; Ladizinsky 1998; Zohary 2004; Zohary *et al.* 2012). Sowing, which is the main mode of reproduction for fruit trees in natural conditions, is not thought to be appropriate for arboriculture because most species are allogamic and highly heterozygous. Consequently, descendents are often very different from their parents. This mode of reproduction thus provides results that are often unpredictable and accentuated by the length of time, several years, between germination of the plant and its first fruiting which makes it possible to evaluate the young tree's quality.

Nevertheless, this theoretical model probably minimises the role of sexual reproduction and is tinted by the present-day scientific and socio-economic context, so it may well attribute an importance to standardising clonal varieties which people cultivating fruit trees in former times did not always aim for. For C. Bonneuil and F. Hochereau (2008), cultivars in the present-day sense did not really exist in agronomic literature until the first third of the nineteenth century, even though subdivisions of rank below the species level are attested in horiticultural treatises.

The role that sexual reproduction may have played in certain arboriculture systems and in the domestication of fruit trees can be divided into three parts, which are to a certain extent inter-dependent: it can be a common mode of reproduction for certain species, a resource for the evolution of varieties – either consciously or unconsciously – of species usually propagated vegetatively, and as a contribution to the production of trees appropriately serving needs to other cultivated varieties.

In Europe, various fruit trees that today are propagated vegetatively were more frequently sown in the past. In Roman times, Columella (*De Re Rustica*, Book V, X) advised multiplying by vegetative propagation for most species but sowing *Amygdalus communis* L. (almond tree), *Castanea sativa* L. (chestnut), *J. regia* (walnut) and *P. pinea* (stone pine). Pliny the Elder states that chestnuts and walnuts can only be reproduced by their seeds (*Natural History* XVII, X). In the twelfth century, the Arabo-Andalusian agronomist Ibn Al-Awwâm notes that sowing is a frequent recourse in propagating numerous fruit trees, especially the pistachio (*Pistacia vera* L.), the walnut, hazelnut, almond tree, chestnut, pear (*Pyrus communis* L.), olive (*Olea europaea* L.), peach (*Persica vulgaris* Miller), plum (*Prunus domestica* L.), stone pine, palm and grapevine (*The Book of Agriculture,* V). M. Chauvet (1999) remarks that almond, walnut and peach trees were propagated by seed until modern times. Walnut may not commonly have been grafted in southwestern France until the late eighteenth century, and more generally in the nineteenth.

Sexual reproduction is utilised in arboriculture in biogeographic and cultural contexts that are quite different from Europe. For instance, in the southern Moroccan oases, the apricot tree (*Prunus armeniaca* L.) is cultivated exclusively by seed. Genetic data suggest that sexual reproduction played a more important role in the past in northern Tunisia, where cultivation of this tree is today carried out by grafting (Khadari *et al.* 2006). In agroforestry systems in Africa, Asia and the Americas, the same species can be propagated by protecting spontaneous seedlings, by intentional sowing and by vegetative propagation (Wiersum 1997), with the first two methods depending on sexual reproduction. The sweet bush mango (*Irvingia gabonensis* A. Lecomte), for example, is mainly propagated by agroforesters in the wet forests of Cameroon by protecting spontaneous seedlings, but transplantation or intentional sowing from selected trees is practiced as well (Ayuk *et al.* 1999). Using several propagation methods for the same species seems to be fairly common in traditional agrarian systems. Pliny clearly states that *Citrus* and *Sorbus* reproduce both vegetatively and sexually and that *Phoenix dactylifera* (date palm) does not degenerate, no matter how it is reproduced (*Natural History*, XVII, XI). In the fourth century CE, the Latin agronomist Palladius (*Treatise of Agriculture*, III, XXV) notes that the hazelnut can be either sown or propagated vegetatively. Many other examples could be added.

The more important role taken on by sexual reproduction within the framework of non-specialised production can be explained by the lesser importance attributed to particular qualities of fruit and the diversity of uses for fruit trees. In southwestern France, E. Leterme (1995) recalls that vigor and fine growth of fruit trees were important criteria of quality, since these plantations were not given much care. The taste quality of fruit was a secondary factor, and alternate ways of valorising them were always available, either by cooking or as livestock fodder.

Other needs can involve trees that were intentionally seeded or selected among those growing spontaneously, which are not directly utilised for their fruit but to promote growth or production of selected varieties associated with them. Trees used as graft stock or pollinators are perhaps the best example of this. The article by Y. Aumeeruddy-Thomas *et al.* (Chapter 4.8) is quite enlightening on this subject. It shows how wild olive trees were commonly protected and grafted in northern Morocco, at times after transplantation, but also that they are intentionally kept in orchards to pollinate cultivated varieties, just as spontaneous fig-trees are used both as stock and at times for pollination of female varieties. In southwestern France, it was common in the last few centuries not to carry out grafting on very young seedlings (often natural), but to let them bear their first fruits and, according to the result, to graft them or not. So, the Corne du Périgord variety of walnut was usually sown then grafted if the tree was not hardy or the fruit not appreciated (Leterme 1995). This practice surely has a long history and must have played a major role in the history of arboriculture, the domestication of fruit trees and the creation of agro-diversity.

Pliny clearly states that sown trees grow slowly, degenerate and must be regenerated by grafting. Even the chestnut tree sometimes needed to be grafted (*Natural History*, XVII, X).

The role of sexual reproduction in the process of fruit tree domestication may have been important depending on time, place and on the species. This is necessarily true when the favoured mode of reproduction is sowing. For example, the guaje trees (*Leucaena esculenta* Sessé and Moc. Benth.) – a leguminous plant whose buds, flowers, seeds and fruit are used as food in Mexico –selectively spared after several cycles of itinerant forest maize cultivation, are phenotypically differentiated from trees growing in an unmanaged forest. These differences can be even more accentuated in trees cultivated locally by sowing in gardens, fields and hedges (Casas and Caballero 1996; Casas *et al.* 2007). Cultivation by sowing is not incompatible with selecting and maintaining local landraces. In southwestern France, varietal types of sown walnuts were distinguished, as the ordinary walnut, the fertile walnut and the late Saint John's walnut. In the case of the plum tree, growing the swine plum or the roussotte until recently relied both on propagating the suckers and on sowing (Leterme 1995).

Sexual reproduction may have been a driving force in the varietal evolution of species multiplied by vegetative means. Before hybridisation was mastered and genetics arrived on the scene, the creation of new clonal varieties relied on the occurrence of somatic mutations, which subsequently became fixed by cloning, and on sexual reproduction: individual plants were thus selected among either natural or intentional seedlings, resulting from diverse possibilities of crossing between cultivated, wild and feral trees, before being maintained by cloning. Another plant that is cultivated by cloning, manioc (*Manihot esculenta* Crantz), provides a good present-day example of such practices. In a traditional system in French Guyana, plants resulting from spontaneous crossing between varieties and with wild forms are often integrated in crops where they are multiplied vegetatively, along with cultivated varieties, and some of the new plants are assigned to known varieties, contributing significantly to the evolution of genetic diversity of the cultivated section (Elias *et al.* 2001; Duputié *et al.* 2007). Thus, the importance of sexual reproduction in populations of clonally propagated plants would often have been underestimated (Duputié *et al.* 2007).

Spatial Context and Cultivation Practices Applied to Fruit Trees

Trees play a special role in the construction and representation of space and landscapes for human societies (Jones and Cloke 2002). The location and distribution of fruit trees in an agrosystem can be far more diverse than the mono-species monotony of present-day European orchards might lead one to believe. Potentially, all transitory formulas can be observed – sometimes at the same time in the same agrosystem – running from primary forest, where fruit trees grow spontaneously without any human action aside from gathering their fruit, to the mono-species orchard just mentioned. Basing our work on the continuum of people-plant interaction established by D. R. Harris (1989) and the literature pertaining to present-day agroforesty systems (Torquebiau 2000; Wiersum 1997; 2004), as well as on European historical texts, we propose to outline a series of situations fitting into a gradient of intensification of production and of growing artificialisation (Fig. 4.1). In each of these stages, the ways fruit trees are grown can be combined with cultivation of herbaceous plants and also or only with pastoral practices.

The first stages include various ways of using and managing the forest, among which simple gathering in a natural forest is but one of the particular cases in point. Even in agricultural societies utilising itinerant processes, the examples provided by present-day agroforestry systems indicate that practices aiming at sparing, promoting or even spreading fruit trees are common. This is especially true of the influence brought to bear on the composition and productivity of fallows and secondary forest that develop after cultivated plots are abandoned (Wiersum 1997; Cotton 1998; Coomes *et al.* 2000; Schroth *et al.* 2004).

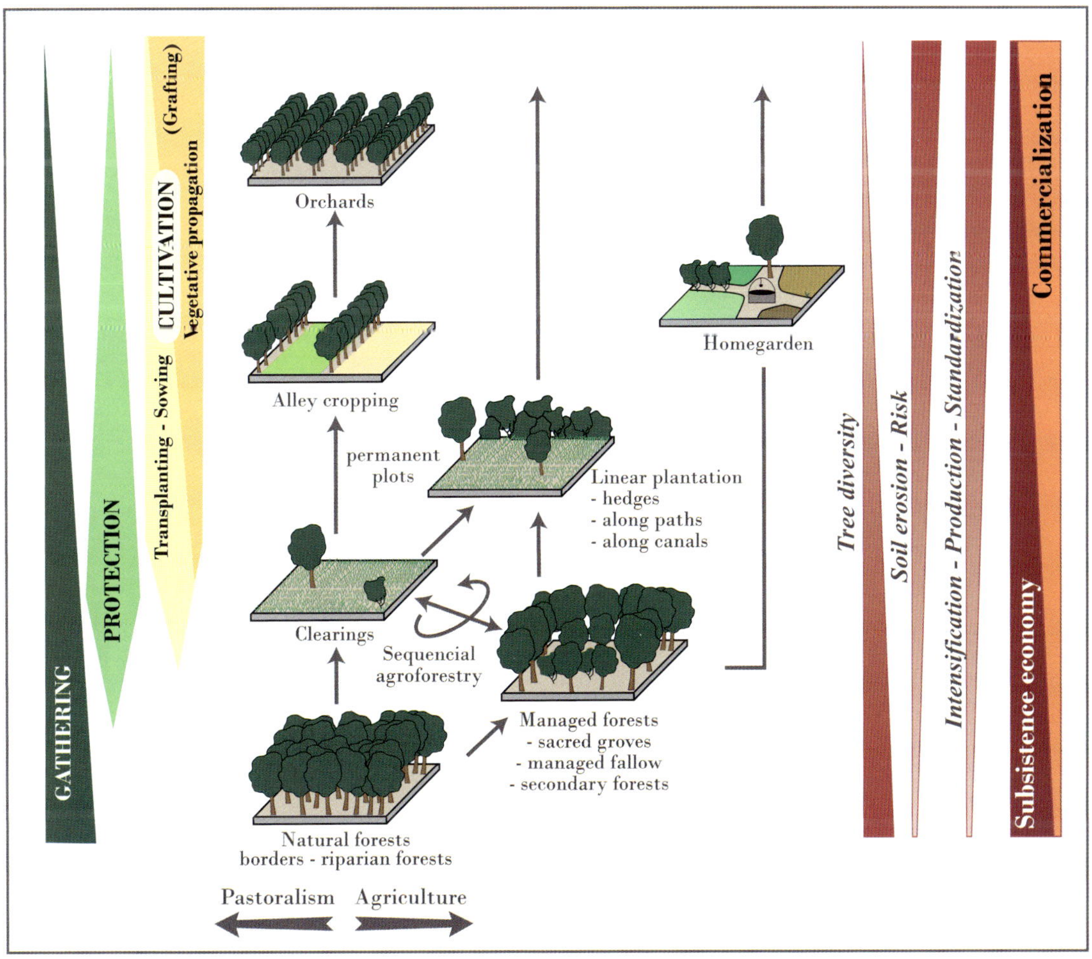

Fig. 4.1. Schematic categorisation of forms of cultivation of fruit trees, ordered in a gradient of intensification and artificialisation of production.

Once agriculture arrived – and doubtless before that in some cases – most forests rapidly became varyingly affected by human action. When the first agriculturalists of the Neolithic settled in southern France, fire systems seem mainly to function in relation to the climate, but human action is perceptible locally as early as this time, to subsequently become preponderant in the proto-historical period (Berger 2006; Rius *et al.* 2009). Several authors have put forth the hypothesis of human impact on European forests as early as the Neolithic, even the Mesolithic, most notably to promote the growth of oak (Mason 2000; Ruffo 2001; Delhon *et al.* 2009). Based on the strong representation of heliophilous taxa in anthroacological records of early Neolithic sites (Linearbandkeramik) in central Europe, A. Kreuz (1992; 1998) proposes the existence of managed forests or even regularly kept hedges around cultivated plots.

Hedges can be seen as a kind of linear plantation. They are interesting in that they represent an arboriculture structure that is not intensive but permanent over many years. Linear plantations – especially hedges – are commonly associated with agrosystems having permanent plots and limited mechanisation, the typical example of which is the hedged farmland or *bocage*. They can nonetheless appear in non-permanent forest agriculture systems along paths (Cotton 1998), and persist in the most intensive of agrosystems.

Of course, the setting up of fixed-plots can be associated with a significant level of intensification in the care given to fruit trees and to their cultivation, which would also be accompanied by the transition from rights of collective use to private rights (Wiersum 1997; 2004). Typical stages we can discern are thus:

- the presence of dispersed trees in plots or permanent pastures,
- crop cultivation associated with trees – sometimes of different species, sometimes on different strata – and herbaceous plants, along the lines of the classic type of Mediterranean *coltura promiscua* (Durand 2003),
- the mono-species orchard.

The gradient outline we have just set out corresponds globally to a gradient of intensification, standardisation and specialisation in production. Without excluding other factors such as increased pressure on forest resources, commercial use of fruits seems to be a major determinant in intensification of production (Den Hertog and Wiersum 2000). The evolution of viticulture in southern France portrayed by L. Bouby *et al.* (Chapter 4.6) on the basis of archaeological and archaeobotanical data provides an example of a possible transition from a relatively extensive cultivation of grapes to the monospecific plantations linked to the development of an intensive wine-production oriented to trade.

This continuum in the modes of operation is linked to the processes of maintenance, selection and multiplication of fruit trees, running from simple gathering, perhaps including practices to protect, promote or even spread species by various techniques, but leaving much room for diversity, to the other extreme of standardisation of production implying exclusive recourse to vegetative propagation and selection of an ever-decreasing number of varieties. In the final analysis, we can say that there is a gradual process of domestication entailing both changes in the ways plants are cultivated and in how fruit trees are reproduced (Wiersum 2008).

For example, the inhabitants of the Tehuacan valley in Mexico eat the fruit of various cacti obtained simultaneously in different ways: the fruit is collected from semi-natural plants, from managed plants, which involves sparing the phenotypes selected when land is cleared, but also from cultivated cacti on agricultural terraces or in gardens. They then use sowing and vegetative multiplication, both from selected plants in the wild population (Otero-Arnaiz *et al.* 2005). Furthermore, this example is revealing about the multitude of interactions and interconnections that can exist between forest and cultivated plots. The very special character of the garden as a cultivated space should stand out. Since it can either be the main locus for agricultural production or accompany all forms of agriculture, the garden represents a site of diversity and interactive experimentation with other species, whether cultivated or not (Kumar 2008). Among agroforestry peoples today as in the European medieval city, the garden is where the most fragile trees are planted, the ones that require the most care and it is there that new varieties are tested, acclimated and developed. The garden could well have played a considerable role in the domestication of fruit trees.

Conclusions

The customs and practices linked to fruit tree cultivation make up a vast panorama of diversity. In this chapter, we have attempted to focus on the examples and debates about diversity that are connected with the use of fruit as food and to the methods and practices involved in cultivating and managing fruit trees. The main point in this introduction is to emphasise the fact that, perhaps in a way even more obvious than for annuals, fruit tree husbandry is organised into a continuum of people-plant interactions. It covers close interrelations between the ways fruit trees are cared for – especially concerning the ways they are organised in space – and the ways they are reproduced, running from wild to cultivated and tending towards growing intensification and specialisation of production and a consequent pauperisation of diversity at all levels. Fruit tree domestication is a process that accompanies this continuum.

By focusing mainly on fruit tree cultivation as a commercial operation and on domestication as the selection of elite clones multiplied vegetatively *ad infinitum*, archaeological, historical and biological literature concentrated on one – certainly crucial – aspect, but which is only one part of the history of fruit cultivation in the Old World. It quite probably had a long history before these important stages, a history that is difficult to relate because it took place in societies without writing and has left but meager archaeological and archaeobotanical traces. However, today there is diverse evidence that enables us to guess at more complex relationships between fruit trees and human beings than simple gathering as early as the Neolithic. Of course, we do not want to oversimplify the situation and to squeeze a whole history into a model that represents a largely theoretical construction. Every fruit species has its own history or singular histories due to the peculiarities of the bioclimatic and socio-cultural contexts they were grown in, and because of their biology, that largely determines the propagation method or methods that must be favored and that weigh in the process of domestication. As a final note, we might recall that the diverse methods and practices in fruit tree husbandry can frequently be associated and interconnected within the same agrosystem and that these diverse situations result from choices and adaptations that can be and could be reversible.

4.2. ACORN USE IN NATIVE CALIFORNIA

Rob Cuthrell

Introduction

In many temperate climates, nut mast resources such as acorns (*Quercus* spp.), hazelnuts (*Corylus* spp.), walnuts (*Juglans* spp.), and buckeye (*Aesculus* spp.), are an abundant and relatively predictable source of nutritious food. The waste products of nuts are often present in hunter-gatherer archaeological sites that have good plant preservation. However, the predominating disciplinary focus of archaeobotanical research on agricultural societies has, with some exceptions (*e.g.* Mason 1992), precluded a full consideration of the myriad ways nut resources might have contributed to ancient foodways.

This summary will focus on ecological, ethnographic, and archaeological research related to the use of acorns by indigenous groups in California. We will briefly review economic modeling of acorn use and discuss how economic modeling has changed the accepted view of acorns as a food source. Finally, we will suggest how research on acorn use in Native California might be relevant to prehistoric European foodways research.

Ecological and Nutritional Aspects of California Oaks

The latest California flora documents nineteen species of native oaks plus several oak subspecies and hybrids (Hickman 1993). Oaks are common over about a third of California's land area, preferring places with 25–40 cm or more of annual precipitation and average temperatures between 14–16°C (Pavlik *et al.* 2002, 52). Though most or all oak species may have been used as food sources by indigenous groups, a subset of six to eight species were preferred, including: black oak (*Quercus kelloggii* Newb.), blue oak (*Quercus douglasii* Hook. & Arn.), white oak (*Quercus lobata* Née), coast live oak (*Quercus agrifolia* Née), and tanoak (*Lithocarpus densiflorus* [Oerst.] Rehder; produces acorns similar to those of *Quercus* spp.) (Baumhoff 1963; Bettinger and Wohlgemuth 2006). Based on the modern distribution of oak species, the majority of native Californian groups would have had access to at least three of the eight most commonly used oak species (Wohlgemuth 2004, 22). However, since the distribution of oak species in the past could have varied with climatic fluctuations – a process which has not yet been well documented – these observations should be treated tentatively when making archaeological interpretations.

Acorns are higher in fat and lower in protein and carbohydrates than modern cultivated grains and the overall caloric value of dried acorns is higher than that of our modern cereals (Fig. 4.2). Although more research is needed on the nutritional content of California acorns, estimates suggest there can be great disparity in nutrition between the acorns of different oak species, most apparent in the relatively high fat contents of *Q. kelloggii* and *Q. agrifolia*. Data from the U.S.D.A. (2008) suggest acorns are a poor source of vitamins and minerals.

All Californian oak species produce acorns containing tannins – bitter and potentially toxic natural compounds that must be removed from foods before consumption (Wohlgemuth 2004, 22). The necessity of reducing tannin content through leaching added significantly to the amount of labor involved in producing acorn-based foods, and may also have

Description	Water	Protein	Fat	Carb.	KCal / 100g1
Lithocarpus densiflorus [Oerst.] Rehder[2]	*	2.9	12.1	54.4	338
Quercus douglasii Hook. & Arn.[2]	*	5.5	8.1	65.5	357
Quercus kelloggii Newb.[2]	*	4.6	18	55.5	402
Quercus lobata Née[2]	*	4.9	5.5	69	345
Quercus agrifolia Née[2]	*	6.3	16.8	54.6	395
Quercus garryana Douglas ex Hook[2]	*	3.9	4.5	68.9	332
Quercus chrysolepis Liebm.[2]	*	4.1	8.7	63.5	349
Acorn nuts, raw[3]	27.9	6.2	23.9	40.8	387
Acorn nuts, dried[3]	5.1	8.1	31.4	53.7	509
Acorn flour, 'full fat'[3]	6	7.5	30.2	54.7	501
Maize (*Zea mays* L.), dried[3]	8.1	9.9	5.2	74.9	386
Barley (*Hordeum* sp.), hulled[3]	9.4	12.5	2.3	73.5	354
Wheat (*Triticum* sp.), hard white[3]	9.6	11.3	1.7	75.9	342

Fig. 4.2. Nutritional data for acorns of California oaks, unspecified oak species, and modern cultigens. Water, protein, fat, and carbohydrate values expressed as percentages of edible portion. [1]Kilocalories per 100g material estimated from protein, fat, and carbohydrate percentages for California acorn species. Exact amounts given for unspecified acorn samples and modern cultigens from U.S.D.A. (2008). [2]Data from Bettinger and Wohlgemuth (2006); modified from Baumhoff (1963, 162); after Wolf (1945). Water percentages not included. [3]Data from U.S.D.A. National Nutrient Database for Standards Reference (2008); oak species unspecified (*Quercus* sp.).

removed nutrients from the acorn material (Basgall 1987; Mason 1992).

Acorn Use in Native California

Before we discuss acorn use in California, we must caution that the present focus on this single resource masks the wondrous diversity of plants used by the native people of California for food, medicine, basketry, shelter, and many other purposes. As some recently published works have demonstrated, simplified characterisations of native Californian groups as eaters of 'nuts, roots, and berries' deny the complexity of indigenous food traditions (Anderson 2005; Timbrook 2007).

California's ecology is characterised by botanical richness. The latest California flora lists over 3400 species of indigenous plants, of which about a quarter are endemic (Hickman 1993). This richness is reflected in the surviving ethnobotanies of California tribes – for example, the Sierra Miwok consumed eight distinct varieties of clover and some tribes recognised subtly different varieties (possibly subspecies) of oak for which we have no taxonomic correlates (Anderson 2005, 269; Goldschmidt 1951). Although indigenous food choices would have always been conditioned by seasonality and ecology, within each context people had to choose from a multitude of plausible food options.

Native Californian groups have been described as semi-sedentary or mostly sedentary, employing a cyclical settlement pattern that responded to changing seasons and available plant and animal resources (Heizer and Elsasser 1980). For groups who made intensive use of the fall acorn harvest, the logistics of acorn procurement – assessing the location and quality of acorn mast; monitoring acorn ripeness; and organising work parties to harvest, transport, process, and store acorns – represented a set of important challenges. Indigenous groups developed many homologous but distinct practices to solve these problems.

Harvesting Acorns

Ownership of oak trees and groves was generally on the tribal or familial rather than the individual level, but it was not uncommon for individuals to have rights to seed-gathering tracts or trees (Bean and Saubel 1972; Mason 1992). A Wintu (Fig. 4.3) man

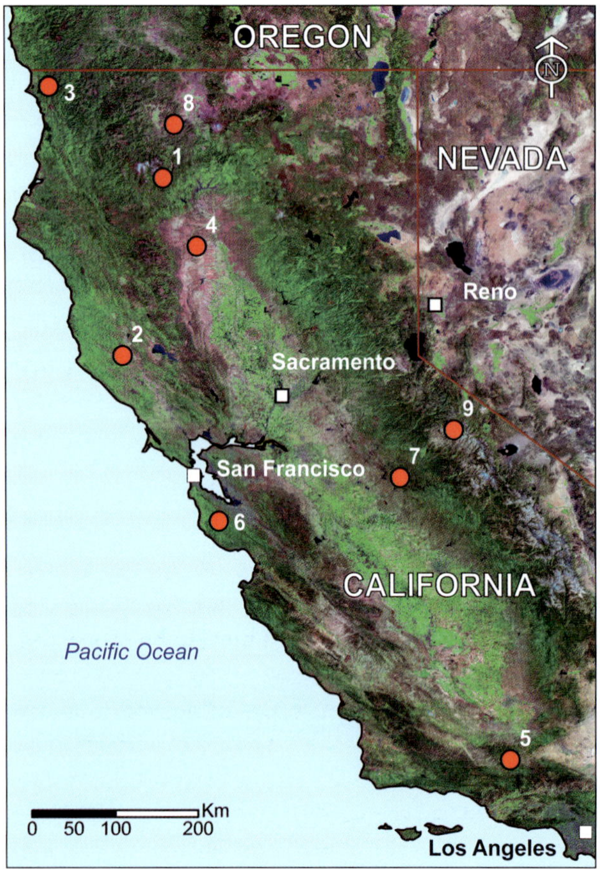

Fig. 4.3. Map of the tribes' settlement areas (only their centres indicated), as mentioned in the text. 1) Wintu tribe; 2) Pomo tribe; 3) Tolowa tribe; 4) Nomlaki; 5) Chumash tribe; 6) Ohlone tribe; 7) Miwok tribe; 8) Shasta tribe; 9) Western Mono tribe. Map: R. Lugon, J.-C. Loubier and A. Chevalier.

could claim an oak tree or even a single branch by leaning sticks diagonally against the trunk (Du Bois 1935, 18). Tribes friendly towards one another would ask, and often receive, permission to make use of plant and animal resources on another's territory (Anderson 2005, 248); however, acorn 'poaching' sometimes occurred when resources were scarce (Du Bois 1935).

Many species of oak only produce a large nut mast every two to five years, and in some cases entire stands can go several years without producing a good crop (Mason 1995). Monitoring oak stands to estimate productivity in the months before harvest would have been most important in years of sparse nut mast, when it would have been critical to make sure the acorn harvest was not eaten by animals, taken by other groups of people, or spoiled by exposure to the weather. Such monitoring would also have been important when deciding where to place acorn camps. These temporary habitation sites, usually occupied only during acorn harvesting and initial processing, were a logistical technology that improved the efficiency of acorn procurement by reducing the amount of time spent walking and transporting acorns between villages and oak groves.

Acorns were usually harvested while still in the trees, though fallen acorns were commonly collected as well. Men and boys climbed oaks and shook limbs by hand or struck them with long poles to knock acorns down (Dixon 1905; Du Bois 1935). Smaller acorn-laden branches might be cut off and thrown to the ground for picking (Mason 1992). Some groups waited for inaccessible acorns near the tops of trees to ripen and collected them after they fell (Goldschmidt 1951). To enhance acorn visibility, Pomo and several other groups kept the ground underneath oak stands clear by regularly burning off understory vegetation (Kniffen 1939).

As the acorns were removed from the trees, women and children collected them in baskets, often biting off the acorn cups to reduce bulk and weight (Du Bois 1935). Women and children usually transported acorns back to acorn camps or villages but, in some cases, men assisted as well (Driver 1936). Transporting acorns could be a difficult task – Tolowa women carried heavy loads of acorn five to fifteen miles between harvesting sites and villages and, in some cases, acorns were harvested from oak groves up to 20 miles from main village sites (Gould 1975; Mason 1992).

Local acorn camps were often the initial processing centres for harvested acorns. At these camps, acorns were commonly sorted to remove spoiled nuts, shelled, and dried. Women usually did this work, but sometimes acorn shelling involved the entire group (Mason 1992). Often a woman would stay behind at the acorn camp during the day to turn and monitor acorns drying in the sun (Du Bois 1935).

Acorn Storage, Processing and Cooking

Acorns could be stored in many forms: whole and unshelled; shelled and dried; as pounded, leached, and dried meal; or underground in streambeds and

Fig. 4.4. Western Mono acorn storage structure, 1918. Photo by Edward Gifford, Catalogue No. 15-6183. Copyright © Phoebe A. Hearst Museum of Anthropology and the Regents of the University of California.

seepages (Goldschmidt 1951; Kniffen 1939). Often different storage and processing methods were used for different types of acorn at different stages of ripeness. While the Nomlaki stored green acorns in the shell, ripe acorns were hulled before storage (Goldschmidt 1951). When stored outdoors, acorns were often kept in circular or conical granaries, sometimes raised off the ground to deter pests. These granaries could be small – woven into a single chaparral bush – or large enough to hold an entire village's acorn supply, up to three metres high and two and a half metres in diameter (Fig. 4.4; Du Bois 1935; Goldschmidt 1951; Mason 1992). Acorns could also be stored in bark-lined pits or shelled and stored exclusively inside houses, as in the case of the Chumash (Du Bois 1935; Timbrook 2007). Acorn meal was commonly stored indoors as well, probably so that it could be monitored more closely for vermin infestation or spoilage (Driver 1936). A final method of storage was in pits dug into streambeds or 'mud seepages,' places where water flowed continually. Shelled acorns placed in seepage pits would slowly be leached over the course of several months (Mason 1992). Acorns processed in this way could be buried in the fall and exhumed for consumption in winter or spring, four to six months after burial (Du Bois 1935). Dried and shelled acorns could reportedly be stored for up to five years, and one report suggests acorns buried in mud seepages could last up to thirty years (Mason 1992; Merriam 1918).

Women prepared acorn meal as needed throughout the year; in rare cases men would also participate (Timbrook 2007). A quantity of dried acorns was removed from storage to the place where acorn meal was made – an important social space for women. If acorns were stored unshelled, the shells would first be removed, usually by women but sometimes also by children using cracking stones and/or teeth (Goldschmidt 1951; Mason 1992). Shelled acorns were ground in one of four ways: in a stone or wooden bowl-shaped mortar with a stone pestle; in a hopper mortar; on a flat rock using a pestle; or with a mano and metate (Mason 1992; Nomland 1935). Coarse particles were separated from fine particles using basketry sieves or by size-sorting particles by shaking them in a shallow basket, causing the small particles to stick to the bottom (Du Bois 1935, 19).

After grinding, the acorn meal was next leached, most commonly in a shallow sand pit, the size determined by the quantity of meal to be processed, and up to over a metre in diameter (Dixon 1905; Merriam 1918). Different types of acorns required different leaching times, from about two hours up to a full day. Sometimes warm or hot water was used to leach acorns, which sped the process but may have removed more nutrients and flavor from the meal. After leaching, sand would have been rinsed off the underside of the acorn mass, which would have had the consistently of a sticky dough (Mason 1992).

The wide variety of unique acorn recipes used by native Californians can be classified into three general categories: acorn mush, a thick soup or stew made of finely ground acorn meal; acorn bread, a baked good usually made from coarser acorn meal; and whole acorns, roasted or boiled with little or no grinding and most closely associated with acorns buried in mud seepages (Du Bois 1935; Goldschmidt 1951; Mason 1992). Most native California groups did not use pottery; acorn mush was cooked in water-tight baskets using heated stones. Acorn bread was formed into loaves after leaching and, sometimes, clay rich in iron oxide was added to neutralise any remaining tannins (Johns and Duquette 1991). The size of acorn bread loaves ranged from 'small cakes' to large masses weighing up to 18kg (Dixon 1907; Goldschmidt 1951). The bread was usually baked for a day in underground pit ovens containing heated rocks. This bread could remain edible for months (Du Bois 1935, 19).

In areas with several varieties of oaks, the acorns of each species were sometimes associated with particular recipes, and less desirable acorn types were used only in lean times (Du Bois 1935, 19–20; Kniffen 1939, 379). In each culture, some types of acorn were preferred over others, likely based on a combination of taste, ease of processing, and cooking properties. For example, an Ohlone group in central coastal California reportedly preferred coast live oak (*Q. agrifolia*) over tanoak (*L. densiflorus*), while Sierra Miwok favored black oak (*Q. kelloggii*) most out of at least seven types of oak they recognised (Barrett and Gifford 1933; Bocek 1984).

Economic Aspects of Acorn Use

Acorns in various forms were one of the most important trade goods for many indigenous groups in California. According to Timbrook (2007, 157), a 'thriving trade in dry, shelled acorns' persisted in Chumash territory into the colonial period. In northern California, Shasta groups traded dry blocks of leached acorn meal to groups in southern Oregon (Dixon 1907). Acorns were also commonly traded for pine nuts in the eastern portion of California, and tribes who controlled land with different varieties of acorns would sometimes exchange them (Mason 1992 after Barrett and Gifford 1933; Steward 1933).

Several researchers have attempted to quantify the productivity of California oaks and to estimate caloric return rates of acorn processing. Baumhoff (1981), while acknowledging the uncertainty of much of his data on acorn productivity, concluded that indigenous population densities were generally more than one order of magnitude below the 'carrying capacity' of the land based on average oak nut mast. He proposed that population densities instead may have reflected the carrying capacity of a minimally productive nut mast year.

Basgall (1987) estimated an overall caloric return rate for acorns (including all steps of processing from collection to consumption) of around 1000kcal/hr, while Bettinger and Wohlgemuth (2006 after McCarthy 1993), working with black oak (*Q. kelloggii*) data, figured returns of just under 800kcal/hr. Wohlgemuth (2004, 24) suggested that for a family of four obtaining half of their calories from acorn products, five hours per day of labor, predominantly by women, would have been required. Heizer and Elsasser (1980, 97–100) estimated that a large Sierra Miwok acorn granary (4.7m^3 with 2.4 metric tons of acorns) might hold enough acorns to supply each member of a family of six with over 5000 calories per day for one year.

In the past two decades, the view of the role of acorns in California foodways has changed from one in which acorns were regarded as a nutritional panacea that would have been adopted as soon as technology allowed (Baumhoff 1963; 1981; Mason 1992), to an alternative consideration of acorns as a somewhat unpredictable and relatively high-cost resource that might have been avoided until its use was necessitated by increased population pressure (Wohlgemuth 2004). Currently, much of the cited nutritional information on the acorns of California oaks comes from a study over 60 years old (Wolf 1945). No research projects have yet comprehensively quantified the relationships between local ecology and acorn productivity for each species of California oak, although some studies have examined productivity of one or a few oak species in specific ecological areas (Mason 1992, Appendix 3). Likewise, no studies have attempted an economic comparison of the many different practices that indigenous peoples employed in the transformation of acorns from natural product into food, leaving much room for speculation regarding variability in caloric return rates. For example, Bettinger and Wohlgemuth (2006) suggested that total acorn processing time might have been halved (doubling caloric return rates) by reducing the intensity of various procedures and achieving economies of scale. Until there is much more complete understanding of all the variables affecting acorn use, attempts to generalise about the economic aspects of acorn production will have a high degree of uncertainty.

Archaeology of Acorns in California

Archaeological detection of acorn use depends on many factors: the location of acorn processing activities, either at village sites or temporary (and archaeologically ephemeral) camps; proclivities to use acorn processing refuse as fuel or to process acorns in proximity to fire; the probability that identifiable food remains would be charred based on the particular cooking method; taphonomic processes such as weathering and bioturbation;[1]

and finally archaeological methods for recovery and analysis of macrobotanical remains (Mason 1992; Wohlgemuth 2004).

In many cases, the ethnographic record can guide our expectations for acorn recovery. For example, Sierra Miwok groups are reported to have preferred acorn shells as fuel (Barrett and Gifford 1933), providing a regular pathway to acorn shell preservation in the archaeological record. When acorns were shelled away from camps and hearths, the chances for preservation would have been much poorer. Likewise, whole or fragmented acorn cotyledons are unlikely to have been preserved except in the less common cases where these were regularly roasted whole.

Wohlgemuth (2004) provided the first comprehensive summary of archaeobotanical evidence for acorn use in much of California over the past 7,000 years. In 940 flotation samples from across central and northern California, Wohlgemuth (2004, 61) found acorn fragments in over 75% of samples examined. Wohlgemuth used standardised ratios of acorn shell remains (mg) to small seeds (counts) to mitigate for variable taphonomic effects in exploring plant intensification trends through time. This project documented the beginnings and trajectory of intensive acorn use in California's geographic sub-regions and set the stage for future studies to explore how the role of acorns in indigenous foodways changed through time in more localised contexts. This study also showed that acorns have a long history of use in California foodways, with a high ubiquity of acorn remains among archaeobotanical samples from the Lower Archaic (*ca.* 12,000–7,000 BP). This finding challenged work by Basgall (1987) which suggested acorns were a high cost resource that would not have been used until high population densities forced the use of less efficient plant resources.

Summary

The preceding brief review highlights some of the ways that indigenous groups in California effectively managed, monitored, procured, stored, and consumed oak nut masts. Stored acorn products were a dependable and energy-rich food source and an important botanical component of the foodways for many of these groups. The technology for long-term acorn storage gave indigenous groups the potential to anticipate future shortages, reducing their susceptibility to ecological and environmental contingencies. Acorn-related work was a central locus for social practice, particularly for women, and acorn products were important trade goods. Despite the many uncertainties in economic models of acorn use, recent archaeological research has demonstrated that the study of acorns can contribute to our understanding of indigenous foodways, particularly on the intensification of human-plant relationships. As more European researchers begin to include botanical remains in their research, particularly in studies of Mesolithic foodways, the archaeological and ethnographic research on acorn use in California may provide a critical data set for comparing and contrasting new European models of changing human-plant relations.

4.3. A WILD SOLUTION TO RESILIENCE AND PROVISION: THE CASE OF *PROSOPIS* SPP. ON THE PERUVIAN NORTH COAST

David J. Goldstein

Introduction

As outlined in this section's introduction, fruit trees are often viewed as additive components to agricultural systems, especially in the case of fruits that are modern cash crops. Archaeological and historical evidence, however, demonstrate that this has not always been the case: fruiting trees were important contributors to the development of agriculture in the Neotropics (Atran 1993; Lentz 1999; Pearsall 1992; 2008; Turner and Miksicek 1984). The importance of fruiting trees is now coming more often to be considered a critical component of agricultural development in the Neotropics (Baleé and Erickson 2006; Clement 2006; Beresford-Jones 2005; Moutarde 2006).

Prosopis spp. (algarrobo) has long been a vital component of subsistence agroecologies – the indigenous agricultural and ecological provisioning system – on the Peruvian Pacific coast (Fig. 4.5). I focus on the use of two of these woody plants, *Prosopis juliaflora* D.C. and *Prosopis pallida* (H. and B. ex Willd.) H.B.K., two sympatric or parapatric species of leguminous trees on the Peruvian coast during the pre-Hispanic and modern periods (Burkhart 1976; Pasecznik *et al.* 2001) and trace these organisms' roles in the development of ancient agricultural systems and in sustaining modern subsistence production. *Prosopis* plays a critical role in the specialised ecosystem of the dry tropical forest, with specific adaptations to the region's climatic cycle. These unique ecological features prove beneficial to local agriculturalists who incorporate *Prosopis* into farming strategies. Archaeological evidence

Fig. 4.5A. Map of Peru with internal political borders and the area of study in the Lambayeque department.

implicates *Prosopis* in the development of coastal agriculture over the past 6000 years: this article provides an archaeological case study from the region, highlighting this genus' use in domestic and craft economies. The long-standing reliance

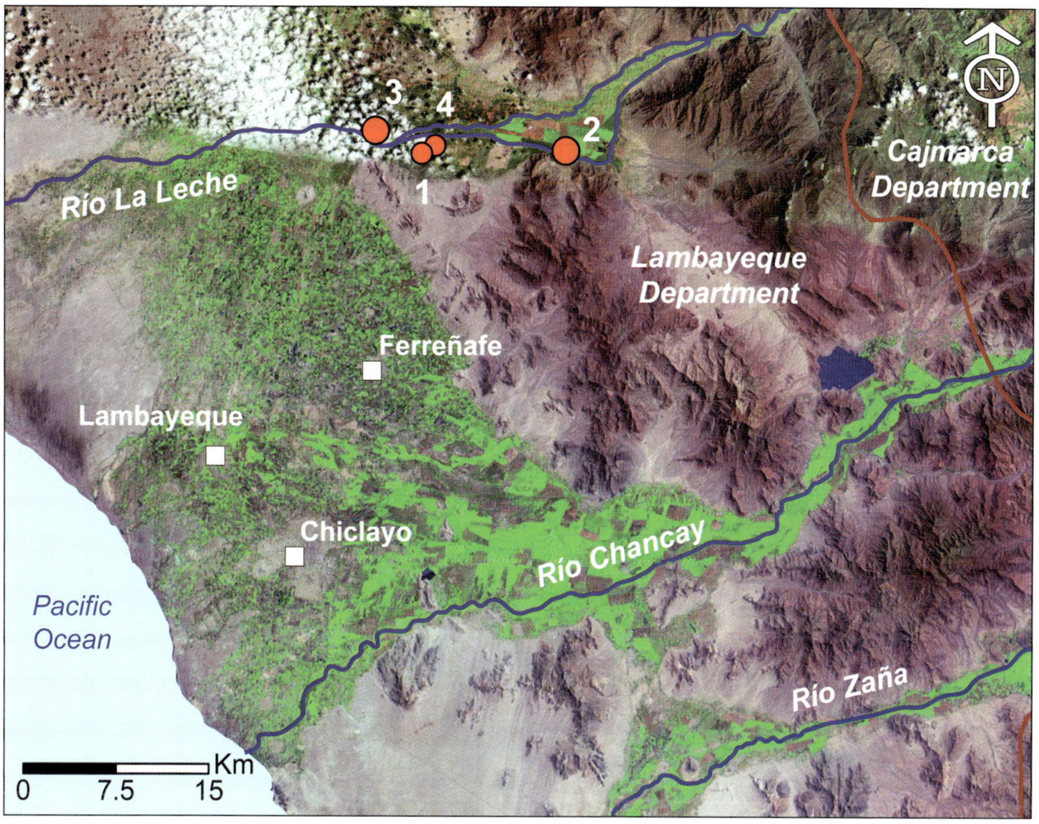

Fig. 4.5A. Map of the Lambayeque department in Peru with the following archaeological sites: 1) La Zaranda; 2) Batán Grande; 3) Sicán Tree National Sanctuary, bosque de Pomac; 4) Milenario Tree, La Zaranda. Maps: R. Lugon, J.-C. Loubier and A. Chevalier.

on *Prosopis* on the Peruvian north coast, however, has not prevented its overexploitation for charcoal production in present-day contexts (Risco 1998; Cuba Salerno *et al.* 1998). Despite the modern threat to dry tropical forests and *Prosopis*, the archaeological data provide potential alternatives to resource exploitation in coastal Peru.

Economic Uses of *Prosopis*

Trees in the genus *Prosopis* are widely known today with the Spanish term *algarrobo* and/or *huarango* throughout the western coast of South America (Brack Egg 1999). *P. juliflora* and *P. pallida*, the focus of this chapter, are economic species endemic to the western coast of South America from Colombia to northern Chile and are currently suffering from severe predation by humans. Most trees are now only found adjacent to or mixed with patches of fallow farmlands, making it difficult to study their long-term growth habits and their interactions with the dry tropical forest ecology. At present, in the coastal Andes, *Prosopis* is sought for charcoal production due to the wood's physical stability at high temperatures. The genus is globally valued as fuels for their savory-smelling secondary metabolites, qualities that provide long-lasting and rich-smelling cooking fires that impart a particular flavor to local delicacies (Pasiecznik 2002; Rogers 2000). These trees are often clear-cut for charcoal production and as a profit windfall from agricultural clearance (Harris *et al.* 1998; Risco 1998). Given such intensive harvesting, people have not actually witnessed forest regeneration and succession here in the modern era (Proyecto Algarrobo 1997; Fig. 4.6). Hence, forest ecologists and local farmers have little experience with the relationship between *Prosopis* growth and regeneration; a large contributing factor to over-harvesting. The propensity to cut these trees beyond regenerative capacity has significantly affected forest restoration throughout the colonial and modern era and has forced the regulation of forest products in the coastal areas of Peru (Ferreyra 1987). At present, rural forest residents often produce *Prosopis* spp. charcoal illegally for additional household income to be sold in local and regional markets and/or domestic use for fuel (Risco 1998).

Fig. 4.6. The view from within a healthy dry tropical forest on the north coast of Peru, Batán Grande, Lambayeque.

Aside from its use as charcoal, *Prosopis* spp. provides a number of economic products. The tree produces 10–20 cm long legume fruits that contain a highly nutritious and stable seed that is a choice animal fodder, as well as a product for refining into fermented beverages, *chicha de algarrobo*, and a sweet syrup, *algarrobina*, that is marketed nationally (Risco 1998). The resin from the trunk is used as an adhesive and as a local remedy to fill dental caries (Brack Egg 1999; Risco 1998). The pods are known regionally as an important famine food for modern humans when resources are scarce after forest fire, flooding or extreme drought (Felger 1979; Felker *et al.* 1984; Schaedel 1988).

The leguminous properties of these trees as nitrogen fixers, sub-soil stabilisers, and water table sustainers, make the arid lands where they grow 'ready-made' prepared soils for farmers. Hence, stands of *Prosopis* spp. create choice locations for farmland clearance and are at risk as increasing modern human migration drives newcomers to the lowland river valleys (Arroyo Baldeon 1997; Pasiecznik 2002). Although *Prosopis* spp. furnishes a number of economic products and is prized by local populations, its utility also endangers its survival in dry tropical forests; as a major player in this ecosystem, its over-exploitation threatens this delicate ecology.

Prosopis spp. and Dry Tropical Forests

In the very few protected areas throughout the north Peruvian coastal departments, the dry tropical forest is the dominant ecosystem (Tosi 1960; Proyecto Algarrobo 1997). The regional woody flora consists of xerophytic trees, low shrubs, and cacti. These tree species are indeed adapted to an extremely dry environment, with a mean precipitation of less than 40 mm/year (Delvaud 1984, 26); most plants in these forests have waxy or small leaves, deep taproots, and diurnally dormant stomatic systems, which keep them from drying out in periods of extreme drought. The biodiversity of woody plants is manifest in two families, the *Capparidaceae* (caper family), and the *Fabaceae* (bean family), among which *Prosopis* species are particularly important (Proyecto Algarrobo 1997). As a dominant species in mature forests, *Prosopis* spp. is the longest-lived, 200–400+ years, and some are in their adolescence for up to at least 60+ years of life (Ocampo 1991; Fig. 4.7). Due to modern exploitation there are currently very few *Prosopis* spp. trees over 40 years of age along the Pacific coast (Arroyo Baldeon 1997; Ocampo 1991).

The scarcity of water in dry tropical forests makes this ecosystem unusually vulnerable: the slow annual growth of dominant tree species, coupled with uncertain periodic water and slow (relative to temperate forest ecology) nutrient recharge, can result in deforestation when the ecosystem is disrupted (Maas 1995; Cuba Salerno *et al.* 1998). When a species is singled out for exploitation, the delicate balance of the ecosystem is significantly disturbed. In the case of *Prosopis* spp., while adapted for periodic climatic disturbance, it is unable to regenerate effectively without protection from human predation in the modern world.

Fig. 4.7. The 'Millenario' tree of La Zaranda, Lambayeque. Algarrobo, *Prosopis pallida* (Humb. & Bonpl. ex Willd.) Kunth

ENSO Dynamics on Peruvian Dry Coastal Forests and *Prosopis*

El Niño/Southern Oscillation (ENSO) weather events have become more and more prominent in the modern discourse about weather and periodic or episodic climate events globally. *Prosopis* is adapted to survive, and even prosper, when ENSO events or heavy rains occur. These trees experience branch and trunk growth spurts following inundation; during dry years *Prosopis* spp. mostly invests in root expansion and growth through its system of deep tap and lateral roots. These adaptations are resilient strategies to cope with weakening soils and rising groundwater, allowing them to survive flooding and benefit from the abundant water. *Prosopis* spp. is equipped to manage water events in other ways to enhance its survival. The drying period following heavy rains tends to force flowering in the trees, and also to accelerate the developmental life cycle for tree pollinators, indigenous flies, wasps and bees, as well as imported honey bees.

The tree is a heavy fruiter: increasing its above-ground biomass by two or three times when fruiting (Ocampo 1991; Risco 1998). In ENSO events during December or January, just as the tree is coming into fruit, trees will often fall, unable to sustain themselves in the wet sandy soils, laden with fruits. While some *Prosopis* spp. have deep taproots, these will not prevent keeling over even in extreme instances, but this does not kill these resilient trees. Instead they immediately begin to put out new branches, reestablishing forest cover, and extending/repairing broken root systems (Fig. 4.8). These unique adaptations permit biological perseverance, encouraging farmers to include these trees in their agrosystems.

Fig. 4.8. The 'Sicán' tree of Pomac, Lambayeque, *Prosopis pallida-juliflora.*

Modern Agroecological Importance

Modern agriculturalists in the research area actively take advantage of these qualities in *Prosopis* growth and resilience, often incorporating them into their different agroecosystems. Despite the periodic and unpredictable nature of ENSO events, farmers demonstrate an attentiveness to their long cycle of seasonality and incorporate agricultural strategies that take advantage of this cyclical nature. In modern farming, the anticipation of ENSO events relies on the incorporation of *Prosopis* spp. into traditional agricultural systems. During non-ENSO event years, modern, non-mechanised agriculture is dominated by extensive, canal-fed field systems along the Peruvian coast. *Prosopis* spp. is a common feature of the landscape as a field border/living fence system and to line canals. Its lateral root system helps maintain soil surface water, and in conjunction with its leaf litter, enhances soil surface nitrogen (Luna-Suarez *et al.* 1998; Nordt *et al.* 2004; Proyecto Algarrobo 1997). The species also plays a significant role in preventing soil erosion and desertification through both the shade and soil cohesion offered by its branch and root systems, respectively (Luna-Suarez *et al.* 1998). The plant's incorporation into the agricultural system highlights a comprehensive understanding of the long-term strategies necessary for resilience to episodic disturbance due to ENSO events, which are often catastrophic.

Additionally, farmers rely on *Prosopis* spp. to maintain floodplains that are only suitable for agriculture in post-ENSO periods. Usufruct landholdings in and around the flood stage riverbanks along the main river systems, as well as along larger canals that overrun their banks at flood stage, provide supplementary farmland for agriculturalists (Nordt *et al.* 2004). These fields are not in production annually: they depend on fertilisation and irrigation as floodwaters recede after ENSO events. Like more formal, irrigated agricultural land, *Prosopis* spp. and other tree crops are tended on these plots even in non-ENSO years, maintaining their fertility and water tables through the benefits of agroforestry practice and the nitrogen fixing qualities of woody Fabaceae (such as *Prosopis*). The plots can be farmed for up to a year and a half once flood stage has been reached. Some smallholders extract, given labor availability, up to three growing seasons off these plots that are often larger than irrigated plots. Undoubtedly, the considerations for using long-lived dry forest woody legumes, such as *Prosopis*, are important and complex (Goldstein 2007; Nordt *et al.* 2004). These 'extra' fields are used as additional income sources; they also mitigate economic stress when existing agricultural resources suffer during ENSO events and reduce dependence on federally regulated and taxed irrigation water. The use of these fields illustrates how local agriculturalists of coastal inland valleys of the Peruvian coast have a broad understanding of the integration of annual wet/dry agriculture with the episodic ENSO cycle.

Archaeological Look at *Prosopis* and Ancient Andean Agroecology

The archaeological signature of tree crops in ancient Andean agriculture over the past 5,000+ years is profound (Pearsall 1992; 2008; Ugent and Ochoa 2006). From the earliest agricultural periods we see the presence of *Annona cherimolia* (chirimoya), *Annona muricata* (guanabana), *Lucuma bifera* (lucuma), *Carica papaya* (papaya), *Persea americana* (avocado), and *Psidium guajava* (guava). In particular, on the Pacific coast, Ugent and Ochoa (2006) report use of a variety of trees from the Fabaceae family, *Acacia macracantha*, *Caesalpinia paipai*, *Erythrina edulis*, *Inga* sp., and *Prosopis chilensis* (algarrobo), all of which are well adapted to living in the region's dry tropical forests.

Although these limited data indicate the longevity of human-*Prosopis* coexistence, little research has been carried out on their interaction. The unsystematic and inconsistent sampling for archaeobotanical remains in Andean archaeology has led to considerable ignorance of how vital tree crops were in antiquity (Bertone *et al.* 2008). A further problem with locating the importance of *Prosopis* spp. in the archaeological record has been with identification. Modern problems exist at the species level of distinction. As a result, evaluating the presence and use of *Prosopis* at the species level in the archaeological record is difficult. This case is especially so when different authors report different species for the wood, seeds, leaves and fruits recovered (Bonavia 1982; Ugent and Ochoa 2006). Investigators should be mindful that species-level distinctions are very difficult in archaeobotany, should be reserved for exceptional instances, and, as seen here, can confound future research (Hather 1994).

The few treatments of the archaeobotanical record that do exist for *Prosopis* spp., however, at least at the generic level, demonstrate that it was present and used consistently throughout Andean prehistory on the Pacific coast from at least 6000 years ago. Uses in these instances include food, fuel, construction and animal fodder; it also played a critical part in indigenous agroecology, still visible today. At present, very few Archaic and Pre-ceramic Period sites have systematic collected archaeobotanical data to report. The best published registry from the Pre-ceramic Period (BCE 8000–1500) comes from Bonavia's (1982) excavations at Los Gavilanes, on the Peruvian central coast. During the period that Engel (1987) and others (Kaulicke 1994; Pearsall 2008; Weir *et al.* 1986; 1988) have named the period of incipient domestication and agriculture, Bonavia's archaeobotanical evidence indicates the use of *Prosopis* spp. over 5000 years ago with dates of 2700 cal. BCE. The tree was utilised as a building material, fuel, and as a primary material associated with subsistence, including human consumption (Bonavia 1982, 330).

More recent work (Haas and Creamer 2004; Shady Solis 1999) confirms that some of the earliest complex societies of coastal Peru were extracting resources from numerous tree crops mentioned above while they were beginning to construct monumental architecture for the first time in the Americas. As at Los Gavilanes (Bonavia 1982), *Prosopis* spp. seeds, fruits, wood, leaves, etc. were recovered from Caral (Shady Solis 1999), Huaricanga and Caballete (Haas and Creamer 2004), a series of 5,000+ year-old sites from the central coast's Supe and Pativilca river valleys. In these cases, they were recovered in conjunction with very early agricultural contexts that include *Zea mays* (maize), *Ipomoea batatas* (sweet potato), *Manihot esculenta* (yuca), and a variety of cucurbits and chili peppers. This cultivation was concurrent with an increasing presence of camelids at coastal settlements; in fact, by the end of the Pre-ceramic, camelids, alpaca or llama (*Llama* sp.) are present on the coast in pastoral contexts (Shimada and Shimada 1985). The presence of this complete domestication complex 6,000 years ago, of crop plants, animals, and tree crops, all correspond to the earliest complex social and economic formations in the region (Smith 1995). *Prosopis* spp. and its part in this development as an agroecological component is just beginning to come to light, and I highlight its importance in the example below.

An Archaeological Account of *Prosopis* use from the Tenth Century

Recent excavation of the metal and ceramic workshop of Huaca Sialupe, a site used from 900–1050 CE pertaining to the Sicán or Lambayeque polity, demonstrates the primary, integrated role *Prosopis* played in ancient agriculture and subsistence. *Prosopis* remains are ubiquitous across the site in all features and floor contexts. The archaeobotanical remains indicate that *algarrobo* served as a vital fuel resource; however, it was also used for construction, fodder and cooking. Fuel remains at the site included a mix of wood fuels, camelid dung from local herds, and organic garbage. Distinct ratios of fuels were apportioned for specific types of fire needs (Goldstein 2007). While wood fuels comprised the largest component of all fuel use, higher proportions of *Prosopis* spp. were limited to meet specific production requirements. In particular, *Prosopis* fuels are primarily used in blackware ceramic kilns, but in no case do they make up more than 20% (count/volume) of the fuel remains. Camelid dung, overall, was used in all burning contexts and consistently comprises nearly 5% of all fuel remains. Organic household and agricultural waste content varied depending on specific combustion needs. The differential apportionment of fuel types in different burning contexts (ceramic, metal or cooking fires) demonstrates artisans' integrated use of resources, which likely prevented the over-exploitation of *Prosopis* spp. (Goldstein and Shimada 2007).

We have abundant evidence at Sialupe that food production took place alongside craftwork. The animal remains, maize cobs and other agricultural products, with residues from large aggregate fires, similar to those used to produce small feasts today, together indicate that artisans were cooking and eating at the site (Goldstein and Shimada 2010). Additionally, the use of camelid dung and agricultural debris demonstrates that an agricultural economy was in close proximity. All hearths used some *algarrobo*, an indication that these resources were common enough to be used in quotidian ways and were not exclusive to craft production. This pattern suggests that in and around Sialupe, the dominant dry tropical forest species, *Prosopis* spp., was in significant abundance for daily use.

While *Prosopis* spp. remains are the most ubiquitous at the site, the most abundant plant organ represented are the pods and seeds, not the woody stem (Goldstein 2007). The fruits from the trees played an important role in the economy, probably as the primary fodder used for camelids kept at the site; it was also likely a part of the human diet. As mentioned above, camelid dung provided a constant fuel resource; in fact, at the northern end of the site, excavations recovered several square metres of burned and unburned camelid dung deposits, six to ten centimetres deep, frequently with durable seed and pod remains in the matrix.

In the case of the Huaca Sialupe workshop, some 30% of the overall fuel assemblage indicates that organic garbage was a primary fuel source (Goldstein 2007; Goldstein and Shimada 2010). Every indication from these fuel remains demonstrates that food garbage from the households that surrounded the workshop contained high levels of *Prosopis* spp. seeds and seed pod remains. Clearly, some of this presence can be attributed to dung used as fuel. The ubiquity, however, demonstrates that on a household level people were processing and consuming *Prosopis* spp. Regional ethnography also supports the use of *Prosopis* spp. for human consumption (Schaedel 1988). Although *Prosopis* spp. is primarily exploited today for charcoal production, fuel remain evidence from Sialupe demonstrates that these trees were of primary importance to coastal animal husbandry practices and an integral part of the local cuisine, as well as an important fuel source (Goldstein 2007).

The decision-making processes reconstructed at Sialupe, based on the archaeobotanical remains, demonstrate a judicious use of fuel resources: based on the specific physical requirements for artifact production, artisans used distinct mixes of fuels (Goldstein and Shimada 2007). The abundance of *Prosopis* remains as charcoal, seeds, seed pods, small branches and wood at the site indicates that this species was of primary importance. *Prosopis* spp. stands clearly were important components to the Sicán era forest, demonstrating the historical depth of human and *Prosopis* ecological interdependence. Additionally, the presence of agricultural production, food preparation, and camelid pasturing at the site demonstrates a high level of integration of *Prosopis* spp. and agriculture prior to European colonisation in the region (Goldstein 2007).

Conclusions

Archaeological research has the potential to identify forest resources, and fruit trees in particular, as historically and culturally important aspects of managed landscapes. Understanding the role that landscape features play in society, like fields that rely on agroforestry practices, is one way to begin to decipher and reclaim histories of land use in regions where these datasets are otherwise elusive. Rival (1998) and Clement (2006) more specifically indicate ways in which trees, especially those that are part of agroecological practices, become touchstones for realising land claims and symbolising long-term investment in land use. The enduring use of *Prosopis* spp. in Coastal Peru justifies adding this venerable tree to the same list of those known to have played significant roles in the development of civilisation in South America.

Only by acknowledging the importance of flowering and fruiting trees in modern and early agricultural systems can we develop sustainable solutions to potential scarcity of resources among human populations. Although looking to the past may not always provide direct solutions to modern ecological problems, it does remind us that alternatives to our present agricultural course exist and offers us the potential to learn from that past.

4.4. BEFORE THE EMPIRE: PREHISTORIC FRUIT GATHERING AND CULTIVATION IN NORTHERN ITALY

Mauro Rottoli

Introduction

The most ancient reference to an orchard is famously to be found in the Odyssey, which contains both an idealised description (*Odyssey* VII, 112–131) and more realistic versions, when Ulysses disembarks in Ithaca and meets his old father Laertes (*Odyssey* XXIV, 246–247 and 340–345). These descriptions – which define an orchard as a garden bounded by stones or thorn bushes where trees (fig, olive, pear, apple, pomegranate) and vines are laid out in rows (see also the *Iliad* XIV, 123) – demonstrate that a system of this type was already known in the Mediterranean area in about the ninth to seventh centuries BCE; it was probably of still greater antiquity, dating at least from Mycenaean times (*i.e.* before 1200 BCE) (Bile 2003).

We do not know if orchards existed in Italy this long ago – still less in northern Italy, which was more marginally involved in contacts with the Mycenaean and Greek worlds. It should be remembered that the orchard recounted in the Odyssey was the property of a prince or king and, therefore, existed in a complex, stratified society; this type of society developed in Tyrrhenian Italy from the ninth century onwards and two centuries later in northern Italy (De Marinis 1987).

In the Mediterranean area and especially in mountainous zones, the presence of trees, vegetables etc. grouped by type and arranged in rows or patches in a circumscribed or fenced area constitutes, in any case, a response to practical needs. Heaps of stones and walls facilitate the conservation of humidity and protect more delicate species from the wind; the construction of dry-stone walls was originally consequential upon the removal of stones from the soil and only in certain situations did they mark property boundaries.

Published and unpublished data from sites located mainly in northern Italy (Fig. 4.9) are gathered and analysed in this paper in an attempt to link the archaeobotanical findings with fundamental questions concerning fruit gathering and cultivation over time (from the Neolithic up to the Iron Age) and across different areas (from northern to southern Italy). In later periods, namely during the Roman Empire, it becomes increasingly difficult to distinguish local from imported products; at the same time, the spectrum of plant species becomes more and more homogeneous from northern to southern Italy (*e.g.* Sadori *et al.* 2009).

The Neolithic

The importance of fruit throughout the history of human nutrition in Italy is not in doubt; archaeobotanical records consistently indicate the existence of a wide variety of fruit, notwithstanding the fact that non-waterlogged archaeological deposits do not readily preserve this material.

In the Mesolithic, only the consumption of hazelnuts (*Corylus avellana* L.) is reliably documented; these have been found in diverse Alpine and Apennine locations (*e.g.* Cornizzolo, near Como, Castelletti *et al.* 1984; Isola Santa, near Lucca, Leoni *et al.* 2002). The consumption of fruit during the Neolithic (*ca.*

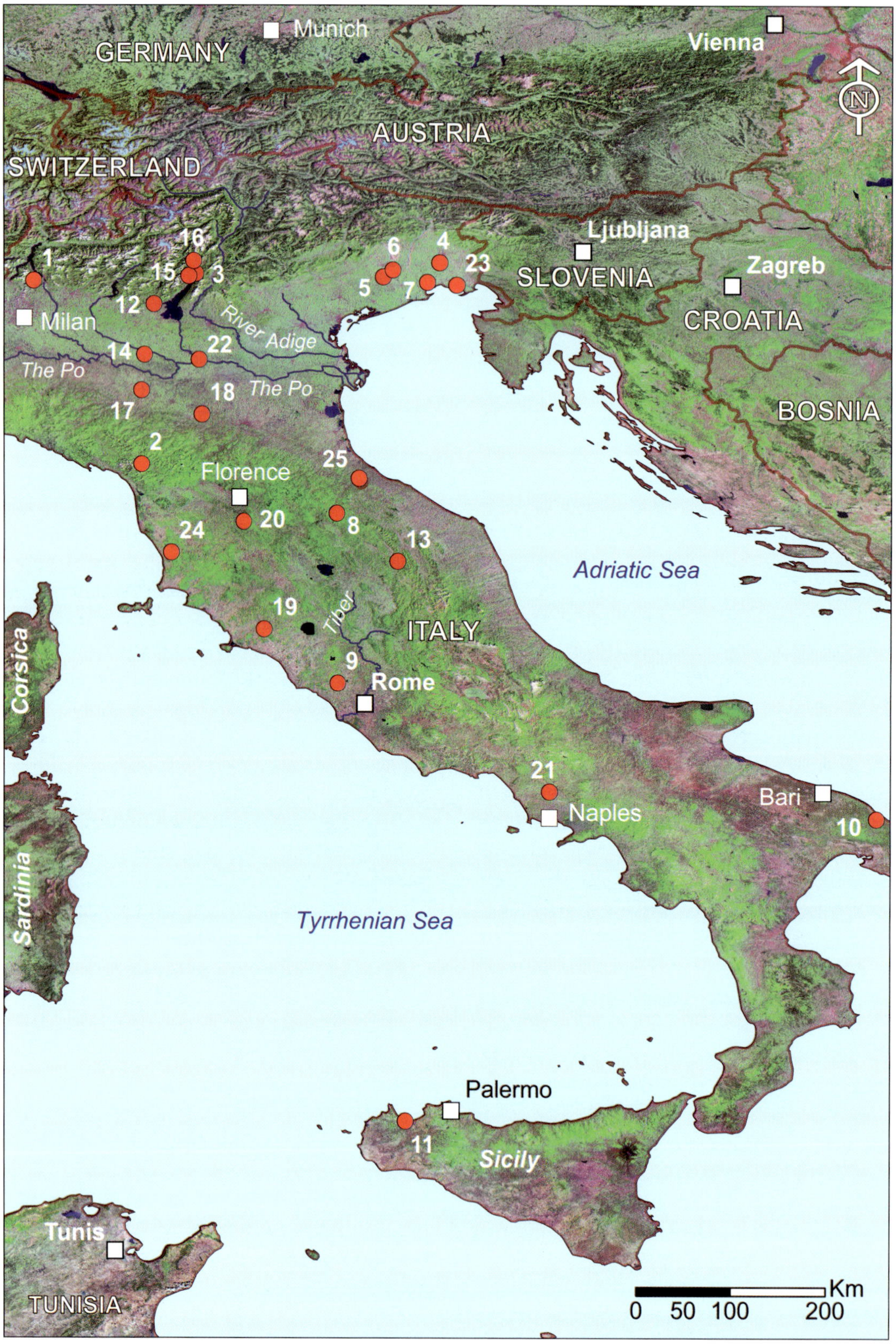

Fig. 4.9. Italian archaeological sites with analysis of macroremains mentioned in the text (type of preservation: c=charred; w=waterlogged; m=mineralised). ***Mesolithic***: 1) Cornizzolo (c); 2) Isola Santa (c). ***Neolithic***: 3) Riva del Garda (c); 4) Sammardenchia (c); 5) Fagnigola (c); 6) Piancada (c); 7) Bannia (c); 8) Pieve S. Stefano (c); 9) La Marmotta (c, w); 10) Torre Canne (c); 11) Grotta dell'Uzzo (c). ***Chalcolithic***: 12) Monte Covolo (c); 13) Meduno (c). ***Bronze Age***: 14) Castellaro del Vhò (c, w, m); 15) Ledro (c, w); 16) Fiavé (c, w); 17) Parma (c, w); 18) Montale (c, w, m); 19) Scarceta di Manciano (c); 20) San Lorenzo a Greve (w); 21) Santa Maria Capua Vetere (w). ***Iron Age***: 22) Forcello (c); 23) Aquileja (c, w); 24) Casale Marittimo (c, m); 25) Verucchio (c, w). Map: R. Lugon, J.-C. Loubier and A. Chevalier.

5600–3500 cal. BCE) appears to have been much greater. In addition to hazelnuts, northern Italian settlements have yielded fruit remains of European cornel (*Cornus mas* L.), common dogwood (*Cornus sanguinea* L.), hawthorn (*Crataegus* sp.), fig (*Ficus carica* L.), strawberry (*Fragaria vesca* L.), apple (*Malus sylvestris* Miller), bladder cherry (*Physalis alkekengi* L.), sloe (*Prunus spinosa* L. agg.), pear (*Pyrus* sp.), acorn (*Quercus* sp.), blackberry and raspberry (*Rubus fruticosus* L. agg., *Rubus idaeus* L.), dwarf elder and elder (*Sambucus ebulus* L., *Sambucus nigra* L.), water chestnut (*Trapa natans* L.) and grape (*Vitis vinifera* subsp. *sylvestris* (Gmelin) Hegi) (Rottoli and Castiglioni 2009).

The same species are also present in peninsular Italy, although in general information is scarcer (Costantini 2002). At La Marmotta, an underwater site in Lake Bracciano (near Rome), where conditions are particularly favourable for preservation, the same fruits listed above for northern Italy are present, with the addition of the fruit of the bay tree (*Laurus nobilis* L.), a plant typical of the local vegetation (*Lauro-Carpinetum betuli*, Lucchese and Pignatti 1991, see Blasi *et al.* 1995), which must have grown to a considerable size since it was also used for making posts in the pile-dwellings (Rottoli and Fugazzola Delpino 2000). The use of fruit is not clear, as a production of oil is possible. At La Marmotta, *Prunus* stones are abundant and the size of some of these (10–12 mm long) suggests that they may derive from still wild forms of damson, *Prunus domestica* L. subsp. *insititia* (L) C.K. Schneider. The Mahaleb plum (*Prunus mahaleb* L.) has been found at Pieve S. Stefano in Tuscany (Castelletti *et al.* 1992).

At sites further south (Torre Canne, Brindisi, Puglia; Grotta dell'Uzzo, Trapani, Sicily) other, typically Mediterranean, species were collected (strawberry tree berries, *Arbutus unedo* L.; almonds, *Prunus dulcis* (Miller) D. A. Webb [syn. *Amygdalus communis* L. and *Prunus amygdalus* Batsch], wild olive, *Olea europea* L.; Costantini 2002).

The list of species occurring in the Neolithic includes, together with fruits generally consumed as food, other fruits which were toxic or at least not particularly pleasant to eat; these were perhaps gathered for use in dyeing or as medicine or oil, such as, for example, dwarf elder (*Sambucus ebulus* L.) or common dogwood (*Cornus sanguinea* L.; *cf.* Maier 1990; Karg and Märkle 2002).

Quantities are often reduced, although clear exceptions are the particularly large amount of hazelnuts found in settlements in the northeast of the country (Sammardenchia, Friuli: Rottoli 1999a; Fagnigola, Friuli: Carugati *et al.* 1996), and the number of grape pips present at La Marmotta, which had not been burnt. At La Marmotta, where numerous fragments (charred and uncharred) of *Vitis* wood were recovered, the occurrence of narrow-necked pottery vessels suggests the possibility that these containers may have been used to serve wine, although further evidence of such a production has not been found to date; the morphological characteristics of the grape seeds (according to Stummer 1911; Mangafa and Kotsakis 1996; Jacquat and Martinoli 1999) fall entirely within the range of the wild form.

There is no proof that cultivation practices were adopted during the Neolithic; fruit was gathered from naturally occurring plants which may, perhaps, have received some kind of human encouragement; there is no concrete evidence that plants were intentionally grown from seed. The analysis of firewood charcoal from several Neolithic settlements in the northeast (Sammardenchia, Fagnigola, Piancada: Carugati *et al.* 1996) suggests the existence of semi-natural hedges of fruit trees (*Pomoideae*) at field margins, with the combined function of keeping animals out and allowing fruit and good firewood to be easily collected. The large amount of charred wood, with a considerable proportion of branches, and the increase in other light-seeking trees (*Corylus*, *Prunus*), suggests that hedge cutting and pruning were practiced, as portrayed on the German Linearbandkeramik (Kreuz 1992).

The Eneolithic and the Bronze Age

A greater abundance of acorns, already present on Neolithic sites but generally in small quantities, is the only feature which distinguishes northern Italian Eneolithic (Copper Age) settlements (*ca.* 3500–2100 cal. BCE). This seems in some way connected with an increase in pastoral activity and animal-rearing in general (Ferrari *et al.* 2002) and the need for increased production of tannin for tanning. In the few Eneolithic sites excavated, the cherry (*Prunus avium* L.) makes its first appearance and there is an increase, with respect to the Neolithic, in the use of cornel (Monte Covolo: Pals and Voorrips 1979; Meduno, Friuli: Castiglioni *et al.* 2003).

True cornel orchards, areas of trees subject to human care or protection, seem to have been present in the Bronze Age, when this species became extraordinarily important in northern Italy, probably for the production of a fermented drink. Wild trees produce a considerable quantity of fruit, but the amount increases greatly when each plant is surrounded by an open area and left to enhance the tree's growth. At present, the cornel is a fairly rare species in lowland woods (Pignatti 1982); to ensure such a plentiful production of fruit, the trees must have been artificially propagated, i.e. grown from seed (Rottoli 2001). In contrast to the carpological evidence, pollen analyses indicate no particular abundance (Ravazzi et al. 2004; Mercuri et al. 2006); this is perhaps due to the tree's modest production of pollen, although the transport of fruit from areas distant from the sites where they were found has been hypothesised (Fiavé, Trentino: Greig 1984). In addition to pits from the fruit, there is a notable record of wood and wood charcoal, although it is not possible to distinguish the wood of *Cornus mas* L. from that of *Cornus sanguinea* L.; fruit of the latter species is, however, quite rare during the Bronze Age.

In addition to cornels, in the Terramare – the Po Plain Middle Bronze Age sites which have yielded most archaeobotanical information – the same fruits listed above for the Neolithic are also found. At Montale, rowan berries (*Sorbus domestica* L.) are also present and cherries (*Prunus avium* L.) (Mercuri *et al.* 2006) are more frequent; roses (*Rosa* sp., Castellaro del Vhò, near Cremona: Rottoli 2001) also occur.

With respect to cereals, fruit is generally plentiful and acorns occasionally present in large quantities (*e.g.* at a pile-dwelling, the Palafitta di Parma: Lancelotti 2005; Casartelli 2007). The occurrence of single-species accumulations of fruit in Bronze Age sites is actually rare; the few exceptions consist of deposits of hazelnuts, cornels and crusts of charred acorns at the bottom of containers. A most unusual case is that of a late Bronze Age (*ca.* tenth century BCE) site at Scarceta di Manciano (Grosseto) in Tuscany, where a building caught fire immediately after a considerable quantity of sloes (*Prunus spinosa* L. agg.) had been collected, perhaps whilst these were being cooked or toasted (Rottoli 1999b). Cornels, hazelnuts, strawberries, bladder cherries, apples, pears, blackberries, raspberries, grapes and acorns are also found in the Adige Valley in the Alps (Fiavé and Ledro, Trentino: Jones and Rowley-Conwy 1984; Karg 1998; Dalla Fior 1940).

It seems reasonable to assume that fruit cultivation in Italy grew up over several centuries, between the end of the Bronze Age and the beginning of the Iron Age (*ca.* 1200–900 BCE), and developed in non-linear fashion generally from the south towards the north; it seems initially to have developed along the Ionic and Adriatic coasts.

In the later Bronze Age (Late and Final Bronze Age, thirteenth-tenth centuries BCE), finds of olive stones become progressively more frequent in southern Italy (Fiorentino *et al.* 2004). The beginning of olive cultivation is a matter of debate; some scholars hold that the information available is not yet sufficient to demonstrate the existence of cultivation prior to the foundation of the first Greek colonies (Costantini 2002; Fiorentino 1998), whereas others maintain that its occurrence in the extreme south of Italy from the Middle Bronze Age onwards is probable, or even certain (Nisbet and Ventura 1994; Fiorentino *et al.* 2004). In the Ligurian mountains, the *Olea* pollen curve is continuous from 3800 BP on, evidencing coastal distribution of the wild plant (Lowe *et al.* 1994). An early cultivation was supposed, but macroremains are however absent up to the Late Iron Age.

Grape pips are found sporadically in central and southern Italian settlements; there is no clear evidence of local wine manufacture, although wine probably circulated in the central Mediterranean in connection with trade exchange for material of Aegean origin (Martinelli *et al.* 2003). The only exception to this lack of positive evidence is the singular find in a well in Santa Maria Capua Vetere (Caserta) (Castiglioni and Rottoli 2001) of numerous pruned pieces of vine shoot, which would seem to indicate the existence of specialised viticulture in the Middle Bronze Age, at least in some areas towards the Tyrrhenian coast. Pollen analyses from the Terramare (Palafitta di Parma: Valsecchi *unpublished*; Montale: Mercuri *et al.* 2006) seem to confirm that interest in vines also increased in northern Italy at the end of the second millennium BCE. At Montale in particular, the pollen diagram shows a growth in the presence of such pollen during the final period of occupation (Late Bronze Age) (Mercuri *et al.* 2006). Wine production appears, though, to catch on much later, with Etruscan expansion in northern Italy.

The increase in vine pollen, which is observed in other sites of the same period in Tuscany, may in part be due to a progressive increase in humidity or in the frequency of poorly-drained areas which offered environments that would have favoured the spread of wild vines (Mariotti Lippi *et al.* 2007), but it might well also be a sign of greater human interest in the species, as recent finds appear to suggest (San Lorenzo a Greve, near Firenze: Bellini *et al.* 2008).

There are very little data regarding the fig, the remaining typically Mediterranean species, but the reason would appear to be the difficulty of recovery of the tiny 'seeds' (see endnote 4).

During the Bronze Age, walnut shells (*Juglans regia* L.) also become common in central and southern Italy (Bellini *et al.* 2008); these are known from the Neolithic in only three sites (in northern Italy, Sammardenchia and Bannia, Friuli: Rottoli and Castiglioni 2009; Riva del Garda, Trentino: Mottes *et al.* 2010), in which they may have arrived from the Balkans as gifts or curiosities. The tree could be indigenous in the Euganei hills (Veneto, northeastern Italy: Kaltenrieder *et al.* 2010) but, at the moment, macroremains are completely absent in this area up to the Romanisation period (second century BCE).

The Iron Age

With respect to the species recorded from preceding periods, the scarcity of excavations in Iron Age wetland sites means that we have a poor idea of the situation, with an apparently low consumption of fruit during this period. However, the introduction of efficient systems of fruit cultivation – which are uncertain and open to debate during the Bronze Age – is known with certainty to have occurred in the Iron Age (*i.e.* the first millennium BCE), on the basis of a combination of historical, archaeological and archaeobotanical information. Although the precise chronology of the various events may not be determined, during this period the techniques of grafting and propagation by cuttings were introduced from Greece and the East, and specialist beekeeping developed; these methods allowed the production of cultivated varieties of fruits, the growth of which from seed had not hitherto guaranteed products of good quality (Zohary *et al.* 2012). Trade in fruit also became established, and 'exotic' (or recently introduced) fruit became an item of prestige amongst food offerings deposited in burials. In Aquileia, a pre-Roman and Roman port on the Adriatic, olives and seeds of white-flowered gourd (*Lagenaria siceraria* (Molina) Standley) were found in a layer from the eighth century BCE settlement (Scotti Maselli and Rottoli 2007). A seventh century BCE princely tomb in Casale Marittimo (Pisa) contained a pomegranate fruit (*Punica granatum* L.), together with hazelnuts and apples. In the same tomb, a honeycomb was preserved as well (Rottoli 1999c). A white-flowered gourd seems also to have been present in a princely grave in Verucchio dating to the eighth-seventh centuries BCE (Gentili 1985).

Olive and vine cultivation were certainly widespread in the sixth-fifth centuries BCE in Etruscan territory (Cristofani 1987), when the export of Etruscan wine to central Europe is attested, but these activities actually began in the early first millennium BCE. Broad-bladed bronze pruning hooks, specialised tools for viticulture, first appear at the end of the Bronze Age (tenth century) in peninsular Italy (Bietti Sestieri 2002). It is probable that at this time vine cultivation started, with the employment of the various methods which were subsequently described by Roman writers (with the plants attached to living supports, such as elm or maple).

Landscape studies, particularly those concerning hill settlements, show the diffusion of villages with scattered houses, adapted to the establishment of market gardens and fruit trees between the buildings; in settlements at higher levels, the villages are compact and the few fruit trees able to survive are planted next to the houses, so as to make use of the protection they offer.

In fact, systems of landscape organisation have recently been detected in more ancient lowland contexts: the use of irrigation associated with methods of division of agricultural land is known in the Campanian plain (near Naples) from the Early Bronze Age, when villages and fields were buried by volcanic ash from an ancient Vesuvian eruption

(Bietti Sestieri 2002). Groups of trees, especially oak, were also present near the houses, and brambles were frequent towards village margins (Castiglioni and Rottoli, *unpublished*). Traces of irrigation systems have been recognised in association with the Middle Bronze Age Terramare in the Po Plain (Bernabò Brea *et al.* 1997), although it is not known whether these structures were connected with intensive fruit production.

During the Bronze Age, the management of trees for pastoral purposes and multiple widespread fires (some of which may have been meant to enlarge grassland) influenced the development of pre-existent hazelnut populations, as recently described in western Apennines (De Pascale *et al.* 2006).

In the course of the first millennium BCE, arboreal cultivation, especially of vines, tended to occupy increasingly large areas of land, and extended further in the Roman period. Notwithstanding evidence of local wine production in many sites, Greek wines continued to be imported for a long time (*e.g.* at Forcello, near Mantova: Castelletti and Rottoli 1986). Although olives (in the peninsula) and vines (in both northern and southern Italy) probably tended to progressively cover the landscape, other extensive cultivations do not seem to have developed until the Roman period.

The cultivation of chestnut trees (*Castanea sativa* Miller), which still today involves a large proportion of the Apennines and the hilly area of northwestern Italy (Cevasco *et al.* 2010), is controversial. Palynological analysis in Liguria (see for example, Cruise *et al.* 2009) and the Euganei hills (Kaltenrieder *et al.* 2010) seems to demonstrate the cultivation during the Neolithic or the Bronze Age, but anthracological, carpological and historical data show that cultivation become established only after CE 100 (Conedera *et al.* 2004; Cottini and Rottoli 2001) in southern Italy, in the northern Apennines (Liguria, Tuscany and Emilia-Romagna) and in the region of Insubri, between Lombardy and Ticino (Switzerland).

Towards the Empire

The accumulated evidence – especially that from archaeobotany – suggests that even before the complete Romanisation of Italian territory, the already wide selection of fruit available since the Neolithic had gradually been extended by the addition of new species, which rapidly entered into local cultivation. A continually increasing circulation of ideas and raw materials, from the flowering of Mycenaean and later Greek culture, led progressively to the knowledge and then the cultivation of new foods, particularly fruit and vegetables, first in coastal areas and Greek colonies and subsequently in the interior and north of the country.

In the Republican period, in the second and first centuries BCE, at least eleven different fruits, either certainly or probably cultivated (grape, fig, apple, pear, cherry, cornel and strawberry), or wild (blackberry, rowan berry, hazelnut and elderberry) were available to the Roman *coloniae* in northern Italy (Bandini Mazzanti *et al.* 2001). As far as is known, though, not all 'Mediterranean' fruits were easily obtainable (almonds and pine nuts are missing from the list, for example, and olives were rare). In the Imperial period, the appearance of new, recently imported species (such as peach *Prunus persica* (L.) Batsch., jujube *Ziziphus jujuba* Miller and plum *Prunus domestica* L. s.s.) was immediate both in northern and southern Italy (Sadori *et al.* 2009); the presence of a powerful unified government and efficient communication systems seems to have led to greater homogeneity of consumption and also, to some extent, of production.

Conclusions

Fruits were crucial for the prehistoric and historical diet. Archaeobotanical and palynological data testify to the gathering of wild plants from the Neolithic on, both as food and for other purposes such as medicine or dyeing. As suggested by the discovery of large amounts of hazelnuts in some Neolithic settlements, some human intervention (or encouragement at least) could be hypothesised for this species as well as for others such as apple, sloe and so on – a more systematic gathering and production, supported by landscape structures, is more evident only during the Bronze Age. Nevertheless, cultivation practices were surely adopted from the Iron Age on (ninth century BCE) and probably during the Late and Final Bronze (thirteenth-tenth centuries BCE), in the context of increasing circulation of ideas and materials, first with the Mycenaean and later with the Greek population.

The spread of fruit cultivation progressed non-linearly inwards – from coasts towards the inland areas – and along a south-north axis. Most species from the Mediterranean were known and available in northern Italy before the Romanisation period, as was grafting and propagation by cutting techniques. Due to this common background, the introduction and cultivation of new fruits coming from the East could take place contemporaneously all over northern and southern Italy during the Roman Empire.

4.5. *CITRUS* (RUTACEAE) WAS PRESENT IN THE WESTERN MEDITERRANEAN IN ANTIQUITY

Bui Thi Mai and Michel Girard

Introduction

Many exotic or indigenous food plants were utilised in the Mediterranean region in Antiquity. Their presence is attested in classical texts, but finds of their remains are relatively rare. Most of the determinations available are based on carpological analyses but, in some cases, palynology has an important role to play in our knowledge of such plants, most especially in analysing the pollen of plants brought in from elsewhere. Thus, it has been possible to provide evidence that a *Citrus* (*cf. Citrus medica* L. = citron) was under cultivation several centuries before the Christian era in the central Mediterranean ports such as Greek Cumae or Punic Carthage (Fig. 4.10). What are we to think of this presence? Might this tree have been cultivated in an urban context without us having any formal proof of the practice?

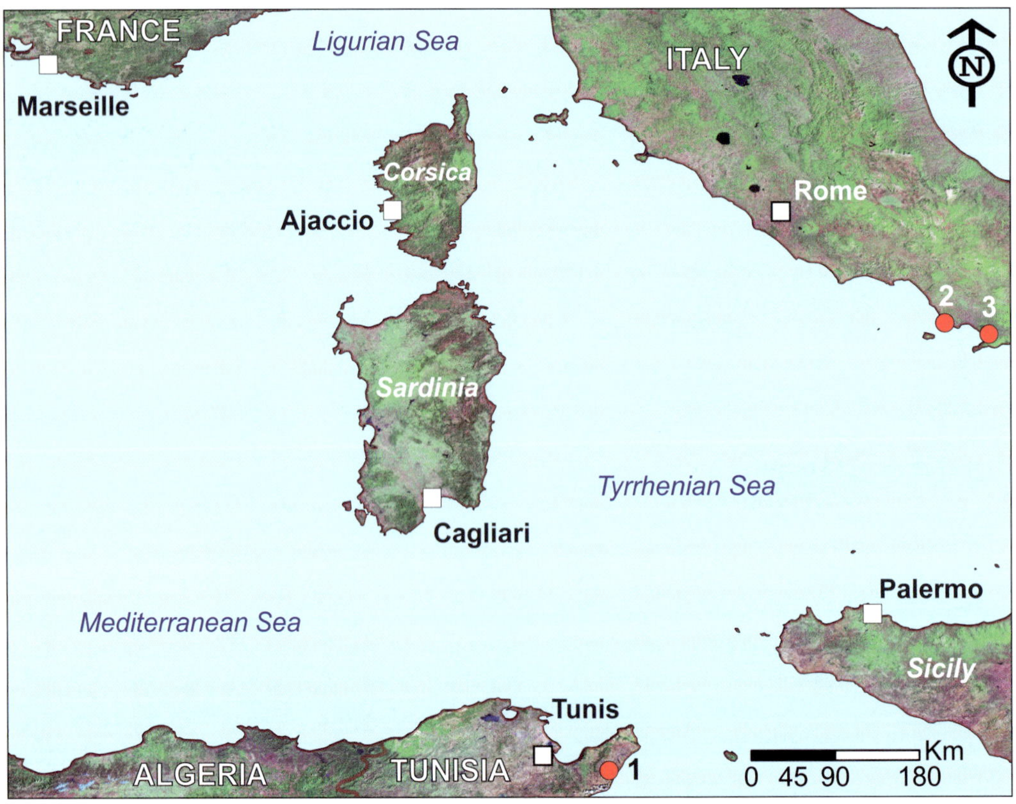

Fig. 4.10. Map of sites in the central Mediterranean with palynological evidence of *Citrus*. 1) Carthage; 2) Cumes; 3) Pompeii. Map: R. Lugon, J.-C. Loubier and A. Chevalier.

Botany

The agrumes (citrus fruit), which belong to the *Rutaceae* family, include the species *Citrus*, *Eremocitrus*, *Microcitrus*, *Clymenia* and *Poncirus* in Eurasia. *Citrus* is represented by species and their hybrids that produce edible fruits such as the lemon, orange, tangerine, bergamot, lime, grapefruit, pink grapefruit (pomelo) and tangerine, while *Poncirus trifoliata* (L.) Raf., which can resist temperatures of -20°C, does not have edible fruit and is essentially utilised as a grafting stock. According to Mabberley (1997), the older name of the species *Fortunella japonica* Thunb. has been included in the genus *Citrus* (*Citrus japonica* Thunb.), but Barkley *et al.* (2006) consider the genus *Fortunella* as a distinct taxon from eight known species giving kumquats. These fruits, whose sizes vary according to the taxon, have an outer fruit coat (exocarp) ranging in color from green to orange, surrounding a green, yellow or orange pulp divided into sections containing the pips.

The nomenclature of agrumes has led to considerable controversy in recent decades. Although the first classifications posited the existence of over 150 species, the latest studies made on the phylogeny of agrumes (Mabberley, 1997) tend to indicate that the original wild species are not very numerous and include only six natural species: citron (*Citrus medica*) (Fig. 4.11), mandarine (*Citrus reticulata* Blanco), true grapefruit (*Citrus maxima* (Burm.) Merr.), kumquat (*Citrus japonica*) and two as yet unknown species. The classification from Scora (1975) and Barret and Rhodes (1976), based on 146 morphological and biochemical characters, suggests that only *Citrus medica*, *C. maxima* and *C. reticulata* should be considered 'true species', while the other cultivated ones more likely represent hybrids occurring from cross-breeding or natural events. Recent phylogenetic[2] work based on SSRs molecular (Barkley *et al.* 2006) and AFLP markers (Pang *et al.* 2007) supports these original suggestions.

The great genetic flexibility of agrumes – highly frequent spontaneous mutations and a remarkable capacity to hybridise – gave rise to an impressive number of varieties represented by hundreds of cultivars. According to Mabberley (2004, 486–487), the yellow lemon tree (*Citrus* x *limon*) is thought to have descended from the citron (*Citrus medica*) and a first unknown species; the lime (*Citrus* x *aurantifolia*) to come from the grapefruit (*Citrus maxima*) and a second unknown species. The sweet orange (*Citrus* x *sinensis*) may come from a group of hybrids (*Citrus* x *aurantium*) arising from the mandarine and the grapefruit. It is not sweet unless the mandarine is dominant, and is sour (*Citrus* x *aurantium*) when the true grapefruit is preponderant. The bergamot orange (*Citrus* x *bergamia*) is thought to have come from hybridisation of the citron and one of the groups of hybrids. This genetic plasticity was taken advantage of very early by peoples in the Far East, who went on to unceasingly improve the plants' sugar and acid content, texture and colors over the last two or three millennia through successive selections and crosses (Bottin 2008).

Fig. 4.11. Flowers and fruit of *Citrus medica*.

Geographical Origin

Some consensus exists about the role of Southeast Asia (southwestern China, northeastern India, Myanmar (Burma) and the Malay archipelago) in the origin and diversification of the genus *Citrus* (Gmitter and Hu 1990).

History

The history of cultivation of agrumes is still uncertain, most especially as regards their introduction into the Occident. The data concerning the zone extending from the Far East to the Middle East have been described by various authors (particularly Bretschneider 1871; de Candolle 1883).

E. Bretschneider notes that the Chinese consider oranges to be among wild fruit, that most of their trees are indigenous and had been cultivated for a long time, since the varieties of fruits have distinct names mentioned in texts. The same author states that around the beginning of the Christian era, the Indians gave the name of *nagrunga* to a fruit lacking a pleasant taste (*cf.* bitter orange). Referring to the works of the historian Josephus, A. de Candolle observed that the Hebrews were familiar with the citron, since they carried it in their festival of Persian apples, *Malum persicum*, one of the Greek names for the citron. Later, the Crusaders (eleventh–thirteenth centuries CE) saw the bitter orange in Palestine.

The European history of the cultivated citron seems to begin only around 300 BCE. The Greeks became familiar with the fruit through the Persians, whence the name *medica* and Theophrastus (*ca.* 372–*ca.* 288 BCE) in fact cites it under the name of 'apple of Media' and 'apple of Persia', without mentioning whether it was cultivated in Greece (Amigues 2007). A. de Candolle (1883, 143) notes that '*the Romans did not have it in their gardens at the beginning of the Christian era*', although it was depicted on frescos and mosaics in Pompeii, and that according to Targioni '*the species was cultivated in Italy in the 3rd or 4th century AD*'. However, according to Palladius, cultivation was well established in the fifth century. By the tenth century, Arab physicians were prescribing its juice and they introduced this shrub into Spain (Ramón-Laca 2003). Hybrids such as the bitter orange and the sweet orange were not introduced into the Mediterranean basin until around the mid-thirteenth century, while the mandarine-tree only arrived in Europe in the nineteenth (Webber 1967). A complete survey of the Greco-Roman written sources and iconography dealing with the description, introduction and uses of the *Citrus* species can be found in C. Pagnoux (2010).

Pollen Morphology (see e. g. Halbritter et al. 2007)

According to various authors (Erdtman 1966; Nair 1961; Grant *et al.* 2000; Mariotti-Lippi 2000, among others) and to the reference slides in our own collection, agrume pollen has tetra- and/or penta-colporate (stephanocolporate)[3] grains, the polar axial length of which varies between 30 and 20µm, depending on the species. Their exine[4] has a reticulum[5] with more or less well developed meshes; they are very slender in *Citrus maxima* and relatively large in *Citrus medica*, for example.

Citrus medica (Fig. 4.12), which is of particular interest to us here, has the following characteristics, as outlined by Grant *et al.* (2000):

- Pollen grains tetra- or tetra-/pentacolporate, some grains with a syncolporate[6] tendency
- Polar axis = 16 (17.9) 19 µm
- Equatorial axis = 18 (18.2) 19 µm[7]
- Mean ratios usually given as 1:10, (oblate spheroidal ± subprolate)[8]
- Polar outline circular, or where angular, plana-perturate
- Equatorial outline elliptical or circular
- Exine thin in cross section
- Nexine[9] with a little or much thickening around endoapertures
- Nexine < 1 µm thick at poles, 1 µ m thick around endoapertures
- Sexine[10] of uniform thickness throughout (1µm). Sexine 1 of short columellae. Sexine 2 a heterobrochate semi-tectate *reticulum*, muri thick, simplicolumellate, smooth, rounded, with no further ornamentation
- Porus (endoporus): 2–3 µm, nearly circular in shape
- Colpus (ectocolpus) very long (11 (13.4) 15 µm) narrow, slightly broader toward the equator (2 (2.5) 3 µm)
- Granules present on ectocolpus membrane

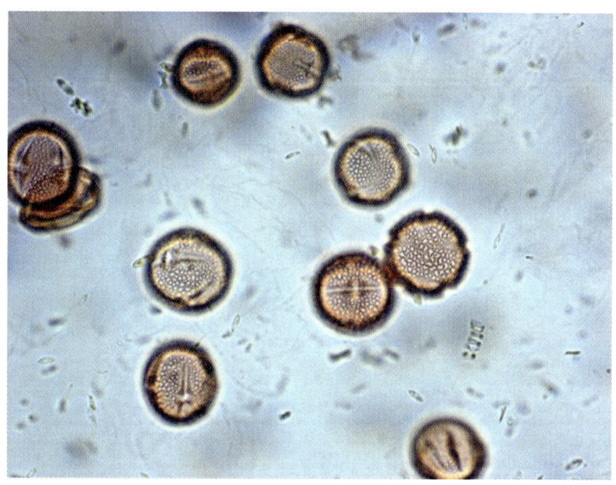

Fig. 4.12. Pollen grains of *Citrus medica*.

According to Maria Mariotti-Lippi (2000), the *reticulum* of *C. limon* and *C. medica* is slightly different. These differences were detectable by Scanning Electron Microscope (SEM), but were less discernible under light microscopy. In *C. limon* the *lumina*[11] were rather more roundish in shape than in *C. medica*. The *reticulum* of *C. medica* pollen was made up of fewer regularly shaped *brochi*; there was a prevalence of large *brochi*. Moreover, as observed by SEM, the ratio between the *muri*[12]/*lumina* in *C. limon* was slightly larger than in *C. medica*. These details are not detectable by optical microscopy; the author refers only to the genus *Citrus* sp. for its fossil grains.

Pollination

Agrumes are characterised by entomophilous pollination (by insects) that disperses very little of their pollen into the air. In order to find relatively large amounts of their pollen in the environment, it is necessary for the trees to be fairly numerous in the area under consideration.

Pollen vs. Seeds

Although this sort of pollen remains strictly in the vicinity of the site from which it is emitted, *Citrus* seeds may come from products that circulate in trade networks linking the western Roman world to Asia Minor and even beyond (the 'Silk Road', for example).[13] Transport of fresh fruit over a longer time appears to be unlikely and, hence, these remains might well indicate dried fruit such as we find in present-day Syria. Some fifteen years ago, we were given 'loomi'[14] (dark-coloured dried citrons) from Syria that are broken up into hot water to prepare an aromatic drink. They are also used in cooked dishes and skin creams. The samples we were able to keep as ethnobotanical examples indicate that they remain unchanged after many years.

The Contributions of Pollen Analysis

Pollen analyses were carried out in the sediment deposits of ports of Carthage in Tunisia, and Cumae and Pompeii (Porta Nocera, Tomb 5, Enclosure 23 of the necropolis) in Italy, and have provided information about the cultivation of agrumes in the mid-Mediterranean over the centuries preceding the Christian era.

Carthage

W. Van Zeist (Van Zeist *et al.* 2001) did, in fact, discover *Citrus* pollen in the sediment of the Punic port of Carthage in the levels contemporary with its period of activity, that is, well before the Punic Wars between 264 and 146 BCE, which led to Carthage being wiped out. The chrono-stratigraphic location of this microscopic botanical testimony is important, because it enabled the author to assert that *Citrus medica* cultivation was being undertaken in this area towards the mid-fourth century, a fact which did not surprise him, seeing that cultivation of the agrume is well attested in the eastern Mediterranean for the same period.

Cumae

Cumae, a city of Magna Graecia located in the Italian region of Campania on the Tyrrhenian Sea coast northwest of Naples, along with the *Phlegraean Fields* just to the west of the city, is today an archaeological zone of the greatest importance, with many and varied remains, among which the Sybil's oracle cave is the most illustrious. Founded by the *Euboeians* around 750 BCE, Cumae was to become one of the Greeks' first commercial settlements in Italy and led to the founding of Naples (Neapolis). It played a decisive role in the Hellenisation of Campania and in transmission of Greek civilisation to the Etruscans, then to the Romans (Bats 2005; Bats *et al.* 2009).

A core sample of sediment 6m long down to -7.95m was taken and dated (Stefaniuk *et al.* 2005). The six dates that resulted located the sequence between the eighth century BCE and the fifteenth–eighteenth centuries CE, which represents a minimum time span of some 2,500 years. Pollen analysis indicates the existence of open landscapes and a broad agricultural use of the hillsides with various crops, including different cereals and linen, the appearance of fruit tree cultivation such as grapevines, olive, walnut, and myrtle, as well as citron.

Discovery of this last taxon is quite important, since 70 pollen grains of *Citrus* were observed throughout the sequence (Fig. 4.13). The first occur at level -641cm after the date 896/657 cal. BCE, mesured at level -656 cm and they can be observed thereafter fairly regularly until a climax in the fifteenth to eighteenth centuries. Thus, regular and relatively consequential occurrence of the pollen enables us to confirm the presence of this fruit-bearing Rutaceae

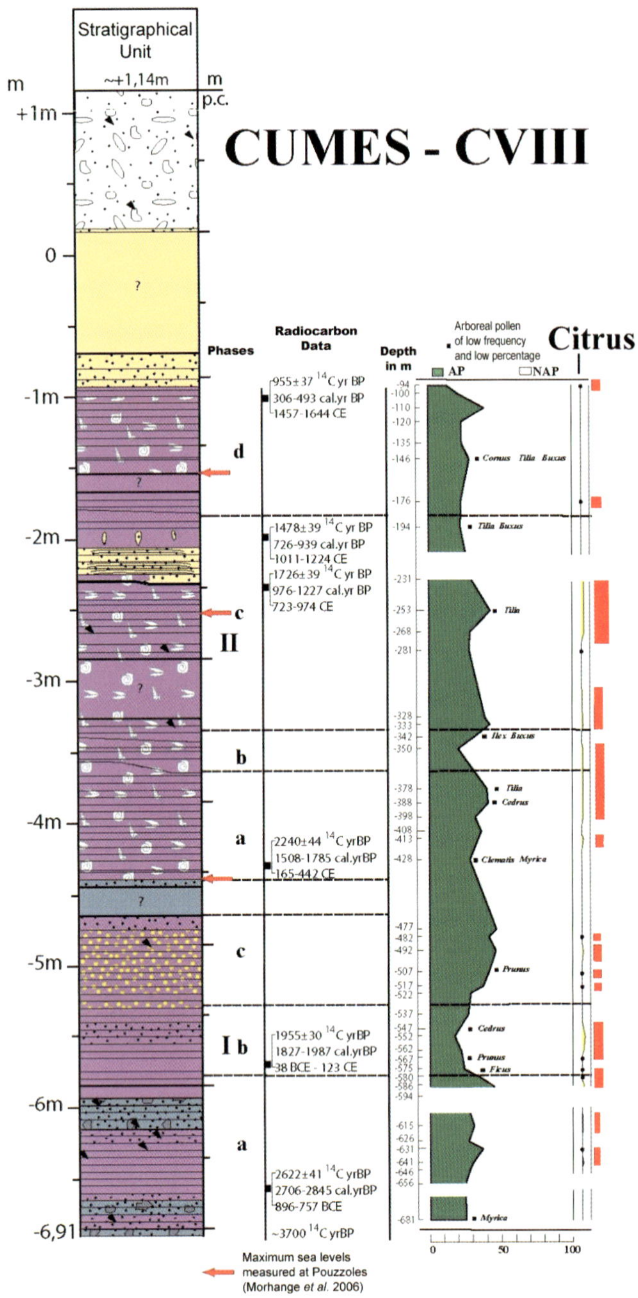

Fig. 4.13. (Partial) pollen diagram of the core sample from CUMAE VIII (curve of trees and herbaceous plants in green and *Citrus* sp. in red). Analyses by Bui Thi Mai.

in the area of Cumae for this time period. It would obviously be important to precisely date the -641 cm level that the first pollen appears in. This horizon is located between two unfortunately rather distant dates: 896/657 cal. BCE at -656 cm and 38 BCE and cal. CE 123 at -575 cm; that is, there is nearly a metre of deposits between the initial occurrence and the later dates.

Pompeii

M. Mariotti-Lippi (2000) discovered pollen in the sediments of Garden VII at Pompeii (destroyed by Vesuvius on the 24th of August, 79 CE). The author attributed these grains of pollen to *Citrus medica* and *Citrus limon*, but points out that precise identification down to species level is very difficult. The pollen grains examined may have belonged to *C. limon*, since the *reticulum* of the Pompeian grains is fairly similar to present-day reference pollen. We also found a *Citrus* pollen grain in Grave n° 5 of the Porta Nocera, Enclosure 23 of the Necropolis (work in progress). The presence of *Citrus* is also confirmed in this site by macro-remains, in particular in the Temple of Venus (third–second centuries BCE) and in the Temple of Fortuna Augusta (10 BCE–3 CE) (Coubray *et al.* 2010). In the 'House of the Orchard' paintings, several trees were portrayed, including recognisable lemon-trees.

Conclusion

The presence of *Citrus* pollen in various sites in the central Mediterranean seems to confirm that this agrume was indeed cultivated in the western part of the ancient world, since these pollen grains imply local existence of the trees that produced them. The fact that this pollen is found in ancient ports may not be fortuitous. Transportation of agrume trees might have been undertaken by sailors who would thus have introduced them to gardeners in the region. If this were indeed the case, these trade ports could be considered the sites through which this fruit tree was introduced. Although the first findings in Cumae were unfortunately not precisely defined chronologically (somewhere between 757 BCE and 38 BCE/123 CE), we can nevertheless propose the hypothesis that *Citrus* was present in southern Italy, probably before the first century of the Christian era.

4.6. FROM SECONDARY TO SPECULATIVE PRODUCTION? THE PROTOHISTORY HISTORY OF VITICULTURE IN SOUTHERN FRANCE

Laurent Bouby, Philippe Marinval and Jean-Frédéric Terral

Introduction

All the traditional grape cultivars have been domesticated from the wild grape (*Vitis vinifera* L. subsp. *sylvestris* C.C. Gmelin). It is a well-known liana climbing on trees and rocks in Mediterranean and mesophilous forests – especially riparian forests – of southwest Asia, North Africa and Europe (Levadoux 1956; Zohary *et al.* 2012; Arnold *et al.* 2005). Wild grape is autochthonous in France where a few small populations are still reported today (Levadoux 1956; Arnold *et al.* 1998; This *et al.* 2001; Lacombe *et al.* 2004). It would have been much more common in the past but would have lost ground due to human impact and American pathogens (phylloxera, mildew...) (Arnold *et al.* 1998; 2005). Palaeobotanical data document the presence of *Vitis* in France all through the Quaternary (*e.g.* Planchais 1973; Renault-Miskovsky 1976; Heinz and Thiébault 1988). According to some pips found in the Early Palaeolithic site of Terra Amata (Nice; Fig. 4.14), fruits would have been collected and eaten as early as around 400,000 BCE (Boone and Renault-Miskovsky 1976). There is much more evidence of such use from the Epipalaeolithic and Mesolithic onwards (Marinval 1997).

The ancient history of viticulture in southern France is closely connected to the development of influences from foreign Mediterranean states and peoples which occur throughout the Iron Age and Antiquity (*e.g.* Py 1993; Roman and Roman 1997; Garcia 2004; Dietler 2007). The beginning of Mediterranean goods importation seems to occur only during the second part of the seventh century BCE. Very soon after that, around 600 BCE, Phoceans (Greeks from western Anatolia) founded the city of Massalia (modern-day Marseilles), the first permanent colonial settlement in southern France and the first Greek city in the whole western Mediterranean. Massalia is a harbour located on the French Mediterranean coast, close to the mouth of the large Rhône River. From the end of the sixth or the fifth century BCE, Massalia created a series of secondary colonial coastal sites. The Massaliotes were to dominate trade in southern France for centuries. Indeed, the situation was only to really change with the rapidly increasing activity of Roman merchants, starting from the third century BCE, and the Roman colonisation at the end of the second century BCE. The transalpine *Provincia* was founded in 121 BCE and the Mediterranean regions of France and the Rhône valley became quickly and deeply Romanised.

The writings by Greco-Roman authors state that Mediterranean people were the first to cultivate grapes in southern France. Especially well known is the story of the foundation of Massalia related by Justin, a Latin author of the second century CE, who was using older sources. The text claims that under Phocean influence, local Celtic people learnt to prune grape and plant olive trees (Justin, *Epitome of the Philippic History of Pompeius Trogus*, L XLIII, 4, 1–2). Beyond the significant, but patchy and Greco-Roman-centred written record, a wealth of archaeological information about viticulture, wine production and consumption has been built up over

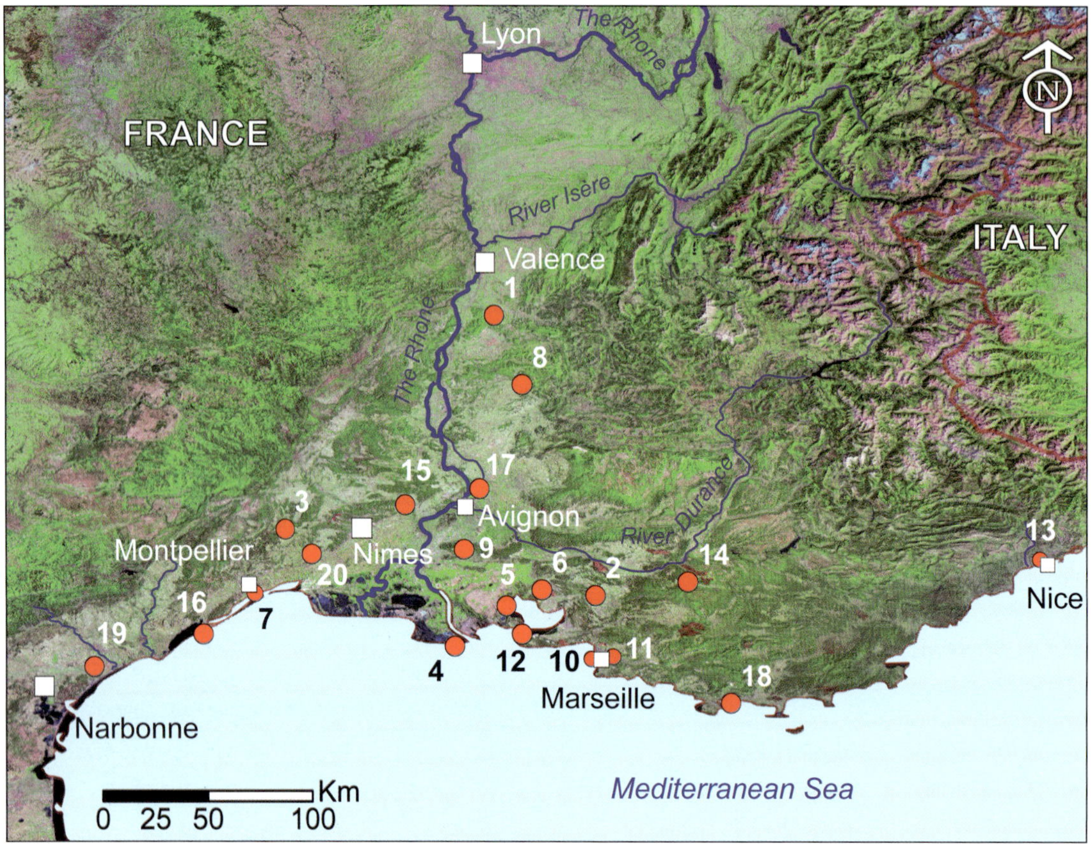

Fig. 4.14. Map of southern France with sites mentioned in the text. 1) La Prairie; 2) Pierredon; 3) Plan de la Tour; 4) La Roque; 5) Castellan; 6) Coudouneu; 7) Lattara; 8) Le Pègue; 9) Les Tremaïe; 10) Massalia; 11) Saint-Jean-du-Désert; 12) L'île; 13) Terra Amata; 14) Les Toulons; 15) Marduel; 16) La Fangade; 17) Mourre de Sève; 18) Besagne; 19) Portal Vielh; 20) Ambrussum. Map: R. Lugon, J.-C. Loubier and A. Chevalier.

recent decades. Thus far, it has not provided any indication of an earlier cultivation than traditionally assumed, but it does raise new questions about the process of development of this first viticulture, at the core of relations between indigenous and colonial peoples. Here we focus on archaeobotanical data, but will refer broadly to other archaeological and textual sources.

Massaliote Viticulture and Indigenous Thirst

The first Mediterranean importations were, above all, composed of ceramics used for the transport and drinking of wine. Wine immediately became the focus of trade between indigenous peoples and colonists (Dietler 2007). It is thought to have played a fundamental role in social, economic and political relations amongst indigenous societies, through hospitality and its ritual significance (Dietler 1990). Besides some Greek ceramics and amphoras of various origins, Etruscan material is largely predominant at the time. It is generally assumed that Etruscan traders were responsible for the import of all these products (*e.g.* Bouloumié 1981; Py 1985). This trade was largely restricted to coastal areas and especially to the lower Rhône Basin (Py 1995; Dietler 2007). However, it has been proposed that Massaliote traders, rather than Etruscans themselves, may have been responsible for the import and distribution of Etruscan wine (Bats 1998).

Archaeological data now clearly shows that Massalia started to produce and export wine soon after its foundation. Grape seeds are common in sixth century BCE layers of the city harbour (Bouby and Marinval 2000). The city began to produce typical amphoras about the middle of the sixth century BCE (Bats 1990; Bertucchi 1992). Hellenistic (fourth–second centuries BCE) vineyards have been evidenced by planting trenches and pits located in

Fig. 4.15. Planting traces of a Hellenistic vineyard in Saint-Jean-du-Désert, near Marseille (Photo: P. Boissinot).

a small valley at Saint-Jean-du-Désert, about four kilometres from the centre of the modern city (Boissinot 1995; 2001). The place is thought to have been part of Massaliote territory at the time, but grapevines could have been planted by indigenous people. The traces are organised into various adjacent fields along a small stream (Fig. 4.15). Grape was the main – if not the only – cultivated plant in these plots, excepting rare trees occurring in one field. In this small valley, grape seems to have occupied most of the cultivated land. Density of plantation is variable from one field to another, but some thousands of grape plants per hectare were cultivated in each vineyard. Such a configuration suggests a rather intensive viticulture. We can assume that it was involved in the production of Massaliote wine. What we can say from these planting traces is in agreement with the description made by the Greek geographer Strabo (first century CE) of a Massaliote territory covered with vines and too poor for grain (Strabo, *Geography*, IV, 1, 5). Other planting traces are reported right at the Greek city's doorstep, dating back at least to the end of the fourth century BCE (Bouiron 2005).

The production of Massaliote amphoras quickly increased and was already important at the end of the sixth century BCE. Their remains have seldom been found in various areas of France and neighbouring countries, but Massaliote wine was mostly sold in the lower Rhône and Mediterranean France (Bats 1990). Amphora sherds are much more common in settlements of all kinds in this area than in any other part of France. Another difference is that imported goods are mainly restricted to modest objects in relation to wine, whereas there is evidence of the import of luxury drinking material but few amphoras in the Hallstatt area. Wine would, therefore, have been much more commonly drunk in the south. Here there would have been a special 'consumer demand' for wine which would have been utilised in feasts and hospitality practices, to mobilise labour and power, in a context of heightened social and political competition (Dietler 1990; 1992).

Grape Seed Morphometrical Features as Evidence of Indigenous Viticulture

Artefacts and archaeological features only rarely provide evidence of viticulture by local Iron Age people. No specific indigenous amphoras were made for the transport of wine. However, imitations of Massaliote amphoras were manufactured on a very small scale in a few native settlements from the fifth and fourthth centuries BCE, in Le Pègue, a hinterland site on the eastern bank of the Rhône River, and in a few other sites close to the Etang de Berre, near Marseille (Lagrand and Thalmann 1973; Verdin 1997; Chausserie-Laprée 2005). Some of the *dolia* (large storage jars) used in Iron Age settlements of the Mediterranean littoral zone during the third–second centuries BCE bear inscriptions and representations related to grapes and wine. This is regarded as an indication of their use for winemaking and storage, along with grain storage (Jannoray 1955; Py and Buxó I Capdevila 2001; Chausserie-Laprée 2005). Such containers were made by indigenous people from the end of the sixth century BCE onwards, perhaps under Greek influence (Garcia 1987; Py and Buxó I Capdevila 2001). Press structures are documented for various sites in Provence and coastal Languedoc from the fourth century BCE onwards. According to their morphology, they are regarded as having been dedicated first to pressing olive oil, but some could have been used to press wine (Brun 2004). *Vitis* planting traces are only reported from the indigenous proto-urban harbour of Lattara, in eastern Languedoc. A few hundred metres from the city, various grape fields, of dates ranging over the third–first centuries BCE, have been uncovered on several thousand square metres. The density of plantation is about 9600 grape plants per hectare (Jung 2000; 2007).

The information about local viticulture obtained from grape seeds (pips) recovered from archae-

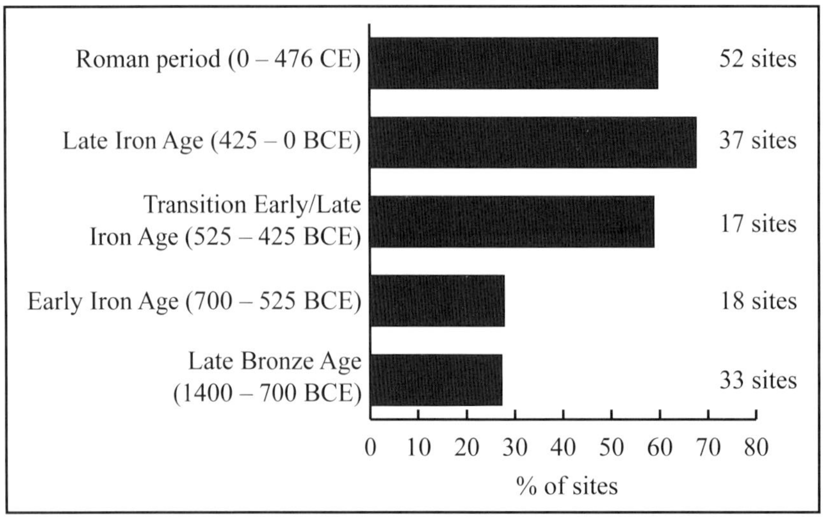

Fig. 4.16. Ubiquity of charred grape (*Vitis vinifera*) pips (percentage of sites) in the Rhône basin and Mediterranean France from the Late Bronze Age to Roman times.

ological layers is much more common (Fig. 4.14). As early as the Late Bronze Age, charred pips were fairly common in sites from the Rhône Basin and Mediterranean France, but their frequency clearly increases about the fifth century BCE, at the transition between the First and Second Iron Age (Fig. 4.16). They then remain frequent (*ca.* 60% of the sites) until the end of Roman times.

Measurements of pips have been used for a long time in archaeobotany to discriminate domesticated from wild grape (Stummer 1911). Pips from *Vitis vinifera* subsp. *sylvestris* are generally small, globular with a short stalk, whereas seeds from *V. vinifera* L. subsp. *vinifera* are larger, elongated, global or pyriform and with a longer stalk (Stummer 1911; Levadoux 1956). Traditional simple index (breadth/length, stalk length/total length) values provided evidence many years ago of the presence of cultivated *Vitis vinifera* seeds in various Iron Age sites (Erroux 1974; Marinval 1988; Buxó I Capdevila 1996), but for a long time such results were based on only tiny comparative values from modern wild and cultivated material, principally on the values given by Stummer (1911) and Schiemann (1953) for the Danube and Rhine areas. We are now able to compare traditional measurement values (total length, length of stalk, total breadth, length from stalk tip to chalaza) of archaeological pips to an important reference collection of wild vines and cultivars originating from various areas of France and Europe (Bouby and Marinval 2001; Bouby *et al.* 2006; Bouby *et al.* 2013). This indeed confirms the possibility of clearly discriminating wild and cultivated pips from simple measurements; especially the length of the stalk is of great help, as stated in earlier studies (Levadoux 1956; Smith and Jones 1990). Experimental work shows that the modifications of size and shape caused by charring have little effect on the possibility of discriminating wild and cultivated grape pips (Mangafa and Kotsakis 1996; Bouby 2010). This is important as most of the pre- and protohistorical pips were preserved by charring.

We had the opportunity to analyse pips from various sites located in southeastern France, ranging from the Late Bronze Age (*ca.* 1400–700 BCE) to the end of the Roman period (*ca.* 0–476 CE) (Bouby *et al.* 2006). As regards the Iron Age, information comes from indigenous settlements as well as from Massalia. Each one of the indigenous sites delivered pips classified both as wild and cultivated compared to our modern reference collection using Discriminant Analysis (Fig. 4.17). Cultivated pips are dominant in most of the sites. Some seeds could not be identified, their measurements being intermediate between modern wild vines and cultivars. We consider such a frequency of cultivated pips as an evidence of local cultivation. It does not seem likely that dried raisins could have been largely imported in all these sites. Moreover, three sites delivered amounts of *Vitis* remains that must be regarded as wine-pressing residues. In these samples, *Vitis* remains are overwhelmingly predominant. A sample composed of grape pips, pedicels and grape skins,

Site	Preserv.	Date	Number of pips	Classification		
				Wild	Cultivated	Unident
Late Bronze Age						
La Prairie, Chabrillan	W	13th–11th c. BCE	65	64	0	1
La Fangade, Sète	W	13th–11th c. BCE	27	20	4	3
Portal Vielh, Vendres	C	13th–8th c. BCE	6	2	4	0
Iron Age indigenous sites						
Mourre de Sève, Sorgues	C	6th c. BCE	9	3	4	2
Marduel, Saint-Bonnet-du-Gard	C	5th c. BCE	51	3	42	6
Plan de la Tour, Gailhan	C	5th c. BCE	75	54	8	13
Lattara, Lattes	W	5th c. BCE	75	13	49	13
Coudouneu, Lançon-de-Provence	C	5th c. BCE	12	1	9	2
L'île, Martigues	C	4th c. BCE	75	31	24	20
Pierredon, Eguilles	C	2nd c. BCE	21	12	4	5
Les Tremaïe, Les Baux-de-Provence	C	2nd–1st c. BCE	9	4	5	0
La Roque, Graveson	W	2nd c. BCE	75	12	46	17
Greek Massalia						
Jules Verne, Marseille	W	6th c. BCE	117	29	77	11
Jules Verne, Marseille	W	5th c. BCE	75	19	34	22
Charles De Gaulle, Marseille	W	4th c. BCE	75	29	38	8
Jules Verne, Marseille	W	3rd–2nd c. BCE	99	20	60	19
Jules Verne, Marseille	W	2nd c. CE	75	25	39	11
Roman sites						
Ambrussum, Villetelle	C	1st c. CE	75	23	38	14
Les Toulons, Rians	C	2nd c. CE	75	48	18	9
Besagne, Toulon	W	2nd–3rd c. CE	75	6	58	11
Les Toulons, Rians	C	4th c. CE	19	4	12	3

Fig. 4.17. Results of the identification of archaeological *Vitis* pips from Bronze to Roman Age sites in southeastern France using the following measurements: total length, length of the stalk, total breadth, length from the tip of the stalk to the chalaza. Identifications are made by means of Discriminant Analysis using statistical threshold of 65% probabilities. Measurements of charred pips have been corrected with coefficients based on the modifications obtained experimentally at 250°C during 45 minutes in oxydising atmosphere (After Bouby *et al.* 2006, modified); preservation: C = charred, W = waterlogged.

some still attached to pips, was taken from the floor of an early fourth century BCE house destroyed by fire in the site of l'Île de Martigues (Fig. 4.18; Marinval, *unpublished*). In Coudouneu, a sample exclusively composed of pips, pedicels and large peduncles was found in a *dolia* dated to the end of the fifth century BCE (Marinval, *in* Verdin 1997). In the oppidum of Le Castellan, another assemblage of pips, grape skins and one pedicel was discovered between a hearth and an oven in a second-century layer (Marty and del Corso 2002; Bouby, 2010). As assessed by ethnographical and experimental work, such assemblages of pips, pedicels and grape skins must be regarded as wine-pressing residues (Margaritis and Jones 2006) and cannot represent remains of dry raisin storage. The three sites are indigenous settlements, located in the vicinity of the Etang de Berre, a few dozen kilometres from Marseille, but outside the Massaliote territory. In coastal Languedoc, important concentrations of pips have been uncovered in Lattara from the end of the 5th century BCE, sometimes including

Fig. 4.18. Charred *Vitis* remains (pips, pedicels, rachis fragments and grape skins) from an early fourth-century BCE wine pressing residue from the site of Île de Martigues (Bouches-du-Rhône) (Archaeobotanical analysis: P. Marinval; Photo: S. Ivorra).

minor proportions of pedicels (Buxó I Capdevila 1996; Py and Buxó I Capdevila 2001). Some at least probably also represent wine-pressing residues. In Lattara, the proportion of grape pips in relation to cereal grains has been used to advocate an increase of viticulture in the end of the third century BCE. Meanwhile the proportion of amphoras decreases whereas *dolia* fragments become more common. This is regarded as evidence of the development of local wine production while imports decrease (Buxó I Capdevila 1996; Py and Buxó I Capdevila 2001).

If we accept the presence of domesticated grape pip morphotypes as evidence of local viticulture, grapes would have been widely cultivated in Mediterranean France, at least from the fifth century BCE onwards. Being already widespread in the coastal area and in the hinterland at that time, cultivation probably started earlier. Unfortunately, archaeobotanical studies of eighth to sixth century BCE sites are rather scarce and, in most cases, only a few samples have been analysed. Consequently, we clearly lack information for that crucial period. A few pips from Late Bronze Age settlements have been identified as domesticated grape but their number is far from being significant enough to prove viticulture. For the sixth century BCE, the only sample analysed comes from Le Mourre de Sève, a hinterland site in Provence. It includes a noticeable proportion of the domesticated morphotype. Existence of viticulture is very likely during the sixth century, although it is impossible to say if it started even before that. If we cannot ascertain that Mediterranean people, and especially Massaliotes, were responsible for the beginnings of viticulture, we must assume that they at least stimulated its development. The frequency of pips increases during the fifth century BCE. There is more evidence of viticulture in the coastal area, where their influence was stronger, and all the winemaking residues known today originate from there.

What was the Role and Status of Indigenous Viticulture during the Iron Age?

Viticulture developed rapidly after the Roman conquest and especially during the second half of the first century BCE. Archaeobotanical evidence is not as straightforward as during Protohistory, but there is plenty of archaeological and textual information proving the widespread and very great importance of speculative viticulture in Mediterranean France, at least until the end of the second century CE. Colonisation by veterans of the Roman armies played a major part in this development. Viticulture was then clearly dedicated to a market economy. Wine was massively sold in northern Gaul, to garrisons of the Rhine area and even in Rome. Over 50 manufactures of amphoras are reported for the first–second centuries CE in Mediterranean France (Brun and Laubenheimer 2001; Brun 2005).

It is not so easy to understand what the role of indigenous viticulture was during the Iron Age. We can assume from the archaeobotanical and archaeological information presented above that it grew in importance, while Massaliote wine was massively and simultaneously sold in the area. Local people, however, were making wine themselves. Local manufacture of amphoras was, at best, rare and seemingly only imitated from Massaliote containers. We do not have any evidence of trade of indigenous wine. Of course, we cannot rule out the possibility that it was transported in perishable containers such as wine skins or wooden barrels. Most especially, we must take into consideration the possibility that indigenous wine was centralised by Massaliotes and perhaps sold in their own amphoras. This could be a reason why the evidence of viticulture and winemaking is so strong in the coastal area, compared to the hinterland (and even to regional Roman dry settlements), but this probably cannot apply to the whole Mediterranean area of France, where most of the Massaliote wine was sold. It is difficult to assume that local people simply bought their own production after it had been stored in amphoras by Massaliotes. More probably, local wine was not of the same nature or did not have the same value and status as the one from Massalia. It could have been consumed in different social situations, perhaps more on a daily basis or by more humble people.

A chronological gap seems to exist between the archaeobotanical evidence, which is already widespread during the fifth century BCE, and the other archaeological sources (inscriptions on *dolia*, press structures and planting traces) which are not so common, appear later and are more specific to the coastal area. This is probably not a matter of simple coincidence and could be related to the form of viticulture initially practised by local people. Specific features and equipment easily identifiable for archaeologists are, to some extent, connected to intensive viticulture. This can be the case for regularly distributed planting trenches or pits of vineyards, for wine cellars equipped with *dolia* and even for press structures. But grape can be cultivated - and even wine can be made - without any of this specific equipment. It could have been cultivated as a secondary crop, for domestic production, leaving hardly any archaeological traces. Grapevines may be planted together with trees to use them as a scaffold, scattered here and there, or located along the tracks, along the borders of fields, as is occasionally still done in the Mediterranean, for example in Portugal. This is an imitation of the way wild grape naturally grows on trees along rivers. Cultivation of grapes on trees was said to have been practised in Gaul according to Columella (*De Re Rustica*, V, 7), but archaeology has thus far been unable to document this technique locally (Boissinot and Puig 2006).

Conclusion

The intensive, commercial viticulture practised in Mediterranean France by Romans and, before them, by Massaliotes, and reported by Latin authors is today documented by many archaeological sources. Paradoxically, the viticulture of indigenous Iron Age people in that area is quite exclusively documented by archaeobotanical information. Besides the specialised viticulture directly or indirectly promoted by Massaliotes and especially Romans, indigenous people seem to have grown grape more as a secondary crop. Many planting and winemaking techniques obviously were introduced into Gaul by Mediterranean people. Was it a one-way relationship? We may especially wonder about the origins of the cultivars that are reported by Latin authors, such as Columella (*De Re Rustica*, III, 2) and Pliny (*H.N.*, XIV, 18, 25–27, 43), to be autochthonous to Gaul.

4.7. FRUIT AS STAPLE FOOD: THE ROLE OF FIG (*FICUS CARICA* L.) DURING THE PRE-HISPANIC PERIOD OF THE CANARY ISLANDS, SPAIN (FROM THE 3RD–2ND CENTURIES BCE TO THE 15TH CENTURY CE)

Jacob Morales and Jaime Gil

Introduction

Most fruit trees were domesticated some time after the initial development of farming and they have been traditionally perceived as secondary products in the human diet. Use of fruits is normally focused on complementing staple food production and as a cash crop. In this last case, fruits are specifically cultivated for exchange rather than for local consumption, and they can be processed to some degree, adding value by creating a more complex product. Olive oil and wine are examples of cash crops that were produced and traded from the Bronze Age on in the Mediterranean zone. Another typical example of a cash crop is the use of cacao (*Theobroma cacao* L.) seeds as coins in Classic Mesoamerica. In this case, the crop itself is the cash (Sherratt 1999).

The use of fruit as a staple food is not common, but there are some examples where some fruits became dominant and indispensable for the survival of human populations. Perhaps the most well-known example is the consumption of dates (*Phoenix dactylifera* L.) by the inhabitants of the Near East and North Africa (Zohary *et al.* 2012). However, there are other examples, such as the consumption of banana (*Musa x paradisiaca* L.) in the wet tropics (Lejju *et al.* 2006) or chestnuts (*Castanea sativa* Mill.) in mountain regions of Europe (Conedera *et al.* 2004).

The aim of this paper is to clarify the role of fruit as a staple food, taking as an example the consumption of figs (*Ficus carica* L.) in the Canary Islands during the pre-Hispanic period (third to second centuries BCE to the fifteenth century CE). We will attempt to reconstruct the pattern in fig consumption by using different sources of information, such as archaeobotany, palaeodiet studies on human bones, ethnohistorical documents, archival records and ethnographic data. Using this multidisciplinary approach, we have tried to identify the role of figs in the diet and discuss the reasons that could have made this fruit a staple food for the first inhabitants of the Canary Islands.

The Fig

Botany of the Fig

The fig is a Mediterranean tree that develops a characteristic and complex reproductive biology. In a strict botanical sense, fig 'fruits' are actually inflorescences called syconia. They are hollow, fleshy structures composed of peduncular tissue, lined on the inside with hundreds of minute flowers. At one end is a small opening (ostiole) lined with dense, overlapping scales. Each tiny flower consists of a five-parted calyx and an ovary with a short style. Following pollination and fertilisation, the ovaries develop into minute one-seeded drupelets with an endocarp surrounding the seed (Condit 1955; Galil 1977). These are the fig remains that archaeobotanists normally retrieve in archaeological digs (Zohary *et al.* 2012).

Most species of the genus *Ficus* in the world are monoecious and have male and female flowers within single syconia. In *Ficus carica*, wild and cultivated forms develop different kinds of inflorescences. Wild fig trees (caprifig) produce male flowers and short-style female flowers, while cultivated fig trees (Smyrna, San Pedro and Common type) only produce long-style female flowers. In fact, they are the two sexual forms of the same species, of which the female is known as the true fig (cultivated) and the male as caprifig (wild).

Caprifigs of *Ficus carica* produce three crops of syconia per year: the *profichi* that ripen in early summer; the *mammoni* that ripen in fall; and the *mamme* that over-winter on the tree and ripen in spring. Fig pollen is transferred from male flowers to female flowers by an insect known as the fig wasp (*Blastophaga psenes*, Hymenoptera). These wasps live, lay their eggs and feed within the fig syconia. Only the winged female wasps emerge and leave the fig, through the ostiole, (wingless males cannot leave the syconia). The female wasps fly to new syconia to oviposit eggs in some of the female flowers, one egg per ovary. However, the *profichi* caprifig has many male flowers near the ostiole, and the pollen from these flowers is carried by the female wasps to the next syconium, in this way pollinating the ovaries. Although caprifig syconia incubate and perpetuate these tiny fig wasps, seeds may also develop in the female flowers. This is especially true of *mammoni* syconia in which pollination results in some ovaries (without wasps) developing seeds with viable embryos (Condit 1955; Galil 1977). It is interesting to note that in some Mediterranean regions, such as Morocco (see Chapter 4.8), caprifigs are also planted to produce fig wasps that are used to pollinate cultivated fig trees.

The reproductive biology of cultivated fig trees is simpler than in wild caprifigs. In the case of fruit setting for Smyrna type figs, these trees produce only two crops of syconia annually, a *breba* crop (*profichi*) that ripens in early summer and the main crop (*mammoni*) that ripens in autumn. For fruit development to occur, the Smyrna fig needs to be pollinated with pollen from the caprifig. This process is called 'caprification' (see Chapter 4.8). The San Pedro fig type produces two crops of figs each year. The *profichi* crop is parthenocarpic (it proceeds directly without pollination and fecundation), but the second crop, like the Smyrna fig, requires caprification. Most extended commercial cultivars of *Ficus carica* in the world belong to the Common type. Cultivars of this type do not require caprification. Unlike the Smyrna and San Pedro types, the syconia remain on the branches and ripen without wasp pollination. The drupelets inside develop parthenocarpically, and the endocarps are generally hollow and without fertile seeds (Condit 1955; Galil 1977).

Distinguishing Wild from Cultivated Figs in the Archaeobotanical Record

This complex reproduction biology allows us to recognise some botanical features of figs that can be preserved in archaeological material, which we can use to distinguish between different types of fig (Fig. 4.19). We consider that it can be possible to distinguish between the 'seeds'[15] of wild and cultivated fig types, at least between Common type and the rest. As we have seen, Common type fig seeds are not pollinated and thus are hollow; while caprifig, Smyrna and San Pedro (only the *mammoni* crop) types require pollination for fruit development. The seeds of these types possess an embryo and endosperm, which can be easily seen through a binocular microscope (Fig. 4.20). It is necessary to bear in mind that there are exceptions to this, such as unpollinated Smyrna seeds and San Pedro *profichi*; however, we regard it a useful criterion to use when samples contain significant number of seeds.

Origin of Figs

Until recently it was thought that figs were domesticated in the Near East during the Bronze Age, together with olives, grapes and dates (Zohary *et al.* 2012). However, recent archaeobotanical data seems to indicate that fig cultivation was practiced in the Near East since the twelfth millennium BP, nearly a thousand years before the domestication of cereals and legumes. The absence of endosperm and embryos in the archaeological seeds suggest that edible fruits were gathered from parthenocarpic trees grown from intentionally planted branches. Hence, fig trees could have been the first domesticated plant of the 'Neolithic Revolution' (Kislev *et al.* 2006).

	fig types	flower morphology	presence of wasp	*profichi* (April to June)	*mamme* (November to April)	*mammoni* (June to November)	seed characteristic
wild plants	Caprifig	**male flower** with five stamens **female flower** with short style	male and female wasp (wasps´ eggs can develop only in short style flowers)	production of pollen in the male flower but no fecundation, fig not edible	not fecundation (pollen is not viable), small fig not edible	fecundation by pollen from profichi (mammoni does not produce pollen), small and hardly edible fig	fertile seed (embryo and endosperm developed)
cultivated plants	Smyrna	**male flower** not presented **female flower** with long style	only female wasp, from a caprifig syconia (wasp can not lay its eggs in long style flowers)	fecundation by pollen from a carprifig, edible fig but it can not be dried	not developed	fecundation by pollen from a caprifig, edible fig (main crop) and it can be dried	fertile seed (embryo and endosperm developed)
	San Pedro	**male flower** not presented **female flower** with long style	only female wasp, from a caprifig syconia (in the mammoni crop)	not fecundation (parthenocarpic), edible fig but it can not be dried	not developed	fecundation by pollen from a carprifig, edible fig (main crop) and it can be dried	infertile seed profichi (hollow seed), fertile seed mammoni (embryo and endosperm developed)
	common type	**male flower** not presented **female flower** with long style	no wasps involved	not fecundation (parthenocarpic), edible fig but it can not be dried	not developed	not fecundation (parthenocarpic), edible fig (main crop) and it can be dried	infertile seed (hollow seed)

Fig. 4.19. Fig types and sexual characteristics.

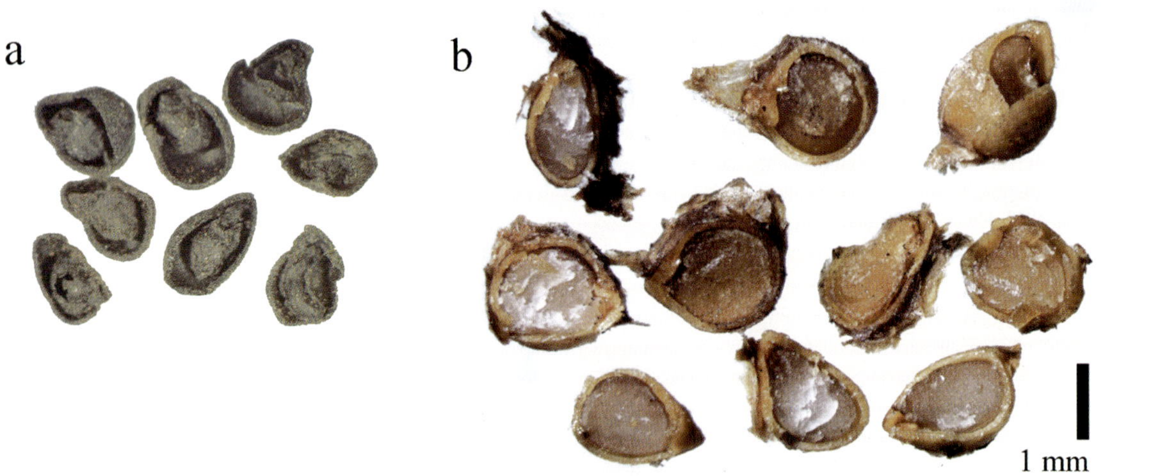

Fig. 4.20. (a) Hollow charred seeds of common type fig, archaeological site of Lomo los Melones (b) Pollinated seeds, with embryo and endosperm, of Smyrna type figs, modern.

The Canary Islands: Geographical and Historical Context

Natural Conditions

The Canary Islands are a volcanic archipelago located in the Atlantic Ocean opposite the coast of Africa, at 28° north latitude and only 100 km from the north-western Sahara (Fig. 4.21). The archipelago consists of seven islands and several islets that are characterised by the existence of high mountains and deep ravines in most of them. The climate varies according to orography. Mean temperature and annual precipitation range from about 21ºC and 100–300 mm, respectively, in coastal zones, to about 13ºC and 500–800 mm, respectively, at higher altitudes. These factors are very significant because they have a direct effect on the vegetation. The flora is clearly distributed as a function of altitude and orientation, including dry and more humid forest habitats. As a consequence of these climatic agents and geographical isolation, the islands possess tremendous botanical and animal biodiversity (Marrero and Pérez 1997).

The Pre-Hispanic Period

The first colonisation of the Canary Islands took place around the third to second centuries BCE and it was carried out by people from northern Africa with a Berber-like cultural and language background. There is not any evidence of contacts between these new islanders and the mainland or even among the different islands. As a consequence, every island developed a singular culture, until Europeans first arrived around the middle of the fourteenth century CE. Pre-Hispanic subsistence depended on food production as well as fishing, hunting, gathering wild plants and marine shellfish.

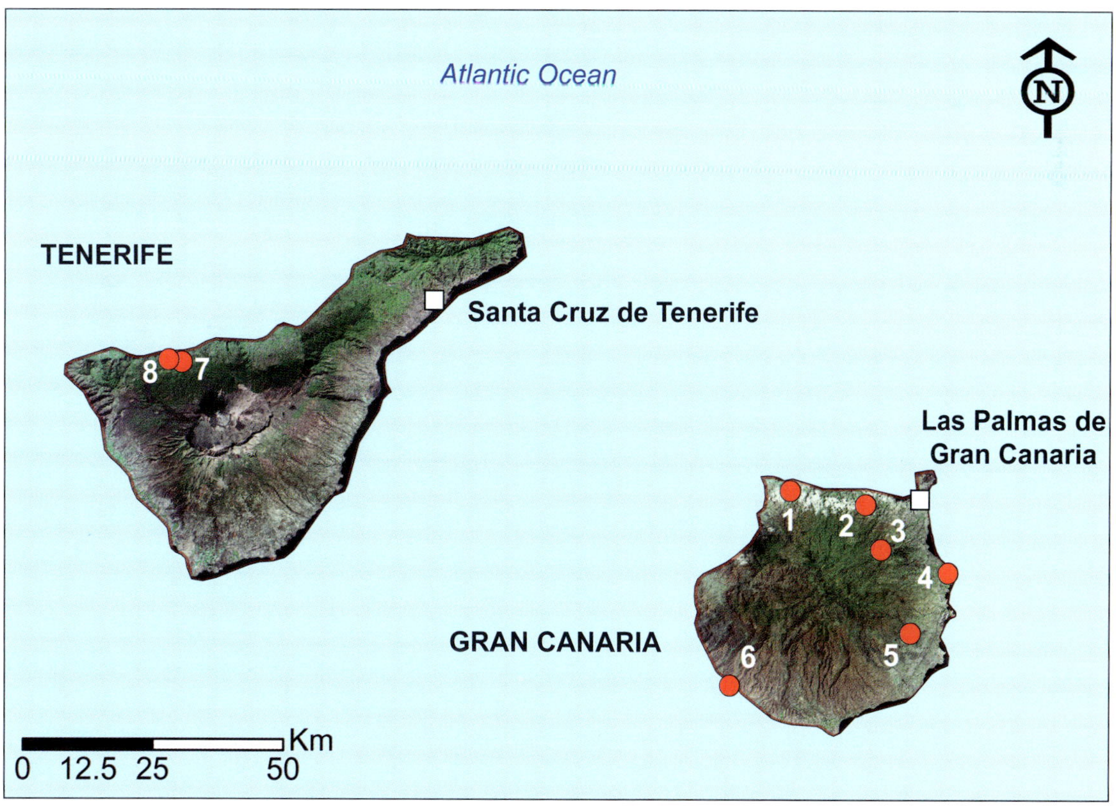

Fig. 4.21. A: Location of the Canary Islands just off the northwest coast of Africa, in relation to the Spanish mainland position. B: The sites on the Canary Islands as mentioned in the text: 1) Cueva Pintada; 2) La Cerera; 3) El Tejar; 4) Lomo los Melones; 5) San Antón; 6) Lomo los Gatos; 7) Las Palomas Cave; 8) Don Gaspar Cave. Map: R. Lugon, J.-C. Loubier and A. Chevalier.

	Archaeological sites										
	La Cerera	El Tejar, context 2	El Tejar, context 1	Ermita San Antón	Lomo los Melones, context 1	Lomo los Melones, context 2	Lomo los Gatos	Cueva Pintada, context 55	Cueva Pintada, context 25	Cueva Pintada, context 52	total
Dates on calibrated centuries AD	4th–15th	7th–12th	13th–14th	11th–15th	13th–15th	13th–15th	15th	13–15th	13–15th	15th	**total**
Volume of sediment, litres	780	426	191	69	405	86	207	300	37	400	**2901**
Seed density per litre	1.2	0.3	0.1	1.4	4.9	8.1	2.4	4.7	17	3.5	**2.7**
Fig remains											
Ficus carica, seed	25	10	22	53	1788	688	108	891	484	206	**4275**
Ficus carica, peduncle							4	1			**5**
Ficus carica, fruit fragment							5				**5**
Cereals											
Hordeum vulgare, grain	880	88	4	17	202	10	245	382	143	940	**2911**
Hordeum vulgare, rachis	3	3		2	4			90			**102**
Hordeum vulgare, glume						1				1	
Triticum aestivum/durum, grain	2	3		2	2		49	18	1	197	**274**
Triticum durum, rachis								1			**1**
indetermined cereal										25	**25**
Pulses											
Lens culinaris	1			21				6		1	**29**
Pisum sativum	1									3	**4**
Vicia faba								1		1	**2**
Wild fruits											
cf. *Adenocarpus foliolosus*, seed								1			**1**
Lauraceae								1			**1**
Neochamaelea pulverulenta, seed				3			3	3			**9**
Phoenix canariensis, seed	1	1			1		21		1		**25**
Phoenix canariensis, perianth	1	2							1		**4**
Pistacia atlantica, seed			1							1	
Plocama pendula, seed							27				**27**
Plocama pendula, complete fruit					1		28				**29**
Retama rhodorhizoides, seed								1			**1**
Rubus sp, seed										1	**1**
Visnea mocanera, seed					2	1	3			4	**10**

Fig. 4.22. Archaeobotanical evidence of figs, cultivated grains and wild edible plant species in pre-Hispanic archaeological sites from Gran Canaria.

The first population group brought with it domestic animals (goat, sheep, pig and dog) and plants (barley, wheat and pulses), which constituted the main component of the diet. They used knapped stone tools because no metals were available and they produced only handmade pottery. At the end of the fifteenth century CE, after 150 years of commercial contacts, slavery and Christianisation, all the islands were conquered by the Spanish Crown (Tejera and González 1987).

The Use of Figs in the Pre-Hispanic Canary Islands

Ficus carica grows wild in the Mediterranean area, but it is not a native tree in the Canary Islands. However, there are several sources of information that indicate the fig was introduced into the Canaries during the first human colonisation and the great importance of this fruit in the diet.

Archaeobotanical Evidence

Archaeological research has confirmed that figs were cultivated during the pre-Hispanic period of the Canary Islands. However, archaeobotanical remains of figs are only present in Gran Canaria and Tenerife islands and they are lacking in the rest of the archipelago. This general absence of data is partially due to the lack of proper recovery techniques that take into account the nature and size of the remains. Fig seeds measure *ca* 1.5 mm, which makes them difficult to detect without sieving and microscopic analysis.

In Tenerife, the only archaeobotanical remains of fig consist of charcoal fragments, recovered from the archaeological sites of Las Palomas Cave and Don Gaspar Cave in levels dated to around 560 CE (Machado *et al.* 1997). Conversely, in Gran Canaria, fig remains are plentiful and we have identified its seeds in all the archaeological sites we have sampled. In total, we have studied six different sites whose levels were dated from the fourth to the sixth centuries CE until the fifteenth century CE (Fig. 4.22) (Morales 2010). Fig remains are the most frequent plant identified in these contexts, where they appear mixed with other crop remains. Hulled barley (*Hordeum vulgare* L. subsp. *vulgare*), together with fig, is the most common species in these samples, but we have also recovered remains of hard wheat (*Triticum durum* Desf.), peas (*Pisum sativum* L.), lentils (*Lens culinaris* Medik.) and broad beans (*Vicia faba* L.), as well as of wild edible plants such as the Canary date palm (*Phoenix canariensis* Hort. ex Chabaud), the mocán (*Visnea mocanera* L.f.), brambles/raspberries (*Rubus* sp.) or the Mt. Atlas mastic tree (*Pistacia atlantica* D.C.) (Morales 2010). However, fig remains are normally more frequent than those from other plants. In some sites such as Lomo los Melones or Cueva Pintada, we found concentrations of more than 100 fig seeds in some samples. Although the number of retrieved fig seeds is rather low, considering that one fig can contain up to 1500 seeds, the constant presence of those seeds in the different domestic structures suggests a high relevance of the fruit in the diet of the inhabitants. In addition, dry figs have been found in silos where pre-Hispanic peoples used to store food, though they have not been excavated properly and we are not sure about the chronological context (Morales 2010).

In all the cases we have studied, the fig seeds were hollow, which means that they were not pollinated (Fig. 4.20b). In the case of the dry figs collected in the silos, they lack male flowers and the short-style female flower. *Blastophaga psenes* insect remains were not found, either, although we were able to identify other fossil insects inside the dry figs. Thus, we conclude that during the pre-Hispanic period, people cultivated the Common type fig. In the Canary Islands, it should be noted that the existence of the caprifig is not specifically mentioned in the ethnohistorical texts, though we have recorded some intriguing references concerning the occurrence of wild figs (Morales and Delgado 2007). In any case, all the fig trees that grow nowadays in the Canary Islands are cultivated and there is not any evidence of wild figs.

Ethnohistorical Evidence

The ethnohistorical sources are documents written between the fourteenth and the seventeenth centuries CE by the first European explorers who arrived in the archipelago. Basically, they are historical chronicles of the conquest of the Canaries or descriptions of the islands, in which some indigenous customs are recorded. The first writings from Italian explorers in 1341 CE had already indicated the presence of figs in the interior of indigenous houses from Gran Canaria (Boccacio 1998). When they refer to the eating habits of the Gran Canaria islanders, most of the authors report on a daily diet composed of goat products, barley and dried figs (Morales 1993). According to these texts, the most common way to eat the fig during the pre-Hispanic period was dried. Figs were first dried in the sun on rush mats, and they were then stored in cylindrical receptacles made of a plant fibre called *carianas*. Finally, they were pressed into loaves or threaded on strings, which allowed the fruits to be kept for long periods and made them more valuable (Abreu 1977). Prehispanic fig production was so important that they even traded them with the European explorers in exchange for metals and other exotic items (Bontier and Le Verrier 1980). At the end of the sixteenth century CE, Abreu documented the different names given to figs by the aboriginal descendants. They were called *arehormaze* when they were fresh and *tehaunenen* when they were already dried and had their sweet flavour (Abreu 1977).

Palaeodiet Evidence

A further, more indirect line of evidence is that of the occurrence of dental cavities or caries in the pre-Hispanic human remains of Gran Canaria. This is a pathology caused by local demineralisation of the dental hard tissues caused by the organic acids formed when bacteria metabolise the carbohydrates from the diet (Hillson 1996). Therefore, diet plays an important role in the presence and frequency of dental caries. The richness of carbohydrate foods in a diet (cereals, fruits like figs, etc.) leads to significant rates of this pathological condition. Oral pathology studies have shown a high rate of cavities among pre-Hispanic communities of Gran Canaria. In particular, this community exhibits a caries frequency of 17.31% (1164 carious teeth/6730 observed teeth). If we consider the origin of this pathology, those results indicate the important role carbohydrate food had in their diet. In addition, the study has shown the presence of fig seeds in four teeth belonging to four individuals (Fig. 4.23). Fig seeds were preserved by desiccation and they were inside the pulpar cavity, exposed because of dental caries in three of the four and because of severe wear in the fourth. This excellent preservation allows us to directly prove the consumption of figs and the role of this fruit in diet. Probably, fig was a staple food as is indicated by the consumption of this fruit by people of both sexes and different age ranges (Morales and Delgado 2007).

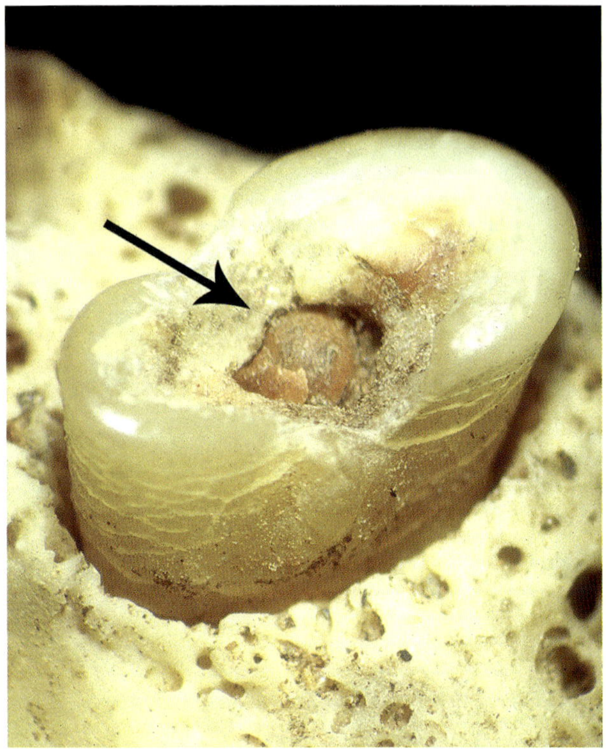

Fig. 4.23. Fig seed inside pulpar cavity. Human skeleton from Temisas, Gran Canaria.

Two traits explain the significant cariogenic role of figs. Firstly, there is their chemical composition, rich in simple or low molecular-weight carbohydrates and therefore rapidly metabolised by oral bacteria. Secondly, the sticky fig seeds adhere longer to teeth than other food (no other food remains have been discovered). Thus, the natural oral cleaning mechanism is less efficient. These characteristics, plus the small size of the seeds, explain why they remained inside the dental pulpar cavity (Morales and Delgado 2007). Some analyses carried out on human groups that had a large intake of fermentable carbohydrates such as dates have shown high frequencies of dental caries as well (Littleton and Frohlich 1989; Nelson *et al.* 1999).

Historical and Ethnographical Evidence

There is not any archaeological or ethnohistorical data about the management of fig cultivation and the system involved during the pre-Hispanic period of the Canaries. In order to fill this gap, we have analysed historical documents and ethnographic records of fig cultivation in the archipelago during the Hispanic period and in modern times. Although this source of information cannot be directly applied to the past, the use of ethnographic and historical data offers new possibilities to interpret the archaeological record.

After the European conquest, the cultivation of the fig tree retained a very important role in the subsistence of the newly-arrived peasant communities of the Canary Islands. Varieties of fig from both the European and African mainland were introduced in all the islands, replacing pre-Hispanic trees, and became a daily component of the diet. In famine times, as is attested in the late eighteenth century CE (Viera 1982; Urtusáustegui 1983), figs used to fill the subsistence gap left by the failure of cereal crops. In addition, leaves and twigs of figs were largely used as fodder for domestic animals.

The study of archival records dated between the eighteenth century and the late nineteenth centuries has made it possibly to identify 15 fig landrace names, most of which are still in use nowadays (Fig. 4.24) (Gil *et al.* 2008a). At the present time, farmers from the archipelago manage nearly 50 fig landrace names, taking into account the colour and the form of the fruit. However, recent molecular characterisation carried out on accessions from La Palma and Tenerife islands have shown that farmers have overestimated the biodiversity of fig trees (Giraldo *et al.* 2008). In any case, the biodiversity of fig trees is still very significant and is similar to some staple crops such as wheat or potatoes (Gil 1997; Gil *et al.* 2008b).

In North Africa, the place of origin of the first Canarians, figs are also a very important resource in the diet at the present time. Although there are no molecular studies of the relationship among figs from the Canaries and North Africa, ethnographic and linguistic studies in Morocco and Algeria have shown several similarities in the naming and managing of fig with the data from the Canary Islands (Chaker 1997; Sabir 2001). It is also interesting to note that in the Canary Islands, fig trees are under private ownership, so are inherited and can be sold or hired out.[16] In El Hierro Island, there are even some archival records of several people who inherited the branches or half of the same fig tree.[17]

Fig landrace names recorded in Historical Archives	Date of the first archival record	Islands where fig landrace names have been recorded at the present time						
		Gran Canaria	La Gomera	La Palma	Tenerife	Fuerteventura	El Hierro	Lanzarote
BICARIÑOS	*ca.* 1810	X	X	X	X		X	X
BLANCOS	1769	X	X	X	X	X	X	X
BLANCA DE INVIERNO	1850	X						
AZAHARILLOS	*ca.* 1810	X						
BOUJASOTE	1622	X	X			X		X
BOBA	circa 1810	X						
BREVAL / BREVERA	1645 / 1778	X	X	X	X	X	X	X
BREVERA BLANCA	1769	X			X			
BREVERA NEGRA	1862	X		X	X			
CASTELLANA	1729	X						
COTIOS	1779	X		X	X		X	
CANARIA	1798	X						
GOMERA	1840			X	X			X
MULATA	1850	X			X			
HARTABELLACOS	*ca* 1810	X						
NEGROS	1772	X	X	X	X	X	X	X
NOGAL	1729	X					X	X
SALVAJE	1523	X						
TARAJALES	1862		X		X			

Fig. 4.24. List of fig landrace names recorded in Historical Archives from the Canary Islands.

Modern fig cultivation takes place in different ecological environments that range from the sea-coast to the high mountains or even inside the lava fields. Fig trees are cultivated at the borders of roads and fields, or confined to enclosures made with stones. These circular structures protect the tree from the wind and livestock. In arid areas, they are cultivated in dry riverbeds to exploit periodic rains.

Conclusion: Fig, a Staple Food?

It is clear from the present evidence that figs were cultivated and consumed during the pre-Hispanic period of the Canaries. However, fig remains are limited to Tenerife and Gran Canaria islands. In Tenerife, only fig charcoals have been identified in archaeological sites, but in Gran Canaria Island, we have found evidence of fig consumption. Archaeobotanical remains, ethnohistorical documents and palaeodiet studies indicate that this fruit played an important role in the diet of the first inhabitants of Gran Canaria. Consequently, this data may suggest that figs could be considered a staple food. Staple foods normally contribute a high caloric input to the diet. They are eaten on a regular basis and are perceived as essential in the nutritional regime of a group. In order to be used on a daily basis, staple foods are normally storable, but they can also include non-storable plants than can be cultivated and harvested all year round, as are some tropical crops (banana [*Musa x paradisiaca*], sweet potato [*Ipomoea batatas* (L.) Lam.] and taro [*Colocasia esculenta* (L.) Schott]), among others. So what are the reasons that could have made figs a staple food for the indigenous people of Gran Canaria?

Firstly, fig was the only fruit crop introduced in the Canaries during the pre-Hispanic period and it has a higher yield that any wild native fruit tree. It is well adapted, as is indicated by modern high biodiversity, and it is easy to cultivate and gather. Secondly, figs have a high sugar content and they can make an important caloric contribution to the diet. Thirdly, figs were dried, which made them sweeter and storable. Dry figs allow people to eat the fruit during the whole year. Finally, figs – especially dry figs – could play an important economic role. They were stored in silos and they could be exchanged with other local products. Additionally, ethnohistorical records note that pre-Hispanic people traded dry figs to European explorers in exchange for imports. This fact suggests that figs could develop an important role as a cash crop as well, which made figs still more essential in their daily lives. According to these considerations, the important role of figs in the diet and the economy suggests that fruits and other vegetal resources can also be considered as staple foods.

4.8. BEYOND THE DIVIDE BETWEEN WILD AND DOMESTICATED: SPATIALITY, DOMESTICITY AND PRACTICES PERTAINING TO FIG (*FICUS CARICA* L.) AND OLIVE (*OLEA EUROPAEA* L.) AGROECOSYSTEMS AMONG JBALA COMMUNITIES IN NORTHERN MOROCCO

Yildiz Aumeeruddy-Thomas, Younes Hmimsa, Mohammed Ater, and Bouchaïb Khadari

Introduction

The distinction between the wild and the domesticated ranges from a stark divide in most European societies between wild and domesticated spaces, animals and plants, to a continuum between the natural world and domesticated spaces, crops and animals in other types of societies (Barrau 1986; Descola 1986).[18] Plant and animal domestication and their level of domesticity bear different meanings with equally different levels of integration between highly transformed crops and their wild relatives (*e.g.* wheat) and less transformed crops that grow side by side with escapees or wild relatives in tropical regions (*e.g.* sago palms) (Barrau 1978). Domesticity relates to the level of integration to domestic life, the definition of which varies from one society to another. In the Mediterranean region, domestication took place over some 2000 years, during which Palaeolithic hunter-gatherers progressively selected wild cereals, 'cultivated the wild relatives' and developed cultivation techniques and tools that radically transformed the wild habitats and the selected crops (Tanno and Willcox 2006; Weiss *et al.* 2006).

This portrays a linear vision of transformation from hunting-gathering to cultivation that led to a high level of transformation of wild cereals into domesticated forms, a period termed the 'Neolithic Revolution', a major civilising step in European and Middle Eastern history (Childe 1958). Tree cultivation in the Mediterranean region has been given a somewhat secondary role, although tree species such as olive and fig are known to have been domesticated at a very early stage (especially the fig) and have very important symbolic and material values (Condit 1947; Roux 2006). The latest archaeological findings of charred fig remains that date back to 11,400 years in the Jordan valley suggest that fig cultivation, through planting of cuttings, could be contemporary with cereal cultivation (Kislev *et al.* 2006). The gesture of planting tree macro-cuttings is a very simple step towards cultivating wild trees and is a well-known technique throughout tropical regions (Aumeeruddy-Thomas and Pinglo 1989). This process of selecting and cultivating wild trees draws from a horticultural approach to plants similar to the social practices described by Haudricourt (1962) and Barrau (*op. cit.*) in a generic model of agriculture called *hortus*. Experience from Indonesian cinnamomum-based agroforests (Aumeeruddy-Thomas 1994) and the extensive work conducted by Michon (2005a) suggest that the *hortus* model as described by Barrau, actually refers to a model of intensified forest-based agriculture where trees originating from forest ecosystems can be incorporated in highly intensified agricultural systems. Such systems call for a new paradigm that Michon (2005b) coined as 'domestic forests' which is a form of agriculture where linkages with forest ecosystems have led to highly productive systems.

Olive and fig trees have been integrated into Mediterranean agricultural systems and create tree formations that are a result of decades and centuries of political, social and commercial contexts. In Roman times, cultivated areas including cereals, vineyards and olive trees, known as *ager*, were distinguished from *saltus* (where animals grazed), perceived as wild areas and the even wilder *silva* (forest), a representation of the wild as opposed to the domestic (Descola 2003). Thus, olive and fig trees and many other fruit trees came to be integrated into spaces side by side with highly domesticated cereals. Moreover, defining *ager* or *saltus* or *silva* was also a means for the Romans to identify different types of land in their cadastral system in close relation to their taxation system. Policies and agrarian rules thus have highly influenced different levels of control of nature as well as the domesticated nature of trees throughout the Mediterranean region, given the high level of control that Romans formerly had in this area.

This paper aims at understanding whether the domesticated and intensive *ager* model of the Romans that fundamentally separates the domesticated from the wild is a reality in mountain farming systems which represent an important proportion of agroecosystems in the Mediterranean region. Indeed, very little has been said about processes of incorporating forest, wild or spontaneous[19] trees into agricultural systems and to what extent these tree components, wild and cultivated, may have both productive and social values that contributed to the social, economic and political development of such societies.

Study Area (Fig. 4.25) and Methods

Our experience since 2006 draws on an interdisciplinary project developed with geneticians about fig and olive domestication in western Mediterranean areas (Aumeeruddy-Thomas and Khadari 2008; Khadari *et al.* 2005). It also draws on the co-supervision with Professor M. Ater of the thesis of Hmimsa on perceptions, practices and uses

Fig. 4.25. Map of the Moroccan Rif with the main cities and places mentioned in the text. Map: R. Lugon, J.-C. Loubier and A. Chevalier.

relating to fig trees in northern Morocco (Hmimsa 2009) and from Hmimsa and Aumeeruddy-Thomas' ethnographic field data. We shall mainly concentrate on ethnographic data from observations made in northern Morocco (Tingitan and Rif region), specifically on fig and olive trees. We shall also refer briefly to the medieval Almohad gardens of Marrakech as well as irrigated olive agroecosystems found in Demenat, a town located east of Marrakech.

The Rif Mountain region of northern Morocco ranges from the western Tingitan peninsula inhabited by the Jbala Arab-speaking groups, through to the easternmost area inhabited by the Berber speaking 'Rifians'. Observations and field data were collected through a survey in villages throughout this region and a more systematic ethnographic survey with the Beni Ahmad, a Jbala tribe, especially in the village of Talandaoued, a hilly area located south of Chefchaouen. The Beni Ahmad and Talandaoued represent communities formed by different families or lineages that trace their origins back to very different areas of Morocco and even from Andalousia. Marriages tend to be confined to the Beni Ahmad tribe, although marrying with other tribes within the Jbala or even with other tribal confederations is also possible, but quite rare. Three major markets were visited in June at the period of the sale of caprifigs (wild and cultivated male fig fruits), in Targuist (Eastern Rif), Taounate (South or Pre-Rif) and Beni Ahmed. Interviews were conducted in the fields and indoors with local farmers, men and women, as well as with local administrators. We have also used Fay's (1972) detailed observations in a neighboring region of the Beni Ahmed, Zomi, in order to acquire a comparative vision between his descriptions of techniques and land use systems and those which we could observe in the Beni Ahmed region. All fig varieties were collected and carefully described with similar pomological criteria, measured and photographed and varieties' vernacular names were discussed through cross interviews with a significant number of persons. An atlas of the fig varieties of the Rif region is under preparation (Hmimsa *et al.* 2012) and leaf samples have been collected for genetic investigations.

Spatiality: Organisation of the Agroecosystem

The Beni Ahmed area is characterised by the extensive mixed olive, fig and cereals (barley, wheat, oat and rye) orchards representing hundreds of hectares surrounding the main villages (*douar* or *dchar*), sub-villages or isolated houses scattered throughout this landscape. Each *douar* is an administrative unit (village) and is also the oldest inhabited area according to oral history. The oldest cultivated fig trees of Talandaoued grow in the *douar* close to the houses and immediate surrounding areas, as well as in old fig orchards scattered in small pockets throughout the territory. On the top of a hill above the village, a multi-century oleaster (*el berri*), locally acknowledged as a spontaneous olive tree, marks the presence of this sanctuary.

Fig and olive trees in the orchards are arranged in a haphazard way (not in rows), the whole landscape looking very much like a dry natural woody savannah. The land between the trees is ploughed and sown as is the case in any cereal production area. However, while ploughing, the farmers make sure not to disturb or destroy the tree root system. Around each tree, tending of the soil is conducted individually. Such areas are termed *nokla* or *nkali* (a group of *nokla*). Bouzidi (2002) in his work on figs in northern Morocco uses the word *nokla*. According to him, *nokla* means 'roots' in Arabic but can also be used to speak of the human or animal race. This *nokla* or extensive orchard of the Beni Ahmad does not show any easily visible enclosures and a foreigner may have difficulties understanding where the limit of each property starts. Goats and sheep roam around in such orchards after the harvest of cereals, and are watched by small children or women and allowed to eat all the lower leaves of the trees. All trees are typically browsed at 'standing' goat-head levels.

Another land use system is represented by the fields lying in the rich alluvial and irrigated areas by the side of the main river (*oued*) where vegetables and pulses (*e.g.* lentils) are planted. These areas called *gharsa* are not planted with trees. Near the river and rivulets, spontaneous vegetation may still persist, including wild olive or escapees of cultivated olive (*el berri*) and fig trees (*nabout*) locally perceived as growing spontaneously. Further away from the main *douar* and beyond the *nkali*, there is another land category, composed of fields with trees but which are differenciated from *nkali* because they are enclosed by the *zarb* or enclosures. Thus, cereal fields relatively far from the village which may have some scattered trees (fig and olive) were protected from grazing with enclosures. Remnants of old shepherd houses are found there and are marked by the presence of very old fig trees. These areas were initially adjacent to *ghaba* or forests where goats and sheep were led for grazing, but all forests in Talandaoued today have been transformed into *nkali*. Inhabitants of Talandaoued consider that there is no more *ghaba*; therefore, no more space for extending the agricultural area.

Cultivating and Managing Wild Oleasters: Transforming Forests into Orchards

The process of transforming forests into *nkali* shows that inhabitants in Talandaoued and in the Rif acknowledge that oleasters (*el berri*) from the forests were usually saved and grafted during periods of extension of agricultural lands. Besides, in Talandaoued a cultivated olive tree is not a tree but a *lokma* which means the 'grafted', although the word *zaytoune,* meaning olive, is also quite commonly used, especially when the tree starts bearing fruit. Thus, the process of acquiring new agricultural lands was fundamentally linked to the preservation of forest oleasters which were ultimately transformed into domestic varieties through grafting. Moreover, large numbers of fig trees were also planted in these previous forest areas as a means of marking access to the land. Although this situation cannot be found in the Talandaoued area, oleaster forests that have been cleared from other vegetation can be found in other areas in the Rif such as in Zomi, as described by Fay (1972). Domestic animals in Zomi currently use these 'oleaster forests' for grazing with particular rules relating to periods during which such areas could be open or closed to grazing. It is also well known in the Rif that the inhabitants produced and still produce oil from oleaster olives which they highly appreciate for its locally perceived medicinal value but also for cooking. The process of transforming oleaster forests into agricultural lands has been described in other North African areas. Brun (1986) states that the technique of grafting wild olive trees had been attested since Homer and was probably very common in Greek Antiquity. He also notes that a differential approach of tax exoneration was applied by the Romans, five years for wild grafted olive stands and ten years for planted olive stands.

Our observations in present *nkali* areas show that the approach to cultivating olive trees is closely linked to this forest-based technique. It is a current practice for farmers in Talandaoued to allow oleasters to grow until they are grafted, thus relying partly on providence, or purposely collect spontaneous *el berri* growing from the *oued* area which they bring back in a lump of soil for planting in their own field. Some of these trees are also maintained to produce oleaster oil. Grafted olive trees distributed by the Spanish colonial agents and later by the Moroccan agricultural services could also be incorporated. Similarly to spontaneous fig trees called *nabout,* farmers may differentiate between *el berri nabout* and *el berri.* The oleaster *el berri,* when transplanted, is thus differentiated from *el berri nabout,* because the conscious act of transforming its destiny makes of *el berri* a non-spontaneous plant; both may become a cultivated olive *zaytoune* through the technical gesture of grafting. These series of categories show the continuity, in the local sense, between what is commonly known as a wild and a domesticated olive tree. Oleasters are also consciously preserved within the agroecosystem to pollinate the cultivated olive trees. The practice of conserving oleasters in cultivated orchards has also been observed in the medieval twelfth-century irrigated Almohad Agdal Gardens in Marrakech, where century-old cultivated varieties grow side by side with a few individuals of *sbouj* (oleaster in that area) of the same period. The *sbouj* improve the pollination of cultivated olives (according to present-day gardeners). Similarly, in the area of Demenat, the very old stands of olive trees, supported by old and sophisticated irrigation systems, host large numbers of spontaneous young oleasters acknowledged locally to be equally important for pollination.

Fig Trees: God's Gift and Man's Control over the Complex Cycle of a Co-Evolved Insect-Plant System.

As far as figs are concerned, a large number of fig varieties are present with a complex classification system (Hmimsa *et al.* 2012). Locally distinct varieties (*kermos*) are cultivated by cuttings and bear specific names such as *el messaria, el kotai, el fassia* etc. and are differentiated from figs growing spontaneously (*nabout*). However, different types of *nabout* are identified according to their affinities with existing varieties. Thus, for instance, a *nabout* may be called a *nabout kotai* because its fruit is very like that of the variety *kotai* (Hmimsa *et al. forthcoming*). In such cases, farmers demonstrate their capacity to identify different varieties of spontaneous fig trees that have originated from seedlings. Some fig varieties have two yields, spring figs or *breba* which are parthenocarpic figs (*bakor*) and autumn figs *(kermos).* The male fig trees or caprifigs (Valdeyron and Lloyd 1979) are known locally as *dokkar*. They can be either cultivated or originate from seedlings. They may be planted by cuttings, grafted on spontaneous *nabout* and have clearly undergone a process of selection. Two types of caprifigs, the *dokkar hlou*, and *dokkar hmer* (early and late) are cultivated on the basis of the period when the insect (*Blastophaga psenes*), the specific pollinating insect of *F. carica,* has achieved its cycle and the relative percentage of *Blastophaga* as compared to parasites. Spontaneous *dokkar* are also present but in small numbers (near wet areas) and are found along the rivers and rivulets. Villages such as Talandaoued are known in the Rif to be highly specialised in cultivating caprifigs, whereas less specialised villages rely on markets or collect from marginal lands such as the *oued* areas. When female figs (*kermous*) need to be pollinated, farmers (mainly children and women) actively bring the male figs in pendulum of two or three caprifigs and hang these in the female trees, a highly socialised activity during which children learn and share knowledge about ecological processes with their parents, close kin or neighbours (Fig. 4.26). The pollinator (*Blastophaga psenes*), locally known as *chenwila,* will then come out of the male figs and pollinate the female figs by entering through the ostiole, which opens up for a limited period of about one or two days during which the female figs display a very specific chemical signal and odour (Grison-Pigé *et al.* 2001) that is recognised locally. The balance between the number of caprifigs placed on a tree and the number of times caprification is repeated is very precise, as some persons consider that the tree may dry out from overproduction if too heavily pollinated. Human intervention here in this highly complex biological cycle is a remarkable illustration of the complex biocultural interactions involved in fig cultivation.

Fig. 4.26. Children preparing caprifigs for the pollination of domestic figs in Talandaoued, in extensive fig, olive and cereal orchards. Photo: Yildiz Aumeeruddy-Thomas.

The relationship between spontaneous fig trees, planted or spontaneous caprifigs, and selected fig varieties forms a continuity and an interconnected system within the same highly domesticated space, *i.e.* the *nkali.* The farmers control over the pollination process of the figs is here closely connected to the natural cycle of the *Blastophaga*, thus consisting of a coupled human and biological process that is at the basis of fig domestication. The technique of caprification dates back very far in history and has been depicted on Mesopotamien bas-reliefs (Marinval 2008).

The *nabout* – although recognised locally to have grown spontaneously and seen as 'a gift of god' – form part of the list of fig types enumerated in that area and lumped together with other fig varieties. Spontaneous tasty *nabout* are also eaten and classified into varieties as described above. Because they do not require any specific techniques to improve their production, they can be eaten by whoever passes by. This same attitude applies to the parthenocarpic *breba* (spring figs) which are also 'a gift of god' because they do not require pollination. However, they are also sold and sometimes dried for feeding animals. Nobody would dare to collect large quantities from somebody else's orchard.

Fig trees have a major importance in local trade and economy, especially dried figs which are extensively used during the Ramadan period, as they replace date palms that do not grow in northern Morocco. Very large quantities have long been traded for making fig alcohol (*mehia*), as well, and the overall fig economy is significantly very important, if not more so than cereals in the household economy in Talandaoued. Caprifigs, both wild and cultivated, are commonly sold on the markets. During sale, the vendor's caprifigs are scrutinised by the buyers who open them to check the proportion of *Blastophaga* and parasites. This is also the occasion of many passionate discussions and endless bargaining, as the vendor possesses the key for an optimum production, but the buyer tries to show in a very ostensible way that he knows better the good caprifigs from the bad ones, a very refined interplay between knowledge and power.

Besides these very concrete uses, both spring figs and dried figs have very strong symbolic values in gift and counter-gift systems. It is not uncommon to see people preparing buckets of *breba* as gifts to

Fig. 4.27. Farmer from the Beni Ahmed area preparing buckets of *breba* (spring figs) as a gift to the Caïd. Photo: Yildiz Aumeeruddy-Thomas.

Fig. 4.28. In the shade of fig trees – Beni Ahmad women sharing lunch after harvesting cereals in the fig and olive orchards of Talandaoued. Photo: Yildiz Aumeeruddy-Thomas.

the Caïd or for more powerful people in neighboring provinces from whom they seek some form of protection (Fig. 4.27). Dried figs also feed the dead, the poor or the angels, as they are deposited in cemeteries on the twenty-sixth night of Ramadan (*Lailatulkadr*), the night during which all heavenly angels are supposed, according to Islamic beliefs, to come down on earth and take away the sins of the poor humans, another relationship of gifts and counter-gifts, here, between human and non-human.

Conclusion and Discussion

Our analysis in northern Jbala mountain farming systems shows that there is no great divide between the wild, spontaneous and cultivated olive trees, nor is there a strict compartment that separates cultivated fig varieties from the spontaneous fig trees. Wild, spontaneous and cultivated trees exist in a common compartment which is the highly domesticated mountain orchards. They have developed with root stock from oleaster stands from the *ghaba* forests on the one hand and the transfer of cuttings from old specimens of cultivated figs from the main village (the *douar*), a double movement to create a highly productive agroecosystem (see Fig. 4.26). Farmers continuously introduce or protect spontaneous oleasters and figs in the core areas of the agricultural systems, *nkali.* The high domesticity of such areas lies in the fact that they are very productive systems that have developed as a continuity over time and space with the *douar* and *sub-douars* which represent the centre of domestic life in that area (Fig. 4.28). They are the basis for a very large array of social arrangements, especially through a variety of *métayage* (sharecropping systems). Major technical gestures such as planting, grafting, and fruit harvesting apply both to wild, spontaneous olive and fig trees in a common attempt at increasing production. Linkages between the various indigenous categories of trees, *el berri* and *zaytoune*, *nabout*, *dokkar* and *kermous*, rely here both upon people's intervention and natural ecological processes. The domestication process applies both at the ecosystem level through conserving entire wild tree populations of oleasters, as well as individual application of horticultural practices such as grafting and planting of cuttings. The maintainance of both clonal and sexual reproduction regarding figs and olive appears here to be part of the domestication and diversification process. Marginal areas, such as the *oued* vegetation, still represent gathering areas especially for caprifigs, although the best varieties are already multiplied by cuttings and are found in an orchard or *nkali.*

Our major conclusion is that on the contrary to cereal domestication in the Mediterranean system which progressed slowly from a wild form to a highly domesticated form, tree domestication could have followed a very distinct path because wild, spontaneous and domesticated forms are still very closely linked, as shown by observations made in the Rif Mountain agroecosystems. Reliance on providence, seedlings growing spontaneously, as well as a general acceptance of what is simply given by god, is an important factor that differentiates approaches to cultivating cereals, as compared to approaches to cultivating trees. The latter draws on a very fine-tuned balance between naturality and humanity. How generalised are such attitudes and such systems today? This situation also suggests a possible continuum in tree domestication processes. Moreover, cereal domestication moved away from natural processes towards highly sophisticated systems, important transformations of soils and habitats where man's input is crucial, while domesticated olive and fig trees analysed in this context rely strongly on natural processes and profound relationships with forests. In mountain agricultural systems, such as the Rif Mountain farming systems, we are thus very far from the Roman *ager* model, which developed very far from the wild *silva* and somehow in a hybrid situation between the highly domesticated *hortus* and the *silva*. Since cereal cultivation and tree cultivation, as well as pastoral systems, coexist in the Rifian orchards, we contend that three different and integrated approaches to controlling and domesticating nature co-exist here in this type of agroecosystem. Rather than a domestic forest, these systems – very similarly to the cork oak *montados* of Portugal or the argan agroecosystem in southern Morocco – are truly agrosylvopastoral systems.

4.9. CONCLUSIONS

Laurent Bouby

Fruits and fruit trees have often received far too little attention in archaeobotanical investigations into prehistoric farming economies. When referring to historical sources, the focus has often been on taxa of high cultural and economic values, such as the olive tree and grapevine in the Mediterranean regions.

It is true that domestication syndromes are generally not very distinct for fruit trees and therefore morphological signs of domestication are difficult for the archaeobotanist to perceive. The adoption of vegetative propagation techniques, the keystone of most fruit tree cultivation, is also difficult to tackle in the field of archaeology.

In this chapter, we bring interdisciplinary contributions to shed light on the role of fruit trees in ancient and modern farming communities. They remind us that fruits can play an important part in human nutrition among hunter-gatherers (Chapter 4.2) as well as in peasant societies at various technical stages of agriculture. For example, archaeobotanical sources help to portray the role of fruits in northern Italy from Neolithic times to the Roman Empire (Chapter 4.4). They also remind us that fruit trees not only provide nutritive and symbolic food but also multiple products and services. This is illustrated in the example of the uses of *Prosopis* in Peruvian pre-Hispanic and modern societies (Chapter 4.3). Various contributions enable us to go beyond the usual image of high commercial and cultural value products of the classical Mediterranean fruits and demonstrate the diversity of their uses and of the cultivation practices applied to these trees in modern and ancient, non-mechanised societies. Figs, for example, were a staple food in the Canary Islands during historical times (Chapter 4.7). The case study from northern Morocco raises the question of the relations between wild and cultivated fruit trees and of the integration of trees coming from seeds into agrosystems (Chapter 4.8). This shows that fruit tree cultivation cannot be reduced to vegetative propagation of elite clonal cultivars and provides evidence of the continuity that may exist in fruit tree husbandry between wild and domesticated forms. In the archaeobotanical study of protohistorical viticulture in southern France (Chapter 4.6) we can see another example of a continuum, from a non-intensive domestic activity to specialised and commercial cultivation. There may be a general relation between the specialisation focusing on elite clonal cultivars and the intensification of fruit production. We hope that these contributions will encourage readers and scholars to take a closer look at the domestication of perennial plants and the history of fruit tree cultivation, which so ably highlights the diversity of uses and practices.

Chapter Acknowledgements

The fieldwork by David John Goldstein (Chapter 4.3) was supported by grants to Dr. Izumi Shimada from the H. John Heinz III Fund for Latin American Studies, Committee for Research and Exploration of the National Geographic Society, and the Wenner-Gren Foundation for Anthropological Research. NSF Dissertation Improvement Grant #0211364, and a Doctoral Fellowship from the Graduate School at Southern Illinois University at Carbondale provided fieldwork support. Robin Coleman Goldstein provided editorial assistance. The author is indebted to members of the Sicán Archaeological Project, the staff at the Museo Nacional Sicán in Ferreñafe, Peru, the Llauce family of La Zaranda, and Don Julio Navarro and Family of Tres Puentes for their consultation and support.

Bui Thi Mai and Michel Girard (Chapter 4.5) would like to thank Laurent Bouby (CNRS-Montpellier) and two anonymous referees for their most helpful comments, as well as Cozette Griffin-Kremer who translated this article into English.

Laurent Bouby *et al.* (Chapter 4.6) thank the 'Viticulture' team in Montpellier (UMR DIAPC, INRA-Montpellier Supagro), and especially T. Lacombe, for their help in constituting their reference collection of pips from modern cultivars. The authors are grateful to P. Boissinot who allowed them to use the photograph of the planting traces of grape in Saint-Jean-du-Désert. Special thanks are due to S. Ivorra and T. Pastor who have been of great help in taking most of the photographs of modern and archaeological pips.

Chapter Notes

1 The reworking of soils and sediments by animals or plants.
2 Research focused on the evolution and systematics of organisms.
3 A colpus is an elongated furrow in the pollen grain, while a porus is smaller and circular. A colporate structure is the combination of the two. A tetracolporate pollen grain has four of these furrow-pore-combinations in its surface, a pentacolporate one has five. Stephanocolporate pollen has an undefined number of these structures.
4 The outer hull of a pollen grain, bearing its characteristical structures.
5 A net-like structure.
6 The colporate structures in syncolporate pollen meet (and fuse) at their ends.
7 Erdtman and Nai propose considerably larger dimensions that probably depend on the sample preparation technique, or even on the age of the reference slides.
8 Spheroidal pollen are sphere-shaped, with the same diametres at their longitudinal (polar) axis and their equatorial region. Prolate pollen is elongated along its polar axis, subprolate pollen only slightly so.
9 The inner layers of the exine.
10 The outer layers of the exine.
11 The inner parts ('holes') of the reticulum.
12 The outer parts ('walls') of the reticulum.
13 The gilt silk tunic of Chinese origin discovered in the Naintré grave is a very fine example of this long-distance trade (Desrosiers, 2000).
14 These were citrons dried either in very warm sand for two or three weeks or in the sun for around two weeks.
15 For ease of reading we use the term 'seeds' when referring to what botanically are 'endocarps'.
16 Historical Archive of the County of Santa Cruz de Tenerife. Signature PSO – 1131: '(...) in order to pay the cost of my wife's funeral I sell a small piece of land with a white fig tree, located in a place called Las Rosas'.
17 Private archive records from Samuel Acosta. Contract of sale in «Costa de Asofa». 10th of May 1903.
18 The word 'wild' hereafter is used in its wider sense,

i.e. an area, a plant or an animal which is assumed not to have been transformed by man. We add 'perceived as wild' when discussed in the context of a particular society.

19 'Spontaneous' relates to the fact of growing spontaneously and is distinct from 'wild' in that it includes plants that grow from seeds originating from cultivated plants (defined by 'feral').

Chapter Bibliography

Abreu, J. (1977) *Historia de la conquista de las siete islas de Canaria*. Goya, Santa Cruz de Tenerife.

Alexandre-Bidon, D. (forthcoming) Du nom des fruits...

Amigues, S. (2007) L'exploitation du monde végétal en Grèce classique et hellénistique. Essai de synthèse. *Topoi* 15 (1), 75–125.

Anderson, K. (2005) *Tending the Wild: Native Knowledge and Management of California's Natural Resources*. University of California Press, Berkeley.

André, J. (1952) Contribution au vocabulaire de la viticulture: les noms de cépages. *Revue d'Etudes Latines* 30, 126–156.

André, J. (1981) *L'alimentation et la cuisine à Rome*. Les Belles Lettres, Paris.

Arnold, C., Gillet, F. and Gobat, M. (1998) Situation de la vigne sauvage *Vitis vinifera* ssp. *silvestris* en Europe. *Vitis* 37 (4), 159–170.

Arnold, C., Schnitzler, A., Douard, A., Peter, R. and Gillet, F. (2005) Is there a future for wild grapevine (*Vitis vinifera* subsp. *silvestris*) in the Rhine valley? *Biodiversity and Conservation* 14, 1507–1523.

Arroyo Baldeon, A. J. (1997) *Consideraciones Ecológicas y Socioeconómicas de la Regeneración Natural del Algarrobo (Prosopis pallida) en el Valle de Olmos (Lambayeque-Perú)*. Facultad de Ciencias Forestales, Universidad Nacional Agraria La Molina, Lima.

Atran, S. (1993) Itza Maya Tropical Agro-forestry. *Current Anthropology* 3, 633–700.

Aumeeruddy, Y. and Pinglo, F. (1989) *Phytopratiques des régions tropicales - recueil préliminaire de techniques traditionnelles d'amélioration des plantes*. UNESCO, MAB, Paris.

Aumeeruddy-Thomas, Y. (1994) Représentations et gestion paysannes des agroforêts en périphérie du Parc National Kerinci Seblat à Sumatra, Indonésie. *People and Plants Working Paper Series*, No 3. UNESCO, Paris.

Aumeeruddy-Thomas, Y. and Khadari, B. (2008) *Domestication et traitement de l'agrodiversité arborée en méditerranée: perpétuation et renouvellement des sociétés. Exemples du figuier et de l'olivier au Maroc*, Conference paper given at « Journée scientifique du Centre d'Ecologie Fonctionnelle et Evolutive », November 13th 2008, Agropolis International, Montpellier.

Ayuk, E. T., Duguma, B., Franzel, S., Kengue, J., Mollet, M., Tiki-Manga, T. and Zenkeng, P. (1999) Uses, management and economic potential of *Irvingia gabonensis* in the humid lowlands of Cameroon. *Forest Ecology and Management* 113, 1–9.

Bakels, C. C. and Jacomet, S. (2003) Access to luxury foods in central Europe during the Roman period: the archaeobotanical evidence. *World Archaeology* 34 (3), 542–557.

Baleé, W. and Erickson, C. L. (eds.) (2006) *Time and Complexity in Historical Ecology*. Columbia University Press, New York.

Bandini Mazzanti, M., Mercuri, A. M., Bosi, G., Marchesini, M. and Accorsi C. A. (2001) The archaeobotanical archive: plants used by man (which, where, how, when?) – What fruits did Romans eat in Emilia Romagna (northern Italy)? Some responses from seeds and fruits. In A. Guirino (ed.) *Science and Technology for the safeguard of Cultural Heritage in The Mediterranean Basin*, 1. CNR-Progetto Finalizzato Beni Culturali, Roma, pp. 318–324.

Barkley, N. A., Roose, M. L., Krueger, R. R. and Federici C. (2006) Assessing genetic diversity and population structure in a Citrus germplasm collection utilizing simple sequence repeat markers (SSRs). *Theoretical and Applied Genetics* 112, 1519–1531.

Barrau, J. (1978) Domesticamento. In *Encyclopedia Einaudi*. Einaudi, Torino, Vol. 5, pp. 49–71.

Barrau, J. (1986) Les hommes et les forêts tropicales d'Afrique et d'Amérique. Mémoires du Museum National d'Histoire Naturelle, Paris, MNHN, 132, pp. 289–286.

Barrett, S. and Gifford E. (1933) Miwok material culture: Indian life of the Yosemite region. *Bulletin of the Public Museum of the City of Milwaukee* 2(4), 127–277.

Barret, H. C. and Rhodes, A. M. (1976) A numerical taxonomic study of affinity relationships in cultivated *Citrus* and its close relatives. *Syst. Bot.* 1, 105–136.

Basgall, M. (1987) Resource intensification among hunter-gatherers: Acorn economies in prehistoric

California. *Research in Economic Anthropology* 9(1), 21–52.

Bats, M. (ed.) (1990) *Les amphores de Marseille grecque, chronologie et diffusion (VIe-Ier s. av. J.-C.).* Etudes Massaliètes 2, ADAM/Université de Provence, Lattes/Aix-en-Provence.

Bats, M. (1998) Marseille archaïque, Etrusques et Phocéens en Méditerranée nord-occidentale. *Mélanges de l'Ecole Française de Rome* 110, 609–633.

Bats, M. (2005) Articles: Agde, Marseille, Olbia de Provence, Cumes, Sybaris, Velia. In J. Leclant (ed.) *Dictionnaire de l'Antiquité.* PUF, Paris.

Bats, M., J-P., Brun, P., Munzi and Pasqualini, M. (eds.) (2009) *Kymè 1: les fouilles du Centre Jean Bérard au sud de l'acropole de Cumes. Rapport 2008-2009.* École Française de Rome.

Baumhoff, M. (1963) Ecological determinants of aboriginal California populations. *University of California Publications in American Archaeology and Ethnology* 49, 155–336.

Baumhoff, M. (1981) The carrying capacity of hunter-gatherers. In S. Koyama and D. Thomas (eds.) *Affluent foragers: Pacific Coasts East and West.* National Museum of Ethnology, Osaka, pp. 77–90.

Bean, L. and Saubel, K. (1972) *Temalpakh (From the Earth): Cahuilla Indian Knowledge and Usage of Plants.* Malki Museum Press, Banning, (CA).

Bellini, C., Mariotti-Lippi, M., Mori Secci, M., Aranguren, B. and Perazzi, P. (2008) Plant gathering and cultivation in prehistoric Tuscany (Italy). *Vegetation History and Archaeobotany* 17 (Suppl. 1), 103–112.

Beresford-Jones, D. (2005) *Pre-Hispanic Prosopis-Human Relationships on the South Coast of Peru: Riparian Forests in the Context of Environmental and Cultural Trajectories of the Lower Ica Valley.* Unpublished Ph.D. thesis, University of Cambridge.

Berger, J. F. (2006) Sédiments, dynamique du peuplement et climat au Néolithique ancien. In J. Guilaine (ed.) *Populations néolithiques et environnements.* Errance, Paris, pp. 155–212.

Bernabò Brea, M., Cardarelli, A. and Cremaschi, M. (eds.) (1997) *Le terramare. La più antica civiltà padana.* Electa, Venezia.

Bertone, G., Bellido Cerda, E. and Li, J. N. (2008) La Arqueobotánica Peruana. In S. Archila, M.Giovanetti and V. Lema (eds.) *Arqueobotánica y Teoría Arqueológica: Discusiones desde Sur América.* Universidad de los Andes, Departamento de Antropología, Bogotá, pp.16–24.

Bertucchi, G. (1992) *Les amphores et le vin à Marseille, VIe avant J.-C. - IIe après J.-C.* 25ème suppl. Revue Archéologique de Narbonnaise, CNRS, Paris.

Bettinger, R. L. and Wohlgemuth, E. (2006) California Plants. In D. Ubelaker (ed.) *Environment, Origins, and Population. Handbook of North American Indians Vol. 3,* W. C. Sturtevant (gen. ed.). Smithsonian Institution, Washington D.C., pp. 274–283.

Bietti Sestieri, A. M. (2002) L'agricoltura in Italia nell'età dei metalli. In G. Forni and A. Marcone (eds.) *Storia dell'agricoltura italiana,* 1, L'età Antica, Preistoria. Accademia dei Georgofili, Edizioni Polistampa, Firenze, pp. 205–217.

Bile, M. (2003) L'agriculture en Crète au Ie millénaire: l'apport de l'épigraphie. *Cretan Studies* 8, 23–40.

Blasi, C., Doegiallo, G., Follieri, M., Lucchese, F., Magri, D., Pignatti, S. and Sadori, L. (1995) La vegetazione naturale potenziale dell'area romana. *Atti dei Convegni Lincei* 115, 423–457.

Boccacio, G. (1998) *De Canarias y de las otras islas nuevamente halladas en el océano allende España.* Ayuntamiento Puerto de la Cruz- Ayuntamiento Villa de Orotava, La Laguna.

Bocek, B. (1984) Ethnobotany of Costanoan Indians, California, based on collections by John P. Harrington. *Economic Botany* 38(2), 240–255.

Boissinot, P. (1995) L'empreinte des paysages hellénistiques dans les formations holocènes de Saint-Jean du Désert (Marseille). *Méditerranée* 3–4, 33–40.

Boissinot, P. (2001) Archéologie des vignobles antiques du sud de la Gaule. *Gallia* 58, 45–68.

Boissinot, P. and Puig, C. (2006) Archéologie du champ et viticulture méridionale. Pourquoi les traces de vignobles sont-elles si peu fréquentes au Moyen Age? *Archéologie du Midi Médiéval* 23–24, 17–26.

Bonavia, D. (1982) *Los Gavilanes.* Editorial Ausonia-Talleres Gráficos S.A., Lima.

Bonneuil, C. and Hochereau, F. (2008) Gouverner le 'progress génétique' Biopolitique et métrologie de la construction d'un standard variétal dans la France agricole d'après guerre. *Annales H.S.S.* novembre-décembre 2008 (6), 1305–1340.

Bontier, P. and Le Verrier, J. (1980) *Le Canarien: Crónicas Francesas de la conquista de Canarias.* Aula de Cultura del Cabildo Insular de Tenerife, Santa Cruz de Tenerife.

Boone, Y. and Renault-Miskovsky, J. (1976) La cueillette. In H. de Lumley (ed.) *La préhistoire française. Les civilisations paléolithiques et mésolithiques de la France,* Vol. 1, CNRS, Paris, pp. 684–687.

Bottin, M. (2008) Les agrumes, la tour de Babel des botanistes et des jardiniers. *Au bout du jardin* 20, 5.

Bouby, L. (2010) *Agriculture dans le bassin du Rhône. du Bronze final à l'Antiquité. Agrobiodiversité, économie, cultures.* Ph. D. thesis, EHESS/Université Toulouse le Mirail. 3 Volumes.

Bouby, L. and Marinval, P. (2000) Ressources végétales à Marseille et dans les sociétés indigènes au Bronze Final et au Premier Âge du Fer: premiers éléments de comparaison. In T. Janin (ed.) *Mailhac et le Premier Âge du Fer en Europe occidentale.* Monographies d'Archéologie Méditerranéenne 7, ARALO, Lattes, pp. 205–214.

Bouby, L. and Marinval, P. (2001) La vigne et les débuts de la viticulture en France: apports de l'archéobotanique. *Gallia* 58, 13–28.

Bouby, L. and Marinval, P. (2004) Fruits and seeds from Roman cremations in Limagne (Massif Central) and the spatial variability of plant offerings in France. *Journal of Archaeological Science* 31, 77–86.

Bouby, L., Figueiral, I., Bouchette, A., Rovira i Buendía, N., Ivorra, S., Lacombe, T., Pastor, T., Picq, S., Marinval, P. and Terral J.-F. (2013) Bioarchaeological insights into the process of domestication of grapevine (*Vitis vinifera* L.) during Roman times in southern France PLoS ONE 8(5), e63195. DOI 10.1371/journal.pone.0063195.

Bouby, L., Terral, J.-F., Ivorra, S., Marinval, P., Pradat, B. and Ruas, M.-P. (2006) Vers une approche bio-archéologique de l'histoire de la vigne cultivée et de la viticulture: problématique, choix méthodologiques et premiers résultats. *Archéologie du Midi Médiéval* 23/24, 61–74.

Bouiron, M. (2005) Secteur de l'Alcazar. In M.-P. Rothe and H. Tréziny (eds.) *Marseille et ses alentours.* Carte Archéologique de la Gaule 13/3, Maison des Sciences de l'Homme, Paris, pp. 581–597.

Bouloumié, B. (1981) Le vin étrusque et la première hellénisation du Midi de la Gaule. *Revue Archéologique de l'Est et du Centre-Est* 32 (125–126), 75–81.

Bouvier, M. (2001) *Les saveurs du vin antique. Vins d'hier, vignerons d'aujourd'hui.* Errance, Paris.

Bouzidi el, S. (2002) Le figuier: histoire, rituel et symbolisme en Afrique du Nord. *Dialogues d'histoire ancienne*, Année 2002, 28(2), 103–120.

Brack Egg, A. (1999) *Diccionario Enciclopédico de Plantas Útiles del Perú.* Centro de Estudios Regionales Andinos "Bartolome de las Casas', Cuzco, Peru.

Bretschneider, E. (1871) *On the Study and value of Chinese Botanical Works, with Notes on the History of Plants and Geographical Botany from Chinese Sources. Illustrated with 8 Chinese Wood cuts.*

Brun, J.-P. (1986) *L'oléiculture antique en provence. Les huileries du département du Var. Revue Archéologique de Narbonaise*, Supplément 15, Editions du CNRS, Paris.

Brun, J.-P. (2004) *Archéologie du vin et de l'huile. De la préhistoire à l'époque hellénistique.* Errance, Paris.

Brun, J.-P. (2005) *Archéologie du vin et de l'huile en Gaule romaine.* Errance, Paris.

Brun, J.-P. and Laubenheimer, F. (2001) Conclusion. *Gallia* 58, 203–219.

Burkhart, A. (1976) A Monograph of the genus *Prosopis* (Leguminosae Subfam. Mimosoideae). *Journal of the Arnold Arboretum* 57, 217–249.

Buxó I Capdevila, R. (1996) Evidence for vines and ancient cultivation from an urban area, Lattes (Hérault), southern France. *Antiquity* 70, 393–407.

Candolle, A de (1883) *Origine des plantes cultivées.* Réédition J. Lafitte, Marseille, 1984.

Carugati, M. G., Castelletti, L. and Rottoli, M. (1996) L'agricoltura nel primo neolitico del Friuli. Le ricerche a Sammardenchia, Fagnigola e Valer. In A Ferrari and A. Pessina (eds.) *Sammardenchia e i primi agricoltori del Friuli.* Credito Cooperativo – Banca di Credito Cooperativo di Basiliano, Udine, pp. 103–112.

Casartelli, S. (2007) *La Palafitta di Parma: analisi archeobotaniche delle unità stratigrafiche 220 e 270.* Università degli Studi di Milano. Master 'Scienze per i Beni Culturali Archeologici: figure professionali di geo-archeologo ed archeometra'. Unpublished thesis.

Casas, A. and Caballero, J. (1996) Traditional Management and Morphological Variation in *Leucaena esculenta* (Fabaceae: Mimosoideae) in the Mixtec Region of Guerrero, Mexico. *Economic Botany* 50(2), 167–181.

Casas, A., Otero-Arnaiz, A., Pérez-Negron, E. and Valiente-Banuet, A. (2007) *In situ* Management and Domestication of Plants in Mesoamerica. *Annals of Botany* 100, 1101–1115.

Castelletti, L. and Rottoli, M. (1986) Resti vegetali macroscopici – rapporto preliminare. In R. De Marinis (ed.) *Gli Etruschi a nord del Po.* Catalogo della mostra, 1, Campanotto Editore, Udine, pp. 177–183.

Castelletti, L., D'Errico, F. and Leoni, L. (1984) Il sito mesolitico del Monte Cornizzolo (Prealpi Lombarde Occidentali). *Preistoria Alpina* 19, 213–220.

Castelletti, L., Martinelli, M. C., Maspero, A. and Moroni, A. (1992) Il sito neolitico della Consuma 1 (Pieve S. Stefano, Arezzo). *Rivista di Scienze Preistoriche* 44, 48–55.

Castiglioni, E. and Rottoli, M. (2001) Capua (Caserta). Località Strepparo e Cento Moggie. Scavi nell'area del CIRA. Resti botanici da un pozzo dell'età del bronzo. *Bollettino di Archeologia* 37–38, (1996), 62–67.

Castiglioni, E., Fontana, A. and Visentini, P. (2003) Recenti dati sulla preistoria del Friuli occidentale. In G. Tasca (ed.) *Giornata di Studio sull'archeologia del medio e basso Tagliamento in ricordo di Giuseppe Cordenos.* San Vito al Tagliamento (PN), pp. 10–35.

Cevasco, R., Menozzi, B. I., Molinari, C., Moreno D., Vaccarezza, C., Guido M. A. and Montanari C. (2010) The historical ecology of Ligurian chestnut groves: archival, documentation and field evidence. In G. Bounous and G. L. Beccaro (eds) *Proceedings of the 1st European Congress on Chestnut Castanea 2009*, Acta Horticulturae, ISHS, n. 866, pp. 43–50.

Chaker, S. (1997) Figue-figuier. In G. Camps (ed.) *Encyclopédie Berbère.* Édisud, Aix-en-Provence, pp. 2825–2833.

Chausserie-Laprée, J. (2005) *Martigues, terre gauloise. Entre Celtique et Méditerranée.* Errance, Paris.

Chauvet, M. (1999) Esquisse historique de la diffusion des espèces fruitières européennes. In M. Chauvet (ed.) *Le patrimoine fruitier. Hier, aujourd'hui, demain. Actes du colloque de La Ferté Bernard (Sarthe), 16–17 octobre 1998.* AFCEV, Paris, pp. 5–14.

Childe, V. G. (1958) *The prehistory of European society.* Penguin, Harmondsworth.

Clement, C. R. (2006) Fruit Trees and the Transition to Food Production in Amazonia. In W. Balée and C. L. Erickson (eds.) *Time and Complexity in Historical Ecology.*

Columbia University Press, New York, pp. 165–186.
Condit, I. J. (1947) *The fig*. Chronica Botanica Co, Waltham Mass, USA.
Condit, I. J. (1955) Fig Varieties: A Monograph. *Hilgardia* 11, 323–538.
Conedera, M., Krebs, P., Tinner, W., Pradella, M. and Torriani, D. (2004) The cultivation of *Castanea sativa* (Mill.) in Europe, from its origin to its diffusion on a continental scale. *Vegetation History and Archaeobotany* 13(3), 161–179.
Coomes, O. T., Grimard, F. and Burt, G. J. (2000) Tropical forestes and shifting cultivation: secondary forest fallow dynamics among traditional farmers of the Peruvian Amazon. *Ecological Economics* 32, 109–124.
Costantini, L. (2002) Italia centro-meridionale. In G. Forni and A. Marcone (eds.) *Storia dell'agricoltura italiana*, 1, L'età Antica, Preistoria. Accademia dei Georgofili, Edizioni Polistampa, Firenze, pp. 221–234.
Cottini M. and Rottoli, M. (2001) Reperti archeobotanici. In C. Varaldo (ed.) *Archeologia urbana a Savona: scavi e ricerche nel complesso monumentale del Priamar. II.2 Palazzo della Loggia (scavi 1969–1989). I materiali.* Istituto Internazionale di Studi Liguri. Collezione di monografie preistoriche ed archeologiche. XI. Bordighera, Savona, pp. 519–528.
Cotton, C. M. (1998) *Ethnobotany. Principles and Applications.* Wiley, Chichester, p. 424.
Coubray, S., Fiorentino, G., Zech-Matterne, V., Bui Thi Mai and Pagnoux C. (2010) New perpectives about the history of *Citrus* in western Mediterrranean: a multidisciplanary investigation. Poster. *15 th Conf. Intern Group for Palaeoethnobotany*. Wilhelmshaven, 31 May–5 June.
Cristofani, M. (1987) Duo sunt liquores. In *L'alimentazione nel mondo antico. Gli Etruschi*, Istituto Poligrafico e Zecca dello Stato, Roma, pp. 37–40.
Cruise, G. M., Macphail, R.I., Linderholm, J., Maggi, R. and Marshall, P. D. (2009) Lago di Bargone, Liguria, N Italy: a reconstruction of Holocene environmental and land-use history. *The Holocene* 19, 987–1003.
Cuba Salerno, A. B., Silva Peralta, A. and Cornejo Flores, C. (1998) *Bosques Secos y Desertificación*. INRENA, Lima.
Dalla Fior, G. (1940) Analisi polliniche di torbe e depositi lacustri della Venezia Tridentina. V. La palafitta di Molina di Ledro, la Torbiera dei Piani di Festa (M. Baldo), la Torbiera di Creèr (M. Baldo), la Palù Longa di Brez (Val di Non), la Palù Longa di Anterivo, Banco di torba dell'alta Val di Martello. Conclusioni generali. *Memorie del Museo di Storia Naturale della Venezia Tridentina* 5, 121–176.
De Marinis, R. (1987) I commerci dell'Etruria con i paesi a nord del Po dal IX al VI secolo a.C. In R. De Marinis (ed.) *Gli Etruschi a nord del Po*. Panini Editore, Mantova, pp. 52–80.
Delhon, C., Thiébault, S. and Berger, J. F. (2009) Environment and landscape management during the Middle Neolithic in southern France: Evidence for agro-sylvo-pastoral systems in the Middle Rhone Valley. *Quaternary International* 200, 50–65.
Delvaud, C. C. (1984) *Las Regiones Costeñas del Perú Septentrional*. CIPCA/Pontificia Universidad Católica del Perú, Lima.
Den Hertog, W. H. and Wiersum, K. F. (2000) Timur (*Zanthoxylum armatum*) production in Nepal. Dynamics in Nontimber Forest Resources Management. *Mountain Research and development* 20 (2), 136–145.
De Pascale, A., Maggi, R., Montanari C. and Moreno, D. (2006) Pollen, herds, jasper and copper mines: economic and environmental changes during the 4th and 3rd millennia BC in Liguria (NW Italy). *Environmental Archaeology* 11, 115–124.
Descola, P. (1986) *La nature domestique, symbolisme et praxis dans l'écologie des Achuar, publié par La Fondation Singer Polignac*. Editions de la Maison des Sciences de l'Homme, Paris.
Descola, P. (2003) Le sauvage et le domestique perception du sauvage et du domestique domestique chez les populations d'Asie orientale, du sous-continent Indien et les populations Européennes. Séminaire en ligne, Archives Audiovisuelles de la Recherche. http://www.archivesaudiovisuelles.fr/FR, 27 Nov. 2003, EHESS, Paris.
Desrosiers, S. (2000) Textiles trouvés dans deux tombes du Bas-Empire à Naintré (Vienne). In D., Cardon and M. Feugère (eds.) *Archéologie des textiles, des origines au Ve siècle: actes du colloque de Lattes, octobre 1999.* Monographies instrumentum 14. Monique Mergoil, Montagnac, pp. 195–208.
Dietler, M. (1990) Driven by Drink: The Role of Drinking in the Political Economy and the Case of Early Iron Age France. *Journal of Anthropological Archaeology* 9, 352–406.
Dietler, M. (1992) Commerce du vin et contacts culturels en Gaule au premier Age du fer. In M. Bats, G. Bertucchi, G. Congès and H. Tréziny (eds.) *Marseille grecque et la Gaule*. Etudes Massaliètes 3, ADAM/Université de Provence, Lattes/Aix-en-Provence, pp. 305–336.
Dietler, M. (2007) The Iron Age in the Western Mediterranean. In W. Scheidel, I. Morris and R. Saller (eds.) *The Cambridge Economic History of the Greco-Roman World.* Cambridge University Press, Cambridge, pp. 242–276.
Dixon, R. (1905) The northern Maidu. *Bulletin of the American Museum of Natural History* 17, 119–381.
Dixon, R. (1907) The Shasta. *Bulletin of the American Museum of Natural History* 17, 381–498.
Driver, H. (1936) Wappo ethnography. *University of California Publications in American Archaeology and Ethnology* 36, 179–220.
Du Bois, C. (1935) Wintu ethnography. *University of California Publications in American Archaeology and Ethnology* 36, 1–148.
Duputié, A., David, P., Debain, C. and McKey, D. (2007) Natural hybridization between a clonally propagated

crop, cassava (*Manihot esculenta* Crantz) and a wild relative in French Guiana. *Molecular Ecology* 16, 3025–3038.

Durand, A. (2003) *Les paysages médiévaux du Languedoc*. Tempus, Presses universitaires du Mirail, Toulouse.

Elias, M., Penet, L., Vindry, P., McKey, D., Panaud, O. and Robert, T. (2001) Unmanaged sexual reproduction and the dynamics of genetic diversity of a vegetatively propagated crop plant, cassava (*Manihot esculenta* Crantz), in a traditional farming system. *Molecular Ecology* 10, 1895–1907.

Engel, F. A. (1987) *De Las Begonias al Maíz: Vida y Producción en el Perú Antiguo*. Lima, Centro de Investigacion de Zonas Áridas, Universidad Nacional Agraria La Molina, La Molina, Perú.

Erdtman, G. (1966) *Pollen morphology and plant taxonomy. Angiosperms (An introduction to Palynology, 1)*. Hafner Publishing Company, New York.

Erroux, J. (1974) Les graines des fouilles de Lattes. In J. Arnal, R. Majurel and H. Prades (eds.) *Le port de Lattara, Lattes, Hérault*. Institut International d'Etudes Ligures, Bordighera/Montpellier, pp. 275–277.

Fall, P. L., Lines, L. and Falconer, S. E. (1998) Seeds of Civilization: Bronze Age Rural Economy and Ecology in the Southern Levant. *Annals of the Association of American Geographers*, 88(1), 107–125.

Fay, G. (1972) *Recherches sur l'organisation de la vie rurale et sur les conditions de la production dans la basse montagne rifaine*, Thèse de IIIème cycle, Université Paris VII, UER Géographie et Sciences de la Société.

Felger, R. (1979) Ancient Crops for the 21st Century. In G. A. Ritchie (ed.) *New Agricultural Crops*. American Association for the Advancement fo Science Symposium, Vol. 38, Westview Press, Boulder, pp. 5–20.

Felker, P., Clark, P. R., Osborn, J. F. and Cannell, G. H. (1984) *Prosopis* Pod Production: Comparison of North American, South American, Hawaiian, and African Germplasm in Young Plantations. *Economic Botany* 38(1), 36–51.

Ferrari, A., Delpino, C., Petrucci, G., Rottoli, M. and Visentini, P. (2002) Introduzione all'ultimo Neolitico dell'Italia padano-alpina e nord adriatica. In A. Ferrari and P. Visentini (eds.) *Il declino del mondo neolitico. Ricerche in Italia centro-settentrionale fra aspetti peninsulari, occidentali e nord-alpini*. Atti del Convegno, Pordenone 5–7 aprile 2001. Quaderni del Museo Archeologico del Friuli Occidentale, 4, pp. 101–121.

Ferreyra, R. (1987) *Estudio Sistemático de los Algarrobos de la Costa Norte del Perú*. Ministerio de Agricultura, Lima.

Fiorentino, G. (1998) Le risorse vegetali. In A. Cinquepalmi and F. Radina (ed.) *Documenti dell'età del Bronzo. Ricerche lungo il versante adriatico pugliese*. Catalogo della Mostra. Fasano di Brindisi, pp. 211–221.

Fiorentino, G., Castiglioni, E., Rottoli, M. and Nisbet, R. (2004) Le colture agricole in Italia nel corso dell'età del Bronzo: sintesi dei dati e linee di tendenza. In D. Cocchi Genick (ed.) *L'età del Bronzo Recente in Italia*. Atti del Congresso Nazionale, Lido di Camaiore, 26–29 ottobre 2000, pp. 219–226.

Galil, J. (1977) Fig Biology. *Endeavour* 1, 52–56.

Garcia, D. (1987) Observations sur la production et le commerce des céréales en Languedoc méditerranéen durant l'âge du Fer : les formes de stockage des grains. *Revue Archéologique de Narbonnaise* 20, 43–98.

Garcia, D. (2004) *La Celtique méditerranéenne*. Errance, Paris.

Gentili, G. V. (1985) *Il Villanoviano verucchiese nella Romagna Orientale ed il sepolcreto Moroni*. Studi e documenti di Archeologia, 1, Bologna.

Gil, J. (1997) *El Cultivo Tradicional de la Papa en la Isla de Tenerife*. Asociación Granate, La Laguna.

Gil, J., González, A. J., Morales, J. and Perera, J. (2008a) Early records of *Ficus carica* diversity in Canary Islands and its permanence as local names until recent times. *Acta Horticulturae* 798, 39–47.

Gil, J., Morales, J., Oliva, O., Delgado, N. T., Mederos, F., Oliva, D. and Alamo, I. (2008b) *Especies y Variedades de Plantas Cultivadas en la Isla de Gran Canaria*. Asociación Insular de Desarrollo Rural de Gran Canaria, Las Palmas de Gran Canaria.

Gilman, A. (1981) The Development of Social Stratification in Bronze Age Europe. *Current Anthropology*, 22(1), 1–23.

Giraldo, E., López-Corrales, M. and Hormaza, J. I. (2008). The use of SSR markers to screen new accessions before their incorporation into Fig Germplasm Collections. *Acta Horticulturae* 798, 165–168.

Gmitter, F. G. and Hu, X. (1990) The possible role of Yunnan, China, in the origin of contemporary *Citrus* species (Rutaceae). *Economic Botany* 44, 267–277.

Goldschmidt, W. (1951) Nomlaki ethnography. *University of California Publications in American Archaeology and Ethnology* 42, 303–443.

Goldstein, D. J. (2007) *Forests and Fires: A Paleoethnobotanical Assessment of the Impact of Middle Sicán Pyrotechnology on the dry tropical forests of the La Leche River Valley, Lambayeque, Peru (950-1050 C.E.)*. Unpublished Ph.D. dissertation, Department of Anthropology, Southern Illinois University.

Goldstein, D. J. and Shimada, I. (2007) Middle Sicán Multi-Craft Production: Resource Management and Labor Organization. In I. Shimada (ed.) *Craft Production in Complex Societies: Multi-Crafting, Sequential Production, and Producers*. University of Utah Press, Salt Lake City, pp. 44–67.

Goldstein, D. J. and Shimada, I. (2010) Feeding the Fire: Food, Metal, and Ceramic Production in the Middle Sicán. In E. Klarich (ed.) *From Subsistence to Social Strategies: New Directions in the Study of Daily Meals and Feasting Events*. University of Colorado Press, Boulder.

Goldstein, D. J., Coleman Goldstein, R. C. and Williams, P. R. (2008) You Are What You Drink. A Sociocultural Reconstruction of Pre-Hispanic Fermented Beverage

Use at Cerro Baul, Moquegua, Peru. In J. Jennings and B. J. Bowser (eds.) *Drink, Power and Society in the Andes.* University Press of Florida, Gainesville, pp. 133–166.

Gould, R. (1975) Ecology and adaptive response among the Tolowa Indians of northwestern California. *Journal of California Anthropology* 2(2), 148–170.

Grant, M. S., Blackmore, S. and Morton, C. (2000) Pollen morphology of the subfamily Aurantioideae (Rutaceae). *Grana* 39 (1), 8–20.

Greig, J. (1984) A preliminary report on the pollen diagrams and some macrofossil results from Palafitta Fiavè. *Patrimonio storico artistico del Trentino*, 8, Trento, pp. 305–322.

Grieco, A. J. (1996) Alimentation et classes socials à la fin du Moyen Âge et à la Renaissance. In J. Flandrin and M. Montanari (eds.) *Histoire de l'Alimentation.* Fayard, Paris, pp. 479–490.

Grison-Pigé, L., Bessière, J. M., Turlings, C. J., Kjellberg, F., Roy, J. and Hossaert-McKey, M. (2001) Limited intersex mimicry of floral odour in *Ficus carica. Functional Ecology* 15, 551–558.

Haas, J. and Creamer, W. (2004) Cultural transformations in the central Andean Late Archaic. In H. Silverman (ed.), *Andean archaeology.* Blackwell Publishing, Malden, pp. 35–50.

Halbritter, H., Weber, M., Zetter, R., Frosch-Radivo, A., Buchner, R. and Hesse, M. (2007) *PalDat – Illustrated Handbook on Pollen Terminology.* PalDat, Society for the Promotion of Palynological Research in Austria: Wien. Website. URL: http://www.paldat.org

Harris, D. R. (1989) An evolutionary continuum of people-plant interaction. In D. R. Harris and G. C. Hillman (eds.) *Foraging and Farming, the Evolution of Plant Exploitation.* Unwin Hyman, London, pp. 11–26.

Harris, P. J. C., Pasiecznik, N. M., Bradbury M. and Ramırez L. (1998) Problems and Potential of *Prosopis.* In H. D. V. Prendergast, N. L. Etkin, D. R. Harris and P. J. Houghton (eds.) *Plants for Food and Medicine.* Royal Botanical Gardens, Kew, pp. 277–93.

Hather, J. G. (1994). Introduction. In J. G. Hather (ed.), *Tropical Archaeobotany.* Routledge, New York, pp. 1–8.

Haudricourt, A. G. (1962) Domestication des animaux, culture des plantes et traitements d'autrui. *L'Homme* 2(1), 40–50.

Heinz, C. and Thiébault, S. (1998) Characterisation and palaeoecological significance of archaeological charcoal assemblages during Late and Postglacial phases in southern France. *Quaternary Research* 50, 56–68.

Heizer, R. and Elsasser A. (1980) *The Natural World of the California Indians.* California Natural History Guides, University of California Press, Berkeley, CA.

Hickman, J. (ed.) (1993) *The Jepson Manual: Higher Plants of California.* University of California Press, Berkeley.

Hillson, S. (1996) *Dental Anthropology.* Cambridge University Press, Cambridge.

Hmimsa, Y. (2009) *L'agrodiversité de l'agroécosystème à l'arbre: Cas du Rif (Nord du Maroc).* PhD, Université Abdelmalek Essaadi, Tétouan.

Hmimsa, Y, Aumeeruddy-Thomas, Y. and Ater, M. (2012) Vernacular taxonomy, classification and varietal diversity of fig (*Ficus carica* L.) among Jbala cultivators in northern Morocco. *Human Ecology* 40, 301–313.

Hmimsa, Y., Ater, M., Achtak, H. Khadari, K., Kjellberg, F. and Aumeeruddy-Thomas, Y. (forthcoming) Atlas of local varieties of Figs (*Ficus carica* L.) of the Rif region in northern Morocco.

Jacquat, C. and Martinoli, D. (1999) *Vitis vinifera* L.: wild or cultivated? Study of the grape pips found at Petra, Jordan; 150 B.C. – A.D. 40. *Vegetetation History and Archaeobotany* 8, 25–30.

Janick, J. (2005) The Origins of Fruits, Fruit Growing, and Fruit Breeding. *Plant Breeding Reviews*, 25, 255–320.

Jannoray, J. (1955) *Ensérune, contribution à l'étude des civilisations préromaines de la Gaule méridionale*, 2 vol. De Boccard, Paris.

Jennings, J. and Bowser, B. J. (2008) *Drink, Power and Society in the Andes.* University Press of Florida, Gainesville.

Johns, T. and Duquette, M. (1991) Traditional detoxification of acorn bread with clay. *Ecology of Food and Nutrition* 25(3), 221–228.

Jones, O. and Cloke, P. (2002) Tree Cultures. The Place of Trees and Trees in their Place. BERG, Oxford.

Jones, G. and Rowley-Conwy, P. (1984) Plant remains from the north Italian lake dwellings of Fiavé (1400–1200 B.C.). In R. Perini (ed.) *Scavi archeologici nella zona palafitticola di Fiavé Carera. Parte I, Campagne 1969–1976. Situazione dei depositi e dei resti strutturali.* Servizio Beni Culturali della Provincia Autonoma di Trento, Trento, pp. 323–359.

Jung, C. (2000) Pérennité et mutations dans l'organisation de paysages agraires: le Tricastin et le terroir de Lattes. *Revue Archéologique de Narbonnaise* 33, 351–380.

Jung, C. (2007) Le vignoble du deuxième âge du Fer et les aménagements agraires républicains. *Lattara* 20, 439–459.

Kaltenrieder, P., Procacci, G., Vannière, B. and Tinner, W. (2010) Vegetation and fire history of the Euganean Hills (Colli Euganei) as recorded by Lateglacial and Holocene sedimentary series from Lago della Costa (northeastern Italy). *The Holocene* 20, 679–695.

Karg, S. (1998) Winter- and Spring-foddering of Sheep/Goat in the Bronze Age Site of Fiavé-Carera, Northern Italy. *Environmental Archaeology* 1, 87–94.

Karg, S. and Märkle, T. (2002) Continuity and changes in plant resources during the neolithic period in western Switzerland. *Vegetation History and Archaeobotany*, 11(1–2), 169–176.

Kaulicke, P. (1994) *Historia General del Perú, Los Orígenes de la Civilización Andina, Tomo I.* Editorial Brasa S.A., Lima.

Khadari, B., Grout, C., Santoni, S. and Kjellberg, F. (2005)

Contrasted genetic diversity and differentiation among Mediterranean populations of *Ficus carica* L.: a study using mtDNA and RFLP, *Genetic Resources and Crop Evolution* 52, 97–109.

Khadari, B., Krichen, L., Lambert, P., Marrakchi, M. and Audergon, J. M. (2006) Genetic structure in Tunisian apricot, *Prunus armeniaca* L., populations propagated by grafting: a signature of bottleneck effects and ancient propagation by seedlings. *Genetic Resources and Crop Evolution* 53, 811–819.

Kislev, M. E., Hartmann, A. and Bar-Yosef, O. (2006) Early Domesticated Fig in the Jordan Valley. *Science* 312, 1372–1374.

Kniffen, F. (1939) Pomo geography. *University of California Publications in American Archaeology and Ethnology* 36, 353–400.

Kreuz, A. (1992) Charcoal from ten early Neolithic settlements in central Europe and its interpretation in terms of woodland management and wildwood resources. *Bulletin de la Société Botanique de France* 139 (2/3/4), 383–394.

Kreuz, A. (1998) Closed forest or open woodland as natural vegetation in the surroundings of Linearbandkeramik settlements? *Vegetation History and Archaeobotany* 17, 51–64.

Kumar, B. M. (2008) Homegarden-based Indigenous Fruit Tree Production in Peninsular India. In Akinnifesi, F. K., Leakey, R. R. B., Ajayi, O. C., Sileshi, G., Tchoundjeu, Z., Matakala, P. and Kwesiga, F. R. (eds.) *Indigenous Fruit Trees in the Tropics. Domestication, Utilization and Commercialization*. Cabi International, Oxfordshire/ Cambridge, pp. 84–99.

Lacombe, T., Di Vecchi, M., Laucou, V., Dechesne, F., Vares, D. and This, P. (2004) Les populations de vignes sauvages du massif de l'Albera. In *Actes del Colloqui L'Albera i el patrimoni en l'espai transfronter*. Consell Comercal del Alt Empordà, Figueres, pp. 313–322.

Ladizinsky, G. (1998) Plant Evolution Under Domestication. Kluwer Academic Press, Dordrecht.

Lagrand, C. and Thalmann, J.-P. (1973) *Les habitats proto-historiques du Pègue, Drôme, le sondage N°8*. Cahier 2, Centre de Documentation de la Préhistoire alpine, Grenoble.

Lancelotti, C. (2005) *Archeobotanica ed aspetti archeologici delle unità US 3 e US 234 della Palafitta di Parma*. Università degli Studi di Milano. Master "Scienze per i Beni Culturali Archeologici: figure professionali di geoarcheologo ed archeometra'. Unpublished thesis.

Lejju, B. J., Robertshaw, P. and Taylor, D. (2006) Africa's earliest bananas? *Journal of Archaeological Science* 33, 102–113.

Lentz, D. (1999) Plant Resources of the Ancient Maya: The Paleoethnobotanical Evidence. In C. D. White (ed.) *Reconstructing Ancient Maya Diet*. University of Utah Press, Salt Lake City, pp. 3–18.

Leoni, L., Castelletti, L. and Castiglioni, E. (2002) I carboni epigravettiani e mesolitici e la dinamica della copertura forestale a Isola Santa. *Rivista di Scienze Preistoriche* 52, 183–195.

Leterme, E. (1995) *Les fruits retrouvés. Histoire et diversité des espèces anciennes du Sud-Ouest*. Editions du Rouergue, Rodez.

Levadoux, L. (1956) Les populations sauvages et cultivées de *Vitis vinifera* L. *Annales de l'Amélioration des Plantes* 6 (1), 59–117.

Littleton, J. and Frohlich, B. (1989) An analysis of dental pathology and diet on historic Bahrain. *Paléorient* 15 (2), 59–84.

Lowe, J., Davite, C., Moreno, D. and Maggi, R. (1994) Holocene pollen stratigraphy and human interference in the woodlands of the northern Apennines, Italy. *The Holocene* 4, 153–164.

Luna-Suarez, S., Luna-Guido, M. L., Frías-Hernández, J. T., Olalde-Portugal, V. and Dendooven, L. (1998) Soil Processes as Affected by Replacement of Natural Mesquite Ecosystem with Maize Crop. *Biology of Fertilized Soils* 27, 274–278.

Maas, J. M. (1995) Conversion of Tropical Dry Forest to Pasture and Agriculture. In S .H. Bullock, H. A. Mooney, and E. Medina (eds.) *Seasonally dry tropical forests*. Cambridge University Press, Cambridge, pp. 399–422.

Mabberley, D. J. (1997) *The Plant Book, A portable dictionary of the vascular plants*. Second edition, Cambridge University Press, Cambridge.

Mabberley, D. J. (2004) *Citrus* (Rutaceae). A review of recent advances in etymology systematics and medical applications. *Blumea* 49, 481–498.

Machado, M. C, Arco, M. C., Vernet, J. L. and Ourcival, J. M. (1997) Man and vegetation in northern Tenerife (Canary Islands, Spain), during the prehispanic period based on charcoal analyses. *Vegetation History and Archaeobotany* 6, 187–195.

Maier, U. (1990) Botanische Untersuchungen in Hornstaad-Hörnle IA. *Bericht der Römisch-Germanischen Kommission* 71 (1), 110–135.

Mane, P. (2006) Images, discours et techniques de la greffe dans l'arboriculture fruitière médiévale. *Archéologie du Midi Médiéval* 23–24, 93–107.

Mangafa, M., and Kotsakis, K. (1996) A new method for the identification of wild and cultivated charred grape seeds. *Journal of Archaeological Science* 23, 409–418.

Margaritis, E. and Jones, M. (2006) Beyond cereals: crop processing and *Vitis vinifera* L. Ethnography, experiment and charred grape remains from Hellenistic Greece. *Journal of Archaeological Science* 33, 784–805.

Marinval, P. (1988). *Cueillette, agriculture et alimentation végétale de l'Epipaléolithique jusqu'au 2ème âge du Fer en France méridionale. Apports palethnographiques de la carpologie*. phD, EHESS, Paris, 2 vol.

Marinval, P. (1993) Etude carpologique d'offrandes alimentaires végétales dans les sépultures gallo-romaines: réflexions préliminaires. In A. Ferdière (ed.) *Monde des morts, monde des vivants en Gaule rurale. Actes du colloque ARCHEA/AGER (Orléans, Conseil Régional, 7-9 février 1992)*. 6ème supplément à la Revue Archéologique du Centre de la France, Tours, pp. 45–65.

Marinval, P. (1997) Vigne sauvage et Vigne cultivée dans le Bassin méditerranéen : Emergence de la viticulture. Contribution archéo-botanique. In P. Frissant (ed.) *L'Histoire du vin, une histoire de rites*. Office International de la Vigne et du Vin, Paris, pp. 137–172.

Marinval, P. (2008) Aliment végétal et plantes utiles de la préhistoire. *In* F. Hallé and P. Lieutaghi (eds.) *Aux origines des plantes. Des plantes et des Hommes*, Fayard, Paris, pp. 20–39.

Mariotti-Lippi, M. (2000) The garden of the 'Casa delle Nozze di Ercole ed Ebe' in Pompeii (Italy): Palynological investigations. *Plant Biosystems* 134(2), 205–211.

Mariotti Lippi, M., Mori Secci, M., Bellini, C. and Gonnelli, T. (2007) Plants in the diet in Prehistoric Tuscany. *Atti Società Naturalisti e Matematici di Modena* 137 (2006), 343–353.

Marrero, A. and Pérez, P. L. (1997) Flora Terrestre de Canarias y su Biodiversidad. In P. L. Pérez, (ed.), *Master en Gestión Ambiental. Ecosistemas Insulares Canarios*. Gobierno de Canarias, Santa Cruz de Tenerife, pp. 177–189.

Martinelli, M. C., Lopes, L., Tindara Calabria, M., Calderoni, G., Fiorentino, G. and Levi, S. T. (2003) Nuove ricerche sul sito dell'età del Bronzo di Portella nell'Isola di Salina. *Atti Istituto Italiano di Preistoria e Protostoria* 35 (2), 883–887.

Marty, F. and Del Corso, M. (2002) L'habitat de hauteur du Castellan (Istres, B.-du-Rh.) à l'âge du Fer. Etude des collections anciennes et recherches récentes. *Documents d'Archéologie Méridionale* 25, 129–169.

Mason, S. (1992) *Acorns in human subsistence*, Ph.D. Thesis. University College, London.

Mason, S. (1995) Acornutopia? Determining the role of acorns in past human subsistence. In J. Wilkins, D. Harvey, and M. Dobson (eds.) *Food in Antiquity*. Exeter University Press, Exeter, pp. 12–24.

Mason, S. (2000) Fire and Mesolithic subsistence – managing oaks for acorns in northwestern Europe? *Palaeogeography, Palaeoclimatology, Palaeoecology* 164, 139–150.

McCarthy, H. (1993) *A Political Economy of Western Mono Acorn Production*, Ph.D. Thesis. University of California, Davis.

Mercuri, A. M., Accorsi, C. A., Bandini Mazzanti, M., Bosi, G., Cardarelli, A., Labate, D., Marchesini, M. and Trevisan Grandi, G. (2006) Economy and environment of Bronze Age settlements – Terramaras – on the Po Plain (northern Italy): first results from the archaeobotanical research at the Terramara di Montale. *Vegetation History and Archaeobotany* 16(1), 43–60.

Merriam, C. H. (1918) The acorn, a possibly neglected source of food. *The National Geographic Magazine* 34(2), 129–137.

Michon, G. (2005a) Ni ager, ni Hortus: la forêt dans tous ses états. Quelques réflexions sur les relations sociétés/forêts inspirées par Jacques Barrau. *Journal d'Agriculture et de Botanique Appliquée* 42(1–2), 147–206.

Michon, G. (2005b) *Domesticating forest. How farmers manage forest resources*. IRD, CIFOR and ICRAF, Bogor, Indonesia.

Morales, F. (1993) *Canarias. Crónicas de su conquista*. Cabildo Insular de Gran Canaria, Las Palmas de Gran Canaria.

Morales, J. (2010) *El uso de las plantas en la prehistoria de Gran Canaria: alimentación, agricultura y ecología*. Cabildo de Gran Canaria, Las Palmas de Gran Canaria.

Morales, J. and Delgado, T. (2007) Figs and their importance in the prehistoric diet in Gran Canaria island (Canary Isles). In R. T. J. Cappers (ed.) *Fields of change. Progress in African Archaeobotany*. Barkhauis, Groningen, pp. 77–85.

Mottes, E., Petrucci, G., Rottoli, M., Visentini, P. (2010) Evolution of the square mouthed pottery culture in Trentino-Alto Adige, Veneto and Friuli: cultural, chronological, palaeoeconomic and environmental aspects. *Gortania* 31, 97–124.

Moutarde, F. (2006) *L'Évolution du Couvert Ligneux et de son Exploitation par l'Homme dans la Vallée du Lurin (Côte Centrale du Pérou), de l'Horizon Ancien (900–100 av. J.-C.) á l'Horizon Tardif (1460–1532 apr. J.-C.) – Approche Anthracologique*. Unpublished Ph. D. Dissertation, Faculty of Archaeology. University of Paris I, Paris.

Nair, P. K. K. (1961) Pollen Morphology of Some Indian Medicinal Plants. *Journal of Scientific & Industrial Research* 20 C(2), 45–50.

Nelson, G. C., Lukacs, J. R. and Yule, P. (1999) Dates, caries, and early tooth loss during the Iron Age of Oman. *American Journal of Physical Anthropology* 108 (3), 333–343.

Nisbet, R. and Ventura, G. (1994) I dati archeobotanici. In E. Peroni and F. Trucco (eds.) *Enotri e Micenei nella Sibaritide*. Istituto per la storia e l'archeologia della Magna Grecia, Taranto, pp. 577–585.

Nomland, G. A. (1935) Sinkyone notes. *University of California Publications in American Archaeology and Ethnology* 36, 149–178.

Nordt, L., Hayashida, F., Hallmark, T. and Crawford, C. (2004) Late Prehistoric Soil Fertility, Irrigation Management, and Agricultural Production in North-west Coastal Peru. *Geoarchaeology* 19(1), 21–46.

Ocampo, M. (1991) *Metodologia para la Determinacion de la Edad y las Tasas de Crecimiento para las Especias Forestales de los Bosques Secos de la Costa Norte del Perú*. Proyecto Algarrobo con Fundacion para el Desarrollo Nacional, Chiclayo, Peru.

Otero-Arnaiz, A., Casas, A., Hamrick, J. L. and Cruse-Sanders, J. (2005) Genetic variation and evolution of *Polaskia chichipe* (Cactaceae) under domestication in the Tehuacan Valley, central Mexico. *Molecular Ecology* 14, 1603–1611.

Pagnoux, C. (2010) *L'introduction des agrumes en Méditerranée occidentale à l'époque gréco-romaine.* Mémoire de master 1 de l'université de Paris I, UFR histoire de l'art et archéologie.

Pals, J. P. and Voorrips, A. (1979) Seeds, fruits and charcoals from two prehistoric sites in northern Italy. *Archaeo-Physica* 8, 217–235.

Pang, X.-M., Hu, C.-G. and Deng, X.-X. (2007) Phylogenetic relationships within *Citrus* and its related genera as inferred from AFLP markers. *Genetic Resources and Crop Evolution* 54 (2), 429–436.

Pasiecznik, N. M. (2002) *Exploiting the Value of Prosopis for Dryland Forestry and Agroforestry Systems.* Henry Doubleday Research Association Brief, Coventry.

Pasiecznik, N. M., Felker, P., Harris, P. J. C., Harsh, L. N., Cruz, G., Tawari, J. C., Cadoret, K., and Maldonado, L. J. (2001) *The Prosopis juliflora- Prosopis pallida Complex: A Monograph.* Henry Doubleday Research Association Brief, Coventry.

Pavlik, B., Muick, P., Johnson, S. and Popper, M. (2002) *Oaks of California.* Cachuma Press and the California Oak Foundation, Olivos, CA.

Pearsall, D. M. (1992) The Origins of Plant Cultivation in South America. In C. W. Cowan and P. J. Watson (eds.) *The Origins of Agriculture: An International Perspective.* Smithsonian Institution Press: Washington D.C., pp. 173–205.

Pearsall, D. M. (2008) Plant Domestication and the Shift to Early Agriculture in the Andes, In H. Silverman and W. H. Isbell (eds.) *Handbook of South American Archaeology.* Springer, New York, pp. 105–120.

Petrucci-Bavaud, M. and Jacomet, S. (1997) Zur Interpretation von Nahrungsbeigaben in römischen Brandgräbern. *Ethnographisch-Archäologische Zeitschrift* 38 (3–4), 567–593.

Pignatti, S. (1982) *Flora d'Italia.* Edagricole, Bologna.

Planchais, N. (1973) Apport de l'analyse pollinique à la connaissance de l'extension de la vigne au Quaternaire. *Naturalia monspeliensia* 23–24, 211–233.

Proyecto Algarrobo (1997) *Inventorio Forestal de los Bosques Secos de Batán Grande.* Universidad Nacional Agraria la Molina, Lima.

Py, M. (1985) Les amphores étrusques de Gaule méridionale. In M. Cristofani, P. Moscati, G. Nardi and M. Pandolfini (eds.) *Il commercio etrusco arcaico.* Consiglio Nazionale delle Ricerche, Roma, pp. 73–94.

Py, M. (1993) *Les gaulois du Midi. De la fin de l'Âge du Bronze à la conquête romaine.* Hachette, Paris.

Py, M. (1995) Les Etrusques, les Grecs et la fondation de Lattes. *Etudes Massaliètes* 4, 261–276.

Py, M. and Buxó i Capdevila, R. (2001) La viticulture en Gaule à l'âge du Fer. *Gallia* 58, 29–43.

Quellier, F. (2006) L'automne horticole du Moyen Âge, permanences médiévales dans les traits de jardinage de la première modernité (1486–1652). *Archéologie du Midi Médiéval* 23–24, 109–117.

Ramón-Laca, L. (2003) The Introduction of cultivated *Citrus* to Europe via Northern Africa and the Iberian Peninsula. *Economic Botany* 57(4), 502–514.

Ravazzi, C., Cremaschi, M. and Forlani, L. (2004) Studio archeopalinologico. Nuovi dati, analisi floristica e sintassonomica della vegetazione nell'età del Bronzo. In M. A. Bernabò Brea and M. Cremaschi (eds.) *Il villaggio piccolo della terramara di Santa Rosa di Poviglio.* Origines, Studi e materiali pubblicati a cura dell'Istituto Italiano di Preistoria e Protostoria, Firenze, pp. 703–737.

Renault-Miskovsky, J. (1976) La végétation quaternaire. La végétation au Pléistocène moyen en Languedoc méditerranéen. In H. de Lumley (ed.) *La préhistoire française. Les civilisations paléolithiques et mésolithiques de la France* Vol. 1, CNRS, Paris, pp.486–488.

Renfrew, C. (1972) *The ermergence of civilisation: The Cyclades and the Aegean in the third millennium B.C.* Methuen, London.

Risco, J. J. (1998) Tecnologías Locales Mejoradas en el Uso de los Rescuros del Bosque Seco en un Contexto Familiar: Cocinas Mejoradas y Algarroberas. In A. B. Cuba Salerno, A. Silva Peralta, and C. Cornejo Flores (eds.) *Bosques Secos y Desertificación,* INRENA, Lima, pp. 129–138.

Rius, D., Vannière, B., Galop, D. (2009) Fire frequency and landscape management in the northwestern Pyrenean piedmont, France, since the early Neolithic (8000 cal. BP). *The Holocene* 19 (6), 847–859.

Rival, L. (1998) *The Social Life of Trees.* Berg, Oxford.

Robert, D., Dumas, C. and Bajon, C. (1998) *La reproduction. Biologie végétale, vol. 3.* Doin, Paris.

Rogers, K. (2000) *The Magnificent Mesquite.* University of Texas Press, Austin.

Roman, D. and Roman, Y. (1997) *Histoire de la Gaule (VIe s. av. J.-C. - Ier s. ap. J.-C.). Une confrontation culturelle.* Fayard, Paris.

Rottoli, M. (1999a) I resti vegetali di Sammardenchia – Cûeis (Udine), insediamento del Neolitico antico. In A Ferrari and A. Pessina (eds.) *Sammardenchia - Cûeis. Contributi per la conoscenza di una comunità del primo neolitico.* Pubblicazione n. 41, Edizioni del Museo Friulano di Storia Naturale, Udine, pp. 307–326.

Rottoli, M. (1999b) Le analisi archeobotaniche. In R. Poggiani Keller (ed.), *Scarceta di Manciano (GR) Un centro abitativo e artigianale dell'età del Bronzo sulle rive del Fiora.* Comune di Manciano, Museo di Preistoria e Protostoria della Valle del Fiume Fiora, Editrice Laurum, Manciano, pp. 151–157.

Rottoli, M. (1999c) Indagini paleobotaniche. In A.M. Esposito (ed.) *Principi guerrieri. La necropoli etrusca di Casale Marittimo,* Electa ed., pp. 87–94.

Rottoli, M. (2001) Analisi archeobotaniche: i macroresti vegetali. In P. Frontini (ed.) *Castellaro del Vhò. Campagne di scavo 1996-1999.* Comune di Milano, Settore

Cultura Musei e Mostre, Raccolte Archeologiche e Numismatiche, pp. 175–195.

Rottoli, M. and Castiglioni, E. (2009) Prehistory of plant growing and collecting in northern Italy, based on seed remains from the Early Neolithic to the Chalcolithic (*c*. 5600–2100 cal B.C.). *Vegetation History and Archaeobotany* 18, 91–103.

Rottoli, M. and Fugazzola Delpino, M. A. (2000) *Lauro-Carpinetum*: a particular forest formation in the Roman countryside; archaeological data and actual diffusion. In *Nouvelles approches méthodologiques, histoire de la végétation et des usages du bois depuis la Préhistoire*, Second Colloque International d'Anthracologie, Paris 13–16 September 2000, Résumés, C.N.R.S., Paris.

Roux, D. (2006) Dire l'olivier en méditerranée, l'homme, la langue et l'arbre, *Forêt méditerranéenne*, 27(4), 303–316.

Ruas, M.-P. (1996) Eléments pour une histoire de la fructiculture en France: données archéobotaniques de l'Antiquité au XVIIe siècle. In M. Colardelle (ed.) *L'homme et la nature au Moyen Age. Paléoenvironnement des sociétés occidentales. Actes du Ve Congrès International d'Archéologie Médiévale, Grenoble (France), 6–9 octobre 1993*. Errance, Paris, pp. 92–105.

Ruffo, S. (ed.) (2001) *Le foreste della Pianura Padana*. Museo Friulano di Storia Naturale, Udine.

Sabir, A. (2001) *Las Canarias preeuropeas y el norte de África. El ejemplo de Marruecos*. Al-Maarif, Rabat.

Sadori, L., Allevato, E., Bosi, G., Caneva, G., Castiglioni, E., Celant, A., Di Pasquale, G., Giardini, M., Mazzanti, M., Rinaldi, R., Rottoli, M. and Susanna, F. (2009) The introduction and diffusion of peach in ancient Italy. In A. M. Mercuri (ed.) *Plants and culture: seeds of the cultural heritage of Europe*. Edipuglia, Bari, pp. 45–61.

Schaedel, R. P. (1988) *La Etnografia Muchik en las Fotografias de H. Bruning: 1886-1925*. Lima, Ediciones COFIDE.

Schiemann, E. (1953) Vitis im Neolithicum der Mark Brandenburg. *Der Züchter* 23(10–11), 318–327.

Schroth, G., Da Mota, M. S. S., Lopes, R. and De Freitas, A. F. (2004) Extractive use, management and in situ domestication of a weedy palm, Astrocaryum tucuma, in the central Amazon. *Forest Ecology and Management* 202(1–3), 161–179.

Scora, R. W. (1975) On the history and origin of *Citrus*. *Bull. Torrey Bot. Club* 102, 369–375.

Scotti Maselli, F. and Rottoli, M. (2007) Indagini archeobotaniche all'ex Essiccatoio Nord di Aquileia: i resti vegetali protostorici e romani. In G Cuscito and C. Zaccaria (eds.) *Aquileia dalle origini alla costituizione del ducato longobardo. Territorio - Economia - Società*. Antichità Altoadriatiche 65, Centro di Antichità Altoadriatiche, Casa Bertoli Aquileja. Editreg, Trieste, pp. 783–816.

Shady Solis, R. (1999) El sustento económico del surgimiento de la civilización en el Perú. *Boletín del Museo de Arqueología y Antropología* 2(11), 2–4.

Sherratt, A. (1999) Cash-crops before cash: organic consumables and trade. In C. Gosden and Hather, J. (eds.) *The Prehistory of food. Appetites for change*. Routledge, London, pp. 13–34.

Shimada, M. and Shimada, I. (1985) Prehistoric Llama Breeding and herding on the North Coast of Peru. *American Antiquity* 50(1), 3–26.

Smith, B. D. (1995). *The Origins of Agriculture*. Scientific American Library, W.H. Freeman Press, New York.

Smith, E. and Jones, G. (1990) Experiments on the Effects of Charring on Cultivated Grape Seeds. *Journal of Archaeological Science* 17, 317–327.

Stefaniuk, L. C., Morhange, P.-F., Blanc, F., Francou, S. and Goiran, J.-P. (2005) Évolution des paysages littoraux dans la dépression sud-ouest de Cumes depuis 4000 ans. La question du port antique. *Méditerranée* 1(2), 49–59.

Steward, J. (1933) Ethnography of the Owens Valley Paiute. *University of California Publications in American Archaeology and Ethnology* 33, 233–350.

Stummer, A. (1911) Zur Urgeschichte der Rebe und des Weinbaues. *Mitteilungen der Anthropologischen Gesellschaft in Wien* 41, 283–296.

Tanno, K. and Willcox, G. (2006) How fast was wild wheat domesticated? *Science*, 311, 1886.

Tejera, A. and González, R. (1987) *Las culturas aborígenes canarias*. Interinsular Canaria, Santa Cruz de Tenerife.

Terral, J.-F (2000) Exploitation and Management of the Olive Tree During Prehistoric Times in Mediterranean France and Spain. *Journal of Archaeological Science* 27, 127–133.

Terral, J.-F. and Arnold-Simard, G. (1996) Beginnings of Olive Cultivation in Eastern Spain in Relation to Holocene Bioclimatic Changes. *Quaternary Research* 46, 176–185.

This, P., Roux, C., Parra, P., Siret, R., Bourse, T., Adam, A.-F., Yvon, M., Lacombe, T., David, J. and Boursiquot, J.-M. (2001) Caractérisation de la diversité d'une population de vignes sauvages du Pic Saint-Loup (Hérault) et relations avec le compartiment cultivé. *Genetic Selection Evolution* 33 (Suppl. 1), 289–304.

Timbrook, J. (2007) *Chumash Ethnobotany*. Publications in Anthropology 1, Santa Barbara Museum of Natural History and Heyday Books, Berkeley, CA.

Torquebiau, E. F. (2000) A renewed perspective on agroforestry concepts and classification. *Comptes Rendus de l'Académie des Sciences Paris, Sciences de la vie* 323, 1009–1017.

Tosi Jr., J. A. (1960) *Zonas de Vida Natural en el Perú: Memoria Explicativa sobre el Mapa Ecológico del Perú*. Instituto Interamericano de Ciencias Agrícolas de OEA, Lima.

Turner, B. L. I. and Miksicek, C. H. (1984) Economic Plant Species Associated with Prehistoric Agriculture in the Maya Lowlands. *Economic Botany* 32(2), 179–193.

Tutin, T. G., Heywood, V. H., Burges, N. A., Valentine, D. H., Walters, S. M. and Webb, D. A. (1964–1980)

Flora Europaea. 5 Vols. Cambridge University Press, Cambridge.

U.S. Department of Agriculture, A. R. S. (2008) *USDA National Nutrient Database for Standard Reference, Release 21.* Nutrient Data Laboratory Home Page, http://www.ars.usda.gove/ba/bhnrc/ndl/.

Ugent, D. and Ochoa, C. M. (2006) *La Etnobotánica del Perú.* Consejo Nacional de Ciencia, Tecnología e Innovación Tecnológica, Lima.

Urtusáustegui, J. A. de (1983) *Diario de viaje a la isla de El Hierro en 1779.* Centro de Estudios Africanos – Colectivo Cultural Valle de Taoro, La Laguna.

Valdeyron, G. and Lloyd, D. G. (1979) Sex Differences and Flowering Phenology in the Common Fig, *Ficus carica* L. *Evolution*, 33(2), 673–685.

van der Veen, M. (2003) When is food a luxury? *World Archaeology* 34 (3), 405–427.

Van Zeist, W. (1991) Economic aspects. In W. Van Zeist, K. Wasylikowa and K.-E. Behre (eds.) *Progress in Old World Palaeoethnobotany.* Balkema, Rotterdam, pp. 109–130.

Van Zeist, W., Bottema, S., van der Veen, M. (2001) *Diet and vegetation at ancient Carthage. The archaeobotanical evidence.* Groningen.

Verdin, F. (1997) Coudouneu (Lançon-de-Provence, Bouches-du-Rhône) : une ferme-grenier et son terroir au Ve s. av. J.-C. *Documents d'Archéologie Méridionale* 19–20, 165–198.

Viera, J. (1982) *Diccionario de Historia natural de las Islas Canarias.* Mancomunidad de Cabildos, Madrid.

Webber, H. J. (1967) History and development of the citrus industry. In W. Reuther, L. D. Batchelor, R. and H. J. Webber (eds.) *The Citrus Industry.* Vol 1. University of California, pp. 1–39.

Weir, G. H. and Eling, H. H. Jr. (1986) Pollen Evidence for Economic Plant Utilization in Prehistoric Agricultural Fields of the Jequetepeque Valley, Northern Peru. In R. Matos, S. Turpin, and H. Eling (eds.) *Andean Archaeology: Papers in Memory of Clifford Evans.* Institute of Archaeology, University of California, Los Angeles, pp. 150–162.

Weir, G. H., Benfer, R. A. and Jones, J. G. (1988) Preceramic to Early Formative Subsistence on the Central Coast. In E. S. Wing and J. C. Wheeler (eds.) *Economic Prehistory of the Central Andes*, Vol. 427, BAR International Series, Oxford, pp. 56–92.

Weiss, E., Kislev, M. E. and Hartmann, A. (2006) Autonomous cultivation before domestication. *Science*, 312, 1608–1610.

Wiersum, K. F. (1997) Indigenous exploitation and management of tropical forest resources: an evolutionary continuum in forest-people interactions. *Agriculture, Ecosystems and Environment* 63, 1–16.

Wiersum, K. F. (2004) Forest gardens as an 'intermediate' land-use system in the nature-culture continuum: Characteristics and future potential. *Agroforestry Systems* 61, 123–134.

Wiersum, K. F. (2008) Domestication of trees or forests: Development pathways for fruit tree production in South-east Asia. In Akinnifesi, F. K., Leakey, R. R. B., Ajayi, O. C., Sileshi, G., Tchoundjeu, Z., Matakala, P. and F. R., Kwesiga (eds.) *Indigenous Fruit Trees in the Tropics. Domestication, Utilization and Commercialization.* Cabi International, Oxfordshire/Cambridge, pp. 70–83.

Wohlgemuth, E. (2004) *The course of plant food intensification in native central California*, Ph. D. Thesis, Department of Anthropology, University of California, Davis.

Wolf, C. (1945) *California Wild Tree Crops: Their Crop Production and Possible Utilization.* Rancho Santa Ana Botanic Garden, Santa Ana Canon, CA.

Zech-Matterne, V. (2010) Le développement de la fructiculture en Gaule du Nord, à l'époque romaine. In Ouzoulias, P. and L., Tranoy (eds.) *Comment les Gaules devinrent romaines.* La Découverte, Paris, pp. 255–266.

Zohary, D. (2004) Unconscious Selection and the Evolution of Domesticated Plants. *Economic Botany* 58 (1), 5–10.

Zohary, D., Hopf, M. and Weiss, E. (2012) *Domestication of Plants in the Old World.* Oxford University Press, Oxford.

Zohary, D. and Spiegel-Roy, P. (1975) Beginnings of fruit growing in the Old World. *Science*, 187, 319–327.

5 Food Plants from the Wild

5.1. INTRODUCTION: WILD FOOD PLANTS IN THE PRESENT AND PAST

Gisella S. Cruz-García and Füsun Ertuğ

Humans have been hunters and gatherers for about 350,000 generations, then mostly agriculturalists for another 600, and quite recently, industrialised agriculture has existed for just two generations[1] (Pretty 2003). Plant gathering is a deeply rooted, shared aspect of human heritage and millions of people in rural areas are still dependent upon gathered plants for the role they play in their diet, medicine, fuel, construction, crafts, animal husbandry, deterrents of pests, as well as in their religious ceremonies and rituals. From the total of 250,000–300,000 known higher plant species, about 5000 have been managed at certain times, and only 20–30 are regarded as staple foods for humanity (Cotton 1996; Heywood 1999). On the other hand, there are thousands of wild and semi-domesticated species used by people, which are cultivated and/or gathered from the wild. This resource has been used by populations in the past and present as shown by multiple archaeobotanical, ethno-historical and ethnobotanical studies.[2]

In order to understand the dynamics of wild food plant consumption and gathering, as well as their importance for rural livelihoods in the present and the past, it is necessary to combine different disciplines and methodological approaches. People-plant relationships in the present are studied by means of ethnobotanical methods (Chapter 2.2), whereas the past use of plants (prehistory and history) is investigated on the basis of several sources of data, ranging from fossilised plant remains (Chapter 2.3), preserved artefacts, graphic representation (Chapter 2.5), to historical documents, folktales and poems (Chapter 2.4, all this volume).

Besides the temporal and methodological scope of wild food plant research, the broad geographical scope must also be taken into consideration. Wild food plants are used across all continents and millions of people depend on this biodiversity of resources for their sustenance. For example, wild food plants are used not only by indigenous societies in Amazonia (Defour and Wilson 1994) but also in northwestern North America (Turner 2003). The use and management of wild edible plants has also been reported in Latin America, for instance, among the Nahua and the Mixtex (Casas *et al.* 1996), in the Bolivian Andes (Vandebroek and Sanca 2007) and Cuba (Volpato and Godinez 2007). The importance of wild vegetables in Africa has been highlighted by Chweya and Eyzaguirre (1999) and their consumption has been reported for the Sambaa in northeast Tanzania (Vainio-Mattila 2000), in the Kingdom of Swaziland (Ogle and Grivetti 1985), and in the Hadejia-Nguru wetlands of Nigeria (Adaya *et al.* 1997), among others. In Asia, the use of wild food plants is also widespread, for instance farmers in northeast Thailand, Laos and Vietnam depend on this resource for their food and nutritional security (Price 1997; Price and Ogle 2008), whereas edible wild plants are also gathered in Turkish Central Anatolia (Ertuğ 2000; 2003b), and by the Rai and Sherpa forager farmers in Nepal (Daniggelis 2003). With respect to Europe, these plants are consumed not only in Sicily (Galt and Galt 1978) and by Arberesh people in Lucania (Pieroni 2003), but also in Spain and Portugal (Pardo-de-Santayana *et al.* 2007; Tardío *et al.* 2006), as well as Poland (Łuczaj 2008; Łuczaj and Szymański 2007). Regarding the vast Australasian area, Tim Low's book '*Wild food plants of Australia*' (1991) presents a detailed description of this resource in the country.

The aim of this chapter is to present an overview of a diverse range of approaches to the study of wild food plants, with a selection of various case studies. Authors make use of ethnobotanical, ethnohistorical and archaeobotanical methodologies, as well as written sources such as documents, oral traditions, ethnographic reports, narrative and poetic material, thereby embracing the challenge of interdisciplinary research. Case studies range from exploratory surveys to in-depth studies of single plants, presenting different insights into the interactions between wild food plants and people. The temporal and geographical range of the studies is also broad, although the latter is not the main focus of the chapter, which is devoted to comparing past and present use of plants across continents and through time.

To start this introductory section, it is necessary to conceptualise 'wild food plants' in relation to human interaction with the 'wild'. Then, four main contemporary issues related to wild food plant use, knowledge and valuation, which come up across the different case studies, are discussed. Firstly, the overlap of the role of wild food plants as food and medicine; secondly, their use during famine and food shortages; thirdly, the role of women; and finally, their stigmatisation versus 'revival'.

Human Interaction with the 'Wild'

Ethnobotanical studies, as well as new data from archaeobotanical research, have caused fundamental revisions of earlier concepts and definitions of 'agrarian' societies. Terms such as 'hunter-gatherer' and 'agriculturalist' were once used by archaeologists as mutually exclusive. However, ethnobotanical research has demonstrated that hunter-gatherers may undertake agricultural practices, and that agriculturalists persist in gathering activities (Adaya *et al.* 1997; Harlan 1975; Ogle and Grivetti 1985). Indeed, wild food plants from agricultural ecosystems provide a critical component to the subsistence system of farmers.[3]

The word 'wild' does not imply the absence of human management. Wild resources are actively managed (including transplanting, promoting, protecting, among other practices) by traditional communities (Cotton 1996). Local management practices are important, not only for the diversity, but also for assuring the long-term availability of plant species, especially in times of seasonal unavailability, famine and stress. Thus, it is possible to cultivate wild plants while cultivated plants are not always domesticated. Furthermore, many of the food plants we grow have not been totally domesticated: for most species the transition from cultivation to domestication does not occur (Harlan 1975). Certainly, Harlan (1975, 63) argues:

> 'Since domestication is an evolutionary process, there will be found all degrees of plant and animal association with man and a range of morphological differentiations from forms identical to wild races to fully domesticated races. A fully domesticated plant or animal is completely dependent upon man for survival.'

In this sense, Harris (1989) and Wiersum (1996) present a continuum model of people and plant interactions at different management intensities: from truly wild to fully managed and cultivated. This continuum model helps us to better conceptualise human-plant interactions, regarding human action with wild, semi-domesticated and domesticated plants. The management of a plant species changes in time and space, but the model is neither unidirectional nor deterministic. Hence, some cultivated wild plants are moving towards domestication, whilst some plants that used to be intensively managed in the past are only tolerated or slightly protected in the present (Harris 1989). Moreover, a plant species can be managed simultaneously in various ways and at different management degrees in some regions, and, at the same time, may not be managed at all in others (González-Insuasti and Caballero 2007; Ogle 2001).

Different degrees of management allow grouping plants in three main categories to facilitate their study: (1a) gathered plants; (1b) plants under incipient management; and (1c) cultivated plants. Furthermore, there is a gradient within incipient management: (2a) tolerance; (2b) protection; (2c) promotion; and (2d) *ex-situ* cultivation. Gathering can also be considered as incipient management by changing the order of gathering locations and restricting harvesting. Furthermore, it is necessary to assess the presence or absence of selectivity when researching the management of a species. Selectivity, aimed at improving the desirable quality of the products – involves higher management intensity and, over a long-term period, could result in the domestication of the species (Casas *et al.* 1997; González-Insuasti and Caballero 2007). On the other hand, in order to better understand management and domestication, it is necessary to recognise the socio-cultural aspects involved in the use and value of the species (Casas *et al.* 1996). The values attributed by local people to the plant species will affect their incentives to manage them (Guijt 1998) and to continue their use (Ogle 2001).

In this chapter, when we refer to wild food plants, we refer to non-domesticated plants, including wild plants that are not managed at all ('truly' wild), wild and semi-domesticated plants that may be tended in some way through encouragement, including clearing surrounding vegetation to reduce competition or selective cutting of perennials. Many of these plants may also be classified as 'weeds' by local people, agronomists and even by scientific literature. While we include in our definition 'naturalised' and 'introduced' plants, domesticated species, namely cultivars, are excluded.

Weeds – defined as 'pioneers of secondary succession' (Bunting 1960) or 'general unwanted organisms that thrive in habitats disturbed by man' (Harlan and de Wet 1965) – have evolved by adapting to disturbed ecosystems. As we are dealing with evolution, we will have many intermediate states: some plants will be weedier than others, some plants will be more resilient to environmental disturbances, and some plants with weedy tendencies may be encouraged while others are despised (Harlan 1975). Many useful wild plants grow as 'weeds' in the context of cultivation. Interestingly, Ogle and Grivetti (1985, 59) in their study in Swaziland, presented the botanical-dietary paradox: although the most intensively cultivated area (Middleveld) experienced the highest species diversity loss, it turned out to have the highest consumption of wild vegetables, mostly weeds. The consumption of weeds as vegetables is illustrated in this chapter by Morales and Gil (Chapter 5.2), who found that, in the Canary Islands, certain weeds are not only considered by local people as wild food plants but also were very likely consumed by the former inhabitants. Likewise, Tardío and Pardo-de-Santayana explain that some edible weeds are collected during hand-weeding of crops, either for human consumption or animal fodder (Chapter 5.3).

Food, Medicine and Famine

Plants' edible parts may include fruits,[4] seeds, flowers, roots, rhizomes, bulbs, tubers, stems and leaves. Wild food plants can be eaten fresh or may need to be prepared or processed before being consumed. There are a myriad of ways to consume these plants; for example, Tardío and Pardo-de-Santayana classified wild plants from Spain in six food-use categories: vegetables, beverages (liqueurs and infusions), fruits, seasonings, preservatives and sweets (Chapter 5.3). Processing, which involves grinding, soaking, drying, heating or parching, is not only important for removing highly toxic secondary compounds or bitter qualities, but also for storage purposes (Cotton 1996). Wild plants are not only consumed alone, but also prepared in numerous combinations of plants and plant-processed products constituting different dishes, which provide an important nutritional benefit (Messer 1972).

The consumption of wild resources as food is essential as a means to ensure food diversity and secure food intake in non-industrialised rural societies, contributing to a balanced diet. Indeed, one of the most important aspects of wild vegetables and fruits is their important role in nutrition: wild food plants are a very important source of vitamins and minerals, as well as secondary metabolites such as alkaloids, phenolics and essential oils (Heywood 1999; Johns 2007). This role was amply documented in Ogle's (2001) publication entitled *'Wild Vegetables and Micronutrient Nutrition'*, which presents a compendium of studies done around the world on dietary intake of micronutrients from wild plants and chemical analysis of their micronutrient content.

The nutritional importance of wild food plants usually overlaps with their frequent use as medicine and their role in disease prevention.[5] In this way, Etkin and Ross (1982) report 'food as medicine and medicine as food' when referring to wild food plants, and Ogle (2003) emphasises the disease-preventing role and presence of health-promoting mechanisms in such plants due to bioactive substances. Some substances, such as flavonoids, tannins, pectins and saponins, have antioxidant activities that stimulate the immune system or exhibit antibacterial, antifungal or antiviral activities. Likewise, Pieroni and Quave (2005) discuss the extraordinary antioxidant properties of these plants. Nevertheless, very little is known about the health benefits of regular consumption of small quantities of medicinal foods and the contribution of wild foods with small quantities of trace minerals and vitamins (Etkin and Ross 1982; 1994; Ogle *et al.* 2003; Price 2005). Furthermore, the overlap of their role as food-medicine is not yet fully understood (*e.g.* Vandebroek and Sanca 2007), even though this role is extremely important as regards their widely documented use as 'famine foods' among poor rural communities throughout the world.

The use of wild plant foods occurs across a spectrum from routine to more irregular use under conditions of mild to extreme stress, holding several positions in the diet (Price 2001). Accordingly, Turner and Davis (1993) distinguish four categories of famine foods: (1) normally eaten foods that become more important during stress periods; (2) less preferred foods that are seldom consumed during normal circumstances; (3) starvation foods that are consumed only in periods of stress; and (4) hunger suppressants and thirst quenchers. Wild food plants are particularly important as 'buffers against shortages' during major stress and scarcity periods (Daniggelis 2003; Heywood 1999; Scoones *et al.* 1992). For instance, when the harvest of staple foods is finished or when cultivated crops fail wild foods can thus complement the seasonal availability of crops (Adaya *et al.* 1997). Hence the frequent reference to them as 'famine foods' or as foods to supplement the diet in periods of shortage or stress (Grivetti and Ogle 2000), as illustrated by Selleger (Chapter 5.4). The role of wild food plants as famine foods in the past is discussed by Scott-Cummings (Chapter 5.6) and Griffin-Kremer (Chapter 5.5, all in this volume), for beeweed and silverweed respectively.

Women and Wild Plants

The role of women is essential in farming and gathering. In many parts of the world, women are the main custodians of the knowledge of wild plant resources, since they are mainly responsible for wild plant food collection, cooking preparation and processing for storage, as well as being home-gardeners, domesticators, herbalists, and seed custodians.[6] Additionally, they are often responsible for household nutrition (Howard 2003a–c; Ogle and Grivetti 1985). Today, rural women are actively engaged in small-scale marketing of wild food and food-medicinal plants in local and sometimes urban markets, which constitutes a significant source of income for them.[7] These 'women's markets' not only provide cash to the gatherers and sellers, but also have important social implications in that they provide women with social networks beyond their immediate communities, as well as access to non-local resources and expertise (*e.g.* Ertuğ 2003b; Kalyoncu 2002). The important role of women in managing wild resources, cultivation and domestication became more visible after female researchers became actively involved in ethnobotanical studies.

Social Stigmatisation Versus Revival

The gathering and consumption of wild food plants in some places is locally stigmatised, increasingly associated with poverty and social marginality (Cruz-García 2006; Malaza 2003; Somnasang and Moreno-Black 2000; also see Chapter 9). Social stigmatisation of wild food plants is also related to their use as famine foods. For example, Selleger states in this volume (Chapter 5.4) that Dogon people feel uncomfortable talking about wild edible grasses, given that their consumption is associated with difficult periods of famine. In addition, Tardío and Pardo-de-Santayana mention that some wild food plants, especially those with a bitter taste, were considered as 'poor people's food' (Chapter 5.3). Price (2005, 85), who conducted research in northeast Thailand explains:

> 'The degree to which 'wild' plant foods are incorporated into the diet of agriculturists depends also in part on whether there are any social status restrictions on the consumption of these foods (or selected species), that is, if they are considered peasant food or foods of poverty and are infrequently consumed by the more prosperous. The greater the social stigma, the more likely these foods will be used as a buffer in times of stress and shortage rather than daily consumption.'

A judgment on the level of acceptance of a wild food plant species is its presence in local markets and its consumption by 'upper' social classes (Pemberton and Lee 1996; Price 2001).

On the other hand, wild plants may participate in new, highly regarded neo-rural cultural identities by appropriation. As Pieroni (2003) pointed out from his research in southern Italy, some wild food plants are nowadays re-valorised by urbanites or highly educated people, as typical local varieties or 'specialty' food products. Likewise, Tardío and Pardo-de-Santayana (Chapter 5.3) affirm that there is a 'revival' in the consumption of some wild food plants in several regions of Spain, related to urban people's need to re-connect with rural life and associated with increasing environmental awareness. It is even possible to suggest that this revival is linked to some Rousseauist nostalgia of an idealised, pre-Neolithic pristine, egalitarian, natural society. Tardío and Pardo-de-Santayana also mention that certain wild plants are considered as delicacies, recognised for their health benefits and even acknowledged as symbols of regional identity (Chapter 5.3). Similarly, Morales and Gil (Chapter 5.2) explain that the Canary Island date palm (*Phoenix canariensis* Hort. ex Chabaud), has been the botanical emblem of the Canary Islands since 1991, constituting a newly created component in Canarians' identity. The recognition of some wild food plants as symbols of local identity has recently come to be a frequent phenomenon in Europe (Pardo-de-Santayana *et al.* 2010). Food movements, such as Slow Food or All Nature shops are the vectors of this new, contemporary, positive perception of 'traditional' food, and therefore also of wild plants. Indeed, recent research on the importance of wild food plants as a source of healthy food, dietary supplements and in medicinal treatments is increasingly fostering their use by contemporary urban people. This is also likely to reflect the trend that 'artisanal' and 'traditional' qualities of foods are important in the marketing and consumption of these products as signifiers of social distinction (Heath and Meneley 2007).

5.2. GATHERING IN A NEW ENVIRONMENT: THE USE OF WILD FOOD PLANTS DURING THE FIRST COLONISATION OF THE CANARY ISLANDS, SPAIN (2ND–3RD CENTURY BCE TO 15TH CENTURY CE)

Jacob Morales and Jaime Gil

The Canary Islands are located opposite the northwestern coast of Africa and they have one of the highest botanical biodiversity rates in the world. Currently, there are 2,594 wild plants, among which 1,893 are native, with 545 of them being endemic (Arechavaleta *et al.* 2010). The archipelago was a pristine environment never occupied by humans until the first colonisation was carried out by Berber-like people from northwestern Africa in the late first millennium BCE (Morales *et al.* 2009). When the first colonists arrived, they introduced crop plants and domestic animals, but they also started to collect the local wild plants. Which plants from the local flora were gathered? Which plants were used as food? What was or were the criteria for choosing these wild foods?

The purpose of the present article is to identify the range of wild food plants used during this early period by the first inhabitants of the archipelago. In order to identify the range of wild species, we make use of recent archaeobotanical data from several sites in the archipelago. We have complemented these data with ethnohistorical documents from the first European colonists and ethnographic records from modern people of the Canaries. Through this multidisciplinary approach, we have tried to show a complete picture of the use of each plant and to explore the potential and limit of each source of information in the study of wild plant consumption in past societies.

Natural Conditions of the Canary Islands

The archipelago consists of seven islands and several islets of volcanic origin (Fig. 5.1). They have been classified into two groups, depending on their topography and climatology: the eastern islands (Fuerteventura and Lanzarote) that lack high mountains and are very dry; and the western islands (Gran Canaria, Tenerife, La Gomera, La Palma and El Hierro), that are uneven in altitude, from 1,501 m in El Hierro to 3,718 m in Tenerife, which generates colder temperatures and higher rainfall. The vegetation in Lanzarote and Fuerteventura is more homogenous than in the rest of the archipelago, where the altitude and orientation towards the trade winds create several bioclimatic layers with many different species. In general terms, five different ecological zones have been categorised in the mountainous islands (Bramwell and Bramwell 2001).

Coastal Vegetation

This is characterised by the influence of the sea. Most of the plants are shrubs between 30 and 100 centimetres high, with succulent small branches or stems and almost no leaves. *Euphorbiaceae* family members, such as leafless spurge (*Euphorbia aphylla*) and balmy spurge (*E. balsamifera*), are dominant in this environment.

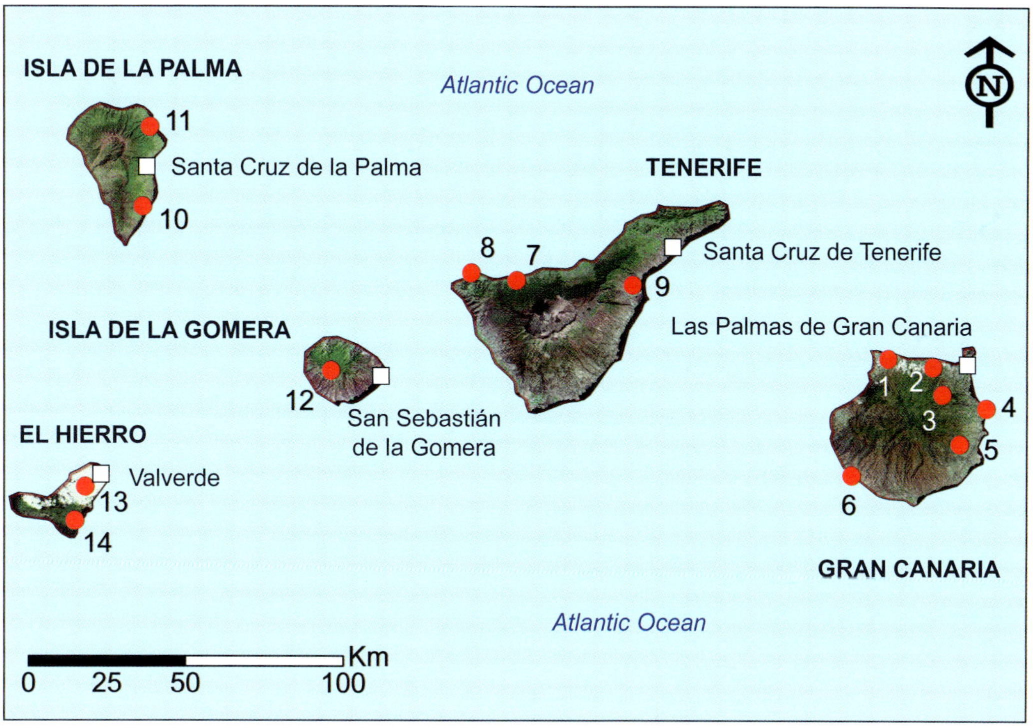

Fig. 5.1. Map of the Canary Islands (not covering Fuerteventura and Lanzarote) indicating the location of the archaeological sites. 1) Cueva Pintada; 2) La Cerera; 3) El Tejar; 4) Lomo los Melones; 5) San Antón; 6) Lomo los Gatos; 7) Don Gaspar; 8) Las Estacas; 9) Chinguaro; 10) Belmaco; 11) El Tendal; 12) Alto del Garajonay; 13) Hoya del Zarzal; 14) La Lajura. Map: R. Lugon, J.-C. Loubier and A. Chevalier.

Thermophile Forest

This ecosystem has developed almost evenly between the coast and the first slopes of the mountains, up to a height of 700 metres. It is characterised by a hot and arid sub-desert climate with rainfall fluctuating between 150 and 250 millimetres per year. Some of the arboreal species of this habitat are mostly the same, or similar to, the species existing in a large part of the Mediterranean area, such as Canarian wild olive (*Olea cerasiformis*), Canarian palm (*Phoenix canariensis*), Canarian juniper (*Juniperus turbinata* subsp. *canariensis*), Mount Atlas pistachio (*Pistacia atlantica*) or white broom (*Retama rhodorhizoides*).

Laurel Forest

The laurel forest consists of abundant groups of evergreen trees such as Canarian laurel (*Laurus novocanariensis*), Canarian avocado tree (*Persea indica*), Canarian khat (*Maytenus canariensis*), and fire tree (*Morella faya*) with laurel-like leaves that develop on the north or northeast-facing mountains in the strip between 500 and 1600 metres high. This forest is regarded as a relict of the sub-tropical humid vegetation, which could be found in large parts of southern Europe and North Africa until the end of the Tertiary period. Most of the plants found in this ecosystem are endemic to the Canaries and are not found nowadays in the Mediterranean area.

Pine Forest

The xerophilous mountain vegetation, developing above the area of laurel forest and in the south-facing mountains, is dominated by the forests of Canarian pine (*Pinus canariensis*). These forests are also characterised by the presence of endemic woody legumes, such as Canarian flatpod (*Adenocarpus foliolosus*) or tagasaste (*Chamaecytisus proliferus*), and also rockrose (*Cistus* spp.).

High Mountain Vegetation

This kind of flora is only present in Tenerife and La Palma islands, which have mountains reaching more than 2000 metres above sea level. These places are characterised by hot and arid summers and very cold winters. Most of the plants that grow there are shrubby or under-shrubby, among which we can highlight the endemic Mount Teide broom (*Spartocytisus supranubius*, Fabaceae) or Mount Teide bugloss (*Echium wildpretii*, Boraginaceae).

The First Inhabitants of the Canary Islands

The first colonists arrived from northwestern Africa around the third to second century BCE and they remained isolated from the mainland until the first Europeans arrived in 1341 CE. During the fifteenth century CE, after a long process of conquest, the Kingdom of Castille (Spain) occupied the archipelago and the Canary Islands became part of European history (Tejera and González 1987).

The economy of the first settlers was focused on agriculture and domestic animal husbandry. In most of the islands, food production depended heavily on animal husbandry, especially on goats and sheep. The bones of these animals have been recovered in large numbers in almost every domestic site on the islands (Fig. 5.2). Dietary studies on human bones have also shown the importance of animal husbandry, emphasising the dependence of the ancient population on the secondary products from the livestock, mainly milk (Delgado *et al.* 2005; González and Arnay 1992; González *et al.* 2001; Velasco *et al.* 1999). All the islands, apart from Gran Canaria, show this pattern of food production. In contrast, on Gran Canaria most of the available food was provided by arable agriculture (Morales 2010; Velasco *et al.* 1999).

Nevertheless, plants were an important source of food in the rest of the archipelago. There is archaeological evidence for the practice of agriculture in all the islands, except Fuerteventura and Lanzarote (Fig. 5.2). Archaeobotanical studies indicate that the most important crop was barley (*Hordeum vulgare*). Wheat (*Triticum durum*), lentils (*Lens culinaris*), fava beans (*Vicia faba*), peas (*Pisum sativum*) and figs (*Ficus carica*) were also introduced to the islands by pre-Hispanic people, though some islands lacked one or more of these species. In addition to cultivated plants, wild plants were gathered as food and, in some cases, became staples, such as fern rhizomes (*Pteridium aquilinum*) in the islands of El Hierro, La Gomera and La Palma (Morales 2010).

The Study: Archaeobotanical, Ethnohistorical and Ethnographic Methods

In order to identify the wild food plants used by the first inhabitants of the Canary Islands, we have analysed the archaeobotanical remains from 14 archaeological sites located in the archipelago (Arco *et al.* 1990; Morales *et al.* 2004; 2007; Morales 2010). Gran Canaria is the best known because it has been

	Fuerteventura	Lanzarote	La Palma	La Gomera	El Hierro	Tenerife	Gran Canaria
Domestic animals							
Canis familiaris Linnaeus, dog	X	X	X	X		X	X
Capra hircus Linnaeus, goat	X	X	X	X	X	X	X
Felis catus Linnaeus, cat			X			X	
Ovis aries Linnaeus, sheep	X	X	X	X	X	X	X
Sus scrofa Linnaeus subsp. *domesticus*, pig	X	X	X	X	X	X	X
Cultivated plants							
Hordeum vulgare L., barley			X	X	X	X	X
Ficus carica L., fig						X	X
Lens culinaris Medik., lentil			X				X
Pisum sativum L., pea							X
Triticum durum Desf., hard wheat			X			X	X
Vicia faba L., faba bean			X			X	X

Fig. 5.2. List of domestic animals and cultivated plants introduced in the Canary Islands during the pre-Hispanic period according to archaeological remains (following Morales *et al.* 2009).

Fig. 5.3. Archaeobotanical evidence of non-cultivated plant species in pre-Hispanic archaeological sites from the Canary Islands.

Archaeological sites	La Lajura	Hoya del Zarzal	Alto del Garajonay	El Tendal	Belmaco	Chinguaro	Las Estacas	Don Gaspar	Cerera	El Tejar	Ermita San Antón	Lomo los Melones	Lomo los Gatos	Cueva Pintada
location, island	El Hierro	El Hierro	La Gomera	La Palma	La Palma	Tenerife	Tenerife	Tenerife	Gran Canaria	Gran Canaria	Gran Canaria	Gran Canaria	Gran Canaria	Gran Canaria
volume of sediment, litres	247	38	337	14	89	192	handpicked	handpicked	780	617	69	491	207	737
nº of samples	21	3	38	2	52	17			71	99	13	41	40	129
non-cultivated species seed density per litre	3.49	0.18	0.65	16.8	3.25	1.65			0.03	0.06	0.85	0.23	0.61	2.24
radiocarbon dates, in centuries AD	3rd–11th	13th–14th	8th–12th	3rd–8th	7th–13th	7th–15th	no dates	3rd–15th	4th–8th	7th–13th	10th–14th	13th–15th	15th	13th–15th
Native species														
Adenocarpus / Spartocytisus sp.	x		x											x
Argyranthemum sp.			x							x				
Euphorbia sp.	x					x								
Hypericum sp.	x			x										
Ilex canariensis Poir.				x										
Juniperus turbinata subsp *canariensis* (A. P. Guyot in Mathou & A. P. Guyot) Rivas-Mart., Wildpret & P. Pérez, seed					x		x							
Juniperus turbinata subsp *canariensis* (A. P. Guyot in Mathou & A. P. Guyot) Rivas-Mart., Wildpret & P. Pérez, fruit					x									
Lauraceae				x										x
Neochamaelea pulverulenta (Vent.) Erdtman						x					x		x	x
Olea cerasiformis Rivas Mart. & del Arco					x									
Phoenix canariensis Chabaud			x						x	x		x	x	x
Pinus canariensis C. Sm. ex DC. in Buch	x													
Pistacia atlantica Desf.							x			x				x
Plocama pendula Aiton, fruit												x	x	
Plocama pendula Aiton, seed						x							x	
Pteridium sp., rhizome fragments	x													
Retama rhodorhizoides Webb & Berthel	x		x	x	x									x
Rubus sp														x
Visnea mocanera L.f.	x	x				x		x				x	x	x
Weed species														
Aizoon canariense L.						x						x		x
Ajuga iva (L.) Schreb.												x		
Amaranthus sp.	x			x	x	x			x		x	x	x	x
Anagallis arvensis L.	x			x	x	x						x		x
Asphodelus sp.														x
Asteraceae	x			x	x	x			x					x
Atriplex sp.												x	x	x
Avena sp.			x											x

Continued over the page

Archaeological sites	La Lajura	Hoya del Zarzal	Alto del Garajonay	El Tendal	Belmaco	Chinguaro	Las Estacas	Don Gaspar	Cerera	El Tejar	Ermita San Antón	Lomo los Melones	Lomo los Gatos	Cueva Pintada
location, island	El Hierro	El Hierro	La Gomera	La Palma	La Palma	Tenerife	Tenerife	Tenerife	Gran Canaria	Gran Canaria	Gran Canaria	Gran Canaria	Gran Canaria	Gran Canaria
volume of sediment, litres	247	38	337	14	89	192	handpicked	handpicked	780	617	69	491	207	737
nº of samples	21	3	38	2	52	17			71	99	13	41	40	129
non-cultivated species seed density per litre	3.49	0.18	0.65	16.8	3.25	1.65			0.03	0.06	0.85	0.23	0.61	2.24
radiocarbon dates, in centuries AD	3rd–11th	13th–14th	8th–12th	3rd–8th	7th–13th	7th–15th	no dates	3rd–15th	4th–8th	7th–13th	10th–14th	13th–15th	15th	13th–15th
Brassicaceae	x													
Bromus sp.	x		x		x					x				
Calendula arvensis L.	x													
Carex sp.				x	x									
Chenopodium sp.														x
Chenopodium murale L.	x			x	x	x				x		x	x	x
Chrysanthemum coronarium L.														x
Convolvulus cf. *siculus*														x
Emex spinosa (L.) Campd.				x						x				
Fabaceae (small seeded legume)	x	x		x	x	x			x		x	x	x	x
Fumaria sp.	x		x		x									
Galium aparine L.	x		x	x	x					x				x
Geraniaceae			x											
Heliotropium ramosissimum (Lehm.) DC.												x		
cf. *Lamarckia aurea* (L.) Moench														x
Lamiaceae	x		x	x	x	x								x
Linum sp.														x
Lolium sp.	x		x											
Malva sp.	x			x	x	x			x	x	x	x	x	x
Malva parviflora L.														x
Mesembryanthemum nodiflorum L.														x
Patellifolia sp.					x							x	x	
Phalaris cf. *canariensis*														x
Phalaris sp.					x				x	x		x		x
Plantago sp.	x													x
Plantago cf. *lagopus*	x			x	x					x		x		x
Poaceae	x		x	x	x									x
Polygonaceae														x
cf. *Ranunculus sardous* Crantz				x										x
Rumex sp.			x	x	x				x	x				x
Scirpus/Cyperus sp.														x
Sherardia arvensis L.			x	x	x				x				x	x
Silene sp.	x			x									x	x
Silene gallica L.	x			x	x	x			x	x		x		x
Silene vulgaris (Moench) Garcke														x
Sisymbrium erysimoides Desf.														x
Solanum nigrum L.	x			x	x					x	x	x		x
Solanaceae	x			x										
Spergularia fallax Lowe	x													
cf. *Veronica hederifolia* L.														x
Vicia sp.														x

extensively sampled (six sites), while in the rest of the islands, we have studied very few sites (three in Tenerife, two in La Palma, two in El Hierro, and one in La Gomera) (Fig. 5.1). A sampling strategy has been employed to obtain a representative assemblage of vegetal macro-remains in most of the sites though, in some cases, we have only worked with handpicked material. As a result, we have recovered and identified a large amount of charred botanical evidence that can be related to the pre-Hispanic uses of wild plants. This data is presented in Fig. 5.3, which summarises the presence of non-cultivated seeds, fruits and rhizomes in pre-Hispanic archaeological sites of the Canary Islands.

In order to complete these data, we have examined the first historical records concerning the contact between the ancient Canarians and the European colonists during the fourteenth and fifteenth centuries CE. Basically, these texts are historical chronicles of the conquest of the Canaries or descriptions of the islands, for which some indigenous customs are recorded. We have examined eleven authors who wrote their manuscripts between the fifteenth and seventeenth centuries CE (Abreu 1977; Bontier and Le Verrier 1980; Frutuoso 1964; Hernández and Delgado 1998; Marín 1986; Morales 1993; Viana 1991). The information gleaned is summarised in Fig. 5.4, where we list the plants, the uses, the island where the plants were consumed and the total number of authors who documented the gathering of these wild foods.

In addition, we have included ethnographic information linked to the use of wild plants in the Canaries during the last century. Certainly we cannot use the ethnographic data as an analogy with the past, but these sources are very useful for understanding the archaeobotanical and ethnohistorical record. In this work we have utilised published data about the use of wild food plants in the Canary Islands (Gil 2005; Gil and Peña 2006; Gil *et al.* 2008; 2009) and current research that is being completed by the authors of this article. This data was obtained through open-

Species	Ethnohistorical records	part used, mode of consumption	Islands						
			H	G	P	T	F	L	C
Allium ampeloprasum L.	1: Morales, ed. 1993	bulbs, cooked in soups							X
Arbutus canariensis Veill	3: Morales, ed. 1993; Núñez 1994; Viana 1991	fruits, raw				X			X
Asphodelus sp.	1: Frutuoso 1964	bulbs, cooked		X					
Canarina canariensis (L.) Varke	5: Abreu 1977; Morales, ed. 1993; Núñez 1994; Viana 1991	fruits, raw	X			X			X
Cistus sp.	2: Abreu 1977; Marín 1986	seeds, roasted and ground into flour			X				
Juncus /Scirpus sp.	1: Hernández and Delgado, eds. 1998	bulbs		X					
Malva sp.	2: Abreu 1977; Marín 1986	roots, cooked with milk			X				
Morella faya (Aiton) Wilbur	1: Viana 1991	fruits, raw				X			
Phoenix canariensis Chabaud	9: Abreu 1977; Bontier and Le Verrier 1980; Frutuoso 1964; Hernández and Delgado, ed. 1998; Marín 1986; Morales, ed. 1993; Viana 1991	sap, cooked as wine or honey; fruits, raw or boiled		X		X	X		X
Pinus canariensis Sweet ex Spreng.	2: Marín 1986; Morales, ed. 1993	seeds, raw or roasted							X
Pteridium aquilinum (L.) Kuhn in Kerst.	4: Abreu 1977; Frutuoso 1964; Marín 1986; Torriani 1978	rhizomes, ground into flour or cooked with milk	X	X	X				
Rubus sp.	2: Núñez 1994; Viana 1991	fruits, raw	X			X			
Visnea mocanera L. f.	8: Abreu 1977; Espinosa 1980; Frutuoso 1964; Marín 1986; Viana 1991; Morales, ed. 1993; Núñez 1994	fruits, raw or boiled to make liqueur	X		X	X			X

Fig. 5.4. List of wild food plants used by the pre-Hispanic population of the Canary Islands according to ethnohistorical sources (H = El Hierro; G = La Gomera; P = La Palma; T = Tenerife; F = Fuerteventura; L = Lanzarote; C = Gran Canaria)

Fig. 5.5. List of wild food plants used in the Canary Islands during the recent past according to ethnographic sources (H = El Hierro; G = La Gomera; P = La Palma; T = Tenerife; F = Fuerteventura; L = Lanzarote; C = Gran Canaria)

Plant species	Part used, mode of consumption	Islands						
		H	G	P	T	F	L	C
Aeonium spp.	immature inflorescences, raw	x	x					
Agave americana L.	nectar							x
Amaranthus sp.	missing data							x
Aizoon canariense L.	seeds, roasted and ground into flour					x		
Aizoon hispanicum L.	seeds, roasted and ground into flour						x	
Allium ampeloprasum L.	bulbs, raw or cooked in soups	x	x	x	x		x	x
Allium roseum L. / *A. subhirsutum* L. / *A. subvillosum* Salzm. ex Schult.	bulbs, raw or as condiment; young leaves, raw or cooked in soups	x		x	x	x	x	
Alternanthera caracasana Humb., Bonpl. & Kunth	missing data				x			
Apium graveolens L.	leaves, raw, as herbal tea, milk substitute or in soups	x			x	x		
Apium nodiflorum (L.) Lag.	young shoots, raw		x		x			x
Arbutus canariensis Veill.	fruits, raw	x			x			
Arisarum simorrhinum Durieu	corms, cooked						x	
Asparagus pastorianus Webb & Berthel.	young shoots, cooked in soups					x		
Asparagus cf. *scoparius* Lowe	young shoots, raw				x			
Astragalus boeticus L.	unripe seeds, raw						x	
Astragalus solandri Lowe.	unripe seeds, raw						x	
Astydamia latifolia (L. f.) Baill.	roots, cooked	x						
Beta macrocarpa Guss.	young leaves, raw or cooked					x		
Borago officinalis L.	young leaves, raw or cooked					x	x	
Calamintha sylvatica Bromf.	leaves, as herbal tea				x			
Canarina canariensis (L.) Vatke	ripe fruits, raw			x	x			x
Chenopodium murale L.	leaves, cooked in soups						x	
Cymodocea nodosa Asch	stems, stewed						x	
Cynara cardunculus var. *ferocissima* Lowe	artichokes and young stems, raw or cooked		x			x	x	x
Cytinus hypocistis (L.) L.	fruits, raw	x		x				
Davallia canariensis (L.) Sm.	shoots, cooked					x		
Dracunculus canariensis Kunth	corms, roasted and ground into flour, boiled and mixed with milk or flour	x	x		x			
Drimia maritima (L.) Stearn	bulbs, cooked					x	x	
Emex spinosa (L.) Campd.	roots, raw; fruits, roasted and ground into flour	x				x	x	
Eruca vesicaria (L.) Cav.	young leaves, in soups	x		x			x	
Ferula lancerottensis Parl.	seeds, raw						x	
Foeniculum vulgare Mill.	roots, in soups; tender stems, raw; seeds, raw	x			x	x	x	x
Fragaria vesca L.	fruits, raw	x						
Glaucium corniculatum (L.) Curtis	ripe seeds, raw						x	
Heberdenia excelsa (Aiton) Banks ex DC.	fruits, raw				x			
Helminthotheca echioides (L.) Holub	young leaves, in soups; tender stems, raw				x			x
Hypochoeris glabra L.	tender basal leaves, raw	x						
Juniperus turbinata ssp. *canariensis* (A. P. Guyot in Mathou & A. P. Guyot) Rivas-Mart., Wildpret & P. Pérez	galbulus, raw		x					x
Lathyrus annuus L.	unripe seeds, raw				x			
Lathyrus cicera L.	unripe seeds, raw	x						
Lathyrus clymenum L.	unripe fruits, in soups; unripe seeds, raw; ripe seeds, roasted and ground into flour	x						
Lathyrus tingitanus L.	unripe seeds, raw; ripe seeds, roasted and ground into flour or coffee substitute, in soups	x	x		x			x
Launaea nudicaulis (L.) Hook. f.	tender basal leaves, raw	x	x		x		x	

Plant species	Part used, mode of consumption	Islands						
		H	G	P	T	F	L	C
Lycium intricatum Boiss.	ripe fruits, raw				x	x	x	
Malva parviflora L.	leaves, raw; unripe fruits, raw; ripe seeds, roasted and ground into flour				x	x	x	x
Mentha x *piperita*	leaves, as herbal tea				x			
Mesembryanthemum crystallinum L.	seeds, roasted and ground into flour				x	x	x	
Mesembryanthemum nodiflorum L.	seeds, roasted and ground into flour				x	x	x	
Morella faya (Aiton) Wilbur	ripe fruits, raw; dry fruits, ground into flour	x	x	x	x			x
Neochamaelea pulverulenta (Vent.) Erdtman	ripe fruits, raw							x
Nicotiana glauca Graham	flowers, sucked						x	
Olea cerasiformis Rivas Mart. & del Arco	fruits, prepared in brine							x
Oxalis pes-caprae L.	tender inflorescences, raw; bulbs, roasted and ground into flour				x		x	x
Papaver sp.	missing data						x	x
Patellifolia patellaris (Moq.) A. J. Scott, Ford-Lloyd & J. T. Williams	tender leaves, in soups; seeds, roasted and ground into flour					x		x
Phalaris coerulescens Desf.	basal internodes, stewed	x					x	
Phoenix canariensis Chabaud	sap, cooked to make wine or honey; young shoots, raw or roasted and ground into flour; tender male inflorescence, raw; unripe fruits, cooked; ripe fruits, raw or dried; seeds, roasted and ground into flour		x		x	x	x	x
Pinus canariensis C. Sm. ex DC. in Buch	seeds, raw and ground, or roasted				x			x
Pleiomeris canariensis (Willd.) A. DC.	ripe fruits, raw				x			
Plocama pendula Aiton	fruits, raw or dried; seeds, to ground into flour				x			x
Portulaca oleracea L.	aerial part, in soups	x	x					x
Pteridium aquilinum (L.) Kuhn	rhizomes, dried and ground into flour	x	x	x	x			x
Quercus ilex L.	fruits (acorn), raw							x
Quercus robur L.	fruits (acorn), raw							x
Quercus suber L.	fruits (acorn), raw				x			
Raphanus raphanistrum L.	young shoots, raw or cooked	x	x	x	x	x	x	x
Rapistrum rugosum L.	tender leaves, in soups				x			
Rorippa nasturtium-aquaticum (L.) Hayek	young shoots, in soups		x	x	x	x		x
Rubus ulmifolius Schott / *R.* cf. *bollei* Focke	tender shoots, raw; ripe fruits, raw or cooked to make wine				x			x
Rumex bipinnatus L. fil.	unripe fruits, raw						x	
Rumex bucephalophorus L.	leaves, raw	x						
Rumex lunaria L.	tender shoots, raw or in soups	x	x		x			
Rumex pulcher L.	leaves, raw	x						x
Rumex vesicarius L.	tender leaves, raw; unripe fruits, raw				x		x	
Rutheopsis herbanica (Bolle) A. Hansen & G. Kunkel	tender leaves, raw or cooked					x		
Salvia aegyptiaca L.	leaves, as herbal tea				x			
Scandix pecten-veneris L.	tender leaves, raw; unripe fruits, raw			x			x	
Scolymus maculatus L. / *S. hispanicus* L.	roots, raw or in soups; tender stems, raw or in soups	x	x		x			
Scorpiurus muricatus L. / *S. sulcatus* L./ *S. vermiculatus* L.	tender leaves, raw or cooked				x	x		
Scorzonera laciniata L.	bottom of unripe inflorescences, raw						x	
Silene vulgaris ssp. *commutata* (Guss.) Hayek	roots, raw; leaves, in soups	x	x		x			
Silybum marianum (L.) P. Gaertn.	unripe seeds, raw						x	
Sinapis alba L.	tender leaves, raw or cooked					x		
Sinapis arvensis L.	tender leaves, in soups		x				x	
Sisymbrium spp.	tender leaves, raw					x	x	
Solanum nigrum L./ *S. luteum* Mill.	ripe fruits, raw	x	x	x	x		x	

Continued over the page

Plant species	Part used, mode of consumption	Islands						
		H	G	P	T	F	L	C
Sonchus acaulis Dum. Cours.	peduncle, raw				x			
Sonchus asper (L.) A. W. Hill	tender leaves, raw				x			
Sonchus bourgeaui Sch. Bip. in Webb & Berthel.	tender leaves, raw						x	
Sonchus oleraceus L.	tender leaves, raw; tender stems, raw	x	x		x	x	x	x
Sonchus pinnatifidus Cav.	young leaves, raw						x	
Sonchus tenerrimus L.	leaves, raw or in soups	x			x			
Spartocytisus supranubius (L.f.) Christ ex G. Kunkel	flowers, raw				x			
Tamus edulis Lowe	tubers, cooked			x	x			
Tetragonia tetragonoides (Pall.) Kuntze	leaves, in soups	x						
Tinguarra cervariefolia (DC.) Parl.	roots, raw				x			
Tolpis spp.	tender basal leaves, raw	x		x				
Tragopogon porrifolius L.	roots, raw; tender leaves, raw				x			
Urtica urens L.	tender leaves, in soups				x			
Vicia benghalensis L.	unripe seeds, raw	x						x
Vicia disperma DC.	unripe seeds, raw				x			
Vicia lutea L.	unripe fruits, in soups; unripe seeds, raw	x		x				x
Visnea mocanera L.f.	ripe fruits, raw or cooked to make wine	x	x	x	x			

oral interviews with 1093 informants from an age-range of 65–90 years old. The overall results are listed in Fig. 5.5, comprising only non-cultivated edible species (feral cultivated species, condiments and aromatics are not included). In any case, this ethnobotanical research is currently in progress and it is still incomplete. The data presented here are preliminary results that cannot be considered as conclusive. Nomenclature of botanical names follows the flora *'Lista de especies silvestres de Canarias'* (Arechavaleta *et al.* 2010) and the list of botanical names including their authors is in Fig. 5.2, 5.3, 5.4 and 5.5. The English common names of most plants are presented in the text. The endemic plants, unique to the Canaries, only have local names in the pre-Hispanic, Hispanic and Portuguese languages and do not have any equivalent in foreign floras. When possible, the local names of these endemic plants were translated into English.

Wild Food Plants of the Canary Islands: Past and Present

It is clear from the results of the archaeobotanical studies that local wild plants were used by the first colonists. Charred remains from these plants have been recovered in all the sites that have been sampled. In total, we have identified 66 different taxa, although in some cases it has not been possible to identify all the seeds to species or even to generic level.

We have separated all the identified species into two ecological groups. In the first group, we included endemic and native plants of the Canary Islands (Arechavaleta *et al.* 2010). Their presence in the sites depends on many factors, but they probably correspond to remains of gathering wild plants from the local flora. The second group is comprised of weeds, which are plants that grow in disturbed places and crop fields. In most cases, these seeds can be considered as remains of the weed flora and they normally appear in archaeological sites as residues of the crop harvest. However, some of them have been identified as wild foods in the ethnographic interviews and it is possible that they could have been used by the ancient population.

In the first group, we have identified seventeen species. Their number is low in comparison to weeds and they are scattered over only a few sites. Some of these species, such as Canarian daisy (*Argyranthemum* sp.) or spurge (*Euphorbia* sp.), are toxic according to the available information and they could be included in the archaeological record as residues of fuel, fodder or other uses. Among the edible plants, the most common are Canarian palm (Fig. 5.6) and *Visnea mocanera*, which are the only species present in six different sites. Seeds from white broom are also abundant and they have been recovered from five different sites. Other important plants are *Neochamaelea pulverulenta* and Mount Atlas pistachio; the remaining species are scarcer and they have been recovered in only one or two sites.

The weed species belonging to the second group are more abundant and they are spread over most of the sites. In total, we have identified 49 different taxa. Small-seeded legumes, amaranth (*Amaranthus* sp.), goosefoot (*Chenopodium murale*) and mallow (*Malva* sp.) are the most common species in the samples. According to ethnographical records, 13 of the weed species identified in the archaeological samples were used as wild foods in the Canary Islands. However, these remains were found associated with crop seeds and pellet dung, so they most possibly represent crop processing by-products or fodder for domestic animals (Morales *et al.* 2004; 2007; Morales 2010). Caution is required when interpreting assemblages of wild plant taxa represented only by their seeds and, consequently, we cannot be certain they should be regarded as wild food plants.

The total number of wild food plants recorded in the ethnohistorical records is rather low, only 13. However, these texts provide very accurate details of the consumption of certain species in some cases. According to the total number of references in these texts (Fig. 5.4), Canarian palm was the most important of the plants gathered by the ancient Canarian population. Native people ate their wild dates and a kind of honey or wine made with the sap of this tree (Frutuoso 1964) (Fig. 5.7). Fruits of *Visnea mocanera* were the second most referenced wild food in the ethnohistorical texts (Fig. 5.8). As indicated by these documents, *Visnea mocanera* was used to produce a beverage called '*chacerquem*' that could be stored and had medicinal properties (Espinosa 1980, 38). Rhizomes of the fern *Pteridium aquilinum* were also eaten abundantly, after pounding them to make flour, or were cooked with milk (Abreu 1977; Torriani 1978). Other significant gathered fruits were *Canarina canariensis*, blackberry (*Rubus* sp.) and those of the Canarian strawberry tree (*Arbutus canariensis*).

In relation to the ethnographic interviews, the number of documented edible wild plants is higher

Fig. 5.6. Canarian palm (*Phoenix canariensis*), Gran Canaria island.

Fig. 5.7. 'Don Miguel Chico' cutting the stem of the Canarian palm to collect the sap, La Gomera island.

Fig. 5.8. *Visnea mocanera*, El Hierro island.

than in the archaeobotanical and ethnographic records. Current data indicates that a minimum of 104 species were used as wild food in the Canary Islands. They include plants that were collected because they provided edible seeds (24 species), fruits (21 species), leaves (45 species), stems (four species), subterranean parts (15 species) and other vegetative parts (11 species) (Fig. 5.9). Some of these wild plants were used as substitutes for the

Fig. 5.9. 'Señora Teresita' collecting the edible roots of the bladder campion (*Silene vulgaris*), Lanzarote island.

staple crops in case of bad harvests and famine, such as slenderleaf ice plant (*Mesembryanthemum nodiflorum*) and *Pteridium aquilinum*. However, it is necessary to emphasise that the most common wild food plants are used as green vegetables or cooked in soups. Leaves from wild radish (*Raphanus raphanistrum*) and *Launaea nudicaulis* are the most referenced wild foods and were consumed regularly. It is also interesting to note that Canarian palm had an important role as a source of wild food since the first human settlement, until very recently. Nowadays, Canarian palm is no longer an important wild food, but remains important in the landscape and to the identity of modern Canarians, because the Government of the Canary Islands declared it the botanical emblem of the archipelago in 1991.

Comparing Methods

The overall results have shown significant differences in the number of wild food plants recorded through each method. In addition, there are clear divergences in the type and accuracy of the data to reconstruct the past use of plants.

Archaeobotanical data represents the most direct information about the use of plants by the inhabitants of each site. This method has provided a significant number of charred seeds and fruits from wild plants that were manipulated for different purposes, including food. However, archaeological charred seeds usually represent a small percentage of the total number of wild food plants consumed by people, since other plant parts such as fruits, leaves, stems or roots, that are consumed raw or processed away from the fire, are usually under-represented. The fact that the archaeological plant remains analysed in this work are charred suggests that these species could be processed by heating, to improve the taste and to remove the toxicity, as some of them are unpalatable when they are eaten fresh (such as white broom and *Pteridium aquilinum*). This indicates that species identified in the archaeological samples primarily represent wild food plants that were cooked with fire, as no other way of cooking with heat has been recorded in the Canaries.

In some cases it is not possible to be sure that the archaeological seeds are food leftovers or whether they really represent residues from other activities, such as fuel gathering, healing activities or sacred practices. Indeed, weed remains normally appear in the archaeological samples associated with crop residues and they have been interpreted as crop processing by-products. But we must emphasise that some of the weeds reported in the archaeological sites, such as slenderleaf ice plant, mallow (*Malva parviflora*) or lesser jack (*Emex spinosa*), were a common wild food until recently (Gil *et al.* 2009).

On the other hand, ethnohistorical records are very significant because they describe, with some detail, the consumption of wild food plants and how they were processed by the ancient Canarians. Additionally, these texts also documented the gathering of fruits, vegetables and roots, which are preserved only with difficulty in charred conditions. However, the number of wild food plants provided by ethnohistorical documents is very limited, if we compare this with the archaeobotanical and ethnographical results. Then again, this type of data was produced by foreigners and strangers who had few contacts with the indigenous population, so they may represent only the edible plants with more visibility.

Finally, ethnographic data has supplied the most abundant and accurate information about the consumption of wild food plants in the Canary Islands during the present and recent past. Oral information has allowed us to document a large range of green vegetables and fruits that are not recorded in the archaeobotanical samples or the ethnohistorical texts. However, the ethnographic information cannot be directly applied to reconstruct the use of plants in the past, because some of the recorded wild food species were introduced through the Spanish occupation and they were not available to the pre-Hispanic inhabitants. Furthermore, after the conquest of the Canaries by the Spaniards, there were significant changes in the distribution and mass of the original flora of the islands (Santana 2001). On one hand, there was massive land clearance for growing staple cereals. On the other hand, sugarcane plantations and 'ingenios' (the Spanish term for the traditional sugar mill and the associated facilities) that needed massive quantities of firewood to produce sugar were established in most of the islands only a few years after the Spanish conquest (Aznar 1983; Lobo 1988; Lobo *et al.* 2007). This high demand for firewood rapidly exhausted the forest of the archipelago and it changed the original flora which had existed during the pre-Hispanic period. In addition, during the Hispanic period, people from Europe, America and Africa re-colonised the islands, bringing with them their botanical knowledge. Conversely, most of the indigenous knowledge about food plants was lost once the native population was almost exterminated by the European conquerors.

Regarding this aspect, it is interesting to note that common food plants in the archaeobotanical samples, such as white broom and Mount Atlas pistachio (Fig. 5.10), have no corresponding ethnohistorical or ethnographical references to their consumption by modern Canarians. This fact may be due to the massive deforestation in past centuries of the thermophile forest where these kinds of plants normally grow. Charcoal analysis in archaeological sites has verified that this type of forest was widely used during the pre-Hispanic period (Machado *et al.* 1997). At the present time, thermophile forests are much reduced in the archipelago, which may explain the limited number of ethnographic records relating to food consumption connected with them. On the other hand, it also suggests that traditional botanical knowledge about the edible properties of these plants was partially lost after the conquest of the Canaries. Finally, we have recovered very few archaeobotanical remains from the laurel forest vegetation; in spite of the high number of plants than can be found in this woodland. It is necessary to note that charcoal from laurel forest trees is common in the archaeological samples (Morales *et al.* 2009; Machado *et al.* 1997), which suggests that these plants were available to the first inhabitants, but that they were not extensively gathered for food. These plants are unique to the archipelago and were unknown to the first colonists during the initial stages. In later stages, after the European conquest of the Canaries, the laurel forests were extensively exploited and these plants became more familiar to people, who gathered some of them as wild food.

Fig. 5.10 Fruits of Mount Atlas pistachio (*Pistacia atlantica*), Gran Canaria island.

Conclusions

The sources of information utilised to elaborate this article are heterogeneous, but altogether show a comprehensive and complete picture of the consumption of wild plants in the distant and recent past of the Canary Islands. The results indicate that the first colonists of the Canary Islands quickly adapted to an unfamiliar and pristine environment, integrating new and unknown plants into their diet. Species such as Canarian palm, Canarian strawberry tree, Canarian pine, among other endemic plants, were extensively consumed and, in some cases, still eaten until recently. Whereas after the arrival of the first Europeans the use of some common pre-Hispanic food plants, such as Mount Atlas pistachio and white broom, was lost, most of the species gathered during the pre-Hispanic period were still collected during the twentieth century CE.

5.3. WILD FOOD PLANTS TRADITIONALLY USED IN SPAIN: REGIONAL ANALYSIS

Javier Tardío and Manuel Pardo-de-Santayana

The geographical location of mainland Spain and its mountainous ranges have helped confer highly varied environmental conditions and, consequently, a very rich flora on the country. The landscape of Spain can be broadly divided into two distinct areas: the Euro-Siberian floristic region and the Mediterranean region. The first, popularly known as 'Green Spain', is a strip of land from the northwest (Galicia) to the Pyrenees and has an oceanic climate. The vegetation consists predominantly of deciduous forest and meadows that remain green throughout the summer. The rest of the country has a Mediterranean climate, with a characteristic drought period and high temperatures during summer. Nevertheless, clear differences exist between different areas in terms of annual rainfall, temperature and the length of the drought period. In general, the climate is more arid in the east and south of the Peninsula, becoming more continental in the centre. The vegetation is mainly evergreen forest, although mountain ranges contain green areas, especially at a certain altitude.

In addition to this high plant biodiversity, a complex history has brought about a great cultural diversity in Spain, with at least three different languages besides Spanish. A rich biocultural heritage lives on as a result of the thorough knowledge of the natural environment (Tardío *et al.* 2006).

In Spain, as in many countries, wild edibles have been used to complement and balance staple agricultural foods, especially during times of scarcity. The aim of this work is to analyse regional differences in the pattern of human consumption of wild plants traditionally used in Spain. Through the application of quantitative and qualitative methods, the paper explores the influence of different factors involved, whether environmental and/or cultural, in selecting wild food species in the various regions.

Ethnobotanical Database of Wild Food Plants and Use Categories

The first step, undertaken over the last ten years, has been collecting wild food plant data from Spanish ethnobotanical and ethnographical literature of the last four decades. Thus, the temporal framework of this research goes back to the last 50–100 years. The exhaustive search of information allowed building a database with more than 4600 records. Part of it has already been published (Tardío *et al.* 2006). Most of the data originate from 67 published works and it also includes some unpublished data. The most recent compilation includes 21 new references from which the majority (17) are bibliographical sources, such as books (*e.g.* ADISAC 2006; Catani *et al.* 2001; Cofradía Extremeña de Gastronomía 1985; Criado *et al.* 2008; Moll 2005; Piera 2006), journals (*e.g.* Hadjichambis *et al.* 2008; Pretel *et al.* 2008) and dissertations (*e.g.* Fajardo 2008; García-Jiménez 2007; Parada 2008). We have also included a few internet references and personal communications (especially those of Daniel Pérez from the Basque Country and Navarre). For taxonomy and plant nomenclature, we follow the *Flora iberica* (Castroviejo *et al.* 1986–2012) for the families included therein, and the *Flora Europaea* (Tutin *et al.* 1964–1980) for the remaining families.

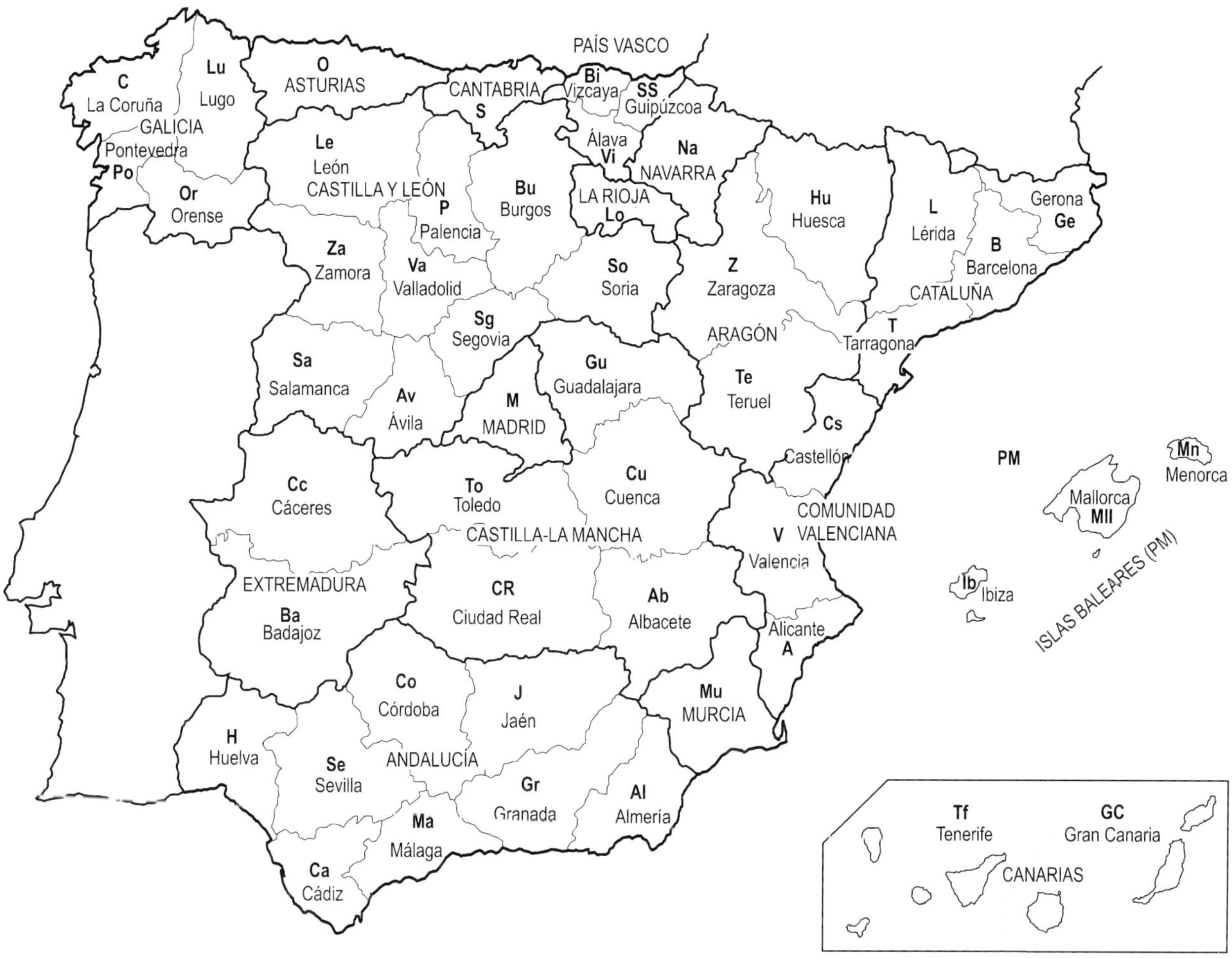

Fig. 5.11. Political map of Spain with the name of all the provinces and Autonomous Regions. Map: J. Tardío and M. Pardo-de-Santayana.

The database contains data from 43 of the 50 Spanish provinces, there are no data from the Canary Islands, and some provinces were scarcely surveyed (*e.g.* ten provinces have less than 20 species registered in the database). Therefore, we selected 28 well-surveyed provinces especially suitable to be explored for regional differences. Fig. 5.11 shows the map of Spain with the codes used for the different provinces. Seven food use-categories were identified: vegetables (or green vegetables), beverages (*e.g.* liqueurs and infusions), fruits, sweets (flowers and underground organs), seasoning, preservatives and other uses.

Because in most of the publications and works quantitative data were missing, it is important to acknowledge some limitations of the database. They relate to the absence of the frequency of citation or use-reports for the different species, which would have assisted in measuring the cultural importance of each use-category in each province (Tardío and Pardo-de-Santayana 2008). In the absence of this type of data, the only way of estimating the relative significance of each use-category was by calculating the number of species assigned to each of the categories selected. Since there were important differences in the overall number of species in each province, the percentages of species in each category were used. Principal Component Analysis (PCA) was carried out for exploring the relationship among the variables and the similarities among the different provinces surveyed. The cases of the data matrix were the different provinces (rows) and the variables (columns) were the seven food-use categories.

Searching for Patterns of Wild Food Plant Use

The dataset consisted of 464 species belonging to 70 families, and covering about 6% of the Iberian flora.

Fig. 5.12. Some wild vegetables traditionally consumed in Spain. A) bundle of asparagus of black bryony (*Tamus communis*); B) peeling the basal leaves of golden thistle (*Scolymus hispanicus*); C) basal leaves of chicory (*Cichorium intybus*).

Among the seven food-use categories considered, vegetables formed the largest group (51%), followed by plants used to prepare beverages (32%). Wild fruits and sweets represented *ca.*17% each, whereas plants used for seasoning only 14% (Fig. 5.12 and 5.13). Species used as preservatives were 6% of the total whereas the category 'other uses' covered 4% of the species. It must be borne in mind that some species were included in more than one category.

Figure 5.14 shows the percentage of species within each food use-category in each province. Northern provinces (*e.g.* Vi, O, Le, or SS) show in general a greater importance of the 'wild fruits' category, whereas in other regions, like the central ones (*e.g.* M, Ba, To), the south (Co, J) and the east (A, V), the 'vegetables' category is of greater importance. Equally, in the northern regions, the percentage of seasoning plants or preservatives is small or almost non-existent, while in the Mediterranean regions (central, southern and eastern Spain) these acquire an important role.

For a better understanding of the relationship among the different food use-categories and the likely similarities in the patterns of wild food consumption in the provinces surveyed, an exploratory multivariate analysis was applied. The correlation matrix analysis (Fig. 5.15) shows various statistically significant ($p<0.05$) correlations among different variables. One of the strongest correlations is that between 'vegetables' and 'fruits' (-0.66), as could be suspected from the analysis of Fig. 5.14. This negative correlation means that those provinces with a larger percentage of the category 'fruits' also have the lowest proportion of the category 'vegetables'. The 'fruits' category is also negatively correlated with 'seasoning' (-0.47) and 'preservatives' (-0.69), while 'vegetables' is positively correlated with 'preservatives' (0.50) and 'other uses' (0.61), and negatively correlated with 'beverages' (-0.57). Fig. 5.16 shows the projection of the cases (provinces) in the plane defined by the two first principal components of PCA with 92% of the total variance. The first component (Factor 1) is highly and negatively correlated (-0.86) with the variable 'fruits'. Therefore, the provinces with a larger proportion of wild fruits are located at the left of the graph. This factor is positively correlated (0.95) as well with 'vegetables' and shows lower

Fig. 5.13. Some wild fruits traditionally used in Spain. A) blackberries (*Rubus ulmifolius*); B) elderberries (*Sambucus nigra*); C) sloe fruits (*Prunus spinosa*).

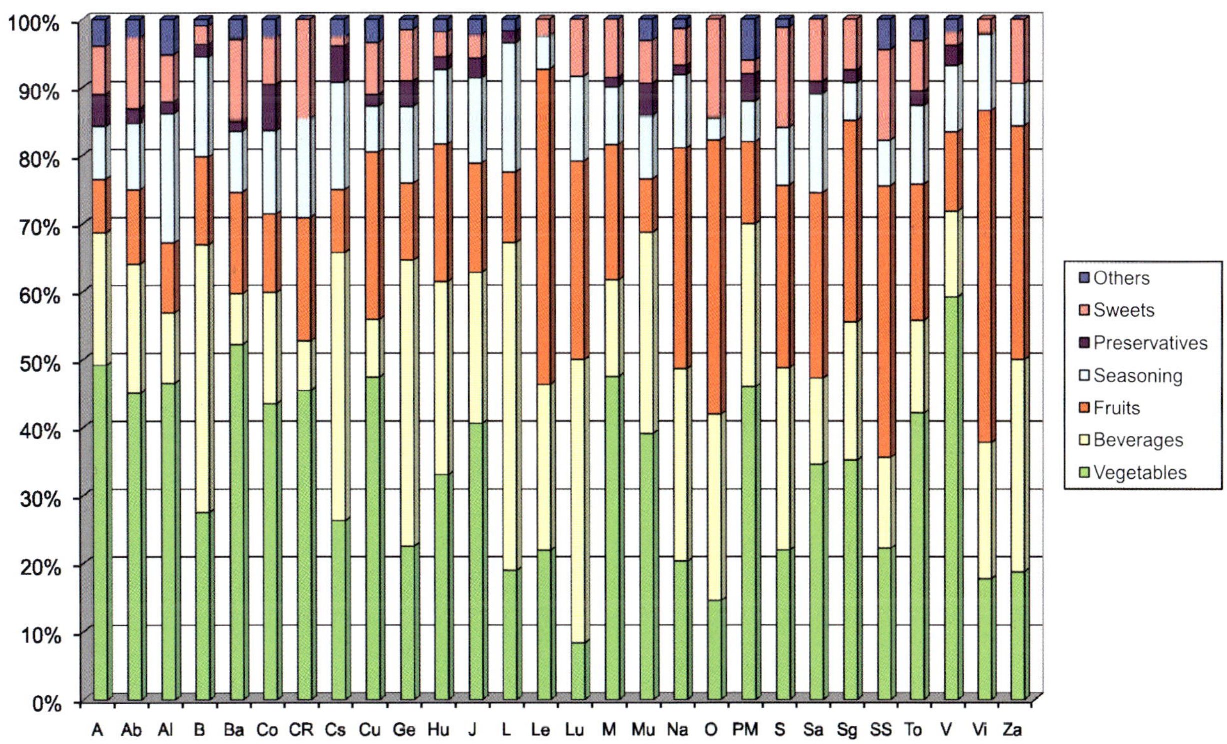

Fig. 5.14. Percentages of species within each food-use category in each of the selected provinces in Spain (for the province codes, see Fig. 5.11).

	Vegetables	Beverages	Fruits	Seasonings	Preservatives	Sweets	Others
Vegetables	1	-0.57 *	-0.66 *	0.05	0.50 *	-0.02	0.61 *
Beverages		1	-0.1	0.35	0.2	-0.39 *	-0.09
Fruits			1	-0.47 *	-0.69 *	0.21	-0.67 *
Seasonings				1	0.28	-0.36	0.21
Preservatives					1	-0.34	0.62 *
Sweets						1	-0.26
Others							1

Fig. 5.15. Correlations among different food-use categories in Spain (with '*' those statistically significant, p<0.05).

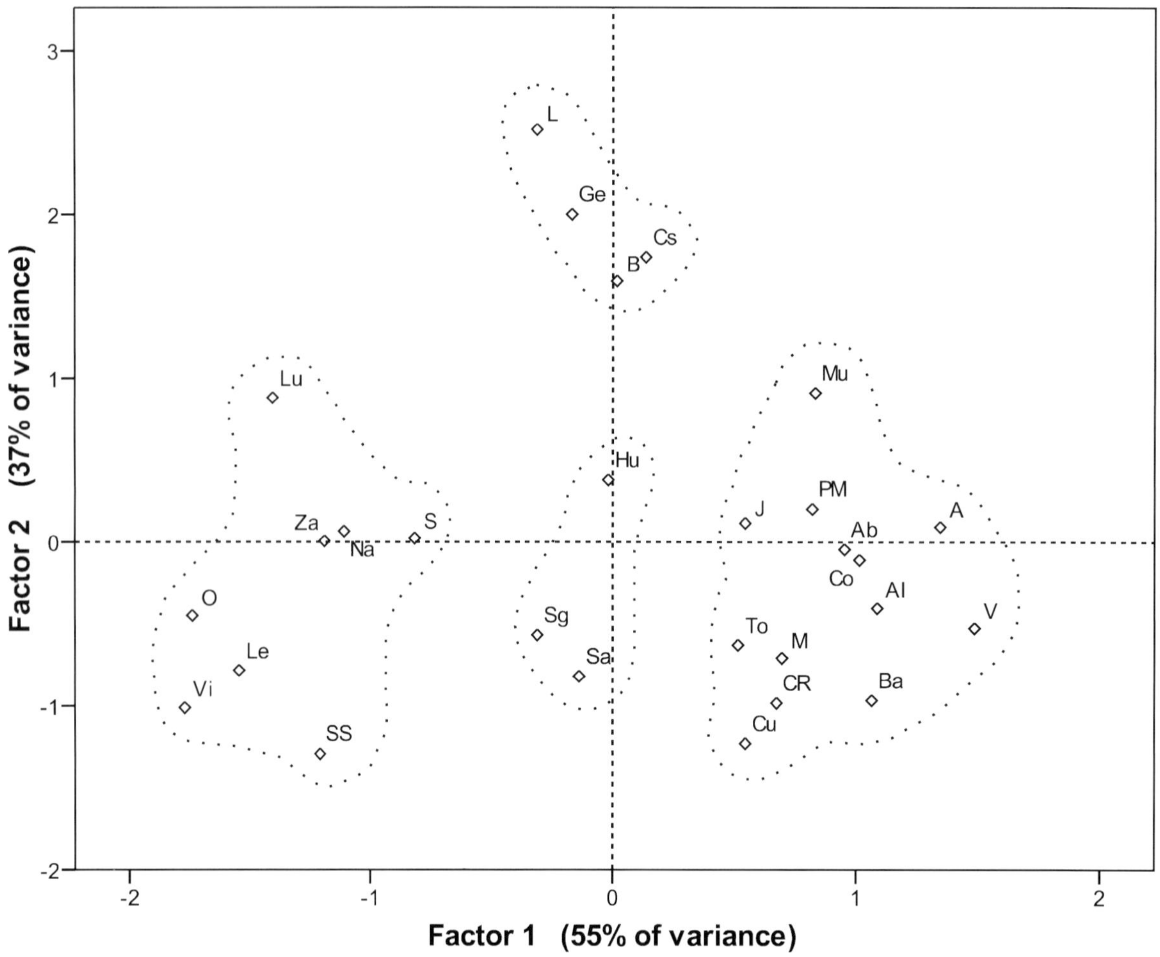

Fig. 5.16. Projection of the Spanish provinces on the plane Factor 1 x Factor 2 that captures 92% of variance. For the province codes, see Fig. 5.11. PCA computed with the covariance matrix (to obtain a clear pattern of loadings, the factors have been rotated by the varimax normalised method).

correlation with 'preservatives' (0.62) and 'other uses' (0.57). Hence, the provinces with a larger influence of those variables are represented at the right side of the graph. The second component (Factor 2) is highly and positively correlated (0.93) with the percentage of plants used for making beverages (and also with seasoning with lower correlation, 0.56).

Summarising, the analysis has evidenced the presence of four different clusters of provinces. A cluster was formed at the left, which includes eight provinces from the north (Vi, O, Le, SS, Lu, Na, Za, S) dominated by a wet climate. In these provinces, there is a high percentage of wild fruits consumption and a low percentage of vegetables. A second cluster is that found at the top of the graph formed by four provinces located in the northeastern part of Spain (L, Ge, B and Cs). In these, a higher percentage of plants used for beverages and a small percentage of wild fruits is the dominant pattern. The third and biggest cluster, located at the right of the graph, consists of thirteen provinces from the south (Co, J and Al), the east (A, V, Ab, Mu, PM) and central Spain (M, Cu, To, Ba, CR,) with high proportions of wild vegetables and little of wild fruits. Finally, there is a cluster of three provinces (Sg, Sa and Hu) occupying an intermediate position.

Though not very clear, further regional differences are found in the overall number of edible wild species encountered. In general, there is a higher diversity in wild plant consumption in the calcareous Mediterranean regions that might be partially explained in terms of species richness in that region. For instance, the southern and eastern regions of Spain, with a warmer climate and predominant calcareous soils have, in general, a rich spectrum of plant species, including a flora with a very high number of aromatic species likely to be used as condiments or herbal beverages. However, this reason does not clearly explain the differences in the use of vegetables. Obviously, a less varied flora offers fewer choices, but apart from environmental constraints, there may be other important factors more in the sphere of cultural traditions, such as culinary habits (in turn, likely influenced by environmental conditions) that need to be considered. There are many examples of species, like black bryony (*Tamus communis* L.) or fiddle dock (*Rumex pulcher* L.), which are commonly consumed in the Mediterranean regions and, despite growing in humid areas, are not traditionally consumed there. The Mediterranean climate, drier in the summer but also milder in winter, has favoured agricultural development, mainly based on cereal and legume cultivation. There is also an increasing development of horticulture as well. Many vegetables and fruits are under cultivation, representing an important contribution to Mediterranean diet. In addition, people have also gathered wild edibles in special periods of the year. For instance, wild vegetables were a valuable resource by the end of the winter and spring seasons when fresh agricultural products were scarce (Tardío 2010). Moreover, the high temperatures reached in the area made the use of condiments and preservatives mandatory for avoiding the decay of agricultural and animal food products. On the contrary, in the north of Iberia, the economy was traditionally based on livestock farming and, consequently, the main elements of people's diet were meat and dairy products; vegetables and condiments, as stated earlier, being less common.

These observations match well with the results shown in the principal component analysis. What is more, despite the exploratory character of this study, comparison with data from other European countries points to the same model. A comprehensive study on wild edibles consumption carried out in Poland (Łuczaj and Szymański 2007; Łuczaj 2008) seems to confirm the same general trend, with a low number of wild edibles consumed, both in overall figures and in the proportions of vegetables and condiments used, when compared to southern regions like the south of Italy (Ghirardini *et al.* 2007; Pieroni *et al.* 2005) or Bosnia-Herzegovina (Redzic 2006). This Polish author (Łuczaj 2008) even uses the term 'herbofilia' when referring to these Mediterranean cultures, as well as to China and Japan, in which the green parts of plants of numerous species are often used and highly prized. As stated earlier, it could be assumed that environmental conditions played a role in determining ways of life and particular practices such as wild plant gathering. However, as we have found in another work (Pardo-de-Santayana *et al.* 2007), many other elements, such as tradition, religion, magical practices, social aspects, cultural or taste preferences may also have influenced people's decisions in using wild plant foods.

Past and Future of Wild Plant Consumption

Wild food plants have played an important role in complementing staple agricultural foods in the past. They balanced the diet by adding variety, new flavours and nutritional value and were, according to many informants, an excellent complement to other foodstuffs in times of scarcity like in the Spanish post-Civil War period. In addition, they were also a buffering element against shortages and crop failures.

The consumption of the vast majority of the wild food plants registered in our database remained steady until about 50 years ago. As previously mentioned, most of them were gathered when no other food was available. For instance, crop hand-weeding in spring offered an excellent opportunity for collecting wild vegetables. Many of them, such as golden thistle (*Scolymus hispanicus* L.), skeleton weed (*Chondrilla juncea* L.) or chicory (*Cichorium intybus* L.), were used for both animal fodder and human consumption.

In rural environments, farmers and shepherds were the main consumers of the wild resources. They were often considered the food of the poor (González Turmo 1997), especially for some species, such as certain vegetables with a bitter taste like lesser burdock (*Arctium minus* Bernh.) and chicory. However, this was not always the case; wealthy people have also consumed some species with better palatability. For instance, golden thistle and bladder campion (*Silene vulgaris* [Moench] Garcke) were collected by poor people and sold to the richer citizens of the villages in the provinces of Albacete (Sánchez López *et al.* 1994) and Jaén (Fernández Ocaña 2000), respectively. Both species, together with others such as wild asparagus (*Asparagus acutifolius* L.) and water-blinks (*Montia fontana* L.), have been sold in local markets or door-to-door and some were even marketed in big cities (González Turmo 1997; Mesa 1996; Tardío *et al.* 2002; Velasco *et al.* 2010).

Although gender issues in the management of wild plant food resources remain understudied in Spain, ethnobotanical references often recall the significant role of women in gathering, processing and cooking wild plants (*e.g.* Oltra 1998; Tardío *et al.* 2002; Verde *et al.* 1998). Women tended to go together for plant collecting; gathering sometimes being part of other activities. For instance, they gathered wild vegetables at the waysides when going for washing or other tasks. In the case of wild fruits, bulbs or flowers, they were mainly consumed as snacks by children or shepherds (*e.g.* Mesa 1996; Pardo-de-Santayana *et al.* 2007). These data are consistent with the information available from other regions (*e.g.* Daniggelis 2003; Ertuğ 2003b; Pieroni 2003).

Nowadays, the expansion of agriculture and the development of supply chains have led to the presence of many kinds of cultivated vegetables and fruits in the markets throughout the year. In the new socio-economic contexts, where cash products occupy a dominant position, the use of wild edibles has enormously decreased and far fewer species are consumed. They are often considered to be old fashioned, unprofitable and too time-consuming. However, over the past few years, some revival in the use of wild plants can be observed. There is a number of interacting drivers for this renewed interest. First, there is an increasing awareness of environmental issues and willingness to reconnect with rural areas and lifestyles in one part of the urban population; *i.e.* city dwellers like walking in the countryside searching for greens or mushrooms. Second, there is growing interest in rural tourism and the re-discovery of local products, including those made with wild plants. Species that were once regarded as poor people's food are now considered delicacies or gourmet food, local specialties, and sometimes they are even a symbol of regional identity (Pardo-de-Santayana *et al.* 2007).

There are many examples of plant food re-discovery which, to some extent, reflect some of the regional differences detected in the PCA. In the central and western part of Spain, the shoots of wild asparagus, the basal parts of wild leek (*Allium ampeloprasum* L.) and the midribs of the golden thistle are prepared and preserved in bottles. It is even possible to purchase them through the internet. But also the less common young shoots or asparagus of black bryony (*Tamus communis* L.), known as 'rabiacanes', can be tasted in luxury restaurants in the north of the province of Cáceres. We are not aware of the marketing of wild greens in the north, where commercialisation of wild plants is more related to homemade jams of wild berries, such as elderberries (*Sambucus nigra* L.), blackberries (*Rubus ulmifolius* Schott.) and bilberries (*Vaccinium myrtillus* L.), and to liqueurs. These are made from different species, the 'patxaran' being the most popular of them. 'Patxaran' is an alcoholic drink, typical of Navarre, made from sloe fruits (*Prunus spinosa* L). Its popularity has led to industrial production, so nowadays it can be found in supermarkets all over the country.

Furthermore, there is also a group of species widely used for preparing herbal teas and liqueurs. Two different species, both known as rock tea (*Jasonia glutinosa* D.C., in the east, and *Sideritis hyssopifolia* L., in the north), are highly valued and herbal teas made with them are even served in restaurants. Both species are considered a feature of local gastronomic identity in many of the areas where they grow. In eastern Spain, several herbal spirits are marketed and considered typical and regional specialties. This is the case of the Catalan 'ratafia' or the 'herbero' from Alicante. The 'ratafia' is the subject of a local feast celebrated in several villages of Gerona and Barcelona provinces.

A final aspect to be considered is the gastronomic interest of urban chefs in discovering 'new' flavours from traditional dishes. Some of the species traditionally consumed and considered food of the poorest have become specialty products and gourmet ingredients. Moreover, the re-discovery of the health benefits related to the consumption of wild fruits and vegetables, the popularity of herbal medicine and also the increasing literature on the nutritional potential of non-cultivated vegetables of the Mediterranean area (Heinrich *et al.* 2006), may also contribute to expanding the interest in wild plants. It is important to note that an excessive increase in their demand could lead to unpredictable consequences to their sustainable use.

Recommendations for Future Research

This chapter has analysed and discussed the regional differences in the traditional consumption of wild plants in Spain. With the available ethnobotanical data, an exploratory analysis has been carried out and some general trends have been pointed out. However, due to the limitations of the method employed, further research is needed for a clear confirmation of this tendency. On one hand, more ethnobotanical research on lesser known areas is necessary, especially to verify the lower number of species that seem to be used in northern regions. On the other hand, the possibility of using another system for evaluating the importance of each food use-category could also help. The use of a cultural importance index for these categories, based on the number of use-reports instead of the number of species, would be more interesting for a more in-depth study of the influence of cultural and ecological factors in wild food plant consumption.

5.4. USE OF WILD FOOD PLANT RESOURCES IN THE DOGON COUNTRY, MALI

Camille Selleger

The transition from gathering to agriculture is a central question in anthropology and archaeology. In sub-Saharan Africa, this transition is blurred and difficult to understand. Recent studies have shown the diversity of subsistence ways during the Holocene. Cereal domestication appeared late, contrasting with the early use of ceramics in Africa. Furthermore, remains (especially of grinding material) attest to the intensive exploitation of wild edible plants, particularly of cereals, on different sites of the region. Some authors have mentioned the abundance of natural vegetation in the African savannah to explain the long-term reliance on wild plant food and the late appearance of domestic agriculture (Clark 1976; Neumann 2003; 2005). The phase of intensive wild plant food exploitation should have been followed by a period of cultivation of wild species, in particular those of cereals. This phase of 'pre-domestic agriculture' (Willcox 1991) is impossible to identify with certainty from the archaeological remains. Indeed, wild and domestic cereals should have coexisted within the primitive fields (Balter 2007).

The aim of our study was to determine which wild plant species are still used for food in the African savannah among local farming populations, given that the study of these plants provides clues to understanding the relation between man and his environment, as well as the key role that these plants have played in the development and implementation of practices among present and past farming populations.

Prehistoric Background

In the savannas of sub-Saharan Africa, there is evidence that ceramics have been used since the tenth millennium BCE in Ounjougou, Dogon country, Mali (Huysecom 2006; Huysecom *et al.* 2006) and since the ninth–tenth millennium BCE in the Sudan (Close 1995; Haaland and Magid 1992; Jesse 2003; Khabir 1987) and in Niger (Roset 1996; 2000). The earliest grinding material, dated to *ca.* 17,000–15,000 BCE (in the late Pleistocene), was found in Wadi Kubbaniya, Egypt, and is associated with the exploitation of edible rhizomes (Rowley-Conwy *et al.* 1997; 1999; Wasylikowa 2001). The earliest evidence of grinding material associated with the use of wild cereals appears during the eighth millennium BCE in Nabta Playa, Egypt (Królik and Schild 2001; Wasylikowa 2001). Grinding material from the early Holocene was also found in Niger (Roset 1996; 2000), in southern Algeria (Camps 1969) and in Mali (Huysecom 2006; Huysecom *et al.* 2006).

According to our recent and scattered information, a gap separates these finds and the appearance of the first domestic cereals on African soil. African domesticated millet (*Pennisetum glaucum* R.Br., also known as pearl millet) appears in the Dhars Tichitt, Mauritania, during the third–second millennium BCE (Amblard 1996). African rice (*Oryza glaberrima* Steud.) was domesticated from its wild form *Oryza barthii* A.Chev. and it appears in the eighth century BCE in Dia, Mali (Murray 2004). Domesticated sorghum (*Sorghum bicolor* (L.) Moench) appears during the first millennium BCE in Kawa, Sudan (Fuller 2004), and later, in the beginning of the first millennium CE, in Qasr Ibrim, Egypt, and Jebel Tomat, Sudan (Barich 1993; Wasilykowa 2001); so almost 6000 years after the first evidence of the exploitation of the wild grass *Pennisetum arundinaceum* Pers. (Barakat and Fahmy 1999; Wasylikowa and Dahlberg 1999; Wasylikowa 2001; Wendorf *et al.* 1992).

This gap between the appearance of ceramics, grinding stones and domesticated cereals can be explained by the intensive exploitation of wild grasses during this period. The discovery of pottery sherds bearing wild grass impressions in several sites (Haaland 1995; Klee *et al.* 2000; 2004; Wasylikowa 2001; Zach and Klee 2003), as well as evidence of wild grasses statigraphically related with ceramic and grinding material, seems to confirm this hypothesis. The heavy reliance on wild plant resources may explain the unusual pattern of plant domestication in Africa.

The Field Study in Dogon Country

The research was conducted in the Dogon country, which is located in the western part of Mali, south of the Niger bend, between the Malian town of Mopti and Burkina Faso (Fig. 5.17). This territory is bisected by the Bandiagara cliff, a sandstone escarpment stretching about 200 km, which forms a natural border between the plateau of Bandiagara and the Seno plain. It is set in the Sahelo-Sudanian area, characterised by an annual mean precipitation of 500 mm. Dogon farmers form a majority of the population and live in villages. A minority of the population is formed by semi-nomadic Fulani cattle herders. Some Fulani cattle herders have been recently converted to agriculture and thus have become sedentary. Most of the farming work is done during the wet season, from June to October. The harvest takes place in October–November. The most common cultivated cereal is pearl millet,

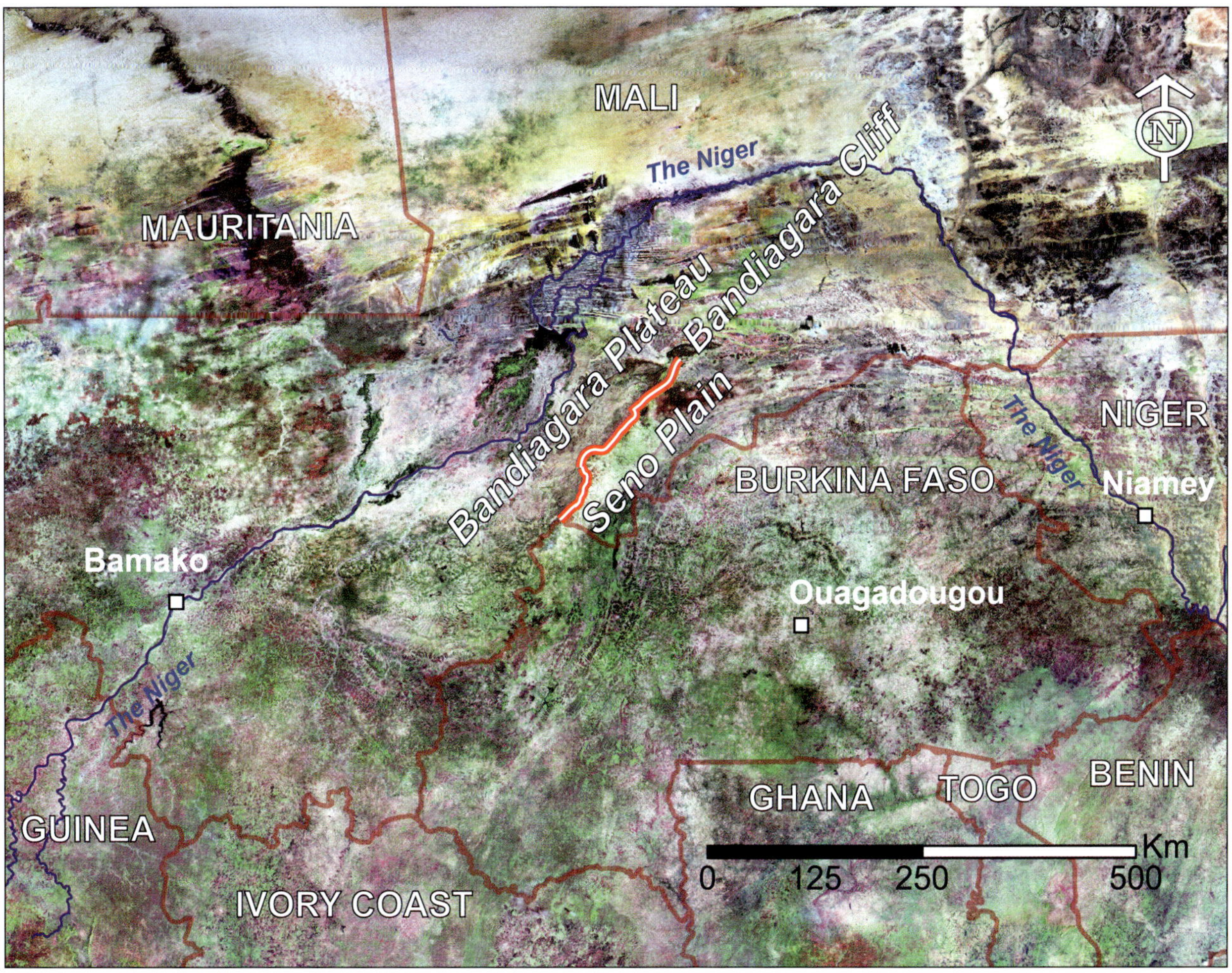

Fig.5.17. Map of Mali showing the Bandiagara cliff, the plateau of Bandiagara and the Seno plain. Map: R. Lugon, J.-C. Loubier and A. Chevalier.

followed by fonio (*Digitaria exilis* Stapf), rice and sorghum. Between August and October, just before the harvest, an annual lean time called 'soudure' (which means 'gap' in French) is characterised by the progressive lack of cultivated cereals. However, the harvest of Bambara groundnut (*Voandzeia subterranea* Thou.) and black-eyed pea (*Vigna unguiculata* subsp. *unguiculata* (L.) Walp.) occurs in this period, providing the population with food resources during the lean time.

The field study was carried out in 2006 in four villages of the Dogon country. These four villages represent different environmental areas: Kobo and Yawa are located along the Bandiagara cliff, Gologou is set on the plateau, while Dimmbal lies in the Seno plain. A total of 45 interviews were carried out. The villagers interviewed were the elderly, because they have kept the best knowledge about wild plants. Both men and women were consulted with during individual and, occasionally, group interviews.

The objective of this research was to determine which plant species are used for food. The use (frequency, context), processing techniques (such as roasting, grinding or cooking) and constraints related to these plants were analysed, in order to evaluate their impact on the subsistence within present-day Dogon agricultural society. Botanical samples of the edible plant species were taken and they are conserved in the Herbaria of the Conservatoire et Jardin botaniques de la Ville de Genève (CJB) in Switzerland, as well as in the Herbarium of the Senkenberg Museum, Frankfurt (FR). Nomenclature of botanical names follows *'Arbres, arbustes et lianes des zones sèches d'Afrique de l'Ouest'* (Arbonnier 2002), *'Flore du Sénégal'* (Berhault 1967), *'Flore illustrée du Sénégal'* (Berhault 1971–1980), *'Flora of East Africa'* (Bruce 1953), *'Flore du Gabon'* (Cavaco 1963), *'Flora of West Tropical Africa'* (Hutchinson *et al.* 1954–1972), *'Les Poaceae de Côte d'Ivoire'* (Poilecot 1995) and *'Les Poaceae du Niger'* (Poilecot 1999). The list of wild food plant species including the authors of the botanical names is presented in Fig. 5.18.

The Dogons' Wild Food Plants

After analysing the botanical samples, we determined 40 species of wild edible plants used in the Dogon country. They were classified in three groups according to their nature and exploitation: the grass grains (wild cereals), the leaves (used as green vegetables) and the roots, rhizomes and fruits (usually used as snack food).

The largest group is made up of the wild cereals and includes such grasses as *Brachiaria lata,* browntop millet (*Brachiaria ramosa*), India sandbur (*Cenchrus biflorus*), *Chloris pilosa,* coast button grass (*Dactyloctenium aegyptium*), crabgrass (*Digitaria ciliaris*), *Digitaria fuscescens, Digitaria horizontalis,* wire crabgrass (*Digitaria longiflora*), jungle ricegrass (*Echinochloa colona*), crowfoot grass (*Eleusine indica*), *Eragrostis tremula*, gray love grass (*Eragrostis cilianensis*), *Microchloa indica, Oryza* sp., elbow buffalo grass (*Panicum subalbidum*), kayasuwa grass (*Pennisetum pedicellatum*), African bristle grass (*Setaria* cf. *sphacelata*) and cattail grass (*Setaria pumila*). Their grains are not exploited every year but only as a famine food. Dogon people are quite uncomfortable speaking about these grasses, because their use is intimately related to hard times. The older generation of Dogon people has a vast knowledge about these plants and knows how to collect, thresh and prepare or process them. The cereals are usually eaten right after their gathering for this reason and are not stored. Furthermore, some other grasses, such as *Oryza* sp., burgu grass (*Echinochloa stagnina* (Retz.) P.Beauv.) and antelope grass (*Echinochloa pyramidalis* (Lam.) Hitchc. and Chase), were exploited regularly and outside of any famine context, but these hydrophilic species have now almost disappeared from the surveyed area due to ongoing aridification.

The wild plants exploited for their leaves are, on the contrary, regularly used as a complementary food in the traditional diet, based on carbohydrates. This group includes spiny amaranth (*Amaranthus spinosus*), red spiderling (*Boerhavia diffusa*), antbush (*Cassia occidentalis*), foetid cassia (*Cassia tora*), false sesame (*Ceratotheca sesamoides*), African spider-flower (*Cleome gynandra*), *Commelina forskaolii, Corchorus tridens,* Cape hibiscus (*Hibiscus diversifolius*), purslane (*Portulaca oleracea*), *Leptadenia hastata, Sesamum alatum,* horse-purslane (*Trianthema portulacastrum*) and caltrop (*Tribulus terrestris*). Generally, the leaves are cooked in a sauce or as a green vegetable. The most popular of them are sun-dried and stored in granaries in order to be eaten during the dry season. The Dogon people particularly appreciate African spider-flower, foetid cassia and *Corchorus tridens*. This

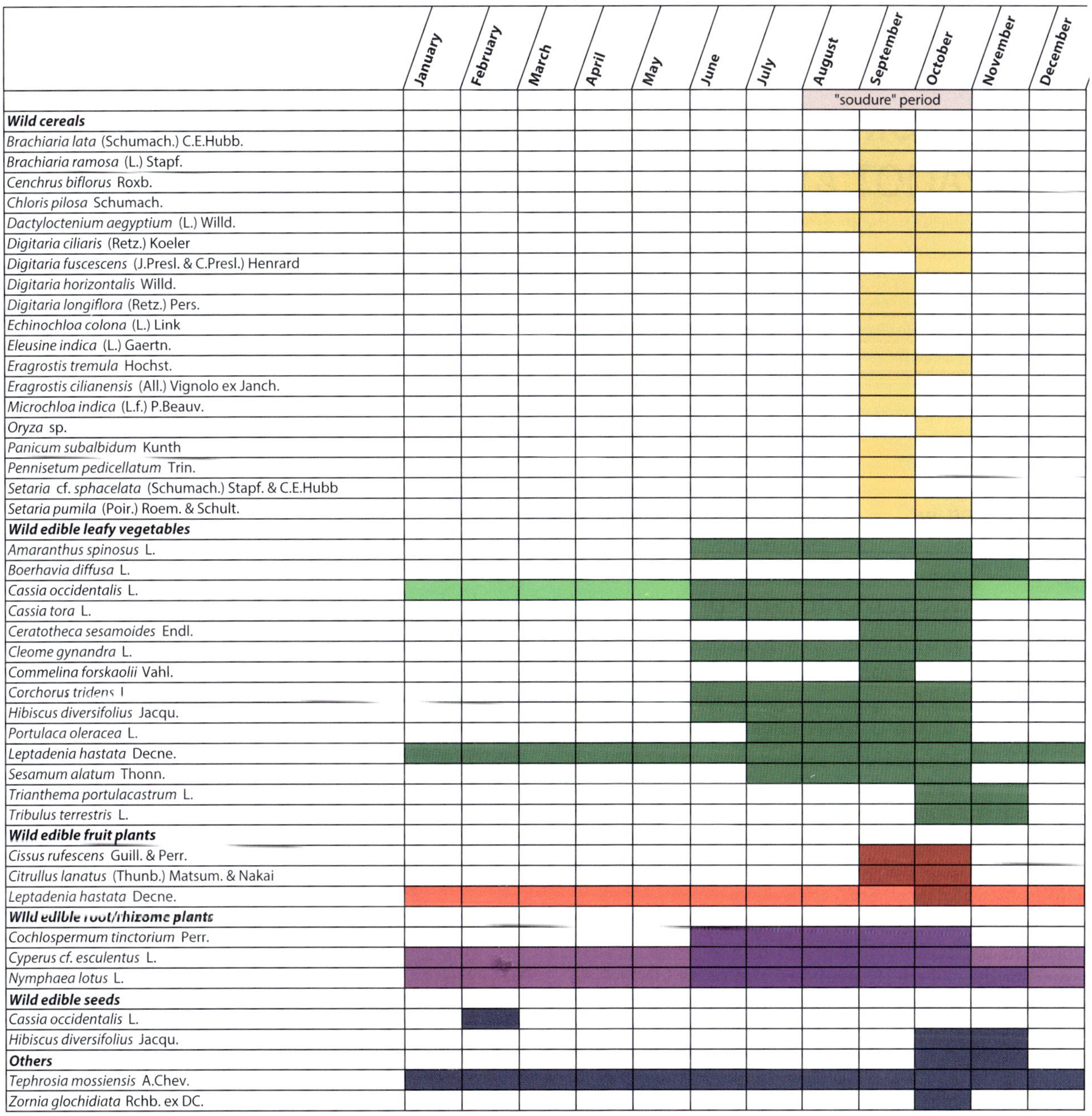

Fig. 5.18. Seasonal distribution of gathered food plants in Dogon country, Mali, based on our field study (a dark colour shows the main gathering period, while a light colour shows a secondary gathering period).

last plant is used as a main ingredient in the famous *fakoy* [fakoi] sauce (Fig. 5.19), which is also popular among Tamasheq people from the northern Niger delta (Jacks *et al.* 1995). To prepare this sauce, dried leaves of *Corchorus tridens* are crushed and cooked with water and spices until the sauce has a dark, thick appearance. Some edible leaves are associated with famine times. This may be linked to the fact that they are less palatable, such as *Leptadenia hastata*, which has a very acid taste, or that they spread on the ground in very ruderal areas, like caltrop.

Some other plants are considered as snacks and are usually consumed on the spot where they are collected, like the underground parts of *Cochlospermum tinctorium*, earth-almond (*Cyperus* cf. *esculentus*) and white lotus (*Nymphaea lotus*), and the fruits of *Cissus rufescens*, citron-melon (*Citrullus lanatus*) and *Leptadenia hastata*. These edible underground organs and fruits are quite rare in the savannah, except for *Leptadenia hastata*, which is very common. These plants are gathered in an opportunistic way, which means that their collection follows no real strategy.

Fig. 5.19. Leaves of *Corchorus tridens* L. are sun-dried at home before being cooked in sauce or sold on markets in Dogon country, Mali.

They are preferentially collected by people who travel in the savannah, such as hunters or shepherds, and also by children. Actually, the Tengu vernacular name for *Cissus rufescens* means 'the grape of the little shepherd', which shows the link between a plant used for snacks and a mobile activity carried on in the savannah.

Other plants and other uses of wild edible plants were also highlighted during our study. *Zornia glochidiata* is a leguminous plant whose seeds are used as a bean substitute during famine times. Some villagers use a decoction of *Tephrosia mossiensis* branches as a sweetener in cereal porridge or drinks. The seeds of antbush can be roasted and boiled to prepare a coffee substitute called *ersatz* [erzatz]. Finally, the seeds of Cape hibiscus are used to prepare a very popular condiment called *datou* [datu], usually made of fermented seeds of domesticated hibiscus (*Hibiscus sabdariffa* L.).

The trading of wild edible plants depends on the type of food. Wild cereals are not traded at all, given that they are immediately consumed after being gathered. Roots, rhizomes and fruits are rarely traded due to their scarcity. The most popular leaves, important in the Dogon diet, are, on the contrary, widely sold in villages and markets. The trade of leaves, as well as their gathering, is predominantly undertaken by women. The gathering and trade of leaves allows the women to contribute to the acquisition of food resources and to promote their knowledge of plants. Furthermore, women can obtain a small independent income in this way (Fig. 5.20). The impact of collecting leaves on the social cohesion and the status of women has already been underlined by Ertuğ (2003) in her study about wild greens collected in Anatolia. Moreover, the predominance of women as plant gatherers has been illustrated by Howard (2003a–c) as well as Price and Ogle (2008).

Leaves, roots, rhizomes and fruits are mainly collected during the wet season between June and October. Gathering of cereals occurs between August and October, during the annual lean time called 'soudure'. The importance of wild plant food during this crucial time is clear; as the reserves of cultivated food decrease, people exploit wild resources more intensively (Fig. 5.18). This phenomenon was emphasised by the study of Toury *et al.* (1961) in Senegal. It shows that, as the wet season progresses, the average ration of cultivated millet decreases and the part of wild greens increases. Other authors have emphasised the importance of wild plant food in Sahelian agricultural societies which must cope every year with a serious shortage of cultivated food resources (de Garine 2005; Tchago and Moupeng 2002).

The consumption of wild plant food provides an important nutritional contribution to the diet of the Dogon population. Generally, leaves are very nutritive and complete the local diet in terms of essential nutrients like protein, minerals and

Fig. 5.20. A woman selling fresh leaves of African spider-flower (*Cleome gynandra* L). in front of her house (second person from the left), Dogon country, Mali.

vitamins. For example, African spider-flower is almost as rich in iron as green amaranth (*Amaranthus hybridus*), a type of cultivated spinach. A moderate ration of *Corchorus tridens*, a plant that is very rich in iron, can also satisfy about half of the protein needed per day by a normal adult (Freiberger *et al.* 1998). Leaves of Cape hibiscus are regularly eaten cooked with peanut balls, while its seeds are used to prepare the condiment *datou*. This popular condiment is very rich in protein (30%) and may be considered as a candidate-vehicle for zinc fortification (Bengaly *et al.* 2007). *Leptadenia hastata* is despised by some people and appreciated by others, but its leaves are an excellent source of β-carotene (176 μg/g), vitamin A and vitamin C (Jacks *et al.* 1995). Thanks to its high level of vitamin A, this green is beneficial for pregnant and breast-feeding women, which was corroborated during our investigation by the testimony of several villagers who emphasised the benefits of *Leptadenia hastata* for young mothers. Leaves of *Sesamum alatum*, caltrop, spiny amaranth and foetid cassia have a good value in terms of protein and calcium, and slender amaranth (*Amaranthus viridis*) is also particularly rich in iron. Furthermore, false sesame provides a good amount of protein (Freiberger *et al.* 1998).

Grass grains have very good nutritional characteristics too. India sandbur has a good energetic value and is also rich in essential minerals like calcium, magnesium, iron and zinc. Surprisingly, this wild cereal has a better nutritional content than the domestic millet (*Pennisetum glaucum*) in terms of protein and essential minerals (Hveem *et al.* 2005). Coast button grass and *Echinochla colona* are rich in protein, zinc and iron. In addition, coast button grass is an excellent source of calcium and magnesium. Nutritional values of the wild cereals are usually higher than those of the cultivated species (Salih *et al.* 1992), except for the lower quality of the protein of India sandbur, and they can be a good diet basis during periods of famine. Until the beginning of the twentieth century CE, wild cereals, in particular India sandbur, *Panicum laetum*, *Eragrostis* spp., *Digitaria* spp., Guinea millet (*Brachiaria deflexa*) and coast button grass were widely collected in the southern Sahara and in the Sahel. These cereals then constituted the diet basis and were more important than cultivated millet. Nowadays, these plants still play a fundamental role in the diet of some Sahelian populations (Bernus 1967; 1992–1993; Maïga 2005).

The composition of white lotus rhizomes is well-balanced in amino acids and this plant constitutes a good source of protein as well as carbohydrate (Nordeide *et al.* 1994; Toury 1961). A recent study has shown that, on the contrary, rhizomes of earth-almond (*Cyperus esculentus*) are of quite poor nutritional value (Glew *et al.* 2006).

Conclusions

Numerous archaeobotanical studies emphasise the importance of wild plant food in the dietary balance of ancient societies. Several of them show that the proportion of wild plant food decreases as the proportion of domestic food increases (Klee *et al.* 2004; Munson 1976). However, these species continued to be exploited after the appearance of agriculture in numerous populations (Grivetti and Ogle 2000).

In the Dogon country, we were able to highlight the key role played by wild edible plants as a dietary complement, as well as a resource in periods of food shortage. In that sense, the exploitation of these plants could be considered as a mainstay of agricultural society in sub-Saharan Africa, permitting agricultural systems to persist.

Furthermore, we were able to observe the exploitation of two wild cereals that are also the progenitors of cultivated species. Crowfoot grass is the progenitor of caracan millet (*Eleusine coracana* Gaertn.), a domestic cereal cultivated in East Africa and in Sri Lanka (Portères 1976), and African bristle grass has led to cultivated varieties (Poilecot 1999). It shows that, if cultivation does not lead automatically to domestication, it is a crucial step in the taming of a vegetal resource.

Ethnobotanical studies help to interpret the archaeobotanical remains found during excavations but they also open onto another perspective: it could become possible to detect the periods of famine in the stratigraphic layers, or at least to point out a heavy reliance on wild plants by identifying the massive presence of remains among agricultural societies. To do so, ethnobotanical data should be correlated with ethnohistorical and palaeoenvironmental data in order to provide a better comprehension of the complex functioning of agricultural societies in sub-Saharan Africa

5.5. SILVERWEED: A FOOD PLANT ON THE ROAD FROM WILD TO CULTIVATED?

Cozette Griffin-Kremer

Verses of praise for the silverweed from Scottish tradition:

> *Honey under ground*
> *Silverweed of spring.*
> *Honey and condiment*
> *Whisked whey of summer*
> *Honey and fruitage*
> *Carrot of autumn.*
> *Honey and crunching*
> *Nuts of winter*
> *Between Feast of Andrew*
> *And Christmastide.*
> (Carmichael 1941, IV, 119)

The silverweed (*Potentilla anserina* L.; PFAF, synonym of *Argentina anserina* (L.) Rydb.) is usually presented as a common wild flower of temperate zones, but its roots were utilised for food well into the eighteenth century and, more recently, as a 'play food' for children (Ary and Gregory 1977; Pankhurst and Mullin 1991; Figs. 5.21, 5.22). Today, the plant seems to be the object of an 'alimentary amnesia' and not even to figure in that happy category of 'old-fashioned' vegetables presently being brought back into fashion (with no trace in Davidson 1999). Earlier assertions that 'tubers leave no traces' or, at least, very few and only indirectly, have been overtaken by finer-honed analytical techniques and, also, an increasing interest in the role of roots and tubers as major food sources around the world (Hather 1993; Kubiak-Martens 2002; Perry 1999). These developments dovetail with the variations on the 'tuber hypothesis' positing that hominins' early fire-use 'tamed' otherwise very difficult-to-digest plant parts, with considerable consequences for cognitive development.

The case of the silverweed is an impetus to compare all the sources we have from written documents, oral traditions and ethnographic reports on the use of what are termed root-plants in vernacular language, especially their underground parts (corms, bulbs, rhizomes, roots, tubers). The emphasis here will be on sources from the British Isles, running from Old Irish lyric poetry to nineteenth-century ethnographic reports on Scotland and Ireland. Indeed, there is no reason to neglect information from narrative or poetic source materials as part of the evidence and casting the net wide enough to do some brief cross-cultural comparison can provide at least some suggestive lines of investigation. The major objective in such an endeavour is simply to remind us that we might always be overlooking something and that a cross-disciplinary approach can be of great value in such instances.

Sources, Tap-Roots, the Silverweed, and the 'Tuber Hypothesis'

Before examining part of the testimony to the use of silverweed as a food plant in Ireland and Scotland, we might first look at root-plants more broadly, since they figure among the most ancient mentions of foodstuffs in a remarkable piece of Irish literature. Its careful textual conservation is considered an indication of the prestige once attached to it. *The 'Quatrains on the Seasons'* (Jackson 1935; Meyer 1884) is an eighth or ninth-century CE Old Irish praise poem in which each group of foods is identified as emblematic of one of the

Fig. 5.21. *Argentina anserina* (synonym: *Potentilla anserina* L.). Copyright Kurt Stueber, 2007. Illustration from Otto Wilhelm Thomé 'Flora von Deutschland, Österreich und der Schweiz, Nur Tafeln' (1885–1905).

Fig. 5.22. A wild stand of silverweed, as photographed in the wetlands near Pulkau, Lower Austria. The yellow flowers of this member of the Rosaceae family show from late spring till late summer and produce tiny nutlets. However, the plant also reproduces by cloning itself – note the reddish stolons (runners) in the picture. The lower sides of the leaves (see Fig. 5.26 and Fig. 5.27) are densely covered by silverish hairs, hence the name silverweed. Caption and photo by Andreas G. Heiss.

four old quarter-days in the calendar system. The foods mentioned include fruits and herbs linked to the summer quarter-day near the beginning of August; meat, ale, mast, tripe, buttermilk and butter associated with November Day; tasting each food in the proper order for the February quarter-day and, first of all, the ale, whey, curds and 'worts' that are associated with May Day.

> The Quatrains on the Seasons:
> *I tell to you, a special festival*
> *The glorious dues of May-day:*
> *Ale, worts, sweet whey,*
> *And fresh curds to the fire.*
>
> *Lammas-day, make known its dues*
> *In each distant year:*
> *Tasting every famous fruit*
> *Food of herbs on Lammas-day.*
>
> *Meat, ale, nut-mast, tripe,*
> *These are the dues of summer's end,*
> *A bonfire on a hill pleasantly,*
> *Buttermilk, a roll of fresh butter.*
>
> *Tasting every food in order,*
> *This is what behooves at Candlemas,*
> *Washing of hand and foot and head,*
> *It is thus I say.*
> (Meyer 1884)

The word used in the May Day verse for 'worts' is *mecon*, also translated as parsnip or carrot. Consensus among scholars has settled on a comfortably safe generic term – tap-root. *Mecon* was cited in Old Irish law as emblematic of land suitable for growing cereal crops, the so-called 'three-wort land' (*tír trí meccon*), which the *Commentaries* explain as meaning the richest soil (DIL 1983, M–76; Kelly 1998; Lucas 1960; Vendryes 1960, M–27). This first note on root-plants from early Medieval Ireland is echoed by a specific mention of the common name for the silverweed (*briscén*) in a satirical text – *'The Vision of Mac Conglinne'* – dated linguistically to the late eleventh century, where it appears among the luxury foods used to tempt a tapeworm to jump out of a king's mouth (Jackson 1990), with no hint about its status as wild or cultivated. It goes without saying here that we are on uncertain ground as regards such common names; although *brisgein*, or its dialect variants, are still the usual term for silverweed in the Gaelic of both Scotland (*briosclan*) (Dwelly 1994) and Ireland (*brisgean*) (Dinneen 1927). Etymologically speaking, the term appears to express crispness or brittleness and is a diminutive, so the name may mean 'little crispy (or brittle) one' (Vendryes *et al.* 1981, B:90–91).

After seeing these attestations in two Irish texts citing 'worts' and silverweed as among foods with a prestige or, at least, a festive image, it is intriguing to find that the most prolific ethnographer of nineteenth-century Scotland also devoted considerable attention to the silverweed (Scottish Gaelic *brisgein*). The plant is the object of two praise poems that speak of it as 'blessed of spring,' as one of the 'seven foods' of the Gael, and as especially nourishing (see opening quote above). The commentary attached to these poems provides further information from the island of North Uist (Fig. 5.23) and might indicate transfer from a primary to a secondary habitat, thus perhaps from a 'wild' to a 'cultivated' status:

> 'The root was much used throughout the Highlands and Islands (Fig. 5.23) before the potato was introduced. It was cultivated, and so grew to a considerable size. As certain places are noted for the cultivation of the potato, so certain places are remembered for the cultivation of silverweed. One of these was Lag nan Tanchasg in Paible, North Uist, where a man could sustain himself on a square of ground of his own length. In dividing *"mòrfhearann,"* common ground, the people lotted their land for *'brisgein'* much as they lotted their fishing-banks at sea and their fish on shore. The poorer people exchanged *"brisgein"* with the richer for corn and meal, quantity for quantity and quality for quality. The *"brisgein"* was sometimes boiled in pots, sometimes roasted on stoves, and sometimes dried and ground into meal for bread and porridge. It was considered palatable and nutritious.' (Carmichael 1941, IV, 118–119).

In earlier documents concerning the Hebrides, use of the silverweed as a food plant is reported several times for the seventeenth and eighteenth centuries (Grant 1961, for Lewis and Harris; Martin 1709, for the islands of Tiree, Skye and St. Kilda in about 1660), when it seems to have been eaten by only some of the inhabitants or by the poor or as a famine food. In his travels in the Islands, Martin stressed the combination of food sources he often found; for example, on Tiree: barley-bread, the dairy triad of butter, milk and cheese, plus fish and 'some eat the roots of the silver-weed' (Martin 1709, 60, 714, 721). He was also quite clear that the silverweed was eaten elsewhere and that there was wide use of, what for him were, wild plants as foodstuffs.

Martin's reports are usually made in passing and never contain the detail of Carmichael's commentary

Fig. 5.23. Location of the Outer Hebrides and the Island of North Uist. Map: R. Lugon, J.-C. Loubier and A. Chevalier.

on the plant and the way it was tended: whether cultivated, hence of large size (*cf.* Hildebrand 2003), the use of a particularly appropriate site or soils, plot size and yield, the exchange rate between silverweed root and other commodities and, lastly, culinary procedures and palatability. However, to his credit, we might recall that Martin's seventeenth-century reports come from a rare human specimen: both a native Gaelic speaker and a highly literate reporter, whose colourful accounts later set the famous Samuel Johnson off on his own tour of the Islands. Martin reports on lands very near to us geographically, whose people have left us few traces of their lives, since they did not write, as we often see in ethnographic reports from farther afield (*cf.* Hildebrand 2003). Could it be that the silverweed as a food plant has been forgotten, because it was mainly used by illiterates and little known to those who wrote?

The period Martin covered so carefully was a full century before adoption of the potato in the area. Does Martin lift the hem of the veil over root-plants to reveal they might be underestimated as components in the diet before the potato took on its role as a staple? Have root-plants in general been given short shrift? The 'tuber hypothesis' refers to the propositions that hominid use of fire and the innovation of cooking tubers might have provided a quantum leap in nutrition, as well as influencing morphological and cognitive development. Such developments would have been most especially beneficial to women and to social restructuration around a hearth. Increased nutritional sources from the cooked root-plants would have led to less dependency on hunting and a posited rise in food security (summarised with bibliography in Wrangham 2009). In any case, the use of root plants should hardly be considered in isolation from the more general spectrum of food plants utilised by any group at a particular time.

Food Assemblages, Culinary Delights, Ownership and Reception of a Newcomer

Suggestion of a broadening spectrum of food resources in very early hominid development recalls David Clarke's (1976) work on the yields of gathered foods in the Mesolithic. He estimated that cooler Mediterranean woodlands would provide, per hectare per year, one ton of hazel nuts, three tons of chestnuts and ten to twenty tons of edible bulbs. He also pointed out that plants tend to grow in assemblages – as we shall see for the case of silverweed and clover – and so would have been gathered together, eaten together and perhaps classified together in popular botanics. Clarke proposed a viewpoint which emphasises the complementarity of various plant resources and includes an energy cost analysis of the activities:

> 'All economies expend energy in the maintenance and control of food supplies on the one hand and the detection and pursuit of food supplies on the other. Perhaps it might be said that husbandry moves into food production when the energy expended in the maintenance and control of food supplies first expands beyond that expended in the detection and pursuit of plant and animal food sources. Such an energy cost threshold might then provide a useful quantitative basis for our conventional division between gathering-hunting-fishing and food-producing systems' (Clarke 1976, 469).

Carmichael's commentary (1941) on the nineteenth-century silverweed praise poems speaks of many things, but not of cost efficiency. Just what might it be like to deal with a silverweed: is it an easy task or a labour-intensive one? Is there some quality – perhaps the taste or the hard-to-get factor – that might explain its apparent prestige in these poems?

It is necessary to look rather farther afield to find people who appreciated the silverweed as much as the Scots of North Uist. There is ample ethnographic testimony that the northwest coast Native Americans (Fig. 5.24) considered the silverweed a delicacy and that it was, what Nancy Turner terms, a 'cultural keystone species' (Turner and Kuhnlein 1982). These Pacific coast peoples noted that the roots varied greatly in taste qualities, desirable texture and keeping properties, and that various tribes valued different nuances. Taste also depended on soil, age of the stands, gathering times and, of course, on the cook's own merits. It could be munched on raw as well, if the gatherer felt a bit peckish (Turner and Davis 1993; Turner and Kuhnlein 1982). Groups that owned fine patches of silverweed (*Potentilla anserina* L. ssp. *pacifica* (T. J. Howell) Rousi, *cf.* Kuhnlein and Turner 1991) and springbank clover (*Trifolium wormskioldii* Lehm.) – forming a common 'assemblage', as in Clarke's terms – shared the roots within their lineage, but also used them to lavishly host visitors, often at the festival of First Fruits and First Shoots held at the beginning and end of summer (Turner and Davis 1993). This recalls that the first mention of 'worts' in the Old Irish praise poem was also connected with a calendar festival, *Bealltaine*, associated today with May Day, which opened the summer season and historically involved large gatherings (*oenach* or 'assembly') of tribes. Carmichael's nineteenth-century North Uist testimony linked the silverweed more vaguely with spring, but he was quite clear that good silverweed growing sites were included in Scottish land allotment traditions.

Pursuing the comparison with the Pacific coast societies, particular tribal groups were said to 'own' – or have sovereignty over – find stands of plants, and the reasoning behind this legal status harks back to Clarke's positing of a threshold in the passage from foraging to food production. These Native American societies were highly complex and had much variation across cultural, linguistic and geographic space, but Salish ethnographic

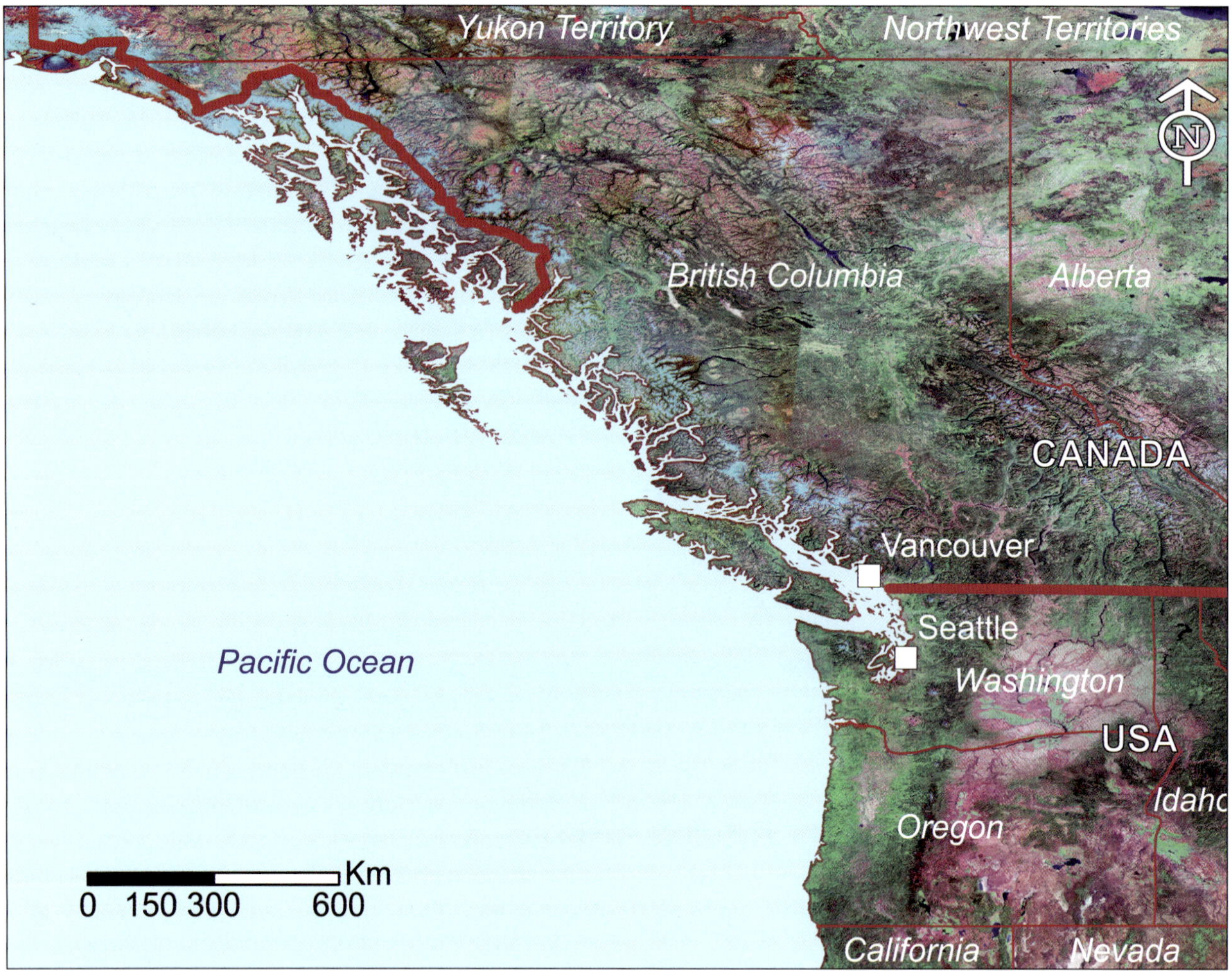

Fig. 5.24. Map of the Pacific northwest coast of Canada and USA. Map: R. Lugon, J.-C. Loubier and A. Chevalier.

descriptions seem to be representative of the general thrust of practices and attitudes: the harvesting of resources which were very localised or that required care or construction and, therefore, a particular expenditure of energy over and above the harvesting itself, was usually subject to some form of control and ownership. This was quite clear to the Native American population, but it proved not to be so clear to others and became a crucial point in relations with Europeans, for whom field agriculture was *the* single criterion for land rights (Turner and Jones 2000).

Ethnographic reporters in the late nineteenth century were dazzled by the wealth in natural resources of the Pacific northwest coast and the capacity of its peoples to produce sufficient food supplies to sustain, what the Europeans considered, a dense population. Much of this depended on land management techniques – selective burning, pruning and replanting – and a broad panoply of storage methods (Turner and Davies 1993; Turner and Jones 2000). Observers were also struck by the very rapid and widespread adoption of the potato, which had never been cultivated this far north in pre-contact times, a remark that recalls signs of the opposite – a certain resistance to adopting the potato – noted for parts of Scotland.

In his classic account of the potato's spread around the world, Redcliffe Salaman (1985) reflected on the rapidity or reluctance with which the potato was accepted and outlined the factors he felt might elucidate its varying reception. There were rare cases in which the plant was outright imposed on growers but, when the potato was willingly adopted, he posited that people already had a strong relation with root plants and practiced appropriate soil preparation techniques. Salaman particularly noted that Highland Scotland and the Hebrides were among the last areas to adopt widespread cultivation of the potato. He tentatively attributed this to a general conservatism in food tastes, as in political organisation, among the Highlanders, although he was quite aware of the silverweed's status as a food plant and did note this (Leach 1997; Salaman 1985; *cf.* particularly J. G. Hawkes' introduction to the 1985 edition for subsequent refinements). Did Salaman put his finger on a sort of 'filiation' between the nigh-forgotten silverweed and the latterly omnipresent potato, based not on any kinship between the species, but on a series of similarities in the way they were managed? This seems to have included the tools used in handling the plants, the notion that particular soils suited them and produced especially fine stands, perhaps even the 'construction' of plots that enabled higher yields and, finally, the way they were stored and processed for consumption.

Cultivation, Complementarity, Waves of Change, Verticality, Coveredness and Spread

Nineteenth-century Scottish testimony spoke of using a tool (translated as a 'small mattock' in English) to dig up carrots and other roots. The same term applies to a stick used to stir dehusked grain in a pot before sifting and grinding in the quern, a pointed instrument for digging, a small spear, and a stick to stir potatoes as well as to plant them in newly reclaimed soil (*sleabhag* in Carmichael 1928, II, 356; *sleaghag* in Dwelly 1994). The Pacific northwest coast peoples still use a digging stick to bring up their silverweed and springbank clover roots, since both are quite brittle and it takes appreciable skill to get them out without breaking or bruising them (Leach 1997; Turner and Kuhnlein 1982). A 'stick' may sound simple, but 'simple' tools may not be simple to use. In twentieth-century Highland Scotland, Isabelle Grant (1961) recalled how an older man asked what she thought rather a high price for a simple-looking tool she wanted for her museum collection. When she remarked on this, he replied that it might well only take a day to find the right piece of wood, but that it took a lifetime to know how to use the implement properly.

This is not to be taken as an argument that people who grew silverweed in the Scottish Highlands or Islands must have used a digging stick like that of the Pacific coast peoples to prepare the soil and gather the roots. However, the Native American informants emphasised that using a digging stick was the only right way to gather delicate roots and that it involved carefully 'rooting around'. This had a considerable impact on the soil, since it was left well aerated. The gatherers took only 'this year's' roots, which were easy to distinguish from older ones. They could be dug up when the plant's leaves had died back for the winter, in October or early November at their nutritional peak, but they were considered good until the next March. Efficient gathering involved considerable experience and skill, as well as familiarity with the location of the best sites (Turner and Kuhnlein 1982) (Figs. 5.25, 5.26, 5.27).

On the Pacific coast, the entire process, from gathering through the various culinary preparations and processing of silverweed for storage, was far more labour-intensive than subsequent potato cultivation. More recently, the era of canned and frozen foods sounded the death knell of this 'cultural keystone species'. In Scotland, perhaps the potato's progress alone might explain disaffection with a hard-to-get root like the silverweed, even if it had come into garden cultivation, as Carmichael's testimony might imply, and the potato was not the only novelty in food production at the time. The potato appears to have arrived during a period of considerable reorganisation of production in both Scotland and Ireland that included massive field turnip plantations to fuel the cattle trade with England and the continent. This major shift from domestic milk production to the international meat trade pushed most of the rural poor onto poorer soils, where only the potato grew well, creating a dangerous dependence. Repeated failures of the potato crop in the late eighteenth and early nineteenth centuries foreshadowed the famine to come (Bourke 1993; Salaman 1985). It took place in the context of a food production system that stands in stark contrast to Clarke's systems, emphasising plant assemblages and their complementarity, which buffered against risk.

Clarke's calculations of the potential tonnage of woodland yields took into account underground resources such as bulbs and roots. The top end of his spectrum included another food source – the acorn – that was significant for many Pacific coast

Fig. 5.25. This pile of silverweed roots (*Argentina anserina* of the interior from beds of wild plants) was harvested in March after the snow melted, but before new top growth had started, which is the ideal time for full flavour and sweetening, but is something of a harvest window that closes after the plants begin growing (in north central Washington, U.S.) Caption and photo by Steve Dupey.

Fig. 5.26. Silverweed (here, *Argentina egedii* subsp. *groenlandica*) plants growing in simple experimental conditions, in three- to five-gallon pots set in a children's wading pool, with a static water level around the bases of the individual plant pots an inch to two inches deep. Some were grown in nearly pure sand, while others were in a medium fairly rich in organic matter. Soluble fertiliser was applied at times. The plants grow quite vigorously in one season, sending out runners profusely, flowering over a long period, and generally overwhelming any available area for spreading within their vicinity. There were some nice-sized roots at season's end, but nothing that was so significantly greater in size than could be found in the larger specimens of the wild growing roots. Notice the difference in leaf colour between the silvery A. *anserina* subsp. *anserina*, and the A. *egedii* on left behind hand. Ongoing experiments with these varieties may shed further light on those reportedly utilised by Northwest Coast native peoples (north central Washington), as well as on tending and gathering techniques. Caption and photo by Steve Dupey.

Fig. 5.27. *Argentina anserina* plants form a thick groundcover, spreading by surface runners and clonal off-set plants. A view of the root and runner system showing the long vertical starch-storage roots from which the next season's growth will commence. Container-grown plants. Caption and photo by Steve Dupey.

peoples and has been posited as once an important food in Europe, including the areas where the silverweed was known to have been consumed (Kelly 1998; Ni Chathain 1979). We have moved from root to treetop, but there was much going on between these strata, since we must also factor in domestic animals' food resources alongside those of the human population.

J. G. Lewthwaite (1982) recommended viewing resources as three-dimensional: tree products occupy volume, while grazing occupies space, and the former often provides 'suspended' food and forage in far greater quantities than the floor cover of forest or meadow can. In his examination of the role of acorns in prehistoric food production in Europe, Lewthwaite commented on an unanticipated integration of dairy pastoralism and acorn collection in the archaeological record for the temperate Mediterranean. The Scottish testimony indicated that the silverweed root, like the acorn, was consumed in the form of a ground meal for both bread and porridge, which could be made with water, whey, milk or even cream. These combinations were unavailable to Native American populations before the European occupation, of course, and hint at some very significant differences in the food systems touched on here. For one thing, milk looms large in Andrew Sherratt's (1981) reasoning about a 'secondary products revolution' that included animal draft and its consequences for European economies and technologies.

A rich spectrum of resources is just one part of food security and reminds us that people often had to cope with food 'insecurity' on a quick and deadly scale. A broadly interpreted sense of the term 'economy' can include 'extractive' activities somewhat beyond gathering or harvesting – simply taking what others have laboured to put together in the form of movable wealth such as cattle (literally a 'capital' transfer) and the often concomitant strategy of destroying what cannot be carried off. Food resources that can be counted on over a good part of the year and that store well in their own soil or in an underground cellar, as is the case of many root crops and the silverweed (Turner and Kuhnlein 1982), may possess a valuable trump card in their 'coveredness' and their seasonal 'spread'. Historical sources for both Scotland and Ireland are replete with references to cattle rustling and cereal-crop-burning. This was the tactic utilised massively by the English forces in the later sixteenth and early seventeenth-century campaigns in Ireland (Lucas 1960), as well as standard practice in the long period covered by the Irish Annals (*e.g.* the *Annals of Ulster AD 431-1201)*, when the principal opponents of the Irish were the Irish and of the Scots, the Scots, with the occasional Viking incursion. It is another parameter to consider when thinking about all the qualities of the foods in a resource spectrum.

Perspectives

As regards finding evidence of the use of the silverweed as a food plant, perhaps we need not wholly adhere to David Clarke's (1973, 10) famous throw-off line about archaeology, that it is: 'the discipline with the theory and practice for the recovery of unobservable hominid behaviour patterns from indirect traces in bad samples'. Advances in analytical capacities have changed the picture for the silverweed and other underground treasures within recent decades. Although earlier conventional archaeobotany analysis of seeds, charcoal remains and pollen proved to be limited in identifying underground starchy food, today starch granule, as well as phytolith analysis, would enable identification of *Potentilla anserina* in some archaeological contexts (Chevalier and Peña-Chocarro *pers. comm.* 2010).

Much may also depend, literally, on a shift in 'perspectives': root-plants become more visible when we take an interest in them and there is ample promise in recourse to other sources, such as we have here, which can be complementary to archaeological investigation in many cases. The silverweed may have been on its way from a 'wild' plant gathered knowledgeably to a 'tended' one in a garden plot, at least in North Uist, in the late eighteenth century. Today, its status is near oblivion, but the intriguing attestations to it as a trustworthy food source have brought up a host of angles from which to view the use of any plant, including those considered 'wild', and especially the – now – far more 'traceable' roots: nutritional value, complementarity, assemblages and filiations, food in three dimensions, cost efficiency, ecological expertise and tool-use skills, seasonality, prestige status and festive uses, trade, warfare, property rights and kinship systems, and the impact of the Columbian exchange.

5.6. *CLEOME*: A WILD PLANT AS COMPLEMENT TO CULTIGENS IN SOUTHWESTERN NORTH AMERICA

Linda Scott Cummings

Use of wild plants as famine foods is difficult to track in the prehistoric record. In fact, designation of a food or foods as 'famine' foods is often made erroneously, as it has been for the genus *Cleome* (beeweed, Fig. 5.28), an important food for Puebloan people. Two species of *Cleome* grow in this area: *Cleome serrulata* Pursch. and *Cleome lutea* Hook. They appear to have been used without distinction between the species. Interpretation of flavour, smell and other parameters that influence opinions about the attractiveness of food, have been made by researchers and people of modern times that have different cultural preferences, experiences and/or tastes than people living either earlier or in a different culture. Harrington (1967) notes that *Cleome* (beeweed) has an unpleasant odour while cooking. This type of observation sometimes leads to speculation on the part of ethnobotanists concerning the use of plants as 'starvation' foods, since it can be hard to understand consistent use of plants with unpleasant characteristics. Elmore (1944, 51) cites Standley (1912, 75, 458) about *Cleome:* 'it is almost incredible to anyone familiar with the beeweed that it could ever be eaten. Its stems and leaves give off a most offensive odour, but this is said to disappear when cooked.' He goes on to describe the Navajo method of cooking *Cleome* as: 'young plants are boiled and pressed out three times, after which they are rolled into balls and eaten, or dried and stored for the winter. When these balls are to be used, they are soaked and boiled with or without tallow. The Navajo say that *guaco* has saved them from starvation on several occasions' (Elmore 1944, 51). It is from published information such as this that *Cleome* has earned (undeservedly) the reputation of having been a 'starvation food'.

Fig. 5.28. Flower of beeweed (*Cleome* sp.) Photo by Linda Scott Cummings.

Cleome in the past

It is usually a matter of luck in sampling sites that have deposits representing both good times and lean times that provide us with relevant samples. A study at Hoy House, located outside the southern boundary of Mesa Verde National Park in southwestern Colorado, provided just such an opportunity in the middle 1970s (Fig. 5.29). Excavation of this cliff dwelling yielded both midden deposits and also coprolites. Examination of the coprolites (palaeofaeces) for pollen by the author (Cummings 1994; Scott 1976) and macrofloral (Stiger 1976) evidence of foods consumed provides a direct

Fig. 5.29 Map of the US Four Corner region indicating the Hoy House site (1) outside the southern boundary of Mesa Verde National Park, southwestern Colorado. Map: R. Lugon, J.-C. Loubier and A. Chevalier.

record of the foods that people were eating. While the coprolites were not dated directly, they are associated with occupation of this site between the mid-1100s and early 1200s CE.

Coprolite evidence of diet at Hoy House includes cultigens such as maize, beans and squash/pumpkin. Evidence for maize was observed in almost every coprolite. However, since there is no evidence as to which portion of the occupation the coprolites date, further information may be available from pollen analysis of stratigraphic samples collected from the midden.

Today, local vegetation in Johnson Canyon, where Hoy House is situated, includes dense growth of trees and shrubs along the canyon bottom. The mesa top above the cliff dwelling supports various scrubby trees and shrubs, yucca, sagebrush, Mormon tea (or ephedra), cactus and sparse grasses.

Excavation of the site revealed evidence of the cultigens maize, beans, and squash, as well as native plants including walnut, reed, prickly pear cactus, and Indian rice grass.

A trash deposit, located downslope from the dwelling area, was sampled stratigraphically to recover samples for pollen analysis by the author (Scott 1976).

Cleome (beeweed) pollen was observed both in samples from coprolites and also the midden. This plant might well have been encouraged or even occasionally cultivated (or at least tolerated) in agricultural fields. It appears to have been important economically to prehistoric residents of the area, as it is also important to Indians of the American southwest in historic times. Whiting (1939, 77–78) discusses the use of beeweed by the Hopi:

> 'The young plants are gathered and boiled for food. Frequently plants are allowed to mature and seed in the corn fields, insuring a supply for the following spring. [Beeweed also was used] in the manufacture of *paho* (prayer-sticks) in *Powamu* ceremony. This plant is used in the preparation of pottery paint among the Tewa of the Rio Grande. At Hano it is named with the three chief cultivated plants – corn, pumpkin, cotton.'

Further, Whiting (1939, 16) describes the importance of *Cleome* to the Hopi when he writes: 'Plants of the Rocky Mountain beeweed are allowed to mature and to disperse their seed in Hopi corn fields providing an abundant supply of young plants for the cook pot the following spring.'

Martin and Byers (1965, 132) also make reference to ethnographic accounts of the use of beeweed:

> 'Elmore, Stevenson, and Whiting record extensive uses of *Cleome* among native agricultural people of the southwest. The Hopi allow it to grow in the corn fields, gathering most of the plants but leaving some to mature and drop their seeds, insuring next year's crop. The plants are boiled and eaten with corn. They also prepare a black residue from *Cleome* which is used to decorate pottery and prayer sticks. The Navajo boil *Cleome* greens and eat them with meat or in a stew. They preserve *Cleome* by boiling and pressing the leaves, storing the dried product for winter use, when it is soaked and eaten plain or mixed with tallow. The "guaco" has saved them from starvation on several occasions.'

Frequent presence of *Cleome* pollen in coprolite samples from archaeological sites in the American southwest, including Hoy House in Johnson Canyon, southwestern Colorado (Cummings 1994; Scott 1976), at Step House in Mesa Verde National Park (Cummings and Puseman 1992) and references to *Cleome* in the ethnobotanical literature, suggest continued use and preference for this plant through time. Although it has been interpreted as having been used as a starvation food, it has a long history that suggests otherwise. In this case, it appears that *Cleome* was not viewed as a starvation food during prehistory, for it was regularly prepared along with maize. It was intentionally grown in agricultural fields, harvested, processed and stored. Recovery of *Cleome* pollen in 95% of the coprolites examined from Hoy House suggests its regular consumption. What has not been established is whether or not the coprolites relate to a period of abandonment and a difficult time, or to earlier, more successful times.

Cleome pollen is present in most of the samples examined from the trash midden. In the lower levels of the midden it is noted in 2% of the total pollen. By the middle of the record, *Cleome* pollen accounts for 42% of the total pollen, persisting in large quantities until the upper portion of the midden, when it drops precipitously to 1% of the pollen. *Zea mays* L. pollen, on the other hand, is noted in low frequencies throughout most of the record, with the exception of large frequencies at the base and top of the record.

The abundance of *Cleome* and scarcity of *Zea mays* pollen in the middle and upper portions of the trash midden might represent a shift in cultural preference or, more probably, represent a decrease in maize harvest and reliance on other foods. Although no direct evidence is available to substantiate a decline or failure in the maize harvest for the Hoy House occupants, a similar situation was observed in pollen frequencies recovered toward the end of the occupation at Long House, also within the Mesa Verde area.

Martin and Byers (1965, 132–133) suggested that *Cleome* might have acted as a starvation plant for a period of at least one decade. Their evidence for this interpretation is a maximum in *Cleome* pollen in samples representing the period just prior to abandonment. Samples examined from a nearby village, Mug House, on the other hand, did not indicate such a rise in *Cleome* pollen. Failure to substantiate other, similar, rises in *Cleome* pollen in samples representing the period prior to abandonment led Martin and Byers to abandon this hypothesis.

At Hoy House it is possible that crop failure in the maize fields resulted in the reduction of evidence for discard of maize in the midden (Scott 1976). One important factor in interpreting this record is the archaeological evidence that a portion of the population abandoned the Hoy House complex near the end of its life, leaving a smaller population of people for a few years. The lower portion of the midden represents occupation during the early building phase of the cliff dwelling. The middle section appears to represent occupation during the later building period. If the population increased during the later building phase, greater demands would have been put on the available food resources. As land was cleared on the mesa top and used for agriculture, soil erosion would have increased, resulting in a decline in productivity. In the face of declining agriculture and increasing population pressure, *Cleome* could have been used to fill the gap between the amount of maize that was being produced and the amount of food necessary to keep the residents of Hoy House alive. The rapid decline in *Cleome* pollen and slight increase in *Zea mays* pollen in the upper samples is consistent with the archaeological interpretation that part of the population moved from this dwelling, leaving a smaller population base. In this case, even dwindling maize supplies might well have been sufficient to feed the smaller population, reducing reliance on the less desirable *Cleome*. This would change the pattern of food discarded in the midden.

Abandonment not only of Hoy House, but the entire area, is signalled by increases in *Pinus* (pine) pollen, representing a re-colonisation of the mesa top by pines after abandonment of agricultural fields. This pollen signal is observed at every site where pollen analysis has included sediments from the post-occupational period. Martin and Byers (1965, 125) were the first to identify this trend: 'The rise in tree pollen is so clear-cut that we have come to expect a similar rise in any late prehistoric profile from Mesa Verde which extends through the thirteenth century.' A similar rise in tree pollen was observed at the top of the deposits at Hoy House, marking abandonment. At the same time, pollen from economically important plants ceases in the deposits.

Discussion and Conclusions

Whether valued as a starvation plant or an important element of the diet that was consumed by choice rather than dictated by economics or starvation, *Cleome* is well represented in palaeofaeces studied from this region. *Cleome* pollen was observed in 95% of the coprolites from Hoy House, whereas it was present in 83% of the coprolites examined from nearby Step House. This plant apparently was an important part of the economy for the Puebloan Indians living in southwestern Colorado. At Hoy House, two rooms identified as storage rooms during excavation yielded different pollen signatures indicating storage of foods. Room 53 appears to have been used to store beeweed. Nearby Room 59, however, appears to have been used to store maize and, perhaps, only a small quantity of *Cleome*.

The positive correlation for the amounts of *Cleome* pollen recovered in samples from the middle of the Pueblo III occupation at both Mug House and Hoy House suggests that, at this time, beeweed played a much more important role in the diet than it had in times prior. It appears that the occupants of both cliff dwellings relied heavily on this plant for food. Although *Cleome* had been part of the economic plant base for previous time periods (for instance during Pueblo I), it did not contribute overwhelming quantities of pollen to the record in habitation settings (Cummings 1998). The importance of this plant in the diet, through time, cannot be ignored, as evidence for its preparation, storage and consumption is ubiquitous at Puebloan sites in the area. It is changes in frequencies of recovery, particularly relative to *Zea mays,* that provide evidence of probable use as a starvation food.

5.7. CONCLUSIONS

Gisella S. Cruz-García and Füsun Ertuğ

This chapter provides a broad overview of different perspectives on the study of wild food plants in the past and present. The case studies demonstrated that, in order to have a holistic view of wild food plant resources, it is necessary to have an interdisciplinary approach in terms of scientific disciplines as well as utilising various quantitative and qualitative research methodologies. The research findings discussed in these case studies are not only the product of a wide temporal and geographical range of ethnobotanical, ethnohistorical and archaeobotanical research, but also draw from the analysis of written documents such as folktales and poems.

This chapter validated, once again, the importance of women as knowledge holders on wild food plant collection and use, as observed across the contributions. Moreover, this chapter argued opposing, but not excluding, concepts attached to certain species in relation to social stigmatisation versus 'revival'; for instance, certain species are regarded as 'poor people's food' whereas others are symbols of regional identity.

The case studies presented in this chapter also re-affirmed the importance of wild food plant gathering as a deeply rooted aspect of our heritage across cultures and time. Although it has been demonstrated that these species constituted a main component of human subsistence in the past, millions of people from around the world still depend on these plants for assuring their basic food needs and health, not only in times of famine and food shortages but also for their daily survival.

Chapter Acknowledgements

Cozette Griffin-Kremer (Chapter 5.5) would like to express special thanks to Steve Dupey and Andreas G. Heiss for the photographs of silverweed roots and of plants growing in the wild or in experimental conditions which they kindly provided to illustrate this article, as well as to Alexandre Chevalier and Leonor Peña-Chocarro for helpful advice on recent developments in techniques for identifying root remains in archaeological contexts.

Chapter Notes

1 Pretty (2003) accepts the dates of 7 million years before present (BP) for human divergence from apes, 12,000 BP for the start of agriculture and 20 years for the average generation length.

2 Further reading on the use of wild food plants is to be found in the book edited by Nina Etkin (1994) *'Eating on the wild side'*, which presents the pharmacological, ecological and social implications of wild food plant consumption by different societies, not only in the present, but also in prehistory. In addition to this, Cunningham's (2000) *'Applied ethnobotany: People, wild plant use and conservation'* as well as Prance and Nesbitt's (2005) *'The cultural history of plants'* are important books on this topic. The book series of PROSEA Foundation (Plant Resources of South-East Asia) documented 6,697 useful plant species from South-East Asia (2010), whereas PROTA (Plant Resources of Tropical Africa) presents information on about 7000 useful plants, including many wild edible plants (2010). The prehistoric use of wild food plants is illustrated in the articles published by Behre (2008), Bouby and Billaud (2005) and Hastorf (1988), among other authors.

3 The consumption of wild plants from agricultural ecosystems is the main point of Scoones *et al.* (1992) publication called *'The hidden harvest: wild foods and agricultural systems: a literature review and annotated bibliography'*, as one of the outputs of the Hidden Harvest international programme, which presents a compilation of literature regarding several aspects of this resource. The importance of wild food plants for farmers is explained by Ertuğ (2000), Guijt *et al.* (1995), Ogle and Grivetti (1985), Bharucha and Pretty (2010), Heywood (1999) and Price (2005, 1997), among other authors.

4 Fruit in this chapter refers either to the botanical part of the plant (Chapters 5.2 and 5.4), or a food-use category (Chapter 5.3).

5 The overlap of wild plant roles as food and medicine, as well as the use of these foods for disease prevention is the focus of the book edited by Pieroni and Price (2005) *'Eating and healing. Traditional food as medicine'*, presenting case studies from around the world. This topic is also interestingly discussed in Johns' (1990) book *'With bitter herbs they shall eat it: Chemical ecology and the origins of human diet and medicine'*. This double role of wild food plants is also central to the proceedings of the joint conference of the Society for Economic Botany and the International Society for Ethnopharmacology *'Plants for food and medicine'* (Prendergast *et al.* 1996).

6 The book edited by Howard, *'Women and plants: gender relations in biodiversity management and conservation'* (2003) discusses and presents several case studies from the world related to the role of women as the main custodians of knowledge on wild plant resources.

7 More information on the role of women in small-scale marketing of wild food and food-medicinal plants was published by Daniggelis (2003), Ertuğ (2003a), Hanlidou *et al.* (2004), Moreno-Black and Price (1993), Putscher and Vogl (2006), among others.

Chapter Bibliography

Abreu, J. (1977) Historia de la conquista de las siete islas de Canaria. Goya: Santa Cruz de Tenerife.

Adaya, A. L., Bdliya, H., Bitrus, H., Danjaji, M., Eaton, D., Gambo, M. B., Goggobe, M., Makinta, A., Okali, D., Omoluabi, A. D., Polet, G., Salisu, M., Sanusi, S. S., Sarch, M. T. and Shuaibu, M. (1997) The Hidden Harvest: the role of wild foods in agricultural systems. Local level assessment of the economic importance of wild resources in the Hadejia-Nguru wetlands, Nigeria. *Hidden Harvest Project Research Series* 3. Institute for Environment and Development (IIED)

ADISAC (2006) *Los guardianes de las semillas*. ADISAC-La Voz, Zamora.

Amblard, S. (1996) Agricultural evidence and its interpretation on the Dhars Tichitt and Oualata, southeastern mauritania. In G. Pwiti and R. Soper (eds.) *Aspects of African archaeology*. Congrès de l'Association panafricaine de préhistoire et disciplines assimilées 1. University of Zimbabwe Publications, Harare, pp. 421–427.

Annals of Ulster AD 431-1201. CELT (*Corpus of Electronic Texts*) www.ucc.ie/celt/published.html

Arbonnier, M. (2002) *Arbres, arbustes et lianes des zones sèches d'Afrique de l'Ouest* (2nd edition). Musée d'Histoire Naturelle, CIRAD, Paris.

Arco, M. C., Arco, M. M., Atiénzar, E. and Hopf, M. (1990) Estudio de los restos vegetales de la cueva de Don Gaspar y algunas anotaciones sobre la agricultura prehistórica de Tenerife. *Investigaciones Arqueológicas en Canarias* 2, 13–25.

Arechavaleta, M., Rodríguez, S., Zurita, N. and García, A. (eds.) (2010) *Lista de especies silvestres de Canarias. Hongos, plantas y animales terrestres. 2009*. Gobierno de Canarias, Tenerife.

Ary, S., Gregory M. and Nicholson B. E. (Illustrations). (1977) *The Oxford Book of Wild Flowers*. Oxford University Press.

Aznar, E. (1983) *La Integración de las Islas Canarias en la Corona de Castilla (1478-1526)*. Universidad de Sevilla-Universidad de La Laguna, Secretariado de Publicaciones: Madrid.

Balter, M. (2007) Seeking agriculture's ancient roots. *Science* 316, 1830–1835.

Barakat, H. and Fahmy, A. G. (1999) Wild grass as 'Neolithic' food resources in the Eastern Sahara. In M. Van de Veen (ed.) *The exploitation of plant resources in Ancient Africa*. Kluwer Academic / Plenum Publishers, New York, pp. 33–46.

Barich, B. E. (1993) Culture and environment between the Sahara and the Nile in the Early Holocene and Mid-Holocene. In L. Krzyżaniak, M. Kobusiewicz, J. Alexander (eds.) *Environmental change and human culture in the Nile Basin and Northern Africa until the second millenium BC*. Polish Academy of sciences, Poznań archaeology museum, Poznań (Studies in African archaeology; 4), pp. 171–183.

Berhault, J. (1967) *Flore du Sénégal* (2nd edition). Clairafrique, Dakar.

Berhault, J. (1971–1980) *Flore illustrée du Sénégal* (27 volumes). Direction des Eaux et Forêts, Ministère du Développement Rural du Gouvernement du Sénégal, Dakar.

Behre, K. E. (2008) Collected seeds and fruits from herbs as prehistoric food. *Vegetation History and Archaeobotany* 17, 65–73.

Bengaly, M. D., Béré, A. B. and Traoré, A. S. (2007) Effects of alkali, pressure-cooking and fermentation on the phytate in roselle seeds during Bi-Kalga production: predicting zinc bioavailability. 3rd African food science and nutrition conference (9–11 April 2007; Ouagadougou). IPICS, Uppsala; OPCW, La Hague (Internet; http://www.fosnna.org/pb2007.pdf).

Bernus, E. (1967) Cueillette et exploitation des ressources spontanées du Sahel nigérien par les Kel Tamasheq. *Cahiers / ORSTOM: Série Sciences humaines* 4, 31–52.

Bernus, E. (1992–1993) Des arbres et des herbes aux marges du Sahara. *Sahara* 5, 17–28.

Bharucha, Z. and Pretty, J. (2010) The roles and values of wild foods in agricultural systems. *Philosophical Transactions of the Royal Society* 365, 2913–2926.

Bontier, P. and Le Verrier, J. (1980) *Le Canarien: Crónicas Francesas de la conquista de Canarias*. Aula de Cultura del Cabildo Insular de Tenerife: Santa Cruz de Tenerife.

Bouby, L. and Billaud, Y. (2005) Identifying prehistoric collected wild plants: A case study from Late Bronze Age settlements in the French Alps (Grésine, Bourget Lake, Savoie). *Economic Botany* 59 (3), 255–267.

Bourke, A. (1993) *The Visitation of God? The potato and the great Irish famine*. Lilliput Press, Dublin.

Bramwell, D. and Bramwell, Z. (2001) *Flores silvestres de las Islas Canarias*. Rueda: Madrid.

Bruce, P. (1953) *Pedaliaceae (Flora of East Africa)*. Crown Agents for the Colonies, London.

Bunting, A. H. (1960) Some reflections on the ecology of weeds. In J. L. Harper (ed.) *The Biology of Weeds*. Blackwell, Oxford, pp. 11–26.

Camps, G. (1969) *Amekni: Néolithique ancien du Hoggar*. Arts et métiers graphiques, Paris. (Mémoires du Centre de Recherches anthropologiques, préhistoriques et ethnographiques; 10).

Carmichael, A. (1928) *Carmina Gadelica* Vol. II, Oliver and Boyd, Edinburgh and London.

Carmichael, A. (1941) *Carmina Gadelica.* Vol. IV, Oliver and Boyd, Edinburgh and London.

Casas, A., Caballero, J., Mapes, C. and Zárate, S. (1997) Manejo de la vegetación, domesticación de plantas y origen de la agricultura en Mesoamérica. *Bol.Soc.Bot. México* 61, 31–47.

Casas, A., Del Carmen Vázquez, M., Viveros, J. L. and Caballero, J. (1996) Plant management among the Nahua and the Mixtec in the Balsas River Basin, Mexico: An ethnobotanical approach to the study of plant domestication. *Human Ecology* 24 (4), 455–478.

Castroviejo, S. (coord.) (1986–2012) *Flora iberica. Plantas vasculares de la Península Ibérica e Islas Baleares*, Vol. I–V, X–XV, XVII, XXI. Real Jardín Botánico-CSIC, Madrid.

Catani, M., Amaya, S. and Díaz-Aguilar, A. L. (2001) Comer en Tentudía: aproximación etnográfica a la comida y a los hábitos de vida de las gentes de la comarca de Tentudía en los últimos setenta años. Centro de Desarrollo Comarcal de Tentudía, Monesterio.

Cavaco, A. (1963) *Polygonacées, Chénopodiacées, Amaranthacées, Nyctaginacées, Phytolaccacées, Aizoacées, Portulacacées, Caryophyllacées (Flore du Gabon, volume 7).* Musée National d'Histoire Naturelle, Paris.

Chweya, J. A. and Eyzaguirre, P. B. (1999) *The biodiversity of traditional leafy vegetables.* International Plant Genetic Resources Institute (IPGRI), Rome.

Clark, J. D. (1976) Prehistoric populations and pressures favoring plant domestication. In J. R. Harlan, J. M .J. de Wet, A. B. L. Stemler (eds.) *Origins of African plant domestication.* Mouton, Paris, The Hague; Aldine, Chicago. (World anthropology), pp. 67–105.

Clarke, D. (1973) The Loss of Innocence. *Antiquity* 47, 100

Clarke, D. (1976) Mesolithic Europe: the economic basis. In G. Sieveking, I. Longworth and K. Wilson (eds.) *Problems in Economic and Social Archaeology.* Duckworth, 449–481.

Close, A. E. (1995) Few and far between: early ceramics in North Africa. In W. K. Barnett and J. W. Hoopes (eds.) *The emergence of pottery: technology and innovation in ancient societies.* Smithsonian Institution Press, London, pp. 23–38.

Cofradía Extremeña de Gastronomía (1985) *Recetario de cocina extremeña. Estudio de sus orígenes.* Universitas Editorial, Badajoz.

Cotton, C. M. (1996) *Ethnobotany.* John Wiley & Sons, London.

Criado, J., Fernández López, J. M., Leocadio, G., Núñez, R. M. and Blanco, E. (2008) *Uso tradicional de las plantas en Toledo.* Diputación de Toledo, Toledo.

Cruz-García, G. S. (2006) The mother – child nexus. Knowledge and valuation of wild food plants in Wayanad, Western Ghats, India. *Journal of Ethnobiology and Ethnomedicine* 2 (39)

Cummings, L. S. (1994) Anasazi Diet: Variety in the Hoy House and Lion House Coprolite Record and Nutritional Analysis. In K. D. Sobolik (ed.) *Paleonutrition: The Diet and Health of Prehistoric Americans.* Visiting Scholar Conference Volumes, Center for Archaeological Investigations, Southern Illinois University at Carbondale, Occasional Paper No. 22. pp. 134–150.

Cummings, L. S. (1998) Sampling Prehistoric Structures for Pollen and Starch Granules. In V. M. Bryant and J. H. Wrenn (eds.) *New Deveopments in Palynomorph Sampling, Extraction, and Analysis*, American Association of Stratigraphic Palynologists Foundation, Contributions Series Number 33, pp. 35–51.

Cummings, L. S. and Puseman, K. (1992) Pollen, Phytolith, Parasite, and Macrofloral Analysis of Coprolites from Room 21 in Step House (5MV1285), Ms. on file with National Park Service, Mesa Verde National Park, Mesa Verde, Colorado. PRI Technical Report 92–01.

Cunningham, A. B. (2000) *Applied Ethnobotany: People, wild plant use and conservation.* Earthscan, London.

Daniggelis, E. E. (2003) Women and 'wild' foods: nutrition and household security among Rai and Sherpa forager-farmers in eastern Nepal. In P. Howard (ed.) *Women and plants: Gender relations in biodiversity management and conservation.* Zed Press and Palgrave Macmillan, London and New York, pp. 83–98.

Davidson, A. (1999) *The Oxford Companion to Food.* Oxford University Press, Oxford.

Defour, D. L. and Wilson, W. M. (1994) Characteristics of 'Wild' Plant Foods used by Indigenous Populations in Amazonia. In N. Etkin (ed.) *Eating on the Wild Side.* University of Arizona Press, Tucson and London.

Delgado, T., Velasco, J., Arnay, M., Martín, E. and González E. (2005) Prevalence of caries among the prehispanic population from Gran Canaria. *American Journal of Physical Anthropology* 128, 560–568.

DIL. (1983) *Dictionary of the Irish Language.* Compact Edition. Royal Irish Academy: Dublin.

Dinneen, P. (1927) *Focloir Gaedhilge agus Bearla/Irish-English Dictionary.* The Educational Company of Ireland for The Irish Texts Society, Cork.

Dwelly, E. (1994) *The Illustrated Gaelic-English Dictionary.* Eleventh Edition by Gairm Publications, Glasgow.

Elmore, F. H. (1944) *Ethnobotany of the Navajo.* Monographs of the School of American Research, No. 8. Santa Fe, New Mexico.

Ertuğ, F. (2000) An ethnobotanical study in Central Anatolia (Turkey). *Economic Botany* 54 (2), 155–182.

Ertuğ, F. (2003a) An Ethnobotanical Research in Friday Markets of Bodrum (Mugla-Turkey). *Third International Congress of Ethnobotany.* DELPINOA 45.

Ertuğ, F. (2003b) Gendering the tradition of plant gathering in Central Anatolia (Turkey). In P. Howard (ed.) *Women and plants. Gender relations in biodiversity management and conservation.* Zed Press and Palgrave Macmillan, London and New York, pp. 183–196.

Espinosa, A. (1980) *Historia de nuestra Señora de Candelaria.* Goya: Santa Cruz de Tenerife.

Etkin, N. (ed.) (1994) *Eating on the wild side.* Arizona Series in Human Ecology The University of Arizona Press, Tucson and London.

Etkin, N. and Ross, P. J. (1982) Food as medicine and medicine as food. An adaptive framework for the interpretation of plant utilization among the Hausa of northern Nigeria. *Social Science and Medicine* 16 (17), 1559–1573.

Etkin, N. and Ross, P. (1994) Pharmacologic implications of 'wild' plants in Hausa diet. In N. Etkin (ed.) *Eating on the wild side.* The University of Arizona Press, Tucson and London, pp. 85–101.

Fajardo, J. (2008) *Estudio etnobiológico de los alimentos locales de la Serranía de Cuenca.* Doctoral Thesis. ETS de Ingenieros Agrónomos. Universidad de Castilla la Mancha.

Fernández Ocaña, A. M. (2000) *Estudio etnobotánico en el Parque Natural de las Sierras de Cazorla, Segura y Las Villas. Investigación química de un grupo de especies interesantes.* Ph.D. thesis. Universidad de Jaén. Facultad de Ciencias Experimentales.

Freiberger, C. E., Vanderjagt, D. J., Pastuszyn, A., Glew, R. S., Mounkaila, G., Millson, M. and Glew, R. H. (1998) Nutrient content of the edible leaves of seven wild plants from Niger. *Plant Foods for Human Nutrition* 53, 57–69.

Frutuoso, G. (1964) *Las Islas Canarias: Saudades da terra.* Fontes Rerum Canariarum XII. Instituto de Estudios Canarios: La Laguna.

Fuller, D. Q. (2004) Early kushite agriculture: archaeobotanical evidence from Kawa. *Sudan & Nubia* 8, 70–74.

Galt, A. H. and Galt, J. W. (1978) Peasant use of some wild plants on the Island of Pantelleria, Sicily. *Economic Botany* 32 (1), 20–26.

García Jiménez, R. (2007) *Etnobotánica Leonesa.* Municipio de Palacios del Sil. Doctoral Thesis. Universidad Complutense de Madrid.

Garine, I. de (2005) Nourriture de brousse chez les Muzey et les Masa du nord Cameroun. In, C. Raimond, E. de Garine, O. Langlois (eds.) *Ressources vivrières et choix alimentaires dans le bassin du Tchad.* Mega-Tchad (11; 2002; Paris). Editions du CNRS, IRD, Prodig, Paris, pp. 47–63.

Ghirardini, M., Carli, M., del Vecchio, N., Rovati, A., Cova, O., Valigi, F., Agnetti, G., Macconi, M., Adamo, D., Traina, M., Laudini, F., Marcheselli, I., Caruso, N., Gedda, T., Donati, F., Marzadro, A., Russi, P., Spaggiari, C., Bianco, M., Binda, R., Barattieri, E., Tognacci, A., Girardo, M., Vaschetti, L., Caprino, P., Sesti, E., Andreozzi, G., Coletto, E., Belzer, G. and Pieroni, A. (2007) The importance of a taste. A comparative study on wild food plant consumption in twenty-one local communities in Italy. *Journal of Ethnobiology and Ethnomedicine* 3, 22.

Gil, J. (2005) Los cultivos tradicionales de la Isla de Lanzarote. Los granos: Diversidad y Ecología. Cabildo Insular de Lanzarote: Arrecife.

Gil, J. and Peña, M. (2006) Contribución al Inventario de Especies y Variedades de Plantas cultivadas tradicionalmente en la isla de El Hierro. *Tenique* 7, 125–154.

Gil, J., Morales, J., Oliva, O., Delgado, N. T., Mederos, F., Oliva, D. and Alamo, I. (2008) *Especies y Variedades de Plantas Cultivadas en la Isla de Gran Canaria.* Asociación Insular de Desarrollo Rural de Gran Canaria: Las Palmas de Gran Canaria.

Gil, J., Peña, M. and Niz, R. (2009). *Usos culturales de las yerbas en los campos de Lanzarote. Bases orales para la reconstrucción del conocimiento etnobotánico tradicional.* Aderlan, Arrecife de Lanzarote.

Glew, R. H., Glew, R. S., Chuang, L. T., Huang, Y. S., Millson, M., Constans, D. and Vanderjagt, D. J. (2006) Amino acid, mineral and fatty acid content of pumpkin seeds (Cucurbita spp) and Cyperus esculentus nuts in the Republic of Niger. *Plant Foods for Human Nutrition* 61, 51–56.

González Turmo, I. (1997) Comida de rico, comida de pobre. Evolución de los hábitos alimentarios en el occidente andaluz (siglo XX). Universidad de Sevilla, Sevilla.

González, E. and Arnay, M. (1992) Ancient skeletal remains of the Canary Islands: Bone histology and chemical analysis. *Anthropologischer Anzeiger* 50, 201–215.

González-Insuasti, M. S. and Caballero, J. (2007) Managing plant resources: How intensive can it be? *Human Ecology* 35 (3), 303–314.

González, E., Velasco, J., Arnay, M., Santolaria, F. and Galindo, L. (2001) Paleonutritional analysis of the pre-Hispanic population from Fuerteventura (Canary Islands). *The Science of the Total Environment* 264, 215–220.

Grant, I. F. (1961) *Highland Folk Ways.* Birlinn, 1997 edition, first printed by Routledge and Kegan Paul, Edinburgh.

Grivetti, L. E. and Ogle, B. M. (2000) Value of traditional foods in meeting macro- and micronutrient needs: The wild plant connection. *Nutrition Research Reviews* 13 (1), 31–46.

Guijt, I. (1998) Valuing wild plants with economics and participatory methods: an overview of the Hidden Harvest methodology. In H. D. V. Prendergast, N. L. Etkin, D. R. Harris and P. J. Houghton (eds.) *Plants for food and medicine.* Royal Botanical Gardens Kew, Kew, pp. 223–235.

Guijt, I., Hinchdiffe, F., Melnyk, M., Bishop, J., Eaton, D., Pimbert, M., Pretty, J. and Scoones, I. (1995) *The hidden harvest. The value of wild resources in agricultural systems. A summary.* Sustainable Agriculture Programme of the International Institute for Environment and Development (IIED), London.

Haaland, R. (1995) Sedentism, cultivation, and plant

domestication in the Middle Nile Region. *Journal of Field Archaeology* 22(2), 157–174.

Haaland, R. and Magid, A. A. (1992) Radiocarbon dates from Mesolithic sites in the Atbara region, Sudan. *Nyame Akuma: a newsletter of African archaeology* 37, 17–27.

Hadjichambis, A. C., Paraskeva-Hadjichambi, D., Della, A., Giusti, M. E., De Pasquale, C., Lenzarini, C., Censorii, E., González-Tejero, M. R., Sanchez-Rojas, C. P., Ramiro-Gutiérrez, J. M., Skoula, M., Johnson, C., Sarpaki, A., Hmamouchi, M., Jorhi, S., El-Demerdash, M., El-Zayat, M. and Pieroni, A. (2008) Wild and semi-domesticated food plant consumption in seven circum-Mediterranean areas. *International Journal of Food Sciences and Nutrition* 59, 383–414.

Hanlidou, E., Karousou, R., Kleftoyanni, V. and Kokkini, S. (2004) The herbal market of Thessaloniki (N Greece) and its relation to the ethnobotanical tradition. *Journal of Ethnopharmacology* 91, 281–299.

Harlan, J. (1975) *Crops and man*. American Society of Agronomy and Crop Science Society of America, Madison, WI.

Harlan, J. R. and de Wet, J. M. J. (1965) Some thoughts about weeds. *Economic Botany* 19, 16–24.

Harrington, H. D. (1967) *Edible Native Plants of the Rocky Mountains*. University of New Mexico Press, Albuquerque.

Harris, D. R. (1989) The evolutionary continuum of people-plant interactions. In D. R. Harris and G. C. Hillman (ed.) *Foraging and farming: the evolution of plant exploitation*. Unwin Hyman, London.

Hastorf, C. (1988) The cultural life of early domestic plant use. *Antiquity* 72, 773–782.

Hather, J. G. (1993). *An Archaeobotanical Guide to Root and Tuber Identification*. Oxbow Monograph 28. Oxford, Oxbow Books.

Heath, D. and Meneley, A. (2007) Techne, Techoscience, and the circulations of comestible commodities: an introduction. *American Anthropologist* 109 (4), 593–602.

Heinrich, M., Müller, W. E. and Galli, C. (eds.) (2006) *Local Mediterranean Food Plants and Nutraceuticals*. Forum of Nutrition, edition I. Elmadfa. Vol. 59., Karger, Basel

Hernández, M. and Delgado, J. A. (eds.) (1998) *De Canaria y de las otras islas nuevamente halladas en el Océano allende España*. Colección a través del Tiempo, 16. Ayuntamiento de Puerto de la Cruz-Ayuntamiento de la Orotava: La Laguna.

Heywood, V. (1999) *Use and potential of wild plants in farm households*. FAO farm systems management series; 15 FAO, Rome.

Hildebrand, E. A. (2003) Motives and opportunities for domestication: an ethnoarchaeological study in southwest Ethiopia. *Journal of Anthropological Archaeology* 22 (2003), 358–375.

Howard, P. (2003a) *The major importance of 'minor' resources: women and the plant biodiversity*. Gatekeeper series 112, IIED, International Institute for Environment and Development, London, pp. 2–22.

Howard, P. (2003b) Women and the plant world: An exploration. In P. Howard (ed.) *Women and plants. Gender relations in biodiversity management and conservation*. Zed Press and Palgrave Macmillan, London and New York, pp. 1–48.

Howard, P. (ed.) (2003c) *Women and plants. Gender relations in biodiversity management and conservation*. Zed Press and Palgrave Macmillan, London and New York.

Hutchinson, J., Dalziel, J. M., Keay, R. W. J. and Hepper, F. N. (revisor) (1954–1972) *Flora of West Tropical Africa: the British West African territories, Liberia, the French and Portuguese territories south of latitude 18° N to Lake Chad, and Fernando Po* (10 volumes, 2nd edition). The Crown Agents for Oversea Governement and Administration, London.

Huysecom, E. (2006) Vers un 'Néolithique africain': concept et apparition. In Programme de recherche international: 'Peuplement humain et évolution paléoclimatique en Afrique de l'Ouest'. *Etudes maliennes* 65 (numéro spécial), 43–67.

Huysecom, E., Ozainne, S., Robion-Brunner, C., Mayor, A., Ballouche, A., Coulibaly, N., Guindo N., Keïta, D., Le Drezen, Y., Lespez, L., Neumann, K., Eichhorn, B., Rasse, M., Schaer K., Selleger, C., Serneels, V., Soriano, S., Terrier, A., Traoré, B. D. and Tribolo C. (2006) Peuplement humain et paléoenvironnement en Afrique de l'Ouest: apports de la huitième année de recherches interdisciplinaires. In *Jahresbericht 2005*. FSLA, Fondation Suisse-Liechtenstein pour les recherches archéologiques à l'étranger, Zürich, Vaduz, pp. 79–160.

Hveem, B., Berge, G., Ag Mahmoud, M., Maïga, A. and Diallo, D. (2005) Utilisation de la céréale Cenchrus biflorus par les Touaregs. In G. Berge, D. Diallo, B. Hveem (eds.) *Les plantes sauvages du Sahel malien: stratégies d'adaptation à la sécheresse des Sahéliens*. Karthala: Paris, pp. 213–230.

Jacks, B., Sall, M. and Petterson, A. (1995) Leptadenia hastata: ecology, use and nutritional value. *Journal d'agriculture traditionnelle et de botanique appliquée* 37, 2, 37–50.

Jackson, K. (1935) *Studies in Early Celtic Nature Poetry*. Cambridge University Press, Cambridge.

Jackson, K. H. (ed.) (1990) *Aislinge Meic Con Glinne*. (*The Vision of Mac Conglinne*) School of Celtic Studies, Dublin Institute for Advanced Studies, Dublin.

Jesse, F. (2003) Early ceramics in the Sahara and the Nile Valley. In: L. Krzyżaniak, K. Kroeper, M. Kobusiewicz (eds.) *Cultural markers in the Later Prehistory of North-eastern Africa and recent research*. Polish Academy of sciences, Poznań archaeology museum : Poznań (Studies in African archaeology; 8), pp. 35–50.

Johns, T. (1990) *With bitter herbs they shall eat it: Chemical ecology and the origins of human diet and medicine*. The University of Arizona Press, Tucson.

Johns, T. (2007) Agrobiodiversity, diet and human health. In D. I. Jarvis, C. Padoch and D. Cooper (ed.) *Managing Biodiversity in Agricultural Ecosystems.* Columbia University Press, New York.

Kalyoncu, V. (2002) Karadeniz Ereglisi'nde Üretici Kadınlar Pazarı (Gala Bazarı/ Gadınla Bazarı) (Women Markets at Karadeniz Eregli in Turkey). In K. Toygar (ed.) *Türk Mutfak Kültürü Üzerine Araştırmalar 2002.* Türk Halk Kültürünü Araştırma ve Tanıtma Vakfı, Ankara, pp. 93–108.

Kelly, F. (1998) *Early Irish Farming.* School of Celtic Studies and Dublin Institute for Advanced Studies, Dublin.

Khabir, M. (1987) New radiocarbon dates for Sarurab 2 and the age of the early Khartoum tradition. *Current Anthropology* 28 (3), 377–380.

Klee, M., Zach, B. and Neumann, K. (2000) Four thousand years of plant exploitation in the Chad Basin of northeast Nigeria, part I: The archaeobotany of Kursakata. *Vegetation History and Archaeobotany* 9, 223–237.

Klee, M., Zach, B. and Stika, P. (2004) Four thousand years of plant exploitation in the Lake Chad Basin (Nigeria), part III: plant impressions in potsherds from the Final Stone Age Gajiganna Culture. *Vegetation History and Archaeobotany* 13, 131–142.

Królik, H. and Schild, R. (2001) Site E-75-6: an El Nabta an Al Jerar village. In F. Wendorf, R. Schild (eds.) *Holocene settlement of the Egyptian Sahara.* Vol. 1: The Archaeology of Nabta Playa. Kluwer Academic / Plenum Publishers, New York, pp. 111–146.

Kubiak-Martens, L. (2002) New evidence for the use of root foods in pre-agrarian subsistence recovered from the late Mesolithic site at Halsskov, Denmark. *Vegetation History and Archaeobotany* 11, 23–31.

Kuhnlein, H. V. and Turner, N. J. (1991) *Traditional Plant Foods of Canadian Indigenous Peoples. Nutrition, Botany and Use.* Gordon and Breach Science Publishers: Philadelphia, Reading, Paris, Montreux, Tokyo and Melbourne.

Leach, H. M. (1997) The terminology of agricultural origins and food production systems – a horticultural perspective. *Antiquity* 71, 135–148.

Lewthwaite, J. G. (1982) Acorns for the ancestors: the prehistoric exploitation of woodlands in the west Mediterranean in M. Bell and S. Limbrey (eds.) *Archaeological Aspects of Woodland Ecology.* B.A.R. International Series 146, Oxford, pp. 217–230.

Lobo, M. (1988) *El comercio canario europeo bajo Felipe II.* Gobierno de Canarias, Gobierno da Madeira: Funchal.

Lobo, M., Santana, G. and Rodríguez, A. L. (2007) *Los usos de la madera: recursos forestales en Gran Canaria durante el siglo XVI.* Cabildo de Gran Canaria: Las Palmas de Gran Canaria.

Low, T. (1991) *Wild food plants of Australia.* Angus and Robertson, Sidney.

Lucas, A. T. (1960) Irish Food before The Potato. *Gwerin* 3, No. 2, pp. 8–43.

Łuczaj, L. (2008) Archival data on wild food plants used in Poland in 1948. *Journal of Ethnobiology and Ethnomedicine* 4, 4

Łuczaj, L. and Szymański, W. (2007) Wild vascular plants gathered for consumption in the Polish countryside: a review. *Journal of Ethnobiology and Ethnomedicine* 3, 17.

Machado, M. C., Arco, M. C., Vernet, J. L. and Ourcival, J. M. (1997) Man and vegetation in northern Tenerife (Canary Islands, Spain), during the prehispanic period based on charcoal analyses. *Vegetation History and Archaeobotany* 6, 187–195.

Maïga, A. (2005) Production et cueillette de grains des céréales sauvages dans le Gourma. In G. Berge, D. Diallo, B. Hveem (eds.) *Les plantes sauvages du Sahel malien: stratégies d'adaptation à la sécheresse des Sahéliens.* Karthala, Paris, pp. 197–211.

Malaza, M. (2003) Modernization and gender dynamics in the loss of agrobiodiversity in Swaziland's food system. In P. Howard (ed.) *Women and plants. Gender relations in biodiversity management and conservation.* Zed Press and Palgrave Macmillan, London and New York, pp. 243–257.

Marín, T. A. (1986) *Historia de las siete islas de Canarias.* Real Sociedad Económica de Amigos del País de Las Palmas de Gran Canaria: Madrid.

Martin, P. S. and Byers, W. (1965) Pollen and Archaeology at Wetherill Mesa. In Contributions of the Wetherill Mesa Archeological Project. *Memoirs of the Society for American Archaeology*, No. 19.

Martin, M. (1709) *A Description of the Western Isles of Scotland.* In J. Pinkerton *Voyages and Travels in All Parts of the World.* First Edition London: Andrew Ball, reprinted by Longman, Hurst, Rees and Orme, 1809, London, pp. 572–730.

Mesa, S. (1996) Estudio Etnobotánico y Agroecológico de la comarca de la Sierra de Mágina (Jaén). Doctoral Thesis. Universidad Complutense de Madrid.

Messer, E. (1972) Patterns of 'wild' plant consumption in Oaxaca, Mexico. *Ecology and Food Nutrition* 1, 325–332.

Meyer, K. (ed.) (1884) The Quatrains on the Seasons in *Hibernica Minora*, Appendix (Rawlinson B 512, p. xxx; fo. 98b2). Oxford, pp. 48–49.

Moll, M. (2005) *Les plantes a Menorca: noms i usos.* Institut Menorquí d' Estudis, Maó.

Morales, F. (ed.) (1993) *Canarias. Crónicas de su conquista.* Cabildo de Gran Canaria: Las Palmas de Gran Canaria.

Morales, J. (2010) El uso de las plantas en la prehistoria de Gran Canaria: alimentación, agricultura y ecología. Cabildo de Gran Canaria: Las Palmas de Gran Canaria.

Morales, J., Marrero, A. and Rodríguez, A. (2004) La socialización de nuevos espacios: transformación del medio y explotación de los recursos vegetales en el yacimiento de El Tendal, La Palma (Islas Canarias). *El Museo Canario* 59, 19–42.

Morales, J., Alberto, V. and Rodríguez, A. (2007) Intervención arqueológica en el yacimiento de

Belmaco (campaña de 2000). Nuevas aportaciones al estudio de macrorestos vegetales en la isla de La Palma. *Revista de Estudios Generales de la isla de La Palma* 3 (2), 135–160.

Morales, J., Rodríguez, A., Alberto, V., Machado, C., Criado, C. and Rando, J. C. (2009) The impact of human activities on the natural environment of the Canary Islands (Spain) during the pre-Hispanic stage (3rd–2nd BC to 15th AD): an overview. *Environmental Archaeology* 14, 27–36.

Moreno-Black, G. and Price, L. L. (1993) The marketing of gathered food as an economic strategy of women in northeastern Thailand. *Human Organization* 52 (4), 398–405.

Munson, P. J. (1976) Archaeological data on the origins of cultivation in the southwestern Sahara and its implications for West Africa. In J. R. Harlan, J. M. J. de Wet, A. B. L. Stemler (eds.) *Origins of African plant domestication.* Mouton, Paris, The Hague; Aldine, Chicago. (World anthropology), pp. 187–209.

Murray, S. S. (2004) Searching for the origins of African rice domestication. *Antiquity* 78, 300.

Neumann, K. (2003) The late emergence of agriculture in Sub-Saharan Africa: archaeological evidence and ecological considerations. In K. Neumann, A. Butler, S. Kahlheber (eds.) *Food, fuel and fields: progress in African archaeobotany.* Heinrich-Barth-Institut: Köln, pp. 71–92.

Neumann, K. (2005) The romance of farming: plant cultivation and domestication in Africa. In A.B. Stahl (ed.) *African archaeology: a critical introduction.* Blackwell, Malden, pp. 249–275.

Ní Chatháin, P. (1979) Swineherds, seers and druids in *Studia Celtica* 14, 200–211.

Nordeide, M. B., Holm, H. and Oshaug, A. (1994) Nutrient composition and protein quality of wild gathered foods from Mali. *International Journal of Food Sciences and Nutrition* 45, 275–286.

Núñez, I. (1994) *Conquista y antigüedades de las islas de la Gran Canaria y su descripción.* Universidad de Las Palmas de Gran Canaria: Las Palmas de Gran Canaria.

Ogle, B. M. (2001) *Wild vegetables and Micronutrient nutrition. Studies on the significance of wild vegetables in women's diets in Vietnam.* Acta Universitatis Upsaliensis. Comprehensive summaries of Upsala dissertations from the Faculty of Medicine 1056. Uppsala.

Ogle, B. M. and Grivetti, L. E. (1985) Legacy of the chameleon: edible wild plants in the Kingdom of Swaziland, Southern Africa. A cultural, ecological, nutritional study. *Ecology of Food and Nutrition* 17, 1–40.

Ogle, B. M., Ho Thi, Tuyet., Hoang Nghia, Duyet. and Nguyen Nhut Xuan, Dung. (2003) Food, feed or medicine: the multiple functions of edible wild plants in Vietnam. *Economic Botany* 57 (1), 103–117.

Oltra, J. E. (1998) *Fer herbes a Quatretonda.* Col·lectiu Cultural Dorresment, Quatretonda.

Pankhurst, R. J. and Mullen, J. M. (1991) *Flora of the Outer Hebrides.* 1994 HMSO Reprint of the 1991 edition by The Natural History Museum, London.

Parada, M. (2008) *Estudi etnobotànic de L'Alt Empordà.* Doctoral Thesis. Facultat de Farmàcia. Universitat de Barcelona.

Pardo-de-Santayana, M., Tardío, J., Blanco, E., Carvalho, A. M., Lastra, J. J., San Miguel, E. and Morales, R. (2007) Traditional knowledge on wild edible plants in the northwest of the Iberian Peninsula (Spain and Portugal): a comparative study. *Journal of Ethnobiology and Ethnomedicine* 3, 27.

Pardo de Santayana, M., Pieroni, A. and Puri, R. K. (2010) The ethnobotany of Europe, past and present. In M. Pardo de Santayana, A. Pieroni and R. K. Puri (eds.) *Ethnobotany in the new Europe. People, health and wild plant resources.* Studies in environmental anthropology and ethnobiology – vol. 14. Berghahn Books, New York, Oxford, pp. 1–15.

Pemberton, R. W. and Lee, N. S. (1996) Wild food plants in South Korea; market presence, new crops and exports to the United States. *Economic Botany* 50 (1), 57–70.

Perry, D. (1999). Vegetative tissues from Mesolithic sites in the northern Netherlands. *Current Anthropology* 40 (2), 231–237

Piera, H. (2006) *Plantas silvestres y setas comestibles en la comarca Valle Ayora-Cofrentes.* Mancomunidad Comarcal Valle de Ayora-Cofrentes, Ayora.

Pieroni, A. (2003) Wild food plants and Arberesh women in Lucania, southern Italy. In P. Howard (ed.) *Women and plants. Gender relations in biodiversity management and conservation.* Zed Press and Palgrave Macmillan, London and New York, pp. 66–82.

Pieroni, A. and Price, L. (eds.) (2005) *Eating and Healing, Traditional Food as Medicine.* Food Products Press, an imprint of The Haworth Press, Inc., New York.

Pieroni, A. and Quave, C. L. (2005) Functional foods or food medicines? On the consumption of wild plants among Albanians and Southern Italians in Lucania. In A. Pieroni and L. Price (ed.) *Eating and Healing, Traditional Food as Medicine.* Food Products Press, an imprint of The Haworth Press, Inc., New York, pp. 101–129.

Pieroni, A., Nebel, S., Santoro, R. F. and Heinrich, M. (2005) Food for two seasons: culinary uses of non-cultivated local vegetables and mushrooms in a south Italian village. *International Journal of Food Sciences and Nutrition* 56, 245–272.

Poilecot, P. (1995) *Les Poaceae de Côte d'Ivoire, Conservatoire et Jardin botaniques, Boissiera.* Mémoires des Conservatoire et Jardin botaniques de la Ville de Genève 50, UICN, CIRAD, Genève.

Poilecot, P. (1999) *Les Poaceae du Niger, Conservatoire et Jardin botaniques, Boissiera.* Mémoires des Conservatoire et Jardin botaniques de la Ville de Genève 56, UICN, CIRAD, Genève.

Portères, R. (1976) African cereals: Eleusine, Fonio, Black

Fonio, Teff, Brachiaria, Paspalum, Pennisetum and African Rice. In J. R. Harlan, J. M. J. De Wet and A. B. L. Stemler (eds.) *Origins of African plant domestication.* Mouton, Paris, The Hague (World anthropology), pp. 409–452.

Prance, G. and Nesbitt, M. (2005) *The Cultural History of Plants.* Routledge, New York, London.

Prendergast, H.D.V., Etkin, N., Harris, D.R. and Houghton, P.J. (eds.) (1996) *Plants for food and medicine. Proceedings of the joint conference of the Society for Economic Botany and the International Society for Ethnopharmacology.* Royal Botanical Gardens Kew, London.

Pretel, M. T., Pérez, V., Sánchez, M. and Obón, C. (2008) Contribución al conocimiento de las plantas comestibles silvestres de la provincia de Alicante. *Agrícola vergel* 314, 58–63.

Pretty, J. (2003) *Agri-Culture, Reconnecting People, Land and Nature.* Earthscan.

Price, L. L. (1997) Wild plant food in agricultural environments: A study of occurrence, management, and gathering rights in northeast Thailand. *Human Organization* 56 (2), 209–221.

Price, L. L. (2001) Women at the center: 'Wild' plant food use, gathering rights and management in disturbed farming environments. *Workshop 'Comparing and Contrasting Different Approaches on Uncultivated Plant Foods'.*

Price, L. (2005) Wild food plants in farming environments with special reference to northeast Thailand, food as functional and medicinal, and the social roles of women. In A. Pieroni and L. Price (eds.) *Eating and Healing, Traditional Food as Medicine.* Food Products Press, an imprint of The Haworth Press, Inc., NY, pp. 406.

Price, L. and Ogle, B. M. (2008) Gathered indigenous vegetables in Mainland Southeast Asia: A Gender Asset. In B. P. Resurreccion, B. P. Resurreccion and R. Elmhirst (eds.) *Gender and Natural Resource Management: Livelihoods, Mobility and Interventions.* Earthscan.

PROSEA Foundation (2006) *Plant Resources of South-East Asia.* http://proseanet.org/prosea/index.php

PROTA Foundation (2010) *Plant Resources of Tropical Africa* http://www.prota.co.ke/en/home

Putscher, J. and Vogl, C. R. (2006) An ethnobotanical survey on the herbal medicine in Quito markets. In F. Ertuğ (ed.) *Proceedings of the IVth International Congress of Ethnobotany.* Ege Yayınları, Istambul.

Redzic, S. (2006) Wild edible plants and their traditional use in the human nutrition in Bosnia-Herzegovina. *Ecology of Food and Nutrition* 45, 189–232.

Roset, J.-P. (1996) La céramique des débuts de l'Holocène au Niger nord-oriental: nouvelles datations, bilan des recherches. In G. Aumassip, J. Desmond Clark, F. Mori (eds.) *The prehistory of Africa.* Section 15. XIIIth International Congress of Prehistoric and Protohistoric Sciences (8–14 sept. 1996; Forlì). ABACO edizioni, Forlì, pp. 175–182.

Roset, J.-P. (2000) Céramique et néolithisation en Afrique saharienne. In J. Guilaine (ed.) *Premiers paysans du monde: naissances des agricultures.* Séminaire du Collège de France (1998–1999 ; Paris). Errance, Paris. (Collection des Hespérides), pp. 263–290.

Rowley-Conwy, P. A., Deakin, W. J. and Shaw, C. H. (1997) Ancient DNA from archaeological sorgum (Sorgum bicolor) from Qasr Ibrim, Nubia: implications for domestication and evolution and a review of the archaeological evidence. *Sahara* 9, 23–34.

Rowley-Conwy, P. A., Deakin, W. J. and Shaw, C. H. (1999) Ancient DNA from Sorghum: the evidence from Qasr Ibrim, Nubia. In M. Van de Veen (ed.) *The exploitation of plant resources in Ancient Africa.* Kluwer Academic / Plenum Publishers, New York, pp. 55–61.

Salaman, R. N. (1985) *The History and Social Influence of the Potato.* Cambridge University Press.

Salih, O. M., Nour, A. M. and Harper, D. B. (1992) Nutritional quality of uncultivated cereal grains utilised as famine foods in Western Sudan as measured by chemical analysis. *Journal of the Sciences of Food and Agriculture* 58, 417–424.

Sánchez López, M. D., García Sanz, J. A., Gómez Merino, A. and Zon Blanco, S. (1994) *Plantas útiles de la comarca de la Manchuela.* Colectivo de Escuelas Rurales de la Manchuela, Albacete.

Santana, A. (2001) *Evolución del paisaje de Gran Canaria (siglos XV-XIX).* Cabildo Insular de Gran Canaria: Madrid.

Scoones, I., Melnyk, M. and Pretty, J. N. (1992) The hidden harvest: wild foods and agricultural systems: a literature review and annotated bibliography, compiled by Ian Scoones, Mary Melnyk, Jules N. Pretty. International Institute for Environment and Development, London.

Scott, L. J. (1976) Hoy House – A Palynological Study. Appendix II in Johnson-Lion Canyon Project Report of Investigation III, Assembled by P. R. Nickens, Mesa Verde Research Center, University of Colorado, Boulder, pp. 8–49.

Sherratt, A. (1981) Plough and pastoralism: aspects of the secondary products revolution. In I. Hodder, G. Isaac and N. Hammond (eds.) *Pattern of the Past: Studies in honour of David Clarke,* Cambridge University Press, 261–305.

Somnasang, P. and Moreno-Black, G. (2000) Knowing, gathering and eating: Knowledge and attitudes about wild food in an Asian village in northeastern Thailand. *Journal of Ethnobiology* 20 (2), 197–216.

Standley, P. C. (1912) Some Useful Native Plants of New Mexico. *Annual Report of the Smithsonian Institution for 1911,* pp. 447–462.

Stiger, M. A. (1976) The Application of Microtechnique to the Study of Hoy House Coprolites. Appendix III in Johnson-Lion Canyon Project Report of Investigation III, Assembled by P. R. Nickens, Mesa Verde Research Center, University of Colorado, Boulder, pp. 50–60.

Stueber, K. (copyright) (2007) *Plants for a Future (PFAF)* http://www.pfaf.org/database/plants.php?Potentilla+anserina

Tardío, J. (2010) Spring is coming: the gathering and consumption of wild vegetables in Spain. In M. Pardo-de-Santayana, A. Pieroni and R. Puri (eds.) *Ethnobotany in the New Europe: people, health and wild plant resources.* Berghahn Books, Oxford-New York, pp. 211–238.

Tardío, J. and Pardo-de-Santayana, M. (2008) Cultural importance indices: a comparative analysis based on the useful wild plants of southern Cantabria (northern Spain). *Economic Botany* 62, 24–39.

Tardío, J., Pascual, H. and Morales, R. (2002) Alimentos silvestres de Madrid. Guía de plantas y setas de uso alimentario tradicional en la Comunidad de Madrid. Ediciones La Librería, Madrid.

Tardío, J., Pardo-de-Santayana, M. and Morales, R. (2006) Ethnobotanical review of wild edible plants in Spain. *Botanical Journal of the Linnean Society* 152, 27–72.

Tchago, B. and Moupeng, B. (2002) *Les produits alimentaires en période de soudure dans la zone sahélo-soudanienne: cas des régions du lac Fitri et des lacs de Léré (Tchad).* Mega-Tchad (11; 2002; Paris). (Internet; http://www.uni-bayreuth.de/afrikanistik/mega-tchad).

Tejera, A. and González, R. (1987) *Las culturas aborígenes canarias.* Interinsular Canaria: Santa Cruz de Tenerife.

Torriani, L. (1978) *Descripción de las Islas Canarias.* Goya: Santa Cruz de Tenerife.

Toury, J., Lunven, P. and Giorgi, R. (1961) Aliments de cueillette et de complément au Sénégal et zone sahélienne. *Plant Foods for Human Nutrition* 8(2), 139–156.

Turner, N. J. (2003) 'Passing on the news': Women's work, traditional knowledge and plant resource management in indigenous societies of north-western North America. In P. Howard (ed.) *Women and plants. Gender relations in biodiversity management and conservation.* Zed Books, London & New York, pp. 101–114.

Turner, N. J. and Davis, A. (1993) When everything was scarce: the role of plants as famine foods in northwestern North America. *Journal of Ethnobiology*, 13(2), pp. 1–28.

Turner, N. J. and Jones, J. T. (2000) Occupying the Land: Traditional Patterns of Land and Resource Ownership among First Peoples of British Columbia on CD ROM Proceedings, IASCP 2000 (Common Property Resources) Conference, Bloomington, Indiana.

Turner, N. J. and Kuhnlein, H. V. (1982) Two Important 'Root' Foods of the Northwest Coast Indians: Springbank Clover (*Trifolium wormskioldii*) and Pacific Silverweed (*Potentilla anserina* ssp. *pacifica*). *Economic Botany*, 36(4), 411–432.

Tutin, T. G., Heywood, V. H., Burges, D. M., Moore, D. H., Valentine, S. M., Walters, S. M. and Webb, D. A. (1964–1980) *Flora Europaea*, Vol. 1–5. The University Press, Cambridge and London.

Vainio-Mattila, K. (2000) Wild vegetables used by the Sambaa in the Usambara Mountains, NE Tanzania. *Annales Botanici Fennici* 37 (1), 57–67.

Vandebroek, I. and Sanca, S. (2007) Food medicines in the Bolivian Andes (Apillapampa, Cochabamba Department). In A. Pieroni and L. L.Price (eds.) *Eating and Healing: Traditional Food as Medicine.* Food Products Press, an imprint of The Haworth Press, Inc., NY, pp. 273–295.

Velasco, J., González, E., Arnay, M., Barros, N., Martín, E. and Santolaria, F. (1999) Bone histology of prehistoric inhabitants of the Canary Islands: comparison between El Hierro and Gran Canaria. *American Journal of Physical Anthropology* 110, 201–214.

Velasco, J. M., Criado, J. and Blanco, E. (2010) *Usos tradicionales de las plantas en la provincia de Salamanca.* Diputación de Salamanca, Salamanca.

Vendryes, J. (1960) *Lexique étymologique de l'irlandais ancien. Lettre M.* Dublin Institute for Advanced Studies and Centre National de la Recherche Scientifique: Dublin and Paris.

Vendryes, J., Bachellery, E. and Lambert, P. Y. (1981) *Lexique étymologique de l'irlandais ancien. Lettre B.* Dublin Institute for Advanced Studies and Centre National de la Recherche Scientifique: Dublin and Paris.

Verde, A., Rivera, D. and Obón, C. (1998) *Etnobotánica en la sierras de Segura y Alcaraz: las plantas y el hombre.* Instituto de Estudios Albacetenses, Albacete.

Viana, A. (1991) *Antigüedades de las Islas Afortunadas.* Gobierno de Canarias: Santa Cruz de Tenerife.

Volpato, G. and Godinez, D. (2007) Medicinal foods in Cuba: Promoting health in the household. In A. Pieroni and L. L. Price (eds.) *Eating and healing: traditional food as medicine.* Food Products Press, an imprint of The Haworth Press, Inc., New York, pp. 213–236.

Wasylikowa, K. (2001) Site E-75-6: vegetation and subsistence of the Early Neolithic at Nabta Playa, Egypt, reconstructed from charred plant remains. In F. Wendorf and R. Schild, (eds.) *Holocene settlement of the Egyptian Sahara.* Vol. 1: The Archaeology of Nabta Playa. Kluwer Academic/Plenum Publishers, New York, pp. 544–608.

Wasylikowa, K. and Dahlberg, J. (1999) Sorghum in the economy of the early Neolithic nomadic tribes at Nabta Playa, Southern Egypt. In M. Van der Veen (ed.) *The exploitation of plant resources in ancient Africa.* Kluwer Academic, Plenum Publishers, New York, pp. 11–31.

Wendorf, F., Close, A. E., Schild, R., Wasylikowa, K., Housley, R. A, Harlan, J. R. and Królik, H. (1992) Saharan exploitation of plants 8'000 years BP. *Nature* 359, 721–724.

Whiting, A. F. (1939) *Ethnobotany of the Hopi.* Museum of Northern Arizona Bulletin No. 15.

Wiersum, K. F. (1996) Indigenous exploitation and management of tropical forest resources: an evolutionary continuum in forest-people interactions. *Agriculture, Ecosystems and Environment* 63, 1–16.

Willcox, G. (1991) Le début de l'agriculture au Proche Orient. In M.-C. Cauvin (ed.) *Rites et rythmes agraires.* Séminaire de recherche. Lyon: GDR-Maison de l'Orient; Paris, De Boccard. (Travaux de la Maison de l'Orient méditerranéen; 20), pp. 9–29.

Wrangham, R. (2009) *Catching Fire. How Cooking Made Us Human.* Basic Books, New York.

Zach, B. and Klee M. (2003) Four thousand years of plant exploitation in the Chad Basin of NE Nigeria, part II: discussion on the morphology of domesticated Pennisetum and complete catalogue of the fruits and seeds of Kursakata. *Vegetation History and Archaeobotany* 12, 187–204.

SECTION 3

Food and Beyond

6 A Versatile World: Examples of Diversity in Plant Use

6.1. INTRODUCTION

Cozette Griffin-Kremer

General Perspectives on Diversity in Plant Use

Thinking first of all in very general terms, plants have a tendency, in people's minds, to remain rather discreetly in the background, unless plant life has been struck by some impressive disaster, such as drought, severe storms or intentional eradication and, hence, the absence of certain plants is felt to have an emotional, aesthetic or economic impact. Of course, for some people, they are very present much of the time and there is a whole spectrum of possible ways for humans to conceive of their relations to plants.

The first uses of plants that would occur too many of us would be as foods, but we will move 'beyond food' in this chapter. Still, it is well to recall how essential to our lives the nourishment of our domestic livestock is (and, at times, even of animals in the wild), in the form of feed or fodder. Plants provide bait and implements for hunting, fishing and trapping – the world's finest fishing poles are of bamboo in longitudinal wedging. Of course, they are a major source of vegetable oils for cooking or conserving, but also for lighting. As a 'support' to food, plants are often a prime source of combustible material for cooking and are used to make cooking or keeping vessels. They are a component in many processing techniques, such as the flexible mesh 'snakes' used to squeeze the toxic elements out of cassava (Bataille-Benguigui and Cousin 1996, pl. 38; Agropolis Museum) or as stirring sticks, sieves, whisks and the supports on which food is dried or frozen. At times, in cooking processes, it is difficult to discern whether plants add flavour, nutritional value, humidify, insulate, contain – or do all of this at one and the same time as a 'package' (Bataille-Benguigui *et al.* 2003).

A second thought about the role of plants might be as medicinals to treat a variety of ailments, but even the least exhaustive of lists can evoke a plethora of other ways they are utilised which do not strike us as strictly 'medicinal'. It is standard practice to employ them as antidotes to plant, animal or mineral poisons, to keep away insects, for ridding oneself of vermin and parasites, as preservatives and other birth control measures, including abortifacients. For example, the sale of rue is still highly regulated in some countries such as France. In some cultures they, or products extracted from them, have been used as poisons to kill or stun animals or people, but also as anaesthetics, hunger suppressants, thirst quenchers, bandages, dressings and poultices – mustard poultices were the bane of our grandparents' existence when they were children. They also serve as cosmetics, mouth fresheners, toilet soap, shampoo, comb, brush, toothbrush and toothpaste, toothpicks, deodorants, toilet paper, wash cloth and disposable diapers, as well as other forms of body care.

Our categories of medicinals, foods and aromatics overlap; in fact, as we can see in the vast family of plant-based fermented or alcoholic beverages (McGovern 2009), teas and coffees, the thickening agents for cosmetics, beverages, soups and stews, when plants are used as flavourings for other foods but not ingested, or are burned or otherwise processed to produce scents. This last utilisation spills over into ceremonial or festive practices, which are dealt with more fully in Chapter 8, but are important in both traditional and industrialised societies – flowering plants alone are among the world's most lucrative commodities and closely associated with both calendar and life-cycle events such as weddings and funerals. Here, we enter the realm of social (or asocial) and symbolic uses – as stimulants or narcotics of various strengths, ingested orally, in liquid or powdered form or by inhalation. They may be intended to produce a state of trance or vision, as aromatics burned like incense, in which instance they are often thought to enhance communication with the dead or the divine and are essential to religious or ritual expression (Goodman *et al.* 1995). Plants and their products are valuable items of exchange, offerings to friends and allies, to divinities or otherworld inhabitants, as well as being used as adornment for the human person, for animals, buildings, tombs and streets.

In a more material vein, plants and their products are used in craftwork as stains and paints, glues and tanning agents, as well as cleaning, scouring or polishing agents. They furnish us with many of the components of clothing and other apparel, dyestuffs and mordants (Cardon 2007), textiles beyond clothing, twine and rope, with their important extension to footwear, saddles, harness, bridges, water transport, boat-building and the construction of shelters and dwellings. In the latter, we use plants as insulation, but also as fuel to keep warm and as effective screening to protect ourselves from a whole gamut of pests. In the context of homely comforts, plant parts can be utilised as wrapping or packaging, containers and tableware, as furniture materials including stuffing for mattresses and pillows, and once had an important role as renewable carpeting in the form of 'strewings' (Fletcher 1998). Of course, this was extended to bedding for animals and the utilisation of plants to catch and hold manure, as well as an essential part of composting for gardens and even fields, a subject dealt with in this chapter.

Broader agricultural uses are legion: ash from spent plant fuels, such as furze from bakers' ovens, was a prized commodity and widely traded upon. One plant can serve as a support for another's growth or to protect it from inclement weather, winds or pests – witness the intertwined fate of maize and beans in the Americas. Aphids on roses are the red light for vineyard-owners to treat their grapevines, so one plant can be the official guardian of another. We even marry plants, in grafting, where one provides the nourishing 'house' and the other brings its own qualities, almost as a dowry. More prosaic uses in horticulture and agriculture spring to mind, as well as the many uses of plant parts (fruit stones, coffee bean shells) in the present-day gardening sector.

In a step beyond use of a plant to its representation, depictions of plants in crafts, art and architecture often call upon rich and polyvalent symbolisms (one need but think of the Green Man in Christian churches or vegetal motifs in other spiritual edifices, garlands on coffins and sarcophagi, motifs in textiles and wearing apparel, among many others). The visual impact of plants alone is well documented and they create 'ambience', even what we have come to understand as 'environment' to a great extent (from restaurants, hotels, homes, on to housing developments, parklands and even 'wilderness'). Coming perhaps full circle on the question of 'uses' of plants, we might note that they definitely, if discreetly, provide a kind of company to human beings and stimulate or enrich a sense of human 'community', as has been widely confirmed by their use in offices, schools, hospitals and nursing homes. People in tomorrow's mega-cities may even depend on plants to produce the oxygen they need in large building complexes, so our ties with plants might become ever more vital to our own survival in unexpected ways. Our diverse uses of plants have affected them deeply in many cases and it is clear that our destinies and development are intimately interwoven in a tapestry of many hues (Wrangham 2009).

Synergies

The discussion undertaken in this chapter touches on periods running from 10,000 BP to contemporary practices from the British Isles across Mediterranean Europe into the Near and Far East. Within each article, various ecosystems are evoked or attentively dealt with, but perhaps the most fruitful aspect of the chapter is the synergy of methodologies, approaches and sources provided by archaeologists or archaeobotanists, ethnography and historical research. The complementarity of these methods most often highlights the notion that the present-day marginality in use of a plant may mislead us about its former importance. This is clear from an examination of furze and nettles by C. Griffin-Kremer (Chapter 6.2) who emphasises first the multiplicity of uses of furze (*Ulex* spp.) in both rural and urban contexts, in part due to its structural characteristics. The plant provided an important fuel, underwrote a broad spectrum of livestock production and its wood was crafted into domestic artefacts. It was also associated with popular traditions of striking diversity and was a clear marker of land values. In many respects, the nettle (*Urtica* spp.) was recognised for the same reasons, being valued as much as a medicinal, as fodder, and as a textile fibre. Like furze, it had associations with ritual and, beyond that, with saintliness.

A recurrent term running through all these discussions of diversity is 'by-product'. True to the etymological implications of the term (Kluge 1999, 92), by-products are made 'around' or 'near' one another when processing a plant, producing material that we sometimes call 'waste' or 'residue'. However, human beings have proven over the millennia that they do *not* 'waste' these in any conceivable way. On the contrary, the ingenuity with which everything is used is comparable to the processing of a pig – 'everything is used, but the squeal'. There is much here to encourage us to pay attention to the notion of collateral assets, as brought out clearly by L. Peña-Chocarro and L. Zapata (Chapter 6.3) in their survey of hulled wheats in Mediterranean mountain areas and their broader archaeological and historical attestations. Einkorn, emmer and spelt (*Triticum monococcum, T. dicoccum* and *T. spelta*) are highly resistant to disease, well adapted to cultivation in mountain soils and fit into an intriguing biodiversity portfolio of rural subsistence strategies. They provide both food and fodder, as well as a rich collection of domestic uses in patterns that vary subtly over time and can be perceived through fruitful confrontation of archaeological evidence with that from other disciplines. At times, what we perceive as a by-product can be determinant in the overall value of crops. Emmanuelle Bonnaire (Chapter 6.4) takes the example of vegetal material produced during threshing of various cereals that has been utilised in tempering earthen structures and thus emphasises a particular aspect of palaeoenvironmental research that helps elucidate the entire sequence of cereal processing. Her work necessitated setting up a reference collection of distinctive criteria in order to identify taxa in situations without preservation of charred, waterlogged or mineralised remains. This has involved elaborating the methodology for sample choice, observation and description of imprints, and identification of elements, using a reference collection from modern cereals. Her results can complement those from seed analysis and also provide indirect information about building techniques and the overall plant economy of a site.

The present-day uses of a grass confirm their value as testimony to past practices, as Patricia Anderson (Chapter 6.5) shows in a review of several years of fieldwork in northwest Tunisia. Her interest in, and encouragement of, an older man well-known for basketry work with the wild grass called *dis* or *diss* (*Ampelodesmos mauritanica*), also used as fodder, took a very positive turn in the promotion and appreciation of a traditional craft artefact. Made with tools well attested from the Neolithic on, these carrying recipients have proved more resilient than their present-day plastic replacements. At a time when motorised transport has taken a down-turn in the face of fuel scarcity, both animal transport and the efficient containers it requires are seen in a new light. Still in the Mediterranean sphere, Bui Thi Mai, Michel Girard and François de Lanfranchi (Chapter 6.6) trace possible ancient uses of the mastic tree (*Pistacia lentiscus L.*), most especially the oil it furnishes, in comparison with recent ethnological inquiries. Experimentation was carried out in a Sardinian community still familiar with the know-how of traditional production practices, an approach that greatly enhances any evaluation of archaeological remains of *Pistacia*. In the final contribution, we move to East Asia, where Bui Thi Mai and Michel Girard (Chapter 6.7) launch onto the trail of the various uses of

Dipterocarpaceae (gurjan family) in Vietnam, most particularly the series of plants that produce an oleoresin suitable for processing into boat caulking. They observed present-day practices in their ethnographic inquiries and discuss the pertinent literature about modern round-boats; as well as the remains and techniques implied by analysis of the late fifteenth-century Brunei wreck. They detail the entire process from tapping to the final product used in caulking, along with the use of other vegetal fibres and animal products, demonstrating that ethnographic inquiry and palynological analysis can dovetail in the same research project.

There are many threads linking these articles and the reader will discover such notions as the cross-over of food and fodder, the significance of fermented products or the highly 'structural' virtues of different products that, literally, mean they hold things together. There are also reminders that, between the 'standing' and the 'cut' crop, it may, at times, be the standing one that most valorises the plant and insures its continued cultivation or care. Whether dealt with explicitly or implicitly, there is much in these analyses to stimulate our discussions of terminology and our conceptions of the spectrum of possible relations between humans and plants. The examples given can join the extensive work already available that nourishes the debate over how we choose and use terms, even terms such as crop or agriculture (see, for example: Brookfield 2007; Brush 2004; Deur and Turner 2005; Estabrook 2006; Hildebrand 1993; Leach 1997; Vázquez-García 2008; Zapata *et al.* 2004).

The authors here also point out some very basic opportunities we must seize, such as the spur of ethnological inquiry and presently-known uses for the careful construction of reference collections. They examine the wealth in variations of patterns of plant use in one area and over time. They speak of risk-buffering and the intricate interweaving of agrarian production and animal management, at times under the shadow of legal restrictions that impose one production and exclude another. There are many examples of replacement and loss, but also of revival, of the wealth of skills and the simplicity of tools. The contributors observe how underutilised and non-commodity products may, suddenly, capture hearts and please the pocketbook or take their place in a scheme of cultural heritage management. Implicit in all of the articles is the conviction that past practices may very well not be taken up again, but that they have much to teach us about the resilience of farming communities, their skills and depth of experience. Above all, these contributions converge to remind us that such practices, and the plants involved, long sustained us, that they had their place in complex and often carefully planned production systems, and that they provided a bulwark against a certain 'loss of sovereignty' (Vázquez-García 2008, 75), which happens when we replace 'artefacts' with 'goods' (Glassie 1999, 77).

6.2. 'HUMBLE PLANTS': USES OF FURZE AND NETTLES IN THE BRITISH ISLES (AND BEYOND)

Cozette Griffin-Kremer

Usefulness can be a rather capricious concept. What was considered highly useful yesterday may no longer find favour today. The positive side of this is the pleasure of finding an often rich heritage of plant lore in ethnographic reports, literature and ancient law, as well as popular wisdom in the form of proverbs, rhymes, songs and fairy tales – sources that are not always sufficiently taken into account. This is definitely the case for common furze (*Ulex europaeus* L., Fig. 6.1) and nettle (*Urtica dioica* L., Fig. 6.2); today often qualified as noxious invasives. Recalling their uses can contribute much to constructing a balanced perspective on notions of diversity. Both have been widely utilised well beyond the European scene, but pride of place here will be given to sources from the British Isles (Fig. 6.3).

Fig. 6.1. Common furze or gorse (*Ulex europaeus*), here as an invasive plant in North America. Image: Forest & Kim Starr, U.S. Geological Survey, Bugwood.org. Inset: drawing from Thomé (1885, plate XVI,3, modified).

Furze

From Fencing to Fuel

A. T. Lucas (1958, 144) comments that: 'cut furze served the purposes today served by fencing wire, barbed wire, netting wire, chain-link fencing and the electric fence' and this was most especially pertinent in the case of Ireland, where observers were often struck by the role of temporary fencing and even non-enclosure of domestic stock (Evans 1956; 1957, 100–118, 201). As temporary fencing, furze could be qualified as a dead hedge, but it was equally important as a live one (a requirement in steeplechase rules even today). From the time of the earliest Norman occupation, it was sown, like the hawthorn, to provide coverts for foxes and rabbits (Humphries and Shaughnessy 1987, 20).

Fig. 6.2. The stinging nettle (*Urtica dioica*), photographed in the backyard of the University of Natural Resources and Life Sciences (BOKU), Vienna. The wind-pollinated (anemophilous) flowers are unisexual and found on separate plants, thus resulting in female (left) and male (right) individuals. Image: A. G. Heiss.

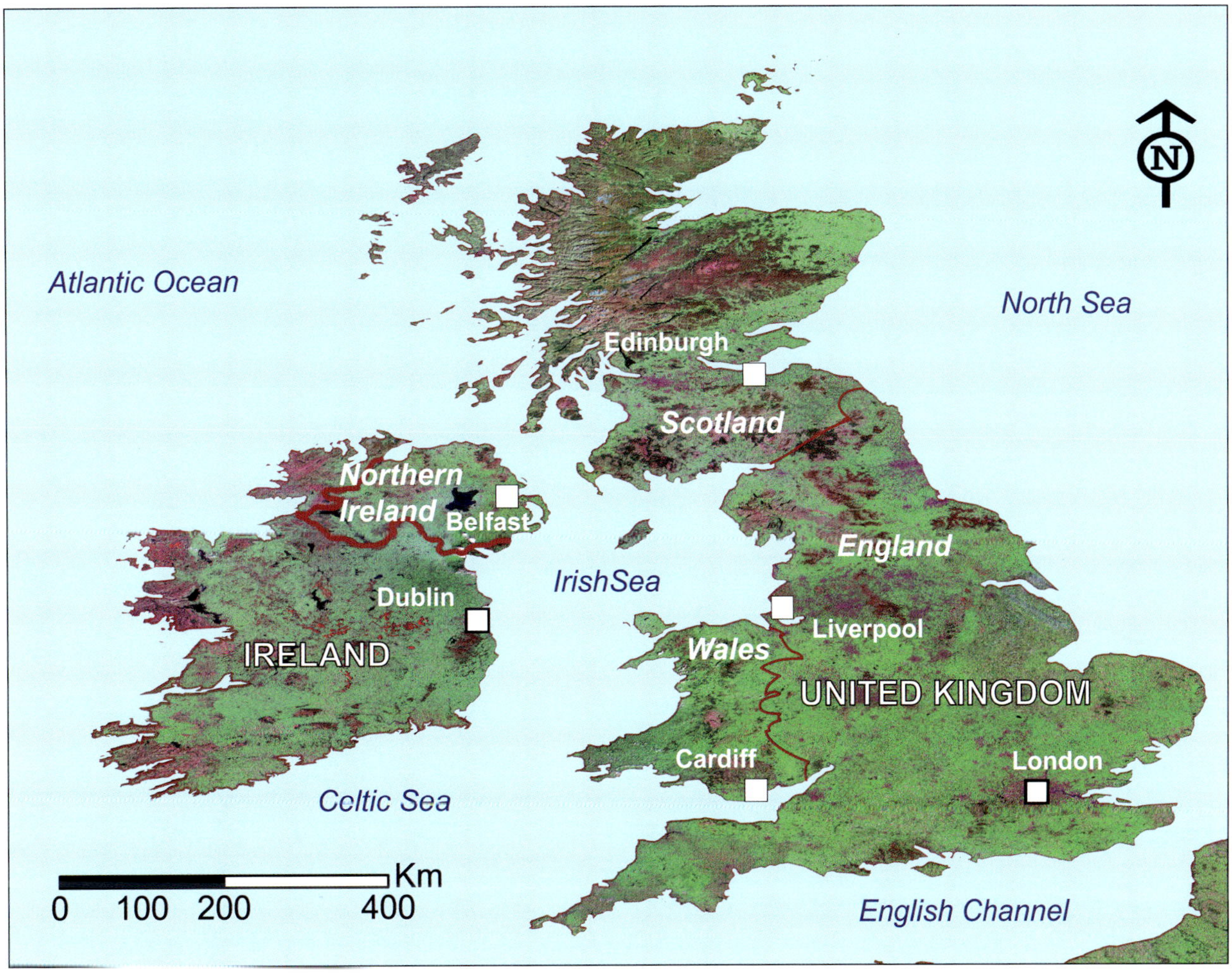

Fig. 6.3. Map of the British Isles (United Kingdom and Ireland). Map: R. Lugon, J.-C. Loubier and A. Chevalier.

Furze also figured as a component in constructions. Its properties of stiffness and prickliness made it an excellent stop-gap in hedges and walls. It was also utilised as wattle in doors and the walls of duck houses, as under-thatch for the roof and as bonding in plaster for the inside and outside coating of houses. Furze could 'construct' what we would ordinarily term the landscape: a furze bush was often planted near the house to dry clothes on after washing – a role shared with the hawthorn – and this protected the laundry from intrusive animals as well.

It was sown with saplings, to guard the young plants against both wind and herbivores, planted with peas to protect them from rodents or cut and laid on top of seed beds to keep birds and rats off. It protected hay ricks and corn stacks from rodents, birds or cattle and attached to ships' ropes to stop rats from boarding. Furze could just as well protect animals, as when it was set out in 'curtains' in brooding boxes to prevent chicks from smothering one another (Lucas 1958, 133–144).

Moving to what is 'under' the landscape; furze was an excellent soil stabiliser and served as fill for reclamation of land. One of its prime roles was in the construction of field drains in a layered plan or allied with stone courses, eventually rotting to leave a hollow drain tunnel (Lucas 1958, 150–151). The same qualities made it an outstanding material for pitting potatoes, turnips and apples for field storage: 'When they pit apples which they have not room to store otherwise they put furze about 18 inches deep in the bottom of a shallow trench and then the apples, piled up like potatoes in a pit, then furze up the sides and top before covering with clay. The apples stored this way do not get the smell of clay as happens when straw is used and they keep fresh and sweet' (Lucas 1958, 146–147).

Furze' role was not limited to rural life. The wood itself was highly prized as a material for domestic articles, running from hurley and walking sticks to umbrella and hammer handles, as pieces for straddles and panniers, as the thole pins for currachs (Irish river boats) and as ribs in lobster pots (Lucas 1958, 165–166). It was an essential part of soil processing for field crops; being used as a 'brush' or harrow for covering seed, to aerate grassland, to spread top dressing (manure) on fields and even to brush the hoarfrost off potato plants. The plant provided a fine 'bedding' material in every sense, for both people and animals, and was also strewn outside to collect manure for dunging (a practice attested even more recently in Portugal for broom, the *estrume*, or composted dung in *giesta*, two *Cytisus* species, Estabrook 2006, 310; Lucas 1958, 124–128, 155).

Furze wood makes fine fuel in limekilns and bakers' ovens, as well as to melt tar for ships. As such, it was a true commodity; subject to intense speculation in the towns and widely traded as a cash crop gathered by the poorest country folk. As a consequence, it was also the object of Church tithes. We know that its function as fuel for bread took on enormous proportions in Ireland from the Norman period on because it is often mentioned as a fire-hazard (Humphries and Shaughnessy 1987, 18; Lucas 1958, 24–25, 55–65). It seems that furze charcoals can be discriminated from other Fabaceae, so this line of investigation might be pursued further in archaeological contexts (Bouby *pers. comm.* 2006, citing Marguerie 1991, 57–58).

Nourishment, Enhancement and Beyond

Furze consumption in bakers' ovens produced a significant by-product – ash – that was sought after as a soil improver and fertiliser, usually placed as a first layer in lazybeds just under the soil to improve the texture and taste of potatoes. There are indications that it was systematically planted in some areas of Ireland, creating 'furze brakes' made by cutting sod rolls to burn and spreading the ash with considerable expertise (Lucas 1958, 38). There may well have been other reasons for planting furze brakes, since they would have been effective nitrogen-fixers, comparable to the broom brakes or *giestals* reported on for the Portuguese Beira Alta area (Estabrook 2006, 307–319).

Furze was given as fodder to rabbits, horses and cattle and was said to give a special lustre to animals' coats in sources all over the British Isles (Humphries and Shaughnessy 1987, 20, 18; Harris 1991, 21). It was added to wine and whiskey as a flavouring, to butter to lend it colour and utilised widely for yellow and green textile dyes. The plant was administered as a medicinal to animals and humans: as a vermicide and a cure for jaundice, kidney stones, obesity, coughing, hiccough, heartburn and various swellings, and even as an insecticide in the garden. Furze 'brew' or tea was thought to be a general tonic and blood-purifier whilst a burning furze twig cured ringworm. Furze bark was even smoked as a tobacco substitute (Humphries and Shaughnessy 1987, 20; Lucas 1958, 186), which hints at properties or older uses of many plants that may have gone unrecorded.

There is also a wealth of attestations to use of furze in ceremonial, ritual or festive contexts. Its yellow colour was associated with the opening of summer (around May Day) and the beginning of the butter season; when furze served as May 'Bush and Bough' because they safeguarded cows and their products. It was burned in the fields to protect the land and cattle were driven between burning bushes for the same reason. The plant loomed large in other holiday practices; for the Wren Hunt on St. Stephen's Day and for making Saint Patrick's crosses in Ireland, and was deemed powerful, at any time of the year, to protect from witches, fairies or 'powers' – so was an essential part of bridal bouquets (Grieve 1931, 367).

If a furze bouquet made a lucky wedding, it was also associated with the last life-cycle event, as we see in attestations from across the British Isles that wicked souls travelling to the land of the dead would be pierced to the bone by furze (Humphries and Shaughnessy 1987, 22). Popular wisdom abounds in etiological tales about the plants – a legend about Saint Patrick explains why furze is prickly and its base so often brown. Saint Patrick (a Briton) told his houseboy he was apt to say impolite things about Ireland while asleep. When he repeatedly cursed the place, the watchful child each time diverted the malediction to the tips of the rushes, the tips of the horns of white cows and to the bottom of the furze, which explains why the end of rushes are withered, the tips of cows' horns are black and the lower parts of the furze are brown (Vickery 1995, 324). Thus is a plant woven into a belief system and used in narrative to elucidate the prickly relationship between a saint and the land he converted.

Most of these examples of popular custom and belief come from nineteenth and, especially, twentieth-century ethnographic sources, but there is no reason to think such practices were not older (Lucas 1958, 149, 158). There is ample evidence to support this proposal, especially as regards the value of furze and the furze brakes. It is in the fourth category ('bushes of the wood') of valuable woods in the Old Irish tree list, furze mountain pasture was rated fifth among the six categories of valuable land and unauthorised cutting of furze was punished by three days of distraint (Kelly 1998, 380–381, 394–395). This is reflected in recent popular sayings about soil qualities, such as 'gold under furze, silver under rushes and famine under heath' (Vickery 1995, 157); whilst a twelfth-century Irish text declares that the reign of a great king is portended by an abundance of furze bushes (Lucas 1958, 11). The earliest sources to mention furze after these Medieval documents are from the seventeenth century and they second the judgement that furze land is considered 'profitable,' noting that furze raises 'wasteland' to the category of 'part profitable' (Lucas 1958, 16).

The purposely-planted broom brakes in Portugal, that could occupy up to one fourth of a land-holding mainly dedicated to rye and fodder crops (Estabrook 2006, 317), have been taken as a hint that furze was more widespread in the British Isles in the past, before the land transformations of the eighteenth to twentieth centuries CE (Humphries and Shaughnessy 1987, 16). The plant was certainly held in high esteem in many areas and taken as emblematic of a homeland, as opposed to a foreign 'intruder': 'on the back side of the Mournes [Mountains in Ulster], the tall coarse *whins* [furze] are termed *Scotch whins,* in contrast to the *wee soft kindly Irish whins*' (Evans 1989, 77).

Nettles

A character in Victor Hugo's *Les Misérables* laments: 'And what does the nettle require? Little earth, no attention, no cultivation.... With a little trouble, the nettle would be useful; it is neglected and becomes harmful' (2005, Vol. 1, 193–194). The same passage notes several of the uses of nettles we also find in the British Isles: as food and fodder, fibre and dye. In spite of its 'neglect', the plant seems to have been highly regarded in some places and at some times. The eighteenth-century Scottish poet Thomas Campbell noted: 'In Scotland, I have eaten nettles, I have slept in nettle sheets and I have dined off a nettle tablecloth. The young and tender nettle is an excellent potherb. The stalks of the old nettle are as good as flax for making cloth. I have heard my mother say that she thought nettle cloth more durable than any other species of linen' (Grieve 1931, 575–579). Nettles have many uses, indeed – in textiles, as food and fodder and, finally, as a support to general well-being.

Textiles

Nettle is usually thought of as 'stinging' but is soft and has great tensile strength when processed into textile fibres. The plant was also utilised as a dyestuff, especially on woollens – a practice taken up again during the Second World War (Vickery 1995, 257). Nettle as a textile plant was still used in the twentieth century in some parts of Europe, especially during World War I, when Germany and Austria ran short of cotton and took up nettle as a substitute, with the quantity gathered rising to 2.7 million kg in 1917 in Germany alone. Due to its extreme fineness, nettle was also utilised to make artificial silk and in the manufacture of gas-mantles and gas-masks (Grieve 1931, 576).

This ancient role as a textile fibre appears to be confirmed by its etymology. 'Nettle' in English and the Irish *nenaid* (modern *neanntog*) are linked to the words for 'netting' and 'knotting' (Kluge 1995, 586; Vendryes 1960, N–9). Perhaps because of the contrast between its stinging aspect and softness as a fabric, nettle cloth is well attested as 'magical' in popular tales or learned renderings of them such as Andersens' (1838) '*The Wild Swans*'. Archaeological remains of both *Urtica dioica* and *Urtica urens* have been found from Gallo-Roman through to early modern sites in France and Germany, but such traces do not provide information on the use of the plant in textiles (Wiethold *per. comm.* 2006). Some early finds of nettle in Scandinavia appear to long have been interpreted as flax (Barber 1991, 19; citing Hald 1942, 34, on the Oseberg Ship fabrics). Nettle seeds are regularly found in waterlogged contexts from the Neolithic on, in France and Switzerland, for example (Bouby *per. comm.* 2006; Jacomet *pers. comm.* 2006), and archaeobotanical analyses are increasingly revealing the possible presence of such textile plants (*e.g.* Dietsch-Sellami *et al.* 2008).

A Humble Food and Fodder

Narrative tales in Germany refer to nettle as a famine food (Heiner 2006) and the plant is often associated with the long-remembered hard times in seventeenth- and eighteenth-century Scotland (Grant 1961, 95), where nettle is still a regular ingredient in culinary traditions and cultivated as a potherb (Grieve 1931, 577; Davidson 1999, 531–532). Nettles may have marked popular memory as a famine food but there is some indication they were once part of the regular diet in Ireland, at least of modest folk, often in combination with charlock (*Sinapis arvensis* L.) (Lucas 1959, 140–141). Older sources mention nettles as a food essential not only to the poor but also to the saintly and, along with dairy products, a vital element for all classes. A tale about the sixth-century Saint Columba (*The Martyrology of Oengus* [CELT]) mentions that he ate only nettles with melted butter. As for Saint Kevin, who died in the early sixth century, he was said to have subsisted for seven years with no food but nettle (or nettle and sorrel). This diet turned his skin green (Plummer 1910); a trait he shared for the same reason with the Tibetan Buddhist sage Milaraspa. Ascetic practices are a use of plants we might rarely think to evoke, but they once motivated socially important minorities in many places in the world – nettles were used for self-flagellation in early Ireland, as elsewhere, which is mentioned in the *Penitentials* and gave rise to the term 'urtication' or 'urtification' (Werblowsky 1968).

Nettles are far more often cited as fodder for poultry, pigs and cattle, and renowned for improving coat sheen and egg production (Grieve 1931, 577, 579). During the 1930s, when turkey production was on the rise in Britain, nettles were regularly used as feed to put weight on young birds quickly, a practice that continued in Suffolk at least into the 1990s. Among its miscellaneous alimentary uses, the plant is still a component in local beer-brewing in Scotland and Lancashire, its juice was used as a rennet substitute and it was a component in making Cheshire cheese (Vickery 1995, 253–255).

From Soil-Food to Sacking to Flogging for Fun

Like furze, nettle also 'feeds' the soil and is highly valued as compost. Nettle leaves were used in packing stone fruits or tomatoes to help retain their bloom and freshness so they would travel, ripen and store better. The plant was also said to keep frogs away from beehives and flies from larders (Grieve 1931, 579). Nettle is a good indicator of garden biodiversity, and its protective properties for orchard and soft fruits are being reassessed (PFAF).

Nettles figure among the plants that were used as medicinals and were considered bracing or tonic in a broader sense, especially at the end of winter (Danaher 1972, 119–120; Baker 1995, 106). The plant was especially renowned as a blood-stauncher and also used as a gargle or to soothe burns. The old practice of 'urtification', or flogging with nettles, was long thought to cure rheumatism and this may be connected with the many customs involving stinging with nettles throughout the British Isles (Danaher 1972, 120; Roud 2006, 167–168, 199).

Perspectives on Diversity

Furze and nettles have many a virtue in common: both are associated with land value and soil improvement, with what we might term bodily improvement, with preserving other plants from unpleasant smells and keeping them fresh, with particular skills and wisdom and, last but not least, with saintly lives. Both plants were used to enliven narrative and explain the ways of the world, even unpleasant aspects of the afterlife. Furze and nettles open a window on plants as part of other ways of looking at the world than our own, and remind us that the plants we focus on today may simply reflect a shift in our attention over time. Anthony T. Lucas (1959, 144) had a most interesting, very tentative, comment on this very subject: 'In the case of both nettles and furze, the writer is inclined to think that this limited medicinal use is the last lingering trace of the tradition of the former widespread use of the plants: of furze for animal food and of nettles for human food'.

A glance at the highly diverse uses of furze and nettles enables us to think again about concepts such as intensification (and, consequently, de-intensification), and about plants that do not fit neatly into our usual categories of wild, cultivated, domesticated and so on. Both furze and nettles seem to be barely tolerated in many present-day recommendations (*e.g.* GISD, USDA), yet furze figured among the valuable woods in older Irish law and nettles were the food of saints. At times, some older practices involving such plants may come back to the battlefield for a surprising new round. This is the case of the 'nettle war' in France, where nettles have hit the headlines in recent years because the production of nettle slurry (*purin d'ortie*) as a home-made fertiliser came into direct conflict with the interests of the agro-chemicals industry, with considerable legal consequences (JO 2009). This incident highlights yet another aspect in the issues of diversity – the intriguing human quest to control by undiversifying.

6.3. VERSATILE HULLED WHEATS: FARMERS' TRADITIONAL USES OF THREE ENDANGERED CROP SPECIES IN THE WESTERN MEDITERRANEAN

Leonor Peña-Chocarro and Lydia Zapata

Introduction

Hulled wheats (einkorn, emmer and spelt) have been the source of food for many cultures across Europe and the Near East. Together with barley, two hulled wheats (einkorn and emmer) were the first domesticated cereals more than 10,000 years ago. From the Neolithic period onwards, einkorn and emmer have always been present in the crop assemblages of many sites in Europe and the Near East. In fact, both species were staples for millennia, supplying nutritional requirements to prehistoric populations. As suggested by Nesbitt and Samuel (1996), their important role in past agricultures contrasts enormously with their modern conception as minor and, in some cases, endangered crops.

The term hulled wheats describes a group of species: *Triticum monococcum* L. (einkorn), *T. dicoccum* Schübl. (emmer) and *T. spelta* L. (spelt) whose tough glumes adhere to the grain after the wheat has been threshed. The final product is, therefore, whole spikelets as opposed to the free-threshing species whose grains easily separate from the chaff. This important characteristic makes them very resistant to fungal diseases and well adapted to poor soils in mountain areas where they still survive.

In Europe, the development of new agricultural systems with more high-yielding varieties and the consequent loss of the traditional agroecosystems, were the causes of the final replacement of the old traditional varieties by new improved ones, thus increasing biodiversity loss. In addition, the difficulties in processing hulled wheats, in contrast with the much easier handling of free-threshing species, also influenced the abandonment of these crops during the last century. Although their final replacement occurred over the twentieth century in most areas, hulled wheats had already started their retreat much earlier. According to the available European archaeobotanical literature, hulled wheats were replaced by free-threshing wheats during the first millennium CE (Nesbitt and Samuel 1996). This replacement was not homogeneous and there were areas where, at particular times, hulled wheats appeared as the dominant species, and others in which they continued to be cultivated until the last century on a smaller scale.

Today, hulled wheats are considered underutilised species. The term can be confusing, since it may include many different categories and situations, as suggested by Padulosi and Hoeschle-Zeledon (2004). In this paper, we have adopted the definition used by these authors: *'non-commodity crops, which are part of a larger biodiversity portfolio, once more popular and today neglected by users' groups for a variety of agronomic, genetic, economic, social and cultural factors'.* Hulled wheats survive today in mountain areas of Europe, the Near East and North Africa, where local farmers still cultivate them by traditional methods. In many cases, these crops are closely linked to rural landscapes where biodiversity and rural livelihoods are extremely threatened.

The Western Mediterranean

In the western Mediterranean, and particularly in Spain, hulled wheats appear from the very beginning of the agricultural expansion (*ca.* 5,500 BCE). In contrast to other European areas, like central Europe or the nearby Italian Peninsula, where hulled wheats became the dominant species when agriculture spread to these areas, Iberia was characterised by the presence of a wide range of crops with no clear specialisation in hulled wheats. Within Spain, there were regions, such as the Basque Country or some areas in the north Castilian plateau, where particularly emmer – and to a lesser extent, einkorn – were the dominant species (Zapata *et al.* 2004). Along the eastern coast, however, both species were seldom major crops. The available data for later periods indicate the presence of these species in the archaeobotanical record with some high peaks in particular periods and areas (Buxó *et al.* 1997). Over time, nevertheless, their importance decreased gradually, although they never disappeared.

The case of spelt is somehow different. The first evidence from the archaeological record comes from the Iron Age from the site of Intxur (Cubero 1996; Fig. 6.4), while a further record is dated to the Roman period (Buxó 1997). In both cases grains and chaff were retrieved. One of the main problems related to the scarce archaeological information on these species from Roman times onwards stems from the rare application of recovery techniques in excavations from historical periods. Hence, most of our data on hulled wheat cultivation in medieval times originates in written sources, both Christian and Muslim. Amongst the numerous monastic and civil documents which contain

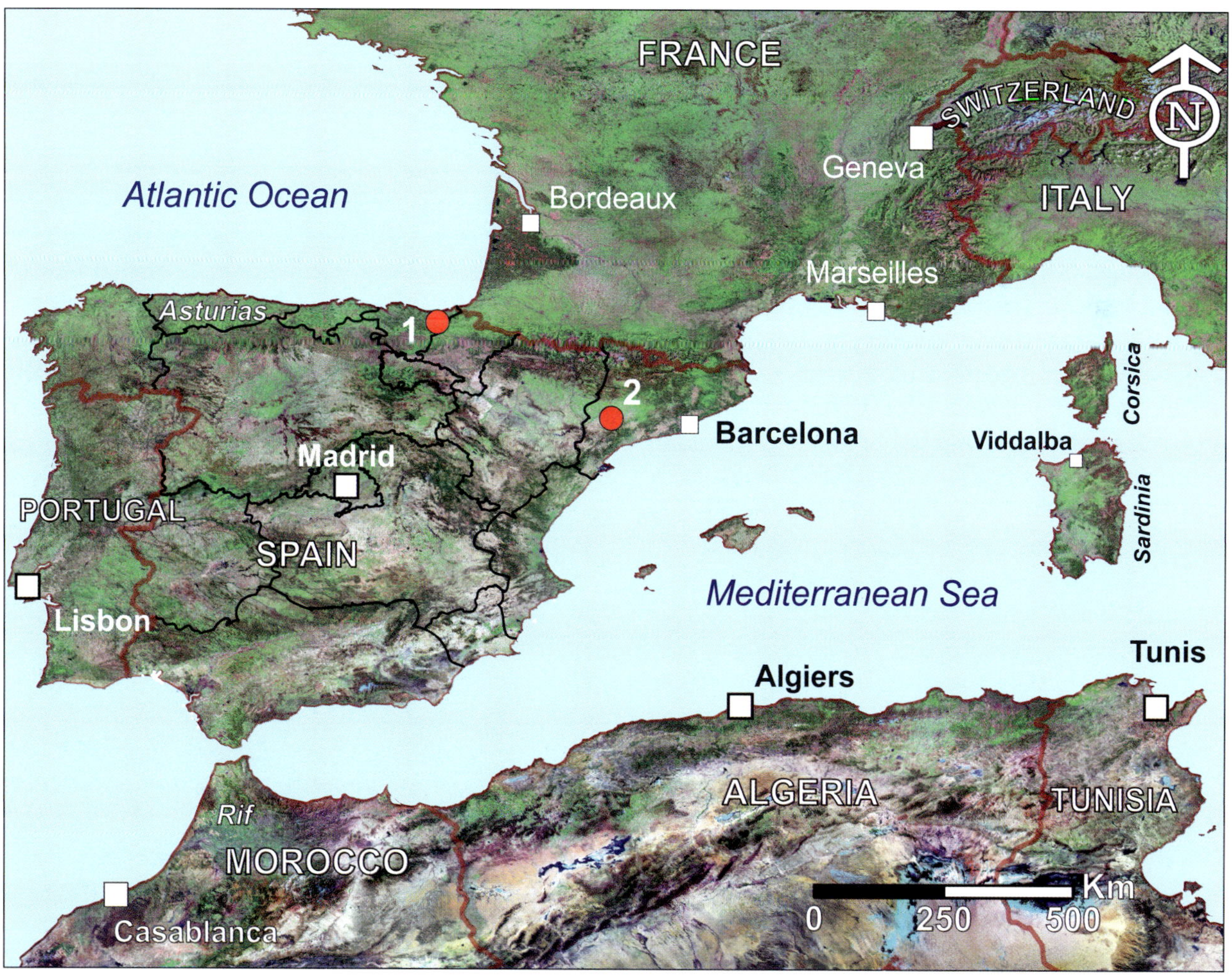

Fig. 6.4. Map of the western Mediterranean Sea with the countries, regions and sites mentioned in the text. 1) Archaeological Iron Age site of Intxur, Basque Country, Spain; 2) Archaeological Bronze Age site of Minferri, Catalonia, Spain. Map: R. Lugon, J.-C. Loubier and A. Chevalier.

insights into agriculture, the *Cronicón Albendense* (833 CE) (Anonymous, twelfth century) contains the first reference to emmer/spelt, indicating that the Asturian *escanda* (generic name for both emmer and spelt) was one of the well-known products of Spain. The Muslim period also witnessed the prolific production of many agrarian treatises in which hulled wheats, in particular, einkorn, are well represented. In his masterpiece, Ibn-al Awwan (1988) refers to several types of hulled wheats, amongst which some could be identified as einkorn. Hernández Bermejo and García Sánchez (2008) mentioned this species in their thorough analyses of the members of the *Gramineae* present in al-Andalus, through the study of Andalusi agronomic texts from the period.

For the twentieth century, the detailed statistical records of the Spanish Ministry of Agriculture show that, even if the numbers of hectares dedicated to the cultivation of hulled wheats decreased with time, einkorn, emmer and spelt were still cultivated on a small scale. This, of course, does not imply that they did not play an important role within the rural economies where their cultivation was maintained. In fact, as this paper will demonstrate, hulled wheats played a crucial role in the rural subsistence strategies of local communities until recently.

Human food, fodder, thatching, crafts, fuel, stuffing, isolating material and food specialties represent some of the different uses of these species in the recent past. These traditional uses are, in some cases, strongly related to regional heritage, and this has sometimes been used to boost their cultivation. In fact, in some European regions (*e.g.* Switzerland, Germany or France), hulled wheats have managed to occupy a specific niche related to organic food and the revival of the now fashionable natural foods. The growing links between food and culture, largely absent in the past discussions on conservation of biodiversity, have led, in some areas, to a revival, accompanied by significant signs of regional identity. This is also the case of emmer in some Italian areas, where its popularity amongst certain communities has led to the revival of its cultivation (Papa 1996; D'Antuono and Bravi 1996), and of spelt in Asturias, where spelt cultivation has recently been promoted. In addition, their specific adaptation to mountain areas, marginal soils and harsh conditions has allowed their continuous cultivation in these rugged regions, filling a specific niche within the rural economies and increasing the agrodiversity of the areas where they grow.

In Spain, these crops were still grown in the 1990s (Peña-Chocarro 1996; 1999; Peña-Chocarro and Zapata 1998; 2003; Peña-Chocarro *et al.* 2009). Emmer and spelt were grown in the central mountain area of Asturias, in northern Spain. Although emmer was once part of the spelt crop, nowadays it has almost vanished from Asturian fields. Instead, spelt is still cultivated in the area, linked to a developing urban market of spelt products. A small-scale association of spelt producers, and a wide range of initiatives promoted by growers and small entrepreneurs interested in their revival, is opening a new avenue for the survival of this crop. Einkorn was grown in the southern part of Spain (Andalucía) where it was cultivated until the early 1990s. In addition to the Iberian Peninsula, einkorn has been documented in the western Rif (northern Morocco), where this species is seriously threatened by the development of modern ways of life and consequent abandonment of traditional activities.

Multipurpose Cereals

Hulled wheats have traditionally provided human and animal food, and this has, perhaps, been the better known aspect of these species. There are many references to the traditional use of their grains to produce human food, in the form of flour, or whole grains consumed as bread, biscuits, soups, or porridge (Hillman 1984b; papers in Devroey and van Mol 1989; papers in Padulosi *et al.* 1996; papers in Duplessy *et al.* 1996). References refer not only to our most recent past, or the present day in some areas, but also to earlier periods. In addition, in many regions, particularly in the Mediterranean, the diet of domestic animals has included hulled wheats, either as green fodder (Peña-Chocarro 1999) or, more commonly, in grain form. Farmers described the advantages of these species as lending particular sheen to animal hair, whiteness to hen eggs and their medicinal properties for cows after giving birth. But, like in modern industry, where cereals are split up into many different products (starch, fibres, bran, etc.) and, in turn, into molecules for supplying new markets and industries with new components, traditional farmers have also taken advantage of the versatile uses of hulled wheat by-products (straw and chaff). At least in the recent past, non-food products have also been an important factor in farmers' decisions to maintain these species.

Ethnographic research carried out in areas of the Iberian Peninsula and northern Morocco, where these three species were still cultivated until recently, has allowed documentation of a wide variety of uses which illustrate the versatility of these crops. The information was collected over several years of research (1991–2000) in different areas and during different research projects (for an updated report see Peña-Chocarro *et al.* 2005).

The different regions where these crops have been studied belong to areas of different cultural traditions, so the patterns of use in each show great variability. As mentioned above, grains have been used both for animal and human consumption, while the straw has been utilised for several purposes, such as thatching, basketry, stuffing, fuel, etc. Some of this information has been published elsewhere (Peña-Chocarro 1996; 1999; Peña-Chocarro and Zapata 1997; 1998; 2003; Peña-Chocarro *et al.* 2005; 2009) and what follows is a review, including references, to other areas for which data is still unpublished.

Einkorn (Triticum monococcum *L.)*

In most areas of einkorn cultivation, this species is generally grown for animal food or other purposes related to its straw. As part of animal diet in Spain, einkorn was fed as whole spikelets to goats, mules, donkeys and hens or, milled and mixed with water and often with other flours, to pigs and cattle. The same applies for areas of Italy (Fig. 6.5) where einkorn was cultivated, as reported by Hammer and Perrino (1984) and Perrino *et al.* (1996). Another interesting area of einkorn cultivation is the Carpathians where the most recent use of einkorn is also for animal food. Several authors (Gunda 1983; Perrino *et al.* 1996) reported its excellent properties in feeding cattle, pigs, sheep and fowl. In Morocco, recent research (Peña-Chocarro *et al.* 2009) indicates that einkorn is commonly fed to animals and, in particular, to hens to produce particularly white eggs.

Although einkorn was part of the human diet from prehistory, its consumption in more recent times seems to have decreased as new generations move away from the traditions of the past. In fact, at least during the last centuries, einkorn has not been a major component of the human diet in Europe. In Spain, nonetheless, farmers still remember its consumption in times of scarcity. In Romania, according to Borza (1945) and Gunda (1983), einkorn was used for making bread. Gunda (1983) notes, however, that its flour is generally mixed with that of wheat, barley or maize. The same author recorded the baking of unleavened einkorn flat bread on hot stones in Transylvania. Einkorn is still an element in the diet of farmers in the western Rif (Morocco). Our recent research in the area (Peña-Chocarro *et al.* 2009) shows that einkorn is dehusked to be consumed in the form of different products (flour, cracked grain, noodles, biscuits, etc.), as is the case of other wheat species.

However, einkorn straw seems to have been the key element that has guaranteed its conservation. In Spain (western Andalucía), Morocco (western Rif) and the Carpathian area, einkorn straw has traditionally been used for thatching houses and huts. In Andalucía there are still examples of what was certainly a long tradition of thatching with einkorn. Einkorn straw was valued for its hardness, durability and ability to protect against rainfall. In order to obtain long stalks suitable for thatching, farmers followed a series of specific techniques for harvesting and threshing. Uprooting or cutting very low in the straw were the two techniques used by farmers, while lashing is the threshing method preferred; in other words, they beat the ears against a fixed object, so the grain was removed while the whole length of the stalk was preserved. Vignet-Zunz (1993) noted, nevertheless, that in some areas of Morocco, the ears were harvested first by sickling very high in the stem and, only later on, the stalks were collected.

Apart from thatching, einkorn straw has been used for many different purposes. Basketry and plaiting of goods such as pan covers and decorative objects, stuffing of mattresses, cushions and saddles are amongst the uses of its straw in the western Mediterranean (Peña-Chocarro 1996; 1999). It has also been used for smaller activities, such as fire-lighting, or as fuel to produce smoke to keep away bees while collecting honey. In Hungary, Gunda (1983) recalled the use of straw for making hats, singeing pigs, tying up vine branches and maize stalks and for litter. Borza (1945) also noted the use

Fig. 6.5. Map of the eastern Mediterranean countries with the main mountain ranges (Alps, Apennines and Carpathians), Tuscany and the Molise regions. Map: R. Lugon, J.-C. Loubier and A. Chevalier.

in Romania (Fig. 6.5) of einkorn straw to tie selected grapevine branches to wires so that growth could be better controlled.

Emmer (T. dicoccum) *and spelt* (T. spelta)

For emmer and spelt, most of the information collected in Spain on their uses relates to the production of bread, either from flour from one of the two species or from mixtures of different flours. Apart from the key product, *escanda* bread, as it is locally known in Asturias, there are other types of foods made with their flour. An example is the *panchón* (a sort of dough slowly baked on the fireplace and consumed with milk and sugar). Other products include a kind of porridge (traditionally consumed with milk) and a variety of pancakes and crêpes (Peña-Chocarro 1996). In this region, the consumption of hulled wheats never spread outside the area of cultivation.

During the 1990s, some regions of Italy witnessed an increase in the popularity of hulled wheat, linked to the revival of traditional foods and the invention of new ones, *e.g.* biscuits, pasta, etc. (Padulosi *et. al.* 1996; Perrino *et al.* 1996). Nowadays, a flourishing industry related to emmer products (pasta, bran, flakes, biscuits, etc.) is established in areas such as Tuscany, as well as in other European regions, such as the Carpathian Mountains (Gunda 1983).

Emmer and spelt are rarely used in Asturias for animal feed. However, in those cases where these species are used for fodder, farmers highly recommended it for feeding animals for sale. Hulled wheat fodder was particularly valued for its properties for increasing hair sheen. In Italy, Hammer and Perrino (1984) reported the same quality, but for pigs. In Navarra, where emmer wheat was cultivated until the 1980s, it was exclusively used as an animal feed for cattle, horses, pigs and sheep. Farmers also recommended it for young horses, either milled or in spikelets or mixed with other grains (Peña-Chocarro and Zapata 1997). In the Carpathians (Gunda 1983), pigs fed with emmer are thought to produce a more tasty bacon.

Animals were also fed with green fodder from hulled wheats. Whilst emmer in Navarra was often harvested in late spring and used for green fodder, farmers in Asturias only used it occasionally. Examples from Italy (D'Antuono *et al.* 1993) show their use in the Molise region and further south (Hammer and Perrino 1984).

Conclusions

Hulled wheats have been an important resource in Mediterranean mountain areas, not only for their multipurpose character, but also for their capacity to cope with harsh conditions and poor soils. They provided human and animal food as well as material for many different purposes beyond nutrition. For example, until recently, einkorn straw – at least – was considered the main product in many of the areas where it was grown. Nowadays, however, hulled wheat cultivation has almost disappeared, with only a very few areas where these crops are still maintained. Factors such as migration, market demands and agricultural policies, amongst others, can be considered the driving forces causing the fragmentation of habitats and the disappearance of many of these valuable agroecosystems and their biodiversity. Hulled wheats had their own role and contributed to the production of food and raw material for different purposes. In addition, traditional farming practices helped to maintain the cultural landscape in which they were embedded. Crop diversity in these areas is an important element, but of equal importance is the set of practices and techniques associated with these crops. Thousands of years of cultivation have allowed the development of specific management practices and uses which explain their presence nowadays in rich ecosystems severely threatened by modern development. The wealth of farmers' traditional knowledge linked to these species is undoubtedly the key to their future conservation.

6.4. USE OF CROP-PROCESSING BY-PRODUCTS FOR TEMPERING IN EARTHEN CONSTRUCTION TECHNIQUES

Emmanuelle Bonnaire

Introduction

Earth is a ubiquitous material found all around the world, both in traditional vernacular housing and also for prestigious buildings in complex societies, where building stone is not available and where climatic conditions allow its use (*e.g.* deserts in the Middle East and South America). This is true even today for some areas in Europe, where these older building traditions have been placed at the cutting edge of sustainable architecture (Cousins 2000; Houben and Guillaud 1994; Jeannet *et al.* 2003). If the occurrence of such techniques is widespread across the world, the technical and material challenges are often highly local, since clay soils often hold only a small proportion of pure clay. The remainder is made up of inert aggregates such as silt, sand and gravel. Clay expands and shrinks according to the ambient humidity, so controlling this expansion and contraction is a prime objective for builders, both ancient and modern. In a few areas, the subsoil can be used just as it comes from the ground, but it generally has to be *tempered*, and the role of straw is paramount in this process. It assists in handling and provides reinforcement, as well as minimising and distributing the fine cracks that form as the earthen material dries (McCann 2004).

Preparation of earth for building often involves the addition of plant temper for both bonding and plasticity. Various parts of cereals are used as temper and, in most cases, these are by-products of crop processing. After a period of time, the plant temper usually disappears, leaving impressions in the clay. In archaeology, these impressions are studied as part of archaeobotanical research providing evidence for past agricultural activities and techniques. Although the study of impressions is done only occasionally, their analysis provides useful insights into the ways crop by-products were used and completes information obtained through traditional seed analysis on the various utilisations of cereals. In many cases, this information has been insufficiently exploited on archaeological sites due to the lack of a reference collection of cereal impressions. Therefore, a reference system has recently been established (and can be used in all temperate regions), enabling the identifications to be made through comparison with modern cereals.

Crop Processing and By-Products

Earth is an ideal building material which can take on various forms, depending on how it is prepared: mud bricks, daub, pisé or plaster. In order to reinforce the structure of the earth used for building, it is necessary to add a plant temper. Also, various techniques improve the properties of the earth.

Most of the plant components used in temper are derived from agriculture, commonly from the waste generated during the threshing of cereals (Willcox 1995; Willcox and Fornite 1999). Naked cereals (bread wheat or durum wheat) seem to be better suited for use as temper, due to their greater ease in threshing, than hulled cereals (*e.g.* einkorn, emmer, spelt, barley) whose glumes, paleas and lemmas (chaff) adhere tightly to the grain. This particular characteristic determines the way naked and hulled cereals are processed, that is, for the hulled cereals, dehusking is needed to separate both grain and chaff, whereas in the naked forms, threshing is enough to separate both (see, for example, van der Veen 1999).

Crop-Processing

Cereal crop-processing includes a series of different steps (winnowing, different sievings, etc.) carried out to clean cereal grain of the different contaminants (chaff, weed seeds, stems, etc.) so it can be used for consumption. The process involves numerous steps which vary according to whether the cereal is hulled or naked. From the 1980s onwards, ethnobotanical data on wheats have been provided by G. Hillman (1984a and b; 1985) after fieldwork in Turkey; by G. Jones (1984), mainly in Greece; by Peña-Chocarro (1999) in Spain; by S. Reddy (2003) in India (millet and sorghum), as well as other researchers working in other parts of the world (Harvey and Fuller 2005). Their observations and studies of traditional agrarian practices in regions where agriculture is not yet mechanised have provided information that is applicable to the reconstruction of past husbandry practices. It has been recognised that the different operations involved in the crop-processing sequence have an important influence on the composition of the archaeobotanical samples. In fact, the proportions of grains, chaff (glumes, rachises, etc.), straw fragments and nodes, weed seeds of each product and by-products generated after each operation vary at each stage (Fig. 6.6).

Our main interest here is in the products and by-products resulting from crop processing. The by-products are often called waste, even if they are not always discarded. On the contrary, when utilised, they become secondary products. In fact, each type of sample resulting from the processing operations may have more than one use: roofing or flooring, fuel, bedding, fertiliser, fodder, human food and temper for building. For this last use, the by-products are collected during the various stages of cereal processing. Most of the cereal by-products used in temper comes from winnowing, during which the threshed grains are separated from their chaff and impurities. Throwing the threshed cereal into the air allows the wind to blow away the lightest fraction (chaff, straw, rachis segments, awns and light weed seeds) or by-product, while the heavier fraction or product (mostly grain) is set aside for further processing.

In the case of hulled wheats, as their glumes adhere tightly to the grain (see above), an additional step is required which makes it possible to free the grain from the chaff. This is done by dehusking, a mechanical action which breaks up the glumes and frees the grain. The mixture then undergoes a second winnowing. The by-products generated in this operation can be incorporated in the building earth, fulfilling a second function as a plant temper for construction material.

Cereals in Building Material

Elements of plant temper

Plant temper is made of cereal by-products generated during crop processing and it includes various parts of the plant (Fig. 6.7). Cereal stems are made up of nodes, where the leaves are inserted, and hollow internodes. The ear (or inflorescence) is a compound spike at the end of the stem. The ear consists of a group of spikelets around a central rachis which is made up of nodes and short internodes. At each node, a spikelet attaches to the rachis. The spikelets have at their bases bracts or glumes and contain the individual flowers (florets in grasses). Each floret, which will produce the grains, is surrounded by bracts called the *lemma* and *palea*. The *lemmas* end in long awns.

Identification: problems and issues

Through time, the organic parts of the plant temper decompose, leaving impressions in the earth that once contained them and providing an opportunity for their study. Attempts to identify the cereal species involved have been made (Willcox and Tengberg 1995; Willcox and Fornite 1999). However, the impressions are often incomplete, making species identification difficult. The major problem is, in fact, the absence of detailed information on the vegetal structure of cereal glumes and inner glumes. This lack of information made it necessary

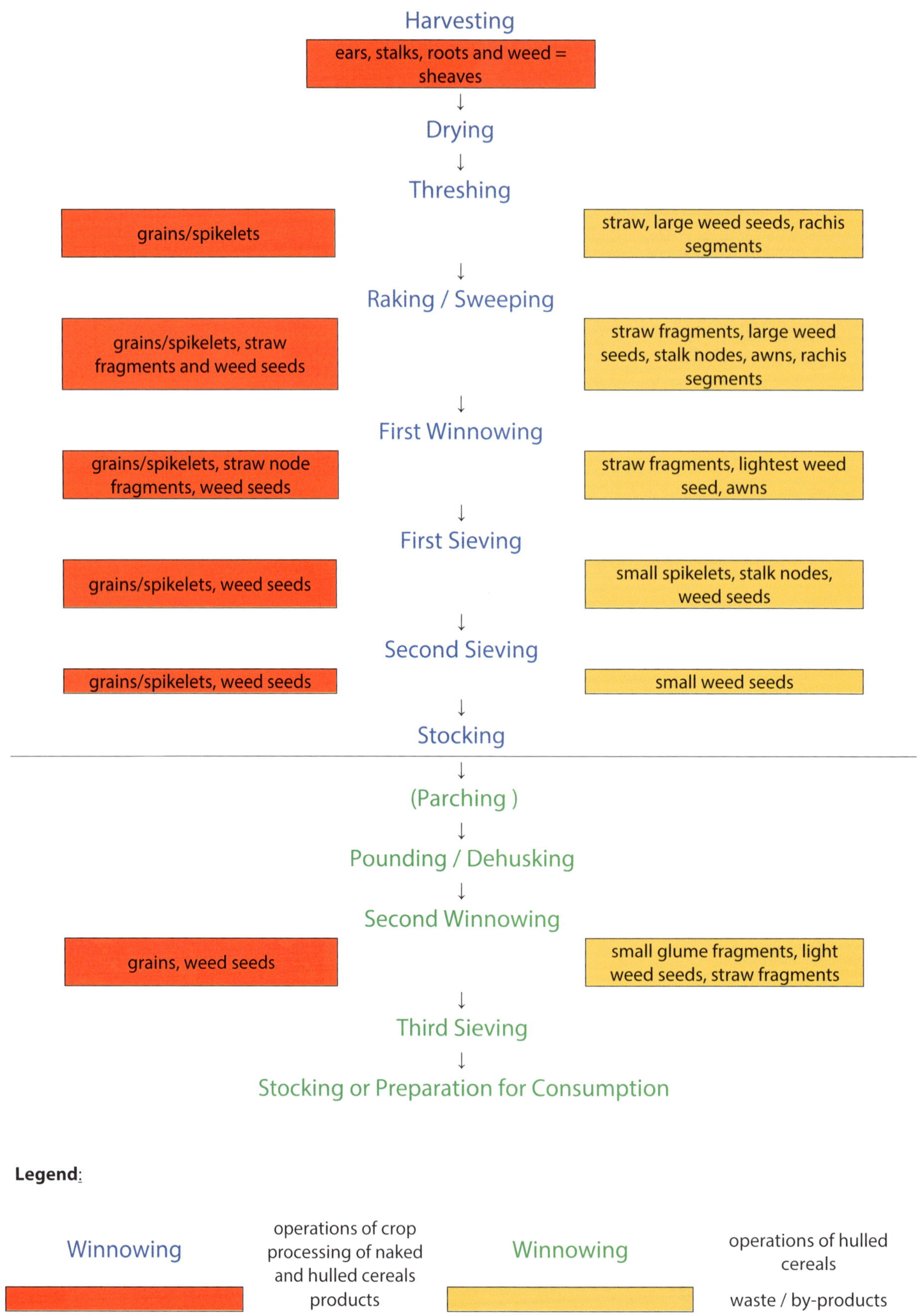

Fig. 6.6. Crop processing for naked cereals and hulled cereals.

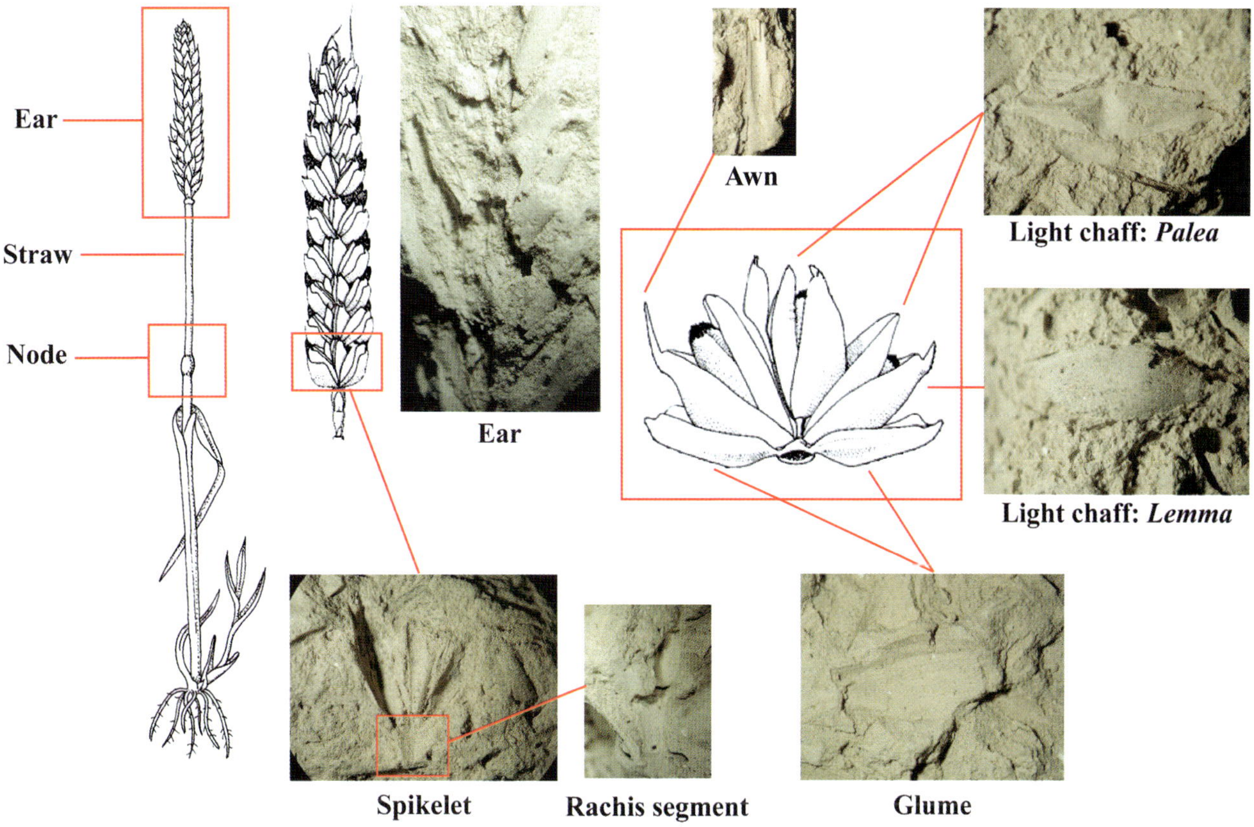

Fig. 6.7. Plant components of tempering. Image: E. Bonnaire.

to constitute a reference system in which distinctive criteria for the different elements involved was established (Bonnaire 2005; 2006). The choice of the descriptive criteria is inspired partially by work by L. Martin (2002), S. Jacomet (2008) and G. Jones *et al.* (2000). The aim was to provide a tool for taxa identification in situations where no macro-remains were available. Apart from species identification, agricultural practices and modes of plant exploitation may be investigated, contributing to a better understanding of past human activities. Evidence is also provided for building techniques.

The reference system was based on the analysis of key characters of modern cereals (various species of wheat, barley, oat and rye). In addition, some sub-species typical of temperate regions were also chosen (*e.g.* emmer, hard wheat, bread wheat, spelt or six-row barley). Most of the diagnostic characteristics were observed in the glumes, *paleas* and *lemmas* at the level of species and sub-species. These elements are extremely fragile and tend to disappear when plant remains are preserved by charring, but they are represented in the majority of impressions found in building material. Following this descriptive work on modern specimens (Bonnaire and Tengberg 2007), particular anatomical characters were attributed to each cereal species and results presented in atlas form with each sub-species described on a card. In addition to the atlas, a summary of all the morphological criteria observed for each taxon was produced in the form of a classic botanical identification key (Bonnaire 2006).

For the archaeological material, samples were chosen and impressions described and analysed using the identification criteria developed. The selection of samples was made by a series of time-consuming sorting procedures. First, the outer surfaces were visually examined in order to find potential material; in other words, impressions with traits likely to be identifiable. Afterwards, sampling was carried out according to the archaeological contexts and the fragments were examined with a microscope (magnification ×10) in order to find the best-preserved impressions. The different vegetal elements were then observed looking at their structural characteristics and comparing them with the reference material (Bonnaire 2008).

Conclusion

In many areas of the world, cereals are the main components of the human diet, providing carbohydrates as well as many other elements important for nutrition. Yet food is not the only use of this plant category. Crop by-products generated during crop-processing (chaff, straw, weeds, etc.) have also been used for many different purposes including tempering clay for building. Evidence on their use comes not only from the plant remains recovered on site, but also from the analysis of plant impressions preserved in mud or clay used for building, therefore providing complementary information on plant use. Data provided by the identification of the elements involved in tempering is of great interest for reconstructing past human activities related to agriculture and plant management, as well as construction techniques.

The precise identification of the crop processing by-products contained in the mineral medium allows us to reconstruct the sequence of steps involved in transforming a crop into an edible product and to explore the different products and by-products generated at each step. Therefore, a more detailed picture of the agricultural activities carried out at the site can be produced. In addition, plant impressions in building material provide insights into techniques used to produce daub, plaster or other types of building material in which plant temper is used. In the absence of any evidence of plants (*e.g.* seeds), impressions provide a unique opportunity for documenting species cultivated in the region and, when material comes from different chronological layers, changes through time can be observed.

6.5. USES OF THE WILD GRASS *AMPELODESMOS MAURITANICA* IN NORTHWESTERN TUNISIA TODAY

Patricia C. Anderson

Introduction

Why are certain wild plants still used today, despite the availability of domesticated plants or even industrially-produced materials that play a similar role? In the largely Muslim society living in rural northwestern Tunisia, where non-mechanised agriculture having a subsistence or mixed commercial/subsistence basis is practiced, a number of wild plants are used. A particular wild grass growing in the mountains in northwestern Tunisia, Mauritanian grass, *Ampelodesmos mauritanica* (Poiret) T. Durand and Schinz, but called in the local vernacular *dis* or *diss*, is the best present-day example of this. As we shall see, it is still used for fodder, although domestic cereals and pulses are used for this as well, and special baskets are made entirely from this plant, despite competition from baskets and bags woven from plastic strips. Men still harvest *dis* for multiple uses at home.

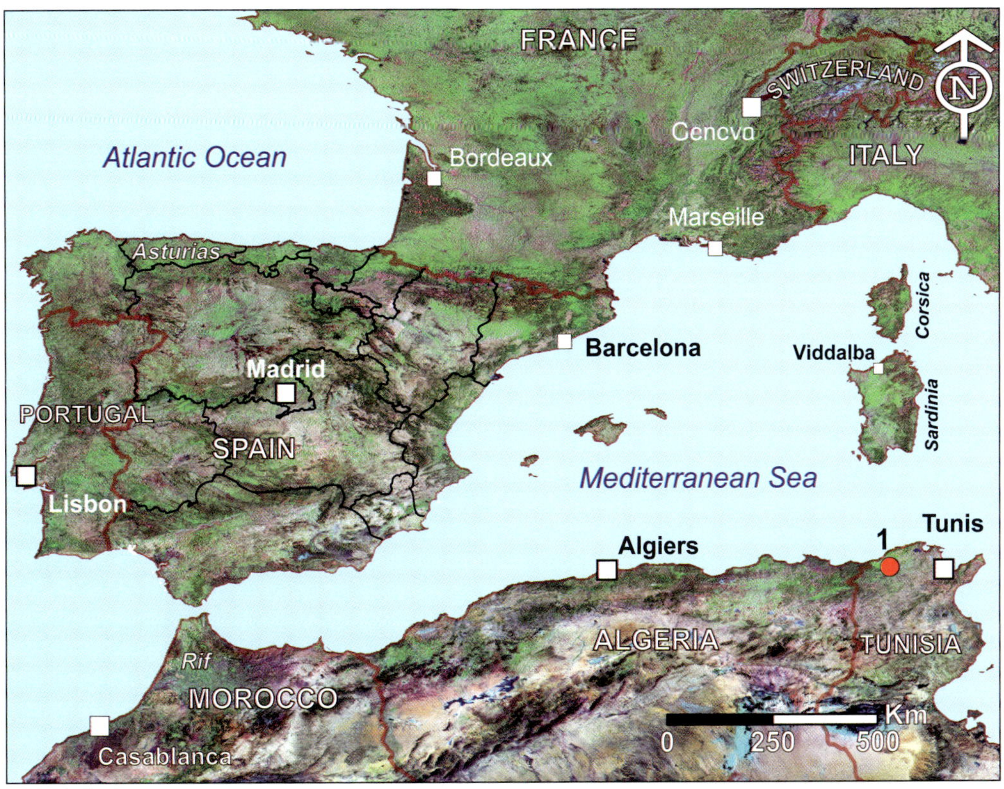

Fig. 6.8. Map of the western Mediterranean Sea with the countries, regions and sites mentioned in the text. 1) Ain Salem in Tunisia. Map: R. Lugon, J.-C. Loubier and A. Chevalier.

Fig. 6.9. Stands of *Ampelodesmos mauritanica*, or *dis*, growing on limestone outcrops in the Atlas Mountains in northwestern Tunisia. Images: P. C. Anderson.

This plant grows at the northeastern end of the Atlas Mountains, at approximately 100–300 m in altitude (Fig. 6.8, Fig. 6.9). In Tunisia, the plant seems to occur only in this area and today its narrow, serrated leaves are still used for fodder, basketry and, more rarely, thatching. Its stems (which bear the inflorescence, although in local memory the seeds are never consumed as food or sown) were, until very recently, used as roofing as well as for a 'snack'. Paradoxically, this plant is attested to in other areas of the Mediterranean, but is rarely, if ever, usually collected or used. In fact it is considered a nuisance that often causes fires, for example, in the Balearic Islands (F. Retamero, *pers. comm.* 2006).

In Italy, where it is called *stramma*, this plant did have multiple uses in the past, such as making rope, whips, small baskets and roofing for huts (Novellino 2005). The plant is protected today in a regional park, and a few small, decorative basketry items made with *stramma*, have been revived for the tourist trade by heritage workers, although the plant is not in use as it was in earlier times (*op cit*). Why was use of this plant stopped in Italy, whereas most of its past uses in northwestern Tunisia still form part of daily life?

Dis *as Animal Fodder and Other Uses*

In northwestern Tunisia today, the most common use of *Ampelodesmos mauritanica,* or *dis*, is of the leaves for green fodder. Dairy cows and milk production have become prevalent here, and the *dis* is said to give a good taste to the milk. Indeed, animals – especially cows, goats and sheep – appreciate this plant and browse its leaves in the wild today (Fig. 6.10), when their owners live in or close to the mountains. This was extremely common in the past. Mountain shepherds who stayed with their animals burnt the plant for heat and pulled up and broke off fresh *dis* plant stems.

Today, there are far fewer shepherds than in the past. In the 1990s, the families living in the mountains in the natural range of *dis* relocated to new houses in the plains, where electricity and running water were available. When the family does not live close to the *dis* plants, men journey the five km or so on animal back from their homes to the mountains where *dis* grows, particularly when dry fodder has run out and the *dis* leaves are tender (from January to March). Today dry animal fodder (wheat, barley and chickpea stalks) are either chopped finely, in the case of wheat, using the threshing sledge or trod by animal feet (barley and chickpea stalks). For green fodder the sources are alfalfa, sorghum, oats, green barley and, of course, *dis*. In the past, that is to say up until the 1980s or 1990s, *dis* was virtually the only source of fodder, other than chopped stems of wheat and barley, and elders think this is far better than the new animal foods. I believe people think that non-mechanised agriculture made food taste better and is healthier for animals and people, as well as for the soil.

Fig. 6.10. Cows grazing *dis* under cover of pine forests. Image: P. C. Anderson.

Fig. 6.11. Bags of leaves of the *dis* plant have been harvested by men and loaded into two large sacks, carried on the back of a donkey. Image: P. C. Anderson.

Fig. 6.12. A dry-stone, one-room structure in the mountains, with a thatched roof made from leaves of *dis*. This resembles how houses all looked in the mountains before the population moved down to the valley in the 1990s. Image: P. C. Anderson.

Fig. 6.13. A dry-stone structure now used as an animal stable. Note that its roof beams are covered by a mat woven from stems of *dis*. Note the inflorescences or panicles hanging over the roof in the foreground. Note also the dung patties drying against the inside wall. These will be used either for fuel or to fertilise the fields. Image: P. C. Anderson.

When men harvest *dis*, they cut its leaves with a sickle (see below), stuff the leaves into sacks which they then load onto their donkey or mule (Fig. 6.11) and take them back home as fresh fodder for their animals. When people lived in the mountains, before they relocated into a valley area in concrete houses, they lived close to this plant, in stone and mud huts with thatched roofs. *Dis* served multiple purposes in daily life, and the leaves, in addition to being made into baskets, harnesses and rope, were used to cover the floor and to thatch roofs and animal stables, etc. The hard stems of *dis* were also woven into mats to cover the floor or the roof (Fig. 6.12, Fig. 6.13).

This plant was – and still is – a male domain, including its harvest and use in basketry, and no one could explain why this was so and why women never worked with this plant. This contrasts with Italy, where only women made objects from *stramma* (Novellino 2005). Interestingly, Novellino reports that the timing of the *stramma* harvest is related to the lunar cycle and that whips made of *stramma* are used in ceremonies to make a cracking noise. No such special instructions or uses were given in relation to *dis* in northwestern Tunisia, according to our informants.

Case Study of a Basket-Maker

My first long encounter with *dis* was with the grandfather of the family with whom I was staying, Ali ben Alaha, in Ain Salem, about 100 km from Tunis (Fig. 6.4). I heard from his grandson that he used to make baskets and was a master specialist in the region. We approached him to make me a basket and he agreed enthusiastically. Some men are specialists in the making of the various basketry items mentioned above. Such skills are largely disappearing today due to the arrival, about twenty years ago, of industrially-made baskets of woven plastic strips, which are sold cheaply in markets and stores and also due to the relocation of inhabitants from the mountain environment to the nearby valley. The demand for certain models of baskets made from wild regional plants (and of some storage silos made from wheat stems) had virtually disappeared by 2007, when I first asked for baskets. The most common baskets made with *dis* are generally referred to as *zembils* (a double basket used to carry grain and other items draped over an animal's back) or the *couffin*, a large, double-handled basket carried by people.

Fig. 6.14. The basket-maker harvesting leaves of *dis*, taking care to select the best ones and never harvesting all of the leaves from any one plant. Image: P. C. Anderson.

Fig. 6.15. After harvest, the leaves are twisted and bent, then bound together by braiding into sheaves. Image: P. C. Anderson.

Fig. 6.16. After the sheaves are brought back home and dried on the roof of the house, the bundles are undone and pounded repeatedly by using a stick, in order to soften and break down the fibres so that the leaves can be woven without breakage. Image: P. C. Anderson.

Fig. 6.17. The leaves are then plaited, extending the plait constantly by adding new leaves. Image: P. C. Anderson.

Fig. 6.18. The plaits are sewn together in a spiral to make the bottom of the basket and continuing up to make the sides. Note that they are sewn together using a rope made from the *dis*, extracted and extended from the plait. Note also the wooden needle. Image: P. C. Anderson.

We will describe here a case study of the process of making the *couffin*, which we had the chance to film and photograph in the spring of 2007, in 2009 and 2010. Ali ben Alaha had not made baskets for several years as the demand had disappeared. He used to make baskets at home, when he had the time and someone in the family needed one or sold baskets to outsiders coming to his home, but never in markets. He was also a builder and a farmer. He had recently been too ill to build or farm and this had left him bored. He has often said he misses earlier times, when better food was produced by completely non-mechanised agriculture, with only natural fertilisers and fodders. These practices still exist in the nearby mountains (Anderson 2006; 2007), but the plains have been largely mechanised. However, the story of his basket-making has evolved over these three years.

The baskets he made were entirely from the *dis* local to the region, not from other plants such as *Juncus* or cereal stems. The first stage consisted of Ali ben Alaha's harvesting of *dis*. The leaves were harvested in April at around 300 m in altitude, where *dis* grows in tufts in mixed forests on limestone bedrock. Unlike the short season preferred for harvesting *dis* for animal fodder, *dis* can be harvested year-round for basketry. He travels to the area, approximately five km from his home but over uneven trails, on a mule or donkey. Ali ben Alaha used a sickle with small teeth (the same one also used to harvest cereals), to harvest the *dis* leaves, (Fig. 6.14), selecting handfuls of strong, green leaves, shaking out the yellowed or short ones. He twisted and braided the bunch as he harvested (Fig. 6.15). Finally, after approximately 45 minutes, five bunches were prepared in this way and transported in a sack to the house, where they were then put on the roof to dry for three days. The bunches were then undone, subdivided and pounded by Ali with a wooden baton, made from a tool handle he had always used for this purpose (Fig. 6.16). He said he has kept this tool for beating *dis* stems and it gets better with use. The stems were then re-braided and left outdoors. The entire process of basket-making took three days and four of the five braided bunches were used.

The first bunch was then undone and the basket-making began as the stems were braided into a thick plait (Fig. 6.17) and constantly extended by the progressive adding of stems which, after three days, had attained about five metres in length. As the plait progressed, Ali wound it around into a coil, first to make the basket bottom and then the sides of the basket. A kind of light rope was made from threads of the plait, by rolling leaf strands between the hands while spitting into them, all the while adding stems to extend the 'rope'. The rope was then threaded into two holes of a special pointed wooden awl he had brought and the plait edges were sewn together (Fig. 6.18). The process of extending the plait, then of making rope to sew it, without ever breaking the chain, occurred several times over the last two days. Finally the two handles were made from this rope, sewn through the basket sides in double thickness, and then covered with more rope wrapped around each handle. The basket is heavy strong and easy to use, and he obviously enjoyed making it.

Before I left with it, other members of his family visited to see and admire the basket as this art was virtually unknown, even to the young adults. The fact that no other artisans of basketry live in the area today, that Ali has no apprentices or helpers and had stopped making the baskets, means the know-how will die with him. The production of the baskets, even if he did train an apprentice, will remain too limited to be commercial and bring money into the family in the short term. Ali said no one would be interested in learning because no one would buy them. As they had some free time and many had no source of income, I hoped that one of his grandsons could learn as seeing my interest made them proud. However, this did not seem to interest them as they too saw no demand.

Three times since then, Ali has obviously enjoyed making me another ten or so baskets, which I pack

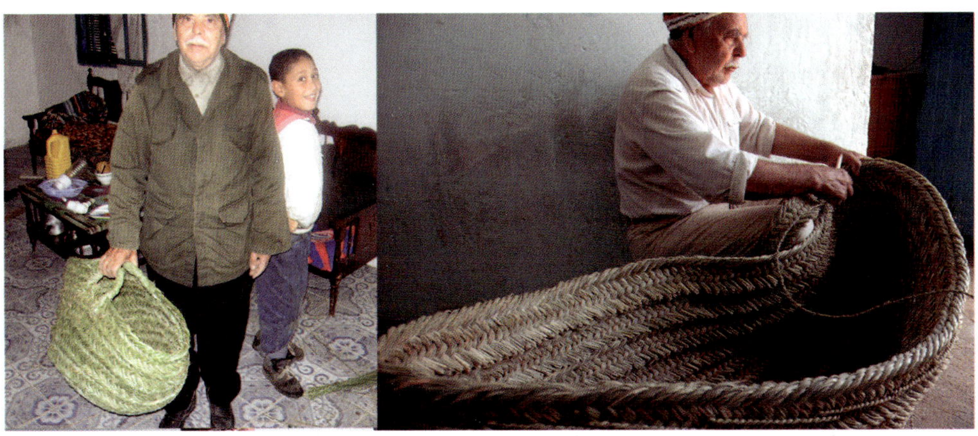

Fig. 6.19. Whereas the basket shown above left has handles attached and is for manual carrying of items, in the right-hand image the basket-maker is shown using the wooden needle to sew together plaits of a large basket designed to be draped over an animal's back for carrying dung, grain, sand, etc. Images: P. C. Anderson.

into my suitcase and sell in Europe for the price I paid him. However, a surprising transformation has occurred since the time he made me the first basket in 2007. First, new generations have discovered Ali ben Alaha's ability to make beautiful, but particularly useful, baskets. Second, there has been disenchantment with the plastic woven baskets and sacks available on the market, as they do not wear well and do not carry heavy loads reliably. Third, animals are again being more widely used for the transport of grain, sand or dung to spread on the fields, instead of the tractors which require expensive fuel. In 2009, I discovered Ali had diversified in the kind of basketry he was making. Rather than only baskets for people to carry by hand (*couffins*), he was also making *zembrils*, a beautiful large basket that drapes over an animal's back for transport plants, dung, etc. (Fig. 6.19). In 2010, he told me that over the year he had made over fifty *couffins* and ten or more *zembrils*, the large baskets used to carry loads draped over animals' backs. He also makes very large baskets for carrying bread. He is selling the baskets to people who order them from him at his home, but he never sells them in markets.

Concluding Comments

This is clearly a case of a revival of basketry due to new awareness of Ali's skills and a preference for traditionally-made baskets over industrially-produced ones. The kind of basket diversifies as the years go by. It is unclear who may carry on this tradition after he is gone. Whilst two of his grandsons have watched him work, neither has taken on the role of apprentice.

The basketry technique as well as the three tools used by Ali ben Alaha resemble those used since the Neolithic: a sickle (albeit the first sickles had a cutting edge of blades of flint or obsidian set in a handle of wood, horn or bone); an awl carved in wood (or often bone, in archaeological examples); and a wooden beating stick. This quick look at *dis* hopefully can assist archaeologists, historians and botanists in understanding the history of techniques, as well as the multiple uses of wild grasses in the past, for example in relation to the origins and and development of agriculture.

6.6. USES OF THE MASTIC TREE (*PISTACIA LENTISCUS* L.) IN THE WEST MEDITERRANEAN REGION: AN EXAMPLE FROM SARDINIA, ITALY

Bui Thi Mai, Michel Girard and François de Lanfranchi

Introduction

Among the vegetal species most commonly utilised to produce oil in Mediterranean Antiquity, there is obviously the olive but, in temperate zones, people had to seek out the oil plants that grew in their environment. This explains the survival of traditions in times of food scarcity when people still have recourse to beechnuts, walnuts, hazelnuts and so on.

One of the preoccupations of human beings since earliest antiquity was production of fatty foods and, more especially, vegetal oils (Lanfranchi and Bui Thi Mai 1998). Every area in the world has oil-producing plants, although there are more in warmer regions, and this makes for a considerable source of oil in Mediterranean and temperate zones. In order to produce this valuable liquid, it was necessary to search through local vegetation to find the species that could produce oil and then select the most productive among them to make extraction 'profitable'. Several plants in the European flora produce oil, but few of them provide an edible product (Dorvault 1938; Paris and Moyse 1971–1981). Oil is usually extracted from a plant's seeds, except for oils produced naturally by tubers or by the rhizomes of some ferns such as the male fern (*Dryopteris filix-mas* (L.) Schott), which have medicinal properties (as antihelminthics). Male fern rhizomes and the base of their fronds contain filicine and phloroglucides, which constitute one of the best remedies against tapeworm in veterinary medicine, and this plant is also used to fight sheep *fascioliasis* (liver fluke; Perrot and Paris 1974; Jahns 1989).

The plants principally utilised to produce oil include the olive, walnut, hazelnut, grapevine, colza, black mustard, beech, Briançon plum tree, linden tree, hornbeam, Indian chestnut, pine, spruce, the rowans, elders, spindle-tree, currant bushes, raspberry bushes and the mastic tree, which is the object of our investigation here (Dorvault 1928; Jouven 1942; Couplan and Styner 1994). It was during ethno-historical inquiries carried out by Corsican/Sardinian teams that François de Lanfranchi found out about the traditional production of mastic oil in the 1960s and this led to further investigations which enable us today to give a general presentation of mastic tree oil production, its extension into the Mediterranean world and, finally, the archaeological evidence for its immemorial use.

The Mastic Tree and its Many Useful Substances

The mastic tree (*Pistacia lentiscus* L.) belongs to the family of the Anacardiaceae. It is a shrub with an average height of three to four metres and contributes largely to making up the *maquis* (on silicious soils), but can also develop in a calcareous milieu in certain *garrigues*. Its distribution area extends on both sides of the Mediterranean from the Iberian Peninsula to the Near East. Here and there, it forms dense populations that are easy to work, as in the island of Chios, for example, where the mastic tree almost exclusively makes up the vegetation in the southern zone of the island.

The mastic tree has twisting branches with dark green, evergreen foliage that secretes a strong acrid smell similar to that of turpentine. Its coreaceous leaves, with a shiny upper side, are composed of three to six pairs of lanceolated folioles that are obtuse and mucronate on top. The staminated and pistillated flowers are born by two different plants (dioicity), which explains the absence of 'seeds' on certain individual plants. Clustered in dense catkins at the base of the leaves, the small male flowers are greenish with red anthers and open from April to June. The female flowers produce small globulous fruit five mm in diameter that ripen in winter (Fig. 6.20, Fig. 6.21). The slightly lumpy, dented seeds, of nearly lenticular shape (4.5 mm in diameter by 2.3 mm thick on average) are covered by a black pulp that is easy to detach.

This vegetal species is mainly exploited for the resin secreted by its branches. However, its leaves, wood and fruit are also used for alimentary, domestic or medicinal uses (Boyer *et al.* 1990). As for the old roots, they are alleged to be luminescent (Rivera-Núñez and Obón de Castro 1991).

Leaves

In southern Spain, mastic tree branches were put into the water of wells disinfected with lime to remove the unpleasant taste. The leaves are reputed to have an anti-parasitic power, so they were put into piles of wheat or barley to ward off weevils and corn-moth and were used in infusions against fleas. In North Africa, a kind of tea was also utilised to counter bad breath and the smell of sweat. In Libya (Fig. 6.5), the tannin-rich leaves were used in tanning (Rivera-Núñez and Obón de Castro 1991). They were also applied, either with or without, the galls that sometimes develop under them, to dye

Fig. 6.20. Flowering mastic tree. Image: M. Girard.

Fig. 6.21. Fruiting mastic tree. Image: A. G. Heiss.

carpet wool black in Macedonia, Tunisia, Morocco and Anatolia (Cardon and Chatenet 1990; Fig. 6.5).

Wood

The especially hard wood of this shrub is used in carpentry and cabinet-making, but also as firewood (Bonnier and Douin 1934). Likewise, it provides excellent charcoal and the ashes can be used as soap (Rivera-Núñez and Obón de Castro 1991).

Fruit

The berries can be eaten raw, but they are usually used in food preparations or, in Sardinia, as fodder. In Arab countries, they are used in making sweets as well as a liquor. In southern Spain, the berries were utilised fresh to whiten teeth, although they produce an indelible ink when they are heated with alum (Rivera-Núñez and Obón de Castro 1991). The berries are mainly used to extract a green-coloured oil, the various steps in the fabrication of which we shall examine later. Dorvault (1928, 983) notes the productivity and the use of this oil: 'The mastic tree is very common in Algeria, the fruit containing 20 to 25% of a fatty green oil that the Arabs use for food and lighting. Its extract is used to make 'Algerian' pills.'

The Resin or 'Mastic' (Putty)

'Mastic' is extracted in summer by making repeated incisions in the branches. In this way, production can reach four to five kg per tree. Light yellow in colour, this transparent, resinous product gives off a fairly strong balsamic scent. It is soluble in turpentine, ether and benzene and was used medicinally in Europe in the early twentieth century as an anti-diarrhetic for children, as an anti-scorbutic and, in the form of cataplasms, for fumigations, in dentistry for the occlusion of cavities and to make varnishes and glues (Dorvault 1928). It was utilised in antiquity as a cosmetic and Dioscorides noted that in his time (the first century CE), Greek women attached false eyelashes to their eyelids with this resin. It was also said to have been one of the substances used in embalming (Hepper 1990).

In North Africa (Morocco), the resin was used as a depilatory and a perfume. In the Orient, the resin is traditionally utilised as a scented chewing gum to protect the gums and refresh the breath. Dissolved in ether, it provides a psychotropic drink employed in some whirling dervish sects (Rivera-Núñez and Obón de Castro 1991). A special drink, likewise called *mastic,* is also made of it. This is produced by still distillation of a mixture made up of aquavit grounds (*marc*), sodium chloride (1%), anise (5%) and mastic resin in teardrop form (4%). The mastic is only put into the still when the alcohol solution, to which salt and anise has been added, has come to a boil. The alcoholic content of the liquor is subsequently brought down to 25° by adding water. It then undergoes another distillation: when it is boiling, 4% anise and 5% mastic resin are again added. The distillate, with the exception of the first litres which are discarded because of their poor taste, is stored temporarily. Finally, a mixture of water containing 3% mastic in teardrops and 5% sugar is distilled separately. The two products are then mixed together, allowed to cool, and their alcohol content is brought down to around 40° (de Cordemoy 1911).

Mastic Oil: Rediscovery of a Traditional Technique Still Used in the Last Century

In the western Mediterranean, there are still areas where traditional harvesting practices for mastic 'seeds' and extraction of the oil are remembered. This is true in Corsica and Sardinia, where a certain number of people remember this old practice, although very few have retained a perfect knowledge of the technical steps in the process. Fortunately, there is a group of people in northern Sardinia (Fig. 6.22) who undertook experimental extraction in the traditional way and this enabled us to record all the stages, from gathering to the finished product. Even though the mastic tree is wild, it represents an economic potential which has resulted in setting precise rules about how it can be worked. Areas covered with mastic pistachio (*a chessa* = the plant) and which produce abundant fruit (*lu baccu* = fruit, berry) indeed have a legal status, because they are on private land (*la tanca*). In order to pick them, it is necessary to have permission from the landowner. This is granted in exchange for a compensation generally made up by one part of the oil produced. So the areas with mastic trees have a real value and are thus part of an economic

Fig. 6.22. A. Sardinia B. Region of Viddalba.

heritage useful for everyone, which explains the fact that they are protected and the vigilance which all the various actors apply to their well-being.

Harvesting time varies from one year to another. There can be differences of a month in harvesting times due to variations in climatic conditions. Obviously, exposure to sunlight influences ripening, which occurs earlier in the sunnier zones of the south-facing slopes and the plateau tops than in the north-facing ones and valley bottoms. Gathering requires appropriate, simple, efficacious materials. Harvesters use the *fricaiola*, a small basket of circular form with a small rim that is a product of local traditional basketwork. A string attached to opposite sides of this basket is placed around the neck of the lady harvester, who thus carries the receptacle horizontally. The basket is held under the mastic tree branches in such as way as to catch the berries. The hands of the gatherers are placed on both sides of the small branches, which they rub with a rapid back-and-forth gesture called *frigare lu listincu* (rubbing the mastic tree). This friction detaches the fruit, which falls into the *fricaiola*. The workers' hands become covered with a very scented, black coating which is hard to remove, even when the hands are washed several times with soap.

After the harvest, the fruit is set aside for three or four days and carefully spread out in a fairly airy place. The berries shrivel due to water loss, which can amount to 25% of the weight. After this, the small branches and leaves are removed and the seeds poured into a riddle (*lu culiri*) to eliminate any dust attached. Then the berries are put into a vessel full of water kept at a boil for around half an hour. A blackish deposit forms on the walls of the receptacle, probably resulting from a mixture made up of the coating on the berries and the black, unstable substance that covers the seeds. In fact, the seeds lose their dark colour after heating and become pale yellow. Once the heating process is over, the berries swell up again in the warm water, and they are taken out with a skimmer and put in a bag called a *baletta*.

The bag and its contents are then placed in the concave part of a special wooden trough (*lu laccu*) that, in Viddalba, was a hollowed-out juniper trunk. Set at an incline, *lu laccu* rests with both of its ends on sturdy supports placed above ground level and a receptacle to catch the product of the berry pressing is put at the base of the trough. A man (*lu calzicadore*), wearing rubber boots, steps onto the bag in the bottom of the *laccu* to tread on (*calpestrare*) the berries (Fig. 6.23). He leans on one foot and uses the other to press the bag, out of which runs a red liquid mixed with water. The pressing process is interrupted from time to time to pour the last grains into the bag, along with a bit of warm water to maintain a constant temperature during all the process.

When pressing is finished, the receptacle and the liquid are placed on a burner and stirred from time to time. When the oil that floats up to the surface is ready to be taken off, some cool water is added to reduce its fluidity and more easily separate it from the bitter residue. The oil is taken out with a spoon and put into another receptacle (*si toglie l'olio dell'amara*). The final product is a green oil called raw oil (*olio crudo*, Fig. 6.24) that is kept 'as is' and can be stored for several years or even a decade. This

Fig. 6.23. Raw extract obtained by pressing. Image: F. de Lanfranchi.

Fig. 6.24. 'Raw' (green) oil and 'purified' (yellow) oil. Image: Bui Thi Mai.

green oil has various uses in the Mediterranean: in Morocco it can be used as an external medication, in the form of an unguent, to treat burns or backache and for domestic lighting. In the Orient, mastic oil is preferred to olive oil for this purpose. It can also be used to make soap, but it is its role as an edible that we will focus on here.

In several countries in the Orient and North Africa, green mastic oil is mixed with flour and almond paste to make a kind of highly-appreciated 'butter.' The Sardinian experiment was undertaken in order to show us how raw oil is transformed, by 'purification', into a yellow liquid whose appearance and taste are considerably modified by the process. Brought to a boil in a pot, the green oil (*olio crudo*) gives off a dark smoke at first. This is when grains of rough salt are added. They play no role in the changes expected, but serve as an indicator of how much water is still there. In fact, a characteristic crackling noise can be heard as long as there is any moisture left. When the smoke turns white, this means that transformation of the green oil into a golden yellow product has been finished: *è torrata giaghiccu* ('yellowish' is taken here in the sense of golden yellow). This yellow oil is kept especially to make a kind of doughnut or traditional pastries. The leftover seeds make up a cake (*ossigheddu* in Sardinian and *a gragnola* in Corsican) that used to be given to swine and chickens.

In the first half of the twentieth century in this rural world, the unit of measure was the cup (*lu cuppa*) and this is how the quantities of berries were expressed. *Lu cuppa* was a wooden receptacle, in the form of the base of a pyramid, with a capacity of 20 kg of wheat and 16 kg of mastic berries, but half-cups and eighth-cups were also used (Fig. 6.25). Some receptacles were made from a plate of cork with the ends folded together to touch and attached with five wooden pegs. The productivity of

Fig. 6.25. Traditional *cuppa* of various dimensions and shapes, Viddalba region. Image: Bui Thi Mai.

mastic trees depends on various factors: their size, the climatic, edaphic (soil) conditions and so on. An evaluation was carried out on individual trees one metre in height and three metres in diameter. Expressed in fresh berry quantity, the average weight of berries obtained per individual tree was about 1300 g, the equivalent of approximately a quarter litre of oil. In the Sardinian experiment, the weight of the gathered fruit dropped from 1.6 kg to 1.2 kg at the end of the operation. This difference of 400 g indicates that the oil and accompanying substances correspond to an extraction of 25%. It is usually thought that the ripe berries in a *cuppa* (16 kg) produce about three litres of oil, that is, a yield close to 18 to 19%.

The chronological and geographical origin of this practice is not yet known with certainty, but we do know that the gathering of mastic tree berries is very old, since many mastic seeds have been recovered in France in Mesolithic, Neolithic and Metal Age sites. The relatively recent discovery of many seeds of *Pistacia lentiscus* in the Minferri (western Catalonia; Fig. 6.4) Bronze Age site confirms this ancient tradition (Alonso i Martínez 2000).

Collection remains have also been observed in Neolithic layers in the Near East where hundreds of seeds of *Pistacia* cf. *atlantica* were found (Van Zeist and Bakker-Heeres 1982). As for the oil, the extraction of which necessitates heating in a fire-resistant vessel, it could theoretically only have been produced from the Neolithic on, when pottery appears. Nonetheless, it is not impossible that this practice is still more ancient, since it appears that heating liquids by immersing hot stones in skin or bark vessels was known to Upper Palaeolithic and Mesolithic hunters.

Founded on harvesting wild berries, the survival of the utilisation of the mastic tree is one of the originalities of the traditional rural economy in Sardinia and Corsica. The process of oil production is relatively complex: if the treading phase is a purely mechanical action that requires only the strength of the operator, purification necessitates a precise knowledge on the part of the worker about physico-chemical reactions of the product in order to achieve an optimum result. Treading is traditionally done by men because of the force they can apply. The second stage, which is more delicate, is entrusted to women, who have the 'skill' revealed by gestures adapted to each phase in the operation.

Conclusion

The current use of different parts of the tree as a fuel (wood), insecticide (foliage) and the practices involved in extraction of resin and oil, fit into the recent living traditions of country people in a particular place and help to elucidate the relationships between people and a wild plant species. Conservation of these techniques is thus a real windfall for archaeological research, since this makes it possible to think anew about certain archaeological remains that were previously unexplained or poorly interpreted. If it should turn out that prehistoric gathering of mastic tree berries was done to produce a fatty oil, these practices – which were still to be observed in the twentieth century in the west Mediterranean region – would thus have their origins more than 7000 years ago.

6.7. ANCIENT AND MODERN BOAT CAULKING: USE OF OLEORESINS IN TROPICAL ASIA

Bui Thi Mai and Michel Girard

Introduction

The process of caulking, or sealing the joints in wooden boats, is paramount to their sea-worthiness and vegetal matter figures among the materials utilised for this purpose. Pollen analysis of a resinous material contained in a receptacle found in the Brunei wreck, a late fifteenth-century vessel transporting around 14,000 pieces of Thai, Chinese and Vietnamese ceramics and discovered in 1997, revealed pollen from *Dipterocarpaceae* (*Dipterocarpus* and *Shorea*) pollen, a family mainly including tropical lowland rainforest trees (Bui Thi Mai and Girard 2003). The *Dipterocarpus* genus has about seventy species, common in southeast Asia and typified, as its name suggests, by 'two-winged fruits', while the *Shorea* occur in alluvial forests of Malaysia, Sumatra and Borneo; the English common name, white meranti, stemming from the native Malaysian term.

The oleoresins (a naturally occurring mixture of an oil and a resin extracted from various plants, such as pine or balsam fir) of these two species have been – and still are – utilised to make caulking joints that are especially flexible and resistant to sea water. The presence of these pollens in ancient resin led us to conduct research in southern Vietnam on how resins are harvested in the forest and applied in traditional boat construction. These resins differ in that *Dipterocarpus* provides a liquid oleoresin collected by deep tapping of the tree trunk, whereas *Shorea* naturally excretes a resin that solidifies upon contact with the air and can thus be directly harvested on the trunk, at branch bases and even from the ground. In Vietnam, these products are still used to caulk large wooden vessels such as junks, as well as smaller craft made of plaited bamboo, such as the *thuyền thûng*, so often to be seen on the beaches.

Wrecks recovered from underwater sites (Bui Thi Mai and Girard 2002) are among the many locations that may provide palynological evidence. In the case of wrecks, organic material may be preserved that was utilised for caulking, carried as repair material or transported for commercial reasons. All of these may furnish pollen evidence pertaining to their use and geographical origins (Pons 1961; Vernet and Leroi-Gourhan 1969) and may, occasionally, even enable us to pinpoint the area in which the vessels were constructed (Diot 1991; Wicha and Girard 2006).

Pine resins were used in Roman times, both to caulk various craft and to seal ceramic containers such as wine amphorae, and there was a well-developed extraction and trade of such resins. For the Mediterranean area, these materials, which may be made of compounds of resins of varying origin, have – thus far – not been the subject of much pollen analysis (Pons 1961; Vernet and Leroi-Gourhan 1969; Arobba 1976; 1983; Arobba *et al.* 1998; Girard 1978; Trevisan Grandi *et al.* 1986; Girard and Maley 1987; 1999; Vogt *et al.* 2002; Wicha and Girard 2006, for example). J. Connan of the Laboratoire de Géochimie Bio-organique (ECPM-Université Louis Pasteur-Strasbourg) is now leading a research programme on ancient resins, pitches and bitumen and this project has been extended to a wider geographical scale with the inclusion of physico-chemical and palynological data.

	D5	D5	D5	308-D10	BA	320	320	G11	G6	G5	BA	D10	D9	BA	D5
Archaeological samples	305 a	305 b	305 c		590–319	H9	H7	317	316	315	625	308	304	318	305
						320	318				F10			307	D
											313				
Dipterocarpus sp.	10.2	12.7	-	-	7.3	-	-	5.3	7.1	6.6	-	10	-	-	-
Shorea sp.	3.4	3 ,6	-	-	12.2	-	-	5.3	-	31.1	-	5	-	2.7	-
TOTAL pollen	88	110	4	22	41	26	11	56	28	45	6	20	36	109	14

Fig. 6.26. Pollen frequencies (%) of *Dipterocarpus* and *Shorea* in the samples of oleoresins from the Brunei wreck.

Palynological Analysis of Archaeological Samples

Within the framework of the above research programme, 15 samples of resinous material discovered in earthenware jars in the early fifteenth-century Brunei wreck studied by Michel L'Hour, have been analysed in the Palynological Laboratory of the CEPAM (Centre d'Etudes Préhistoire, Antiquité, Moyen Age). Although the samples are only a few cubic millimetres in volume, they have provided enough characteristic pollen to orientate research on the origins of the substances concerned. Half of these (8/15) contain grains of *Dipterocarpus* and *Shorea* (Dipterocarpaceae) pollen. Both of these trees secrete usable hydrofugal oleoresins (Fig. 6.26). Both of these taxa can be found in monsoon Asia from India to Indo-Malaysia and the Indochinese Peninsula. Some *Dipterocarpus* trees can reach 40 metres in height with a trunk of one metre diameter, at a metre above ground level (Fig. 6.27). Since their tubular flowers (Fig. 6.28) are pollinated by insects, pollen diffusion in the atmosphere remains sparse (Fig. 6.29). However, the relatively high frequencies observed in the Brunei samples suggest their pollen grains may have been caught in their own resins. This hypothesis had to be verified by measuring the present pollen levels in oleoresins of this taxon. In contrast, *Shorea* has flowers in small clusters with air-borne pollen, so they are more abundantly diffused.

Several species in these genera provide oleoresins that can be used for caulking (de Cordemoy 1911; Pham Hoâng Hô 1999). Their flexibility, resistance to sea water and to drying when boats are beached, as well as their repellant action on xylophagic organisms (such as wood-eating insects), has underwritten their use in traditional boat-building in southeast Asia until recent times (Aubaile-Sallenave 1987; Amos 1989; 1997; Tallec

Fig. 6.27. *Dipterocarpus costatus* grove.

Fig. 6.28. *Dipterocarpus obtusifolius* (Gurjan oil tree) flower.

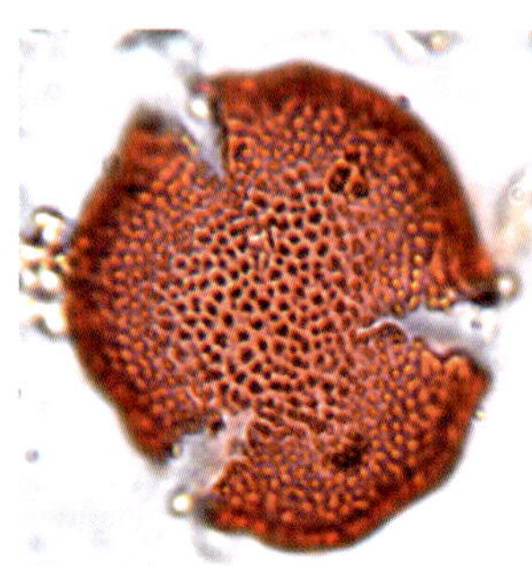

Fig. 6.29. *Dipterocarpus obtusifolius* (Gurjan oil tree) pollen.

1997). Vietnamese botanists and ethnologists confirmed they were still being used in present-day construction. These resins were collected from both trees and commercial products to establish the necessary biochemical and palynological referentials within the framework of a Franco-Vietnamese project (Bui Thi Mai *et al.* 2005).

Oleoresins and Tapping

An enquiry undertaken in the biosphere reserve of Lô Gô Xa Mat (Fig. 6.30) enabled us to observe the resin harvest and see those species tapped in the region, amounting to only a few, since most Dipterocarpaceae do not produce enough to be profitable. In Lô Gô Xa Mat, resins are mainly harvested from *Dipterocarpus alatus* Roxb., *D. costatus* Gaertn., *D. dyeri* Pierre and *Shorea guiso* BF.

Liquid resins are collected by making a rectangular incision up to 20 cm deep into the tree trunk about 1.50 m from the base. The cut is slanted towards the inside so that the resin will not spill out. Older documents (Aubaile-Sallenave 1987) indicate that these incisions could be much larger, depending on the tree diameter, perhaps up to 80 cm wide and 40 cm high.

The tapper uses a curved, lance-headed tool with a wooden handle 50–60 cm long to scrape the resin into a square tin set against the trunk (Fig. 6.31). After harvesting the resin, the tapper moistens a piece of coconut floss with the resin and sets it on fire with a lighter, then puts this into the cut and allows it to burn for about one minute until the trunk nearby is covered by soot (Fig. 6.32). The fire is put out by pressing a few large flat leaves over the opening and this process, which is repeated at each tapping, quickens the cut and makes the resin

Fig. 6.30. Map of the areas in Vietnam as mentioned in the text. 1) Biosphere reserve of Lô Gô Xa Mat, Tây Ninh region; 2) An Khânh, Mekong Delta; 3) Đâ Năng region. Map: R. Lugon, J.-C. Loubier and A. Chevalier.

ooze out. It is common to see a tree with several cuts, but they do not seem to weaken the trees, since we observed gradual scarring on all the trees no longer used. In Lô Gô Xa Mat, tapping carried out on 300 trees provides 30 litres of resin, about 100 ml of resin per tree, a harvest that necessitates three days' work. Harvesting from several species of Dipterocarpus entails a blend of resins which must be taken into account in analysing older product remains.

Shorea guiso secretes a resin that hardens rapidly in the air. This oozes out on the tree trunks (Fig. 6.33) and forms stalactites up to 20 cm long under the branches. These concretions gradually become heavier and break off, so they can be harvested by scraping them off the tree or picking them up from the ground, and provide a dry product that can be powdered and mixed with liquid resin to make a kind of putty-like substance.

Fig. 6.31. Tapping *Dipterocarpus costatus*.

Caulking

A visit to the boat construction yard in An Khânh (Fig. 6.30) enabled us to observe how these oleoresins are used in caulking junks. This meticulous operation consists of stuffing vegetal fibres (bamboo floss or other material) soaked in liquid resin between the strakes (planking), then the final touches are added with the putty mentioned above. It takes some 200 litres of resin to caulk a 25- to 30-metre junk.

Fig. 6.32. Burning.

Fig. 6.33. Natural excretion of *Shorea guiso* oleoresin.

Fig. 6.34. Rubbing the *Dipterocarpus* resin into the woven structure.

Fig. 6.35. Red *Shorea* resin.

Operating on a smaller craft scale, the people who make round boats of plaited bamboo strips (*thuyên thûng*) in the Đâ Nãng region (Fig. 6.30) caulk the hulls with these natural products harvested in the nearby mountain forests. The boats are waterproofed in two steps: first, by filling the gaps between strips with cow (essentially zebu) dung, letting it dry and then applying several coats of liquid *Dipterocarpus* sp. resin (Fig. 6.34) (Bui Thi Mai and Girard 2007). *Shorea* resin is also used (heated) in liquid form to treat the upper part of the mat that will be fitted into the hoop (Fig. 6.35).

	Dipterocarpus (no.)							*Shorea* (no.)		
Sample	B	D	C	H	E	F	I	A	A′	G
Dipterocarpus	-	1	1	12	1	2	-	-	-	2
Shorea	-	1	-	2	13	-	-	5	44	204
TOTAL pollen	34	34	4	53	191	9	15	1651	623	247

Fig. 6.36. Number of pollen grains of *Dipterocarpus* and *Shorea* observed in the oleoresins collected for these two taxa.

	Dipterocarpus (%)							*Shorea* (%)		
Sample	B	D	C	H	E	F	I	A	A′	G
Dipterocarpus	-	3	(25)	22.6	0.5	(22)	-	-	-	0.8
Shorea	-	3		3.8	6.8		-	0.3	7	82.6

Values noted in parentheses are indicative only, since pollen counts are far too low to calculate a percentage

Fig. 6.37. Pollen frequencies (%) of *Dipterocarpus* and *Shorea* in these oleoresins.

Pollen Analysis of the Oleoresins Collected from Trees

Pollen analysis of the resins collected from different *Dipterocarpus* (Fig. 6.36) and *Shorea guiso* (Fig. 6.37) indicate the presence of pollen from these taxa at frequencies varying according to the place they were collected and on how long tapping had been going on. As would be expected, direct collection from trees (March and April, 2003) indicated that pollen content varied from one sample to another and that these adhesive resins essentially capture ambient pollen rain. In the case of frequent collecting, the interval between harvests is too short for pollen to settle on the resin. Furthermore, the flowering period was over when the sampling was carried out.

Pollen Analysis of Oleoresins Utilised in Small Boats

There are no pollen traces in the liquid *Dipterocarpus* resin utilised in the two boatyards we visited. As in the case of the samples taken from trees, this absence may be due to the frequency of harvesting, as mentioned above. In the dried-out *Dipterocarpus* resin fragment from a *thuyên thûng*, we found only a few *Shorea* pollen, which might be due to pollen

	Gerrycan of oleoresin of *Dipterocarpus* sp. (boat-building sites)		Caulking of TT	*Shorea* powder	Commercial resin	Old caulking
Site Sample	Anh Kanh J1	Da Nang J2	Da Nang K	Anh Kanh L	Ba Tri M	An Khanh N
Dipterocarpus	–	–	–	–	5	38
Shorea	–	–	2	6	4	18
Pollen sum	4	6	21	52	200	625

Fig. 6.38. Pollen data for oleoresins utilised in boat-building.

pollution from this taxon. The dispersion capacities of the pollen might explain why we find traces of them in most of the samples analysed, whether they belong to this taxon or to *Dipterocarpus*.[1] Complementary analyses were undertaken on the paste-like commercial resin, as well as on an older piece of caulking recovered from the An Khanh boatyard. In the hardware store product, the content for both *Dipterocarpus* and *Shorea* is around 2% and in the older caulking, there is 6% *Dipterocarpus* and 2.9% *Shorea*. The older caulking thus has a *Dipterocarpus* content that is fairly near that of the Brunei resins (5 to 12%), but this may be simply a coincidence.

In spite of the considerable variation already mentioned, the pollen frequencies for *Dipterocarpus* and *Shorea* observed in the present-day and archaeological resin samples seem to justify proposing a botanical relationship between these two materials, as is possible from the pollen data for present-day pine resins compared to resins from Mediterranean Antiquity.

In the case of the Brunei vessel, recent biochemical analyses confirm the nature of these oleoresins, since these products come from Dipterocarpaceae and most probably from Dipterocarpus (Fig. 6.38; Burger *et al.* 2009). Thus, the presence of Dipterocarpaceae oleoresins in the fifteenth-century high-seas junk confirms that they were collected at that time and we may assume that knowledge of their hydrofugal properties was, most probably, even older. We may thus hope that the discovery of wrecks earlier than the fifteenth century will make it possible to trace the use of this material in ship-building even farther back. Hence, archaeopalynology, with its many facets, both initiated this enquiry and enabled us to rediscover one of the many links that have united humans and plants for so long.

6.8. CONCLUSIONS

Cozette Griffin-Kremer

A brief sequence of articles on the subject of diversity can only hint at the many ways in which humans utilise plants, not to mention animals and the environment around all them, and how this affects what we term the landscape, which today is ever more anthropogenic, even when we are standing in what we deem to be a wilderness. However, the other chapters in the book, although they are set under different headings, will complement the examples here, just as they will dovetail to provide more than an inkling of the benefits of interdisciplinary cooperation among historians, archaeologists, archaeobotanists, field biologists, ethnographers and yet other disciplines. What is most fruitful in this chapter may also be the most invisible, although the authors are quite aware of the process that this 'meeting' represents – essentially, focusing their efforts to communicate to one another and to an interested public *in the same language*. This exercise alone highlights disciplinary blind spots and breaks new ground for further research, often thanks to questions from colleagues that might appear surprisingly ingenuous at first hearing.

Perhaps the second, equally fruitful outcome of this effort to communicate both expertise and enthusiasm arises from the juxtapositions within articles and across the entire chapter. This can apply to the widely differing time frames, running from excavation results from several disciplines in archaeology to recent ethnography and present-day investigation of living practice. It explicitly implies faith in the value of comparison of similar techniques and product uses across cultural, geographical and temporal boundaries, however aware the authors remain of the limitations these similarities also impose, and of the real differences in their disciplinary tools and methodologies.

The third point that stands out in these articles is about people – about farmers, house-builders, urban bakers, shipwrights, forest denizens out gathering and many others showing resilience, ingenuity, the capacity to utilise resources exhaustively, often without incurring lasting impact on the land. These articles are broad enough in scope to at least imply both the continuity and the mutability we divine emerging from the way products reflect human needs and desires, and remind us of the 'patterns' evoked in several of the articles and their illustrations – the shapes and shades of plants, where colour and texture can be as important as taste; the sudden appearance under a microscope of fragments that attest to particular threshing techniques; the silhouette of a thatched roof or the weave of a craftsman's basket. That is, this sampling reveals threads that cross societies and link people, far away and long ago, to us and to our own enquiries into possible fact, artefact, meaning and our present-day strategies for survival.

Chapter Acknowledgements

The work by Leonor Peña-Chocarro and Lydia Zapata (Chapter 6.3) has been carried out within the framework of the projects AGRIWESTMED (ERC-AdG-230561), HAR2008-03976/HIST, HAR2008-01920/HIST and HAR2011-23716 *Nuevos cultivos, nuevos paisajes* and Programa Consolider TCP-CSD2007-00058 from the Spanish Government. Lydia Zapata is part of the Research Group of the Basque Government IT-622-13 and UPV/EHU UFI 11-09.

Patricia Anderson acknowledges support for ethno-archaeological research from the GDR 2517 ('Regards interdisciplinaires sur les activités et techniques agricoles anciennes et préindustrielles') of the Centre for National Scientific Research, France. Emmanuelle Bonnaire (Chapter 6.4) wishes to thank Michael Ilett for translating her text.

Chapter Note

1 This sampling was carried out within the framework of a Franco-Vietnamese programme to set up a pollen referential for southeast Asia (Bui Thi Mai *et al.* 2005).

Chapter Bibliography

Agropolis Museum. Presse ou 'Couleuvre' à Manioc, Acara, Brazil, Inv. N° 98, 1.175, http://www.museum.agropolis.fr/pages/expos/aliments/racines/objets/couleuvre.htm

Alonso i Martínez, N. (2000) Registro arqueobotánico de Cataluña occidental durante el II et I milenio A.N.E. *Complutum* n° 11, p. 221–238.

Amos, J.-C. (1989) Les bateaux paniers du Viet Nam, *Neptunia* 176, 42–44.

Amos, J.-C. (1997) Typologie des bateaux-paniers vietnamiens. Chasse-Marée. *Histoire et Ethnologie Maritime* 105, 24–27.

Andersen, H. C. (1838) The Wild Swans in *Fairy Tales Told for Children. New Collection,* C. A. Reitzel, Copenhagen. Based on 'The Eleven Swans' in Matthias Winther's *Danish Folktales* (1823).

Anderson, P. C. (2006) Non-mechanised processing and storage of cereals, grasses and pulses used for fodder, fuel, food and crafts: Examples from N. Tunisia, Atlas Region, Northwestern Tell, In Z. Füsun ERTUG (ed.) *Proceedings of the IVth International Congress of Ethnobotany (ICEB 2005),* ISBN: 975-807-153-X, Ege Yaynlar, 223–231.

Anderson, P. C. (2007) Le travail à l'araire aujourd'hui en Tunisie, le point de vue d'une ethnoarchéologue. In R. Bourrigaud and F. Sigaut (eds.) *Nous Labourons. Actes du Colloque, Techniques de travail de la terre, hier et aujourd'hui, ici et là-bas. Nantes, Nozay, Chateaubriant, 25–29 octobre 2006.* Centre d'histoire du travail, Nantes, pp. 247–258.

Anonymous (12th century) Cronicon Albendense. Manuscript. *Corpus Pelagianum et alia scripta minora,* Biblioteca de El Escorial, Madrid.

Arobba, D. (1976) Analisi pollinica di una resina fossile rinvenuta in un dolio romano. *Pollen et Spores* 18 (3), 385–393.

Arobba, D. (1983) Analisi palinologica spettrofotometrica di resine in reperti ceramici del I/III secolo d.C. provenienti dai fondali dell'Isola del Giglio (Grossetto, Italia), *Giorn. Bot. Ital.* 117 (suppl. 1), 50–51.

Arobba, D., Bandini Mazzanti, M., Bertolani Marchetti, D., Galasso, M., Giardini, G. and Mannoni, T. (1983) *Studio pluridisciplinare del materiale proveniente da un carico navale del I–III secolo d.C. scoperto sui fondali dell'Isola del Giglio (Grosseto-Italia).* Forma Maris Antiqui, XI–XII (1975–1981), 117–144.

Arobba, D., Caramiello, R. and Martino, G. P. (1997–1998) Analisi paleobotaniche de resine dal relitto navale romano del golfo Dianese. *Rivista di Studi Liguri* 63–64, 339–355.

Aubaile-Sallenave, F. (1987) *Bois et Bateaux du Viêt-Nam.* Ethnosciences 3. Selaf, Paris.

Baker, M. (1995) *The Folklore of Plants.* Shire Publications, Pinces Risborough, Buckinghamshire.

Barber, E. J. W. (1991) *Prehistoric Textiles.* Princeton University Press, Princeton, New Jersey.

Bataille-Benguigui, M.-C. and Cousin, F. (1996) *Cuisines, reflets des sociétés.* Editions Sépia/Musée de l'Homme, Paris.

Bataille-Benguigui, M.-C., Cousin, F. and Monzon, S. (2003) *Les mets en petits paquets.* Alimentarium, Musée de l'Alimentation, Montpellier, 2008, http://www.alimentarium.ch/fr/expositions-temporaires/precedentes/pot-pourri/les-mets-en-petits-paquets/envelopper-et-cuire.html

Bonnaire, E. (2005) *Les empreintes végétales dans les briques crues de construction. Identification des céréales utilisées comme dégraissant.* Unpublished diploma thesis, Université de Paris I.

Bonnaire, E. (2006) *Etude des empreintes et des phytolithes des céréales utilisées comme dégraissant. Applications sur du torchis de Kovacevo (Bulgarie).* Unpublished diploma thesis, Université de Paris I.

Bonnaire, E. (2008) Le torchis comme témoin ethnographique. Eléments sur les pratiques architecturales et agricoles au VSG ancien : Le site des Ouches à Sours (Eure et Loir). *Internéo* 7, 23–30.

Bonnaire, E. and M. Tengberg (2007) Eléments pour l'identification des empreintes végétales dans la terre à bâtir. In B. Van Doosselaere and C. Oberweiler (eds) *Dégraissants organiques : identifications, nomenclatures et référentiels.* Actes de la table-ronde. Cahier des thèmes transversaux de l'UMR 7041 n°7, Nanterre, pp. 81–85.

Bonnier, G. and Douin R. (1934) – *Flore complète illustrée en couleurs de France, Suisse et Belgique.* Librairie Générale de l'Enseignement, Paris.

Borza, A. (1945) *L'alac (engrain, Triticum monococcum) chez les roumains. Étude botanique et d'histoire culturelle.* Centru de studi si cercetări privitoare la Transilvania, Sibiu.

Boyer, R., Barrandon, J.-N., Binant, C., Bui Thi Mai, Girard, M, Gratuze, B. and Guineau, B. (1990) Les collyres. In R. Boyer (ed.) Découverte de la tombe d'un oculiste à Lyon (fin du IIe siècle après J.C.). Instruments et coffret avec collyres. *Gallia* 47, 215–249.

Brookfield, H. (2007) Intensification, and alternative approaches to agricultural change. *Asia Pacific Viewpoint* 42 (2/3), 181–192.

Brush, S. B. (2004) *Farmer's Bounty. Locating Crop Diversity in the Contemporary World.* Yale University Press, New Haven and London.

Bui Thi Mai. and Girard, M. (2002) Contenu végétal. Spores et grains de pollen. In J.-C. Miskovsky (ed.) *Géologie de la Préhistoire. Méthodes, techniques, applications.* GEOPRE. Editions Maison de la Géologie, Paris. pp. 683–703.

Bui Thi Mai. and Girard, M. (2003) *Contribution de la palynologie à l'étude du site de Oc Eo et de l'épave de Brunei. Résultats préliminaires. 1er Congrès du Réseau Asie, Paris, September, 2003.* Website. http://www.reseau-asie.com

Bui Thi Mai. and Girard, M. (2007) Les thuyền thúng, bateaux-paniers ronds en bambou tressé du Centre Việt Nam. *Neptunia, Revue des Amis du Musée national de la Marine* 248, 17–23.

Bui Thi Mai., Trinh Thi Lâm, Hoàng Việt, and Girard, M. (2005) Banque de pollens de référence de la péninsule indochinoise. La 'civilisation du végétal' en question. 2ème Congrès du Réseau Asie, Paris 28–30 September, 2005. Website. http://www.reseau-asie.com

Burger, P., Charrié-Duhaut, A., Connan, J., Flecker, M. and Albrecht, P. (2009) Archaeological resinous samples from Asian wrecks: Taxonomic characterization by GC-MS. *Analytical Chimica Acta* 648, 85–97.

Buxó. R. (1997) *Arqueología de las plantas.* Barcelona, Crítica.

Buxó i Capdevila, R., Alonso, N., Canal, D., Echave, C. and González, I. (1997) Archaeobotanical remains of hulled and naked cereals in the Iberian Peninsula. *Vegetation History and Archaeobotany* 6, 15–23.

Cardon, D. (2007) *Natural Dyes.* Archetype Publications, London.

Cardon, D. and du Chatenet, G. (1990) *Guide des teintures naturelles. Plantes – Lichens, Champignons, Mollusques et Insectes.* Les Guides du Naturaliste. Delachaux and Niestlé, Neuchâtel.

CELT Corpus of Electronic Texts. The Free Digital Humanities Resource for Irish history, literature and politics. http://www.ucc.ie/celt/

de Cordemoy, J. (1911) *Les plantes à gommes et à résines.* O. Doin, Paris.

Couplan, F. and Styner, E. (1994) *Guide des plantes sauvages comestibles et toxiques.* Delachaux et Niestlé, Neuchâtel.

Cousins, R. (2000) *Lincolnshire Building in the Mud and Stud Tradition.* Heritage Lincolnshire, Sealford, Lincolnshire.

Cubero, C. (1996) Estudio paleocarpológico de muestras de Intxur (Albistur-Tolosa, Guipúzcoa). In P. Ramil, C. Fernández Rodríguez, and M. Rodríguez Guitián (eds.) *Biogeografía Pleistocena de la Península Ibérica.* Xunta de Galicia, Santiago de Compostela, pp. 297–315.

Danaher, K. (1972) *The Year in Ireland. Irish Calendar Customs.* Mercier Press, Cork and Dublin.

D'Antuono, L. F. and Bravi, R. (1996) The hulled wheats industry: present developments and impact on genetic resources conservation. In S. Padulosi, K. Hammer and J. Heller (eds.) *Hulled Wheats. Promoting the conservation and use of underutilized and neglected crops* 4. Proceedings of the First International Workshop on Hulled Wheats. IPGRI, Rome, pp. 221–233.

D'Antuono, F., Bignami, C. and Mastronardi, E. (1993) Traditional agroecosystems and genotypes of cultivated plants inventory and conservation: case study in a Mediterranean highland (Alto Molise, central Italy). In M. G. Paoletti, T. Napier, O. Ferro, B. Stinner and D. Stinner (eds.) *Socio-economic and policy issues for sustainable farming systems.* Cooperativa Amicizia, Padova, pp. 269–286.

Davidson, A. (1999) *The Oxford Companion to Food.* Oxford University Press: Oxford.

Deur, D. and Turner, N. J. (2005) *Keeping it living: traditions of plant use and cultivation on the northwest coast of North America.* University of Washington Press, Seattle.

Devroey, J. P. and van Mol, J. J. (1989) *L'épeautre (Triticum spelta). Histoire et ethnologie.* Editions Dire, Treignes.

Dietsch-Sellami, M.-F., Durand, S. and Verdin, P. (2008) Traitement des plantes textiles à Maurecourt 'la Croix de Choisy' (Yvelines) au Néolithique ancien. Questionnement et premiers résultats. *Les Nouvelles de l'archéologie* 114, 28–32.

Diot, F. (1991) Analyse pollinique des mousses de calfatage du bateau fluvial à Godefroy, commune de Bouliac (Gironde). *Aquitania* 9, 225–235.

Dorvault, F.-L.-M. (1928) *L'officine ou répertoire général de pharmacie pratique.* 17ème édition. Vigot frères, Paris.

Duplessy, B., Gabert, A. Valabrègue, J.P. Vermorel, M.C and Dalmasso, L. (1996) *Le livre de l'épeautre.* Edisud, Aix-en-Provence.

Estabrook, G. F. (2006) Neither Wild nor Planted: Essential Role of Giesta (*Cytisus*, Fabaceae) in Traditional Agriculture of Beira Alta, Portugal. *Economic Botany* 60 (4), 307–320.

Evans, E. E. (1956) Fields, Fences and Gates. *Ulster Folklife* 2, 14–18.

Evans, E. E. (1957) *Irish Folk Ways.* Reprinted 1988. Routledge, London and New York.

Evans, E. E. (1989) *Mourne Country. Landscape and Life in South Down.* Dundalgan Press, Dundalk.

Fletcher, L. (1998) 'Strewings'. *Folk Life* 36, 66–71.

Girard, M. (1978) Résultats préliminaires de l'analyse pollinique. L'épave de la Madrague de Giens (Var). *Gallia* 34 (suppl.), 112–116.

Girard, M. and Maley, J. (1987) Autopsie d'une momie égyptienne du Muséum de Lyon: E-5 étude palynologique. *Nouv. Archiv. Mus. Hist. Nat. Lyon* 25, 103–110.

Girard, M. and Maley, J. (1999) La sépulture féminine du cercueil en plomb du quartier Trion-Gerlier de Lyon (IV e siècle après J. C.) Analyses polliniques. *Revue Archéologique de l'Est et du Centre Est* 50, 397–410.

Glassie, H. (1999) *Material Culture.* Indiana University Press, Bloomington and Indianapolis.

Goodman, J., Lovejoy, P. E. and Sherratt, A. (1995) *Consuming Habits. Drugs in History and Anthropology.* Routledge, London.

Grant, I. F. (1961) *Highland Folk Ways,* 1997 reprint. Birlinn, Edinburgh.

Grieve, M. (1931) *A Modern Herbal.* Volumes I and II. 1971 Reprint Edition, Dover, New York.

Gunda, B. (1983) Cultural ecology of old cultivated plants in the Carpathian area. *Ethnologia Europaea* 13 (2), 145–179.

Hald, M. (1942) The Nettle as a Culture Plant. *Folk-Liv* 6, 28–49.

Hammer, K. and Perrino, P. (1984) Further information on farro (*Triticum monococcum* L. and *T. dicoccon* (Schrank) in south Italy. *Die Kulturpflanze* 32:143–151.

Harris, A. (1991) Gorse in the East Riding of Yorkshire. *Folk Life,* 30 [1991–1992], 17–29.

Harvey, E. L. and Fuller, D. Q. (2005) Investigating crop processing using phytolith analysis: the example of rice and millets. *Journal of Archaeological Science* 32, 739–752.

Heiner, H. A. (2006) Maid Maleen (online) at *SurLaLune Fairy Tales.* Accessed November 2011: http://www.surlalunefairytales.com/maidmaleen/index.html

Hepper F. N. (1990) *Pharao's Flowers. The Botanical Treasures of Tutankhamun.* HSMO, London.

Hernández Bermejo, J. E. and García Sánchez, E. (2008) Las Gramíneas en al-Andalus. In García Sánchez, E. and Álvarez del Morales, C. (eds.) *Ciencias de la naturaleza en al-Andalus. Textos y Estudios VIII.* CSIC, Escuela de Estudios Árabes, Granada, pp. 235–287.

Hildebrand, E. A. (2003) Motives and opportunities for domestication: an ethnoarchaeological study in southwest Ethiopia. *Journal of Anthropological Archaeology* 22, 358–375.

Hillman, G. C. (1984a) Interpretation of archaeological plant remains: the application of ethnographical methods from Turkey. In W. van Zeist and W. A. Casparie (eds) *Plants and Ancient Man. Studies in palaeoethnobotany.* Proceedings of the sixth symposium of the IWGP 1983. A. A. Balkema, Rotterdam, pp. 1–41.

Hillman, G. C. (1984b) Traditional husbandry and processing of archaic cereals in recent times: the operations, products and equipment which might feature in Sumerian texts. Part I: the glume wheats. *Bulletin of Sumerian Agriculture* 1, 114–152.

Hillman, G. C. (1985) Traditional husbandry and processing of archaic cereals in recent times: the operations, products and equipment that might feature in Sumerian textes. Part II: The free-threshing cereals. *Bulletin on Sumerian Agriculture* 2, 1–31.

Houben, H. and Guillaud, H. (1994) *Earth Construction – a comprehensive guide,* CRATerre-EAG, Intermediate Technology Publications, London.

Hugo, V. (2005) *Les Misérables,* Editions de la Seine, MaxiPoche, Tôme 1.

Humphries, C. J. and Shaughnessy, E. (1987) *Gorse*. Shire Natural History Series, Shire Publications, Princes Risborough, Buckinghamshire.

Ibn al-'Awwām (1988) *Kitāb al-Filāha. Libro de agricultura*, edition and translation J.A. Banqueri, 2 vols. Madrid, 1802. Ministerio de Agricultura, Madrid.

Jacomet S. [and collaborators] (2008) *Identification of cereal remains from archaeological sites*. 3rd edition. IPAS/IPNA, Universität Basel, Basel.

Jahns H. M. (1989) *Guide des fougères, mousses et lichens d'Europe*. Delachaux and Niestlé, Paris.

Jeannet, J., Pignal, B., Pollet, G. and Scarato, P. P. (2003) *Le Pisé. Patrimoine, Restauration, Technique d'avenir*. 3rd Edition, Editions CREER, Collections Matérieaux, techniques et tours de mains, Nonette.

JO: Journal Officiel (25 Juin 2009) *Décret du Ministère de l'Agriculture et de la Pêche*.

Jones, G. (1984) Interpretation of archaeological plant remains. Ethnographic models from Greece. In W. van Zeist and W. A. Casparie (eds.) *Plants and Ancient Man. Studies in palaeoethnobotany*. Proceedings of the sixth symposium of the IWGP 1983. A. A. Balkema, Rotterdam, pp. 43–61.

Jones, G., Valamoti, S. and Charles, M. (2000) Early crop diversity: A 'new' glume wheat from northern Greece. *Vegetation History and Archaeobotany* 9 (3), 133–146.

Jouven M. (1942) *Les plantes à huile*. Coll. Rustica. Montsouris, Paris.

Kelly, F. (1998) *Early Irish Farming*. School of Celtic Studies and Dublin Institute for Advanced Studies, Dublin.

Kluge, F. (1995) *Etymologisches Wörterbuch der deutschen Sprache*. Bearbeitet von Elmar Seebold, 23. erweiterte Auflage. Walter de Gruyter, Berlin, New York.

Kluge, F. (1999) *Etymologisches Wörterbuch der deutschen Sprache*, entry *bei*. De Gruyter, Berlin, New York.

Lanfranchi, F. and Bui Thi Mai. (1998) L'oléastre et le lentisque, plantes oléagineuses sauvages dans l'économie néolithique en Corse et en Sardaigne. In V. M. Balmuth and R. H. Tykot (eds.) *Sardinian and Aegean Chronology: Towards the Resolution of Relative and Absolute Dating in the Mediterranean*. Studies in Sardinian Archaeology. Oxbow Books, Oxford, pp. 103–110.

Leach, H. M. (1997) The terminology of agricultural origins and food production systems – a horticultural perspective. *Antiquity* 71, 135–148.

Lucas, A. T. (1958) Furze. A Survey and History of its Uses in Ireland. *Béaloideas, The Journal of the Folklore of Ireland Society* 26 [1958–60], 1–204.

Lucas, A. T. (1959) Nettles and Charlock as Famine Food. *Breifne, Journal of the Breifne Historical Society* 1 (2), 137–146.

Marguerie, D. (1992) *Evolution de la végétation sous l'impact humain en Armorique du Néolithique aux périodes historiques*, Association des Travaux du Laboratoire d'Anthropologie, Préhistoire, Protohistoire et Quaternaire Armoricains.

Martin, L. (2002) *Les céréales de l'âge du Bronze à Ouroux-sur-Sâone (Sâone-et-Loire): essai pour une nouvelle approche de la morphométrie des blés vêtus*. Master thesis, Université de Besançon.

McCann, J. (2004) *Clay and Cob Buildings*. 3rd edition. Shire Publications, Princes Risborough.

McGovern, P. E. (2009) *Uncorking the Past*. Berkeley/ University of California Press, Los Angeles, London.

Nesbitt, M. and Samuel, D. (1996) From staple crop to extinction? The archaeology and history of the hulled wheats. In S. Padulosi, K. Hammer and J. Heller (eds.). *Hulled Wheats. Promoting the conservation and use of underutilized and neglected crops* 4. Proceedings of the First International Workshop on Hulled Wheats. IPGRI, Rome, pp. 4–41.

Novellino, D. (2006) An account of basket weaving and use of fibre plants in the Mount Aurunci Regional Park (central Italy) In Z. F. Ertuğ (ed.) *Proceedings of the IVth International Congress of Ethnobotany (ICEB 2005)*. Ege Yaynlar, 317–326

Padulosi, S. and Hoeschle-Zeledon. I. (2004) Underutilized species. What are they? *Leisa Magazine*, March 2004, 5–6.

Padulosi, S., Hammer, K. and Heller, J. (eds.) (1996) *Hulled Wheats. Promoting the conservation and use of underutilized and neglected crops* 4. Proceedings of the First International Workshop on Hulled Wheats. IPGRI, Rome

Papa, C. (1996) The 'farre de Montelione': landrace and representation. In S. Padulosi, K. Hammer and J. Heller (eds.). *Hulled Wheats. Promoting the conservation and use of underutilized and neglected crops* 4, Proceedings of the First International Workshop on Hulled Wheats. IPGRI, Rome, pp. 154–168.

Paris, R. R. and Moyse H. (1971–1981) *Matière médicale*. 4 volumes. Masson, Paris.

Peña-Chocarro, L. (1996) In-situ conservation of hulled-wheat species: the case of Spain. In S. Padulosi, K. Hammer and J. Heller (eds.). *Hulled Wheats. Promoting the conservation and use of underutilized and neglected crops 4*. Proceedings of the First International Workshop on Hulled Wheats. IPGRI, Rome, pp. 129–146.

Peña-Chocarro, L. (1999) *Prehistoric agriculture in southern Spain during the Neolithic and the Bronze Age; the application of ethnographic models*. BAR International Series 818. Archaeopress, Oxford.

Peña-Chocarro, L. and Zapata, L. (1997) El *Triticum dicoccum* (ezkandia) en Navarra: de la agricultura prehistórica a la extinción de un trigo arcaico. *Zainak* 14 249–262.

Peña-Chocarro, L. and Zapata, L. (1998) Hulled wheats in Spain: history of minor cereals. In J. Jaradat (ed.). *Triticeae III. Proceedings of the Third International Triticeae Symposium, Aleppo, Syria, 4–8 May 1997*. Science Publishers Inc., New Hampshire, pp. 45–52.

Peña-Chocarro, L. and Zapata, L. (2003) Post-harvesting processing of hulled wheats. An ethnoarchaeological approach. In P. C. Anderson, L.S. Cummings, T. K. Schippers and B. Simonel. (eds.). *Le traitement des récoltes: un regard sur la diversité, du Néolithique au présent.* Actes des XXIIIe rencontres internationales d'archéologie et d'histoire d'Antibes, 17–19 octobre 2002, Éditions APDCA, Antibes, pp. 99-113

Peña-Chocarro, L., Zapata, L., González Urquijo, J. E., Ibáñez Estévez, J. J. and Moreno García M. (2005) An ethnoarchaeological project in the Western Rif (northern Morocco): First results. *Quaderni del Laboratorio di Archeobiologia* 1, 21–32

Peña-Chocarro, L., Zapata, L., González Urquijo, J. E. and Ibáñez Estévez, J. (2009) Einkorn (*Triticum monococcum* L.) cultivation in mountain communities of the Westren Rif (Morocco): An ethnoarchaeological project. In E. Weiss and A. Fairbairn (eds.) *From Foragers to Farmers.* Oxbow Books, Oxford, 103–111.

Perrino, P., Laghetti, G., D'Antuono, L. F., al Ajlouni, M., Kanbertay, M., Szabó, A. T. and Hammer, K. (1996) Ecogeographical distribution of hulled wheat species. In S. Padulosi, K. Hammer. and J. Heller (eds.) *Hulled Wheats. Promoting the conservation and use of underutilized and neglected crops* 4. Proceedings of the First International Workshop on Hulled Wheats. IPGRI, Rome, pp. 101–121.

Perrot, E. and Paris R. (1974) *Les plantes médicinales.* 2 vol. PUF (éd.), Paris.

PFAF *Plants for a Future.* Database: http://www.pfaf.org/index.php

Phạm Hoàng Hộ (1999) *Cây Cỏ Việt Nam, an illustrated Flora of Vietnam* (in Vietnamese) Nhà Xuất Bản Trẻ. 3 volumes.

Plummer, C. (1910) *Vitae sanctorum Hiberniae*, vol. 1, Bethada i 126 § viii (13); trans ii 122, Oxford, Clarendon.

Pons, A. (1961) Analyse botanique du contenu d'une amphore antique découverte au large du Cap d'Agde. *Naturalia monspeliensia*, série botanique, f. 13, 43–50.

Reddy, S. N. (2003) *Discerning Palates of the Past. An Ethnoarchaeological Study of Crop Cultivation and Plant Usage in India.* International Monographs in Prehistory. Ethnoarchaeological Series 5.

Rivera Núñez, D. and Obón de Castro, C. (1991) *La guía de las plantas útiles y venenosas de la Península Ibérica y Baleares (excluídas medicinales).* Incafo, Madrid.

Roud, S. (2006) *The English Year.* Penguin, London.

Tallec, J.-L. (1997) Viêt-Nam. Les bateaux-paniers. *Chasse-Marée. Histoire et Ethnologie Maritime* 105, 10–23.

Thomé, O. F. (1885) Flora von Deutschland Österreich und der Schweiz. Gera-Untermhaus. In: K. Stüber (ed.) www.biolib.de. A collection of historic and modern biology books.

Trevisan Grandi, G., Dallai, D. and Mercuri, A. M. (1986) Contributo alla conoscenza dei contenuti pollinici in resine naturali e tentativi di interpretazione. Atti del 2° Congresso Nazionale dell'Associazione Italiana di Aerobiologia, pp. 368–373.

USDA gorse (*Ulex europaeus* L.) http://plants.usda.gov/java/profile?symbol=ULEU

van der Veen, M. (1999) The economic value of chaff and straw in arid and temperate zones. *Vegetation History and Archaeobotany* 8, 211–234.

Van Zeist, W. and Bakker-Heeres, J. A. H. (1982) Archaeobotanical studies in the Levant. 1: Neolithic sites in the Damascus Basin: Aswad, Ghoraifé, Ramad. *Palaeohistoria* 24, 165–256.

Vázquez-García, H. (2008) Gender, ethnicity, and economic status in plant management: uncultivated edible plants among the Nahuas and Popolucas of Veracruz, Mexico. *Agriculture and Human Values* 25, 65–77.

Vendryes, J. (1960) *Lexique étymologique de l'irlandais ancien. Lettres MNOP.* Dublin Institute for Advanced Studies, Dublin, and CNRS, Paris.

Vernet, J.-L. and Leroi-Gourhan, A. (1969) Étude botanique de résines de l'étang de Mauguio (Hérault). Ann. Soc. Hort. Hist. Nat. Hérault 109 (1), 32–36.

Vickery, R. (1995). *Dictionary of Plant-Lore.* Oxford University Press, Oxford.

Vignet-Zunz, J. (1993) Manifestations insolites dans le patrimoine technique des Jbala (Maroc). In P. Barraca (coord.). *Primeras Jornadas Internacionales sobre Tecnología Agraria Tradicional.* Museo Nacional del Pueblo Español, Madrid, pp. 137–146.

Vogt, C., Gouin, P., Bourgeois, G., Girard, M., Schvoerer, M., and Thiébault, S. (2002) Notes on Some of the Abbasid Amphorae of Istabl 'Antar-Fustat (Egypt). *Bulletin of the American Schools of Oriental Research* 326, 65–80.

Werblowsky, R. J. Z. (1968) Flagellation. *Encyclopaedia Britannica.* Vol. 9, pp. 410–411.

Wicha, S. and Girard, M. (2006) Archaeobotanical characterisation of three, ancient, sewn, Mediterranean shipwrecks. Connected by See. *Proc. of Tenth Intern. Symp. on Boat and Ship Archaeology, Roskilde 2003.* ISBSA 10. Chap 19. Oxbow Books, Oxford, pp. 111–116.

Willcox, G. (1995) Some plant impressions from Umm an-Nar Island. In K. Frifelt (ed.), *The Island of Umm an-Nar*, vol. 2. The third Millennium Settlement. Jutland Archaeological Publications XXVI/2, pp. 257–259.

Willcox, G. and Fornite, S. (1999) Impressions of wild cereal chaff in pisé from the 10th millennium uncal. B.P. at Jerf el Ahmar and Mureybet: northern Syria. *Vegetation History and Archaeobotany* 8 (1–2), 21–24.

Willcox, G. and Tengberg, M. (1995) Preliminary report on the archaeobotanical investigations at Tell Abraq, with special attention to the chaff impressions in mud brick. *Arabian archaelogzy and epigraphy* 6, 129–138.

Wrangham, R. (2009) *Catching Fire. How Cooking Made Us Human.* Basic Books, New York.

Zapata, L., Peña-Chocarro, L., Pérez Jordà, G. and Stika, H.-P. (2004) Early Neolithic agriculture in the Iberian Peninsula. *Journal of World Prehistory* 18 (4), 283–325.

7 Plants Used in Ritual Offerings and in Festive Contexts

7.1. INTRODUCTION

Ann-Marie Hansson and Andreas G. Heiss

Plants have always played an integral role in human life, reaching far beyond being just an arbitrary part of the available natural resources. Especially in pre-industrial societies, man's dependence on the growth of wild and cultivated plants is reflected by their significance in social and religious life.

We have chosen here to highlight two essential contexts where plants and plant products played – or still play – a role due to their social significance, allowing researchers to trace plant-related activities using a wide range of methods and across a broad time-scale (also refer to Chapter 9 in Book 2). Ritual contexts usually involve rites – ceremonies – performed in a prescribed manner. Festive contexts, in contrast, might sometimes hold a ritual character, but should always be joyous and mirthful. They can either have a profane or sacral background, or both (also see Chapter 7.4).

Ritual and festive contexts have in common that, from a 'modern' point of view, they do not seem to serve an 'objective' purpose, but instead strive after goals like communication with invisible beings, or maintaining natural life cycles by the use of spiritual forces. It is, however, crucial to note that any differentiations like 'material vs. transcendent', 'objective vs. irrational' or 'religious vs. secular' must not be applied uncritically to cultural contexts other than a modern European one, as they are very recent inventions only dating back to the Enlightenment of seventeenth-eighteenth century Europe. Past or present cultures which have not adopted Enlightenment ideas need to be investigated with caution regarding such categorisation: even the most 'secular' activities, such as sowing, may be steeped in transcendental thoughts (Brück 1999). At any rate, there is one promising common denominator for many ritual actions, past and present: magical, that is, animistic thinking. This is described as a belief that inanimate objects have a mind of their own, as do plants and animals (Subbotsky 2010, 7–8). Embedded in the persuasion that there is an invisible transcendent world, they are thought to become agents involved in the communication with this 'otherworld'.

Focusing on plants and plant products, we intend to present a broad survey of their various uses within the framework of the two contexts, ritual and festive. By no means can this survey be exhaustive; it is rather a mere sample of different applications to which plants and plant products were put in various parts of the world, most often in relation to agriculture. After this introductory section, five in-depth studies from different parts of the globe may serve to further illustrate these different approaches.

Ritual Plant Offerings

Written and oral traditions as well as archaeological and ethnological data show that, apparently in all times and all over the world, people have believed in supernatural forces, be they gods, the spirits of the forefathers, or elemental beings – some good, some evil. Man has turned to these invisible beings in various situations of decisive importance or when problems threatened an insecure existence. The various ways in which communication with these superior forces was established depended on social and economic circumstances, as well as on the environmental situation. Consequently, they have changed over the ages.

One very common way of communicating with, or paying worship to, the spiritual world was by means of an offering; here considered as a part of a sacral or cultic act in a ritual context. The word 'offering' derives from the Latin word *offerre* (to bring, to carry). An offering can be seen as the removal of objects or living things from the material world by depositing them in the ground or in the water, by consuming them during ritual festivals or by burning them (Lang 2002; Schwager 2002; see Fig. 7.1). Thus deprived of their material form, the offerings are thought to be transferred to a transcendental otherworld, or afterlife, there serving to placate the revered (or dreaded) spiritual beings (Schwager 2002). Among the various types of offerings, a broad variety of plants has been used by people in association with ritual purity, fertility, good health, prosperity and life, on the one hand, or with ritual impurity, infertility, sickness, ill-fate and death on the other (Simoons 1998, 3).

Fig. 7.1. Pyre of the altar of Olympian Zeus, as inspired by Pausanias' descriptions. Image from Herrmann (1984), modified.

Similar beliefs can be found in a broad variety of cultures and religions over time – although these cultures may be as different from each other as the Classical and Hellenistic Greeks in the east Mediterranean (see below; also *e.g.* Naiden 2006), the Iron Age populations of the Alps (Chapter 7.3), the Abrahamic religions from the Near East (see below; also *e.g.* Douglas 1999; Dafni *et al.* 2006; Naiden 2006), modern Thai farmers (see below), the ancient Chavín in north Peru (Chapter 7.6; also *e.g.* Miller and Burger 1995), the Zulu in South Africa (Lambert 1993), the Scandinavian peoples (see below, and in Chapter 7.2), or the East Polynesians (Leach 2003).

Circumstances which Affect our Capabilities to Gain Knowledge about Plant Offerings

Ethnological studies provide an excellent picture of present ritual activities and ancient texts or epigraphic sources tell us something about the past. But written records were not kept by prehistoric cultures. Those that exist concerning these people are usually provided by their literate (and often not too objective) contemporaries, sometimes also their own non-descriptive notation-writings (such as ogham and runic inscriptions) are preserved, but which are of very limited use to researchers.

Secondly, written sources from proto-historical and historical times seldom provide an exact record of actual daily life (be it the more 'secular' or the more 'ritual' aspects). A good example of this is the ritual cremation of the dead in the Roman provinces: archaeobotanical analyses have shown deliberate selection of the firewood used (Kreuz 2000; Heiss *et al.* 2008a; see below), while there is no mention of this in the written sources. Likewise, Roman written texts disdain barley as food for slaves and animals. Yet scientific data demonstrate that barley consumption was indeed frequent – and possibly essential – among Roman soldiers (Kreuz 2006; Britton and Huntley 2010). One reason for this may be that those who wrote the ancient texts usually belonged to the higher (literate) social stratum. Thus, environmental archaeology and bioarchaeology are necessary and important tools for shedding light on ceremonial activities of the past, even when dealing with ancient literate civilisations. In archaeobiological (regarding the role of plants: archaeobotanical) analysis, site formation and preservation conditions (taphonomy) play an important role. However, what we find and what has been lost depends on many factors (also see Chapter 2.2).

Without favourable preservation conditions, plant remains usually decay within weeks or months and are lost to analysis. Generally speaking, these are conditions which cause a reduced activity of microorganisms, as for example the acidic and anoxic conditions in peat bogs, permafrost in arctic or high-Alpine regions, the dryness in desert sands or the presence of toxic substances near decomposing copper or bronze objects. But even under unfavourable conditions, plant remains can be preserved for millennia if in a charred state. Charring takes place during exposure to fire under special conditions, such as low and constant temperatures and low availability of oxygen. However, various plant species, as well as plant parts, differ in their combustion and charring properties leading to varying probabilities of preservation. Even the probability for certain plants (or plant parts) of being exposed to fire influences preservation and clearly depends on the way the respective plants are used (for food, tools, construction, medical or other purposes) and on their role in ceremonial contexts. Mineralised plant remains, often found in cesspits, are also preserved very well. However, their occurrence is limited to special archaeological contexts or environmental conditions. The same is true for plant impressions in clay (either in pottery or daub) or for deposits of pollen or phytoliths (see Chapter 2.2 for detailed information on these finds).

Taking into account these considerations on environmental setting and archaeological context, it seems clear that the question of what we can find depends, to a great extent, on the particular region, its climate and topography. For instance, the desiccated finds from Egyptian Pharaonic graves are in an excellent state of preservation and may appear almost recent, while the organic material from much younger Scandinavian Iron Age cremation graves has almost totally disintegrated.

Excavation, sampling and analysis may also play important roles. In contrast to most artefacts and skeletal materials, plant remains are hardly ever visible during excavations. Tiny black concretions in the soil may represent the remains of grave goods, such as bread or other food, but are often overlooked or are not sampled. Only rarely do these minute remains find a place in excavation reports, or get stored in a museum. Also, due to the considerable amount of time required for archaeobotanical analyses, funding is often not available, although these plant remains are of great potential value.

So what do we know, and what can we know about the plant material used as offerings? What we find and interpret as offerings is based on our current knowledge of the botanical material itself as well as of the social context in present, historical, and prehistoric times. This social context can only be reconstructed in an approximate way from oral and written traditions, from archaeological contexts and ethnographic comparison. Still, even though every effort is taken to be as objective as possible, to some extent the resulting interpretations will always reflect the state of our own knowledge. Analysis of ritual contexts requires a high degree of caution, and conclusions need to be rooted in the observation of structural similarities and generally observable phenomena. The article on the festive use of plants on May Day (Chapter 7.4) illustrates the complexity of present-day ritual conceptions, which would be virtually imperceptible if they had to be explored via archaeological remains.

As already mentioned, the range of plant use in ritual and festive contexts is very wide, and only some of the more prominent examples can be dealt with in this chapter. Thus, while a few examples of other ritual and festive uses for plants are provided, our main focus has been applied to nothing less than the two most fundamental aspects of existence: life and death. As we shall see, many religious perceptions bear witness to a strong linkage between these antitheses, reflected in bonds between fertility and funeral. It is from the dead that new life springs, and it is the offering which brings energy to the force of transformation and creation.

Sustaining Life: Rituals in Agricultural Processes

In agricultural communities, it was common to use rituals or ceremonies involving plants or plant products in various forms associated with the different stages of the agricultural process as a reflection of human beings' dependence on the forces of nature. This included sowing, harvesting or winnowing as well as the subsequent stage of plant processing, whether from cultivated crops or gathered plants, berries, nuts and fruit. It must be taken into consideration that these and many other activities of daily life were usually influenced by ritual, or had ritual meaning in themselves. In the past, archaeologists have sometimes too easily qualified apparently non-functional or simply inexplicable finds and find situations as 'ritual' (Brück 1999; Insoll 2004, 10–12). Renfrew and Bahn (2008, 412–414) address this issue by offering a comprehensive 'check-list' of how to detect ritual activity in the archaeological record. While bearing this in mind, it is still challenging for us to try to understand how people from past times might have thought and felt, in this case about plants.

Fig. 7.2. An ear of modern hulled barley (*Hordeum vulgare* L.) as photographed in a field at Termen in the Valais, Switzerland. Image: A. G. Heiss.

The Fertility Cult of Demeter and Persephone/Kore

The first in our series of case studies leads us to the east Mediterranean. For Ancient Greek culture, ample evidence from multiple sources is available: written and iconographic records, as well as artefactual and biological remains, can tell us us about past rituals of the region and about some of the thoughts people may have connected to these rituals.

A large number of annual festivals in Ancient Greece were devoted to Demeter and Persephone, concentrating ritual attention on processes critical to the crops, and thus to nutrition itself. Demeter ruled over growth, especially of the crops, and her influence was greatest over barley (*Hordeum* sp.; Fig. 7.2), the most important cereal species at the time. She was also the symbol for the origin of bread (Buxton 2007, 72) and was seen as having given the knowledge of agriculture to man (Fig. 7.3), thus laying the foundation for nutritional security and for civilisation itself (Taylor-Perry 2003). Snakes and piglets, both symbols of fertility, were also often associated with Demeter.

Fig. 7.3. Demeter (to the left) teaching the secrets of agriculture to the mythical prince Triptolemos of Eleusis, handing him a sheaf of grain. Persephone (to the right) offers him her blessings. Replica of a *ca.* 430 BCE relief at the Museum Schloss Hohentübingen, Germany (original at the National Archaeological Museum of Athens). Image: A. G. Heiss.

Fig. 7.4. A Hellenistic tetradrachm from Syracuse, Sicily (*ca.* 310–304 BCE), showing a female goddess' head adorned with an ear of barley, a typical way to depict Persephone. Image: Numismatica Ars Classica NAC AG, Zürich.

Demeter's daughter Persephone (also called Kore, the 'virgin' or 'young maiden') was worshipped as the goddess of birth and death. In the form of Kore, Persephone was also believed to symbolise the power of growth within the seed or cereal kernel (Day 2007, 72), as depicted on many Greek coins from the Hellenistic period (Fig. 7.4). The concept of worshipping the growth power of seeds seems to be an archetype, having already been of vital importance and widespread at a far earlier date than Greek Antiquity. We can find, for instance, the same idea of the growing seeds painted on pottery as far away as the Ukraine and dating back as far as the fourth millennium BCE (Fig. 7.5).

The fertility rituals in Classical Attica were surrounded by mysticism and ceremonial secrecy. A large amount of today's knowledge about what was going on during these ceremonies is derived from archaeological finds rather than written sources, which may seem unusual for Antiquity. One of the reasons for the lack of written texts may well be that many of these rituals were exclusively carried out by women while it was mainly men who recorded

Fig. 7.5. Germinating seeds commonly occur in vase paintings of the Cucuteni-Tripolye culture. On (3), sprouting seeds (lower part) are associated with a three-fingered figure enclosed in a seed (the Goddess with bird's feet instead of human hands), surrounded by streams: note the snakes on the upper end central bands. Painted black on red. (1) and (2) Cucuteni AB (Verem'e, W. Ukraine: *ca.* 4000 BCE). (3) Cucuteni B1 (Luka Vrublevetskaya, W. Ukraine: early fourth millennia BCE). (1) H. 24.2 cm. (2) H. 22.7 cm. (3) H. 38 cm. Image from Gimbutas (1989, Figure 165), modified.

history. Women performed the central rituals because of their culturally-determined value as bearers of the secrets of fertility (Brumfield 1981, 239, 240). The most prominent of these festivities were the Mysteries of Eleusis and the *Thesmophoria* (see below). However, there were also other festivals dedicated to Demeter and Kore open to all, not only to women (Bookidis 1990, 86–87). Surveys of these more than twenty annual festivals can be found in Harrison (1908) or Taylor-Perry (2003), among others.

One of the most important written sources giving information on the ceremonies is the Homeric hymn to Demeter (Lawton 1898, 154–179; Björkeson 2004). The principal theme deals with women re-enacting aspects of the myth of Demeter searching for her daughter after Kore's abduction by Hades. This period of searching and mourning for her daughter took one third of a year, during which Demeter abstained from her role as Goddess of the harvest and growth. So while Persephone was in Hades' realm, vegetation died.

While temples for Greek Gods were usually situated on high, open places (especially sanctuaries for Zeus); Demeter was very often worshipped in sanctuaries under the earth or in flat *megaron*-type temples, corresponding to the chthonic (earth-bound) nature of this goddess. An impressive example can be found in Selinous/Selinunte, Sicily (Fig. 7.6). There, the fifth to sixth-century BCE temple of Demeter Malophoros ('apple-bearing Demeter'; see Fig. 7.7 and Miles 1998) forms an ample sanctuary with close links to the underworld – represented by the proximity to the Selinuntine necropolis – and the integration of the temples of two other chthonic deities, Hekate and Zeus Meilichios ('Zeus as sweet as honey'; see also Hermann 1847; Rose 1921).

Fig. 7.6. Map of the central Mediterranean sea with the following archaeological sites mentioned in the text: 1) Selinunte; 2) Sanctuary of Monte Papalucio; 3) Eleusis; 4) Acrocorinth. Map: R. Lugon, J.-C. Loubier and A. Chevalier.

Fig. 7.7. Part of the temenos (temple area) of Demeter Malophoros in Selinous, Sicily, as seen from the east. a: propylon (entrance hall), b: great altar, c: megaron. Image: A. G. Heiss

The Offerings

Offerings for Demeter were often thrown into pits (*megara* or into natural chasms, as described for the *kathodos* (the 'lowering of the offerings') during the festivities of the *Thesmophoria* (see below). These offerings usually consisted of biscuits or bread in different shapes associated with fertility, such as snakes or phalli (Brumfield 1981, 79, 232; Taylor-Perry 2003, 52). The snake symbol seems to represent some more archaic traits of the cult, possibly originating from some Stone Age mother goddess as suggested by M. Gimbutas (1989, 147). Aside from written sources and iconography, archaeological finds also tell us about the things offered: for example, charred biscuits resembling coiled snakes were found at the Archaic/Hellenistic sanctuary of Monte Papalucio in Apulia, southern Italy (Ciarialdi 1999; Fig. 7.6). Charred remains of various food plants were also discovered at this site: broad bean (*Vicia faba* L.) regarded as associated with the underworld in general (see below; Hanelt 1972, 209) and pomegranate (*Punica granatum* L.; Fig. 7.8), a symbol of fertility, as well as one of the key elements in the Persephone myth: Persephone had eaten seeds of pomegranate during her stay in the underworld which obligated her to return there every year for a certain time. This was regarded as the reason for the recurring seasons, the barren season being caused by her absence.

Opium poppy (*Papaver somniferum* L.; Fig. 7.9) was likewise closely associated with Demeter and Persephone. Seed impressions of poppy were found in the pastry from Monte Papalucio, along with the remains of cultivated fruits such as grapevine, fig and olive. In general, depictions of pomegranate

Fig. 7.8. Flower (above left) and fruit (above right) of modern pomegranate (*Punica granatum* L.) as photographed in the gardens of Schloss Trauttmansdorff at Merano/Meran, Italy. Image: A. G. Heiss.

Fig. 7.9. Flowers and capsules (right) of modern opium poppy (*Papaver somniferum* L.) as photographed in a field at Rappottenstein, Austria. Image: M. Kohler-Schneider.

and poppy are commonly observed in Persephone/Kore sculptures, such as those found in a sanctuary of Herakleia near Policoro, southern Italy (Kurz 2007; 2008; Fig. 7.6). One of the fragments of painted plates from Acrocorinth displays the lower part of a female head (Demeter/Kore) in profile with a poppy capsule held close to her mouth (Stroud 1968, 302).

As mentioned above, the animal elements in the sacrifices were mainly represented by pigs and piglets or terracotta figurines of them (Burkert 1985, 242). A terracotta *kalathiskos* bearing representations of acorns was found in Policoro, thus documenting one main food source for those pigs (Kurz 2007; 2008).

A Bean from the Underworld

As indicated above, the broad bean, or field bean (*Vicia faba* L.; Fig. 7.10), was closely connected to Demeter and Kore. In general, although being an important staple food, the plant was regarded as a death symbol in the Greek world, and it was seen as associated with the souls of deceased people (Ciarialdi 1999). This chthonic connotation of a food plant may be related to the effects of 'favism', a syndrome induced by the ingestion of broad bean (or inhalation of its pollen), together with an inherited enzyme deficiency in some humans, leading to a sometimes lethal haemolytic anaemia (Flint-Hamilton 1999). Another reason may be that the cultivars in Antiquity generally contained more anti-nutritional substances than we know from modern broad beans (Leopold and Ardrey 1972, 512).

Fig. 7.10. Flowering broad bean (*Vicia faba* L.) in the reconstructed Neolithic field in the Urgeschichtemuseum Niederösterreich. Image: A. G. Heiss.

Perhaps for the same reasons, other Mediterranean cultures also treated the broad bean with dread and reverence, or at least with caution (Hanelt 1972, 209). If we believe Herodotus' reports, Egyptians regarded *Vicia faba* as unclean (*Historiae*, Book II 'Euterpe', 37; Macaulay 1890). In Ancient Rome, there was no explicit negative connotation related to *V. faba*, yet the plant was of symbolic and sacral importance: in the annual ritual of the *Fabrariae Calendae* (the 'calends of the beans') in the beginning of June, a sacrifice of mashed broad beans with lard was offered to *Dea Carna*, the Goddess of death and life (von Stokar 1951, 521). Porridge made of *lomentum* (faba bean meal) was considered an acceptable offering in general, termed *refriva*, or *referiva* (Pliny the Elder, *Naturalis Historia* 18, 117–119: Bostock 1855).

The Thesmophoria

One prominent example of the festivities tied to the myth of Demeter and Kore are the Athenian festivals of the *Thesmophoria*, which were celebrated in autumn and aimed at magically and ritually influencing the course of events surrounding the important autumn sowing. The word derives from θεσμοφόρος, 'bearing the ordinance' (Taylor-Perry 2003), because Demeter had given the knowledge of agriculture and the laws of civilisation to humans, as already mentioned. Beginning with the eleventh of the month of *pyanepsion* (which corresponds to the transition from October to November), the absence of Persephone and Demeter and the resulting withering of vegetation was commemorated for three days. Each day of the festival had its own name, depending on the activities during the day; *e.g. anodos*, the ascent to the temple and *kathodos*, the lowering of the offerings into the offering pits, both on the first day of the *Thesmophoria*. The festival ended with *kalligenia*, the '(re)birth of the beautiful Kore', the celebration of the reunion of the two goddesses and of the beginning of new life for the germinating crop.

One important part of the annual ritual activities was also to remove (and to reuse!) the decayed matter: the decayed corpses of pigs that had been thrown into the caverns or pits during the *Thesmophoria*, together with seeds, were collected and placed on the altars. Later they were sprinkled over the fields to promote the fertility of the crops (Brumfield 1981; Gimbutas 1984, 214f).

Carrying the Liknon

An interesting recurring element in the fertility rituals of Ancient Greece is the *liknon* (λίκνον meaning 'winnowing-fan'), in some contexts also synonymised as *kernos* or *ptuon* (Harrison 1908). The winnowing-fan is an oblong, scoop-like basket usually made of wickerwork (Nilsson 1957, 21) used for crop cleaning: it is rotated and shaken in both hands to separate the lighter chaff from the heavier grains (for more information see Chapter 6.6).

This winnowing-fan was a common element of both the Eleusinian mysteries of Demeter and Kore (Brumfield 1997, 148) and the Dionysian mysteries (Harrison 1908, 158; Nilsson 1957, 30). In processions (the *Liknophoria/Kernophoria*), the *likna* were carried containing fruits of the earth and often also shrouded phalli (Fig. 7.11; Nilsson 1957, 21), thus emphasising the close connection between the growth of food plants and human fertility. Athenaios, an author active during the third century CE, gives a list of the foodstuffs usually contained in a *liknon/kernos*, mentioning: 'sage, white poppies, wheat, barley, pulse, vetch, *ochroi*, lentils, beans, spelt, oats, a cake, honey, oil, wine, milk, sheep's wool unwashed' (Harrison 1908, 159).

Fig. 7.11. Campana relief dating to the late first/early second century CE, depicting an initiation scene during the Dionysian mysteries: a satyr approaching from the left carries a liknon filled with fruits and a phallus. A maenad leads forward the initiate with shrouded head. Another woman bears a tympanon. Image: Friedrich-Schiller-Universität Jena, Archäologische Sammlungen (Inv. no. SAK T 200).

However, according to Athenaios' description, these things were put in a more elaborate type of

Fig. 7.12. Proto-Geometric clay kernos from Kourtes, Crete (*ca.* tenth century BCE), kept at the Herakleion Archaeological Museum. The individual vessels are connected. Image: Robert H. Consoli / SquinchPix (2009).

Fig. 7.13. Early Hellenistic votive plaque (mid-fourth century BCE) from Eleusis, held at the National Archaeological Museum, Athens. Displayed is a scene of the Eleusinian mysteries.Ttop right: Demeter enthroned. Two other deities, Persephone and Iakchos, bear torches, leading the initiates forward. A sacred omphalos stone (in white) lies in the middle of the scene. Note the four female figures who wear kernoi in their hair. Image: marsyas@wikimedia.org (2005).

kernos, different from the agricultural implement, 'a vessel made of earthenware' (Harrison 1908, 159), and often divided into many compartments. These written records are complemented by the objects found at several archaeological sites, such as Melos, Crete (Fig. 7.12), Eleusis (Fig. 7.13) and Acrocorinth. As an example, at the sanctuary of Demeter and Persephone at Acrocorinth, numerous clay objects were found resembling winnowing fans containing cakes. There were also smaller empty vessels, which were probably once filled with real kernels of wheat (Stroud 1965, 23). Votive *likna* were present in the sanctuary from around the early sixth until the second century BCE, usually in secondary fills all over Acrocorinth, possibly originating from periodic cleanings of the sanctuary – none was found *in situ* (Brumfield 1997, 147).

The role of *likna/kernoi* in the rituals seems to have had three main aspects: firstly, the functionality of cleaning the grain was analogous to a mystical purification, important for the Eleusinian initiates. Secondly, the symbolism of human fertility: aside from the Dionysian symbol of a phallus associated with the *liknon*, winnow-fans were also said to have served as cradles for Zeus and for Dionysos, according to certain myths. Brumfield (1997, 149) sums this up as 'the divine infant in the *liknon* was a potent symbol of the renewal of life that agriculture made possible.' And thirdly, there was an explicit focus on agriculture in the *Liknophoria/Kernophoria* rituals, a 'late and elaborate form of the offering of first fruits', as Harrison (1908, 160) put it.

Rome: Ceres and Tellus

In the Roman world, many fertility-related rituals were in existence. Most of them are well-documented, especially in the cereal-cultivating areas of the Mediterranean. A good number of them bear strong resemblance to older Greek traditions (see above). The goddess Ceres, for example, is characterised by many aspects of the Greek Demeter and also the Carthaginian Tanit (Währen 1980), such as the polarity of life-giving and death-related powers. Likewise, she was worshipped in the course of the agricultural year, such as during the *Cerealia* which took place at the time of spring sowing (Lurker 2004). At the end of January, during the *Feriae Sementivae*, the ending of the winter sowing was celebrated with offering cakes and pigs to Tellus (the 'Mother Earth') and Ceres (Bailey 1907).

Other Agricultural Contexts and their Offerings

There are many publications dealing with early agriculture in different parts of the world: how it has developed through time, which crops have been cultivated and on the origins and morphological and ecological changes of the various crops. As regards early agricultural societies' relations to the invisible forces, gods and goddesses, which were believed to impact the crop in a positive or negative way, we will only mention a few here, as it is difficult to trace rituals connected to cultivation. It is even more difficult to find out what the purpose of these offerings or other rituals might have been. Thus, evidence within this field is scarce.

Finds from Ukrainian Hill Forts

Open settlement type villages of the North Carpathian culture (*ca.* 500–200 BCE) are found on elevated flat river terraces, such as in the district of Kharkov/Charkov. There were also many large

Fig. 7.14. Sites in Ukraine and Russia as mentioned in the text. Map: R. Lugon, J.-C. Loubier and A. Chevalier.

hill forts that were centres of administrative, political and religious power. Whole complexes of archaeological monuments dating from the same period are usually found around these hill forts. In both early- and proto-Slavic periods, the peaks of these hill forts often were the site of temples. In such cult places, starting from Scythian times, offerings of acorns (*Quercus* sp.), cereal grains and bread were found, together with miniature and normal-sized pots, iron ornaments and other objects (Gimbutas 1971, 49, 50, 158).

Ceramic idols of bread loaves or biscuits were also found in the hill fort of Karavan in the Kharkov (Charkov) district (Fig. 7.14), dating to the fifth to sixth centuries BCE. The structure at the hill fort's highest point was interpreted as a small sanctuary consisting of a clay sacrificial basin and three oval pits. In and around the sanctuary, offerings of all kinds were found: some miniature pots, a miniature crucible and miniature bread loaves made of specially prepared clay. The clay used for the production of the pots was very fine-grained and without any sand admixtures, while the clay for the production of the small loaves was tempered with cereal grains, chaff and leaves. Some of the biscuits had straw imprints (Gimbutas 1971, 51). There were also clay representations of grains of food plants: wheat, barley, rye, pea, chick pea, cow pea and millet. A similar find was made in the hill fort at Paterskoe in the district of Kirovograd (Fig. 7.14), where the miniature biscuits were baked of actual flour and grains of millet (Petrov 1948, 79; Monah 2002, 94).

Iron Age Europe

During recent years, much research has been focused on the Iron Age cultures of Europe, usually subsumed as the 'Celts': all written documents from that time attesting to their daily life, warfare, or rituals, were in fact written by Greek or Roman authors (Hofeneder 2005) and this introduced a strong cultural bias in the perception of these peoples. Direct evidence on ritual plant use by the Celts, however, is scarce – not least due to methodological reasons: the most famous Iron Age sanctuaries in France, such as Gournay-sur-Aronde, Ribemont-sur-Ancre, or Corent, have never been investigated for botanical remains...

A Celtic farmstead at Fellbach-Schmiden (south-western Germany), probably belonging to the local elite, held a rich inventory of botanical finds drowned in a former well (Körber-Grohne 1999). The more than 20 m deep shaft had been filled with manure, soil and rubble after the building had burned down – covering three animal statues made of oak wood: two he-goats and a stag. The he-goats had belonged to a statue, probably some 'Lord/Lady of the Animals', while the stag seems to display a relation to the Celtic god Cernunnos (Hoppe 2009). For a showy, yet highly speculative combination in reconstructing the respective finds see Fig. 7.15.

a

b

Fig. 7.15. a) The he-goats found in the well of Fellbach-Schmiden, Baden-Württemberg, Germany. Image: Landesmuseum Württemberg, Stuttgart (2011). b) Replica of the two he-goats in an interpretative reconstruction, combining them with the sitting statue of the Celtic god Cernunnos as inspired by the depiction on the famous Gundestrup cauldron. Reconstruction done by Wolfgang Lobisser, VIAS, and kept at the Urgeschichtemuseum Niederösterreich, Asparn/Zaya. Image: Urgeschichtemuseum Niederösterreich (2009).

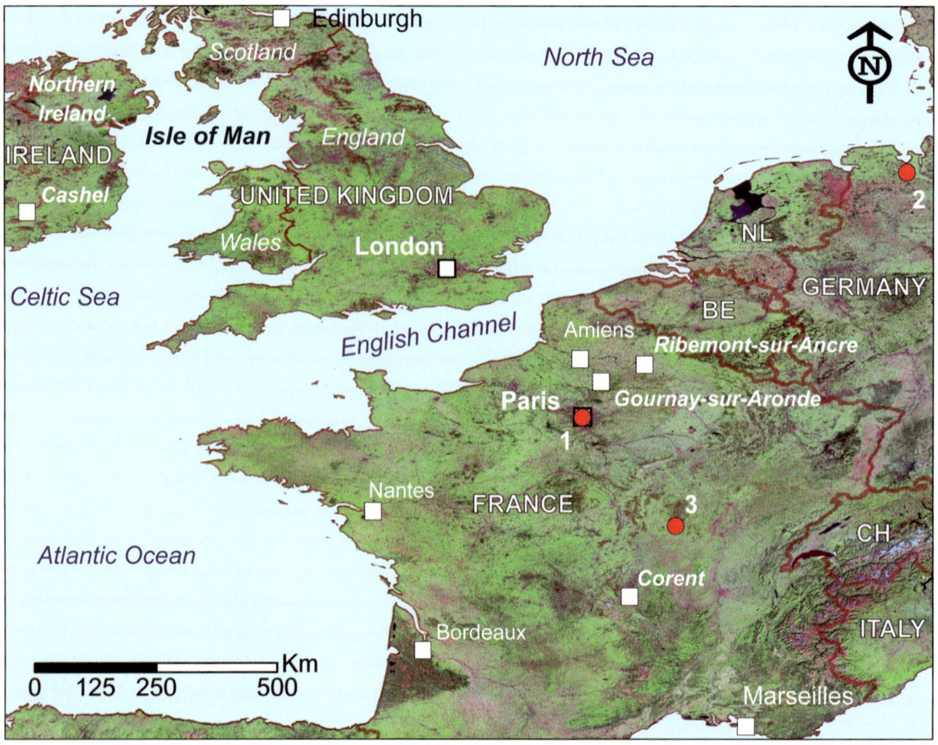

Fig. 7.16. Map of the British Isles and northwestern Europe with the locations mentioned in the text. The Loire-Atlantique region is centred around the city of Nantes in France. The following archaeological sites are mentioned in the text: 1) Le Louvre; 2) Ipwege bog; 3) Bibracte. Map: R. Lugon, J.-C. Loubier and A. Chevalier.

In the sanctuary '1' of Roseldorf (Lower Austria, *ca.* 250–130/110 BCE), a site located in the remote east of Celtic cultures but roughly corresponding to French sanctuaries like Gournay-sur-Aronde or Corent (Poux *et al.* 2007; Holzer 2009), animal and plant remains seem to derive from ritual feasting. Also, possible ritual significance of some of the discovered plants, such as rowan (*Sorbus aucuparia*) and hawthorn (*Crataegus laevigata*), is suggested (Caneppele *et al.* 2010). Similar, if poorer, remains from feasting also seem to be indicated at Frauenberg, another Late Iron Age sanctuary in Austria (Styria; Popovtschak 2005).

In the Celtic oppidum of Bibracte (Beuvray, central France; Fig. 7.16), the remains of a wooden box found near a sanctuary contained 'a strange mixture of burned twigs, carbonised straw, single culm nodes, chaff, cereal grains and small pieces of burned bones' (Wiethold 1996, 107). The major part of the charred plant remains were identified as emmer (*Triticum dicoccum*) chaff, accompanied by remains of other hulled wheats, pulses, and wild plants. The question of whether the plants inside the box were connected with some ritual activity or were just animal fodder is, as yet, unanswered. Not

Fig. 7.17. Map of the southern Scandinavian peninsula and neighbouring countries with the sites mentioned in the text: 1) Birka; 2) Lovö; 3) Helgö; 4) Ovelgönne; 5) Boberget; 6) Oseberg; 7) Odensala; 8) Runsa. Map: J.-C. Loubier and A. Chevalier.

strictly part of the peoples today known as 'Celtic', the Alpine population also had their own way of offering. Their offering sites (*Brandopferplätze*), reflecting agricultural aspects of the prehistoric Alpine societies, are discussed in Chapter 7.3.

One of Sweden's most famous archaeological sites is Helgö, Ekerö parish, east central Sweden, at that time an island in Lake Mälaren (Fig. 7.17). From 1954 to 1978, regular archaeological excavations were carried out at the site (Melin and Sigvallius 2001), as well as archaeological and environmental fieldwork (Miller and Hedin 1988). The site comprises seven building groups and six cemeteries; in three of the cemeteries bread was found. Many house foundations, some indicating they belonged to workshops, were found at the site. One of them has lately been interpreted as an offering terrace used from *ca.* 200 CE up to the Viking Age (Hansson 2011; Bergström 2007). Charred bread buns were retrieved from the archaeological layers at the terrace. Now this terrace is believed to relate to rituals associated with agricultural processes and with influences from the Classical world (Hansson 2011).

A Selfless Divinity: the Rice Mother

In southeast Asia, rice is the most important cereal, and there are many rice-growing countries where a Rice Goddess was and is worshipped. As an example, we will mention Thailand, where evidence of rice cultivation can be dated as far back as *ca.* 3000 BCE (Glover and Higham 1996, 436). Thailand is, in fact, a predominantly Buddhist country, but with strong undercurrents of Hinduism and animism. In present-day Thailand, traditional rituals involving the Rice Goddess are still important for rice farmers (Sanyal 2008). The Rice Goddess (โพสพ, *poh-sòp*), also called the Rice Mother (แม่โพสพ, *mâe poh-sòp*; Fig. 7.18), is a benevolent deity related to the goddesses Dewi S(h)ri in Indonesia, Saning Sari on Sumatra and other deities found on Borneo, in Malaysia, or Burma (Frazer 2009, 972f). Quite comparable to Greek Persephone and Egyptian Osiris, the Rice Mother reigns over the main cereal of the region and is responsible for the fertility of the rice fields. Furthermore, the rice plant is imagined to represent her body and the grains her children, whom she selflessly gives away to ensure human subsistence. Accompanied by offerings of fruit and rice, her spirit is invited to the rice fields during sowing and harvest and, during her 'pregnancy'

Fig. 7.18. Depiction of mâe poh-sòp (Rice Mother) on a modern lucky charm amulet, as found in countless numbers on Thai amulet markets. Image: Ajarn Spencer.

with the rice grains, she is nourished with offerings of bitter and sour fruit (reflecting the peculiar food cravings attributed to pregnant women). The major part of the rituals is exclusively performed by women (Rajadhon 1955; 1968; Hubert 1981). Again pointing out the parallels to Persephone and Osiris, the connection of the main cereal of a region to a guardian goddess or god seems to form an archetype in several – and quite unrelated – religions.

Flesh and Grain, Blood and Beer: A Few African Examples

A drastic example similar to *mâe poh-sòp*'s motif – the nourishing deity giving away part of his/her body as food – is found in the tales of the Bulsa (or Builsa) (Fig. 7.19) of Ghana: their God (*Wen* or *Naawen*) is said to have initially lived on earth, providing humans with a paradisiacal state of life without work. They nourished themselves with pieces of flesh which the god offered to let them cut off of him and these would then become food (Schott 1997, 1). This is, however, an unusual example, as many other African peoples initially did not follow the idea of a deity associated with nature and the cycle of life and fertility. Much more frequent is the concept of the ancestors being worshipped in

Fig. 7.19. Map of Africa showing the four regions mentioned in the text: the Gauteng and Kwazulu-Natal provinces in South Africa (Zulu); the Arusha province in Tanzania (Iraqw), and the the Bulsa (or Builsa) district in Ghana. Map: R. Lugon, J.-C. Loubier and A. Chevalier.

these contexts. Fertility-related rituals of the Zulu in South Africa once featured the sacrifice of cattle and of grains. Lambert (1993, 304) sees a parallel to the rituals in the *Thesmophoria* (see above), as one part of the mixture of grains with the sacrificed animal's blood was used for sowing in order to maintain fertility of the fields. Today, beer libations seem to have taken the place of the cereals in Zulu harvest offerings, as is also reported from their neighbours, the Lovedu (Krige 1968, 183). Likewise, the sorghum beer (*buura*) prepared by the Iraqw in Tanzania (Arusha region; Fig. 7.19) plays a crucial role in their offerings to the ancestors, and in the *geetla/angw* ritual, marking the transition between two agricultural cycles (Rekdal 1996, 371).

Burial Rituals

Burial rituals may be seen as an expression of resurrection and reproduction, symbolising the end of the old life and, at the same time, bringing the vital force back to a new life. In this way they express perceptions about fertility and the deceased's continued existence (Kaliff 1997), especially in prehistoric agricultural communities. Such a synthesis of burial cults and fertility cults is commonly supposed to have existed. Cultivation and the cult of the dead are phenomena linked together in many ways, not least by their affinity to the soil. The dead are placed into the earth in one form or another in the same way grains are planted.

Both the grain and the dead thus enter another dimension, hidden from other spheres of existence. In the soil, the plants will grow to a new life and, it is imagined, that also the dead have a new life. In such an earth-centred and life-revering worldview, the soil is the place for the cycle of creation's eternally recurring fertility (Gimbutas 1989, 321). In various religions, these perceptions are mirrored by similar recurring motifs, such as the self-sacrifice of a deity in order to sustain human life. This basic concept is found in the Greek Persephone/Kore (see above), in the Egyptian Osiris (see below), and in the Thai *mâe poh-sòp* (see above). Certainly, they all make use of quite different expressions of how to support or sustain human life and are not always strictly linked to fertility. Often these sacrifices are also connected with the motif of resurrection.

Taking extant religions as modern analogies, we might suppose that, in prehistoric societies, the eschatological conceptions about what the afterlife was like must have influenced the way in which vegetable materials were used as symbols in their rituals, what was offered, and in which condition. It is thought that in many prehistoric communities, existence after death was believed to resemble life on earth to a certain extent, as deduced from findings of expensive grave goods. These gifts varied depending on when and where the burial took place and, of course, on the social position of the dead. For civilisations of Antiquity, we have written evidence and also scientific data indicating that the burial ritual was highly formalised and that many of the functional aspects were dictated by traditions.

Plants from Egyptian Tombs: Flowers, Fruit, and Cereal Sprouts

An enormously wide range of rituals, many of them involving food and plants, is documented by archaeological evidence from Ancient Egyptian culture. The large variety of finds from these contexts usually owes its preservation to the special environmental conditions in tombs, where dry and hot air desiccates biological material (see also Chapter 2). However, any account claiming to comprehensively summarise them inevitably has to fail. Thus, in this section only a few of the most prominent examples are highlighted.

Finds of cereals as grave goods in Egyptian contexts date back as early as Predynastic times: cereal grains were found on and around the deceased in burials of *ca.* 5000 BCE. The concentration of cereals was especially high near the corpse's mouth, which might indicate the grains were meant as symbolic provisions for the afterlife (Scharff 1947, cited in Centrone 2005). A Predynastic (*ca.* 3600 BCE) woman's grave from Hierakonpolis (Fig. 7.20) provided a variety of botanical grave goods, among them sedge tubers (*Cyperus* sp.), fruits of dill (*Anethum graveolens* L.), and balanites (*Balanites aegyptiaca* L.), interpreted as indicators of the deceased's high status (Fahmy 2005). Figure 7.21 shows desiccated barley grains from an 18th-dynasty graveyard (*ca.* 1450 BC).

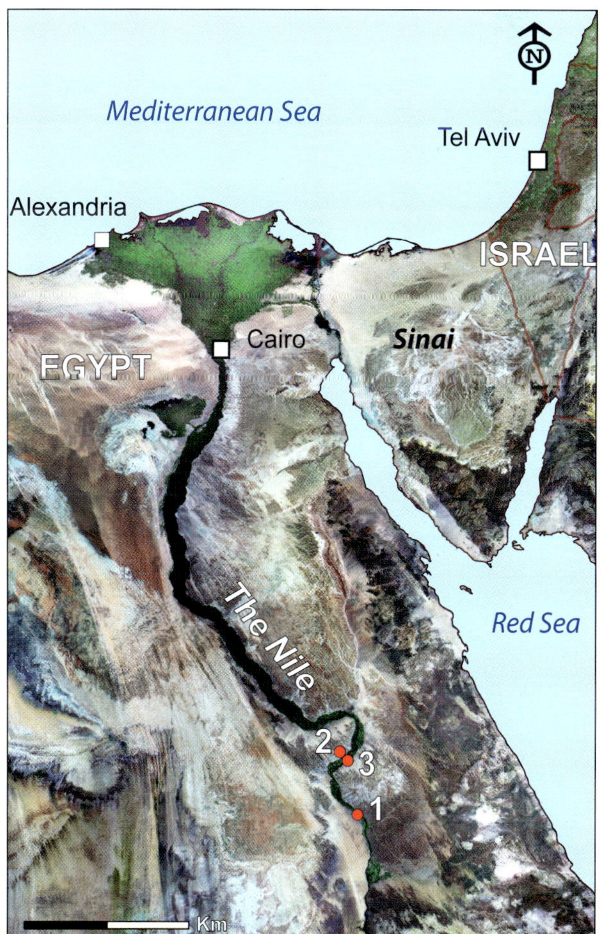

Fig. 7.20. Egyptian sites as mentioned in the text. 1) Hierakonpolis; 2) Valley of the Kings; 3) Thebes/Luxor. Map: R. Lugon, J.-C. Loubier and A. Chevalier.

Fig. 7.21. Desiccated grains of six-rowed barley (*Hordeum vulgare* convar. *hexastichon*) from the 18th dynasty Western Graveyard of Qurnet Murai, south of Sheikh Abd el-Qurna, Egypt. Material stored at the Musée du Louvre, Paris. Image: A.G. Heiss.

Sprouting for of Osiris: Bricks and Mummies

The river Nile has always been extremely important for agriculture in Egypt. Every year, during the dry season between January and June, the land dried up until it seemed dead during the warmest season. Then, in the middle of July, the annual flooding of the Nile gave the hope of new life, providing the crop fields with both moisture and nutrient-rich silt. So, after its 'death', the soil was reborn to a new life every year. According to the myth, the god Osiris also apparently died and was reborn: he was murdered and dismembered by his own brother Seth, then restored and reanimated by Isis, his sister and wife (Lurker 2004). In the Ancient Egyptian belief system, both phenomena – the annual life cycle in the Nile valley and the myth of Osiris – were closely associated with each other. Osiris was an ambivalent deity, reigning over the realm of the dead but also causing plants to grow, especially barley – these are aspects found in the character of the Greek goddess Persephone/Kore, as well (see above).

The hope of rebirth and growth, personified by Osiris, is embodied in the sprouting 'Osiris beds' ('Osiris bricks' according to Tooley 1996), which were created as grave goods for nobles in the New Kingdom (*ca.* 1600–1000 BCE) and, as part of the ritual activites to worship Osiris (Adrario 2002; Raven 1982), a hollow form resembling Osiris was crafted from clay or from stone. This form was filled with a mixture of barley grains (*Hordeum vulgare* L.), Nile mud or sand and water. The barley was left to sprout until the shoots reached about 16 cm in length. The object was then covered with a lid: a 'dead Osiris' lying in his grave surrounded by a living green sheet, and with new life inside and outside – a powerful symbol of rebirth. One of these objects was also found in the tomb of Tutankhamun (see below; Fig. 7.20; Hepper 1990, 54), while the oldest finds of sprouting cereal kernels in tombs come from the Middle Kingdom and are dated to *ca.* 2000–1700 BCE (Adrario 2002, 55).

During the Late Period (*ca.* 700–300 BCE), more complex rites lead to the production of 'corn-mummies', mummiform and often ithyphallic packages of about 35–50 cm in length, again consisting of soil or mud and grains of barley and usually endowed with attributes of the two gods Osiris and Seker. Ninety-two of these objects are known today, according to M. Centrone (2007), originating from five necropoleis. Just like 'regular' human or animal mummies, they were wrapped in resin-soaked linen bandages, and placed in wooden sarcophagi (Fig. 7.22, Fig. 7.23; Centrone 2005, 11, 19). Archaeobotanical analyses have been carried out on corn mummies from the Archaeological Museum in Kraków, Poland (Wasylikowa and Jankun 1997) and from the Museum der Brotkultur in Ulm, Germany (Adrario 2002; Sokiranski *et al.* 2005). They confirmed only the use of barley in the corn mummies (Fig. 7.24), although emmer wheat (*Triticum dicoccum* Schübl ex Schrank) had previously been postulated as a possible component (Centrone 2005). Variation in coffin types and shape of corn-mummies suggests continuity in the use of these objects over a considerable period (until the third century BCE; Centrone 2005, 22).

It is interesting to note that the Greek 'Adonis Gardens' are of a very similar structure, use and symbolism: cereal grains were put in potted soil in order to prematurely germinate after the harvest and to be used in the festivities for Adonis as symbols of death and rebirth (Servais-Soyez 1983; Baudy 1995). A hint at the fundamental character of the underlying conceptions of these rites might be their widespread distribution over the world: the *powamu* ceremony of the Hopi (Chapter 7.5) also refers to sprouts as symbols of life, fertility and resurrection.

Fig. 7.22. The wooden coffin of the corn mummy stored at the Museum der Brotkultur at Ulm, Germany, resembling the Egyptian death god Seker. The object dates to *ca.* 600 BCE. Image: W. Lutz (from: Sokiranski *et al.* 2005).

Fig. 7.23. Inside the coffin the wrapped corn mummy is visible, wearing an Osiris wax mask. Image: W. Lutz (from: Sokiranski *et al.* 2005).

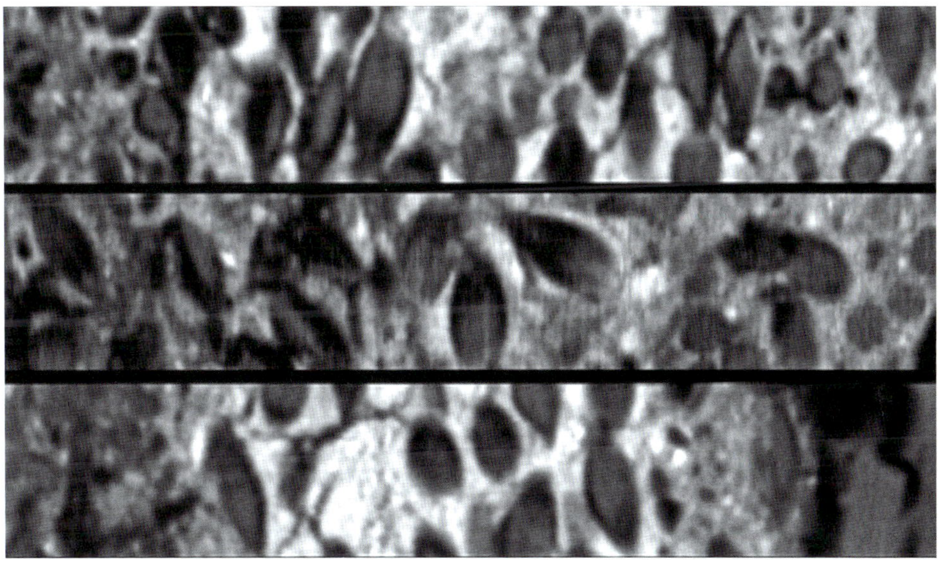

Fig. 7.24. CT image of the corn mummy from Ulm, showing sprouted barley (*Hordeum vulgare* L.) grains embedded in their soil matrix. Image from Sokiranski *et al.* (2005).

The Tomb of Tutankhamun

Well-known examples for perhaps the richest and most elaborate grave goods known are from the Pharaonic tombs in the Valley of the Kings. Egyptian noble graves in general show a multitude of offerings, presented here as concrete examples of how people expressed belief in and, the wish for, a life after death. These grave goods consisted of both 'real' plants and 'created' plant-based objects, as well as their symbolic representations. Various types of bread were offered, as well.

The tomb containing Tutankhamun's mummy (*ca.* 1323 BCE), which was not disturbed by tomb-raiders until its discovery by scientists, provided large amounts of perfectly preserved plant-based grave goods. An impressive masterpiece is the nine-row floral collar which was found lying on the golden death mask (Hepper 1990, 9–10). It was composed of strips of date palm leaves (*Phoenix dactylifera* L.) and papyrus pith (*Cyperus papyrus* L.), with beaded fruits of withania nightshade (*Withania somnifera* [L.] Dunal) and persea tree (*Mimusops laurifolia* [Forssk.] Friis), colourful flowers and petals of Egyptian blue lily (*Nymphaea caerulea* Savign.), oxtongue (*Picris radicata* Less.) and low cornflower (*Centaurea depressa* M. Bieb.), as well as leaves of pomegranate (*Punica granatum* L.) and olive (*Olea europaea* L.).

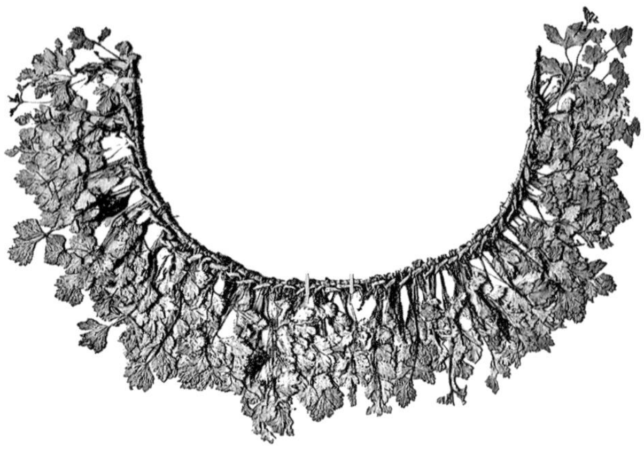

Fig. 7.25. A wreath made of celery leaves (*Apium graveolens* L.) mixed with petals from the Egyptian blue lily as found in the tomb of the noble man Kent in Thebes. In Schweinfurth's herbarium. Image from Laurent-Täckholm (1951, 110). Photo from the Agricultural museum (in Cairo).

Fig. 7.26. Celery leaves and inflorescences (flowering sprigs) from the wreath, as in G. A. Schweinfurth's herbarium. Image: The Agriculture Museum in Cairo (Laurent-Täckholm 1951, 110).

Additional collars and wreaths were found in other parts of the tomb including the plants already mentioned, as well as leaves of Safsaf willow (*Salix subserrata* Willd.) and wild celery (*Apium graveolens* L.). A further chest decoration or wreath composed of sprigs of wild celery mixed with petals of the Egyptian blue lily was found in the tomb of the noble man Kent in Thebes (Fig. 7.20, 7.25, 7.26). His tomb was dated to the twentieth dynasty, *ca.* 1200 BCE.

The symbolic language of all these plant species has not been revealed, but we do know that celery was considered a symbol of sorrow in Ancient Egypt (Laurent-Täckholm 1951, 111). Likewise, it was regarded as a chtonic plant by the Greeks, according to written sources (see, for example, Plutarchus' *De Timoleonte*; Lenz 1859, 183) – though, the Greek word σέλινον (*selinon*) can either be translated as celery or parsley (*e.g.* in Stewart and Long 1894). For general discussion on the problem of transferring ancient concepts, and names, of plants to their modern scientific equivalent, see Box 7.4a, p. 361.

Offering traditions changed over time. In the oldest graves from Egypt, we see that the symbols seem to have focused on food: provisions for the journey to the other side or food to consume in the afterlife. Later, the rituals and symbols became more elaborate, expensive and complex, and accentuated the forces of life and resurrection symbolism, and plants as a source of life. Here also, the relationship between the religious and cosmological perceptions of other Mediterranean areas, for instance Greece, is visible.

Other Grave Finds

As already discussed in the beginning of this chapter, preservation conditions play a crucial role in all research applying archaeobiological methods. This has been accounted for in the following section by dividing it into **inhumation** and **cremation** graves, as these two main forms of burial bear their own taphonomical (preservation-related) characteristics. Preservation of uncharred plant remains in inhumation graves strongly depends on favourable conditions provided by the surrounding soils and local climate. As these favourable conditions are rare (see above and in Chapter 2.2), hardly any plant material is preserved from most burial sites, if no heating or burning was applied prior to or during the funeral. However, if conditions are favourable, these inhumation graves can be the source for a wide range of detailed archaeobiological information. This may also include palynological (pollen-based) evidence indicating details of the funerary rituals or providing hints about the environment at the time of the burial, such as the season (see Chapter 6.1).

Cremation graves, in contrast, more often provide plant remains, but usually only a small part of the spectrum is preserved depending on the conditions during the fire and also on the plant remains' resistance to combustion. This is the reason why charred wood (charcoal) and grains – the most robust parts found in plants – are the most frequently recovered plant remains.

Inhumation Graves

One of the most famous grave finds from 'Celtic' Europe is certainly the **princely grave (*Fürstengrab*) of Hochdorf** (Baden-Württemberg, Germany) dating to the Early Iron Age (*ca.* 530 BCE). Excavation of the approximately 57 metre wide burial mound revealed the remains of a richly furnished burial chamber. The grave goods included, among others, a bronze couch, a wagon, a birch bark hat or headgear, badger fur, the finest textiles dyed red and blue and a huge cauldron crowned by lions, originally crafted in the Mediterranean. In the cauldron, palynological and chemical analyses for pollen and beeswax were carried out on what must once have been roughly 350 litres of mead (Körber-Grohne 1985; Stika 2009). The grave goods, including those of plant origin, were to a great extent luxury items which also could have been used in the chieftain's lifetime, and were items indicative of power and richness.

Separated from the Hochdorf grave both in distance and time is the **Oseberg ship grave** in Vestfold, Norway (Fig. 7.17), dating to the ninth century CE. This was excavated in 1904–1905. The ship served as a massive burial chamber for a high status Viking lady and her supposed bondwoman, and it was filled with provisions as if for a long journey. The oak wood constituting the *ca.* 22 metre long ship was perfectly preserved by the surrounding clay (Fig. 7.27; Rosenqvist 1959) and so were the botanical remains contained in it (Holmboe 1927). Among the organic finds in the grave were a number of horses, dogs, one ox, weaving equipment and very fine textiles of silk and wool. Among the botanical matter were cereals, seeds from flax and hemp and also seeds from plants used in textile dying – just as if the women had planned to go on with their earthly life in the afterlife.

Fig. 7.27. The Oseberg ship during excavation in 1904. Image from Rosenqvist (1959, 13).

Cremation Graves

Cremation of the deceased has a very long history and has been carried out all over the world. The oldest archaeological evidence, for instance, comes from Australia (Lake Mungo, New South Wales, Fig. 7.28) and dates to approximately 40,000 BP (see Becker *et al.* 2005).

Contemporary and prehistoric cultures traditionally using cremation in their funerary rites, such as many East Asian cultures, see the act of burning the body as a ceremony of transformation and transcendence, thus liberating the soul (Newall 1985; for a comment on the Hindu *rigveda* texts see Kaliff 1997, 79). In many cremation rites, grave goods are to be burnt in the same way as the dead body. As in many past cultures, fire and the hearth were considered divine and Kaliff (1997, 79–81) suggests that worship of the sun may lay behind some of these traditions, establishing fire as a sun symbol aiding a human in reaching a superior existence.

Cultural groups who believe in the resurrection of the body, on the other hand, are quite dismissive of the idea of burning a corpse to ashes, as seen in Christianity, Islam or in Ancient Egypt. Over time, in different regions of the world, the balance between inhumation and cremation burial has often shifted, mainly in relation to social changes (Saunders 1925; Childe 1945; Newall 1985; Becker *et al.* 2005). These shifts can, for instance, be observed in the changing funeral habits during Roman times, but also in today's 'western' civilisations. Beginning with initiatives such as 'cremationist' movements in nineteenth-century France and the United Kingdom

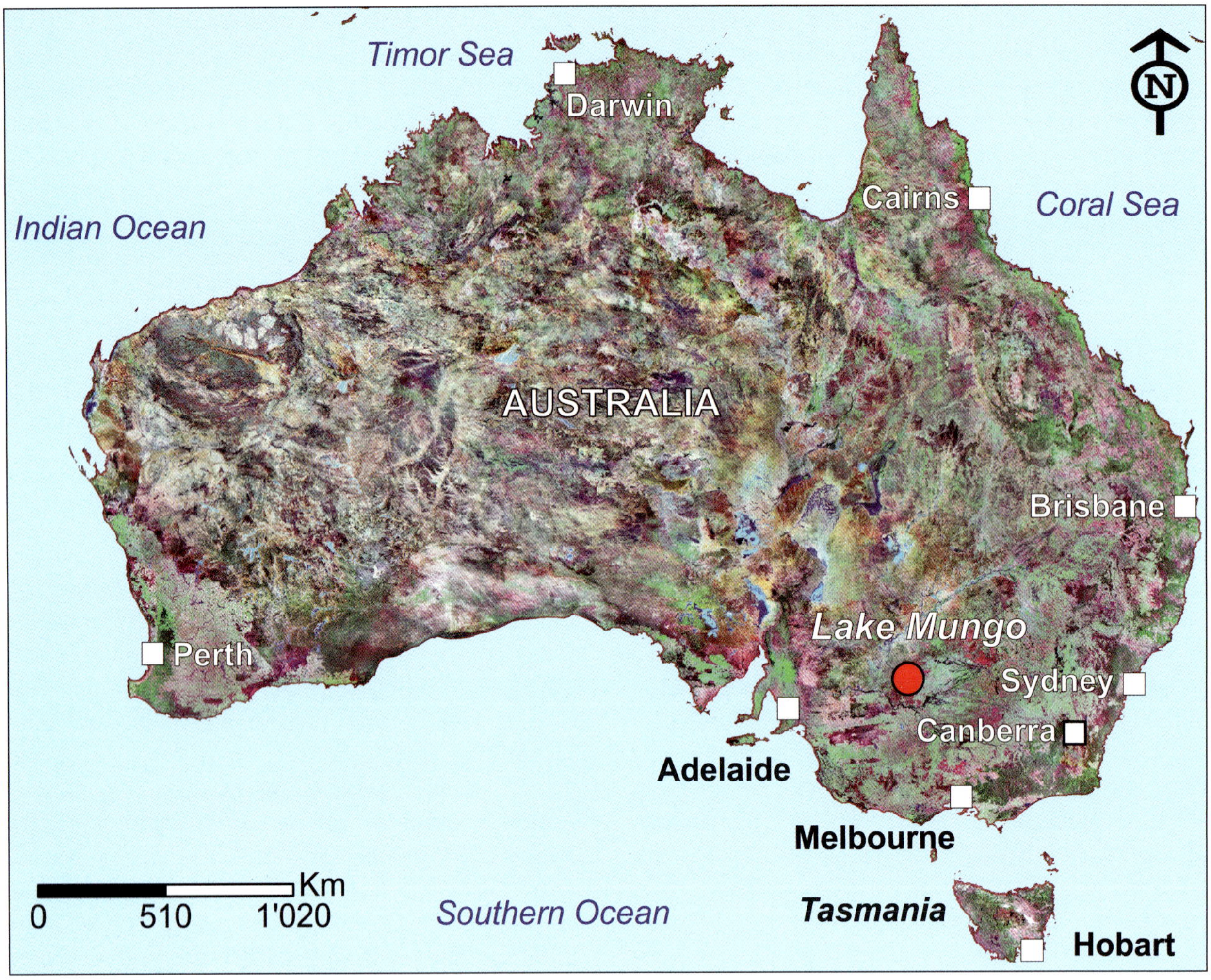

Fig. 7.28. The Lake Mungo site in New South Wales, Australia. Map: R. Lugon, J.-C. Loubier and A. Chevalier.

(Pasteur 1997; Bradbury 1999, 14), cremation is now increasingly common, mostly for practical and hygienic reasons.

From the Ancient Greek world, Homer's *Iliad* tells us in detail about the cremation of the hero Patroklos (Book 23; Voß 1990). Still, in the Archaic and Classical periods, cremation remained 'of secondary importance' until the fourth century BCE (Gualtieri and Becker 1982, 475). From this period, botanical finds are recorded for **incineration graves at Thasos** (Greece) in the north Aegean: charred remains of wheat (*Triticum aestivum* L./*T. durum* Desf.), olive (*Olea europaea* L.) and grape (*Vitis vinifera* L.), walnut (*Juglans regia* L.) and pomegranate (*Punica granatum* L.) were recovered from the pyres, as well as bread fragments (Megaloudi *et al.* 2007). The offering of pomegranate is especially worth pointing out, as with its many seeds it was a symbol of fertility, but also had strong mythological associations with the underworld (see the myth of Persephone/Kore, as mentioned above). The charred cloves of garlic (*Allium sativum* L.) from the same archaeological context might represent an aspect of protection: a panacea in most cultures, the plant has also been attributed protective powers against evil spirits (Megaloudi *et al.* 2007). When found in a grave, the question has to be asked of whether such an agent was meant to shield the deceased from evil forces, or rather to protect the living from the dead (see below, and also Chapter 7.2).

Cremation graves dating to the **Late Bronze Age/ Early Iron Age in Odensala** (central Sweden, Fig. 7.17) contained numerous seeds of two oil plants: gold-of-pleasure (*Camelina sativa* [L.] Cranz) and flax (*Linum usitatissimum* L.) (Hansson 1995) – both produce seeds that hardly ever survive charring due to their high oil content which makes these finds extraordinary.

The results from **multidisciplinary research on Roman cremations** is rich: analyses of burials from throughout Europe (Germany, Belgium, France, Austria) resulted in various plant-based luxury foods utilised as grave goods: condiments, cereals, pulses, imported fruits and bread were found (*e.g.* Lannoy *et al.* 2002; Bakels and Jacomet 2003; Bouby and Marinval 2004; Heiss 2008b). We find no information in written Roman sources about the firewood used in the pyres, although the texts do provide plenty of information on the ritual activities surrounding the cremation and the burial, and even the use of fire accelerants (tar) in the pyre is attested (Becker 1882, 529). There is one single record explicitly mentioning fuel wood selection in cremations, available from Tacitus. Unfortunately, the text does not deal with Roman customs at all, but only with those of Germanic tribes and Tacitus only reports the use of *certis lignis* ('certain woods') for the cremation of Germanic nobles (*Germania*, XXVII; Fuhrmann 1971). Charcoal analyses have however shown that the species used for firewood were deliberately chosen for their purpose: oak (*Quercus* sp.), hornbeam (*Carpinus* sp.) and beech (*Fagus* sp.) were the main species that were used (De Groote 1999/2000; Kreuz 2000; Heiss *et al.* 2008a).

Early Medieval cremation graves in Scandinavia (Migration Period to Viking Period) have produced burial gifts comprising bread, cereals, seeds from oilplants, root tubers of dropwort (*Filipendula vulgaris* Moench), and stem tubers of oat-grass (*Arrhenaterum elatius* (L.) P.Beauv. ssp. *bulbosum* [Willd.] Schübl. & G.Martens) (see Engelmark 1984; Hjelmqvist 1984, 1990; Hansson 1996; 1997; 2005b; Hansson and Bergström 2002; Bergström 2007; Viklund 1997; also see Chapter 7.2). We also find bread offered in the cremation graves at Helgö in eastern central Sweden. It has been suggested that this practise is a Roman influence, originating from Greece and from the Fertile Crescent (Hansson 2011).

Ibn Fadlan's account of the Rus (Rūsiyyah) people in Russia provides a description of a chieftain's burial at about 921 CE, mentioning plant grave goods such as vegetables and condiments (Wikander 1978). Such descriptions complement the often difficult situation of charred plant remains in cremation graves, as they may not survive in a sufficient amount to document the full spectrum of plant-material offerings. It is known today that the Rus were connected to Scandinavia and to the eastern central region of Sweden in particular, with a Scandinavian impact in the Old Russian Kingdom traceable both in graves and settlements (Melnikova and Petrochin 1990). In the necropoleis of the Russian cities of Kiev and Cernigov (Fig. 7.14), the Scandinavian assemblages are the richest (Petrukhin 2007, 66). One facet of this interrelation may also be the offering of bread loaves in cremation graves. While the small bread loaves on metal strings found in cremation graves at the Viking Age town of Birka, in Sweden, are to be considered unique for the whole region (Hansson 1996 and literature cited therein), there is one important exception: a cremation grave near Moscow which contains the same kind of offering, a small loaf of bread on a metal string (Fekhner 1963), indicating a strong cultural interaction between east and west.

Plants as Adornment of Graves

A modern account of funeral-associated plants is provided for Muslim graveyards in Israel. The investigations showed a wide range of traditions leading to today's choice of ornamental plants (Dafni *et al.* 2006): the first group – aromatic herbs, among them sage (*Salvia fruticosa* Mill.), rosemary (*Rosmarinus officinalis* L.), and basil (*Ocimum basilicum* L.) – are traditional health-bringers in many cultures and they are also often regarded as warding off the influences of death and decay. A second group of ornamental plants are all kinds of white-flowered plants: in the study, for instance, narcissus (*Narcissus tazetta* L.), sea squill (*Urginea maritima* [L.] Baker) and day cestrum (*Cestrum diurnum* L.) were observed. The white colour symbolises purity in many cultures and, in Islamic tradition, it is also a colour of mourning. The various shrubby or treelike species found in the cemeteries refer to a multitude of symbolic meanings, either because of their being mentioned in the Qur'an (such as olive, *Olea europaea* L., Arabian jujube, *Paliurus spina-christi* Mill., or date palm, *Phoenix dactylifera* L.), because their evergreen leaves are symbols of immortality (for instance cypress, *Cupressus sempervirens* L., myrtle, *Myrtus communis* L., or laurel, *Laurus nobilis* L.), or have other symbolic meanings. Altogether, the results show how ancient traditions from the Middle East blended with influences from the Greco-Roman world and from western Europe.

Plants as Protective Agents Against the Deceased

Finally, it should also be pointed out that by no means every object (or plant) found in a grave is necessarily to be interpreted as a gift to the deceased. As was pointed out for the garlic cloves in the Greek Thasos graves, objects buried in graves may also be some kind of protection for or even from the dead (see also Chapter 7.2). English, German, Swedish, and Italian medieval and early modern folk tales concerning the 'undead' mention various ways to protect the living from the dead (Thiselton-Dyer 1889; Nordén, 1928). They tell us, for example, about the dangers of placing flowers near the mouth of the deceased before interring them as they would chew on them in the grave, turning into *nachzehrer*, living corpses who drain the strength – and eventually life – from the living while still lying in their graves (Thiselton-Dyer 1889; Schürmann 1990). These creatures and the related revenants (walking corpses), were thought to be kept from their deeds by throwing cereal grains or flax seeds into the coffins before the funeral: as the dead could only proceed with their deeds after having counted all the seeds but, unable to count further than two or three, they were thought to be helplessly locked in their graves forever (Nordén 1928, 346; Hagberg 1937, 627), a tradition also known from other European countries. So we see that in some (if, however, rare) cases, archaeological finds of food plants might indeed signify something quite different from provisions for the afterlife.

The Ritual of Everyday Life

This section highlights a few of the offerings not primarily related to 'matters of life and death', but rather concerning everyday ritual activities as mentioned in the beginning of this chapter. Some of these aspects are dealt with in individual contributions, such as the important role of plants in the May Day festivities in the British Isles and France, an ancient but still living tradition (see Chapter 7.4). For the New World, the importance of plants in agricultural and fertility rites and, especially, of the various colours of maize kernels, is given for the Hopi in southwestern North America (see Chapter 7.5). For pre-Columbian South America, the ritual use of hallucinogenic plants is documented in Chapter 7.6.

Bread as a part of (and a symbol for) daily nutrition has already been mentioned as an offering in agricultural contexts. However, cereal preparations have been recognised as playing a major role in many prehistoric, historic and present-day rituals.

Greek and Roman Antiquity

Many festivals and religious events in Ancient Greece involved the offering of bread, and of bread-like cereal products. The **festival of *Elaphebolia***, dedicated to Artemis, goddess of hunting, involved stag-shaped cakes made from unbaked wheat dough, honey and sesame, according to Athenaios (Marcolin 2008, 45). There are written records of a variety of offerings where cakes (the *popana* and the *pemmata*) play an important role, such as in the cult of Asklepios, god of healing, on the island of Kos, or as a donation to Hekate or Hermes. In many cases, just as in the *Elaphebolia* mentioned above, the cakes are meant as a replacement, or at least a complement, for blood (human or animal) sacrifices (Marcolin 2008).

In Ancient Rome, offerings of bread or other cereal products were very common, such as the rather pottage-like *puls fitilla* (Pliny, *Naturalis Historia* 18, 84: Bostock 1855), the spelt cake (*farreum libum*) offering involved in the nuptial ceremony (*confarreatio*) or the *mola salsa* (a mixture of grain-paste and salt water) which was used in most **Roman offering rites** (Marquardt 1864). However, we can also see that at least some people in these diverse communities had a rather secularised attitude toward these altars and the offerings. An example of this is the story of the emperor Vitellius (around CE 69), who was said to have been so gluttonous that he did not even refrain from stealing and consuming sacrificial bread loaves from the altars (Suetonius, *de vita Caesarum* 7,13: Rolfe 1928).

Bread from Swedish Hill Forts

During the late Roman Iron Age and the Migration Period, the construction of hill forts increased in eastern central Sweden (Olausson 1987, 399). One of these structures, the **Runsa hill fort**, Ed parish, county of Stockholm, is situated on a peninsula, on the northern shore of Lake Mälaren (Fig. 7.17). During the most recent excavation in 1992, two finds of bread were made, one from the shaft through the southern inner wall and one from the shaft in the largest house terrace inside the hill fort (Bergström 2007, 44–46; also see Olausson 1996, 18). Bread was also found in 1908 at **Boberget hill fort**, Konungsund parish in the eastern part of the province of Östergötland (Fig. 7.17): a bun was found in a rich cultural layer (Statens Historiska Museum och K. Myntkabinettet 1909, 273; Schnittger 1912, 2–3). Analysis of the bread was carried out by H. V. Rosendahl (1909). Dating of the buns from these two hill forts shows that they originate from the same period (Roman Iron Age/Migration Period) as the oldest bread loaves found at the site of Helgö (eastern central Sweden, see above), where they were probably offered on a sacrificial terrace. The buns from the two hill forts may also have been offerings: future research will hopefully result in more information about the role of bread in hill forts.

Sometimes we have brief, on-the-spot accounts relating to specific offerings in specific ancient places. For instance, the **Saga of King Olav**, one of the Old Norse kings' sagas in Snorri Sturluson's *Heimskringla* (*ca.* 1230 CE; see translations in Johansson 1992; Simrock 1876) contains an episode about Gudbrand, who believed that the hollow wooden idol of the Asa god Thor was more powerful than the invisible Christian god in which king Olav believed. The wooden statue of Thor had gold and silver as offerings and also meat and four pieces of bread loaves every day. When one of King Olav's men struck the idol down, Gudbrand and his men saw that nothing happened, except that rats, snakes and lizards were crawling away from it, and so had to admit the Christian god was more powerful than Thor (Johansson 1992, 154). Of course, this part of the saga is mainly a vehicle for proselyte propaganda, but the bread offerings mentioned might probably have been made in real life. There have, thus far, been no archaeological finds of bread in Norway but this saga provides us with a good reason to look for it in archaeological contexts.

Offerings in Bogs

This type of offering is still rather mysterious. A bread-shaped object was found beneath a plank roadway, excavated at the Ipwege bog ('Ipweger Moor') near Oldenburg, Germany (Fig. 7.16). Dendrochronological dating of the overlying wood planks resulted in an age of *ca.* 713 BCE. Analysis of the object revealed that, aside from some cereal compounds (broomcorn millet, *Panicum miliaceum* L., and hulled barley, *Hordeum vulgare* L.; see Behre 1991), it mainly consisted of beeswax and thus cannot be regarded as palatable bread, but rather as a bread-shaped symbol (Brockner and Mitchell 1994).

Comments on Ritual Plant Use in 'Modern Societies'

The case studies in this section, just as the major part of the individual authors' contributions to this chapter, deal predominantly with past phenomena of ritual plant use or with rituals in extant pre-industrial or aboriginal cultures (see Chapter 7.5 and Chapter 7.6). This selection was intentional as, in these cases, archaeological and written records, together with ethnographical data, help to trace long-term developments of belief systems and to find common traits in societies separated by hundreds of kilometres or by centuries.

This set of case studies might also evoke the impression that these forms of 'non-secular' (or even 'irrational' as some might put it) plant use are just a phenomenon of past times or of remote and aboriginal cultures. However, if we take a closer look, we find that we ourselves engage in many ritual activities involving plants – right in the heart of an 'enlightened' post-industrial world.

Ritual Meals

A wonderful example of the symbolic meanings of plants passed down to modern times is found in Judaism. The traditional ritual ***seder* (רדס) dinner** on the first night of Passover is a commemoration of the exodus from Egypt. It includes, among others, unleavened *matzo* (הָצַּמ) bread recalling the haste in which the Israelites left Egypt and thus did not have enough time to let the dough rise. The bitterness of Egyptian slavery is symbolised by bitter herbs and vegetables (רוֹרָמ, *maror*), typically horseradish (*Armoracia rusticana* Gaertn., Mey. & Scherb), romaine lettuce (*Lactuca sativa* L. var. *longifolia*), or endive (*Cichorium* sp.). Parsley (*Petroselinum crispum* [Mill.] Fuss), as a symbol of spring, is eaten after the Kiddush blessings, dipped in salt water (Cohn-Sherbok 2005). It has to be emphasised that the *seder* ritual is not at all based on magical thinking (see Subbotsky 2010) but that the plants involved serve their purpose as symbols, or mnemonics, in the ritual.

Ritual food consumption is also a central characteristic in Christian religion, employing wine and sacramental bread as central elements in the **Eucharist** (Douglas 1999). The analogies therein – wine with blood, bread with flesh – form an archetype which is found in a great variety of cultures (see the notes in the sections on the *Thesmophoria* and on African rites above). However, in contrast to the food in the Jewish *seder* serving as a mere mnemonic, contrasting beliefs in bread and wine as *symbols* of flesh and blood versus, in the true transubstantiation, *into* flesh and blood in the Eucharist, have lead to severe doctrinal conflicts since the Middle Ages (Herbermann 1913). In general, Christian traditions are often characterised by a special role for bread, such as the many different sorts of Christmas, Lent or Easter breads found in different regions of the world (Toussaint-Samat 2009).

Festive Plants

The **Christmas Tree**, allegedly originating in Celtic and/or Germanic beliefs, was first documented in the sixteenth century and became well-established in the nineteenth century throughout Europe and North America (Watts 2007). Although originally a focus of Christmas festivities, it has become a more and more secularised symbol in the past decades and is also getting ever more popular among Muslim, Jewish, or Buddhist people, admittedly also often causing conflicts (Plath 1963; Simel 1996; Gayasaddin 2007).

Another annual ritual employing plants which is still performed across Europe is **May Day** (see Chapter 7.4) and so are the well-known festivities at **midsummer night** (summer solstice), especially in northern and northwestern Europe, such as the dances around a maypole (*midsommarstång*) or bonfires during Swedish *midsommar*. Many of the beliefs surrounding midsummer night are regarded as predating the Roman period, although they were later strongly influenced by Christianity in most regions and associated with Saint John the Baptist (Grimm 1844). Several records not only document the continuity in the use of flowers in the festivities (Reynolds and Sawyer 1959; Sillasoo 2009), but also illustrate some of the past attributions of magical properties to flowers gathered at the time of the summer solstice, such as St. John's wort (*Hypericum perforatum* L.), mugworts (*Artemisia* sp.), holly (*Ilex aquifolium* L.) or the mysterious (and fictive) 'fern-seed' (Grimm 1844; Thiselton-Dyer 1889; Heiss and Kohler-Schneider 2011).

Who has not heard of the 'magical properties' of **four-leaf clover** (usually *Trifolium pratense* L. or *T. purpureum* Loisel.) to bring its bearer luck? In British folk-lore, it is also a powerful agent against witches, and it helps reveal fairies or unveil deceptive magic and treachery (Watts 2007; Thiselton-Dyer 1889). However, the modern flower industry no longer relies on the rare fortune of *Trifolium* growing a fourth leaflet: every New Year, millions of pots with the lucky leaf (*Oxalis tetraphylla*), a four-leaf species not related to clover, are sold as lucky charms for the coming year (Watts 2007).

Festive use of flowers is an on-going tradition that continues to develop all around the world: in present-day Thailand, the traditional reverence for the **Rice Goddess** in rural areas (see above) has recently been officially revitalised by Queen Sirikit, in order to increase the population's awareness of the importance of rice for daily nutrition (Rashmisrisethi 2008).

7.2. HIDDEN STONE – A UNIQUE BREAD OFFERING FROM AN EARLY MEDIEVAL CREMATION GRAVE AT LOVÖ, SWEDEN

Ann-Marie Hansson

Early Bread in Sweden

Charred prehistoric bread finds from Sweden have usually been classified as 'buns' or 'biscuits' depending on their appearance, as their identification was originally based only on morphology. Most of these objects resemble small biscuits, and they are often very thin. These small black lumps of bread are often difficult to recognise during excavation because they resemble charcoal. During later analysis of retrieved charcoal from cremation graves, bread fragments were found among the charcoal. The majority of finds comes from cremation graves in eastern central Sweden, dating mainly to the first millennium CE. The oldest finds are dated to the Swedish Roman Iron Age and the youngest from the period of Christianisation (around CE 1000), when cremation ceased to be the common burial rite. The bread finds increase during the Swedish Migration Period with a climax during the Vendel Period. After that period, the finds decrease, with the exception of the many bread finds in Viking Age Birka (Bergström 2007, 69, 70; Fig. 7.17). Later than this, any possible deposition of bread in what are now always uncremated burials in graves can no longer be traced archaeologically due to the low chances of preservation of uncharred bread. A number of the charred finds have now been analysed.

With regard to written sources in Sweden, there is a wide range of ethnological evidence for various bread types, baking methods, etc. For instance Nils Keyland (1919) very thoroughly described Swedish rural fare, province by province. Bread and porridge were given a special place in his research. He divided his research into two parts, one for vegetal foods and the other for animal foods. Åke Campbell (1950; 1952) also concentrated his research on bread, mainly different bread types from the 1880s, which he calls 'the Swedish folk culture's most central cultural element, the home-baked everyday bread'. He discussed bread from geographical, social and technical baking approaches, including the evidence from comparative linguistics. Campbell also makes interesting digressions into the field of medieval bread and the bread types mentioned in the Icelandic Sagas, as well as touching upon the subject of military bread.

In Campbell's research, professional bakeries are not mentioned, but Gösta Berg (1960) filled this gap by working on this topic. He found that many bread types are described in the town and guild records and, especially, the apprentices' masterpieces are carefully listed. Many of these bread loaves had German names and seem to be rather archaic, with a strange morphology indicative of strong influences from Germany and central Europe. There is a great difference between prehistoric bread and bread mentioned in written sources. We still have the actual bread loaf preserved from prehistoric times, even if it tends to be from ritual or burial contexts surviving as small black pieces in a charred state, while the bread mentioned in written sources was meant to be eaten.

In What Context is Bread-Like Organic Material Found?

Carbonised bread loaves are found mainly in cremation graves, but occasionally also in domestic settlements and in hill-forts. In ingredients and morphology, the bread varies considerably from locality to locality and even between different graves within the same cemetery where the time factor may be an important variable. There exists what might be termed a 'standard bun morphology', that is, a flat circular bun with a slightly domed upper side, but many other shapes and sizes also occur. In addition, there seems to be considerable variety in the ingredients used.

In the Lake Mälaren area in eastern central Sweden, which until *ca.* CE 1000 was a bay of the Baltic Sea, many bread loaves have been unearthed from cremation graves and settlements dating predominantly to the mid-first millennium CE. For example, more than 70 buns have been recorded from the settlement and cemeteries of Birka, the Viking Age proto-town situated on the island of Björkö (Hjelmqvist 1984; Hansson 1996; Ahlsén 2003/2004; Bergström 2007). On another island in Lake Mälaren – namely Helgö, in Ekerö parish in the province of Uppland (Holmqvist 1963; Holmqvist and Arrhenius 1964) – a large number of buns were found (Lindeberg 1961), some in a very fragmentary state (Hansson 2003; Bergström 2007). Helgö has produced the oldest charred buns found in Sweden, dating to the Roman Iron Age (*ca.* CE 200). Other finds of bread on Helgö come from the later settlements and graves. Bread has also been found on Lovö, yet another island in Lake Mälaren (Fig. 7.17) with early settlements and graves (Petré 1978, 1984a–c, 1999a and b; Bergström 2007, 261–263).

The Lovö Cremation Grave *ca.* CE 600

In 2004, Bo Petré of the Archaeological Institution at Stockholm University led a seminar excavation of a cemetery (Raä 28C) in Söderby, Lovö parish, in the province of Södermanland. Lovö island is in Lake Mälaren and contains settlements and graves dating mainly to the mid-first millennium CE. (Fig. 7.17) (also see Petré 1984, 4, 150, 151). One of the graves (A20) was a low stone setting *ca.* 4 m in diameter (Fig. 7.29) in the middle of the cemetery.

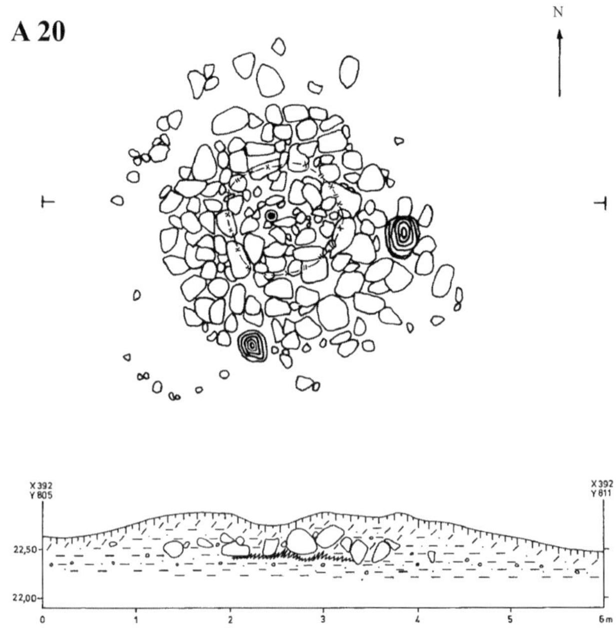

Fig. 7.29. Plan of the stone setting A20. Image: B. Petré.

Fig. 7.30. Stone setting A20 during excavation. Image: B. Petré.

The grave had a pit in the centre, showing that it unfortunately had been plundered. Centrally placed in the lowest layer of the grave (Fig. 7.30) was an inner ring of stones *ca.* 1 m in diameter encircling a cremation layer containing a fragmentary ceramic vessel with burnt bones. According to preliminary analysis of the skeletal material, the age and sex of the human individual is indeterminable. Grave goods or gifts placed with the deceased that ended up in the cremation layer and in the filling included scattered burnt bones of a dog, sheep/goat and a bird. Other grave goods consisted of iron and bronze fragments, tiny textile fragments, the remains of a whetstone, comb fragments, rivets and a burnt flint for a 'strike-a-light'. An important find was a fragment of a shield that indicated that this was a male burial (Petré 2011). These finds were rather normal grave gifts and the same grave field contained other graves that produced remains of weapons. Surprisingly, however, a small burnt bun of bread was also found in grave A20.

The Lovö Bun

A small bun of bread, *ca.* five cm in diameter and *ca.* three cm thick, was discovered in a cremation grave dating to *ca.* CE 600 (Lovö, Raä 28C, grave A20). The bun had broken into three pieces and the fractures were impregnated with ash and clay, as was the outer surface of the bun. This indicates that the bun had already split open when it was deposited in the grave. Tiny bone fragments, remains from the funeral pyre, also adhered to the surrounding layer of ash and clay. Most surprisingly, a small oblong stone with a worked pointed end extended out of the centre of the bun (Fig. 7.31). This is a unique find in Sweden.

During this stage of the medieval period we often find graves supplied with very luxurious and expensive gifts, as a deliberate choice of the best grave goods one could offer. Why then hide a stone in a bun, rendering it inedible? What ideas or reasons can lie behind such a 'destroyed' gift? Do any other prehistoric bread finds with inorganic content exist elsewhere in Europe? Is the dough in this bun standard or does it consists of something more than cereal flour? Microscopic analyses can tell us if the bun was made from standard ancient bread dough of cereal flour or something else. X-rays can show us the extent and shape of the remaining part of the stone, now hidden inside the bun.

Fig. 7.31. The bun from Lovö is 5 cm in diameter and has a thickness of *ca.* 3 cm with two of the loose pieces drawn out, so that the stone is visible. Image: L. Sundkvist.

The Lovö bun is charred on the outside, which is to be expected, since it was found in a cremation grave. The interior also seems very dark and burnt. It is difficult to decide if this is because the stone was hot when it was placed in the dough, which would mean it played a part in the baking process. Contradicting this interpretation, however, and disregarding the fact that the stone is worked into shape, is the smooth, even upper side of the bun which, most likely, was formed through first kneading the dough carefully and then spinning the lump of dough smooth with the hot (?) stone inside it, after which it was pressed down to give a flat underside. The easiest way to create this form would probably be if the stone were not heated. The bun is unusually thick (max. three cm) for a small grave bread. Very few prehistoric buns or biscuits have the form and thickness of the Lovö bun – most are much thinner. In Viking Age Birka, the usual bun/biscuit thickness is *ca.* 0.5 cm with some exceptions.

Analyses Performed on Ancient Swedish Bread

Cell-structure analyses using a light microscope to discern diagnostic cell patterns and cell morphology in charred bread remains were carried out very early in Sweden by H. V. Rosendahl, a pharmacist who identified the presence of cellular structures of tissues from caryopses from various cereal species and also cell structures from peas (Rosendahl 1909; 1912a and b). Later, Hakon Hjelmqvist applied light microscopy using a different preparation technique to analyse a large number of bread and biscuit samples from the Iron Age/early medieval period (Hjelmqvist 1984; 1990). Cell-structure analysis by light microscopy is now just one of a number of techniques used on bread at the Stockholm University Archaeological Research Laboratory (Hansson 1996; 2005a; Hansson and Bergström 2002; Bergström 2007). Results show that early bread often contains several different species of cereals and sometimes includes peas, flax and vetches and, occasionally, seeds from weeds. The cell tissues of the cereal caryopses found in the bread originate from the husks, the pericarp (fruit coat) and the testa (seed coat), surrounding an inner endosperm (starch-rich storage tissue) (Fig.

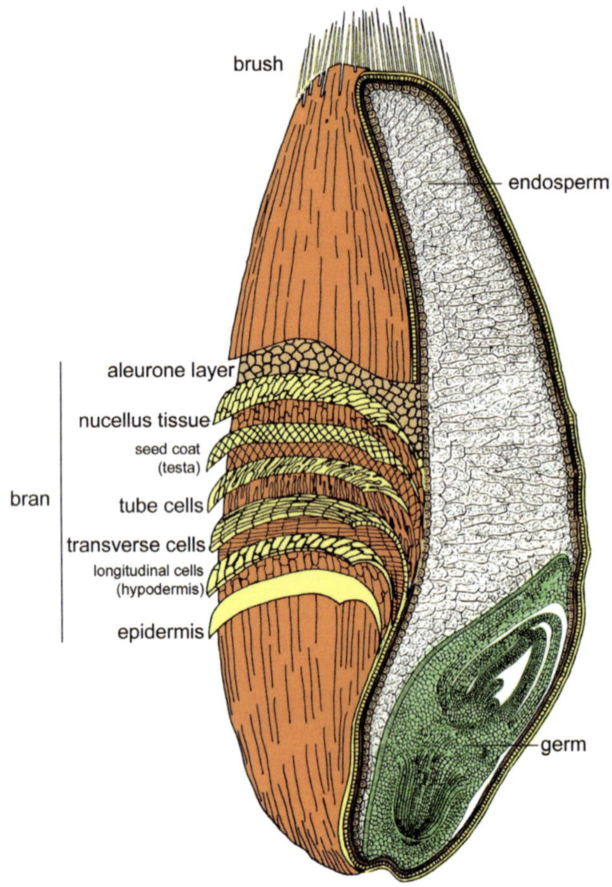

Fig. 7.32. Diagram of a typical cereal grain (bread wheat, *Triticum aestivum* L.), depicting the main components: bran (the sum of all outer layers), endosperm and germ.

7.32). In some species, for example hulled barley, the husks are tightly attached to the pericarp. The endosperm consists of parenchyma cells where most of the nutrition is stored as starch. The outermost cell-layers of the endosperm contain protein- and vitamin-rich aleurone cells. The pericarp is built up of several layers of protective cells. The main diagnostic parts of the caryopsis are the aleurone cells in the endosperm and the transverse cell layer in the pericarp. Starch grains are stored in amyloplasts, forming the inner region of the cells in the endosperm, and occur in different sizes within the same cell (Winton and Winton 1932, 25).

Scanning Electron Microscopy (SEM) in Sweden has also been applied for cell-structure analyses of bread in the last decade (Hansson and Bergström 2002; Bergström 2007). The aim of this analysis method is to reveal the species of the cereal or other plants contained in the bread. Many of the SEM-analyses have been performed on very small charred fragments. In recent years, chemical analyses have been applied to a number of bread fragments and organic foodstuffs (Hansson 2001; 2003; Ahlsén 2003/2004; Schierman 2005/2006; 2007), although most chemical analyses are performed on food residues attached to pottery (Isaksson 1999; 2000). These chemical analyses serve as a corroboration of the morphological identification of the organic material as bread or biscuits. As a consequence, the number of confirmed prehistoric carbonised bread fragments in Sweden has increased considerably. This creates a wider basis for comparative studies with regard to ingredients and morphology. It also allows for new information concerning deposition rituals.

Analysis of the Lovö Bun Using X-ray and Light Microscopy

The X-ray photo of the Lovö bun shows that there is a small pocket of air surrounding the stone. Small gravel/sand grains are also visible around the end hidden within the bun. This end narrows to a rather blunt point while the exposed end is even more pointed. Both ends seem to have been deliberately worked into shape (Fig. 7.33). A small piece from the bun was dissolved in ammonia (25% NH_3), bleached in hydrogen peroxide (H_2O_2), and studied under a light microscope at a magnification of ×400. This allowed the identification of cells from the husks, and the rather small aleurone cells from the endosperm of a caryopsis from a cereal species, probably barley (*Hordeum vulgare*), and starch grains were also identified in polarised light (Fig. 7.34). The organic material was very hard and difficult to dissolve, so there might also be other cereal types or other seeds in the bun.

Other Unusually Thick Swedish Circular Buns, Probably Without Stone Content

Excavation of the hill-fort of Boberget, in Konungsund parish in the province of Östergötland, produced a bun which resembles the morphology of the Lovö bun, although it is somewhat larger: seven cm in diameter and with a thickness of four cm (Fig. 7.35; Schnittger 1912, 2). Rosendahl (1909) analysed the ingredients of the bun microscopically and found that the dough was baked from a coarse flour of barley mixed with mineral particles, which were probably contamination from the grinding stone.

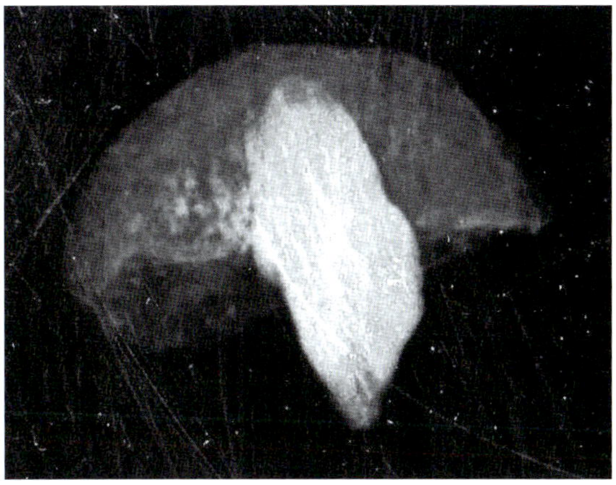

Fig. 7.33. X-ray of the bread loaf from Lovö, where the stone is stuck to the largest lump of the bun. Image: J. Storå, Osteological Research Laboratory, Stockholm University.

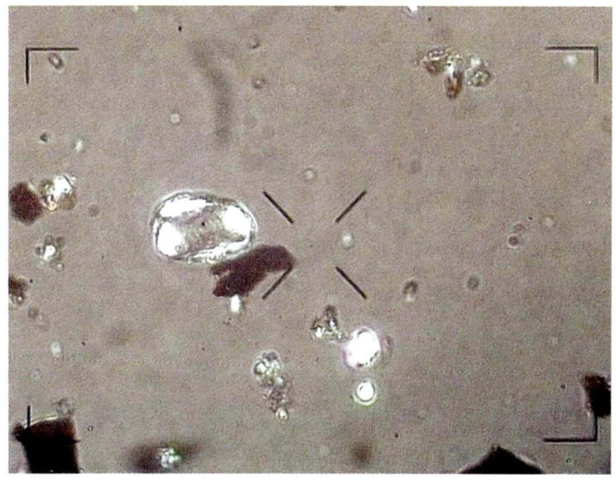

Fig. 7.34. Starch grain showing a centric cross in polarised light, indicating Cerealia (magnification: 400×). Identification and image: L. S. Cummings.

The Boberget bun is whole, so we do not know if anything is concealed inside it. A refuse layer in house-group III at Helgö produced a fragmented thick bun of bread (SHM 29435: 11566) (Bergström 2007, 259, 260), and in this case, we know that no stone or any other inorganic material was ever hidden within it.

Fig. 7.35. The bun from the hill-fort Boberget. From Schnittger (1912, 2).

The Use of Stones in Baking Bread or Cooking Porridge

The association of a stone together with bread immediately conjures up the image of baking or cooking with hot stones as, for example, described by Währen (2000b). Such a use of fire-heated stones for food preparation predominated in the Stone and Bronze Ages (Heer 1866, 9). In such a process, coarse porridge or grain paste is laid on top of or around heated spherical stones. Finds of such prehistoric food preparations have been made in Switzerland for instance (Fig. 7.36).

Fig. 7.36. Grain paste/porridge from Twann 3600–3500 BCE. (The stone, which has been pushed into the organic material, is not the original one). From Währen (2000a, 106).

This procedure is quite different from that used to bake the bun from Lovö. Whenever the stone's function was to assist baking, the stone was of naturally spherical form (Fig. 7.37). There is also a chronological difference between the bun from Lovö and the 'stone-baked buns' described by Währen. As mentioned previously, the tradition of baking with hot stones occurs predominantly during the Neolithic and Bronze Ages, and the bun from Lovö dates to the early medieval period. While we do not know what function the stone served at Lovö, we have established that the stone is worked and rather

Fig. 7.37. A stone, which was used to process cereal-based food. Measurement 54 × 52 × 40 mm, found in Muntelier, Switzerland. From Währen (2000a, 104).

sharp at both ends. It thus has quite a different morphology to the naturally spherical stones used for baking or cooking in these earlier periods.

X-ray Analysis of Bread from Ovelgönne

The Swiss bread researcher Max Währen X-rayed many of the prehistoric bread buns or biscuits which he analysed, and was able to show that some of these contained sand or gravel from the grinding stone (*cf.* Währen 2000b, Fig. 15). One of Währen's analyses concerned a find from Ovelgönne in Germany (formerly Kr. Harburg, now Kr. Stade; Fig. 7.17) (Wegewitz 1955), which he identified as half (3.5 cm) of a bread originally pointed at both ends which he termed a '*Spitzwecken*' (pointed roll). It was found in 1952 in an irregular loamy clay pit that measured 1.6 m wide and 1.5 m deep. The pit contained numerous pottery sherds, charcoal, burnt clay and brittle burnt stones, and this charred fragment of bread lying on a baking slab. The irregular form of the pit indicated that it was the result of clay extraction with fill of pottery sherds and other cultural material. The pottery sherds represented almost all known pottery types from the period 900–600 BCE, which dated the span of deposition. Similar deposits are known from the Urnfield culture.

The underside of the Ovelgönne bread was somewhat irregular (Fig. 7.38), which shows that the baking slab was not entirely smooth. This side also showed traces of charcoal and occasional mineral particles. The remaining half is shaped into a protruding point at one end, which is somewhat broken. There is a short cut from the pointed end along the length of the loaf. When X-rayed (Fig. 7.39), two metal pieces became visible inside the bread, reminiscent of the Lovö bun. Mindful of a find of Ukrainian bread that is formed in the shape of cereal grains, the Ovelgönne fragment could possibly be reconstructed to resemble a cereal grain with its distinctive ventral furrow. The Ukrainian bread was found in a hill-fort which has been dated to the fifth–fourth centuries BCE (Petrov 1948; Gimbutas 1971; Monah 2002).

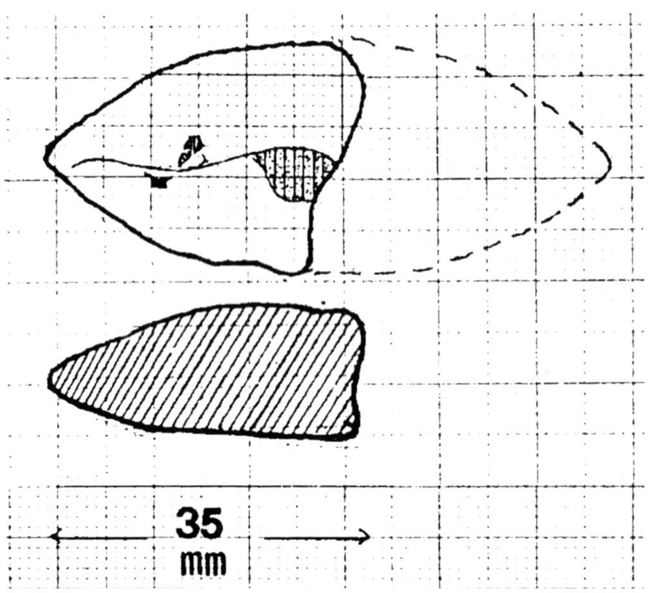

Fig. 7.38. Bread loaf (right) from Ovelgönne, max. length 32 mm, max. width 22.94 mm, max. thickness 11.44 mm, weight 2,5 g. From Währen (1995, 210).

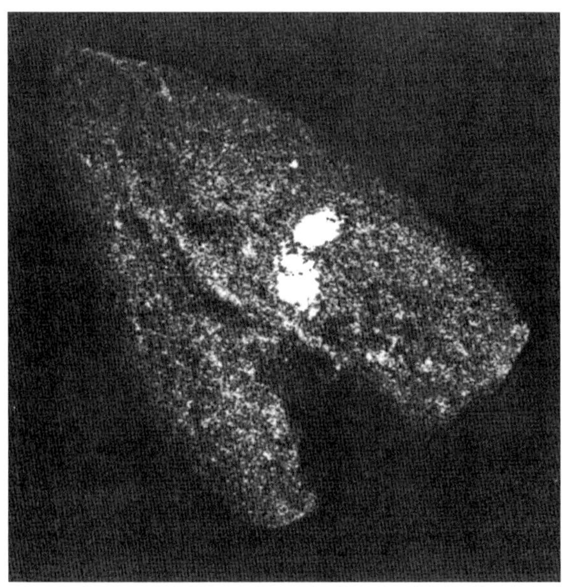

Fig. 7.39. X-ray of the loaf from Ovelgönne. Dr. H. Cames, Bern, who performed the X-ray, considers that the two light spots are metal pieces. From Währen (1995, 213).

Discussion

Bread can be far more varied than one might first think. It is a very complex food item, to judge from comparative studies. The find from Ovelgönne can partly be compared with the Lovö bun, since both contain inorganic material. The bun from Lovö with its hidden stone can hardly be considered as 'normal', not even when considering the context of an offering. The same is true for the German Ovelgönne bread-roll with its metal content, whether from a ritual deposition or from an ordinary settlement. No explanation is given for the very unusual feature of two metal pieces hidden inside the Ovelgönne bread and, to the best of my knowledge, the ingredients have not been analysed. However, the Lovö bun appears to have been baked from the cereals which are quite normal for cremation bread. It is only its stone content that makes the Lovö bun extraordinary. Thus, these two bread finds from quite different contexts – one a male cremation, the other a settlement – are quite intriguing. They also vary greatly in date.

Sacrificial gifts may be considered a form of communication with the non-visible world. The bread offerings might in these two situations signal different types of messages whose meaning is now difficult to interpret. In the very earliest phases of agricultural cultivation, we find evidence for the custom of offering bread sacrifices. The cultivation of cereals was of vital importance and therefore surrounded by countless rituals. Cereals were offered and sacred bread loaves were prepared for the goddess of crops, of vegetation and even of fertility in general. The tradition of offering sacred bread loaves has left its mark within the Christian Holy Communion rite concerning unleavened bread, where the sacred sacramental wafer is baked only of the finest wheaten flour.

The central question surrounding the Lovö bread is: towards whom was the sacrifice directed in this grave?

a) The deceased? – To use as food on his journey to another world? Or to harm him in the afterlife and prevent him from harming the living?
b) Some things or beings which were believed to populate the prehistoric living world?
c) Someone or something which was considered to exist in the afterlife?

To understand for whom – or for what – a certain piece of bread had been baked or what a stone-cored bun represented at the time of its deposition, one must try to understand the beliefs and cosmologies of the people who were living when this bread was prepared. What concepts did they have about their surrounding world, what invisible beings did they believe in, populating not only the living world but also the world of the dead – the afterlife?

The grave is dated to the early seventh century CE and is thus earlier than the time reportedly described in the Icelandic Sagas where the religion of the Viking society was that of the Asa (or Aesir), one of the two clans of deities composing the Nordic pantheon. Neil Price, in his treatment of Viking Age religion in Scandinavia, discusses the mythology of the period. He also stresses that the religion was 'far from static, and changed both regionally and over time' (Price 2002, 55). He shows that there seems to be a swarm of beings beyond the gods and goddesses forming their own invisible world. Probably some of the Asa-religion already existed before the Viking Age.

Comparative ethnological research also enables us to get some grasp on prehistoric concepts, though we must of course take into consideration the altering impact Christianity had on maintaining such ideas. Surprisingly, however, many heathen traits have survived into recent times in popular belief. Probably some of these beings from the myths have survived both belief in the Asa and Christianity. For instance, Granberg (1935, 160) considers the female wood-spirit (*skogsrå*) of Swedish (and other countries') folklore to be an element of folk tradition that goes back to a time long before the introduction of Christianity. According to this belief, the wood-spirit needed to be placated as it could be a danger to humans, especially to men, in certain situations. However, this was just one of many beings in the forest and around the farm that humans had to be wary of and felt subjected to.

Older farming societies provided greater scope than nowadays for a belief in the existence of an invisible world that interfered with the life of the living. Traditional popular belief provides an important cultural resource with its clear rules for all phases in life (Schön 2002, 9, 16). If one did not follow traditional taboos, invisible beings could interfere with one's daily actions, and one could be heading for disaster. Most likely there was similar reasoning for such beliefs during prehistoric times. In the Icelandic Sagas, we find many references to persons relying on soothsayers, on dream interpretation, etc. (Price 2002), which was probably valid for previous times but for which we lack any record. To a certain extent, individuals themselves could govern daily life through magical ritual actions. It was often difficult to define the border between beliefs in magic and religion (Schön 2002, 22).

If the placing of the stone in the bun from Lovö was a magical or ritual action, did it serve good or evil? If serving evil – was it meant to be food for the deceased and to harm him as it was being consumed? Was he considered an evil and hostile person? In that case it would not have been necessary to use a worked stone, which was not sharp enough to really harm him. Any stone of the same size could have been used. Perhaps this stone was meant to prevent him from harming living persons? If the stone was serving 'good' – did it provide a measure of protection? In Swedish folk tradition, steel and in particular steel objects, preferably knife blades, which were broken during manufacture (so-called *ångerstål*) had the power to protect against all elemental beings (Granberg 1935, 161). The sharp steel shielded against being spirited away into the mountain, against the seductive tricks of the wood-spirit, and other potential misfortunes (Schön 2002, 36). During historical times, it was the metal itself that was considered to hold protective powers, and among iron objects it was those with an edge or in some way sharp, which were preferably chosen (Hagberg 1937, 622; Carlie 2004, 147ff).

In historical times, metal/iron was often hidden under the floor or inside the threshold of a newly built house (Schön 1998). In various prehistoric cultures, there is evidence of cases of objects buried inside dwellings to protect the house and those who lived there. Could the sharp stone have been a substitute for such a steel or sharp object? Did it signify that the deceased man or even the whole grave was to be protected, as in the case of a house sacrifice? If we see the grave as a sort of dwelling – the house of the deceased – this interpretation could make the bread with its stone understandable up to a point. In a cremation grave, it might be difficult to find a suitable place to hide the stone – the magical protecting object – in such a way that it is connected to the grave but remains hidden. So why not place it inside a ritual offering: once inside the bread, the object becomes invisible. The bun from Lovö lacks iron, but the stone placed in it had been worked into some sort of points. Perhaps we should understand the concealing of this stone as the taking of suitable measures against evil forces that might otherwise encroach on the dead person in the grave?

If we are dealing with the notion of protection and grave offerings, was this belief necessary protection against invisible evil beings in the prehistoric living world that could invade the grave and injure the dead? Or is it more conceivable that the hidden stone was meant for something, someone, or several beings located in the afterlife or otherworld, for instance a god/goddess? If so, it could be an offering which gave a sort of indirect protection.

The above presentation of the possible significance of the strange deed of placing a stone inside a bun of bread no doubt only examines one of several possible interpretations of this rare phenomenon. Most likely we shall have to await further finds of 'stoned-' bread before we can begin to understand the full implications of this strange sacrifice. As the bun itself can be seen as a ritual grave gift, and its worked stone content as a further ritual gift, then we might say that the Lovö bun is a rare case of an offering inside an offering.

7.3. CEREMONIAL FOODSTUFFS FROM PREHISTORIC BURNT-OFFERING PLACES IN THE ALPINE REGION

Andreas G. Heiss

Introduction: Sacrificial Fires in the Alps

Burning is a very common and simple way of transforming an offering into something immaterial and at the same time transporting it into an otherworld by sending the gifts up into the heavens as smoke (Lang 2002; Schwager 2002; see Fig. 7.1). In the following article, one particular group of archaeological sites featuring burnt offerings will be discussed: the *Alpine Brandopferplätze* (Alpine burnt-offering places). The term was coined by the German archaeologist W. Krämer in the 1960s, in a publication suggesting the re-interpretation of some prehistoric sites in the Alpine area, then known as '*Wallburgen*' (hillforts; Krämer 1966). These structures – comprising ramparts, pavings and levellings, or stone circles – had previously been seen as the remains of ancient fortifications, but he demonstrated that the unusual deposits they contained were to be interpreted rather as the remains of burnt sacrifices: soil layers blackened by charcoal, masses of calcinated and highly fragmented animal bones, and artefacts displaying traces of fire and of intentional destruction. The following decades brought intensified research into these offering places, leading to the discovery of more than 200 sites in the Alpine and circum-Alpine area, although some of them are of uncertain classification (Weiss 1997). Their temporal range covers the Middle Bronze Age to the Roman Iron Age. Recent results suggest even older roots, reaching back to the Copper Age (Oberrauch 2005). Spatial extension of this cult phenomenon has recently been expanded to the region of Alpes-de-Haute-Provence, France, in the west (Garcia *et al.* 2006), and to the Sölk valley in Styria, Austria, in the east (Hebert *et al.* 2003). There is wide variety in the complexity of the built structures of burnt-offering places: some consist simply of a stone pavement, or one single stone circle, whereas others have several altars, multiple stone circles (see Fig. 7.40), and often also buildings associated with the offering site.

The gods worshipped in these offering places are still unknown, although for the Early Iron Age there are several hints pointing towards the northern Italian goddess 'Reitia' (as an example in Kossack 2002). As Alpine burnt-offering places share a prominent common feature with the cult of Demeter, *i.e.* sacrificing meat still attached to the bones (see Chapter 7.1), Ardovino (2002) suggested the possibility of – albeit very remote – links between the two types of rituals. It must however be stressed that the hypothesis of direct relations between Alpine burnt-offerings and the Olympian rite, as originally formulated by Krämer (1966), has long been abandoned by archaeologists.

The Question of Plant-Based Food in the Offerings

Plants, while representing the most important part of everyday nutrition, have long been neglected in *Brandopferplatz* research, as is evident in W. Zanier's (1999) review of diagnostic traits for burnt-offering places: these had always remained exclusively focused on archaeological features, artefact finds and animal remains. One reason for this might be the low visibility of individual plant remains in

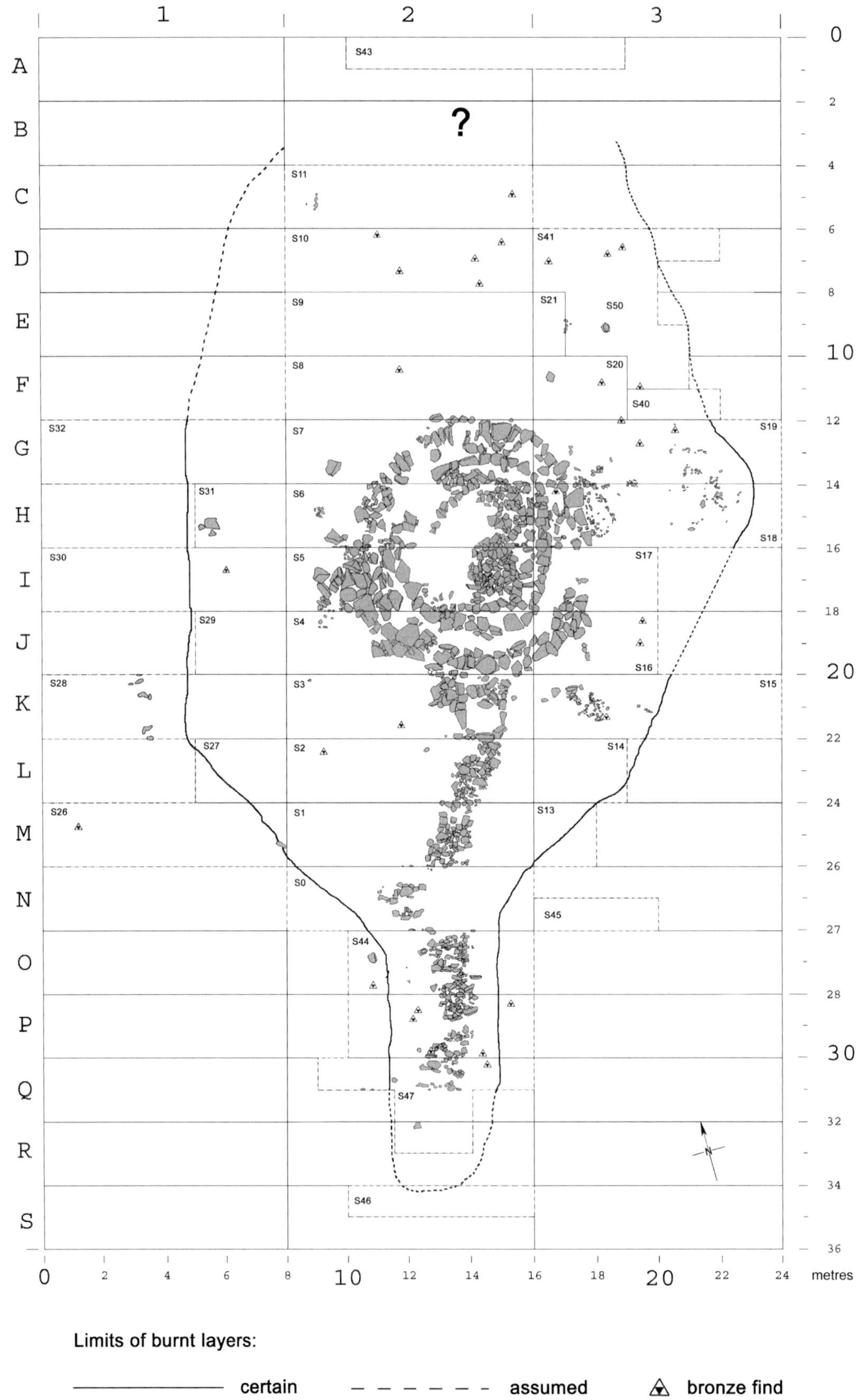

Fig. 7.40. The Late Bronze Age Alpine burnt-offering place of Altenstadt, Grütze, in Feldkirch (Austria, Federal State of Vorarlberg). The plan displays site dimensions, location of built stone structures, and extent of the ashy layers. In the centre lies a double stone circle of ca. eight metres in diameter, with a clay accumulation in its centre, the supposed position of the former altar. Extending to the south lies a row of fireplaces. Drawing from B. Heeb (2009)

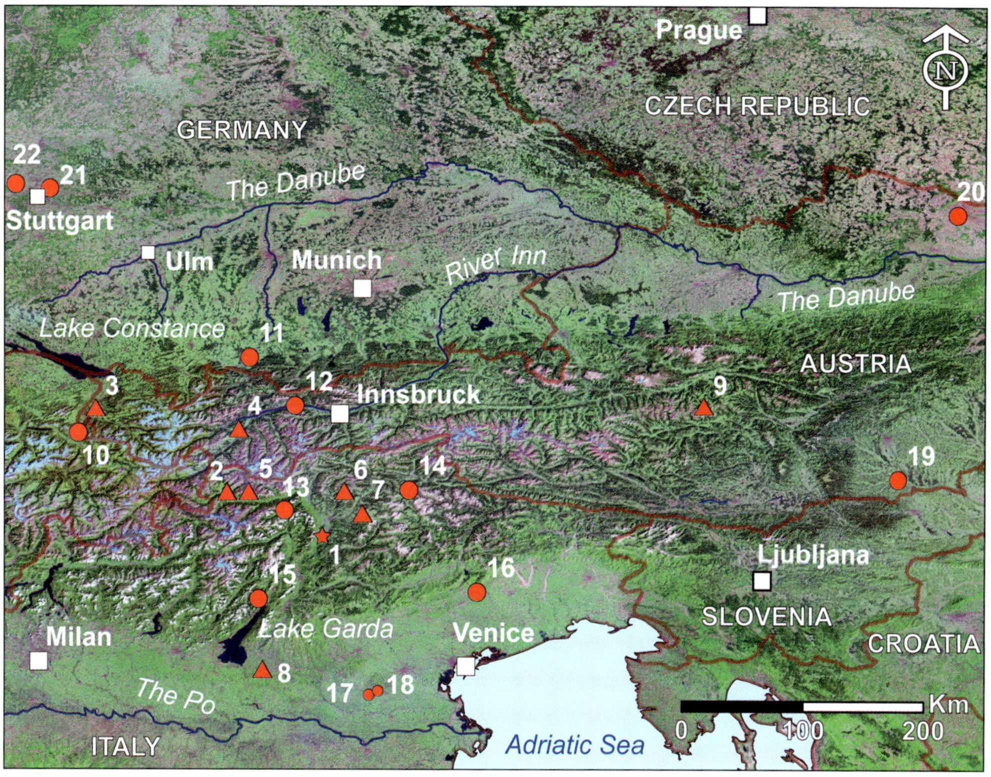

Fig. 7.41. Map of the investigated area. Numbers correspond to the sites in Fig. 7.42. Map: J.-C. Loubier and A. Chevalier.

Site no.	Site name	Period	Municipality	Region/province	Country	Altitude [m a.s.l.]	References
1	Pigloner Kopf	CA	Vadena/Pfatten	Bolzano/Bozen	Italy	550	Gattringer (2006)
2	Ganglegg, Hahnehütterbödele	MBA–LBA	Sluderno/Schluderns	Bolzano/Bozen	Italy	1,100	Heiss (2008)
3	Altenstadt, Grütze	LBA	Feldkirch	Vorarlberg	Austria	444/445	Heiss (2008)
4	Pillerhöhe	LBA–IA; RIA	Fließ im Oberinntal	Tyrol	Austria	1,559	Oeggl (1993), Heiss (2008)
5	Maneidtal, Grubensee	LBA–LIA; LIA–RIA	Silandro/Schlanders	Bolzano/Bozen	Italy	2,435	Heiss (2008)
6	Seeberg, Lago Nero/ Schwarzsee	LBA	Villandro/Villanders	Bolzano/Bozen	Italy	2,035	Castiglioni & Cottini (2000), Oeggl in Niederwanger & Tecchiati (2000)
7	Schlern, Burgstall	LBA; RIA	Siusi/Seis	Bolzano/Bozen	Italy	2,510	Heiss (2008)
8	Sommacampagna	LBA	Custoza	Verona	Italy	124	Nisbet (1996/1997)
9	Sölkpass	LBA	St. Nikolai	Styria	Austria	1,788	Heiss (2008)
10	Ochsenberg	LIA	Wartau	St. Gall	Switzerland	660	Heiss (2008)
11	Forggensee	LIA–RIA	Schwangau	Bavaria	Germany	773	Küster (1999), Tegel (1999)
12	Trappeleacker	LIA	Pfaffenhofen	Tyrol	Austria	700	Heiss (2008)
13	S. Valburga/St. Walburg	LIA	Ultimo/Ulten	Bolzano/Bozen	Italy	1,200	Oeggl (1992), Rösch (2002), Heiss (2008)
14	Prati del Putia	LIA	S. Martino/St. Martin	Bolzano/Bozen	Italy	2,170	Cottini *et al.* (2007)
15	Campi, Monte S. Martino	LIA	Riva del Garda	Veneto	Italy	1,450	Castiglioni (2007)
16	Villa di Villa	LIA-RIA	Cordignano	Treviso	Italy	160	Boaro *et al.* (2005)
17	Fondo Baratella	LIA	Este	Padova	Italy	15	Pasternak (2005)
18	Meggiaro	LIA	Este	Padova	Italy	15	Motella de Carlo (2002)
19	Frauenberg	LIA–RIA	Leibnitz	Styria	Austria	380	Popovtschak (2005)
20	Sandberg, Heiligtum 1	LIA	Roseldorf	Lower Austria	Austria	340	Caneppele *et al.* (2010)

Fig. 7.42. Investigated sites as indicated in Fig. 7.41. Abbreviations: CA: Copper Age, MBA: Middle Bronze Age, LBA: Late Bronze Age, LIA: Late Iron Age, RIA: Roman Iron Age.

archaeological sites: although blackened layers may seem very conspicuous, the charred plant remains they contain usually only become visible after thorough processing of soil samples (by wet-sieving or flotation; see Chapter 7.1).

However, written and iconographic sources on Roman and Greek burnt offerings had already suggested a multitude of plant-based offerings, and consequently these had also to be taken into consideration for the prehistoric Alpine rituals: cereals and their products (porridge, bread, cakes, pastry), fruits, flowers, leaves, roots, wine and beer, oil, and incense are mentioned as sacrifices (Zanier 1999). Just like animal products (meat, wool, fur, honey, and dairy products), they were the results of either hard farm labour or of tedious foraging in the wild, making them precious goods. Thus, there must have been important reasons for sharing them with the gods (see Chapter 7.1). A certain agricultural character of the burnt offerings has been assumed for several years (Weiss 1997) and only recently it was confirmed by the dominance of domesticated animals over game in the archaeozoological record (Rizzi 2000; Leitner 2002; Hebert *et al.* 2003; Schmitzberger 2007; Heeb 2010). Thus, the hypothesis of a linkage between offering rituals and events of agricultural importance, *e.g.* sowing or harvesting (Steiner 2007; 2010), was to be tested for plant-based offerings as well.

Archaeobotanical Analyses: Past Research in a Nutshell and Recent Investigations

The first published data on plant remains from Alpine burnt-offering places date back to the early 1990s, when K. Oeggl (1992) analysed a few samples from the site at St. Valburga/St. Walburg in the Ultimo/Ulten valley, northern Italy. Subsequent occasional analyses were carried out on several other sites. Only recently a comprehensive archaeobotanical study on nine Alpine burnt-offering sites was carried out comprising a synopsis of extant analyses from *Brandopferplätze* and comparable sites in and around the Alps, excluding only sites with a clear connection to funerary rituals (Heiss 2008a; 2010a). The sanctuaries included extend over a large spatial and temporal range, from Switzerland through northern Italy to eastern Austria, and from the Eneolithic (Copper Age) up to Roman times (see Fig. 7.41, 7.42).

The methods applied in the current study mainly followed standard procedures (Jacomet and Kreuz 1999), such as flotation to extract charred plant remains from the surrounding soil, and macrofossil analysis of seeds, fruits, wood charcoal and other plant parts. However, these methods were supplemented by detailed studies of irregularly shaped objects of unknown composition, corresponding to S. Jacomet's 'AOV' category (*Amorphe Objekte Verkohlt*, meaning 'amorphous charred objects'; Jacomet *et al.*

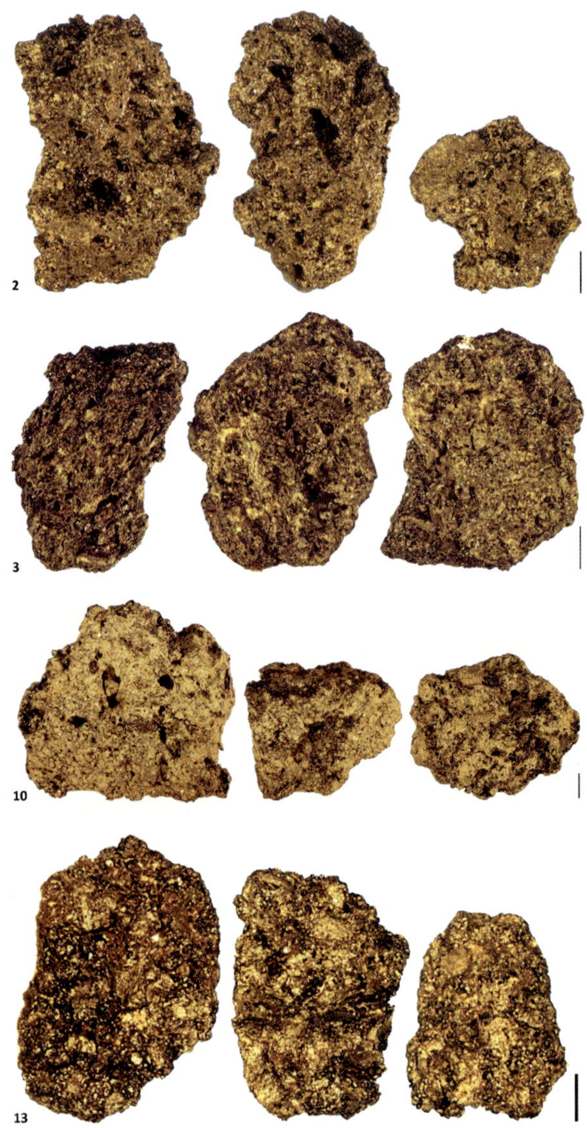

Fig. 7.43. Amorphous charred objects, later identified as cereal products. Scale bar length 1 mm. Site numbers correspond to the sites in Fig. 7.42. 2: Ganglegg, Hahnehütterbödele, 3: Altenstadt, Grütze, 10: Ochsenberg, 13: S. Valburga/St. Walburg. Image: Andreas G. Heiss.

Site no.	2	3	4	5	7	9	10	12	13
Amorphous charred objects (AOV)									
- amount	395	898	418	311	237	2	105	1	4,025
- weight	2,04 g	11,95 g	3,49 g	4,58 g	4,81 g	< 0,01 g	1,01 g	< 0,01 g	42,60 g

Fig. 7.44. Amorphous charred objects (Amorphe Objekte Verkohlt, AOV) as found in the most recent archaeobotanical study of Alpine burnt-offering places. Numbers correspond to the sites in Fig. 7.42. 2: Ganglegg, Hahnehütterbödele, 3: Altenstadt, Grütze, 4: Pillerhöhe, 5: Maneidtal, Grubensee, 7: Schlern, Burgstall, 9: Sölkpass, 10: Ochsenberg, 12: Trappeleacker, 13: S. Valburga/St. Walburg.

2006). Aside from wood charcoal, these amorphous fragments represented the largest proportion of charred finds in the recently investigated Alpine offering sites (Fig. 7.43). Their regular and numerous occurrence (up to several thousand fragments per site, see Fig. 7.44) had suggested a certain importance in the offering rituals and thus required a more thorough investigation. In archaeobotany, charred amorphous remains are usually classified as food remains, and considered as either originating from fruit pulp, or being a cereal product of some kind. So they were searched for potentially identifiable tissue remains with the aim of finding hints about the origin of the material. The methods applied included reflected light microscopy at high magnifications (up to 1000x), Scanning Electron Microscope (SEM) imaging, and chemical digestion of the charred materials following the methodology by Hansson and Isaksson (1994; for details, see Heiss 2008, 26–30, 131–137).

New Insights

The data of the current analyses, together with the synopsis of previous work, resulted in a broad spectrum of plants, of which only food plants shall be discussed here (for the other finds, refer to Heiss 2008a; 2010a). The amorphous objects, for instance, turned out to contain tissue fragments from cereal bran. The term 'bran' describes the sum of all outer layers of a cereal grain, such as the pericarp (the tissues forming the fruit), the testa (the seed coat), and some of the seed's more inward tissue layers. Even in a charred state, the characteristic cell patterns of the transverse cell layer from the pericarp allow the identification of cereals. Also, the structure of the thick-walled aleurone layer which stores the main part of the grain's protein is diagnostic for certain cereal species (Fig. 7.45; see also Chapter 7.2).

For sites 3, 5, 7, 10, and 13 (numbers corresponding to Fig. 7.42), wheat (*Triticum* sp.) could be identified as a component of the charred mass via the transverse cells (Fig. 7.46). The fragments from site No. 3 (Altenstadt, Grütze, western Austria) also contained barley (*Hordeum vulgare*), which was recognised via its multi-layered aleurone layer (Fig. 7.47; Heiss 2010b). Many other tissue remains, mainly fragments of single-layered aleurone layers, were not attributable to particular cereal species.

The tissue remains are clear evidence of cereals in the amorphous objects, but no indication was found for fruit pulp. This suggested the conclusion that the mass was at least predominantly cereal-based. All observed portions of cereal tissue were of minute size, not exceeding 300 µm. Such a high degree of fragmentation indicates grinding of the

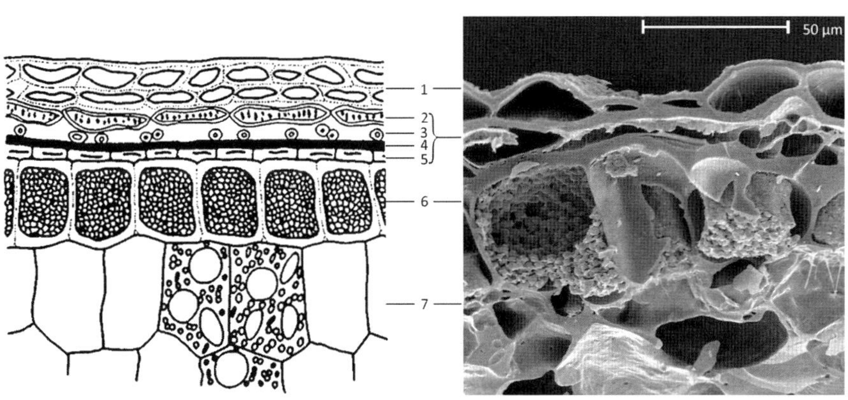

Fig. 7.45. Composition of the seed/fruit layers of a cereal grain, illustrated for bread wheat (*Triticum aestivum* L.). Left: schematic drawing (Hahn and Michaelsen 1996, modified), Right: Scanning Electron Microscope image of an experimentally charred grain. 1: longitudinal cells, 2: transversal cells, 3: tubular cells, 4: testa (seed coat), 5: nucellus remains, 6: aleurone layer, 7: endosperm. Layers 2–6 are collapsed in the experimentally charred specimen. Image: Andreas G. Heiss

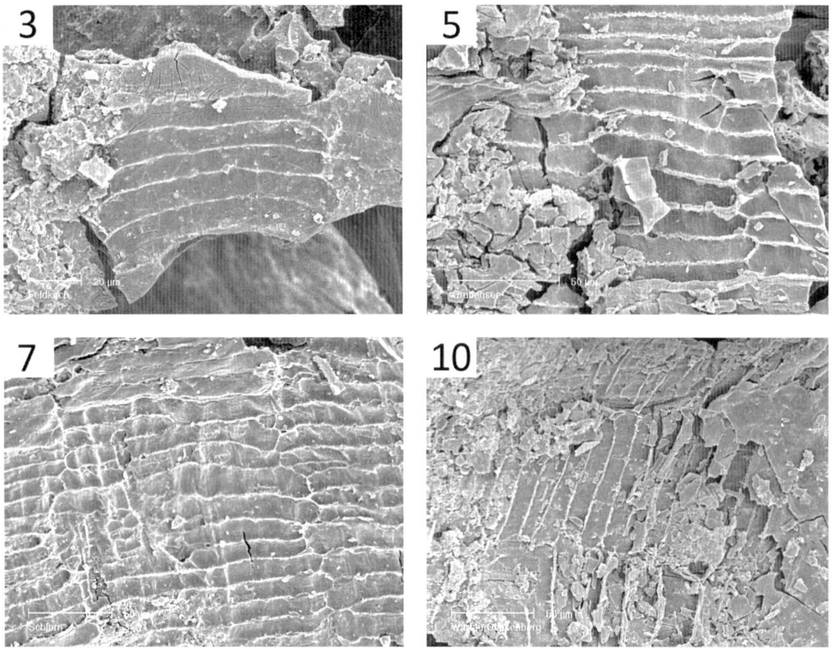

Fig. 7.46. Transverse cell layers from the cereal products. Numbers correspond to the sites in Fig. 7.42: Altenstadt, Grütze (smooth and thin-walled cells indicating barley, *Hordeum vulgare* L.), 5: Maneidtal, Grubensee, 7: Schlern, Burgstall, 10: Ochsenberg (the latter three with regularly thickened and pitted cell walls indicating wheat species, *Triticum* sp.). Image: Andreas G. Heiss.

cereals, with grain sizes comparable to dunst or fine semolina. This was corroborated by comparative structural studies of the prehistoric material from burnt-offering places with the data and SEM images from ancient charred bread finds (Fig. 7.48), such as the bun from a Celtic salt mine in Bad Nauheim, Germany (Heiss and Kreuz 2007), and bread from grave goods in the Roman cemetery in St. Memmie, France (Heiss 2008b; Heiss *et al.* 2008b), and the Viking Age bread from Birka on Björkö island in Sweden (Hansson and Bergström 2002). Eventually, the amorphous objects were identified as the remains of some prehistoric flour-based food. Due to the high fragmentation, it was however not possible to discern between certain cereal products, such as grain-paste, porridge/gruel, or bread (Hansson 1994; Lannoy *et al.* 2002).

Figure 7.49 shows the total cereal remains covered in the study. It is clear that in addition to the cereal product, cereal grains also played a major role in the offerings, both in terms of diversity and numbers of finds. Still, the amounts of fragments of the cereal product suggest that this material was of much higher importance in the sacrificial rites than the actual grains. Among the cereals identified, hulled barley (*Hordeum vulgare* L.) and broomcorn millet (*Panicum miliaceum* L.) show the highest frequency in the sites, whereas wheat species (*Triticum* spp.) dominate in terms of numbers of finds. Chaff is virtually absent. Comparison of these spectra with archaeobotanical data from settlements in the region shows that barley, although the main cereal crop in the prehistoric Alps, is not at all well represented in the offerings during the Late Bronze Age. One factor causing this seeming discrepancy might be the less dense data for ritual sites of this period. But even at sites rich in botanical finds (such as No. 3 Altenstadt, Grütze), barley is barely represented. In the Late Iron Age sites, the remains from the rituals reflect the actual situation of daily nutrition more closely. The cereal product is however well represented during both periods. The presence of rye (*Secale cereale* L.) is only documented for one Late Iron Age site. This goes very well with the rather late cultivation of rye in central Europe, supposedly dating back to the Early Iron Age. However, utilisation of this cereal did not rise to a position of economic importance until Roman times (Behre 1992; Körber-Grohne 1995).

Seeds of legumes (pulses) are preserved less well in a carbonised state than cereal grains (Willerding 1971; Stika 1996; Jacomet, *pers. comm.*), and are thus found somewhat more rarely in settlements. Faba bean (*Vicia faba*), also called broad bean, or field bean, is the most frequent pulse identified from the ritual sites' burnt layers (see Fig. 7.50). The second most frequent legume seeds in burnt offerings are garden pea (*Pisum sativum*) and lentil (*Lens culinaris*). Judging from the data available, pulses are poorly represented in the older sites, but possibly more

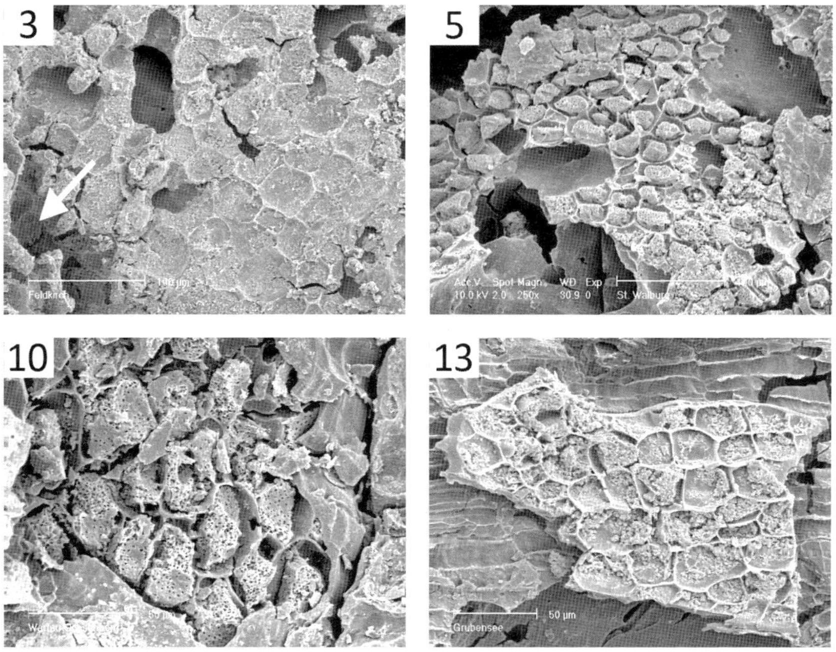

Fig. 7.47. Aleurone cell layers from the cereal products. Numbers correspond to the sites in Fig. 7.42: Altenstadt, Grütze (multi-layered aleurone layer indicating barley, Hordeum vulgare L.), 5: Maneidtal, Grubensee, 10: Ochsenberg, 13: S. Valburga/St. Walburg (the latter three with single-layered aleurone as found in most cereals). Image: Andreas G. Heiss.

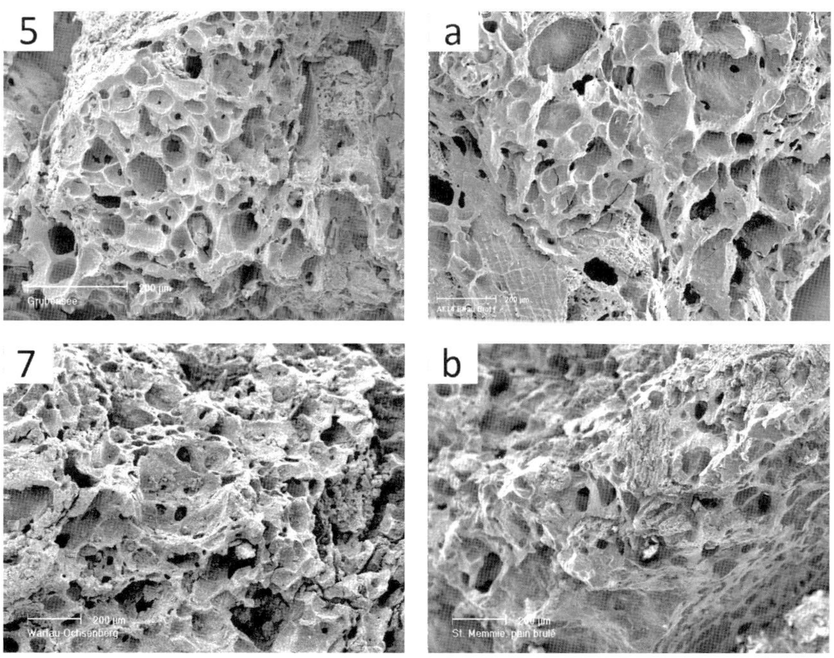

Fig. 7.48. Comparative microstructure of the cereal products from the burnt-offering places and of archaeological bread buns. a: the Late Iron Age saline of Bad Nauheim, Germany (Heiss and Kreuz 2007), and b: the Roman cemetery at St. Memmie, France (Heiss 2008b). Numbers correspond to the sites in Fig. 7.42. 5. Maneidtal, Grubensee, 7: Schlern, Burgstall. Image: Andreas G. Heiss.

related to lower preservability than to rarer use. The spatial distribution of sites with legume finds, however, seems to show a focus in northern Italy: only two sites out of eight from outside this area (No. 11, Forggensee, Bavaria and No. 20, Sandberg, Lower Austria, *cf.* Fig. 7.51) resulted in finds of charred pulse seeds. In contrast, seven out of twelve north Italian sites provided evidence of pulses. This may suggest a local characteristic in the burnt-offering cult practice and is reminiscent of the high symbolic significance of faba bean in the Mediterranean during Antiquity (see Chapter 7.1). Postulating an actual connection would however be purely speculative.

Evidence of oilseeds (such as flax/linseed, *Linum usitatissimum* L., opium poppy, *Papaver somniferum* L., hemp, *Cannabis sativa* L., or gold-of-pleasure, *Camelina sativa* [L.] Crantz) is often difficult to obtain in archaeobotany. Under the influence of

	CA	Middle to Late Bronze Age								Late Iron Age to Roman Age											Ubiquity (percentage of sites)
	1	2	3	4	5	6	7	8	9	10	11	12	13	14	15	16	17	18	19	20	
Cereals																					100%
Hordeum vulgare L. (hulled barley)	+	•	•	•	–	–	–	–	–	•	+	–	+	+	+	–	++	–	+	+++	60%
Triticum monococcum L., *T. dicoccum* Schübl., *T. spelta* L. (species of hulled wheat, total)	++	–	++	–	–	–	–	–	–	–	–	–	++	+	+++	–	++	–	++	++++	40%
Triticum aestivum/durum/turgidum (species of naked wheat)	•	–	–	–	–	–	–	–	–	–	–	–	+	–	++	–	+	•	–	+	30%
Triticum sp. (unidentified wheat species)	•	–	+	•	–	+	–	–	–	•	–	–	+	–	–	–	–	–	+	++	40%
Panicum miliaceum L. (broomcorn millet)	–	+	+	–	•	+	–	+++(+)	–	–	–	–	++	–	++	–	+	–	++	++	50%
Setaria italica (L.) P. Beauv. (foxtail millet)	–	–	•	–	–	–	–	–	–	–	–	–	•	–	–	–	++++	–	++	–	20%
Secale cereale L. (rye)	–	–	–	–	–	–	–	–	–	–	–	–	–	–	–	–	–	–	–	++	5%
indet. (unidentified cereals)	++	•	+	•	–	–	–	–	–	+	–	–	++	–	++++	–	–	–	++	++++	45%
porridge/bread fragments	–	+++	++++	+++	+++	–	+++	–	+	+++	–	•	++++	–	–	+(+++)	–	–	–	–	50%

Fig. 7.49. Semi-quantitative list of charred cereal remains from Alpine and circum- Alpine sites with burnt offerings. Uncertain identifications ('cf.') are included in the respective taxa. Site numbers correspond to the sites in Fig. 7.42. CA... Copper Age. Categories of find numbers: • ... single find, + ... 2–9, ++ ... 10–99, +++ ... 100–499, ++++... 500 and more.

	CA	Middle to Late Bronze Age								Late Iron Age to Roman Age											Ubiquity (percentage of sites)
	1	2	3	4	5	6	7	8	9	10	11	12	13	14	15	16	17	18	19	20	
Pulses/legumes																					45%
Lathyrus sativus L. (white pea)	–	–	–	–	–	–	–	–	–	–	–	–	–	–	–	–	+	–	–	–	5%
Lens culinaris Medik. (lentil)	–	–	–	–	–	–	–	–	–	–	–	–	•	+	–	–	++	–	–	++	20%
Pisum sativum L. (garden pea)	–	–	–	–	–	–	–	–	–	–	+	–	+	++	–	–	•	–	–	+	25%
Vicia ervilia (L.) Willd. (bitter vetch)	–	–	–	–	–	–	–	–	–	–	–	–	–	–	•	–	•	–	–	–	10%
Vicia faba L. (broad bean)	–	–	–	–	•	–	–	•	–	–	+	–	+	+	•	–	–	–	–	–	30%
indet. (unidentified legumes)	–	•	–	–	+	–	–	–	–	–	++	–	++	+	•	–	++	–	–	–	35%
Oilseeds																					20%
Camelina sativa (L.) Crantz (gold-of-pleasure)	–	–	–	–	–	–	–	–	–	–	–	–	–	–	–	–	++	–	–	++	10%
Linum usitatissimum L. (linseed)	+	–	–	–	–	–	–	–	–	–	–	–	+	–	–	–	–	–	–	–	5%
Papaver somniferum L. (opium poppy)	–	–	–	–	–	–	–	–	–	–	–	–	•	–	–	–	+	–	–	+	15%

Fig. 7.50. Semi-quantitative list of charred legume and oilseed remains from Alpine and circum-Alpine sites with burnt offerings. Uncertain identifications ('cf.') are included in the respective taxa. Site numbers correspond to the sites in Fig. 7.42. CA... Copper Age. Categories of find numbers: • ... single find, + ... 2–9, ++ ... 10–99, +++ ... 100–499, ++++... 500 and more.

	CA	Middle to Late Bronze Age								Late Iron Age to Roman Age											Ubiquity (percentage of sites)
	1	2	3	4	5	6	7	8	9	10	11	12	13	14	15	16	17	18	19	20	
Gathered fruits																					65%
Cornus mas L. (Cornelian cherry)	+	–	–	–	–	–	–	•	–	–	–	–	–	–	+	–	–	–	–	–	15%
Cornus sanguinea L. (common dogwood)	–	–	–	–	–	–	–	–	–	–	–	–	–	–	–	–	+	+++	–	–	10%
Corylus avellana L. (hazel)	++	–	+	++	–	+	–	–	–	–	–	+	+	++	–	–	++	++	++	•	55%
Crataegus laevigata (Poir.) DC. (hawthorn)	–	–	–	–	–	–	–	–	–	–	–	–	–	–	–	–	–	–	–	•	5%
Fragaria spp. (strawberry)	–	–	–	–	–	–	•	–	–	–	–	–	•	–	–	–	–	–	•	–	15%
Quercus sp. (oak)	+	–	–	–	–	–	–	–	–	–	–	–	–	–	–	–	•	+	–	–	15%
Pinus cembra L. (Swiss stone pine)	–	–	–	–	–	–	–	–	–	–	–	–	–	++	–	–	–	–	–	–	5%
Prunus spinosa L. (blackthorn)	+	–	–	•	–	–	–	–	–	–	–	–	–	–	–	–	–	–	–	–	10%
Rosa sp. (rose)	–	–	–	–	–	–	–	–	–	–	–	–	•	–	–	–	–	–	–	–	5%
Rubus spp. (brambles, raspberry etc.)	–	–	•	–	–	–	–	–	–	–	–	–	+	–	–	–	+	+++	–	–	20%
Sambucus spp. (elder)	++	–	++	–	–	–	–	–	–	–	–	–	+	–	–	–	+	++	+	–	30%
Solanum nigrum L. agg. (black nightshade)	•	–	•	–	–	–	–	–	–	–	–	–	–	–	–	–	++	–	–	–	15%
Sorbus aucuparia L. (rowan)	–	–	–	–	–	–	–	–	–	–	–	–	–	–	–	–	–	–	–	+	5%
Vitis vinifera L. ssp. *sylvestris* (C.C.Gmel.) Hegi (wild grapevine)	+	–	–	–	–	–	–	–	–	–	–	–	–	–	–	–	++	+++	–	–	15%
Cultivated/introduced fruits																					20%
Ficus carica L. (fig)	–	–	–	–	–	–	–	–	–	–	–	–	–	–	–	–	•	–	–	–	5%
Juglans regia L. (walnut)	–	–	–	–	–	–	–	–	–	–	–	–	–	–	–	–	+	–	+++	–	10%
Olea europaea L. (olive)	–	–	–	–	–	–	–	–	–	–	–	–	–	–	–	–	+	–	•	–	10%
Vitis vinifera L. ssp. *vinifera* (cultivated grapevine)	–	–	–	–	–	–	–	–	–	–	–	–	–	–	–	–	++	+++	+	•	20%

Fig. 7.51. Semi-quantitative list of charred fruit remains from Alpine and circum-Alpine sites with burnt offerings. Uncertain identifications ('cf.') are included in the respective taxa. Site numbers correspond to the sites in Fig. 7.42. CA... Copper Age. Categories of find numbers: • ... single find, + ... 2–9, ++ ... 10–99, +++ ... 100–499, ++++... 500 and more.

heat, the fat contained in their storage tissues usually starts to boil, developing flammable and expansive gases. Their escape causes ruptures in the tissue as well as favouring complete combustion of the seeds. As a consequence, oilseeds survive charring with difficulty and thus are only rarely preserved in dry soils (see Chapter 7.1). Their chances for preservation in a context of intentional combustion (as can be assumed for burnt offerings) can be imagined to be even lower. In contrast, impressive find numbers sometimes reaching tens of thousands are documented for some waterlogged sediments, as for instance in lakeshore settlements in Germany and Switzerland (Jacomet 2009). Against this background, the very few grains of oilseeds recovered from burnt offerings are more inspiring than disappointing: their presence in four sites (Fig. 7.50) is the proof of their utilisation in the Alpine (sites No. 1, 13) and circum-Alpine (No. 17, 20) offering rituals. Their absence in the other sites however is of no significance.

Edible fruits from trees or shrubs (and rarely from forbs, such as strawberry) are commonly regarded as having been gathered in the wild on purpose. Their seeds constantly occur in areas of human activity throughout the archaeological record. Besides their vital importance for hunter-gatherers, even in agricultural societies they made a good complementary food rich in vitamins (and probably were a welcome relief in regards of taste). Yet, in spite of their common occurrence in human settlements, they are not well represented in the burnt-offerings (Fig. 7.51): after a high diversity of remains of gathered fruits found at the Copper Age site of Pigloner Kopf in South Tyrol/Alto Adige (No. 1), they are only sporadically present in the Late Bronze Age and Late Iron Age sites. The only conspicuous exceptions during these periods are Altenstadt, Grütze (No. 3), and S. Valburga/St. Walburg (No. 13). Later, under Roman influence (sites No. 17–19), the diversity of gathered fruits rises again and is complemented by finds of imported fruits such as fig (*Ficus carica* L.), grapevine (*Vitis vinifera* L.) or walnut (*Juglans regia* L.), some of which might also have been cultivated locally. Hazel (*Corylus avellana* L.) is represented during all periods. Preservability of the respective finds is not a likely reason for these changing patterns, as the seeds/fruits of gathered fruit plants easily 'survive' charring due to their morphology: they either have a hard-shelled pericarp such as nuts (as in hazel, acorn, or sweet chestnut) or nutlets (in strawberry and rosehip), or they are thick-walled fruit stones (as in raspberry, elder, or Cornelian cherry). For the Late Bronze Age and Late Iron Age sites (prior to Roman influence), a low importance of gathered fruits for the offering rituals seems to be a likely explanation for their low presence in the charred layers.

The remaining wild plants discovered in the charred record were usually segetal (crop weeds) and ruderal (wayside and fallow) plants: plants belonging to these groups need nutrient-rich soils and plenty of light, easily tolerate disturbance (such as repeated damage by being trampled on), and their reproductive life cycles are very short. These plants – weeds – thus are perfectly adapted to man-made ecosystems like waysides, fallow land and crop fields. From the archaeological record, however, most of these plants cannot be clearly attributed to either ruderals or segetals: they might have been transported to the sacrificial fires with the crops, or might have been part of the on-site vegetation developing during times the pyres were not in use. Using the flowering times (and consequently, the beginning of the fruiting times) of the weeds found in the nine most recently investigated sites (No. 2, 3, 4, 5, 7, 9, 10, 12, 13), a *terminus post quem* for the offerings may be suggested for August when most of the plants found would already bear fruit (Fig. 7.52). However, this hypothesis can only be maintained if the weeds were indeed growing on-site. If they were brought to the pyres together with the crops (which might have been stored for several months before the sacrifice) or inadvertently transported there in the sacrificial animals' fur (and thus could have been in a stable for months), we must dismiss it.

Conclusions and Perspectives

The spectra of charred food plant remains recovered from Alpine burnt-offering places (*Brandopferplätze*) and from circum-Alpine sacrificial sites with burnt offerings reflect an important part of daily nutrition, although the Late Bronze Age sites show an under-representation of barley. The common and constant representation of cereals (mainly represented by a cereal product) and other cultivated crops in the burnt plant remains goes

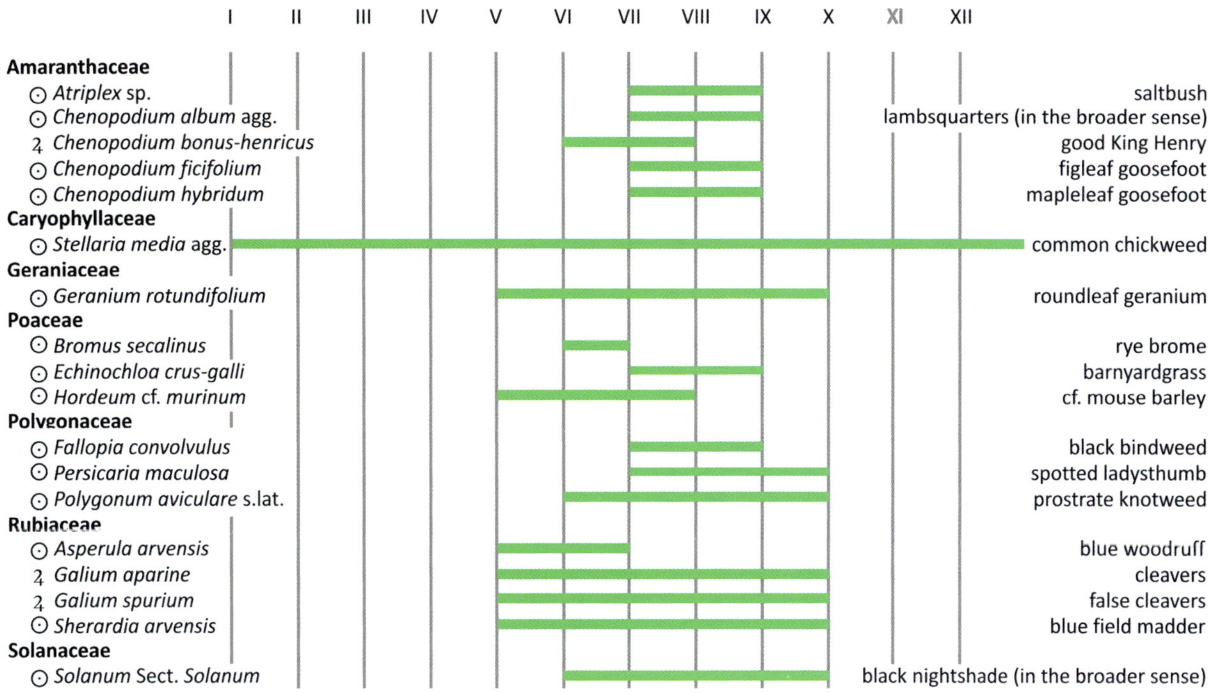

Fig. 7.52. Weeds found in the nine most recently investigated Alpine burnt-offering sites (numbers 2, 3, 4, 5, 7, 9, 10, 12 and 13 according to Fig. 7.42) with their flowering times according to Lauber and Wagner (1996). Roman numerals indicate months, symbols indicate annual (⊙) and perennial (♃) plants.

well with the clear dominance of domesticated animals in the archaeozoological record, and with the obvious lack of importance of gathered fruits. All lead to the conclusion of a close link to agricultural practices, in contrast to gathering and hunting. The oldest Alpine far burnt-offering place found thus, the Pigloner Kopf (Fig. 7.41, No. 1), shows the same correlations in an opposite pattern: moderate amounts of cereals are opposed by a more numerous and more varied record of gathered fruits, and the bulk of animal bones originates from wild animals (Riedel and Tecchiati 2005). So this Copper Age site seems to display not only an older, but also a different aspect of burnt offerings.

The initial question of whether the most recently investigated Alpine sites could have been linked to certain agricultural events such as sowing and harvesting must remain unanswered for now: indeed there is no evidence to argue against annual post-harvest sacrifices in autumn, perhaps even at the equinox, whereas the evidence supporting this hypothesis is still markedly weak. Currently, multidisciplinary research on a larger scale is being carried out at one of the sites investigated, the high plateau of Mount Schlern/Sciliar in northern Italy, by the University of Mainz (Haupt 2009), and this promises to provide further insights into a complex matter.

7.4. FESTIVE USE OF PLANTS: A DIACHRONIC GLIMPSE OF MAY DAY IN THE BRITISH ISLES, FRANCE AND SLIGHTLY BEYOND

Cozette Griffin-Kremer

'I tell to you, a special festival
The glorious dues of May-day:
Ale, worts, sweet whey,
And fresh curds to the fire.'
The Quatrains on the Seasons (Old Irish, Meyer 1894)

Of Plants and Calendars

The use of plants in festive practices is a commonplace for us today, whether they contribute their part to life-cycle events such as weddings or funerals, or to calendar festivities around the world. The two threads may cross over, as when life-cycle events such as graduations, circumcisions and marriage are tied more or less formally to the yearly calendar. Plants have their own life cycle, of course, and this in itself often underlies particular calendar events such as the harvest home celebrations well known in European traditions and widely attested for ancient Middle Eastern cultures, which may be found on seal stones and in other iconography. It is indeed well to look 'abroad' and into the past as a literally timely reminder that our own calendar is a cultural artifact reposing uneasily on one of the most puzzling, and deeply motivating of phenomena: the movement of celestial bodies and their complex relation with the unfolding of vegetal and animal fecundity that underlie the sustenance – both material and spiritual – of humankind.

Calendar events change, even over the span of a lifetime, but calendars change as well and this can deeply affect the way plants are used. The Julian calendar reform was intended, among other objectives, to do away with the confusion and lack of cohesion that makes managing an expanding empire a bother. More pointedly related to the use of plants, the Gregorian calendar was not adopted in Czarist Russia until 1917. Although it had been promulgated in Rome in 1582 after years of debate, it was only gradually adopted in regions under Protestant sway over the course of the following two centuries. By the time Britain accepted the proposition in 1752, nearly eleven days had to be dropped to align with the newcomer, which created a deeply felt dissatisfaction and a long-lasting calendar schizophrenia. This is why the compendia of *British Calendar Customs* take into account holidays in both the 'old style' or OS and 'new style' or NS and list many practices that were carried out OS well into the twentieth century, as well as those prudently undertaken twice and an entirely new category of propitious 'time' for observing some practices – the period *between* NS and OS.

Of May Day and Plants: Food, Fuel, Protection…

In many cases, this had considerable impact on the use of plants, especially flowering plants such as the hawthorn, often closely associated with May Day and maying customs, the opening of the summer season. The widespread conception of a year made up essentially of summer and winter (Nilssen 1920) opens onto a calendar system that is still interwoven with our own. This is a division of the year into halves, fourths and eighths, most probably based on recurrent stellar cycles in which the heliacal rising of the Pleiades (or Vergiliae) is a major event announcing the opening of the summer half-year. This sort of 'peasant' calendar is mentioned by both Hesiod (*Works and Days* ll. 383–4; Evelyn-White 1914) and Pliny (*N.H.* XVIII, 59; Bostock 1855), as well as being engraved in the Gaulish calendars of the very early Christian era (Duval and Pinault 1986).

A forerunner to May Day in Gaelic-language traditions was the festival of *Bealltaine*, the term as we find it today and in the oldest Irish testimony (Beltane). It appears along with the names for the other 'quarter-days' in the Quatrains on the Seasons, dated linguistically to the Old Irish of the eighth to ninth centuries (Meyer 1894). This poem itself is of interest, since it cites the foods associated with each of the quarter-days, highlighting the connection between the calendar system and an agro-pastoral cycle of production. The *Bealltaine* name, along with those of the other three feast days, indicates they were inherited from a time well enough removed from their first analyst that they had become philologically opaque to a competent scholar. That was the bishop-king of Cashel, Cormac MacCuillenan, writing about 900 (Kelly 1997, 537). Cormac provides a contextual etymology, noting that the word proceeds from the two fires made by druids for protecting cattle from disease (Meyer 1912; Corm 2.122; Vendryes *et al.* 1981, B–31). Another name for the holiday was *céattamain* in Irish, *cyntefin* in Welsh, which attest to a common Celtic heritage that would predate the emergence of Goidelic and Brittonic before the Christian era (Vendryes *et al.* 1987, C–58).

The practice of running cattle between two fires or casting burning branches into the four corners of their pen is well attested in ethnographic records across Europe as a method to purify the animals and keep disease, particularly cattle murrain, away from them. There is a plethora of such accounts from Ireland from the eighteenth to the twentieth century (Danaher 1972, 96), some of which mention the fuel utilised. Furze was a standard, both because of its burning qualities and its portability (Lucas 1960, 183–4). It was well known for its protective virtues, as were many other plants, which were inserted in cattle's tails, often tied with a red thread (another frequent prophylactic) to keep them safe from sunset on May Eve to sunrise on May Day. This was the most dangerous time for both people and cattle, either of which could be struck by elf-shot (a fairy dart, usually held responsible for a sudden illness or unexpected death).

One of the most spectacular 'uses' of furze went on until the late eighteenth century on the Isle of Man, where the population was said to burn all the furze bushes in the island on May Eve, because that is where witches and fairies took shelter after sunset. The 1837 account recalling the older 'conflagration' practice mentions that the islanders still gathered particular herbs to place at the doors and inside the house to prevent the entrance of witches, without specifying which plants were sought out (Moore 1891, 111). It is also a reminder that plants used as fuel were generally thought to share in the protective or prophylactic properties attributed to May Day flowers and greenery on the whole.

The hawthorn or 'may' was among the floral royalty of the holiday, laden with special protective powers, and was usually placed at the strategic points of entrance to dwelling, byre, stables, farm or even parish. It presents a special case, in that it was often forbidden entry to the house, since the triethylamines it contains give off an odor reminiscent of putrefaction and hence death, but the notion that it was unlucky to bring other mayflowers beyond the doorway or window sill and into the house was widespread. One thing that is certain is that people in the same township often had diametrically opposite opinions on the subject (Danaher 1972, 89).

Bringing Luck and Health

Either green branch or flowering branch, or simple flower, could be utilised to bring luck to the house or to protect its openings onto the wide world, the paths leading to it and to the well or spring that served the household. The plants included could be named from this use only as 'mayflowers' or specified, as they are in nineteenth and twentieth century surveys: birch, blackthorn, bracken, broom, buttercups, chestnut, cowslips, daisy, elder, fern, furze blossoms, hawthorn, hazel, herb-of-grace, holly, iris, ivy, juniper, lily-of-the-valley (Fig. 7.53),

Fig. 7.53. May Day lily-of-the-valley as porte-bonheur (luck-bringer), postcard, private collection of the author. Image: C. Griffin-Kremer.

marsh marigold, marshmallow, mugwort, nettle, primrose, rosemary, the powerful rowan, savory, sycamore, vervain, willow, woodbine and yarrow, among others. Even serious ethnographic reports rarely attempt to go beyond a generic name such as these, and local accounts sometimes use *very* local names. Grigson (1958, 166) proposes twenty-six of these for the hawthorn alone, not counting the thirty-nine names given for its berries, which is delightful for dialectologists, but perhaps not too satisfying for botanists. (On the difficulties of attributing historical common names to actual plant taxa, see Box 7.4a on page 361)

It is important to add that nothing is ever as simple as we might imagine and that it was often thought paramount to gather flowers or branches laden with the May Day morning dew, a powerful luck- and fertility-bringer. Of course, people also went out to roll naked in the dew-laden grass, which adds grass to the list! In fact, grass still wet with the morning dew was called the 'flower' of the grass. The 'flower' of the well was the 'top' or 'cream' of the well water, which also had to be 'gathered' by a person of pure intent, from the household, usually a young girl, herself still with 'flower'.

Gathering real plants could entail special precautions, such as never touching a plant with iron, hence only uprooting the smaller ones, breaking or twisting branches off the sturdier ones, walking around them three times sunwise (clockwise) before gathering, and so on. Considering the time of year, many of these plants had yellow blossoms – the major epithet of Bealltaine in Irish and Scottish Gaelic is 'yellow' – and quite blanketed the sites they favored, producing visual and olfactory effects which we should not underrate when thinking about the uses of flowers. (Blossoms with a heady or musky scent such as lily-of-the-valley or woodruff may be widely used in perfumery and are currently having these very properties investigated to verify a reputation in popular traditions for a connection with sexual arousal.)

Many of these plants also had a double life as may-flowers and ingredients in traditional medicines, as was the case of the marsh marigold (Grieve 1931, 519) or the furze as both a horse and human tonic, said to be especially effective at stopping hiccough in children (Lucas 1960, 185–186). Health-bringing or 'purifying' decoctions were made with nettles or hawthorn, and nettle-stinging – striking other children or even grown-ups with the plant – was thought to be quite salutary, as was striking with marshmallows (Danaher 1972, 119–120).

Moving and Mediating

The morphology of some of these plants was especially significant, since they were the opposite of wilting violets. The sturdy bushes lent themselves to being carried in processions or circumambulations, as well as being set up at certain points either to protect the home or as the focus of a social entity, such as a cluster of houses, a neighborhood or village. In this case, they were themselves decorated with more may-flowers and, when called a bush or maypole, were often thought to hold the luck of the new season – hardly a negligible quality – and so be well worth stealing. The kind of riotous behavior which resulted in this back-and-forth, intercommunity light warfare was just the kind of thing that particularly irritated Puritans and hence has left a very generous paper trail of persecution and popular riposte. This is the case of Philip Stubbes in the mid-sixteenth century, a marvelous and quite early source of information about the popular customs which he loathed (Furnivall 1882). There is no reason to think that such older territorial attachments to 'our' maytree or pole have lost their power, witness the 1997 Günzberg, Bavaria, incident of maypole theft that saw seven young men condemned to pay a fine of 600 to 1800 Deutschmarks each. This served to highlight in the press the fact that most people in the area found the arrest and prosecution absurd, because they had done or continued to do the same thing themselves every May Day (Anonymous 1997).

The role of plants in festive practice as a mediator in inter- or intra-community relations is especially rich in the realm of reproduction, that is, in the itinerary of young people towards marriage. The Irish May ball games, which were eventually forced out of fashion due to their rowdiness and one or two ensuing deaths, were inaugurated by the latest-married young couple of the parish giving a flower-decorated ball to the group of young players (Danaher 1972, 103–107). Outside our floral considerations, we must not forget that violent clashes such as faction fights were regarded as insuring the success of both crops and human reproduction and generally ended with a kiss between bruised or bleeding warriors. (As it was also said 'no luck at a wake without blood,' but that is a life cycle story before the days of flowers at funerals...) Most customs involving plants and hopes of romance fell into the categories of social signals or divination practices. A Welsh custom cited for the mid-nineteenth century involved young men placing a bunch of rosemary with white ribbons at the bedroom window of an admired young woman (Trevelyan 1909, 25), but the flowers left expressed the whole range of judgments of men on young women, from a 'desirable catch' to a 'slut', a practice called 'birching' on the Welsh borders (Simpson 1976, 148–149).

In the wealth of divinatory custom associated with marriage prospects, we can take an example from Scottish practice of a girl picking a sprig of yarrow, to which she addressed a three-part rhyme, then put it under her pillow to dream of her lover-to-come (Banks 1946, 216). May Eve and Day as a time-warp, a kind of fold in the regular surface of the year, made divination customs rife then and we occasionally find a trace somewhat farther back than the avid collecting of the nineteenth century, usually because some action was deemed unseemly or associated with evil intentions towards the social body. In a Scottish account of May of 1597, a woman was indicted for 'passing to the green growing corn', that is, she sat before sunrise May Day to peel the blades of corn in order to see whether they were growing withersuns (counterclockwise), which would mean a poor year, or sunwise (clockwise), which meant a fine, prosperous season (Banks 1946, 205).

Other Conceptions and Other Times

The list of plants used in May Day and maying celebrations concerned rural, village, town and urban life and still do today. The plants mentioned here are but a 'short list' among the many cited in the British Isles and across Europe in nineteenth and twentieth century reporting. We may think too little about our own attachment to flowers and plants, or our aversions, and how they were formed, but it helps to listen to others' thoughts. An Irish saying noted that the flowers were the stars of the day, reflecting the stars in the sky at night. There is often impressive conceptual cohesion in popular tradition, as well as aesthetic sensibilities that we have moved away from, such as the obvious compliment referring either to prominent pubic or axillary hair which set the epithet 'furze-pelt' on a blond woman in an Old Irish epic. Many references to plants or their effects remain opaque for us, but appeared obvious to people in the country even a century and a half ago; for example, the notion that cows' urine possessed valuable properties as a disinfectant and hence a prophylactic, most especially because it was the quintessence of distilled vegetal virtues and called 'all-flower water'.

One of the most important facts about flowers especially is that they are 'as welcome as the flowers in May' and they come at what was not so long ago a hard time, the memories of which we have now pretty well buried in our ambient wealth. May Day and May were once a time of dying, most usually from what is termed 'spring hunger', when the stocks of meal had run out in the most careful households and the absolutely vital whitemeats season that would bridge over to the first grain harvest was only at a hesitant beginning. This is when poor people ate grass or clover, while they watched their infants, pregnant women and elderly die.

Protective customs took a great blow during the hard times accompanying the 1840s famine in Ireland, since they rarely insured even a semblance of aid in dire distress. The resulting decline in population added to a whole range of economic and social changes across the British Isles and Europe, such as the dispersal and deportation of rural communities in Scotland, which also impinged on festive practices like May Day. By the early twentieth century, improved access to human and veterinary medicine, or the standardisation and hygienisation of milk production, as a particular example, did much to erode faith in practices that had aimed at insuring the well-being of people, their crops and animals. Much remained, of course, by force of habit or out of the pleasures involved. As in the Isle of Man, massive movements such as the furze-burning declined, but the window dressing remained and even increased, while overall, many play and game events, such as those around the maypoles or dancing, were even reinvested with interest in the context of intense community-building, often highly commercialised operations.

What is perforce missing in a superficial overview such as this are the many nuances or outright regional variations in customs of plant use. The association of mayflowers with the Virgin Mary in Ireland ensured continued support from the Church for house altars as an element in personal piety, although officialdom, of all kinds, had done much to counter the more boisterous practices of stealing hawthorn bushes from great houses or playing May Ball games until the blood flowed. Many of the fine shades of custom are lost to us through lack of more precise reporting. It seems that the territories where the maypole dominated and those where the May Bush was more prevalent in Ireland often interpenetrated, but were quite distinct and this is certainly true of the may branch and may flowers in Brittany. Many of the events such as processions that were carried out in towns as early as the eighteenth century remained popular and were underwritten by the municipal authorities. Plants continued to be associated with the on-going wassailing traditions in Wales and the bringing-in-of-summer songs of Ireland, both often mentioning the plants involved, as in the Munster chant 'holly and hazel, elder and rowan, and bright ash from beside the ford' (Danaher 1972, 89).

Novelty and Continuity

The advent of folklore or folkdance societies converged on the festival in force from the end of the nineteenth century and provided the sanction of learned approval, especially when festivities were decorous and good tourist attractions. The formerly neighborhood-based children's 'garland' processions in England that involved carrying flower-laden objects of various shapes, singing and collecting money, were increased in scale and complexity through adult intervention and organisation during the nineteenth century, and even enshrined in literary efforts such as autobiographies and novels with a bent towards picturesque regionalism (Simpson and Roud 2000, 227–228). The interweaving of literary or at least literate and oral sources, as well as the far from 'innocent' implication of reporters (to this day, for that matter) is much too subtle a subject to take up here (Judge 1993), but must be kept in mind, nonetheless, and the time of reports can be of paramount import. Many speak of events 'within living memory,' so even when they are accurate and can be extensively cross-checked, they are perforce veiled by the passage of time and the application of much thinking on popular traditions that saw them transformed and re-invented in the course of a widespread search for meaning in industrialising countries.

On top of that, festive events are products of their time just like we are, and styles change. Decorous, effective tourist events that often simultaneously underwrite local or group identitary quests in which the use of plants is important are thriving. The fact that May Eve revels in Edinburgh today involve wearing little more than an ivy leaf (if that) and rave parties, as well as the occasional clash with police, has proven just how reinventable the holiday can be. That is just one avatar in a broad contemporary spectrum (as one can see by trying 'Beltane' in a search machine...). Present-day festivities, on the whole, are rarely riotous and there is often considerable insistence on at least a semblance of research into past events to provide 'authentic' models for the use of plants in celebrations, all the

Fig. 7.54. 2007 Lily-of-the-Valley Festival, Rambouillet, France. King Kong float, one of 14 lily-of-the-valley floats in the mid-May festival. Image: C. Griffin-Kremer.

more because such uses seem to faithfully espouse many of the tenets of New Age lifestyles.

Plants are often emblematic of particular festive events and can be on the leading edge of commercial tsunamis that suddenly flood markets and introduce novelties – think of the Halloween pumpkin and its reception in recent years or the tulip speculation bubble in the seventeenth-century Netherlands. Just as interesting as uniformised waves of consumer products are the marked differences that subsist in the ways plants are used in neighboring areas or countries, even today. A good example of this is the present use of the lily-of-the-valley (*Convallaria majalis* L.) in May Day and maying customs in France (Fig. 7.54). This comes as a surprise to visitors from the British Isles, where the plant is worn in a procession in one town event, but has generally kept a reputation (like the hawthorn, because of its scent) of being unlucky, if brought indoors (Vickery 1995, 220–221).

Quite to the contrary in France, the lily-of-the-valley is a ubiquitous emblem of good luck, happiness and May Day, and widely given as a gift at that time. It is consequently the object of a massive commercial production on the part of market-gardeners and shipped throughout the country to regular florist outlets, as well as supermarket chains. Its omnipresence occasionally arouses local reactions and it has stimulated a revival of the May branch customs in parts of Brittany, for instance. It is also an indicator that territorialities can be awakened – in a 2008 incident, the caretaker of a private estate shot and wounded one of the members of a wildcat lily-gathering group in the Ile-de-France (Allezy and Persidat 2008). Generally speaking, turf friction is limited to the conflict of interests between professional florists and 'free' sellers in this brief and intense multi-million-euro holiday event (€12,000,000 in 2006 for the Loire-Atlantique market-garden consortia alone, according to *Terra Economica*).

This rather extreme example of commercial hyper-ventilation brings up an interesting aspect of the use of plants in festivities, be they life-cycle events such as weddings and funerals or calendar customs: just what traces of their use might remain afterwards in any taphonomic context? This highlights several issues. Firstly, the dates of the merry-making are determined by our own calendar and not a roving lunar calendar appointment, so the materials available, especially as regards the flowering plants, are limited to a particular time frame. Whether those available are actually the ones used is a different question and one for which reports often leave researchers relatively unsatisfied. What was left of a lily-of-the-valley or a primrose uprooted to bind into a nosegay to give to older neighbors in an Irish hamlet in 1932, once wilted and discarded? May Day garlands in village and town processions could include quite an accumulation of flowers attached to something like a light wooden frame in the shape of a wreath, bell, and so on, but they still had to be carried by children, so there is a limit to the amount of material involved. Some of the festive objects mentioned above may have been burned (the maytrees were in some cases), but this fate is not frequently cited and we know that maypoles in towns were often kept over many years. Perhaps the occasional one was thrown into a river by a Puritan, but hoping to find identifiable remains of such an event would be overstraining the normal potential of serendipity.

This brief survey promised to be diachronic, but the dates noted obviously do not give anything like blanket coverage until the heyday of the newspaper and magazine, which corresponds largely with a period of reinvention and reappropriation of many rural or popular-class festive activities, if not the outright invention of holidays to meet new needs – Mothers' Day is an obvious case in point. There is nothing to indicate that rural or urban festivities organised by illiterate tradition-bearers were in any way static, though we are often faced with an absence of evidence. When it is possible to follow a 'popular' holiday like May Day in any depth at all for even a few decades, the very unstatic, adaptive and opportunistic nature of the beast is usually the most striking aspect (Phythian-Adams 1983). At times, one might be tempted to say that a particular recorded event is celebrated on or around May Day, rather than posit it is a continuation of an earlier festivity (Cawte 1983). We are lucky that people like Philip Stubbes hated popular traditions and every conceivable form of sentimentality enough to write extensively describing their lurid details. (His wife confessed to him on her deathbed that she had once, as a small child, sinningly lavished love on a puppy, or so the story goes.) Otherwise, our luck on attestations to practices before the eighteenth century is generally circumscribed by the hope that someone was injured (preferably killed) in a playful fight or that there was an accusation of witchcraft, actual theft of dew-covered mayflowers or hexing of milch cows that made it into court records.

This is no reason to be unduly pessimistic, however. We have pan-European attestations fairly early, if thin on the ground, of festivities associated with May Day, old style, by the medieval period. Bringing in the summer in processions, preceded by gathering green boughs and flowers could prompt a comment, if it interfered with something important, such as arriving in time for a battle. This may be the case in a thirteenth-century Spanish scoffing song (Gonzalez and Mele 1944, 9). Bringing in the May occasioned one of the earliest complaints about priests in England joining in the activity, in about 1240, although the Church often officially sponsored the processions, in principle as part of Marian devotion (Simpson and Roud 2000, 226). This may certainly have reinforced some of the modern events centring on the virginally colored and uterinely shaped lily-of-the-valley, abundantly reflected in art from at least late medieval times on (Sillasoo 2006), as the flower of choice for May Day or maying celebration. All was not virginal bliss as we see in the language of flowers, which could be quite brutal before its Romantic transformations; witness a 1367 attestation from Amiens in France, because a young woman filed a legal suit. Someone had put a branch of elder on her house, which meant that she stank and she resented this deeply (Belmont 1978, 20). When the maypole in front of the Louvre (Fig. 7.16) fell over by itself in 1610, it was taken as a bad sign and later thought to have predicted the assassination of Henry IV on 14th May (Hoffmann-Krayer and Bächtold-Stäubli 1927, 1523). This is about the time when iconographic representation of plants often becomes discriminatory, giving us an accurate enough depiction to enable identification of a species like the lily-of-the-valley, again usually associated with Marian devotion and only later attested in French maying customs (Schmidt 2000, 114, 130).

A Closing Remark

Festive uses of plants for this one 'holiday' have been variously described as prophylactic, medicinal, decorative or devotional, and are themselves set within the far broader scope of the diversity of plant use, some of which has probably been underrated due either to lack of remains, lack of ethnographic coverage or lack of imagination (Sherratt 1998). The plants mentioned here have definitely been the out-of-garden variety, on the whole, so fit well with the classic popular categorisation of May Day and maying activity as 'bringing in' the summer, as though the transportation of greenery in its most material form towards the home had something compelling to do with the unfolding of a season.

Certainly, ceremonial or festive uses often highlight the importance of non-agricultural plants, and the diversity in the ways plants are mobilised for consumption or even for company. Plants, after all, are part of our daily lives and have quietly kept a power over our minds, because of their combination of the ephemeral and the stubbornly recurrent. They are 'timely' in ways we often overlook. Do we choose them for our festivals simply because they are there at the right time, a matter of pleasure and expediency, or has their 'being there' helped us to shape our notions of time? These are part of the elective affinities between humankind and the environment, the depths of which we cannot even pretend to plumb. In any case – be it festive event, proverb, etiological tale, fable or song – people have often chosen to say what they mean by using or evoking plants, as though they were a measure of man. As an example, we can seek a closing remark in an Irish account, in which a poor man could reply to the rich farmer, who chided him for cutting furze as fuel for his family's fire, in this way (Lucas 1960, 186):

> *'You and I will pass from this place*
> *And the furze will be growing when we are gone.'*

Common Plant Names, Now and Then – The Botanical Viewpoint

Cozette Griffin-Kremer and Andreas G. Heiss

The botanical identifications given in Chapter 7.4 for common names are no more than a best approximation. They are intended to aid readers in identifying a probable species or group. Use of the term 'fern' suffices to discourage the meekest of affirmations, even when all the texts cited refer to the British Isles, so the identification as *Pteridium aquilinum* (L.) Kuhn must be taken with a grain of salt as large as the word. The term 'bracken' is often synonymous with 'fern' in regional use, and this is true for many plant names, such as the series gorse-broom-whin, terms used interchangeably by some reporters. Equally daunting (even if they do not intend to be) are herb-of-grace, elder (perhaps referring here to ground elder, *cf.* Grigson 1958, 216; Grieve 1931, 368), ivy (*Hedera helix*) or ground ivy (*Glechoma hederacea*, *cf.* Grigson 1958, 327–328) – terms which may however also refer to larger bindweed (*Calystegia sepium*, *cf.* Grigson 1958, 288), woodbine (probably a synonym for honeysuckle *Lonicera*, in Grigson 1958, 355), rosemary (*Rosmarinus officinalis* or *Andromeda polifolia*, *cf.* Ary and Gregory 1960, 118) and savory (perhaps the *Satureja montana* referred to in Grieve 1931, 718–719). 'Herb-of-grace' may refer to rue (*Ruta graveolens*), but in all likelihood – and in the logics of May Day celebrations – is simply another name for vervain (Grieve 1931, 831). 'Mountain ash' is usually taken as a synonym for the rowan, but this appears unlikely, when the words occur together, as in the quote above 'and rowan, and bright ash from beside the ford' (*cf.* Grieve 1931, 69–70, who proposes *Sorbus aucuparia*). The flora consulted are Ary and Gregory (1960), Grieve (1931), Grigson (1958), Pankhurst and Mullin (1991), Vickery (1995), all helpful; Grieve (1931), Grigson (1958) and Vickery (1995) being especially rich in providing long lists of common names and/or remarkable folklore information to trawl through for those bold enough to be tempted.

The source for all this mischief is the huge gap between modern concepts of plant taxonomy and the historical (and, most probably, also prehistoric) perception of what a plant 'species' is. In a nutshell, these differences can be explained as follows: a scientific plant species name refers to all plant individuals sharing common constant features with an actual individual in a herbarium, gathered by its author (the 'holotype' of the plant species). These features comprise morphological, ecological, reproductive, chemical and genetic characters. Using the definition of the species as a basis, other taxonomic levels are defined, for instance genus, family, order or class above species level, and subspecies, tribe, or variety below species level – each depending on how many traits are shared by the respective plants. And finally, any botanical taxon name may only be assigned once – so, in spite of the (nerve-racking, to some) diversity of synonymous names for the same plant, homonyms (*i.e.* the same name for different kinds of plant) must never occur (IAPT 2006).

In contrast to this, older concepts for 'classifying nature' as they are found in historical sources have a completely different philosophical background, usually based on single aspects of a plant such as the plant use, its seed colour, basic appearance, or even its place of growth. Thus, even plants with a completely different phenotype could share the same name due to having, for example, the same use. A good example for this is *Nigella sativa* L. (black cumin) and the genus *Nigella* in general (Heiss and Oeggl 2005): already widely known and used in Antiquity, it is mentioned in Ancient Greek written sources as *gith/git* and *melanthion*. But in Hippocrates' writings (von Grot 1887), the latter also seems to have designated ergot (*Claviceps purpurea* [Fr.] Tul.), a highly toxic parasitic fungus on grasses and cereals, blackening their kernels. And the name *gith* as well as the more recent Latin term *nigella* were also applied to *Agrostemma githago* (corn cockle), as both plants share the feature of black seeds, regardless of their completely different shapes and flower colours (huge and pink vs. small and whitish-blue) or leaves (entire vs. pinnately dissected). Even worse, due to the toxicity of both *A. githago* and *Lolium temulentum* (darnel), the name *gith* was passed on to the latter species in the Middle Ages (for instance, in the works of Hildegard of Bingen, see Daremberg and von Reuss 1855).

Going back now to the common plant names from Chapter 7.4, Fig. 7.55 gives a few hints about their possible interpretation, and tries to assign them to scientific taxa known to occur in the British Isles today. Flowering times of the respective plants are indicated by the horizontal bars (data from Fitter and Peat 1994). The period from March to May is highlighted, indicating that plants flowering in these months are likelier to have played a part in the maying festivities, simply as evergreen plants due to their leaves. The time of non-native plants' introduction to the British Isles has to be considered as well, as it influences the time span available for a certain plant to become part of the festive rites and to enter folklore. Moreover, even the flowering times given must be considered with great caution, as they are based on modern climate. Major shifts in the flowering time might have occurred in the past, caused by climatic changes such as the warmer Medieval Climate Optimum *ca.* 700–1200 CE, or the cooler Little Ice Age occurring in about the 16th to the 19th centuries CE (*e.g.* Lamb 1982, Grove 1988) and, as mentioned in the article, the Gregorian calendar reform played havoc with both official and traditional timing. Thus, even plants not falling into the 'correct' season today might have done so in the past, when their traditional use began.

Fig. 7.55. List of common plant names from Chapter 8.4, together with their most likely counterparts in plant taxonomy. The plants listed here occur in the British Isles according to Clapham *et al.* (1987) and Stace (1991). Horizontal bars correspond to their flowering times according to Fitter and Peat (1994), periods important for maying customs are highlighted in red.

Common name	Locally occurring plants most probably referred to	Flowering time (I II III IV V VI VII VIII IX X XI XII)	Remarks
birch	*Betula pendula* Roth, *B. pubescens* Ehrh.		
	Betula nana L.		dwarf shrub, thus very unlikely to have been perceived as a (tree-like) birch in folk-lore
blackthorn	*Prunus spinosa*		
bracken	*Pteridium aquilinum* (L.) Kuhn and any other fern species!		not a flowering plant
broom	*Cytisus scoparius*		
buttercup	*Ranunculus* subgen. *Ranunculus* (e.g. *R. acris* L., *R. bulbosus* L., *R. repens* L., …)		flowering times taken from the most common species (as named left)
chestnut	*Castanea sativa* Mill. [a]		
cowslips	*Primula veris* L.		
daisy	*Bellis perennis* L.		
	Leucanthemum vulgare Lam. agg.		
elder	*Sambucus nigra* L.		
	Aegopodium podagraria L.		
fern	see bracken		
furze	*Ulex europaeus* L. *		
	U. gallii Planch., *U. minor* Roth *		
gorse	see furze		
ground elder	see elder		
ground ivy	see ivy		
hawthorn	*Crataegus laevigata* (Poir.) DC., *C. monogyna* Jacq.		
hazel	*Corylus avellana* L.		

Common name	Locally occurring plants most probably referred to	Flowering time (I II III IV V VI VII VIII IX X XI XII)	Remarks
herb-of-grace	*Verbena officinalis* L.		also important when not in bloom?
holly	*Ilex aquifolium* L. *		
iris	*Iris foetidissima* L.		
	Iris pseudacorus L.		
ivy	*Hedera helix* L. *		
	Calystegia sepium (L.) R.Br.		
	Glechoma hederacea L.		
juniper	*Juniperus communis* L. *		
lily-of-the-valley	*Convallaria majalis* L.		
marsh marigold	*Caltha palustris* L.		
marshmallow	*Althaea officinalis* L.		
mugwort	*Artemisia campestris* L.		
	A. norvegica Fr., *A.vulgaris* L.		
nettle	*Urtica dioica* L.		
	U. urens L.		
	Lamiaceae?		Many members of Lamiaceae family have nettle-like leaves and showy flowers (e.g. the genera *Lamiastrum*, *Lamium*, *Galeopsis* ...)
primrose	*Primula vulgaris* Huds.		
rosemary	*Andromeda polifolia* L.		
	Rosmarinus officinalis L. [a]		
rowan	*Sorbus aucuparia* L.		
	S. aria (L.) Crantz agg., *S. intermedia* (Ehrh.) Pers. agg., *S. latifolia* (Lam.) Pers. agg., *S. pseudofennica* E.F.Warb.		
savory	*Satureja hortensis* L. [a], *S. montana* L. [a]		
sycamore	*Acer pseudoplatanus* L. [b]		
	Platanus orientalis L. [c]		
vervain	see **herb-of-grace**		
willow	*Salix* sp. (e.g. *S. alba* L., *S. caprea* L., *S. viminalis* L., ...)		flowering times taken from the species most common on the British Isles (as named to the left)
woodbine	*Lonicera periclymenum* L.		
yarrow	*Achillea millefolium* L. agg.		

* evergreen species (ritual importance is not necessarily tied to their flowers); [a] introduced in Roman times; [b] introduced in Late Middle Ages/Early Modern Times; [c] introduced in Modern Times

7.5. CEREMONIAL PLANTS AMONG THE HOPI IN NORTH AMERICA

Linda Scott Cummings

Understanding ritual or ceremony in the past relies upon understanding modern and historic references to ceremony and ritual and interpreting past botanic, archaeological and historic records. Representation is important in ritual. For instance, using the Hopi as an example of documented ceremonial and ritual plant use by a native group in the American southwest (Fig. 7.56) is a great help to understanding the archaeological record. The *kiva* is a ceremonial structure that is partly or wholly underground. As such, it represents the underworld from which the ancestors emerged. Climbing in and out of the underworld on a ladder of pine is important to ritual entry into the structure and hence, the underworld (Fig. 7.57). Tradition indicates that people climbed out of the underworld on a ladder made of pine, so *kiva* ladders have traditionally been made of pine. Firewood used in the *kivas* is prescribed by custom to include four fuels: *Atriplex* sp. (saltbush), *Chrysothamnus* sp. (rabbitbrush), *Rhus trilobata* Nutt. (sumac), and *Sarcobatus* sp. (greasewood). It is important to note that these woods are all from shrubs, not trees.

Presents or ritual distribution of foods is made in the form of gifts from the kachinas, richly decorated wooden dolls representing links to the spirit world (Fig. 7.58, Fig. 7.59). Loaves of home-baked wheat bread and fruits are given to older people, while sticks of *piki* bread, which is a very thin wafer made from maize, are given to all. Blue or red maize is often used in the making of *piki* bread, making a grey/blue or pink *piki*. Some of the *piki* also may be dyed a bright yellow (Whiting 1939, 38).

Fig. 7.56. Map of the US Four Corner Region and 1) the Hopi National Reservation. Map: R. Lugon, J.-C. Loubier and A. Chevalier.

The departure of the *Niman* kachinas, usually just before mid-summer, marks the end of the planting season. Before they leave, they bring the first maize of the season to the village. This is usually sweet

Fig. 7.57. Outside and inside views of two reconstructed kivas (top: at Bandelier National Monumen; bottom: at Mesa Verde National Park). Top image: Brian0918@wikimedia.org, bottom image: BenFrantzDale@wikimedia.org (modified).

Fig. 7.59. Wolf Kachina, Hopi. Image: L. Scott Cummings.

Fig. 7.58. Multiple Kachinas. Left to right: A. Ogre Kachina, B. Police Kachina (by Sammie Walker), C. Cold bringing woman (front, First Mesa, Arizona), D. Hopi Female Shalako (back, by Stacy Talahytewa), E. Snake Dancer Kachina, F. Hopi Crow Mother (by Milton Takala), G. Snake Dancer Kachina, H. Cloud Kachina. Image: L. Scott Cummings.

maize that has been planted early in a protected location and might have been irrigated. These 'first fruits' of the harvest are often accompanied by gifts of hand-carved dolls made from cottonwood roots. Traditionally, all kachina dolls are carved from cottonwood roots, whether they are of the cradle-board variety, which are given to young children, or are the more artistic variety. Another present that is often given at this time is *Typha* sp. (cattail) stalks. The children chew on the stems, which are sweet. When mixed with tallow or fat, the tops of these stalks make an excellent chewing gum (Whiting 1939, 39).

Zea mays L. is an important component of ritual for the Indians of the southwestern United States. Maize pollen is imbued with ritual power and is combined with paints to paint murals on *kiva* walls. In some instances maize pollen may be mixed with maize starch. Examination of paint from *kivas* used approximately 1000 years ago resulted in recovery of *Zea mays* L. pollen (Cummings *et al.* 2009, 168, 177), substantiating this practice as one of antiquity. Examination of 'maize pollen' from a modern Native American group indicated that sometimes what is called maize pollen is, in fact, maize starch, probably derived from very finely ground maize kernels. Not only was *Zea mays* pollen recovered in plaster samples, but *Cleome* (beeweed) pollen also was common in some of the layers, suggesting that the black pigment used on the walls was made from boiling *Cleome* leaves, stems, and flowers into a paste that was then used as paint. Mythical beings and natural objects are also associated with each of the cardinal directions. *Zea mays* L., for instance, is often arranged by colour and aligned with the directions (Fig. 7.60).

Phaseolus sp. (beans) have an important ritual significance to the Hopi. White beans are the first food consumed by priests after a fast. February is the month of the great *Powamu* or bean ceremony. The *kivas* are heated to provide the correct temperature to sprout the beans at this cold time of year. The beans are planted in boxes. Larger beans, such as lima beans, are prized over smaller beans, which have smaller cotyledons (first leaves). The Hopi believe that the germ god of the underworld fashions crops below ground, gradually pushing them through the earth in the place the farmer has planted seed (mind the strikingly similar motif in the Osiris myth, see Chapter 7.1). God rolls purple string beans between his palms, leaving the marks of his fingers on the pods. Imitations of the pods are made from finely ground sweet maize meal that has been colored with dye from purple maize. The artificial pods are attached to stalks of the sprouted beans as part of the ceremony, thus involving maize in this bean ritual. Maize also may be sprouted in the *kivas* at the time of the bean ceremony. The success of the *kiva* crops is meant to foreshadow the larger harvest of the coming season (Whiting 1939, 40–41).

Planting beans in February in a dark, subterranean room (a *kiva* with ceremonial functions), which is protected from the cold, allows earlier consumption of nutritious fresh food that would not normally be available for several more months. The bean ceremony, held in February, comes towards the end of the winter or during early spring, which are often considered to be 'starvation' months with food being rather scarce. Fresh vegetable foods are at a minimum at this time of year (*cf.* the 'spring hunger' mentioned in Chapter 7.4). People rely upon stored foods during the winter and early spring in most of the temperate areas. By fashioning sprouting ceremonies, people are able to introduce much-needed nutrition into the diet, if even only as a brief, ceremonial consumption. Deficiencies in the B vitamin complex

Direction	Color	Plant
Northwest	Yellow	yellow maize (*Zea mays* L.) Mariposa lily (*Calochortus* sp.) Douglas fir (*Pseudotsuga menziesii*) rabbitbrush (*Ericameria* sp.)
Southwest	Blue or green	blue maize (*Zea mays* L.) larkspur (*Delphinium* sp.) white fir (*Abies concolor*) sand sagebrush (*Artemisia filifolia*)
Southeast	Red	red maize (*Zea mays* L.) painted cup (*Castilleja* sp.) red willow (*Salix* sp.) cliff rose (*Purshia* sp.)
Northeast	White	white maize (*Zea mays* L.) whitest evening primrose (*Oenothera albicaulis* var. *runcinata*) aspen (*Populus tremuloides*) rabbitbrush (*Ericameria* sp.)
Zenith	Black	purple maize (*Zea mays* L.)
Nadir	All colors or gray	sweet maize (*Zea mays* L.)

Fig. 7.60. Colours are important for ritual and are associated with directions. Many colours also are associated with particular plants, as may be seen in this chart.

are expected to be showing up at this time of year and sprouting seeds for consumption is a good way of introducing these and other important nutrients into the diet. Ceremonies take on a special importance. Ceremonial consumption might be a means to obtain critical nutrients. However, the ceremony itself imparts a special meaning to the consumption. Beans sprouted in *kivas* were grown for a longer period of time than the five days that we are used to today. Bean sprouts frequently grow to lengths of 12–18 inches or more in the *kivas*. Sprouts experience increases in vitamin C (500%) and vitamin E (300%) over the quantities present in the seeds. The sprouting process also converts starches from the beans into simple sugars, making the sprouts more easily digested than the beans. The B vitamins originally contained in the beans remain available in the sprouts. Hence, bean sprouts provided valuable water-soluble vitamins B and C that were often lacking in the winter diets of pre-modern Puebloan peoples of the American southwest.

Gourds are used to make horns, trumpets, flutes, elements of masks such as noses, flowers, water containers and rattles. These items have ceremonial value and are considered to be part of the complex of ceremonial equipment used by native people. Artificial squash blossoms are painted orange with paint from *Castilleja* (painted cup) flowers. The long spines of the fruits of devil's claw (*Proboscidea* sp.) are used to make artificial squash and jimson weed flowers and other pieces of ceremonial equipment.

Yucca roots make good suds and are compared to clouds ritually. Washing hair with yucca suds is a means of ritual purification and is important before every ceremony. Whips used in purification rites also are made from yucca leaves.

Solanum triflorum Nutt. (cutleaf nightshade) is a fast-growing weed that looks somewhat like a miniature watermelon plant. Watermelon is a late introduction to the Hopi culture during the historic era (after CE 1500), so the cultural significance of this weedy plant is also relatively recent. Planting this weed in the same location as the watermelon seeds indicates continued introduction of ritual activities into daily life. Planting this weedy vine along with the watermelon vines is done to transfer the property of the fast-growing weedy vine to the watermelon, although the quantity of the weed should be limited so that the vines do not grow so fast that they do not bear fruit (Whiting 1939, 42).

The Hopi are concerned with signs and omens. Branches of *Pseudotsuga* sp. (Douglas fir) are important in rituals. If, when they are collected in the spring, the branches are bright green, it is taken as a sign that plenty of rain will fall and there will not be bad winds. Dull boughs, on the other hand, bode evil for the coming summer (Whiting 1939, 42).

This brief review of some of the Hopi rituals illustrates not only the importance that ritual plays in daily and yearly life, but also the rich fabric available for interpretation of prehistoric records. Understanding symbols and their meanings provides an opportunity to understand prehistoric people in a new light. They were not people solely focused on collecting food and finding shelter. Their lives, like those of more recent people, were a rich tapestry of activities, some of which were for the purpose of nourishing the body and others for nourishing the spirit. And for many people of the world (past and present), there was not a hard line between nourishing the body and spirit.

7.6. CEREMONIAL PLANTS IN THE ANDEAN REGION

Matthew Sayre

The consumption of plants for uses other than mere sustenance is common throughout the entire world. Moreover, humans through the course of history have been quite familiar with the psychoactive properties of many of them. Yet one continent, South America, has a greater proportion of the world's psychoactive plants than any other (Schultes *et al.* 2001). This essay will examine the use of ceremonial plants in the Andean region, a distinct mountainous region that stretches from Colombia to Argentina. This ecological region also borders the Amazonian jungle, where many of the plants discussed here may have been utilised for the first time by humans.

Ceremonial plants are often difficult to classify as crops or domesticates. This does not mean that these plants were not actively cared for or nurtured by humans, but it is often the case that these plants are able to survive on their own without human care (Hastorf 2006). Although not always actively cultivated or intensively domesticated, those with the special ability to temporarily alter the human psychological state have acquired particular resonance and power in human society. In many ancient Andean societies, such as the Incas and the Moche, psychoactive plants were commonly used in medicinal or religious practices. These plants could be considered as conduits to the gods, or even gods in their own right. Some theorists (Furst 1972) believed that religion in the Americas was derived from Palaeolithic shamanic practices brought over from Asia. So, while it has been suggested that mind-altering plants may have led to the first religions (La Barre 1972), this does not appear to be likely, as religion is a ubiquitous aspect of human life even in regions with limited amounts of mind-altering biotica. Regardless, religious activities often do include practices that alter the mind without the use of pharmacopeia. For example, the physical processes of fasting, praying and chanting over extended periods of time can lead to altered states capable of producing visions (Goody 1982).

This paper focuses on the use of ceremonial plants in the Andes at the site of Chavín de Huántar, located in central Peru (Fig. 7.61). Chavín has often been portrayed as the emblematic site of the Early Horizon/Formative Period (1000–200 BCE) in the central Andes. The elaborate iconography of the site has led to many inferences about resource use at this site, particularly in regards to plants. Chavín long attracted the gaze of archaeobotanists drawn to its iconic imagery of lowland plants prominently featured on the Tello Obelisk (Fig. 7.62). This obelisk appeared to depict cassava (*Manihot esculenta* Crantz), *achira* (*Canna indica* L.), peanuts (*Arachis hypogaea* L.), chilli peppers (*Capsicum* sp.) and bottle gourds (*Lagenaria* sp.) (Lathrap 1973; Burger and Merwe 1990; Burger 1992). However, alternatively, some scholars have suggested that the iconography of this obelisk depicted sacred plants/hallucinogens used at the site.

It has been postulated that *San Pedro* cactus (*Trichocereus pachanoi* Britton & Rose), *huacacachu* (*Brugmansia* sp.) *ayahuasca* (*Banisteriopsis caapi* (Spruce ex Griseb.) Morton, along with other ingredients), *cebíl* or *vilca* (*Anadenanthera colubrine* (Vell.) Brenan), *epená* (*Virola theiodora* (Spruce ex Benth.) Warb.), and tobacco (*Nicotiana tabacum* L.), among others, are psychoactive plants that were possibly ritually consumed at the site (Cordy-Collins and Stern 1977; Mulvany de Peñaloza

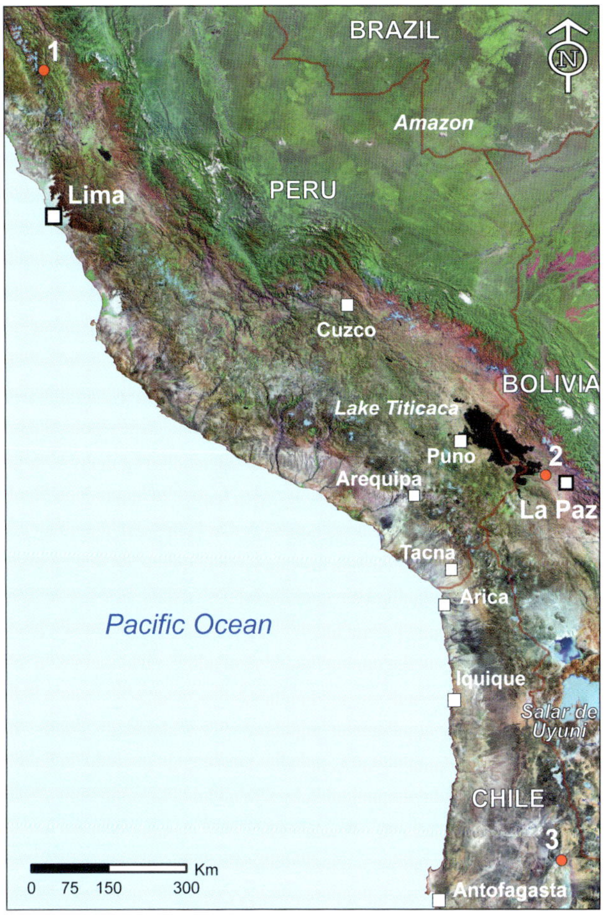

Fig. 7.61. Map of south central Andes with the archaeological sites mentioned in the text: 1) Chavín de Huántar (Peru); 2) Chiripa (Bolivia); 3) San Pedro de Atacama (Chile). Map: R. Lugon, J.-C. Loubier and A. Chevalier.

1984; Burger 1995). Of these, the only one whose botanical ingredient is unequivocally depicted in the iconography is the San Pedro cactus, a plant of highland origins. This plant was depicted on a carving in the circular plaza; the other worked stone that contains possible evidence of ceremonial plants are the Tello Obelisk and the tenon heads which used to be attached to the exterior of the temple.

In their discussions of iconography, many scholars refer to the pioneering work of Drs. Cordy-Collins (1976) and Sharon (2000). Cordy-Collins' (1976) analysis of the shamanic textiles revealed that many of the elements of Chavín iconography were repeated on coastal textiles that depict shamans on trancelike voyages. This work emphasised the broad-scale use of San Pedro cactus on the Peruvian coast, and illustrated how this practice appears to be pan-regional. Another scholar who approached this iconography was Eleonora Mulvany de Peñaloza (1984) who completed a unique interpretation of the Tello Obelisk, in which she stated that all of the plants depicted on the stela are hallucinogens used in rituals at the site. While it has often been noted that there are images of shamans and San Pedro cactus on Chavín's iconography, this article opens up the question of whether or not Lathrap and others misidentified some of the botanical elements on the Tello Obelisk. This point has to be considered and debated more thoroughly, as it is clear that the Chavínos were not obsessed with realism in their depictions of the natural world. The images of birds and other creatures are not always easy to identify to the species level (Burger 1995). Finally, while I do not agree with the emphasis on *ayahuasca*, a compound whose ingredients are from the lowland, it is probable that San Pedro cactus and *vilca* were commonly consumed at the site. It should be noted that these two plants produce vastly different bodily experiences that can last for either short or extended periods of time (Schultes *et al.* 2001). Future work should examine these bodily possibilities separately in conjunction with spatial, architectural, and societal data in order to see how practitioners of rites would have experienced their visions throughout the course of a treatment or ritual.

Recent work by Drs. John Rick (2008) and Constantino Torres (2008) has suggested that shamanic (understood as ecstatic religious experience) practices were employed at the site, although the precise term 'shamanism' is debated, as it is usually reserved for societies without monumental architecture (Moore 2005). These scholars run through the debate of hallucinogens but also mention other psychoactive botanicals such as tobacco, which has been found in later time periods, and *coca*, which is identifiable in macrobotanical samples but has not been recovered as of yet at Chavín (Sayre 2010). Although a few archaeobotanical analyses have been carried out at Chavín (Miller and Burger 1995; Sayre 2006), thus far none of these have provided any evidence. These conjectures and the sculptural evidence of shamanic transformation have ensured that discussions of sacred plant use in the Americas almost always reference Chavín without acknowledging that, as of yet, there has been no actual botanical evidence of these practices.

Fig. 7.62. Tello Obelisk.

Plants and Mixed Compounds in the Past and Present

The great diversity of ceremonial plants of the Andes means that it is difficult to fully examine all of them. Even a cursory examination of mind-altering ceremonial plants reveals the distinct experiences and possibilities involved in their consumption. The following is a non-exhaustive list of the major mind-altering plants that were used in the ancient Andes. Additionally, the most widespread mind-altering product on the planet – alcohol – is a product that can be produced from almost all plant foodstuffs. As such, the use of alcoholic beverages will not be discussed below, but in many instances (such as *vilca*) it is possible that ceremonial plants may have been consumed as an ingredient in an alcoholic beverage.

Although there is a great diversity of plants in South America, the most widely used psychoactive plants today are not necessarily from South America. Tobacco is one of the most widely consumed plants in the modern world and the utilisation of this plant in the past was vastly different from its use in the modern world. Coca's original use has also been vastly transformed. What was once a plant chewed for its mild stimulus and energy-enhancing characteristics is now chemically processed to produce a highly potent and addictive drug (Schultes *et al.* 2001). The other plants described in this chapter are still used in medicinal practice, but have not reached anything near industrial-scale production.

Direct Data

The archaeobotanical identification of these psychoactive plants is quite difficult. Some of the plants leave chemical signatures that are quite similar, such as *epená* and *vilca* (Torres *et al.* 1991; Ogalde *et al.* 2010), while other plants leave distinct chemical signatures. Although some plants may be difficult to identify, there is ample macrobotanical evidence for the use of tobacco in post-temple times at Chavín (Sayre 2006). Tobacco seeds dominate one sample found outside of the area known as the Caracoles gallery. This intriguing find is not symptomatic of widespread tobacco use at the site, as the sample comes from a post-temple context. However, it does raise the possibility that tobacco was found at the site and the diversity of uses of tobacco in the South American past was much wider than current industrial consumption practices (Wilbert 1987).

Macrobotanical evidence would appear to be the clearest indication of sacred plant use, yet this must be questioned, as certain plants native to the region, notably San Pedro cactus, may enter the archaeological record as environmental indicators rather than indicators of cultural activities. It also seems likely that the hallucinogenic plants such as *vilca* or San Pedro at Chavín may have been imbibed rather than snorted, which would make the identification of this practice even more difficult. Various projects associated with Dr. Hastorf's work at Chiripa, Bolivia (Fig. 7.61), have attempted to discern whether or not phytoliths (plant silica bodies that can preserve for millennia) can be used to investigate the trade and use of sacred plants. This work has determined that there may be diagnostic phytoliths capable of revealing these practices but, as of yet, they have not revealed the presence of these plants (Logan 2006).

In an effort to recover direct botanical evidence, in my own work at Chavín I collected samples for GCMS (Gas Chromatography Mass Spectometry) analysis, in order to recover chemical signatures of different plants. Unfortunately, these samples were insufficiently large for analysis. While Torres *et al.* (1991) found ten grammes of botanical material, the samples I have gathered from the interiors of possible snuff tubes were not found in conjunction with large pieces of botanical material that could have been easily subjected to chemical analysis. The preservation conditions at arid San Pedro de Atacama (Fig. 7.61) were clearly superior to those of the wet/dry seasonal climate of Chavín. Future work is needed to truly assess what size the samples need to be in order to search for the key alkaloids present in *vilca* and other sacred plants. However, the interdisciplinary collaborative work conducted at San Pedro de Atacama is clearly a model for future research.

Indirect Evidence: Snuff Tubes and Iconography

Bone tubes and snuff trays are carved implements that may have been used for snorting ceremonial plants or they could have been decorative elements with multiple uses that we have not yet considered. The Chavín snuff trays discussed by Burger (1995) appear to be constructed from whale bone and it is difficult to see them as having some use other than for grinding and consuming plants. These obviously exotic goods were brought to the site from the coast. Recent evidence from the La Banda region (Sayre 2006) near the main temple has made us question whether or not these materials were brought to the site fully constructed, or if the materials were worked into their particular forms at the site. Finally, there is the possibility that snuff tubes were made from reeds which would most likely not preserve in the botanical record.

The iconographic evidence for San Pedro cactus use has been further clarified by our research project (Rick 2008). The lower remnant of the shamanic figure carrying the cactus removed any doubts that remained about whether or not the image depicted a cactus or staff (an image of power recurrent in Chavín iconography).

Psychoactive Plants and Ceremonial Plant Use at Chavín

The difficulty in identifying these ceremonial plants in archaeological contexts does not limit the questions that their possible uses provoke, such as: How were these practices experienced spatially? Who was allowed to use these plants? Where and when were they provided and who carried them to Chavín?

The use of ceremonial plants was a social activity that would have been constrained and shaped by local practices, many of which are not clearly defined. Yet, further analysis of the architecture of the site reveals that it is possible to learn how space contained and structured the events that occurred at the site. The confined labyrinths of the chambers in the main temple likely would have provided a claustrophobic environment where the

Fig. 7.63. The temple of Chavín de Huántar, view from above. Image: M. Sayre.

Common Name	Scientific Name	Use and History
San Pedro cactus	*Trichocereus pachanoi* Britton & Rose	San Pedro and other cacti are still used in divination rite in Andean South America although they have never become common recreational drugs. The San Pedro cactus is consumed by processing the central stalk into a beverage. This means that the presence of the seeds in macrobotanical samples may not indicate ceremonial consumption of this plant as the seeds and flower are not used in creating the drink. The cactus is common throughout the Andean Region (Sharon 2000).
Vilca	*Anadenanthera colubrina* (Vell.) Brenan	This plant may have been one the most widely consumed plants in Andean South America, and yet it is not commonly discussed in the literature. The beans from the tree are gathered and separated from their pod. Then the beans are ground and snuffed through the nose. In Inka times vilca was sometimes consumed in alcoholic maize *chicha* (*aqha*) (Schultes, Hofmann *et al.* 2001).
Ayahuasca	*Banisteriopsis caapi* (Spruce ex Griseb.) Morton	There is little direct evidence for the use of this beverage in the highland Andes during pre-conquest times but the fame of this product and its thorough documentation by modern ethnologists and botanists have lead to its prominent place in the literature. Additionally, this beverage is known to produce some of the strongest hallucinations of any of the substances produced in South America. The beverage is generally made by mixing leaves from the plant with ingredients from other tropical plants (Schultes, Hofmann *et al.* 2001).
Tobacco	*Nicotiana* sp.	This is the only listed plant that spread widely across South America and then became widely used in Central America and North America. The diversity of use meant that many early colonialists encountered it and it spread into Europe quickly after the European invasions. Although most scholars are familiar with the modern means of consuming tobacco there was a greater diversity of practice in the Andean past. In the lowlands and the jungle regions, tobacco is used in many shamanic rites and is known as a strong and potent plant that can be consumed in liquid form, in this instance it is capable of producing visions. The strength of the plant depends on the variety or species consumed (Wilbert 1987).
Coca	*Erythroxylum coca* Lam.	This plant did spread broadly across South America but it did not cross Panama. While this plant does not produce visions when consumed during chewing, its prevalence throughout the region necessitated its inclusion in this section. The coca plant is well-noted in the chronicles and has been found in many ritual contexts, including Capa Cocha sacrifices. It was also used in divination rites (Plowman 1984).

Fig. 7.64. Ceremonial Plants of South America.

Scientific name	Common name	Family	Parts tested
Anadenanthera colubrina (Vell.) Brenan	*vilca*	Fabaceae – Mimosoideae	seed, pod
Banisteriopsis caapi (Spruce ex Griseb.) Morton	*ayahuasca*	Malpighiaceae	bark, leaf
Datura innoxia Mill.	pricklyburr	Solanaceae	seed, leaf

Fig. 7.65. Hallucinogenic plants with diagnostic phytoliths (Logan 2006, 64).

music provided by *pututus* (shell trumpets) would have reverberated solidly off the walls and minds of the participants. The galleries themselves are capable of producing altered states (Lumbreras 1989). The mazes while not impossible to navigate without light are quite difficult to explore without the assistance of non-natural light. The interaction between the use of psychoactive plants and the experience of enhanced visual imagery, increased feelings of claustrophobia, distinct audial input and different experiences of light and darkness would all have combined for an overwhelming event.

The issue of who was allowed to consume these ceremonial crops is one that is particularly difficult to address through material culture. While the iconography depicts some individuals of unclear gender, some overtly male and female figures, and some animals, there does not appear to be clear depictions of children. The tenon heads which were attached to the exterior of the central monument (Fig. 7.63) have been described as illustrating a shamanic march through different stages of hallucinogenic use (Burger 1995). There are heads that vary from clearly human to those that appear to only depict animals. Additionally, some of these heads show anthropomorphic figures with mucus flowing from their noses which has been interpreted as a side effect from the consumption of *ayahuasca* or some other ceremonial plant. It is possible that only shamans or some other form of religious specialists were allowed to consume these ceremonial plants, but the iconography is not a clear indicator in this area. This is another region where further research would be revealing, as the presence of these ceremonial plants in burials or in areas outside of the temple would assist in the analysis of who was allowed to participate in the rites at Chavín.

Some of the plants could have been locally gathered (see Fig. 7.64 and Fig. 7.65), such as San Pedro cactus, and are still used locally by a few medicinal practitioners. However, the general assumption about the use of ceremonial plants at Chavín is that these taxa were brought to the site by pilgrims interested in partaking in the rites at the temple (Burger 1995). *Vilca* grows in the highlands but many of the centres of ancient use of this plant are in highland areas south of Chavín. If *ayahuasca* was consumed at the site, some of its ingredients would have had to come from lowland regions. Local farmers could have provided coca as it could have been produced in nearby regions; however, coca is generally grown in lowland environments. In the case of tobacco it is worth noting that this is one of the few Native South American plants to have spread across the entire hemisphere. This stands in contrast to domesticated animals such as the llama, which never spread north of Colombia before the conquest. Tobacco in ancient South America would never have been assumed to be a habitual product consumed twenty plus times a day, nor would hallucinogenic plants be removed from religious contexts (Wilbert 1987). The experience of consuming sacred plants would presumably have been part of broader rituals that could have lasted several days and invoked multiple realms of religious practice.

Conclusion

A specifically scientific focus on identifying uses of sacred plants will lead us to a more situated understanding of how these plants were moved across the landscape, where and perhaps even when they were prepared and consumed, and who was allowed to consume them. This archaeobotanical research will provide the contextual data that we need to broaden our understanding of the past, rather than inferring much behaviour from stylised depictions of plants in art.

Chavín is an emblematic site, yet much is still unknown even after roughly eighty years of investigation. It appears that chemical analysis may be our best hope for enlivening our conception of when and where sacred plants were used at Chavín. Once there is more actual botanical data it will be possible to further elucidate the social events that necessitated and structured how these plants were used and indicate how society may have been structured through their use.

The question remains, why is Andean South America so diverse in mind-altering ceremonial plants? The diverse regional ecology, in particular the remarkable biota of the Amazonian lowlands, certainly meant that there was greater natural abundance of plants with the chemical components necessary to alter human daily experience. However, there was also the cultural practice of ecstatic shamanism and ritual priests who valued and searched for means of expanding their visions and dream states. These forms of contact with mental states outside of the norm were not always encouraged in many other regions of the world. The question of whether the psychoactive plants were the catalysts for behaviours, or the behaviours were the impetus of psychoactive plant use remains a conjectural issue. Regardless, it is clear that ceremonial plant use was an important aspect of Andean ritual life, and this merits special attention both in approaches to the botanical record and models of Andean society.

7.7. CONCLUSIONS

Andreas G. Heiss and Ann-Marie Hansson

The last decades have brought a great deal of change to historical and archaeological work. New methods have been invented, and old ones have been improved. Interdisciplinary research has brought new and sometimes unexpected insights by combining written sources, archaeological finds and natural scientific data, with ethnographic records as modern analogues. In addition, many paradigms, often concerning human social life, have shifted or have made way for new ones, just like the theories arising from feminist studies and queer theory have begun to broaden our perspectives on past societies in general (see *e.g.* Hays-Gilpin 2000; Perry and Joyce 2001; Schaan 2006). Our knowledge of the past can thus be expected not only to expand in the future, but also to keep changing and shifting.

In terms of the investigation of rituals, we have seen the changing views on what can be regarded as ritual, and thus the question of whether our modern concepts of what 'ritual' means might be applicable to the past at all. When trying to find a common denominator for ritual actions in the past and the present, at least magical (*i.e.* animistic) thinking may be a promising candidate (Subbotsky 2010, 7–8): it is defined by the belief that inanimate objects as well as plants and animals have a mind of their own, either set to help or to harm people. In the dialogue with an invisible world – in the ritual – these objects or life forms are imagined to be valued or dreaded agents in favour of, or against, human will.

In the beginning of this chapter, we outlined various concepts of ritual plant use, illustrating the strong desire in humans to explain their world, and to control their own fate. Among these concepts, especially the connection between the annual cycle of seasons, essential for agricultural societies, and the image of a cycle of life and death in humans could be a recurring theme in almost all human cultures (Chapter 7.1). This connection seems to be reflected by the frequent attestations, be they archaeological, written, or ethnographic, of agricultural offerings in ritual contexts (Chapter 7.1, 7.2 and 7.3), as gifts to invisible (divine) beings or in order to nourish, to placate, or even to banish the spirits of the deceased (Chapter 7.2).

The mirthful aspects of plant symbolism were discussed in Chapter 7.4, using several examples to illustrate the continuity of plants serving as symbols for good luck and fertility from historical till present times, while explaining some of the difficulties researchers encounter when following plant traces in written sources (Box 7.4a, p. 361).

Turning from the Old World to the Americas, Chapter 7.5 showed that the connection between plants and fertility formed a recurrent theme in most past and present cultures, as already pointed out. Chapter 7.6 focused more on the use of hallucinogenic plants as a means of communication between the material world and the otherworld, reminding us of the fact that some magical plants do not require us to enlist our imagination to have magical effects.

Despite illumination and secularism, modern rituals in which we separate our profane and sacral life are still being documented and investigated. Just like the people from societies which are sometimes dismissively (and unjustifiably) called 'primitive', 'modern' people also share the same human desire to influence their fellow human beings, their environment, and their very own lives through means exceeding their immediate human action – by using rituals.

Chapter Acknowledgements

Ann-Marie Hansson's and Andreas G. Heiss' (Chapter 7.1 and Chapter 7.7) warmest thanks go to R. Sokiranski and P. Zwaka (Klinikum Traunstein), C. Adrario (Basel), H.-P. Stika (University of Hohenheim, Stuttgart) and W. Pirsig (Ulm) for allowing them to use their unpublished manuscript and photos on the 'Ulmer Kornmumie'. They also thank T. Hoppe (Landesmuseum Baden-Württemberg, Stuttgart) for providing original imagery of the finds from Fellbach-Schmiden, and I. Hein and U. Thanheiser (both VIAS, University of Vienna) for their support with additional literature on corn-mummies. Additional thanks are due to Kai Fechner (Université Libre de Bruxelles and Université d'Europe, Bruxelles), Delwen Samuel (Kings College, London) and Marina Ciaraldi (University of Birmingham) for so kindly sharing their knowledge of archaeological bread finds.

Ann-Marie Hansson (Chapter 7.1 and Chapter 7.2) is most grateful to B. Petré, who kindly allowed her to study the bread bun found during the excavations on Lovö under his leadership. She would also like to specially thank the international network of EARTH, U. Miller and U. O'Meadhra (both University of Stockholm) for encouraging her to write this article, and L. Scott Cummings (PaleoResearch Institute, Golden, CO) for sharing her knowledge of starch analysis with her. She is deeply indebted to the Berit Wallenberg Foundation for financial support over the years, furthering the analyses of archaeological bread samples and studies into the history of bread.

Part of the research by Andreas G. Heiss (Chapter 7.3) was funded by the Austrian Science Foundation (FWF) under the project number P16714, which the author acknowledges with special thanks.

For their most useful comments to improve the manuscript, all authors in the production of this chapter are greatly indebted to J. A. Blain (Asana, San Francisco), M. Kohler-Schneider (University of Natural Resources and Life Sciences, Vienna), U. Thanheiser (VIAS, University of Vienna) and two anonymous referees.

Chapter Bibliography

Adrario, C. (2002) *'Ta', Getreide und Brot im Alten Ägypten. Ausstellung im Museum der Brotkultur Ulm vom 15.5.-10.11. 2002.* Deutsches Brotmuseum, Ulm.

Ahlsén, M. (2003/2004) *Brödet i Birka. Analys av organiskt material från Garnisonen och Svarta Jorden på Björkö.* Diploma thesis, Archaeological Research Laboratory, Stockholm University.

Allezy, L. and M. Persidat (2008) Le gardien d'une propriété a tiré hier sur un groupe de jeunes qui dérobaient des brins de muguet. *Le Parisien*, 28 April 2008 (online).

Anonymous (1997). Geldstrafen für Diebstahl eines Maibaums. Unsigned article. *FAZ (Frankfurter Allgemeine Zeitung)*, June 17, 1997.

Ardovino, A. M. (2002) Il cammino di Herakles – Ideologia greca ed ideologia settentrionale. In L. Zemmer-Plank (ed.) *Kult der Vorzeit in den Alpen. Opfergaben - Opferplätze - Opferbrauchtum.* Schriftenreihe der ARGE ALP. Athesia, Bolzano/Bozen, pp. 211–229.

Ary, S. M. Gregory (text) and Nicholson, B. E. (illustrations) (1960) *The Oxford Book of Wild Flowers, reprinted and corrected* 1977. Oxford University Press, Oxford.

Bailey, C. (1907) *The Religion of Ancient Rome.* Archibald Constable: London.

Bakels, C. and S. Jacomet (2003) Access to luxury foods in central Europe during the Roman period: the archaeobotanical evidence. *World Archaeology* 34(3), 542–557.

Banks, M. M. (1946) *British Calendar Customs for Scotland.* William Glaisher for The Folk-Lore Society, London.

Baudy, G. (1995) Cereal Diet and the Origin of Man. Myths of the *Eleusinia* in the Context of Ancient

Mediterranean Harvest Festivals. In J. Wilkins, D. Harvey and M. Dobson (eds.) *Food in Antiquity*. University of Exeter Press, Exeter, pp. 177–195.

Becker, M., H.-J. Döhle, M. Hellmund, R. Leineweber and R. Schafberg (2005) Nach dem großen Brand. Verbrennung auf dem Scheiterhaufen — ein interdisziplinärer Ansatz. *Berichte der Römisch-Germanischen Kommission* 86, 61–195.

Behre, K.-E. (1991) Zum Brotfund aus dem Ipweger Moor, Ldkr. Wesermarsch. *Berichte zur Denkmalpflege in Niedersachsen* 1991(1), 9.

Behre, K.-E. (1992) The history of rye cultivation in Europe. *Vegetation History and Archaeobotany* 1(3), 141–156.

Becker, W. A. (1882) *Gallus oder Römische Scenen aus der Zeit Augusts zur genaueren Kenntnis des römischen Privatlebens*. Readaptation by H. Göll. 3rd volume. S. Calvary & Co., Berlin.

Belmont, N. (1978) Le joli mois de mai. *L'Histoire* 1, 16–25.

Berg, G. (1960) Några äldre brödtyper i Sverige. In B. Stoklund (ed.) *Folkeliv og kulturlevn. Studier tilegnet Kai Uldall 14 september 1960*. Nationalmuseet, København, pp. 181–200.

Bergström, L. (2007) *Gräddat. Brödkultur under järnåldern i östra Mälardalen*. Ph.D. thesis, Stockholm University. Theses and Papers in Scientific Archaeology 9, Stockholm.

Björkeson, I. (2004) *De homeriska hymnerna tolkade och kommenterade av Ingvar Björkeson*. Tryckort förlag, Stockholm.

Bookidis, N. (1990) Ritual Dining in the Sanctuary of Demeter and Kore at Corinth: Some Questions. In M. Oswyn (ed.) *Sympotica. A symposium on the Symposion*. Oxford University Press, Oxford, pp. 86–94.

Bostock J. (transl.) (1855) *Pliny the Elder. The Natural History*. In G. R. Crane (ed.) Perseus Digital Library Project. Tufts University website. http://www.perseus.tufts.edu/cgi-bin/ptext?lookup=Plin.+Nat.+toc.

Bouby, L. and P. Marinval (2004) Fruits and seeds from Roman cremations in Limagne (Massif Central) and the spatial variability of plant offerings in France. *Journal of Archaeological Science* 31, 77–86.

Bradbury, M. (1999) *Representations of Death. A Social Psychological Perspective*. Routledge, London.

Brockner, W. and Mitchell, B. (1994) Der sog. Brotfund vom Ipweger Moor – Ein Brotimitat auf Bienenwachsbasis. *Archäologisches Korrenspondenzblatt* 24, 399–401.

Britton, K. and Huntley, J. (2010) New evidence for the consumption of barley at Romano-British military and civilian sites, from the analysis of cereal bran fragments in faecal material. *Vegetation History and Archaeobotany* 20 (1), 41–52.

Brück, J. (1999) Ritual and Rationality: Some Problems of Interpretaton in European Archaeology. *European Journal of Archaeology* 2 (3), 313–344.

Brumfield, A. C. (1981) *The Attic festivals of Demeter and their relation to the agricultural year*. Ph.D. thesis. The Ayer Company, Salem.

Brumfield, A. C. (1997) Cakes in the *liknon*. *Hesperia* 66 (1), 147–72.

Burger, R. L. (1995) Chavín and the origins of Andean civilization. Thames and Hudson, London.

Burger, R. L. and Merwe, N. J. V. D. (1990) Maize and the Origin of Highland Chavín Civilization: An Isotopic Perspective. *American Anthropologist* 92(1), 85–95.

Burkert, W. (1985) *Greek Religion*. Harvard University Press.

Buxton, R. (2007) *Den grekiska mytologins värld*. Prisma Bokförlag, Stockholm.

Campbell, Å. (1950) *Det svenska brödet. En jämförande etnologisk-historisk undersökning*. Svensk bageritidskrift, Stockholm.

Campbell, Å. (1952) Nordvästeuropeisk brödkultur i Rígsþula. In Sahlgren J (ed.) *Saga och Sed. Kungl. Gustav Adolfs Akademiens årsbok* 1951. Uppsala, pp. 5–19.

Caneppele, A., Heiss, A. G. and Kohler-Schneider, M. (2010) Weinstock, Dill und Eberesche: Pflanzenreste aus der latènezeitlichen Siedlung Sandberg/Roseldorf. *Archäologie Österreichs* 21 (1), 13–25.

Carlie, A. (2004) Forntida byggnadskult. *Skrifter* 57. Riksantikvarieämbetet Avdelningen för arkeologiska undersökningar, Stockholm.

Cawte, E. C. (1983) It's an Ancient Custom – but how ancient? In T. Buckland and J. Wood (eds.) *Aspects of British Calendar Customs*. Mistletoe Series 22. Sheffield Academic Press for The Folklore Society, Sheffield, pp. 37–56.

Centrone, M. C. (2005) Behind the Corn-mummies. Current Research in Egyptology 2003. In K. Piquette and S. Love (eds.) *Proceedings of the Fourth Annual Symposium. Institute of Archaeology, University College London 18–19 January 2003*. Oxbow Books, London, pp. 11–28.

Centrone, M. C. (2007) Corn-Mummies: A Case of 'Figuring It Out'. In J.-C. Goyon and C. Cardin (eds.) *Proceedings of the Ninth International Congress of Egyptologists. Grenoble, 6–12 September 2004*. Volume I. Uitgevereij Peeters en Departement Oosterse Studies, Leuven/Paris/Dudley.

Childe, V. G. (1945) Directional Changes in Funerary Practices During 50,000 Years. *Man* 45, 13–19.

Ciaraldi, M. (1999) *Food Offerings at the Archaic/Hellenistic sanctuary of Demeter and Persephone at Monte Papalucio (Oria, Apulia, southern Italy)*. Accordia Research Papers 7, 75–91.

Clapham A. R., Tutin T. G. and Moore, D. M. (1987) *Flora of the British Isles*. 3rd edition. Cambridge University Press, Cambridge.

Cohn-Sherbok, D. (2005) *Judaism: History, Belief and Practice*. Taylor & Francis, Ltd., London/New York.

Cordy-Collins, A. (1976) *An iconographic study of Chavín textiles from the south coast of Peru: discovery of a pre-Columbian catechism*: xxi, 307 leaves.

Cordy-Collins, A. and J. Stern (1977) *Pre-Columbian art*

history: selected readings. Peek Publications, Palo Alto, California.

Cummings, L. S., Puseman, K. and Dexter, J. (2009) Appendix 4: Pollen, Macrofloral, and Coprolite Analysis of Samples from the Bluff Great House Site, 42SA22674, southeast Utah. In C. M. Cameron *Chaco and After in the Northern San Juan: Excavations at the Bluff Great House*. University of Arizona Press.

Dafni, A., Lev, E., Beckmann, S. and Eichberger, C. (2006) Ritual plants of Muslim graveyards in northern Israel. *Journal of Ethnobiology and Ethnomedicine* 2(38). DOI: 10.1186/1746-4269-2-38.

Danaher, K. (1972) *The Year In Ireland: Irish Calendar Customs*. Mercier Press, Cork/Dublin.

Daremberg, C. and von Reuss, F. A. (1855) *S. Hildegardis Abbatissae opera omnia, ad optimorum librorum fidem edita*. Patrologia Latina 197. Jacques-Paul Migne, Paris.

Day, M. (2007) *Gudar och gudinnor i antik mytologi*. Wahlström & Widstrand, Stockholm.

De Groote, K., Bastiaens, J., Clercq, D., Deforce, K. and Vandenbruaene, M. (1999/2000) Gallo-Romeinse graven te Huise't Peerdeken (Zingem, provo Oost-Vlaanderen). Een multidisciplinaire analyse. *Archeologie in Vlaanderen* 7, 31–64.

Douglas, M. (1999) The Eucharist: Its Continuity with the Bread Sacrifice of Leviticus. *Modern Theology* 15(2), 209–224.

Duval, P.-M. and Pinault, G. (1986) *Les Calendriers de Coligny et Villards d'Héria. Recueil des inscriptions gauloises*. Volume 3. Centre National de Recherche Scientifique, Paris.

Engelmark, R. (1984). Two useful plants from Iron Age graves in central Sweden. *Archaeology and Environment* 2, 87–91.

Evelyn-White H. G. (transl.) (1914) Hesiod. Works and Days. In Hare J. B. (ed.) Sacred Texts. Website. http://www.sacred-texts.com/cla/hesiod/works.htm

Fahmy, A. G.-E.-D. (2005) Missing plant macro remains as indicators of plant exploitation in Predynastic Egypt. *Vegetation History and Archaeobotany* 14, 287–294.

Fekhner, M. V. (1963). Timerevskij mogi'lnik. In A. P. Smirnov (ed.) *Jaroslavskoe Povolzhe* X–XI. Gosudarstvennyj Istoricheskij muzei, Moscow, pp. 5–19.

Fitter, A. H. and Peat, H. J. (1994) The Ecological Flora Database. *Journal of Ecology* 82, 415–425. http://www.ecoflora.co.uk

Flint-Hamilton, K. B. (1999) Legumes in Ancient Greece and Rome: Food, Medicine, or Poison? *Hesperia* 68(3), 371–385.

Frazer, J. G. (2009) The Golden Bough. A Study of Magic and Religion (reprint). The Floating Press, Waiheke Island.

Fuhrmann, M. (1971) *Tacitus. Germania*. Reclam, Stuttgart.

Furnivall, F. J. (1882) Phillip Stubbes's *Anatomy of abuses in England in Shakspere's youth, A.D. 1583*. New Shakespeare Society, London. http://www.archive.org/details/phillipstubbescorru00stubuoft

Furst, P. T. (1972) *Flesh of the gods; the ritual use of hallucinogens*. Praeger Publishers, New York.

Garcia, D., Mocci, F., Tzortzis, S. and Walsh, K. (2006) Archéologie de la vallée de l'Ubaye (Alpes-de-Haute-Provence, France): présentation des premiers résultats d'un Projet Collectif de Recherche. *Studi Trentini di Scienze Naturali, Preistoria Alpina* 42, 5–30.

Gayasaddin, R. (2007) 'Keeping the Faith or Not Keeping the Faith?' is the Question in a Western Secular Society. *Macalester Islam Journal* 2 (3), 43–51.

Gimbutas, M. (1971) *The Slavs*. Ancient Peoples and Places 74. Thames & Hudson, London

Gimbutas, M. (1984) *The Goddesses and Gods of Old Europe: Myths and Cult Images*. University of California Press, Berkeley.

Gimbutas, M. (1989) *The Language of the Goddess*. Thames & Hudson, London.

Glover, I. C. and Higham, C. F. W. (1996) New evidence for early rice cultivation in South, Southeast and East Asia. In D. R. Harris (ed.) *The origins and spread of agriculture and pastoralism in Eurasia*. UCL Press, London, pp. 413–440.

González Palencia, A. and Mele, E. (1944) *La Maya: notas para su estudio en España*. Consejo Superior de Investigaciones Cientificas, Instituto Antonio de Nebrija, Madrid.

Goody, J. (1982) *Cooking, cuisine, and class: a study in comparative sociology*. Cambridge University Press, Cambridge/New York.

Granberg, G. (1935) *Skogsrået i yngre nordisk folktradition*. Gustav Adolfs Akademien för folklivsforskning, Uppsala.

Grieve, M. (1931) *A Modern Herbal*. 2 Volumes. Harcourt Brace: New York. Reprints 1971 and 1982, Dover Publications, New York. Available at http://www.botanical.com/botanical/mgmh/mgmh.html

Grigson, G. (1958) *The Englishman's Flora*. Phoenix House, London.

Grimm, J. L. K. (1844) *Deutsche Mythologie*. Vol. 1. 2nd edition. Dieterichsche Buchhandlung, Göttingen.

Grove, J. (1988) *The Little Ice Age*. Methuen, London.

Gualtieri, M. and Becker, M. J. (1982) Cremation among the Lucanians. *American Journal of Archaeology* 86(4), 475–481.

Hagberg, L. (1937) *När döden gästar. Svenska Folkseder och Svensk Folktro i samband med Död och Begravning*. Wahlström & Widstrand, Stockholm.

Hanelt, P. (1972) Zur Geschichte des Anbaues von *Vicia faba* L. und ihrer verschiedenen Formen. *Die Kulturpflanze* 20, 209–223.

Hansson, A.-M. (1994) Grain-paste, porridge and bread. Ancient cereal-based food. *Laborativ Arkeologi* 7, 5–20.

Hansson, A.-M. (1995) Bilaga IV: Analys av fossilt växtmaterial från RAÄ 235, Odensala Prästgård, Odensala socken. In M. Olausson (ed.) Det inneslutna rummet – om kultiska hägnader, fornborgar och befästa gårdar i Uppland från 1300 f Kr till Kristi födelse. Riksantikvarieambetet, Stockholm, pp. 262–269.

Hansson, A.-M. (1996) Bread in Birka and on Björkö. *Laborativ Arkeologi* 9, 61–78.

Hansson, A.-M. (1997) *On Plant Food in the Scandinavian Peninsula in Early Medieval Times.* Ph.D. thesis, Stockholm University. Theses and Papers in Archaeology B:5.

Hansson, A.-M. (2001) *Om förekomst och analyser av bröd och brödliknande organiskt material från yngre järnåldern. Slutrapport för forskning inom ramen för stipendium från Anna Ahlström och Ellen Terserus stipendiefond.* Unpublished report. Stockholm Archaeological Laboratory, Stockholm University.

Hansson, A.-M. (2003) *Helgöbröd från boplats och gravar – kemisk analys av organiskt material.* Unpublished report. Stockholm Archaeological Laboratory, Stockholm University.

Hansson, A.-M. (2005a) *Cellstrukturanalys på bröd från Helgö med hjälp av ljusmikroskop.* Unpublished report. Stockholm Archaeological Laboratory, Stockholm University.

Hansson, A.-M. (2005b) Buried plants: Fossil plant remains from two Early Medieval burial mounds in east-central Sweden. *Journal of Nordic Archaeological Science* 15, 39–57.

Hansson, A.-M. (2011) Bread offerings at Foundation IV, Helgö – Further influence from the Classical World? In O. M. Uaininn and B. Arrhenius (ed.) *Helgö Excavations XVIII.* Kungl. Vitterhets Historie och Antikvitets Akademien, Stockholm, pp. 53–77.

Hansson A.-M. and L. Bergström (2002) Archaeobotany in prehistoric graves – concepts and methods. *Journal of Nordic Archaeological Science* 13, 43–58.

Hansson, A.-M. and Isaksson, S. (1994) Analyses of charred organic remains. *Laborativ Arkeologi* 7, 21–29.

Harrison, J. E. (1908) *Prolegomena to the Study of Greek Religion.* 2nd edition. Cambridge University Press, Cambridge.

Hastorf, C. A. (2006) Domesticated Food and Society in Early Coastal Peru. In W. L. Balée and C. L. Erickson (eds.) *Time and Complexity in Historical Ecology: Studies in the Neotropical Lowlands.* Columbia University Press, New York, pp. 87–126.

Haupt, P. (2009) Archäologische und geophysikalische Untersuchungen im Bereich der bronzezeitlichen (Brandopferplatz) und römischen (Heiligtum) Fundstelle 'Burgstall' auf dem Schlern (Südtirol, Italien). Excavation report. Universität Mainz. Available at http://www.archaeologie.geschichte.uni-mainz.de/Forschung/Schlern/Schlern.htm.

Hays-Gilpin, K. (2000) Feminist Scholarship in Archaeology. *Annals of the American Academy of Political and Social Science* 571, 89–106.

Hebert, B., Einwögerer, T., Christandl, G., Schachinger, U. and Windholz-Konrad, M. (2003) Archäologische Untersuchungen auf dem Sölkpass. Altwege, ein hochalpiner urgeschichtlicher Brandopferplatz und weitere Funde von der Steinzeit bis in die Moderne. In F. Mandl (ed.) *Sölkpass – Ein 6000 Jahre alter Saumpfad über die Alpen.* ANISA, Verein für alpine Forschung, Haus im Ennstal, pp. 89–112.

Heeb, B. (2010) *Feldkirch, Altenstadt-Grütze. Ein urnenfelderzeitlicher Brandopferplatz in Vorarlberg.* Frankfurter Archäologische Schriften 13. Dr. Rudolf Habelt GmbH, Bonn.

Heer, O. (1866) Die Pflanzen der Pfahlbauten. *Neujahrsblatt der Naturforsch. Gesellschaft* 68, 1–54

Heiss, A. G. (2008a) *Weizen, Linsen, Opferbrote – Archäobotanische Analysen bronze- und eisenzeitlicher Brandopferplätze im mittleren Alpenraum.* Ph.D. thesis, University of Innsbruck. Südwestdeutscher Verlag für Hochschulschriften, Saarbrücken.

Heiss, A. G. (2008b) *Analyse archéobotanique d'une galette carbonisée trouvée dans la cimétière Romaine de 'La Trussonnerie', St. Memmie (France, Champagne-Ardenne).* Unpublished report, University of Natural Resources and Applied Life Sciences, Vienna.

Heiss, A. G. (2010a) Speisen, Holz und Räucherwerk. Die verkohlten Pflanzenreste aus dem jüngereisenzeitlichen Heiligtum von Ulten, St. Walburg, im Vergleich mit weiteren alpinen Brandopferplätzen. In H. Steiner (ed.) *Alpine Brandopferplätze. Archäologische und naturwissenschaftliche Untersuchungen / Roghi votivi alpini. Archeologia e scienze naturali* Forschungen zur Denkmalpflege in Südtirol / Beni culturali in Alto Adige : studi e ricerche 5. TEMI, Trento, pp. 781–815.

Heiss, A. G. (2010b) Nur Schutt und Asche...? Verkohlte Pflanzengroßreste aus dem spätbronzezeitlichen Brandopferplatz 'Grütze' (Feldkirch, Altenstadt, Österreich). In B. S. Heeb (ed.) *Feldkirch, Altenstadt-Grütze. Ein urnenfelderzeitlicher Brandopferplatz in Vorarlberg.* Frankfurter Archäologische Schriften 13. Dr. Rudolf Habelt GmbH, Bonn, pp. 339–352.

Heiss, A. G. and Kohler-Schneider, M. (2011) Galgenmännlein und Wunderblumen — Eine kurze (Ur) Geschichte der Zauberpflanzen in Niederösterreich und anderswo. In E. Lauermann and S. Sam (eds.) *Drei Farben — Magie. Zauber. Geheimnis. Bedeutung der Farben über Jahrtausende. Begleitpublikation zur Sonderausstellung im Urgeschichtemuseum Niederösterreich Asparn an der Zaya vom 2. April bis 30. November 2011* Urgeschichtemuseum Niederösterreich, Asparn an der Zaya, pp. 44–83.

Heiss, A. G. and Kreuz, A. (2007) Brot für die Salinenarbeiter – das Keltenbrot von Bad Nauheim aus archäobotanischer Sicht. Untersuchung zur latènezeitlichen Ernährung in Bad Nauheim, Wetteraukreis. *hessenArchäologie* 2006, 70–73.

Heiss, A. G. and Oeggl, K. (2005) The oldest evidence of *Nigella damascena* L. (Ranunculaceae) and its possible introduction to central Europe. *Vegetation History and Archaeobotany* 14 (4), 562–570.

Heiss, A. G., Thanheiser, U., Scherrer, P. and Risy, R. (2008a) Bau und Brand – Aspekte der Holznutzung im römischen *Aelium Cetium* (St. Pölten). *Römisches Österreich* 31, 11–32.

Heiss, A. G., Wiethold, J. Pouget, N., Le Goff, I. and Délor-Ahu, A. (2008b) Le site funéraire gallo-romain de Saint Memmie (Marne): évidence de galettes brûlées et des autres d'offrandes alimentaires: aspects méthodologiques, taphonomiques et résultats scientifiques. Rencontres d'Archéobotanique 2008 (Carcassonne, Frankreich, 19.–20. September 2008).

Hepper, F. N. (1990) *Pharaoh's Flowers. The Botanical Treasures of Tutankhamun*. Royal Botanic Gardens Kew & HMSO, London.

Herbermann, C. (ed.) (1913) 'Fourth Lateran Council (1215)'. Catholic Encyclopedia. Robert Appleton Company.

Hermann, K. F. (1847) Ueber die attischen Diasien und die Verehrung des Zeus Meilichios zu Athen. *Philologus* 2, 1–11.

Herrmann, H.-V. (1984) *Olympia. Heiligtum und Wettkampfstätte*. Hirmer Verlag, München.

Hjelmqvist, H. (1984) Botanische Analyse einiger Brote. In G. Arwidsson (ed.) *Birka II: 1. Systematische Analysen der Gräberfunde*. K. Vitterhets Historie och Antikvitets Akademien, Stockholm, pp. 261–272.

Hjelmqvist, H. (1990) Über die Zusammensetzung einiger prähistorischer Brote. *Fornvännen* 85, 9–21.

Hofeneder, A. (2005) *Die Religion der Kelten in den antiken literarischen Zeugnissen. Band I: Von den Anfängen bis Caesar*. Mitteilungen der Prähistorischen Kommission 59. Verlag der Österreichischen Akademie der Wissenschaften, Wien.

Hoffmann-Krayer, E. and Bächtold-Stäubli H. (eds.) (1927) *Handwörterbuch des deutschen Aberglaubens*. De Gruyter, Berlin.

Holmboe, J. (1927) Nytteplanter og ugræs i Osebergfundet. In A. W. Brøgger and H. Schetelig (eds.) *Osebergfundet* 5. Oslo.

Holmqvist, W. (1963) Forntida bröd och bakugnar från Helgö. Från Helgö till Svalöv, 1700 år svenskt bröd. *Information från brödinstitutet* 6(5), 3–8.

Holmqvist, W. and B. Arrhenius (eds.) (1964) *Excavations at Helgö II. Report for 1957–1959*. Kungl. Vitterhets Historie och Antikvitets Akademien, Stockholm.

Holzer, V. (2009) *Roseldorf. Interdisziplinäre Forschungen zur größten keltischen Zentralsiedlung Österreichs*. Forschungen im Verbund 102. Österreichische Elektrizitätswirtschafts-Aktiengesellschaft (Verbundgesellschaft), Wien.

Hoppe, T. (2009) Ein Ensemble aus Meisterhand: Das Kultbild von Fellbach-Schmiden, 127 v. Chr. In F. Müller (ed.) Kunst der Kelten 700 v. Chr. – 700 n. Chr. Belser Verlag, Stuttgart, pp. 238–241.

Hubert, A. (1981) Rites à Mae Posop en Thaïlande du Centre. *Péninsule* 2–3, 5–16

IAPT (2006) *International Code of Botanical Nomenclature (Vienna Code)*. Regnum Vegetabile 146. A. R. Gantner Verlag, Ruggell.

Insoll, T. (2004) *Archaeology, Ritual, Religion*. Themes in Archaeology. Routledge, London/New York.

Isaksson, S. (1999) Guided by light – the swift characterization of ancient organic matter. *Laborativ Arkeologi* 12, 35–43.

Isaksson, S. (2000) *Food and Rank in Early Medieval Time*. Ph.D. thesis, Stockholm University. Theses and Papers in scientific Archaeology 3, Stockholm.

Jacomet, S. (2009) Plant economy and village life in Neolithic lake dwellings at the time of the Alpine Iceman. *Vegetation History and Archaeobotany* 18(1), 47–59.

Jacomet, S. and Kreuz, A. (1999) *Archäobotanik. Aufgaben, Methoden und Ergebnisse vegetations- und agrargeschichtlicher Forschung*. Verlag Eugen Ulmer, Stuttgart.

Jacomet S., Petrucci-Bavaud, M. and Kühn, M. (2006) Samen und Früchte. In C. Schucany (ed.) *Die römische Villa von Biberist-Spitalhof/SO (Grabungen 1982, 1983, 1986–1989). Untersuchungen im Wirtschaftsteil und Überlegungen zum Umland*. Verlag Bernhard Albert Greiner, Remshalden, pp. 579–624.

Johansson, K. G. (transl.) (1992) *Nordiska kungasagor II (Kringla heimsins): Olav den heliges saga*. Fabel Bokförlag, Stockholm.

Judge, R. (1993) Fact and Fancy in Tennyson's 'May Queen' and in Flora Thompson's 'May Day'. In T. Buckland and J. Wood (eds.) *Aspects of British Calendar Customs*. Mistletoe Series 22. Sheffield Academic Press for The Folklore Society, Sheffield, pp. 167–183.

Kaliff, A. (1997) *Grav och kultplats. Eskatologiska föreställningar under yngre bronsålder och äldre järnålder i Östergötland*. Ph.D. thesis, University of Uppsala. Aun 24. Uppsala.

Kelly, F. (1997) *Early Irish Farming*. School of Celtic Studies and Dublin Institute for Advanced Studies, Dublin. Reprint 1998.

Keyland, N. (1919) *Svensk Allmogekost*. A.-B. Svenska teknologföreningens förlag, Stockholm. Reprint 1989, Carlssons i samarbete med Institutet för folklivsforskning och Nordiska museet, Stockholm.

Körber-Grohne, U. (1985) Die biologischen Reste aus dem hallstattzeitlichen Fürstengrab von Hochdorf, Gemeinde Eberdingen (Kreis Ludwigsburg). *Hochdorf I*. Konrad Theiss, Stuttgart, pp. 87–265.

Körber-Grohne, U. (1995) *Nutzpflanzen in Deutschland von der Vorgeschichte bis heute*. Verlag Konrad Theiss, Stuttgart. Lizenzausgabe Nikol Verlag, Hamburg.

Körber-Grohne, U. (1999) Der Schacht in der keltischen Viereckschanze von Fellbach-Schmiden (Rems-Murr-Kreis) in botanischer und stratigraphischer Sicht. In G. Wieland (ed.) *Die keltischen Viereckschanzen von Fellbach-Schmiden (Rems-Murr-Kreis) und Ehningen (Kreis*

Böblingen). Forschungen und Berichte zur Vor- und Frühgeschichte in Baden-Württemberg 80. Verlag Konrad Theiss, Stuttgart, pp. 85–149.

Kossack, G. (2002) Der zentrale Alpenraum während der Bronze- und vorrömischen Eisenzeit – Institutionen, Sachbesitz und religiöse Ausdrucksformen. In L. Zemmer-Plank (ed.) *Kult der Vorzeit in den Alpen. Opfergaben - Opferplätze - Opferbrauchtum*. Schriftenreihe der ARGE ALP. Athesia, Bolzano/Bozen, pp. 285–340.

Krämer, W. (1966) Prähistorische Brandopferplätze. *Helvetia Antiqua* (Festschrift Emil Vogt), 111–122.

Kreuz, A. (2000) Functional and conceptual archaeobotanical data from Roman cremations. In J. Pearce, M. Millet and M. Struck (eds.) *Burial, society and context in the Roman world [Roman Burial Archaeology Conference, Durham]*. Oxbow Books, Oxford, pp. 45–51.

Kreuz, A. (2006) Erste Hinweise zur Ernährung der römischen Soldaten und Arbeitstiere in der Kapersburg. *hessenArchäologie* 2005, 83–85.

Krige, E. J. (1968) Girls' Puberty Songs and Their Relation to Fertility, Health, Morality and Religion among the Zulu. *Africa: Journal of the International African Institute* 38(2), 173–198.

Kurz, U. C. (2007) Kalathiskos con ghiande: sul significato di un ex voto fittile dal santuario di Demetra a Herakleia. *Bollettino Storico della Basilicata* 23, 79–103.

Kurz, U. C. (2008) Früchte und Opferkuchen in der Koroplastik des Demeter- und Kore/Persephonekultes von Herakleia/Policoro. In C. Franek, S. Lamm, T. Neuhauser, B. Porod and K. Zöhrer (eds.) *Thiasos. Festschrift für Erwin Pochmarski zum 65. Geburtstag.* Phoibos-Verlag, Wien, pp. 513–536.

La Barre, W. (1972) *The Ghost Dance: Origins of Religion.* Allen and Unwin, London.

Lamb, H. H. (1982) *Climate, History and the Modern World.* Methuen, London.

Lambert, M. (1993) Ancient Greek and Zulu Sacrificial Ritual: A Comparative Analysis. *Numen* 40(3), 293–318.

Lang, A. (2002) Speise- und Trankopfer. In L. Zemmer-Plank (ed.) *Kult der Vorzeit in den Alpen. Opfergaben - Opfer-plätze - Opferbrauchtum*. Schriftenreihe der ARGE ALP. Athesia, Bolzano/Bozen, pp. 917–934.

Lannoy, S., Marinval, P., Buleon, A., Chiron, H., Mejanelle, P., Pin, S., Rech, J. And Tchapla, A. (2002) Études de 'pains et galettes' archéologiques français. *Civilisations* 49(1–2), 119–160.

Lathrap, D. W. (1973). Gifts of the cayman: Some thoughts on the subsistence basis of Chavín. In D. W. Lathrap and J. Douglas (eds.) *Variation in Anthropology: Essays in Honor of John McGregor*. Illinois Archaeological Survey, Urbana, pp. 91–105.

Lauber, K. and Wagner, G. (1996) *Flora Helvetica*. Verlag Paul Haupt, Bern/Stuttgart/Wien.

Laurent-Täckholm, V. (1951) *Faraos blomster. En kulturhistorisk-botanisk skildring av livet i Gamla Egypten byggd på verklighetens grundval och med bilder från de sista årens grävningar*. Natur och Kultur, Stockholm.

Lawton, W. C. (1898) *The Successors of Homer*. Macmillan, New York.

Leach, H. M. (2003) Did East Polynesians have a concept of luxury foods? *World Archaeology* 34 (3), 442–457.

Leitner, W. (2002) Feldkirch-Altenstadt, Grütze. In L. Zemmer-Plank (ed.) *Kult der Vorzeit in den Alpen. Opfergaben - Opferplätze - Opferbrauchtum*. Schriftenreihe der ARGE ALP. Athesia, Bolzano/Bozen, pp. 801–805.

Lenz, H. O. (1859) *Botanik der alten Griechen und Römer, deutsch in Auszügen aus deren Schriften, nebst Anmerkungen*. E. F. Thienemann, Gotha.

Leopold, A.C. and Ardrey, R. (1972) Toxic Substances in Plants and the Food Habits of Early Man. – *Science* 176 (4034), 512–514.

Lindeberg, I. (1961) Bread. In W. Holmqvist and B. Arrhenius (eds.) *Excavations at Helgö II. Report for 1957-1959*. Kungl. Vitterhets Historie och Antikvitets Akademien, Stockholm, pp.141–143.

Logan, A. (2006) *The Application of Phytolith and Starch Grain Analysis to Understanding Formative Period Subsistence, Ritual, and Trade on the Taraco Peninsula, Highland Bolivia*. Master thesis, University of Missouri.

Lucas, A. T. (1960) *Furze, a survey and history of its uses in Ireland*. Béaloideas 26(1958). Folklore of Ireland Society, Dublin.

Lumbreras, L. G. (1989) *Chavín de Huántar en el nacimiento de la civilización andina*. INDEA-CONCYTEC, Lima.

Lurker, M. (2004) *The Routledge Dictionary of Gods, Goddesses, Devils & Demons*. Routledge, London/New York.

Macaulay, G. C. (transl.) (1890) *The History of Herodotus*. Macmillan, London. Project Gutenberg e-book #2707. URL: http://www.gutenberg.org/etext/2707.

Marcolin, R. (2008) Gâteaux et galettes : offrandes aux Dieux dans le monde Grec. In P. Marinval (ed.) *Histoire de pain du Néolithique au Moyen Age*. Archéo-Plantes 4. AITAE/AEP/CRPPM, Toulouse, 43–60.

Marquardt, J. (1864) *Römische Privataltherthümer.* Handbuch der Römischen Alterthümer 5. S. Hirzel, Leipzig.

Megaloudi, F., Papadopoulos, S. and Sgourou, M. (2007) Plant offerings from the classical necropolis of Limenas, Thasos, northern Greece. *Antiquity* 81, 933–943.

Melin, J. and Sigvallius, B. (2001) *Excavations at Helgö 14: Cemetery 118 and Building Group 7 and Cemetery 115*. Kungliga Vitterhets Historie och Antikvitets Akademien, Stockholm.

Melnikova, E. A. and Petrochin, V. (1992) The origin and evolution of the name Ru's. The Scandinavians in Eastern-European ethno-political processes before the 11th century. *Tor* 23(1990–91), 203–34.

Meyer, K. (1894) *The Quatrains on the Seasons in Hibernica Minora* (Appendix). Supplement to Mr. Whitley Stokes' descriptions of the Contents of Rawlinson B 512 (p. xxx, 1° 9862). Oxford University Press, Oxford, pp. 48–49.

Meyer, K. (1912) Sanas Cormaic (Corm.2, 122). In O. J. Bergin, E. I. Best, K. Meyer and J. G. O'Keeffe (eds.) *Anecdota from Irish Manuscripts*. Vol. IV. Max Niemeyer and Hodges, Digois & Co, Halle/Dublin.

Miles, M. M. (1998) The Propylon to the Sanctuary of Demeter Malophoros at Selinous. *American Journal of Archaeology* 102(1), 35–57.

Miller, G. R. and Burger, R. L. (1995). Our Father the Cayman, Our Dinner the Llama: Animal Utilization at Chavín de Huántar, Peru. *American Antiquity* 60(3), 421–458.

Miller, U. and Hedin, K. (1988) *Excavations at Helgö 11: The Holocene development of landscape and environment in the south-east Mälaren Valley, with special reference to Helgö.* Kungliga Vitterhets Historie och Antikvitets Akademien, Stockholm.

Monah, D. (2002) Découvertes de pains et de restes d'aliments céréaliers en Europe de l'est et centrale. Essai de synthèse. *Civilisations* 49(1–2), 77–99.

Moore, A. W. (1891) *The Folk-Lore of the Isle of Man.* D. Nutt, London. In J. B. Hare (ed.) Sacred Texts. Website. http://www.sacred-texts.com/neu/celt/fim/fim00.htm.

Moore, J. D. (2005). *Cultural landscapes in the ancient Andes: archaeologies of place.* University Press of Florida, Gainesville.

Mulvany de Peñaloza, E. M. (1984) Motivos fitomorfos de alucinógenos en Chavín. *Chungará* 12, 57–80.

Naiden, F. S. (2006) Rejected sacrifice in Greek and Hebrew religion. *Journal of Ancient Near Eastern Religions* 6, 189–223.

Newall, V. (1985) Folklore and Cremation. *Folklore* 96(2), 139–155.

Nicholson, B. E. and Nilssen, M. P. (1920) *Primitive Time-Reckoning.* Acta Societatis Humanorum Litterarum Lundensis I. Lund.

Nilsson, M. P. (1957) *The Dionysiac Mysteries of the Hellenistic and Roman Age.* Skrifter utg. av Svenska institutet i Athen, 8-5. C. W. K. Gleerup, Lund. Reprint 1975, Arno Press, New York.

Oberrauch, H. (2005) Kupferzeitliche Brandopfer am Pigloner Kopf (Pfatten). *Palyno-Bulletin* 1(3–4), 39–45.

Oeggl, K. (1992) Zur Besiedlung des mittleren Alpenraumes während der Bronze- und Eisenzeit: Die Vegetationsverhältnisse. In J. Kovar-Eder (ed.) *Palaeovegetational Development in Europe and Regions Relevant To Its Palaeofloristic Evolution. Proceedings of the Pan-European Palaeobotanical Conference (PEPC), Vienna, September 19–23, 1991.* Naturhistorisches Museum Wien, pp. 47–57.

Ogalde, J. P., Arriaza, B. T. and Soto, E. C. (2010) Uso de plantas psicoactivas en el north de Chile: evidencia química del consumo de ayahuasca durante el periodo medio (500–1000 d. C.). *Latin American Antiquity* 21(4), 441–450.

Olausson, M. (1987) Hillforts, subsistence and economic centralisation A.D. 300–500 in eastern Middle Sweden. In G. Burenhult, A. Carlsson, Å. Hyenstrand and T. Sjøvold (eds.), *Theoretical Approaches to Artefacts, settlement and Society. Studies in honour of Mats P. Malmer.* BAR International Series 366, 399–417.

Olausson, M. (1996) *Runsa fornborg. En befäst stormannagård från folkvandringstid.* Skrifter om Väsby 5. Upplands, Väsby.

Pankhurst, R. J. and J. M. Mullin (1991) *Flora of the Outer Hebrides.* Natural History Museum, London.

Pasteur, P. (1997) Les débuts de la crémation moderne en France. *Le Mouvement Social* 179, 59–80.

Perry, E. M. and Joyce, R. A. (2001) Providing a Past for 'Bodies That Matter': Judith Butler's Impact on the Archaeology of Gender. *International Journal of Sexuality and Gender Studies* 6 (1/2), 63–76.

Petré, B. (1978) Delundersökning av fornl. 56, Hemmet, Lovö sn, Uppland. *Arkeologiska rapporter och meddelanden från Institutionen för arkeologi vid Stockholms universitet.* Nr 5 1978. Stockholm.

Petré, B. (1984a) Arkeologiska undersökningar på Lovö. Del 2. Fornlämning Raä 27, Lunda. *Studies in North-European Archaeology* 8. Stockholm.

Petré, B. (1984b) Arkeologiska undersökningar på Lovö. Del 3. Gravar, gravfält och boplatser på Lovö. *Studies in North-European Archaeology* 9. Stockholm.

Petré, B. (1984c) Arkeologiska undersökningar på Lovö. Del 4. Bebyggelsearkeologisk analys. *Studies in North-European Archaeology* 10. Stockholm.

Petré, B. (1999a) *Gravfältet Raä 13, Söderby, Lovö sn, Uppland. Ett gravfält med två familjer från yngre järnålder. Rapport, analys, tolkning.* Lovö Archaeological Reports and Studies 6. Stockholm University, Stockholm.

Petré, B. (1999b) *Gravfältet Raä 16, Söderby, Lovö sn, Uppland. Ett familjegravfält från yngre järnålder. Rapport, analys, tolkning.* Lovö Archaeological Reports and Studies 7. Stockholm University, Stockholm.

Petré, B. (2011) *Arkeologiska undersökningar på fornlämning Raä 28, Söderby, Lovö sn, Uppland. Gravfält från vendeltid och vikingatid samt några gravar och boplatslämningar från bronsålder.* Lovö Archaeological Reports and Studies 10. Stockholm University, Stockholm.

Petrov, V. (1948) *Kharchovi reshtki z Patirskogo gorodischa.* Arkheologija II, Kiev, pp. 79–86.

Petrukhin, V. I. (2007) Viking Women in Rus': Wives, Slaves or 'Valkyries'? In U. Fransson, M. Svedin, S. Bergerbrant and F. Androshchuk (eds) *Cultural interaction between east and west. Archaeology, artifacts and human contacts in northern Europe.* Stockholm Studies in Archaeology 44. Stockholm University, Stockholm, pp. 66–69.

Phythian-Adams, C. (1983) Milk and Soot: The Changing Vocabulary of a Popular Ritual in Stuart and Hanoverian London. In D. Fraser and A. Sutcliff (eds.) *The Pursuit of Urban History.* Edward Arnold, London, pp. 83–104.

Plath, D. W. (1963) The Japanese Popular Christmas:

Coping with Modernity. *The Journal of American Folklore* 76 (302), 309–317.

Popovtschak, M. (2005) Teil V. Verkohlte Pflanzenreste aus dem Bereich des Kultplatzes Frauenberg/Leibnitz, Steiermark (Grabungen 2002–2004). In S. Groh and H. Sedlmayer (eds.) *Der norisch-römische Kultplatz am Frauenberg (Österreich)*. Protohistoire Européenne 9, pp. 181–194.

Poux, M., Demierre, M., Foucras, S., Garcia, M. and Guichon Pou, R. (2007) Blutige Opfer und Weinspenden in Gallien am Beispiel des spätkeltisch-römischen Heiligtums von Corent (Frankreich). In S. Groh and H. Sedlmayer (ed.) *Blut und Wein — Keltisch-römische Kultpraktiken. Akten des vom Österreichischen Archäologischen Institut und vom Archäologischen Verein Flavia Solva veranstalteten Kolloquiums am Frauenberg bei Leibnitz (Österreich), Mai 2006* Protohistoire Européenne 10. éditions monique mergoil, Montagnac, pp. 11–34.

Price, N. S. (2002) *The Viking Way. Religion and War in Late Iron Age Scandinavia*. Ph.D. thesis, University of Uppsala. Aun 31. Uppsala.

Rajadhon, P. A. (1955) Me Posop, the Rice Mother. *Journal of Siam Society* 43(1), 55–61.

Rajadhon, P. A. (1968) *Essays on Thai Folklore*. The Social Science Association Press of Thailand, Bangkok.

Rashmisrisethi, A. (2008) HM the Queen to observe ceremony to invite Rice Goddess to rice fields. *Thaindian News* August 9th, 2008 – 9:50 am ICT. http://www.thaindian.com/newsportal/thailand/hm-the-queen-to-observe-ceremony-to-invite-rice-goddess-to-rice fields_10081713.html

Raven, M. J. (1982) Corn-Mummies. *Oudheidkundig Mededelingen uit het Rijksmuseum van Oudheden te Leiden* 63, 7–38.

Rekdal, O. B. (1996) Money, Milk and Sorghum Beer: Change and Continuity among the Iraqw of Tanzania. *Africa: Journal of the International African Institute* 66(3), 367–385.

Renfrew, C. and Bahn, P. (2008) *Archaeology: Theory, Methods and Practice*. 5th edition. Thames & Hudson, London.

Reynolds, L. A. and Sawyer, P. (1959) Folk Medicine and the Four Fairies of A Midsummer-Night's Dream. *Shakespeare Quarterly* 10 (4), 513–521.

Rick, J. W. (2008) Context, Construction, and Ritual in the Development of Authority at Chavín de Huántar. In W. J. Conklin and J. Quilter (eds.) *Chavín. Art, Architecture and Culture*. UCLA Monographs 61. Cotsen Institute of Archaeology, University of California, Los Angeles, pp. 32–47.

Riedel, A. And Tecchiati, U. (2005) Die Fauna des kupferzeitlichen Opferplatzes am Pigloner Kopf. Ergebnisse der archäologischen Ausgrabungen des Jahres 1998 am Fundplatz bei Pfatten. *Der Schlern* 79(2), 4–19.

Rizzi, J. (2000) Die Fauna des Brandopferplatzes vom Seeberg – Schwarzsee. In G. Niederwanger and U. Tecchiati (eds.) *Wasser – Feuer – Himmel. Ein Brandopferplatz spätbronzezeitlicher Bergknappen*. Folio, Bolzano/Bozen, pp. 37–38.

Rolfe, J. C. (transl.) (1928) *Suetonius*. Vols. 1 and 2. Loeb Classical Library. G. P. Putnam's Sons, New York. In B. Thayer (ed.) *LacusCurtius: Into the Roman World*. University of Chicago website. URL: http://penelope.uchicago.edu/Thayer/E/Roman/Texts/Suetonius/12Caesars/home.html

Rose, H. J. (1921) Zeus Meilichios, Zeus Agamemnon, and Zanes. *The Classical Review* 35(7/8): 147–149.

Rosendahl, H. V. (1909) Mikroskopisk analys af brödfynd från 400-500-talen. *Svensk Botanisk Tidskrift* 3(1), 41–46.

Rosendahl, H. V. (1912a) Mikroskopisk analys af brödfynd från 400-500-talen. *Svensk Farmaceutisk Tidskrift* 1, 1–3.

Rosendahl, H. V. (1912b) Ärtbröd från vikingatiden (800–1050 e.Kr.), *Svensk Botanisk Tidskrift* 5(4), 432–433.

Rosenqvist, A. M. (1959). The Stabilizing of Wood Found in the Viking Ship of Oseberg: Part I. *Studies in Conservation* 4(1), 13–22.

Sanyal, A. (2008) Farmers in Thailand grapple with theft of rice crops. *channelnewsasia.com* April 10th, 2008 – 21:42.

Saunders, C. (1925) Cremation and Inhumation in the Aeneid. *The American Journal of Philology* 46(4), 352–357.

Sayre, M. (2006) New Data on Interregional Exchange at Chavín. Presented at 'Searching for Patterns in the Diversity of the Central Andean Formative'. Roundtable Symposium, held at Stanford University 3–4 March 2006.

Schaan, D. P. (2006) Is There a Need to (Un)Gender the Past? *Advances in Gender Research* 10, 45–60.

Schierman, C. (2005/2006) *Bröd vid död i Kalvshälla. Analys av förhistoriskt organiskt grav- och boplatsmaterial från Barkarby i Järfälla socken, Uppland*. Diploma thesis, Archaeological Research Laboratory, Stockholm University.

Schierman, C. (2007) *Analyses of organic finds from Museum der Brotkultur in Ulm, Germany. Find number O 2967, O 2969, O 4513 and O 4517*. Unpublished report, Archaeological Research Laboratory, Stockholm University.

Schmidt, M. (2000) *Warum ein Apfel, Eva? Die Bildsprache von Baum, Frucht und Blume*. Schnell & Steiner, Regensburg.

Schmitzberger, M. (2007) Archäozoologische Untersuchungen an den bronze-, eisen- und römerzeitlichen Tierknochen vom Ganglegg und vom Tartscher Bichl. In H. Steiner (ed.) *Die befestigte Siedlung am Ganglegg im Vinschgau – Südtirol. Ergebnisse der Ausgrabungen 1997–2001 (Bronze-/Urnenfelderzeit) und naturwissenschaftliche Beiträge*. Forschungen zur Denkmalpflege in Südtirol 3. TEMI, Trento, pp. 617–742.

Schnittger, B. (1912) Några förhistoriska brödfynd. *Fornvännen* 7, 1–6.

Schön, E. (1998) *Svensk folktro A - Ö. Hur vi tänkt, trott och trollat.* Prisma Bokförlag, Stockholm.

Schön, E. (2002) *Folktro om ödet och lyckan.* Natur och Kultur, Stockholm.

Schott, R. (1997) *Bulsa Sunsuelima. Folktales of the Bulsa in Northern Ghana.* Series S: Folktales of the Supernatural. Volume 1: Tales of the Sky-God (*Wen, Naawen*) Part II–III. Litt Verlag, Münster.

Schürmann, T. (1990) *Nachzehrerglauben in Mitteleuropa.* Elwert, Marburg.

Schultes, R. E., Hofmann, A. and Rätsch, C. (2001) *Plants of the Gods: Their Sacred, Healing and Hallucinogenic Powers.* Healing Arts Press, Rochester.

Schwager R. (2002) Gottheit/Gott und Opfer. In L. Zemmer-Plank (ed.) *Kult der Vorzeit in den Alpen. Opfergaben - Opferplätze - Opferbrauchtum.* Schriftenreihe der ARGE ALP. Athesia, Bolzano/Bozen, pp. 43–59.

Servez-Soyez, B. (1983) Lits d'Osiris et jardins d'Adonis. In A. Théodoridés, P. Naster and J. Ries (eds.) Vie et survie dans les civilisations orientales. Acta Orientalia Belgica 3. Peeters, Leuven, pp. 219–226.

Sharon, D. (2000). *Shamanism & the sacred cactus: ethnoarchaeological evidence for San Pedro use in northern Peru / Shamanismo & el cacto sagrado: evidencia etnoarqueológica sobre el uso del cacto San Pedro en el norte del Peru.* Museum of Man, San Diego.

Sherratt, A. (1998) *Sacred Weeds.* Four part television series. Channel 4, London.

Sillasoo, Ü. (2006) Medieval plant depictions as a source for archaeobotanical research. *Vegetation History and Archaeobotany* 16, 61–70.

Sillasoo, Ü. (2009) Plants in Late Medieval festivals and customs in written and pictorial sources from southern central Europe. *Environmental Archaeology* 14 (1), 76–89.

Simel, D. L. (1996) Exclusionary Christian Civil Religion for Jewish and Islamic Students in Bavarian Schools. *Comparative Education Review* 40 (1), 28–46.

Simoons, F. J. (1998) *Plants of Life, Plants of Death.* University of Wisconsin Press, Madison.

Simpson, J. (1976) *The Folklore of the Welsh Border.* B.T. Batsford, London

Simpson, J. and S. Roud (2000) *A Dictionary of English Folklore.* Oxford University Press.

Simrock, K. (1876) *Die Edda, die ältere und jüngere, nebst den mythischen Erzählungen der Skalda.* Verlag der J. G. Cotta'schen Buchhandlung, Stuttgart.

Sokiranski, R., Zwaka, P., Adrario, C., Stika, H.-P. and Pirsig, W. (2005) *Three-dimensional spiral CT reconstruction of an ancient Egyptian corn-mummy.* Unpublished manuscript, Klinikum Traunstein and others.

Stace, C. (1991) *New Flora of the British Isles.* Cambridge University Press.

Statens Historiska Museum och K. Myntkabinettet (1909) Tillväxten under år 1909. *Fornvännen* 1909: 15–322.

Steiner, H. (2007) *Die befestigte Siedlung am Ganglegg im Vinschgau - Südtirol. Ergebnisse der Ausgrabungen 1997–2001 (Bronze-/Urnenfelderzeit) und naturwissenschaftliche Beiträge.* Forschungen zur Denkmalpflege in Südtirol 3. TEMI, Trento.

Steiner, H. (2010) *Alpine Brandopferplätze. Archäologische und naturwissenschaftliche Untersuchungen / Roghi votivi alpini. Archeologia e scienze naturali.* Forschungen zur Denkmalpflege in Südtirol / Beni culturali in Alto Adige: studi e ricerche 5. TEMI, Trento.

Stewart A. and Long, G. (1894) *Plutarch's Lives. Translated from the Greek with Notes and a Life of Plutarch.* Vol. I. George Bell & Sons, London/New York.

Stika, H.-P. (1996) *Römerzeitliche Pflanzenreste aus Baden-Württemberg. Beiträge zu Landwirtschaft, Ernährung und Umwelt in den römischen Provinzen Obergermanien und Rätien.* Materialhefte zur Archäologie in Baden-Württemberg 36. Verlag Konrad Theiss, Stuttgart.

Stika H.-P. (2009) Landwirtschaft der späten Hallstatt- und frühen Latènezeit im mittleren Neckarland — Ergebnisse von pflanzlichen Großrestuntersuchungen. In K. Schatz and H.-P. Stika (eds.) *Hochdorf VII. Archäobiologische Untersuchungen zur frühen Eisenzeit im mittleren Neckarraum.* Forschungen und Berichte zur Vor- und Frühgeschichte in Baden-Württemberg 107. Konrad Theiss, Stuttgart, pp. 125–293.

Stroud, R. S. (1965) The Sanctuary of Demeter and Kore on Acrocorinth. Preliminary Report I: 1961–1962. *Hesperia* 34(1), 1–24.

Stroud, R. S. (1968) The Sanctuary of Demeter and Kore on Acrocorinth. Preliminary Report II: 1964–1965. *Hesperia* 37(3), 299–330.

Subbotsky, E. (2010) *Magic and the Mind. Mechanisms, Functions, and Development of Magical Thinking and Behavior.* Oxford University Press, Oxford/New York.

Taylor-Perry, R. (2003) *The God Who Comes: Dionysian Mysteries Reclaimed.* Algora Publishing, New York.

Thiselton-Dyer, T. F. (1889) *The Folk-Lore of Plants.* Project Gutenberg e-book #10118. URL: http://www.gutenberg.org/etext/10118.

Tooley, A. M. J. (1996) Osiris Bricks. *The Journal of Egyptian Archaeology* 82, 167–179.

Torres, C. M. (2008). Chavín's Psychoactive Pharmacopoeia: The Iconographic Evidence. In W. J. Conklin and J. Quilter (eds.) *Chavín. Art, Architecture and Culture.* UCLA Monographs 61. Cotsen Institute of Archaeology, University of California, Los Angeles, pp. 239–260.

Torres, C. M., Repke, D. B., Chan, K., Mc Kenna, D., Llagostera, A. and Schultes, R. E. (1991) Snuff Powders from Pre-Hispanic San Pedro de Atacama: Chemical and Contextual Analysis. *Current Anthropology* 32(5), 640–649.

Toussaint-Samat, M. (2009) *A History of Food.* 2nd edition. John Wiley & Sons, Chichester.

Trevelyan, M. (1909) *Folk-Lore and Folk-Stories of Wales.* Elliot Stock, London.

Vendryes, J., Bachellery, E. And Lambert, P.-Y. (1981) *Lexique étymologique de l'irlandais ancien. Lettre B.* Centre

National de la Recherche Scientifique and Dublin Institute for Advanced Studies, Paris/Dublin.

Vendryes, J., Bachellery, E. And Lambert, P.-Y. (1987) *Lexique étymologique de l'irlandais ancien. Lettre C.* Centre National de la Recherche Scientifique and Dublin Institute for Advanced Studies, Paris/Dublin.

Vickery, R. (1995) *Dictionary of Plant-Lore.* Oxford University Press.

Viklund, K. (1997) Mat och magi. Om forntida människors förhållande till växterna. In Dialekt-, ortnamns- och folkminnesarkivet i Umeå (ed.) Aspekter på växtnamn. Serie E. Växtnamn 1997(1), 117–125.

von Grot, R. (1887) *Ueber die in der hippokratischen Schriftensammlung enthaltenen pharmakologischen Kenntnisse.* Schnakenburg's Buchdruckerei, Dorpat.

von Stokar, W. (1951) *Die Urgeschichte des Hausbrotes.* Barth, Leipzig.

Voß, J. H. (1990) *Ilias. Odyssee.* insel taschenbuch, Frankfurt am Main.

Währen, M. (1980) Tanit et son pays de Carthage. *Le Boulanger - Pâtissier.* Paris, H.7–12, H. 1/1981, pp. 477–494.

Währen, M. (1995) Brot und Gebäck in der Urgeschichte mit ausführlicher Berücksichtigung des Weck-Feingebäcks von Ovelgönne und des Haferbrotes von Langenrehm. In H. Eiselen (ed.) *Max Währen. Gesammelte Aufsätze zur Brot- und Gebäckkunde und -geschichte. 1940–1999.* Deutsches Brotmuseum, Ulm, pp. 191–222.

Währen, M. (2000a) Haushaltskonserven der Jungsteinzeit. Mit Neuidentifizierungen. In H. Eiselen (ed.) *Max Währen. Gesammelte Aufsätze zur Brot- und Gebäckkunde und -geschichte. 1940–1999.* Deutsches Brotmuseum, Ulm, pp. 29–50.

Währen, M. (2000b) Vorgeschichtliche Brotreste aus der nördlichen Lüneburger Heide. In H. Eiselen (ed.) *Max Währen. Gesammelte Aufsätze zur Brot- und Gebäckkunde und -geschichte. 1940–1999.* Deutsches Brotmuseum, Ulm.

Wasylikowa, K. and Jankun, A. (1997) Identification of Barley from the Ancient Egyptian Corn-Mummies in ther Archaeological Museum in Cracow. *Materiały Archeologiczne* 30, 13–17.

Watts, D. C. (2007) *Dictionary of Plant Lore.* Academic Press, Burlington/San Diego/London.

Wegewitz, W. (1955) Arbeitsbericht des Helms-Museums für die Zeit vom 1. April 1950 bis zum 31. März 1954. *Harburger Jahrbuch 5,* pp. 163–204, 191.

Weiss, R.-M. (1997) *Prähistorische Brandopferplätze in Bayern.* Internationale Archäologie 35. Verlag Maria Leidorf, Rahden.

Whiting, A. F. (1939) *Ethnobotany of the Hopi.* Museum of Northern Arizona Bulletin 15. Northern Arizona Society of Science and Art, Flagstaff.

Wiethold, J. (1996) Late Celtic and early Roman plant remains from the oppidum of Bibracte, Mont Beuvray (Burgundy, France). *Vegetation History and Archaeobotany* 5, 105–116.

Wikander, S. (1978) *Araber, vikingar, väringar.* Svenska Humanistiska Förbundet, Lund.

Wilbert, J. (1987) *Tobacco and shamanism in South America.* Psychoactive Plants of the World Series. Yale University Press, New Haven.

Willerding, U. (1971) Methodische Probleme bei der Untersuchung und Auswertung von Pflanzenfunden in vor- und frühgeschichtlichen Siedlungen. *Nachrichten aus Niedersachsens Urgeschichte* 40, 180–198.

Winton, A. L. and Winton, K. B. (1932) *The structure and composition of foods. Volume 1: Cereals, starch, oil seeds, nuts, oils, forage plants.* John Wiley & Sons and Chapman & Hall, New York/London.

Zanier, W. (1999) *Der spätlatène- und römerzeitliche Brandopferplatz im Forggensee (Gde. Schwangau).* Münchner Beiträge zur Vor- und Frühgeschichte 52. C. H. Beck'sche Verlagsbuchhandlung, München.

8 Social Status, Identity and Contexts

8.1. INTRODUCTION

Alexandre Chevalier

Following Brillat-Savarin's famous quote: 'Tell me what you eat, and I will tell you who you are' (Brillat-Savarin 2009, Aphorism 4), we argue that looking at the plants eaten in a specific social context allows us to identify socially, culturally, economically, symbolically and politically both present and past eaters. Indeed, among the potential plants and animals available in nature, humans eat only some of them. Technical and environmental constraints may explain some of the choices, but humans definitely select what they eat for cultural reasons as well (Farb and Armelagos 1983). For example, chillies are praised in Peru and Mexico but not in Germany or in Iceland; cocoa was drunk very spicy in Mesoamerica up to the Spanish conquest and reserved to a social elite, whereas it is now a sweet treat to most children in the western hemisphere; wild mushrooms are a common dish in most rural areas but feared by city-dwellers...

In the same way, it is necessary to remember that the so typical southern France ratatouille, when it is not taken archetypically as the most refined dish of 'connoisseurs' by Hollywood studios, was just a mix of any vegetables and, most probably, including potatoes, up to the mid-end nineteenth century, when local regionalism movements and cooking formalisation by chefs included the American zucchini, bell pepper and tomato as key ingredients. Some Western countries' most 'national' dishes, such as gratin Dauphinois in France, 'French' fries in Belgium, *knödel* in central Europe, *gnocchi* in northern Italy, *Rösti* in Switzerland or mashed potatoes in the USA, would still be be prepared with any other sort of root vegetables such as parsnips, turnips, Swedes or Jerusalem artichokes, if the 'purple flower' of the Andes had not being carried all over the world by Spaniards as early as the sixteenth century.

These examples can be multiplied indefinitely, transposed to the different regions of every country, down to every village within a given region and, even, among households of the same town. There are no two identical dishes prepared by two people: people everywhere seem to have developed their own way of preparing the very same plant by cooking it in a different way, by changing the order of the steps, by adding or subtracting essential spices and condiments. Thus, there are not two people who eat the same plant in the same way. In other words, beyond the innate preference for sweet food shared with other mammals, individuals seem to choose, value and eat different food plants across time and space according to their cultural identity and social status, as well as the context for which and in which this plant is prepared and ingested.

The different examples given in this volume, both historically (mainly Chapters 3, 4 and 7) and across regions and cultures (mainly Chapters 5 and 6), seem to support this perception of a complete individual freedom in the choices made by humans, if not an infinite multiplicity of their choices. Our Western world's modern urban way of living reinforces this feeling, through the availability of food from all over the world and the choices offered to the consumer; if only for coffee, for example, with five different brands, normal or decaf, with brown or white sugar, sweetener, and three to four different kinds of milk, which adds up to at least 120 different possibilities for a cup of coffee! In this perspective of multiplicity of choices, it would be pointless to try to understand food choices made by people, for they would too variable and individual. Moreover it would be highly hypothetical, if not simply impossible, to interpret plant remains from archaeological contexts, for we would lack the individual's intentionality and food choice finality.

Nonetheless, everyone acts and chooses according to specific social and cultural rules that change according to the physical and social context. It is indeed commonly considered that food preference is regarded as a personal process and food intake as a social act. Thanks to the anthropology of food and social psychology studies, we know that this is absolutely wrong and that our 'personal' choices are, in fact, completely shaped by our identity, social status and by the contexts in which food is ingested. It is, therefore, possible to relate food choices to cultural and social status, as is highlighted by all the contributions in this chapter.

When people refer to taste, health or cost to explain their choices and use attributes such as likes-dislikes, healthy-unhealthy, or cheap-expensive to justify their choices, they relate in fact to a whole set of cognitive and symbolic associations generated by an object and, in our case, by a food plant. In other words, every human being has representations of plants (Lahlou 1998) that allow her or him to include them – or to reject them – in that person's social and cultural realms, as stated in the introductory chapter. These representations are rooted in cultural and social settings that create a whole taxonomy of plants that people are allowed to eat, or prohibited from using (Douglas 1975; Mead 1997). However, following Sahlins (1978; 2008) and Descola (2005; 2011), we would propose that physical nature is not an objective truth but, from the very beginning, a social and cultural construction. Otherwise, how can we explain that humans learn to eat bitter plants that mammals tend to avoid because they are perceived as poisonous, or the infinite ways of preparing food plants such as potatoes or rice around the world; if not because of the variety of cultural and social perceptions of these plants that lead to specific ways of including them in our everyday, festive meals or rituals, as is brought out in the contributions to this chapter.

When this process is common to several people and when these people are aware that they share a common associational process, we can refer to a social representation, or social identity. The social psychologists John Turner and Henri Tajfel have developed theories that help us to understand that social influences on individual choices are, in fact, determined by one's membership in a social group. Turner's 'self-categorisation theory' (Turner 1989) explains the formation of one's identity across life events and time through an alternation of self-evaluations (personal identity level) and assessments of appropriateness of one's personality and behaviour to a given social situation (social identity level). Hence, how starting from an 'individual' preference for some food plants, one tends to align to some food rules related to his/her social group, or learns to enjoy specific food intake in specific social contexts. When we teach children how and what to eat, we induce them to learn social norms about food according to their age and gender. These social norms change throughout their lives: if vegetable purees are appreciated by young children, it is because they are allowed to like it by their parents and their social group, apart from the smooth texture or the ease of eating plants without having to chew, but the same food tends to be rejected by teenagers as being 'childish' food, because adolescents align to the social expectations for their age – to become adults – and not because they fundamentally do not like a food texture any more. Up until very recently, Western women were not encouraged to eat very spicy, mustard or peppery foods, for these food plants could have a direct impact on their sexual behaviour, while men would be praised for their taste for spicy foods as an anticipation of their pugnacity and resistance to pain (Flandrin and Montanari 2000). Even if this is no longer the case, probably both because of the broad diffusion of Asian and north African cuisine in the Western hemisphere and internal societal changes in perception, an aura of being 'aphrodisiac' still clings to spicy and peppery meals. These ingested symbolic attributes are well known and described (Fischler 1994), but they are socially codified in terms of gender and power and not just anyone can consume them.

In her contribution, Cruz-García (Chapter 8.5) shows that wild food gathering and consumption in northern India diminishes with age; children being the ones who harvest and eat such plants most, because the social stigma linked with wild plants does not affect them in the same way as adults. Being of a low social status already, or enjoying more social neutrality, they do not need to make their tastes fit a given social group and refrain from or hide the fact that they are eating wild plants, if they individually like them.

Alcohol consumption (Jennings and Bowser 2008; Scholliers 2001) provides another example of food ingestion related to social inclusion or exclusion. Children are not allowed to drink alcohol but, up until very recently, in Western hemisphere societies, teenagers were encouraged do it as a sign of adulthood, although usually only men were involved. Currently discouraged if not prohibited, pre-adulthood alcohol consumption still constitutes a sign of supposed adulthood, as the *botellones* – binge drinking carried on in the street by teenagers and young adults – amply proves. The kind and quantity of alcohol that is ingested is, therefore, very normative and socially controlled, as is shown by several contributions (Chevalier and Dulanto, González Reyero, Durand and Wiethold) in this chapter.

Two consequences follow on from the self-categorisation processes. On the one hand, there is an accentuation of the differences between groups and of the similarities within a group, which eventually become stereotypical. On the other hand, there is a depersonalisation in favour of the group: individuals renounce their personal preferences and adapt to the group's norms and rules. These consequences are, perhaps, more obvious nowadays regarding clothing codes among teens, business people, or rich versus poor people. But they apply perfectly to food and, specifically, to food plants, even if they are more obvious at national(istic) levels, many times in a negative, insulting way. In a more positive way, nations often claim a specific plant or product based on plants as their own, such as the many national dishes based on the potato which we have already mentioned, by prohibiting others from growing specific plants or making and commercialising plant products under a specific name through the legal status of *controlled designation of origin.*

At another level, if it seems obvious to parents that members of a brother/sisterhood hardly like the same food; the latter would still have to learn to ingest the very same plant all their life in order to share a common social identity, whether externally imposed or internally selected and projected toward others as the marker of their social membership. It would be a definite socio-political statement for a member of a high social status group, aware of his/her membership, to refuse to eat or praise truffles, even if he/she initially did not consider them as a tasty plant. In the opposite way, it is acceptable to serve something as common as rice in a formal *bourgeois* dinner if there is some saffron on it, or out-of-season vegetables, such as asparagus or strawberries in winter.

Social identity concepts are defined as one's awareness of being part of a group and of sharing common values and emotions with this group (Tajfel and Turner 1979; 1986). These concepts predict that if a person's group is involved in a conflicting issue, cognitive representation of his own group, as well as of the other group, which is stereotypical in both cases, is modified in order to strictly adhere to the individual's group stereotypes in terms of beliefs, attitudes and norms (Turner 1982). Social identity implies that the individuals and the groups do have an absolute need to keep up and emphasise the positive social identities in order to stimulate their respective self-esteem. In the case of intergroup interactions, this self-esteem stimulation is achieved through positive, not negative, differentiation and can be expressed in various ways, such as plant choices, or in the way plants are used. The more frequent the intergroup interactions are, the more intense the positive differentiation will be. These concepts mirror the anthropology of food theories. For instance, according to Mennell (1997), the more a society is divided into subgroups – in other words, the more hierarchical a society is – the more its *cuisines* will be different from one group to another. Goody (1982) and Gumerman (1997) state the same: complex societies are supposed to have differentiated food, both regarding quality and diversity, according to each social status. Both Goldstein and Hageman

(Chapter 8.8) and Chevalier and Dulanto (Chapter 8.2) show how different social groups differentiate themselves from others by selecting specific plants, while affirming their identity. In addition, González Reyero (Chapter 8.3) demonstrates how Iberian groups, living close to each other, not only symbolically delimited their territory through plant representation in worshipping places located at their territory's borders, but also how high status classes utilised wild plant representations to affirm their identity and power over other groups within their own ingroup, and toward the outgroups.

Two different processes of food differentiation have been proposed by Mintz (1986). Firstly, when usage and values linked to food are transmitted downward socially, Mintz uses the word *intensification*; that is, poor people will adopt a high social class product in order to be considered rich by mimicry. We use the concept of *reappropriation* when this process allows the *intensification* to move upward; when 'old', low social class products are adopted by high social class people and bring back past attitudes and (supposed) significance. Mintz uses the word *extensification,* when a food product is spread among several social groups but its signification and value are modified, as well as the context of its use.

Intensification examples can be found in the use of exotic fruits in Christmas desserts, for instance; what was initially reserved to wealthy people is now widespread among all social classes. *Heimweh* (homesickness) syndrome is a good example of *intensification*, very often with *reappropriation.* It leads migrants to *reappropriate* food they would neglect, or despise in their own country of origin. Through the emphasis put on food products from their group of origin, or food preparations that would taste or smell in the same way as in their home country, some migrants mobilise all their senses to recreate a lost world and try to keep a separate positive identity from that of their social surroundings (Sutton 2001). If this is not always the case of the first generation because of assimilation issues, the second generation very often feels the need to renew the link with their cultural roots through a few but crucial foods that are considered as a 'family tradition'. A later generation of migrants will definitely reinterpret this legacy, blending fantasised cultural roots, family habits and local social needs, along with a few family dishes, that would be transmitted and prepared for very specific family celebrations or holidays according to each family 'tradition'.

Following Mintz' exemplification of *extensification*, we would say that, even today, tea still constitutes a good example of *extensification.* The current trend of organic green tea used by young upper class people infused in glass teapots and mugs, while lower social classes may still brew with teabags in ordinary mugs and upper classes rely on tea of exclusive origins, produced in very low quantities, is quite meaningful. Potatoes are another good example of *extensification*, since they are eaten in all social groups, but a basic potato puree will be shaped in rolls, fried and called potato *Duchesse* in a high social class context. In the same way, onions are not well considered among high social class groups, unless they are prepared in the *Soubise* way (a puree with a *béchamel*).

The current *Slow Food* movement partakes of both *extensification* and *intensification.* Old, peasant, neglected and despised products, such as chestnut flour used to elaborate gnocchis, are now praised for being natural and local by new educated upper-class people. High class people *reappropriate* the 'simple' peasant way of living in the hope of appearing uncomplicated and close to nature. In turn, sundried tomatoes are an example of *extensification.* Being initially a way to preserve and store tomatoes during the winter by peasants, they are now used in very elaborate salads and dishes by members of wealthy social classes, because they are associated with a way of living and good health and are eaten in winter, as well as in the summertime, in place of fresh tomatoes. In their contribution, Kirleis and Klooß (Chapter 8.6) suggest that the over-representation of wild fruits in a tumulus, in comparison with domesticates, would symbolically indicate a *rite de passage*, while their under-representation in a settlement context would indicate food habits. Wild plants are present in both contexts, but their respective meanings would be different, in particular considering the high social status of the dead. If the context could be meaningful, as we will see developed further, the issue here is that someone of a high social status uses the same category of plants – wild plants – with *another signification.*

Personal and social identities are closely related to social status and vice-versa. The anthropological and historical literature on food is very extensive on this subject and has already shown the influence of social status on food, and therefore crop choices, and brought to light some of the mechanisms to get and/or keep socio-political power through plant products. However, social status is often confused with wealth. Although social status may well rely on the accumulation of wealth, this is not always the case. Otherwise how could we explain the differences between young girls and boys in most cultural groups around the world, who enjoy definitely different privileges, but represent the same initial wealth? Social status is, in fact, more closely related to the social potential for accessing goods (opposition empowered/ underprivileged), than to materially possessing the goods (opposition rich/poor). In this definition, gender, age and social class based on wealth are like many social statuses of empowered or underprivileged people and imply a hierarchy between people and groups. Plant choices are made to express and identify this hierarchy: they may relate to the wealth of their owner through their expensiveness and exclusiveness, or relate to the power status of those who use them through their symbolic signification of empowerment. They may, however, also relate to underprivileged people or groups as being disdained by others or because those plants are left unused by others.

Mingote's (Chapter 8.7) contribution shows how this hierarchy was expressed in Spain during the Middle Ages through agricultural regulations forcing farmers to grow specific plants. Most often, however, these social statuses are not put into the form of written regulations but expressed in implicit codes and may consist of a few local, but symbolically important, plant use differences, as Chevalier and Dulanto show is the case for the site of Pampa Chica in Peru (Chapter 8.2).

In contrast, in their contribution, Durand and Wiethold explore (Chapter 8.4) the social differences related to wealth among the inhabitants of Gaulish Bibracte (Mont Beuvray in present-day Burgundy, France) and show that exotic foods and, therefore, expensive ones – such as olives and coriander – were controlled by the aristocratic elite of the Gaulish tribes. The diversity of plant products seems also to be directly related to social status. Durand and Wiethold demonstrate that, in fact, at Bibracte social status was more related to the diversity of plant foods and less to the presence of 'exotic' new species. As for González Reyero (Chapter 8.3), she mentions the representations of wild plants as a symbol of empowerment of those who display them in material culture and on funerary monuments.

At a lower level, it is obvious that – in any society – family members play different roles and have different obligations and constraints, both within their own family unit and towards other social groups. Therefore, the question is not of knowing whether age and gender have an influence on plant choices, but how this influence is expressed. If the relationship between gender status and food has been explored in the literature, mostly from women's perspective (*e.g.* Balakrishnan 2005; Counihan and Kaplan 1998; Eckman 1994; Greenfield and Southgate 2003; Gurung 2002; Koopman 1997; Ruiz-Arranz *et al.* 2002; Vázquez-García 2008), Cruz-García (Chapter 8.5) has to be given credit for being one of the few social scientists to focus her interest on the relationship between food and children, in particular regarding wild food plants, which are related to both children and underprivileged social groups.

Both social and physical contexts are crucial to understanding the plant choices made by social groups in order to state and maintain their social status and identity. We already mentioned above the influence of social context for the expression of one's identity that must fit with this, and the reinforcement of one's ingroup stereotypes in case of conflictive interaction with an outgroup individual. For instance, community celebrations – usually referred to in the literature as feasts (*e.g.* Dietler and Hayden 2001; Hastorf 2003; Hayden 1998; Joannes 2000) – are specific social contexts where social status relates highly to plant choices. It may be a means of carrying out an economic redistribution and, therefore, producing social cohesion, as well as of showing the individual's own social status. Plants used during these social occasions may be everyday food plants, but displayed either in huge quantities or prepared in very different ways, or luxury, exotic (in)-edible plants. Goldstein and Hageman (Chapter 8.8) explore these plant choices and cultural codes in two Late Classic Maya settlements in northwestern Belize by comparing feasting and household plant composition. They show that some of the fallow and wild plants were used exclusively for feasting and were, therefore, involved in negotiating the status of the lineage head, others for domestic consumption, and still others were utilised in both contexts as a statement of community identity toward other Maya groups. In contrast, Chevalier and Dulanto (Chapter 8.2) show that everyday food plants have been used in an egalitarian Peruvian coastal society, but with clear differences in plant composition and plant uses, to worship ancestors of the group in order to maintain the social and cultural community ties between the dead and the living.

Physical contexts, such as shrines, temples or tombs, may directly influence the choice of plants for their symbolic associations and specific ritual uses (see also Chapter 7 for examples of specific plants associated with offerings). In this chapter, Kirleis and Klooß (Chapter 8.6) and González Reyero (Chapter 8.3) present examples of plant presence and representations in various graves in Spain and Germany. Curiously, wild plants seem to play an important role in rituals associated with death and afterlife in very different societies. It is uncertain whether it is a mere coincidence, whether it is due to the social status of the dead in both cultures or to the specific location of the graves within the social space.

Food is indeed a very complex cultural element, in which social status, identity and contexts are intertwined to lead individuals to eat what they eat. However, because the link between the individual and social representation of plants and their choice depends upon the social and physical context in which this representation is elaborated and the choice will be made, it is very difficult to interpret food plant remains in archaeological contexts: we usually lack the precise social setting in which the representations were elaborated. Even in the case where social attitudes toward food are depicted in texts, the specific events which led to the creation of the archaeobotanical remains are often unknown. For instance, plant remains in a tomb can be interpreted as much as food for the journey of the dead in the afterlife, a symbol of how the dead functioned when he/she was still alive or offerings with specific unknown ritual or religious functions. Also, in the case of a prehistoric site, we very often do not know the exact function(s) of the archaeological context in which the action took place: 'public' space, private precinct, storage room, dormitory, etc? Examples of these issues are to be found in Chevalier and Dulanto (Chapter 8.2), Durand and Wiethold (Chapter 8.4), Kirleis and Klooß (Chapter 8.6) and Goldstein and Hageman (Chapter 8.8).

Nevertheless, because the food choices human beings make are social and cultural constructions, they can be analysed and the social processes behind these choices can be explored. Food is the Maussian *fait social total* that structuralism favored, together with family structures and sexuality, as a metaphor of the whole society and a way of understanding how these societies were organised (Lévi-Strauss 1964).

In this chapter we do not intend to cover all aspects of the social processes behind plant choices made by humans described above, nor to propose new theories to understand how humans make food choices: we attempt only to illustrate some of the diversity found in humankind's behaviour and to relate it to the known social processes that shape our choices.

8.2. PLANTS FOR THE ANCESTORS: PERPETUATION OF SOCIAL STATUS AND JUSTIFICATION OF POWER IN A LATE FORMATIVE (400–100 BCE) ANDEAN GROUP

Alexandre Chevalier and Jalh Dulanto

Introduction

The specific social status of every individual within a given human group, and of every group itself, can be identified through its respective food choices, as well as through the ways these foods are secured, prepared, eaten or used in rituals (see Chapter 8.1, as well Chapter 7 on food in rituals). As a consequence, food remains uncovered in archaeological contexts should reflect these humans' attitudes toward food and, therefore, the social status of the people who produced, transformed, ate these foods or used them in rituals may be identified; but with some limitations, however, as we have seen in the introduction.

The pre-Columbian site Pampa Chica, located on the central Peruvian coast, is of special interest to this issue, since part of its architecture is characterised by spaces that are duplicated in a mirror-like way. It is very doubtful that this specific spatial organisation is due to an accretion of spaces (a new part being built to replace an old one), since the C^{14} data indicate that they are roughly contemporaneous. This particular spatial organisation, therefore, results from a specific building plan that corresponds to a specific social need to proceed in this way. Based on the material culture uncovered, there is no indication that these duplicated spaces were used for very different purposes. On the contrary, they seem to have sheltered the same kind of activities (Dulanto 2002a and b). In the same way, there is no indication that different cultural groups, in a broad sense, were using these buildings. Our assumption is, therefore, that these areas were used by two sub-groups with different social status within their ethnic group. The composition of vegetal remains from each of these spaces was analysed in order to identify what may have been the respective social status of these two sub-groups.

A Site on the Edge of a Valley

Pampa Chica is located on the central Peruvian coast, ~12°10'00"S 76°52'06"W, at 180 m asl, twelve kilometres away from the Pacific shoreline, on the northern bank of the Lurín river, in a lateral *quebrada* or ravine that is 500 m away from the valley thalweg[1] (Fig. 8.1). The site overlooks the valley at a height of 50 m, where no direct water is available from the river or through canalisation, making a permanent housing settlement there unlikely. The buildings are not orientated toward the valley but face the rocky hill on the opposite side, that marks the beginning of the *quebrada* de Manchay and narrows its access. Pampa Chica seems to be set back from the farming economy of the valley and its everyday activity, facing some natural element not yet identified, while being visible from the valley (Fig. 8.2). If, as Bastien (1978b) points out, the physical geographical space in the Andean world is both a reproduction and the symbol of social space, Pampa Chica is then embedded in a structured spatial network of relationships with the surrounding natural landmarks. *Huacas* (usually designating a natural location, but they can also be

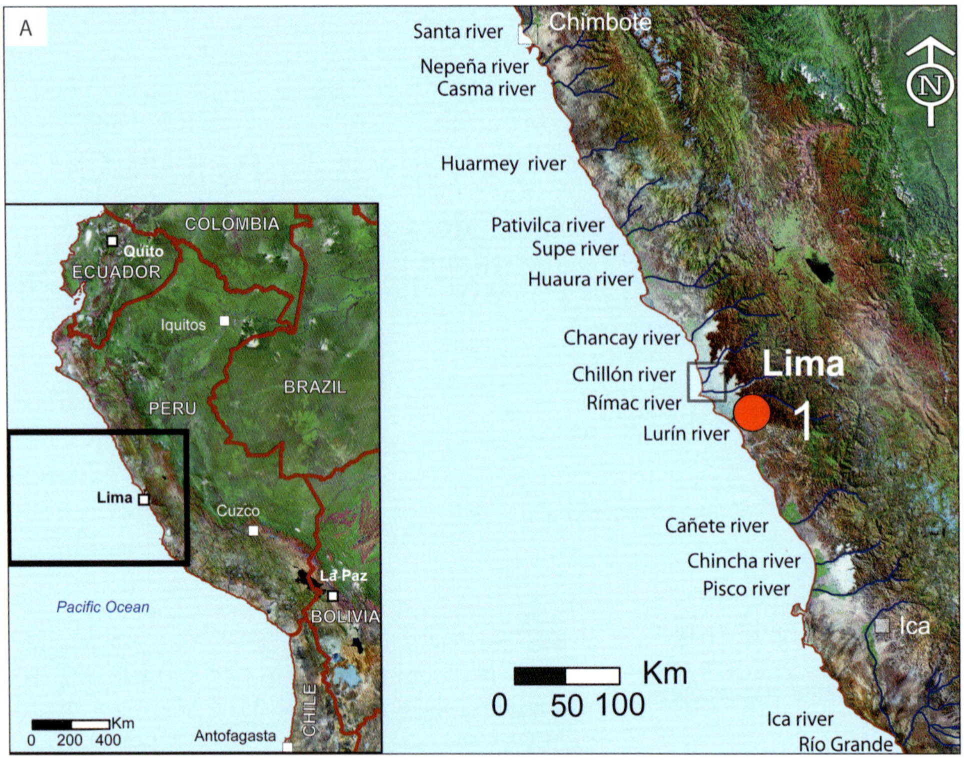

Fig. 8.1. A: Map of central Peru and B: of the Lurín valley, both indicating 1) the Pampa Chica site. Map A: R. Lugon, J.-C. Loubier and A. Chevalier. Map B: A. Chevalier and Y. Reymond.

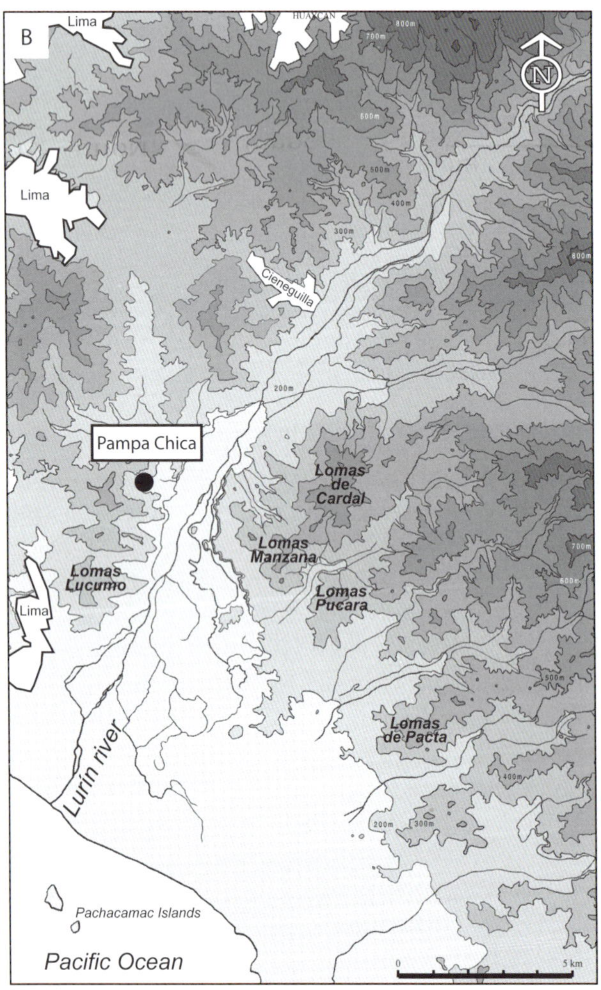

a place where mummies of a group's ancestors are located and venerated) and *apus* (usually mountains) are anchor points representing, respectively, the most direct ancestors of a community and the place of origin of the mythical founders. They constitute as many social elements with a specific status within the society and its history; they forge kinship, justify ancestry and locate the place of origin of a specific group in the operational geographical environment (Bourdieu 1980; Solomon 1995). In this general Andean symbolic framework, it is clear that the location of Pampa Chica was not chosen randomly, but corresponds to specific social and symbolic needs of a community, as well as being the representation of them.

The site comprises two different structures that are 140 m away from each other,[2] extending over two hectares. A big complex of different areas with public plazas, called Structure 1 dominates another isolated, simpler, internally-subdivided rectangular structure, termed Structure 2. It seems to have been built quickly and at one time, following precise planning, since very few additions or juxtapositions have been observed. Nevertheless, four different phases have been identified, dated between 800/400 and 150 BCE.[3] The first two phases (1A1 and 1A2) would correspond to the building phase itself;

Fig. 8.2. View of the Lurín valley from the Pampa Chica structure 1. Image: A. Chevalier.

almost all the walls and surfaces were constructed but without any finishing. The material culture attributed to these first two phases would reflect the building activities more than the peculiar use of the spaces. It is assumed that this first period is quite short. The two following phases (1B1 and 1B2), at the transition between the Early Horizon and the Early Intermediate Period, would correspond to the actual use of the buildings and would have lasted longer than the first phase. We shall focus on this period, since these phases are assumed to better reflect the activities carried out in the buildings. Two other occupation phases are recorded for the Late Early Intermediate Period and between the Middle Horizon and the Colonial Period, but they are reoccupations of the site without any link with the initial buildings.

Structure 1 is comprised of some 20 spaces in five sectors, among which four are organised mirror-like in two pairs (Fig. 8.3). Each pair is constituted by a big open space with many accesses that is located downhill (sectors 1 and 3), and a smaller, more restricted space that is uphill (sectors 2 and 4).

The first compound, located in the northern half of the site, includes two sectors. Sector 1 is a big patio (R20) downhill, surrounded on three sides by upraised terraces (only two have been excavated, R15 and R16), but open on its lowest side. Material culture is scarce, with mostly fragments of ceramics whose functional shapes are related to food

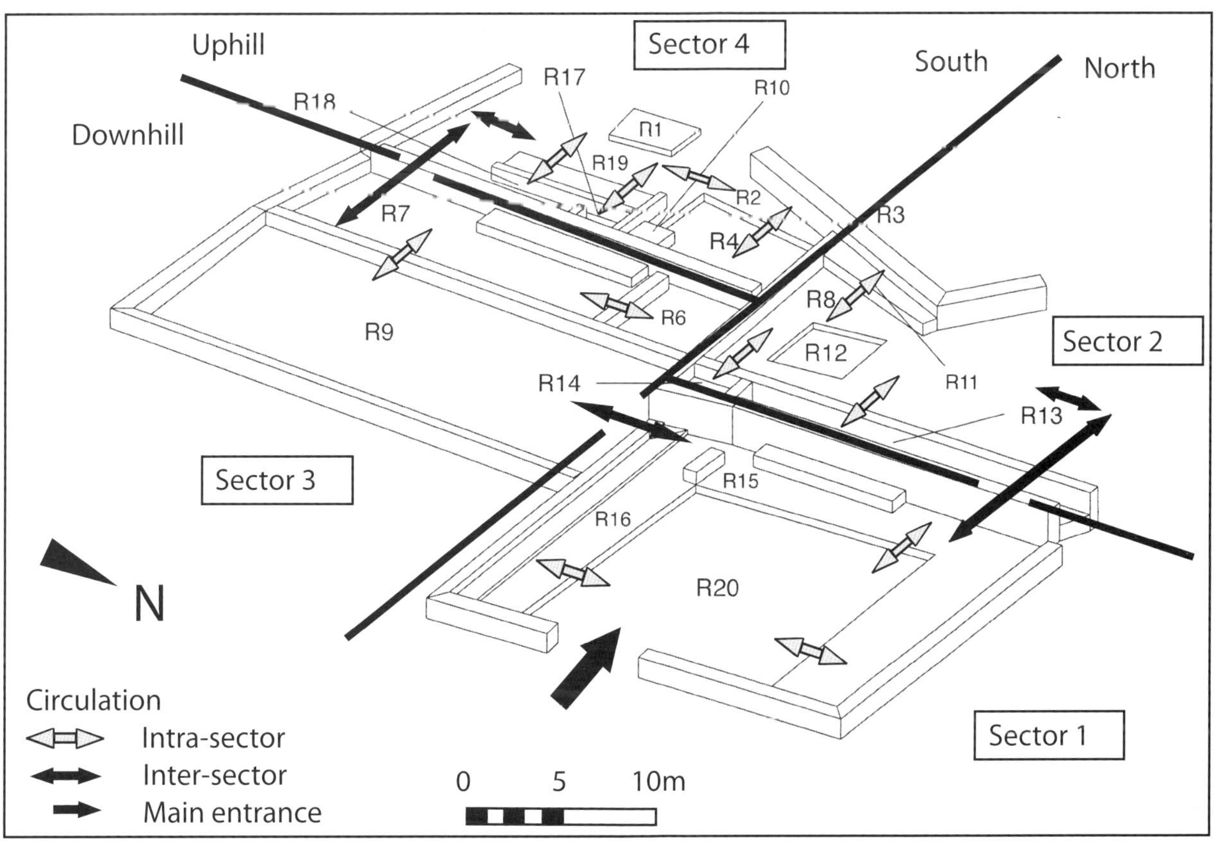

Fig. 8.3. Synthetic isometry of Pampa Chica structure 1 with the double division in Upper-Lower and Southern-Northern parts (redrawn from Dulanto 2002).

consumption ('plates') and liquids ('bowls'); only a hearth pit has been found in the patio itself, while the terraces held no remains. Sector 2 is organised around a small sunken patio (R12), surrounded by a banquette (R8), probably roofed, and flanked downhill by two enclosed spaces (R13 and R14) and another raised terrace (R11) uphill. Many human remains have been found in theses spaces, whether full skeletons in primary context or disarticulated human bones in secondary or tertiary context (removed from their original deposition site), as well as fragments of ceramics related to transport, storing or preparing liquids ('bottles') or food ('pots'/'bowls'), and to the consumption of food ('plates') and liquids ('jugs').

The second compound, located in the southern half, also contains two sectors. Sector 3 is again a

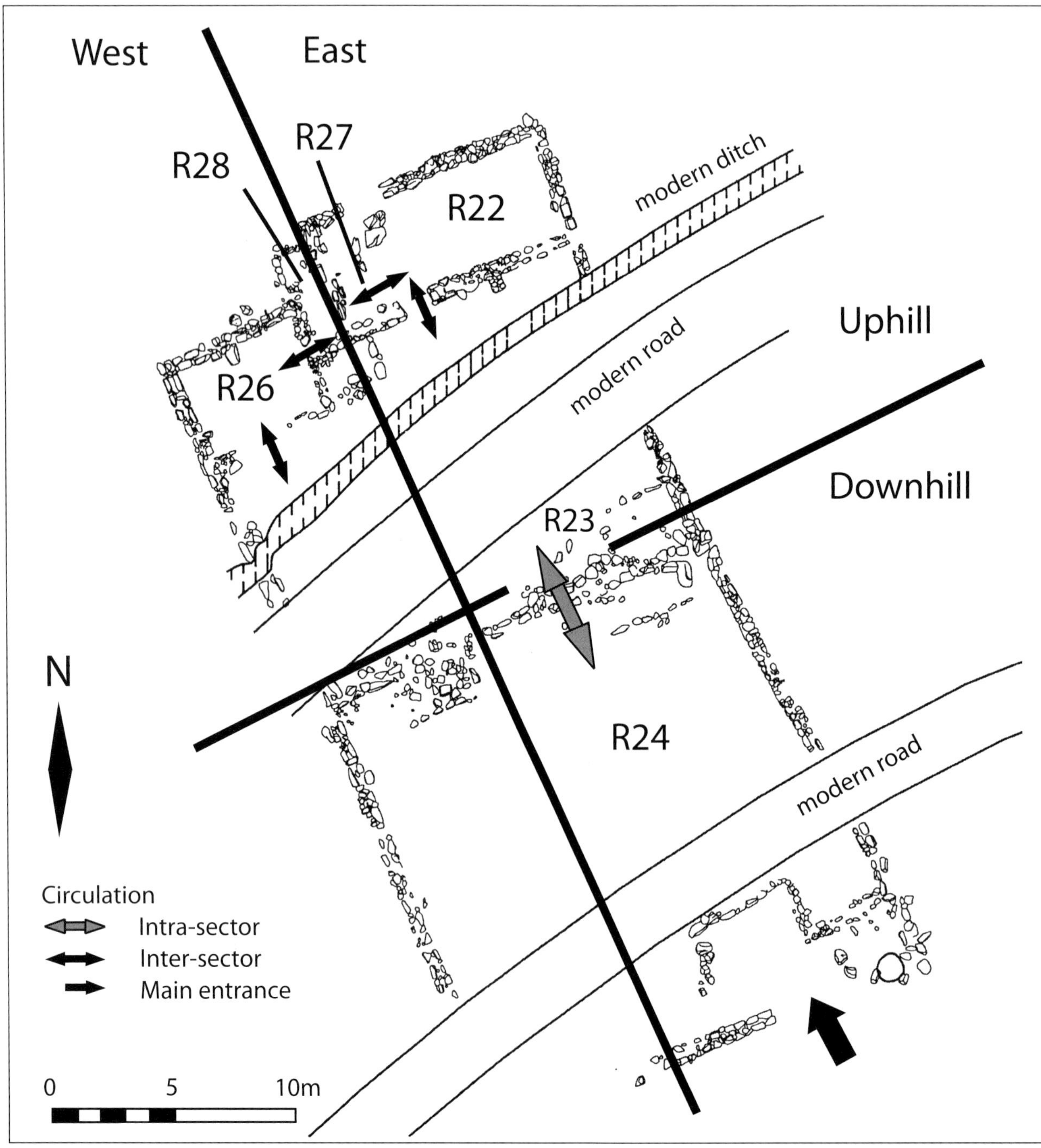

Fig. 8.4. Archaeological map of Pampa Chica structure 2, with the double division in Upper-Lower and Eastern-Western parts (redrawn from Dulanto 2002).

big open-air patio (R9), which communicates with the other patio (Sector 1) through a small opening, dominated by an elevated terrace in its uphill part, that comprises a banquette facing the patio (R7) and a small enclosed space (R6). The patio itself did not contain many remains, while the banquette space contained many organic remains, and the enclosed space had a hearth pit in its first occupation phase and many fragments of 'bowls' in its second phase. Sector 4 shares similarities with Sector 2, although it is organised a little differently. The sunken patio (R4) is off-centre, but also delimited by an elevated roofed banquette (R2) and a podium (R1). Two other enclosed spaces (R17, R18) complete this sector. Again, human remains have been uncovered, as well as fragments of ceramics related to transport, keeping or preparing liquids and to their consumption, though their number is lower than in Sector 2.

Structure 2 comprises six precincts distributed in two sectors (Fig. 8.4). Sector 7 is an open space located in the lower part of structure 2, with a banquette in its northern side, and its organisation looks like that of the open sectors 1 and 3 of structure 1. Sector 8 includes an open space that seems to have been roofed (R23) and two pairs of spaces organised mirror-like – a medium one (R22 and R26) that connects to the roofed space R23, and a small one (R27 and R28). Only parts of R22, R23 and R24 have been excavated and only precinct R22, uphill, had specific structures during the several occupational phases: during phase 1A1, four pits aligned next to a big hearth pit, among which three had fragments of big neckless jars together with maize cobs and other vegetal remains, and the fourth one had huge quantities of wood charcoal; this organisation is repeated for the phase 1A2. During phase 1B1, nine hearth pits have been uncovered, together with lumps of mollusc shells, maize cobs and other plant remains. Finally, five hearth pits, as well as lumps of organic remains, have been dated to phase 1B2. In the R23 precinct, a large fragmented neckless jar, as well as remains of two bottle gourd containers, have been excavated.

Andean Duality and the Ayllu Concept

Both the analysis of the space sizes and their organisation, as well as of the circulation between the spaces, point toward a differentiated use and access to these four sectors: the big open patios (sectors 1, 3 and 7) clearly could hold large-scale activities and bring together quite a number of people, whereas their restricted small counterparts (sectors 2, 4 and 8) would have seen more private activities with fewer people. This asymmetry would reflect the different social status of the people accessing these spaces and performing activities there: a small social sub-group with more specialised skills and privileges for those who would have had access to sectors 2, 4, and 8 and a larger social sub-group attending, but not performing, some sort of activities in the open areas 1, 3 and 7 – thus people with less influence and hold on their community.

At the same time, the replication of the spaces in the two structures suggests that two subsets of the same community were using those spaces for the same purposes but in a separate way. It is of no real importance to determine whether these activities would, in fact, have been exactly the same, somehow replicated by the other subset of the community, or if this community's subset actually performed different activities. The simple fact that they were using different spaces implies that they had a different social status within the community.

This double dual organisation of space is not uncommon in the Andean world and has been identified archaeologically on earlier sites, such as Cardal in the same Lurín valley (Burger and Salazar-Burger 1991; 1993), or in later sites such as Incaic Cuzco (Zuidema 1964), as well as ethnohistorically and ethnographically (Orlove 1974; Rostworowski de Diez Canseco 1986; Wachtel 1966).

Some scholars refer to the concept of *ayllu* (see Burger and Salazar-Burger 1993) to describe this kind of dual social organisation. In spite of the fact that there are many different definitions of an *ayllu*, depending on the period considered, the nature of the description and whether from the viewpoints of ethnography, social anthropology or archaeology, the *ayllu* can be considered as both a social and geographical unit defined spatially, composed of extended families who share the same mythical ancestors, even if strong variations in social organisation of communities across time and space are documented (Rivière 1983; Platt 1978). Every *ayllu* reveres their own minor god, or *huaca*, usually a natural landmark within the space or at the limits of the *ayllu* territory. The *ayllus* are divided in two symbolic parts or halves: the upper one (*hanan*) and the lower one (*hurin*), which may also have their own *huaca* and refer to their own ancestors. In this socio-political organisation, the means of production are private, but the land is common property. Whenever possible, *ayllus* try to exploit agricultural fields in different ecologies to secure their food sources, otherwise they may elaborate their own economic solution to overcome ecological constraints, whether through exchanges or by setting up colonies in other ecological zones. Halves may exploit different ecological zones, though it is not always clear if one half has better fields to exploit than the other one (Orlove 1974). *Ayllus*, and their subdivisions (the halves), are said to have the power to create and transform reality by competing with others toward the same goal (Bastien 1978a). This competition is expressed symbolically in the *huacas*, and economically through material culture such as plant products. However, it is unclear to what extent this specific notion can be applied to archaeological cultures.

Everyday Food for the Ancestors?

The material studied consisted of 251 soil samples composed of three litres taken randomly in every square metre of the excavated surface and by archaeological layer. In addition, 178 macroremain samples were taken *in situ*. Presence-absence and ubiquities (number of contexts where a specific taxon is represented) analyses have been applied to both kinds of samples once merged together, while quantitative analyses (distribution of percentages and ratios) have been applied only to the sedimentary samples. Standard extraction, separation and identification procedures have been applied (Chevalier 2002). Results are quite contrasting among sectors: strong differences can be observed between the big open patios and the small, enclosed areas, but also between the open patios themselves, as well as between the small restricted sectors (Fig. 8.5).

Sector 1

This sector does not contain many remains. Most of them are located in the patio (R20) and consist of – for almost half of them – wood charcoal. A couple of charred coprolites (palaeofaeces), as well as low quantities of very diverse wild charred and uncharred flora, have also been recovered. The southern terrace (R16) shows an even greater diversity of charred and uncharred wild plants. In addition, it is in this context that we identified almost all of the uncharred food plants, in particular peanuts, but also chilli peppers, maize cobs and squash.

Sector 2

Wood charcoals charocal constitutes about one third of all the botanical remains found in the restricted spaces of Sector 2 and reflect, without any doubt, burning activities, but they are limited mostly to the terraced banquette (R11) facing the central sunken patio, hence its ubiquity is very low. This same terraced banquette also contained most of the charred cotton seed, charred maize cob fragments and charred beans that are unique to this Sector 2. However, uncharred food plants were the most numerous remains in this context, with chilli peppers, peanuts and different kinds of squashes, among which the *zapallo loche* (*C. ficifolia* Bouché) is unique to Sector 2. (See list of plants mentioned under their common and Latin names at the end of the article). The central sunken patio (R12) did not contain any charcoal, but did contain burnt coprolites and a lot of different charred wild plants, as well as medicinal/hallucinogenic taxa (such as charred soapberry that is unique to this Sector 2) and dried tobacco-related seeds. Food plants are scarce, with only charred maize kernels and dried passion fruit seeds. The space just around the sunken patio (R8) contained a lot of uncharred food plants, such as peanut, squashes and *lucuma,* as well as *caña brava* that probably come from a collapsed roof. Interestingly enough, the few remains of containers made of bottle gourd have been found in the terraced-banquette and around the sunken patio.

Sector 3

Only the terrace (R7) and the small, enclosed space (R6) have been sampled. The distribution of the densities is quite contrasted: there are only charred remains in the enclosed space R6, with a predominance of wood charcoal, quite a lot of charred maize cob fragments, a few charred chilli pepper remains and charred cotton seeds. In turn, terrace R7 shows very few wood charcoal and other charred materials but contains a lot of dried food plant remains, mostly peanut shells, squashes and chilli peppers. There are few wild plants except for the *heliotropes*. In addition, fragments of uncharred bottle gourds have been uncovered. Finally, and above all, it is in this Sector 3 that we have found charred wild tomato.

Sector 4

Here again, the distribution of the plant remains varies from space to space. For instance, most of the remains are found in the storage rooms (R17, R18) and on the terraced banquette R2; but very few come from the central sunken patio R4 and there are not any from the circulation and access space R19. The central sunken patio R4 contains most of the wood charcoal but almost no burnt coprolites or charred cotton seeds. This space also contains most of the charred wild plants but almost no food plants. It is also in the sunken patio that we have found remains of Peruvian wild tomato. In turn, the terraced banquette R2 contains most of the food plants (mostly peanut shells, followed by chilli peppers, squashes and maize), as well as wood charcoals and wild plants, but in lesser diversity and quantity than in the sunken patio R4. Uncharred passion fruit and *lucuma* fragments have been found in both R4 and R2 spaces, but not in the other three spaces of this sector. The two spaces R17 and R18 have the highest densities of plants, mostly uncharred peanut shells, but also a few chilli peppers and squashes. They also have some specific charred and uncharred wild plants in high quantities, in particular *heliotropes* and charred *nolanas* in R17. As for the area around the sunken patio (space R19), it contains almost no remains aside from a few wood charcoal and uncharred *Cactaceae* and *heliotropes*.

Sector 7

No soil samples have been taken from this sector. However, botanical remains were collected during the excavation. They consist mainly of uncharred maize cobs (~25% of all the remains with 39 whole specimens), peanut shells, squashes and bottle gourd, as well as elements of a building structure, such as wood fragments and stems of *caña brava* and sedges.

Sector 8

Again, there are no soil samples from the roofed precinct R23 and the information on plant remains comes from *in situ* collections only, which consist solely of maize cob fragments. In fact, most of the organic remains come from precinct R22, with both soil samples and *in situ* materials collected during the excavation. Wood charcoal is ubiquitous in almost all the contexts reflecting intensive burning activities. Charrred and uncharred wild plants (Fig. 8.5) show an important taxonomic diversity. As in Sector 7, *in situ* remains consist mainly of uncharred maize cobs, peanut shells, squashes and bottle gourd, as well as a few other fruits such as *lucuma* and *cansaboca*.

Double Duality

Clear botanical assemblage differences thus exist between the open sectors 1 and 3 and the restricted sectors 2 and 4, both in terms of taxa presence and their quantities, as well as regarding their respective ubiquities. Sectors 7 and 8 show no real differences due to sampling issues (see above); moreover, only part of the eastern side of the structures has been excavated and none of the western part of these structures has been analysed.

Public vs. Restricted Spaces

For instance, the ubiquity of charred remains is higher in the open areas 1 and 3 but their quantities and densities are higher in the enclosed areas 2 and 4, due to a completely different pattern of concentration of the charred remains in the open areas as opposed to the restricted spaces. However, when looking at specific categories of use, carbonised food plants indeed have a higher ubiquity in the restricted sectors 2 and 4. In the same way, the highest ubiquity of uncharred food plant remains is in the restricted spaces 2 and 4, comprising chilli peppers, squashes, peanut and maize. Again, carbonised wild plants are more ubiquitous in the two small enclosed areas 2 and 4 than in the open areas 1 and 3. And although uncharred wild plants are quite abundant everywhere, the restricted sectors show the highest taxonomical diversity. In addition, uncharred bottle gourds are more ubiquitous in the restricted spaces 2 and 4.

Table a)	Ubiquities per sector					Frequencies (%) per sector				
Plant taxa (charred)	I	II	III	IV	VIII	I	II	III	IV	VIII
Food plants										
cf. *Lepidium meyenii* Walp.	-	1	-	-	-	-	12.5	-	-	-
Fabaceae	-	-	-	-	1	-	-	-	-	0.7
Arachis hypogaea L.	-	-	-	-	1	-	-	-	-	0.7
Inga feuilli DC.	-	-	-	-	2	-	-	-	-	1.41
Phaseolus cf. *vulgaris* L.	-	2	-	-	5	-	25	-	-	3.52
Phaseolus lunatus L.	-	-	-	-	2	-	-	-	-	1.41
Psidium guajava L.	2	-	1	5	1	14.29	-	16.67	7.35	0.7
Passiflora sp.		-	-	1	4	-	-	-	1.47	2.82
Pouteria lucuma (R&P) Kuntze	1	1	1	3	6	7.14	12.5	16.67	4.41	4.23
Capsicum sp.	-	3	1	6	16	-	37.5	16.67	8.82	11.27
cf. *Lycopersicon peruvianum* (L.) Miller	-	-	1	1	1	-	-	16.67	1.47	0.7
Physalis sp.	-	1	-	4	1	-	12.5	-	5.88	0.7
Zea mays L., kernel	1	2	-	4	46	7.14	25	-	5.88	32.39
Zea mays L., cupule	-	1	1	2	40	-	12.5	16.67	2.94	28.17
Medicinal plants										
Sapindus saponaria L.	-	1	-	-	-	-	12.5	-	-	-
Nicotiana sp.	-	-	-	1	5	-	-	-	1.47	3.52
Technically useful plants										
Lagenaria siceraria (Mol.) Standl	-	-	-	-	7	-	-	-	-	4.93
Gossypium barbadense L.	1	3	1	5	44	7.14	37.5	16.67	7.35	30.99
Cyperus sp.	-	-	-	-	2	-	-	-	-	1.41
Gynerium sagittatum (Aubl.) Beauv.	-	-	-	-	1	-	-	-	-	0.7
Wild plants										
Sesuvium portulacastrum (L.) L.	3	2	-	6	17	21.43	25	-	8.82	11.97
Trianthema portulacastrum L.	1	-	-	6	8	7.14	-	-	8.82	5.63
Amaranthus sp.	-	1	-	-	4	-	12.5	-	-	2.82
Apiaceae	-	1	-	-	-	-	12.5	-	-	-
Asteraceae	-	-	-	2	-	-	-	-	2.94	-
Heliotropium sp.	2	1	-	2	-	14.29	12.5	-	2.94	-
Brassicaceae	1	-	-	-	-	7.14	-	-	-	-
Cactaceae#5	-	1	-	-	1	-	12.5	-	-	0.7
Cactaceae#6	-	-	-	-	1	-	-	-	-	0.7
Haageocereus sp.	-	1	-	1	2	-	12.5	-	1.47	1.41
Cheno-Amaranthaceae	1	1	-	7	2	7.14	12.5	-	10.29	1.41
Chenopodium sp.	1	2	1	7	-	7.14	25	16.67	10.29	-
Euphorbia sp.	1	2	1	1	4	7.14	25	16.67	1.47	2.82
Fabaceae	1	1	1	3	-	7.14	12.5	16.67	4.41	-
Lamiaceae	-	-	-	1	1	-	-	-	1.47	0.7
Malvaceae	2	1	-	1	-	14.29	12.5	-	1.47	-
Melochia lupulina Swartz.	-	-	-	2	3	-	-	-	2.94	2.11
Urocarpidium sp.	-	1	-	-	-	-	12.5	-	-	-
Sida cf. *rhombifolia* L.	-	-	-	1	-	-	-	-	1.47	-
Nolana sp.	4	3	1	28	47	28.57	37.5	16.67	41.18	33.1
cf. *Oxalis* sp.	-	-	-	-	-	-	-	-	1.47	-
Argemone subfusiformis Ownbee	-	1	-		-	-	12.5	-	-	-
Phytolacca dioica L.	-	-	-	1	-	-	-	-	1.47	-
cf. *Gilia* sp.	1	-	-	-	-	7.14	-	-	-	-
Polygonum sp.	-	-	-	2	1	-	-	-	2.94	0.7
Calandrinia sp.	-	-	-	-	1	-	-	-	-	0.7
Portulaca oleracea L.	-	-	-	2	2	-	-	-	2.94	1.41
Scrophulariaceae	-	1	1	3	6	-	12.5	16.67	4.41	4.23
Solanaceae	1	-	-	1	4	7.14	-	-	1.47	2.82
Browallia americana L.	1	1	-	1	-	7.14	12.5	-	1.47	-
Solanum americanum L.	3	2	-	4	13	21.43	25	-	5.88	9.15
Solanum montanum L.	8	3	2	29	37	57.14	37.5	33.33	42.65	26.06
Verbena sp.	-	-	-	5	1	-	-	-	7.35	0.7
Cyperaceae	1	2	-	3	-	7.14	25	-	4.41	-
cf. Sisyrinchium sp.	-	-	-	1	1	-	-	-	1.47	0.7
Juncaceae	-	1	-	2	1	-	12.5	-	2.94	0.7
Poaceae	-	1	2	1	5	-	12.5	33.33	1.47	3.52
other charred finds	-	-	-	-	-	-	-	-	-	-
unknown seed	4	6	3	27	21	28.57	75	5	39.71	14.79
unknown epidermis	1	1	-	-	1	7.14	12.5	-	-	0.7
dense porous	1	2	1	2	21	7.14	25	16.67	2.94	14.79
unknown others	-	-	-	1	6	-	-	-	1.47	4.23
wood charcoal	12	7	5	63	140	85.71	87.5	83.33	92.65	98.59
charred coprolites	2	2	-	1	5	14.29	25	-	1.47	3.52

Table b)	Ubiquities per sector					Frequencies (%) per sector				
Plant taxa (uncharred)	I	II	III	IV	VIII	I	II	III	IV	VIII
Food plants										
Annona cherimolia Mill.	-	-	-	-	1	-	-	-	-	0.7
Cucurbita sp.	6	7	3	54	58	42.86	87.5	5	79.41	40.85
Cucurbita ficifolia Bouché	-	3	-	-	7	-	37.5	-	-	4.93
Cucurbita maxima Duch.	-	2	1	3	11	-	25	16.67	4.41	7.75
Cucurbita moschata Duch.	1	5	2	6	22	7.14	62.5	33.33	8.82	15.49
Arachis hypogaea L.	4	5	4	30	88	28.57	62.5	66.67	44.12	61.97
Bunchosia armeniaca (Cav.) DC.	-	-	-	1	2	-	-	-	1.47	1.41
Psidium guajava L.	-	-	-	1	2	-	-	-	1.47	1.41
Passiflora sp.	-	1	-	3	-	-	12.5	-	4.41	-
Pouteria lucuma (R&P) Kuntze	-	2	-	4	3	-	25	-	5.88	2.11
Capsicum sp.	4	5	3	40	66	28.57	62.5	5	58.82	46.48
Liliaceae	-	-	-	-	1	-	-	-	-	0.7
Zea mays L., cupule	1	4	1	17	36	7.14	50	16.67	25	25.35
Zea mays L., husk unch.	-	-	-	1	1	-	-	-	1.47	0.7
Medicinal plants										
Nicotiana sp.	-	1	-	3	6	-	12.5	-	4.41	4.23
Technically useful plants										
Lagenaria siceraria (Mol.) Standl. pepo	-	-	1	9	34	-	-	16.67	13.24	23.94
Lagenaria siceraria (Mol.) Standl. seed	1	3	2	10	24	7.14	37.5	33.33	14.71	16.9
Gossypium barbadense L.	-	-	1	7	23	-	-	16.67	10.29	16.2
Gynerium sagittatum (Aubl.) Beauv.	1	1	-	2	21	7.14	12.5		2.94	14.79
Wild plants										
Sesuvium portulacastrum (L.) L.	1	-	-	-	-	7.14	-	-	-	-
Trianthema portulacastrum L.	-	1	-	-	-	-	12.5	-	-	-
Amaranthus sp.	2	1	2	3	20	14.29	12.5	33.33	4.41	14.08
Asteraceae	8	4	3	37	38	57.14	50	5	54.41	26.76
Heliotropium sp.	10	6	4	41	31	71.43	75	66.67	60.29	21.83
Brassicaceae	2	1	3	6	7	14.29	12.5	5	8.82	4.93
Cactaceae#3	-	-	-	2	6	-	-	-	2.94	4.23
Armatocereus sp.	1	-	-	-	1	7.14	-		-	0.7
Cactaceae#5	1	-	-	13	4	7.14	-	-	19.12	2.82
Cactaceae#6	1	-	-	2	5	7.14	-	-	2.94	3.52
Haageocereus sp.	-	2	-	33	20	-	25	-	48.53	14.08
Atriplex sp.	-	-	-	-	1	-	-	-	-	0.7
Chenopodium sp.	2	1	1	13	9	14.29	12.5	16.67	19.12	6.34
Convolvulaceae		-	-	1	-	-	-	-	1.47	-
cf. *Geranium* sp.	-	1	-	-	-	-	12.5	-	-	
Euphorbia sp.	1	-	1	-	-	7.14	-	16.67	-	-
Fabaceae	-	-	1	-	-	-	-	16.67	-	-
Lamiaceae	-	-	-	1	1	-	-		1.47	0.7
Sida cf. *rhombifolia* L.	1	1	-	-	-	7.14	12.5	-	-	-
Urocarpidium sp.	-	-	-	1	1	-	-	-	1.47	0.7
Nolana sp.	-	-	-	2	6	-	-	-	2.94	4.23
Polygonum sp.	-	1	-	-	1	-	12.5	-	-	0.7
Calandrinia sp.	-	-	-	3	-	-	-	-	4.41	-
Portulaca olerацea L.	2	1	-	2	2	14.29	12.5	-	2.94	1.41
Diodia sp.	4	2	-	21	7	28.57	25	-	30.88	4.93
Solanaceae	1	1	-	3	6	7.14	12.5	-	4.41	4.23
Browallia americana L.	-	1	1	6	2	-	12.5	16.67	8.82	1.41
Solanum sp.	-	-	1	-	1	-	-	16.67	-	0.7
Valeriana sp.	-	1	-	1	-	-	12.5	-	1.47	-
Lippia nodiflora (L.) Michaux	-	1	-	-	-	-	12.5	-		-
Verbena sp.	-	-	-	2	-	-	-	-	2.94	-
Eleocharis sp.	-	-	-	-	1	-	-	-	-	0.7
cf. *Juncus* sp.	-	-	1	-	-	-	-	16.67	-	-
Poaceae	-	2	-	1	2	-	25	-	1.47	1.41
Other uncharred finds	-	-	-	-	-	-	-	-	-	-
unknown seed	9	5	3	46	43	64.29	62.5	5	67.65	30.28
unknown epidermis	2	1	1	9	15	14.29	12.5	16.67	13.24	10.56
unknown others	6	3	1	21	44	42.86	37.5	16.67	30.88	30.99
uncharred wood	5	5	2	41	76	35.71	62.5	33.33	60.29	53.52
uncharred coprolite	1	2	-	5	8	7.14	25	-	7.35	5.63

Fig. 8.5. Pampa Chica: Ubiquities and frequencies of plants by sector and structure. a) charred, and b) uncharred remains.

But there are still more differences between open and restricted spaces. Charred *lucuma* are unique to the restricted spaces, as are the following food plants: Cape gooseberry, passion fruit, thin vermicelli pumpkin and the Peruvian prune. Finally, medicinal/hallucinogenic plants are found only in the enclosed small sectors 2 and 4 with tobacco-related seeds and soapberries, not to mention prickly poppies.

Northern vs. Southern Halves

The existing differences between the northern (sectors 1 and 2) and the southern (sectors 3 and 4) halves of structure 1 of the site are to be found mostly between the restricted spaces (2 vs. 4) rather than between the open areas (1 vs. 3). For instance, the use of coprolites as fuel is more a characteristic of the northern half, and mainly in the restricted Sector 2; charred cotton seeds are found in all sectors, but with a specific ubiquity in the enclosed Sector 2. In the same way, the highest ubiquity of uncharred food plant remains, and in particular of chilli peppers, squashes, peanut and maize, is to be found in Sector 2. In addition, a couple of taxa are unique to the restricted northern Sector 2: the thin vermicelli pumpkin, as well as the potentially medicinal/hallucinogenic prickly poppy seeds and the soapberry stones.

In turn, charcoal is the most ubiquitous fuel of the southern half, in particular of the restricted Sector 4. Wild plants, whether charred or uncharred, are ubiquitous all over the site but they reach their highest ubiquity and diversity in the southern Sector 4. The same Sector 4 also has the highest taxonomic diversity. Some taxa are only present in the southern half, such as the wild tomato and uncharred cotton seeds. And again, a couple of taxa are only present in the southern restricted space 4, such as the charred passionfruit seeds, the charred Cape gooseberry seeds and the Peruvian prune.

From the Dead to the Living: Social Status and Identity in Pampa Chica

Based on the numerous human remains that have been uncovered in the restricted spaces 2 and 4, as well as on the overall material culture, it has been suggested that Pampa Chica played a specific symbolic function within the group who built it and used its premises, that is, for ancestor worship (Dulanto 2002a and b). Indeed, in comparison with other sites of the same period, the architecture of the different premises and the internal distributions of spaces at Pampa Chica clearly show that this site did not serve as an economic centre with a redistributive role, such as the ceremonial centres of the previous period (Chevalier 2002), nor as a human settlement, since Pampa Chica does not have domestic housing structures. The only domestic activities that can be observed are the food preparation areas of Structure 2, and in particular it's Sector 8, but they are clearly related to the main structure 1.

Indeed, we know through ethnohistorical sources (Salomon 1995) that ancestor mummies were kept, manipulated and worshipped in enclosed spaces by a small group within the society who obviously had more social power, and also displayed in niches or on terraces, for public adoration and rituals aimed at creating a strong sense of group affiliation by repeating the group's founding myths, through oral or theatrical performance. Identification with a group through a common ancestor is at the core of Andean pre-Columbian beliefs (Ossio 1992; Rostworowski de Diez Canseco 1986). We also know that food offerings were made to the deities, in particular meat and animal fat and fermented drinks, and that ritual feasting activities also took place (Moore 1989; Kuznar 2001; Hastorf 2003; Goldstein 2003; Goldstein *et al.* 2008).

Plant remains and their distribution within the different areas of Pampa Chica correspond to this fundamental nature of ancestor worship activities; with an open plaza for public gatherings facing terraced banquettes where mummies were publicly displayed and restricted spaces with burning activities reflecting specific rituals performed by small groups. The two open spaces R20 and R9 of sectors 1 and 3 contained few plants, consisting mostly of uncharred wild plants. Because of their nature and quantities, we think that these plants were brought in accidentally by people gathered there or stuck to the fur of animals brought there, since camelid dung has been found in the site. With the exception of the enclosed space R6, facing the patio R9 in sector 3, where most of the charred plant remains have been uncovered (in particular maize cob fragments), no burning activity seems to have taken place publicly. Also, it may be noted that these two open Sectors 1 and 3 do not show differences in plant composition or in plant quantities. In contrast, the terraces surrounding and above the open spaces contained food plant remains, such as – mostly uncharred – peanuts, squashes and chilli peppers and fragments of bottle-gourd containers, but again in low quantities. These plants were either put there and left as offerings when mummies were publicly displayed on the terraces or they are the remains of ritual feasts. Bottle-gourd containers may have been used for liquids such as fermented beverages for ritual libations or intoxication.

In contrast, burning activities seemed to have been performed quite often in the restricted sectors 2 and 4; both sunken patios R12 and R4 contained different kinds of fuels and the only hallucinogenic plants found in the whole site. They were, therefore, 'central' to rituals, whether intentionally burnt to cause a ritualistic intoxication among participants or manipulated in association with the mummies. Food plant remains and containers have been found all around these sunken patios, as well as in the two small spaces R17 and R18, together with human remains and huge quantities of *heliotropes* and charred *nolanas*, either linked to specific food and liquid offerings or rituals associated with worship of the mummies.

Finally the material culture and plant remains uncovered in structure 2, with numerous hearth pits, wood charcoal, fragments of large neckless jars within pits aligned in rows, and many larger plant remains such as squash seeds and whole maize cobs, as well as the only occurrence of common and lima beans, point toward the preparation of food and probably also maize-based fermented beverages, because of the numerous maize cobs found there in association with the jars and hearth pits, although no direct evidence can definitely prove this (Chevalier 1999).

It is highly significant that the same plant species remains have been found in all the terraced banquettes of the open sectors 1 and 3, in the restricted-access sectors 2 and 4, as well as in the cooking sector 8. As Hastorf noted (2003, 546): 'food illuminates any political setting, as its production is directly tied to the blessing of the ancestors and thus to the justification of social existence and political difference'. González Reyero (Chapter 8.3) refers to this social justification of the exercise of power though feasting representations in the Iberian world. In the Andes, 'a major component of ancestral recognition, group identity and elite formation [seems to be] food presentation' (ibid). In the case of Pampa Chica, these different sectors are all dedicated in different ways to ancestor worship; through preparation of foods and probably fermented maize drinks in sector 8, through to the public displays of the group's ancestor mummies, food offerings, libations and potential feasts in the terraced banquettes, and through a very ritualistic process carried out by a few people in the group, involving specific food, drinks, fuel and hallucinogenic plants in the restricted spaces where the mummies were kept.

We know that the more power is equally shared among members of a given society, the more food products will be equally distributed, which implies a greater similarity of food products used by all segments of this society and a lesser difference between the products used in everyday life and feasting products (Mennell 1997). Only their quantities and their preparation will differ. In turn, complex societies are assumed to create and maintain elaborate alimentary systems and, in particular, use different foods and prepare them in a different way according to the social status of their members (Goody 1982; Gumerman 1997). In the case of Pampa Chica, the two segments of the society that used the premises seem to be socially equal and have the same kind of power. The material culture uncovered does not show anything that would commonly be called 'luxury' items; that is objects or products that are scarce or prized, such as gold, or items from some distant ecological setting, such as the *Spondylus* shells, or with a specific ritualistic or social class attribution, such as deposits in high-class or royalty inhumations that are found in later Peruvian Moche or pre-Columbian Inca cultures (Isbell and McEwan 1991; Alva and Donnan 1993; Dillehay 1995). Following Mennell (1997), Goody (1982) and Gumerman (1997), plants used in rituals at Pampa Chica would, therefore, not differ from the usual food plants and there would not be drastic differences of taxa between the two halves, which is actually the case. In the same way, both human remains and material culture should be considered as equivalent, even though it is impossible to assess the symbolic or economic value of objects two thousand years later. Therefore, social status differences are mostly expressed through quantities, a very few local taxa differences and 'setting, timing, preparation and accessories' (Hastorf 2003, 545).

Moreover, if the goal in Pampa Chica was ancestor worship – in other words, literally nurturing the link with the ancestors in order to re-create the social community between the dead and the living and justify the present social and political order – we should not be surprised that everyday food plants are used by the living to maintain the connection with the ancestors, since there is no noticeable social hierarchy within Pampa Chica architecture, material culture and plant remains. Why would they choose exotic or luxury plants when their lives did not include them and this did not relate to the ancestors? It may be the case of empires or hierarchical societies, since their rulers may have resorted to these sorts of products during feasts to show and maintain their power as well as to worship important deities, but this is obviously not the case at Pampa Chica.

Conclusions

Activities carried out in Pampa Chica were aimed at assuring protection from the group founder members, through the performance of specific rituals that also involved specific plants or specific preparations of plants, as well as justifying and maintaining social status through the keeping, manipulation and display of their ancestors' mummies, as indicated by the numerous human remains found. The splitting of the architecture, mirror-like, replicating the same kind of spaces that had the same overall functions, indicates that two different sub-groups of the same society were performing these religious activities in parallel, for the material culture is identical with the exception of plant remains. The specific composition of plant remains in every half of structure 1 of Pampa Chica – in particular, within spaces in which mummies seemed to have been manipulated and kept and access to which was probably restricted to a few members of every sub-group – point toward the affirmation of identity within a socially equal world. Therefore, social status does not always refer to social or economical power (in other words, to social hierarchy), but may be related to social and cultural identity through ancestor worship and filiation.

We are, however, reluctant to call the specific social organisation of Pampa Chica '*ayllu*', for this term refers to contemporary and sub-contemporary groups, and there is no clear indication that the Pampa Chica builders and users were organised exactly the same way and that their social status would be identical. Not to mention the fact that there are multiple definitions of what an *ayllu* is. Nevertheless, it is evident that two socially equal sub-groups within the same society occupied and performed rituals in their own precinct. In this perspective, the choice of specific plants within a religious context expresses a symbolic competition between the two sub-groups and, therefore, a different social status. But this was within a socially and culturally equal world: the plants found grow all over the valley and no exotic products, that would express specific economic wealth and social power, are present in our plant remains in Pampa Chica.

8.3. PLANTS IN THE EASTERN IBERIAN IRON AGE: FROM DAILY WORK TO THE IDEOLOGICAL CONSTRUCTION OF THE COMMUNITY

Susana González Reyero

The relationships human societies established with nature are crucial to helping us carry out effective historical analyses on social formations of the past. We will describe below the case of the Iberian societies of the Second Iron Age around the Mediterranean area of the Iberian Peninsula (Fig. 8.6). Our goal is to demonstrate that nature plays a key role, not only within economic processes and in generating activities that aim at upholding the structures elaborated within Iberian societies, but also in shaping the social relationships and institutions within those societies. It is fundamental that nature not only has to do with the purely productive, but is central to the ideology and the legitimisation of inequality. Its role was central in asserting or achieving a certain social status and developing forms of identity. Hence, to be protagonists in the fight with a gryphon or in deer-hunting provided social legitimacy. That is, nature – both the representation of fantastic beings and the appropriation of crops – appears linked to the ideology and social differentiation processes experienced by these peoples of the Iberian Iron Age. What analytical tools we can use to dissect these aspects of ancient societies? Several disciplines study these human-nature relationships. Among them, palaeoenvironmental analyses have gained outstanding importance in this field, even if they are not applied systematically to all case studies for the Iberian Peninsula. Nevertheless, a better integration of all these methods and analytical techniques, and above all a greater awareness of their usefulness, should provide us with better knowledge of various aspects of the activities of everyday life in these communities and of the use they made of natural resources (Rodríguez-Ariza and Esquivel 2007).

The diversity of remains from the archaeological record help us to reconstruct aspects of production and work carried out by these societies, by providing palaeoecological as well as palaeoeconomic information on natural resources they were able to exploit, on how they transformed them and on how they implemented new means of resource exploitation and transformation for food or trade.

Furthermore, our view of human-nature relationships will definitely be enhanced by bringing together several sources of information. We specifically refer here to the Iberian epigraphic and iconographic expressions that have been found, in addition to the palaeoenvironemental and archaeological records. These varying sources allow us to emphasise the diversity of the aspects of Iberian social organisation in which nature is present. Indeed, only an integral analysis gives us clues about the relationships every Iberian community created and maintained with nature and, more specifically, with plants. The interaction between the information we find expressed in iconography and epigraphy, and the imprints these societies' activities life left on the landscape, may change the ways we regard this plant-human relationship, which, until now, have given us only a quite partial and restrictive view.

Iberian peoples were affected by an extensive process of population concentration and urbanisation that

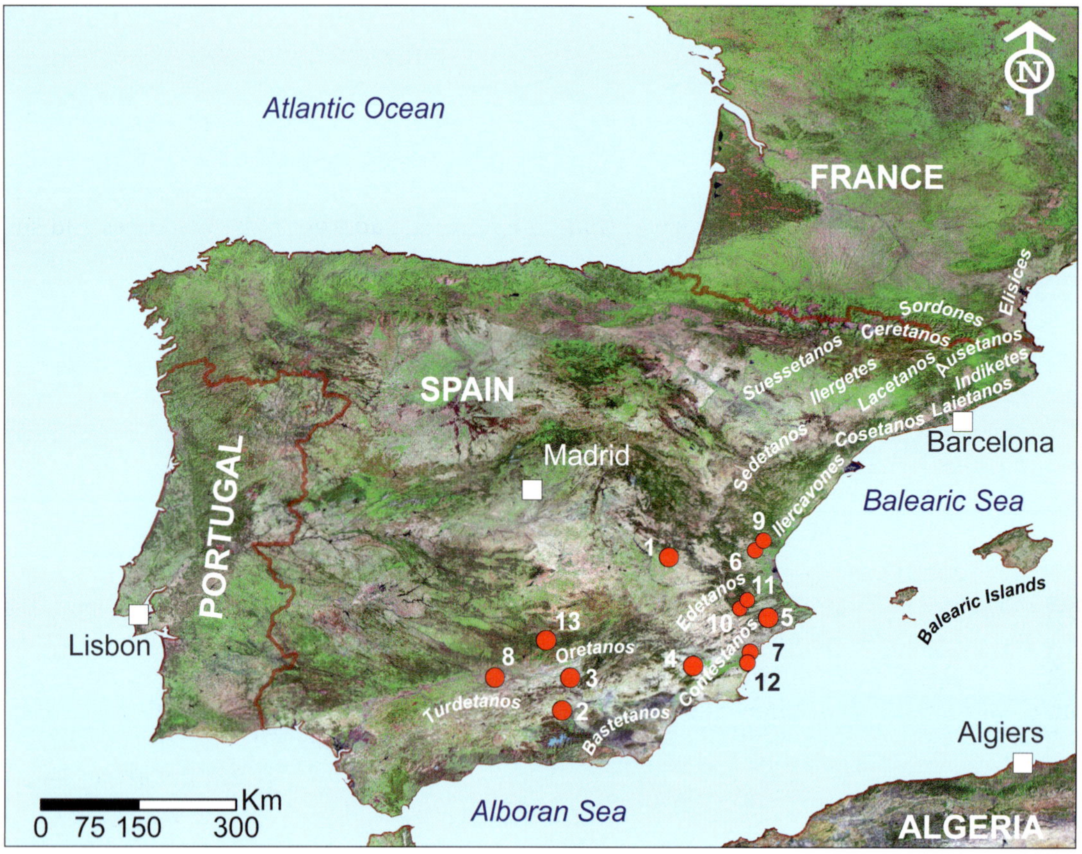

Fig. 8.6. Map of the Iberian Peninsula with the location of the Iberian peoples according to classical sources, and sites mentioned in the text. 1) Cerro Gil (Iniesta, Cuenca); 2) Castellones de Ceal (Jaén); 3) Toya (Jaén); 4) El Cigarralejo (Murcia); 5) La Serreta de Alcoi (Alicante); 6) Sant Miquel de Llíria (Valencia); 7) La Alcudia (Elche, Alicante); 8) Cerrillo Blanco de Porcuna (Jaén); 9) Puntal dels Llops (Olocau, Valencia); 10) La Bastida de les Alcusses (Moixent, Valencia); 11) Corral de Saus (Moixent, Valencia); 12) Cabezo Lucero (Guardamar del Segura, Alicante); 13) Collado de los Jardines (Santa Elena, Jaén). Map: R. Lugon, J.-C. Loubier and A. Chevalier.

occurred in the first half of the first millennium BCE in the Mediterranean basin, as well as in some other zones of temperate Europe. There was a mosaic of heterogeneous communities that shared some characteristics, especially those peoples living in the Mediterranean area of the Iberian Peninsula and southeastern France, during an extensive period from the sixth century BCE up to the second century BCE, when the Roman conquest brought an end to their political independence.

These Iberian communities were organised in hierarchical societies, lived in an increasingly urban landscape, encouraged long-distance trade and developed specialised ceramic and metallurgical products. They also underwent considerable transformations induced by lengthy cultural contact with other Mediterranean cultures such as the Phoenicians and Greeks (Olmos and Rouillard 1997; Ruiz and Molinos 1999). This context of extensive multidirectional, asymmetrical and transforming cultural contacts affected local societies in very different ways (Gosden 2008; Delgado and Ferrer 2007; Cañete and Vives 2011). The Iberian world appears, therefore, as a multilingual mosaic and as a product of complex exchanges among different cultural traditions.

Over recent years, investigation into various Iberian groups through archaeological remains has placed greater emphasis on the spatial expression of the socio-political system, and this has led to an increasing emphasis on analysis of the landscape and settlement organisation (Ruiz 1998; Ruiz *et al.* 2001; Ruiz and Sanmartí 2003; Sanmartí 2004; Sanmartí and Belarte 2001; Grau 2002). Throughout the long historical process that affected them, the Iberian communities experienced transformations enabling them to exercise power, legitimise it and to subject others to that power. In some areas, archaeology has detected various practices of social hierarchy and shed light on how a stratified and aristocratic society was built up on the basis of fictitious kinship ties (Grau 2007).

What seems to be happening, over the time period examined here, is that Iberian social relationships move from a more egalitarian organisation to a client aristocracy system, involving recognition of a few individuals who establish themselves as the bosses or the leaders of the majority of the population, who then become their clients. The Iberian client aristocracy system implies new social relationships and social cohesion, as well as new ways of accessing and appropriating the means of production and surpluses that are essentially of an agricultural nature (Ruiz and Molinos 1999; Grau 2010, 263). In other words, during the Iberian period, we can observe changes, peculiar to every community, in the ways dominant social groups draw from the population the surpluses they need to secure the social reproduction of these elites and to maintain their status quo (Chapa and Mayoral 1998, 64).

These new relationships in the client aristocracy system shape a specific social landscape in which the *oppidum* becomes the urban heart and the organisational centre of the Iberian territorial units and political spaces (Fig. 8.7). The rise of the *oppidum* as urban heart is linked to a social process and is crucial to the subsequent territorial reorganisation. The situation evolves from a rural landscape with scattered hamlets composed of socially and economically undifferentiated populations during the Bronze Age, to urban areas differentiated from the farming countryside, where political, social and economic functions become highly specialised during the Iberian period. Social inequality was thus consolidated and new leaders with hereditary power emerged. Specifically, the presence of the *oppidum* reflects the successes of some families, who managed to concentrate the population, to develop complex and diversified economies and to play a key role in appropriating resources and surpluses, which allowed them to accumulate power, prestige and wealth. Differences observed in necropoleis, the variability of the domestic spaces, as well as the unequal distribution of material culture such as objects of everyday life, imports, jewellery or clothing, all point toward the existence of strong social complexity with unequal access to wealth (Grau 2010, 263).

Fig. 8.7. Aerial view of the oppidum of Puente Tablas (Jaén). Centro Andaluz de Arqueología Ibérica.

Recent investigations have given us more insight into Iberian farming resources management, in particular in Catalonia and in Valencia. It remains difficult to accurately assess the respective importance of plants and animals in the Iberian diet and, in order to do this, we call upon information coming from other cultural settings, but with a similar way of life. From these, we infer that cereals constitute the main dietary element amounting to 65–70%, while fruits, vegetables and greens would account for 20–25%, and finally, wine, oil, meat and fish would cover 5–15% of the diet (Gallant 1991, 68). These general proportions are considered as good approximations, even if we must take into account differences among social groups and geographic location (Iborra *et al.* 2010, 99).

As in any other pre-industrial society, natural resources provided the conditions for survival or the accumulation of wealth, and played a key role in the production processes that set the rhythm of life events and celebrations. Natural resources, and the way nature was perceived, therefore formed not only a fundamental part of material reality but, above all, effected a cultural modelling of reality that would allow it to be understood by these communities. Today, we are able to understand this cultural materiality and give sense to this cultural modelling of nature through information left by these societies. In the case of the Iberian peoples, we can examine the images they produced, since written sources are scarce, not to mention the fact that not all of their languages have been deciphered yet.

It is crucial to keep in mind several characteristics of the imagery involved. Firstly, an image is contextualised: it can only be understood within its specific context. We cannot isolate it from its archaeological context since this same context is part of its signification. Secondly, the image is a message involving the specific materiality of its support. This is both an object where social relationships coincide and an active subject of the social processes. An image can undergo significant changes after being produced and placed within a social context: it can be destroyed, mutilated, etc. In each of these states, an image can have different significations, due to its polysemous[4] character, and can, therefore, be integrated in specific historic discourses. All of the above allows us to characterise the image as a cultural artefact and make possible its contextualisation, through an historical analysis, within the social structure that produced it and the identification of the precise moment in which the image appears.

We can, therefore, state that Iberian peoples did not intend to reproduce reality when producing images. For instance, what is really shown when nature is depicted is a social construction that aims at communicating an understandable message within a given cultural setting. Representations of nature do not mirror nature but constitute a cultural construction. This is why some natural elements are selected and favoured, shapes and attributes of plants may be exaggerated, and some elements are invented (Fig. 8.8), in accordance with precise iconographic rules of representation.

Iberians depict fundamental aspects of their social institutions through their images of the natural world (Olmos 1992; 1996b; 1998; 2004; 2008; 2010; Tortosa and Santos 2003; Tortosa 2006; González Reyero 2008; González Reyero and Rueda 2010). They describe how social groups configured specific visions of the world and of themselves within this same world. In these iconographic representations, nature is a central element that defines the social position of the person represented together with it.

We will, therefore, look at the ways Iberian peoples legitimated the social position their own group held within this hierarchic and aristocratic society through three key aspects: the ways they related to nature, to other members of the community and to gods and the ideational realm, by using representations of plant life.

Fig. 8.8. Representing the lushness of nature on a vase from the Iberian oppidum of Sant Miquel de Liria (Valencia), third century BCE.

Defining Social Place: How to be Socially Different and How to Maintain this Difference

Deeds or epic tales concerning a man appear frequently in Iberian iconography, following a very schematic composition: a man fights alone against another man, a wolf or a hybrid being (gryphon, sphinx), or hunts alone or together with a group, such as in the sculpture of Porcuna (Jaén), created during the second half of the fifth century BCE (Fig. 8.9). This specific representation is one of the legitimisation stories of the Porcuna compound, an iconographic programme explaining the history of the lineage of an aristocratic family (Olmos 2002; Ruiz and Molinos 2007). There is another example with the exceptional Braganza fibula, crafted for a distinguished person of the third century BCE (Perea 2011), depicting a young man, naked and barefoot, fighting off an extremely fierce wolf with only a sword, a shield and a helmet (Fig. 8.10). This same topic of single combat has also been found on a jeweller's mould discovered in an inhumation in Cabezo Lucero (Alicante) and dated to the fourth

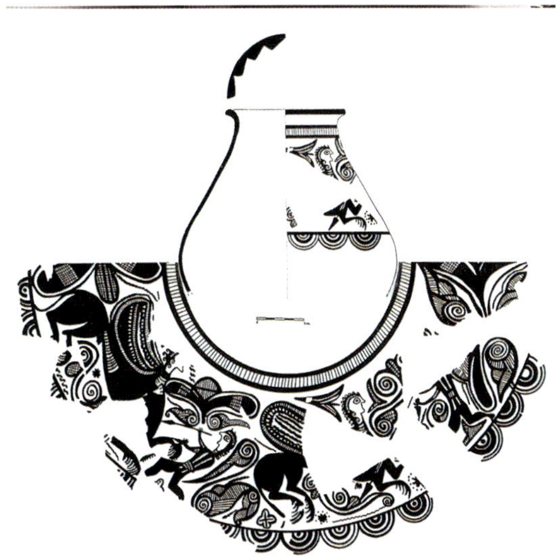

Fig. 8.9. Confrontation between the man and the griffin in the sculpture of Cerrillo Blanco de Porcuna (Jaén): a man and a sphinx fighting in a vessel from the necropolis of the Corral de Saus (Valencia) and a man and a griffin in a die from Cabezo Lucero (Alicante).

century (BCE Perea and Armbruster 2011), in which a man is shown spearing a gryphon (Fig. 8.9). From the third century BCE on, Iberian peoples were to produce large ceramic pieces with elaborate decorations, depicting deeds, which are probably custom-made products used and stored in specific spaces. In La Serreta de Alcoi, the Vase of the Warriors (Fig. 8.11) may depict the initiation labours of a young man (Grau and Olmos 2005; Grau *et al.* 2008) who takes part, successively, in a deer hunt, a fight against a wolf and, finally, in a confrontation with another warrior (Fig. 8.12). According to the plants depicted, all these actions seem to happen in the wilderness, in a space removed from agricultural fields or from the immediate vicinity of the *oppidum*. Thus, the wilderness seems to be the space of combat and confrontation. The same kind of representation – a young man fighting against a sphynx – is found on another narrative vase from the necropolis of Corral de Saus (Valencia). The representation of fights and hunting are recurrent in Iberian iconography: to be victorious, to be successful in a venture through courage or ruse, is the Iberian way to claim one's social difference and legitimate the peculiar role which may be exerted on society by a family, whose members are real or imaginary descendants of this outstanding man who fought against a gryphon, a sphinx, etc.

Archaeological and palaeoenvironmental data indicate other forms of social differentiation, among which wine plays a major role, through its production, procurement, distribution and consumption. Indeed, wild vines grew in the Iberian Peninsula well before the domestication of the grape vine, and their use to produce wine is

Fig. 8.10. The Braganza Brooch, third century BCE. Archivo Au, CCHS, CSIC.

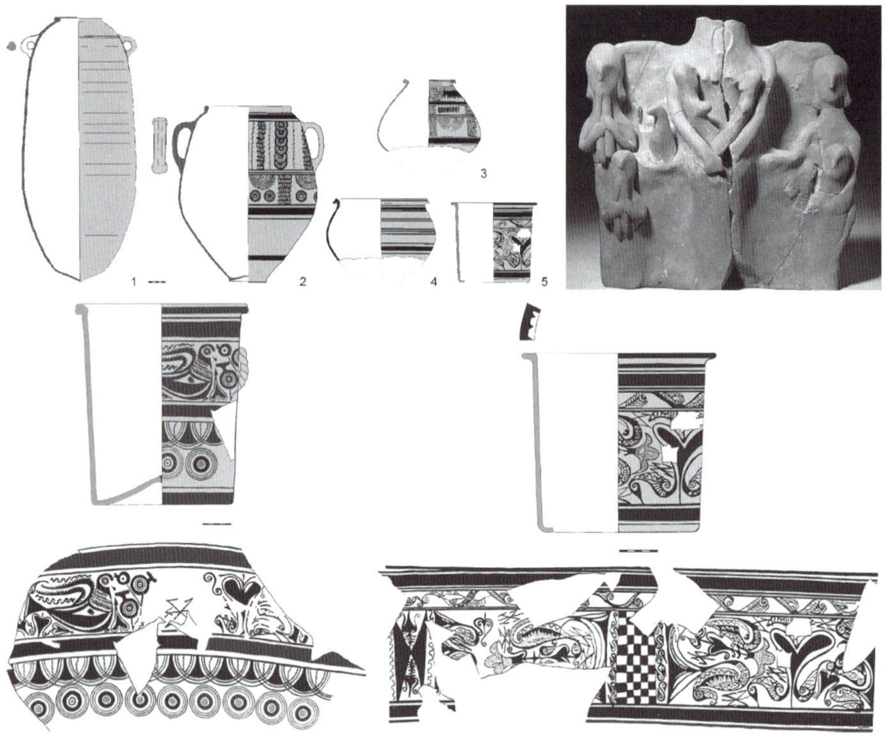

Fig. 8.11. Material culture from structure F1 in La Serreta (Alcoi, Alicante) (Grau *et al.* 2008, figs. 1, 3, 6, 9 and 10).

Fig. 8.12. The hero's exploits take place in rich and luxuriant nature, in the 'Vessel of the Warriors' from La Serreta de Alcoy, third century BCE.

very well documented during the Iberian period: presses and cellars are documented from the end of the seventh century BCE in the eastern part of the Iberian Peninsula, at the site of L'Alt de Benimaquia (Alicante), until the first quarter of the second century BCE, in the regions of Edeta (Sant Miquel de Llíria, Valencia), Kelin en Caudete de las Fuentes (Valencia) and Illeta dels Banyets (Alicante) (Pérez Jordà 2000; Mata *et al.* 2010, 32). The role played by wine, the development of vine-growing and of olive tree agriculture, as well as the gradual development of new agricultural systems, is due to the Phoenicians, who developed trade with extended areas around the western Mediterranean Sea (Buxó 2008).

The appearance of differentiated residences, associated with new local leaders, is linked to evidence of feasts in which wine, among other products, was collectively drunk (Buxó *et al.* 2010, 83). This collective consumption is part of key activities involved in the gradual social differentiation that takes place between the seventh and the sixth centuries BCE. During these feasts, wine, along with imported dishes and prestige goods that are exposed, would generate as much social cohesion as social differentiation. Indeed, the group's identity and capacity to share the same values would be reinforced during this kind of communal meeting, while differentiated participation in the feasts would create social inequalities (Grau 2010, 269). In fact, these feasts are the place where social interdependences characteristic of the Iberian client aristocracy system were created and maintained, since the elites provided access to Greek vases and imported products, the ownership of which was a means to enhance prestige and social status (Ruiz 2008, 807). Potential rivalry during a feast would eventually reinforce leadership positions. Feasts also made it possible to veil the fact that both the control and the distribution of prestige goods were in the hands of the aristocracy, perpetuating in this way the social inequality that characterised the system (Dietler 1996, 69; Grau 2010, 269). Finally, feasts would have provided the opportunity to sanction and to maintain monopolies on the appropriation of farming surpluses.

Defining Social Place: Winning the Favour of the Gods

What is the place of gods among Iberians? Are they linked to nature? In Early Mediterranean societies, depictions and narrations of an inexhaustible fertile nature are plentiful, as demonstrated by the Greek concept of *physis*. Plants are linked with a mythological world of immense wealth: wheat ears full of grains are a gift from the Goddess Demeter who offers them periodically to humans; pomegranate is the fruit that gives access to hell; vine and ivy are attributes of Dionysos, etc. (Olmos *et al.* 2005; Jäger 2006).

Some social sectors promote the representation of an unrealistically portrayed natural world, in which unusual hybrid beings appear, such as the winged goddess, always surrounded by specific vegetation. This composition, which links plants together with birds and a winged deity, is widely distributed across Iberian territories, such as the marble bas-reliefs of Medellín (Badajoz) or of Cortijo Colorao (Granada), the *baetylus* (or sacred stone) of Paterna del Campo (Huelva), the mosaic of the tumulus at Cerro Gil (Iniesta, Cuenca), the reliefs of the Pozo Moro monument (Albacete) or the sculptures of a tumulus at the Elche theme park (Alicante). We argue here that the elaboration of such a winged deity, and its exclusive appropriation by some social groups, contributes to social inequality. For instance, in the necropolis of Cerro Gil in Iniesta (Cuenca) the winged goddess is represented on a collective tumulus, but not inside the grave as an element of the paraphernalia that would reflect a belief in the power of the dead (Fig. 8.13). It is the opposite: on a huge mosaic made of colored pebbles that surrounds the tumulus, the goddess is portrayed with lotus flowers in her hands; birds and canids, as well as exhuberant vegetation, are portrayed around her symmetrically. Very clearly, the people inhumed there – a family of four members, incinerated – publicly claim a specific link with this goddess, differentiating themselves from the remaining inhumations of this necropolis dated to the fifth to the fourth century BCE.

The same phenomenon occurs in Pozo Moro (Albacete), a complex, isolated monument in the Albacete plain on which an almost identical goddess as the one in Cerro Gil is represented (Almagro-Gorbea 1983; Olmos 1996a; Prieto 2000; López Pardo 2006; García Cardiel 2009). However, in the case of Pozo Moro, it is important to highlight the presence of a man who seems to be authorised to enter the goddess' space, and who is involved in deeds depicted all over the monument. The purpose of the monument and its narrative are, presumably, to

Fig. 8.13. The mosaic of a winged goddess of nature is surrounding a burial mound at Cerro Gil (Cuenca), fourth century BCE.

claim a specific social status and rights over harvests and the workforce of the communities living in this area, based on the ascendancy of this man who fought against multi-headed monsters and had the right to be placed next to the goddess. Thus, myths found and justify social inequalities.

Alongside this broader juxtaposition of a hero figure and images of divinities, we observe the frequent presence of plant life as a particular aspect of these iconographies. While an 'unreal' and lush vegetation is a constituent of the goddess' space, specific plants are used in offerings to deities and the dead, such as in the case of the Iberian village of El Amarejo (Bonete, Albacete), where among other offerings, seeds of vine,[5] almonds, walnuts, pine nuts, rye and broad beans have been discovered in a pit excavated in the rock (Broncano 1989) within a ceremonial complex related to a feminine deity (Blánquez 1996). Different plants have been found in funerary contexts, such as grapes in inhumations 95 and 298-B of the El Cigarralejo necropolis (Page 2003); olives in graves of El Cigarralejo (Murcia) y La Vital (Valencia); figs in tombs of Cabezo del Tío Pío (Murcia) and in Burial VII of Casa del Monte (Albacete), in which charred bread has also been discovered (Mata *et al.* 2010, 33–34, 156), as well as almonds, walnut and pine nuts in the necropolis of El Cigarralejo (Murcia). Pine nuts have been also found in inhumations of Cabezo del Tío Pío (Mata *et al.* 2010, 76).

In a funerary context, the representation of vegetal fecundity, sprouting plants and continuous regeneration, like the spikes represented in some chamber tombs in Castellanes de Ceal, Jaén, could allude to an idea of regeneration which resembles the over-representation of wild fruits in a tumulus, as argued by Kirleis and Klooß (Chapter 8.6) as a possible *rite de passage*.

Defining Community Territory: the Symbolic Appropriation of the Border

From the fourth century BCE on, the Iberian world began to be organised into small territorial units similar to the Mediterranean city-state model in which a central town is surrounded by hamlets, farmhouses, craft centres, work houses, defensive sites, etc. The various Iberian political spaces make up a grid of nature and inhabited centres, in which ceremonial centres and sanctuaries play a key role in marking the boundaries of every community territory. However, the whole territory is controlled politically from the *oppidum*. The case of Pajarillo is characterstic: it is a place of worship located at the southern limits of the Úbeda la Vieja *oppidum* territory, which is the city of origin of the family who financed this project at the beginning of the fourth century BCE. The bas-reliefs of El Pajarillo relate the deeds of a man, presumably the ancestor of the family, who fights off a huge wolf. This story and its commemoration during rites performed in the sanctuary aimed at underwriting the expansion of the Úbeda la Vieja political territory toward the valley of the Jandulilla River (Molinos *et al.* 1997). From the second half of the fourth century BCE, territorial sanctuaries start to appear in some communites, such as in Alta Andalucia, where the Los Altos del Sotillo en Castellar (Jaén) and Collado de los Jardines (Jaén) are related to expansion projects directed from the big *oppidum* of Cástulo and serve as milestones defining the limits of political territory (Ruiz *et al.* 2001; Rueda 2011). Rituals performed in these sanctuaries aimed at promoting the cohesion between aristocrats and businessmen. Among the thousands of bronze ex-votos found, some depict bread or cereal cakes held in hands (Fig. 8.14), which probably reflect rituals carried out there; in other words, offerings of bread and cereals aimed at requesting abundance and

Fig. 8.14. Offerings of bread in the sanctuary of Collado de los Jardines (Jaén), third–first century BCE.

prosperity from the gods. The palaeoenvironmental record indicates that cereals, together with pulses and fruits, were the Iron Age staple foods: the bread offered in the sanctuaries was therefore a very common food (Buxó 1997; Alonso 2000). Indeed, wheat (*Triticum* spp.), barley (*Hordeum* spp.), common millet (*Panicum miliaceum* L.), and foxtail millet (*Setaria italica* (L.) P. Beauv.) were cultivated in the whole Iberian area and were used in many different ways, such as breads, soups, or fermented beverages. Among these cereals, barley (*Hordeum vulgare* L.) and free-threshing wheat (*Triticum aestivum* L.) played a major role.

Of course, ovens played an important role in food transformation. In villages, such ovens could be either in a house and, therefore, be for domestic use, as in the El Oral (Alicante) (Abad and Sala 1993, Fig. 138 and 147) and El Zoquete (Valencia) cases (Pérez Jordà *et al.* 2007), or located in a public space for collective use, as in the departments 43 and 42 of Sant Miquel de Llíria (Bonet 1995, 367–369, Fig. 5.2). But even in this latter case, some families may have controlled the community's access to the ovens (Iborra *et al.* 2010). Cooking bread in ovens may, therefore, not have been generalised, or represented a daily activity for the whole community, but may have been limited to some wealthy families, and/or specific festivals (Iborra *et al.* 2010, 107–108). We can argue, therefore, the existence of a certain control over cooking methods, which would have enabled some social groups to differentiate themselves from others and to construct their social identity. Food, but above all, the ways of preparing food, is thus likely to highlight social status and economic power, and constitute a cultural construction used by particular actors in their social and political strategies (Delgado 2010, 27). Cooking methods have been ways to assert their identity, as well as the selection of certain plants, as stressed also by Goldstein and Hageman (Chapter 8.8) and Chevalier and Dulanto (Chapter 8.2).

Conclusions: Towards an Integral Analysis of the Natural World in Ancient Societies

An integrative analysis of the archaeological record is the most advisable way to understand how past societies carried on their lives within a physical space and how they provided it with meaning in order to make it a culturally understandable reality. The integration of the palaeoenvironmental record has enabled us to reconstruct ways of life and economic strategies, and to understand the diversity, or not, of plants found and the reasoning that may have been involved in these choices.

However, in considering only the palaeoenvironmental record, how could we imagine the world of gryphons and monsters created by Iberians and the social consequences of the confrontation between the hero and these figures? The analysis of images in their archaeological context allows us to propose some cultural interpretations for the presence of food and plants, and to understand the key role played by lush nature within the human world. We have explained how these narratives contributed to legitimising a differentiated social space, the appropriation of part of the harvest and of farming products. We have also indicated that the acquiring and maintaining of any high social position implies playing a leading role in deeds or mythical narratives and fulfilling the same social role within society by demonstrating the same character of courage and ruse.

This seems the most appropriate approach in any effort to explain social inequality among Iberians, since no image of coercive action based on physical violence, no representation of war, no victorious or defeated people were ever produced by them, which is not the case of the other cultures around the prehistoric Mediterranean Sea. For Iberian peoples, the image provided the necessary ideological explanations to legitimise and make acceptable the appropriation of wealth by a few, through deeds that are carried out in the wild, as indicated by the lushness of nature depicted in the iconography, but this does not involve violence that occurs in the social, domesticated and productive space.

8.4. SOCIAL STATUS AND PLANT FOOD DIET IN BIBRACTE, MORVAN (BURGUNDY, FRANCE)

Frédérique Durand and Julian Wiethold

Introduction

According to Montanari (2010), food and food practices are the first mark of difference between people and social classes. Consequently, food is a way to express identity and to assert social hierarchy, plant choice being used to assert this hierarchy (Chapter 8.1). The review of M. van der Veen (2003) on different aspects of luxury food highlights the universality of this common phenomenon in every hierarchical society through time and space. She exposes the point of view of A. Appardurai (1986) who defines luxury food as restricted to elites, difficult to access, serving as a social symbol, requiring appropriate consumption and being linked to people who consume it. Luxury food is different in each society and it changes over time. Luxury could either be expressed by the consumption of bigger quantities of the same product or by the choice of different food or different ways of cooking. Van der Veen (2003) underlines the consumption of exotic food by upper classes as the evidence easiest to observe in archaeology. During the Late Iron Age and the early Gallo-Roman period in Gaul, the exchange of agricultural products, especially the trade in wine, olive oil and other long-distance imports, was controlled by the aristocratic elite of the Gaulish tribes, being an important part of the elites' power. Exotic food is easy to recognise in archaeobotanical assemblages because of its foreign origin. The acquisition of these rare and in most cases also expensive import products required well developed exchange networks and long-distance trading contacts.

Based on these remarks, we wonder if archaeobotanical assemblages can reveal the social status of the consumers through variations in vegetal diet. In this study, our approach is to test this hypothesis by applying both quantitative and qualitative criteria to the archaeobotanical assemblages from an archaeologically well documented settlement of proto-urban character. Quantitative criteria provide data on food diversity by evaluating the number of recorded taxa and the number and proportion of the different groups of plants consumed (cereals, pulses, oilseeds, herbs, cultivated fruits and gathered fruits).

The number of species in these different groups of consumed plants can be used to investigate whether the predominance of one of them could be related to a social class. The presence of common species or, on the contrary, of exotic species, provides valuable information on food choice.

Those criteria are applied to six archaeological contexts from the *oppidum* of Bibracte (Fig. 8.15 and Fig. 8.16), located on the Mont Beuvray (842 metres altitude) in the Morvan, a small mountainous area in western Burgundy (France). As noted by S. González Reyero (Chapter 8.3), the rise of *oppida* implies a process involving a hierarchical structurisation of society. In addition, we can deduce from ancient texts as well as from archaeological data that Gaulish societies were strongly contrasted hierarchical systems (Perrin 2002). Bibracte was the capital of the *Aedui*, one of the richest and most powerful Gaulish tribes. The Romans considered the *Aedui* as allies and brothers. In his *De Bello Gallico*, Caesar reports his close relationship with Diviacus, a powerful leader of the *Aedui*. The power and wealth of this tribe were built on exchange of many imported products like potteries, food and drinks held in amphorae from Greece, from many places in Italy, from Massalia or from Spain (Goudineau and Peyre 1993; Olmer 2003, 223–230).

The main occupation of the Bibracte site dates from the end of the second century BCE to the Augustan period (end of the first century BCE–14 CE). The *oppidum* was organised in hierarchical areas (Guillaumet *et al.* 2002, Map 1). A market place was situated on the central plateau. The residential area, in the centre of the *oppidum*, has been thoroughly studied in archaeological investigations. It has urban equipment such as a water basin and fountains and at the Pâture du Couvent, a forum and basilica complex was identified as well (Szabo *et al.* 2007; Bessière and Guichard 2010). Four different excavation sites in this district are archaeobotanically well documented: the occupation at the Pâture du Couvent is attributed to the Celtic and the early Roman period (Vitali and Wiethold 1996). Archaeobotanical studies also included two waterlogged plant assemblages recovered from the water basins of the fountains, La Fontaine Saint Pierre and La Fontaine de l'Ecluse (Wiethold 2009). Craft areas are located next to the gates of the *oppidum*. The botanical data from the craft areas were obtained from La Porte du Rebout (Wiethold 1999a) and from the site of Le Verger au-dessus des Grandes Portes (Wiethold 2011).

Results

Concerning results from the residential area, in the Late La Tène levels at La Pâture du Couvent, all the seeds are charred (Wiethold 1996a). Five cereals, three pulses and three gathered fruits were recovered. In the Augustan levels, seven cereals were found together with three cultivated pulses and seeds from probably cultivated pear (*Pyrus communis* L.), six wild fruits and one olive (*Olea europaea* L.) stone for oilseeds were also determined (Vitali and Wiethold 1996). Whereas naked wheat (*Triticum aestivum* s.l./*durum*/*turgidum*) is the most abundant cereal in the late Celtic levels, emmer (*Triticum dicoccum* Schübl.) predominates in the assemblages of the early Roman period. The fountain and sanctuary of La Fontaine Saint Pierre are located on the slope of the main residential area. The sediments of their huge water basin are dated to the period between La Tène D2 and the Augustan period. Archaeological results suggest a *terminus post quem* of 60/50 BCE and a *terminus ante quem* of 10/5 BCE (Barral and Richard 2009). The clayey and waterlogged sediments enabled excellent conditions for uncarbonised waterlogged plant remains and numerous registered taxa, particularly wild species of wet and disturbed areas (Wiethold 1993; 2009). The sample provided evidence of five cereals. Glume bases of spelt (*Triticum spelta)* and the inner glumes of broomcorn millet (*Panicum miliaceum* L.) were the most frequent cereal remains. Two oilseeds, precociously known in northern Europe (Zohary *et al.* 2012), were recorded: opium poppy (*Papaver somniferum* L.) and turnip (*Brassica rapa* L.). A fruit stone fragment determined as *Prunus domestica* L. cf. subsp *insititia* (L.) C. K. Schneider, nine gathered wild species and two aromatic plants were also revealed. No pulses were found, probably due to the waterlogged preservation conditions. Pulses have very soft seed coats and protein-rich storage tissue, and thus even decompose easily under waterlogged conditions (see *e.g.* Jacomet and Kreuz 1999). Finally, the fountain, La Fontaine de l'Ecluse, also situated at the border of the residential district, again provided waterlogged macro-remains, dated to the transition La Tène D2/early Gallo-Roman, 60–30 BCE (Wiethold 1999b, Fig. 38, 2009). Only a few cereals were identified. The abundance of the inner glumes of broomcorn millet (*Panicum miliaceum* L.) was favoured by the waterlogged preservation conditions. Furthermore, seeds of turnip (*Brassica rapa)* and three fruits of coriander (*Coriandrum sativum* L.) were also recorded. Cultivated fruits like fig (*Ficus carica* L.) and grape (*Vitis vinifera* subsp. *vinifera* L.) and some gathered fruits must also be mentioned. Again we have to suppose that the waterlogged conditions did not allow the preservation of pulses.

From the craft areas, four samples were studied from a craft workshop at La Porte du Rebout, but seeds were recovered in only two of them (Wiethold 1999a). In these samples, all the seeds are charred. The first sample has provided some naked wheat grains (*Triticum aestivum* s.l./*durum*/*turgidum)* and indeterminable cereals. More species were recovered from the second. In this excavation,

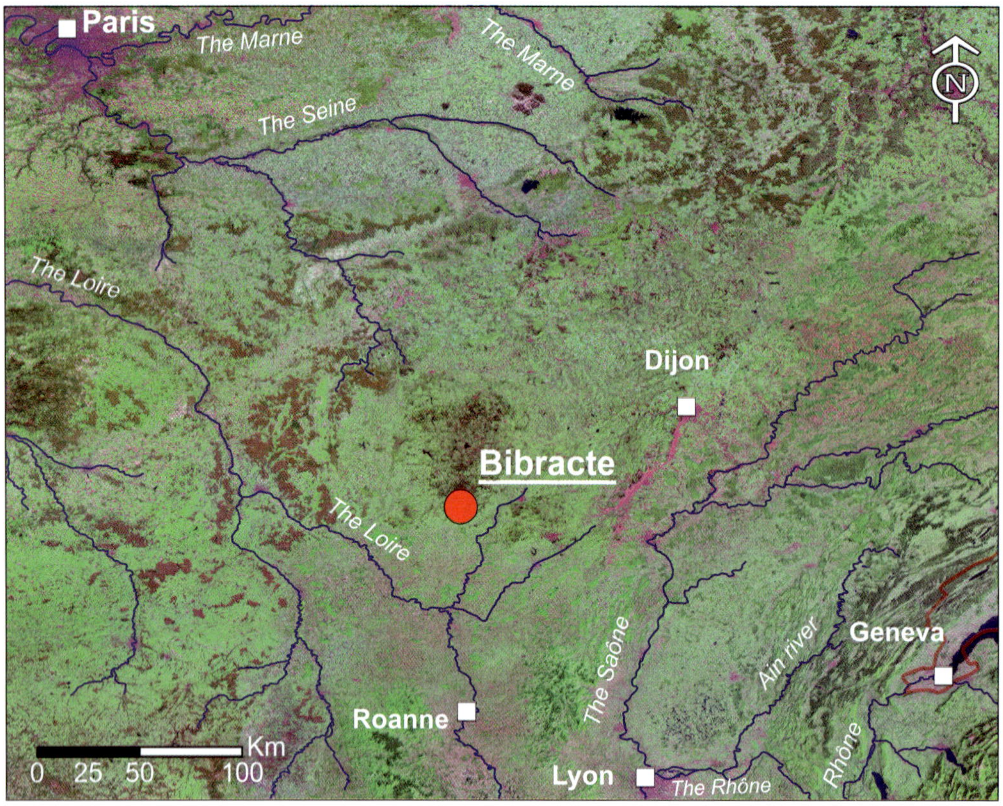

Fig. 8.15. Map of eastern France with the archaeological site of Bibracte. Map: R. Lugon, J.-C. Loubier and A. Chevalier.

four cereals, especially emmer (*Triticum dicoccum* Schübl.) and hulled barley (*Hordeum vulgare* L.), and two undeterminable pulses are present. Five archaeobotanical samples were analysed at the site of Le Verger au-dessus des Grandes Portes: four were taken from the filling of two small pits, US 194 and US 196, which provided remains of carbonised cereal storage, mainly emmer wheat and four-rowed hulled barley (Wiethold 1996b). In total, a great diversity of cereals was found in the samples, but spelt (*Triticum spelta* L.), naked wheat, and broomcorn millet (*Panicum miliaceum* L.) were recorded in smaller quantities. A few caryopses of foxtail millet (*Setaria italica* (L.) P. Beauv.) and a single rye grain (*Secale cereale* L.) were minor weedy contaminations. A single seed evidenced the cultivation of pea (*Pisum sativum* L.).

Regarding Fig. 8.17, which summarises the groupings from each archaeobotanical assemblage, those from the residential area revealed a higher number of species than those from the craft areas. They also provided a higher number of vegetal foods. The evidence of the diet in the craft areas is scarcer and only based on two kinds of food: cereals and gathered fruits at La Porte du Rebout and cereals and pulses at Le Verger au-dessus des grandes Portes. On the contrary, there are five groups of plants utilised that were recorded from the Roman contexts of La Pâture du Couvent, of La Fontaine de l'Ecluse and from the site of La Fontaine Saint Pierre. Three groups of these plants were brought to light in the Celtic levels at La Pâture du Couvent. Consequently, we can conclude that both the numbers of species and of groups of plants used indicate a greater diversity in the residential area. This result presents a general tendency, but we have to keep in mind that differences in the number of sampled structures, the sample volumes dealt with, the absence of waterlogged structures in the craft areas and also the storage finds may have biased this approach.

Observing the representation of each group of plants consumed in archaeological contexts (Fig. 8.18), it appears that cereals are always well represented due to better chances of becoming carbonised. The occurrence of cultivated pulses is strongly influenced by preservation conditions; there is no evidence of pulses in the waterlogged contexts.

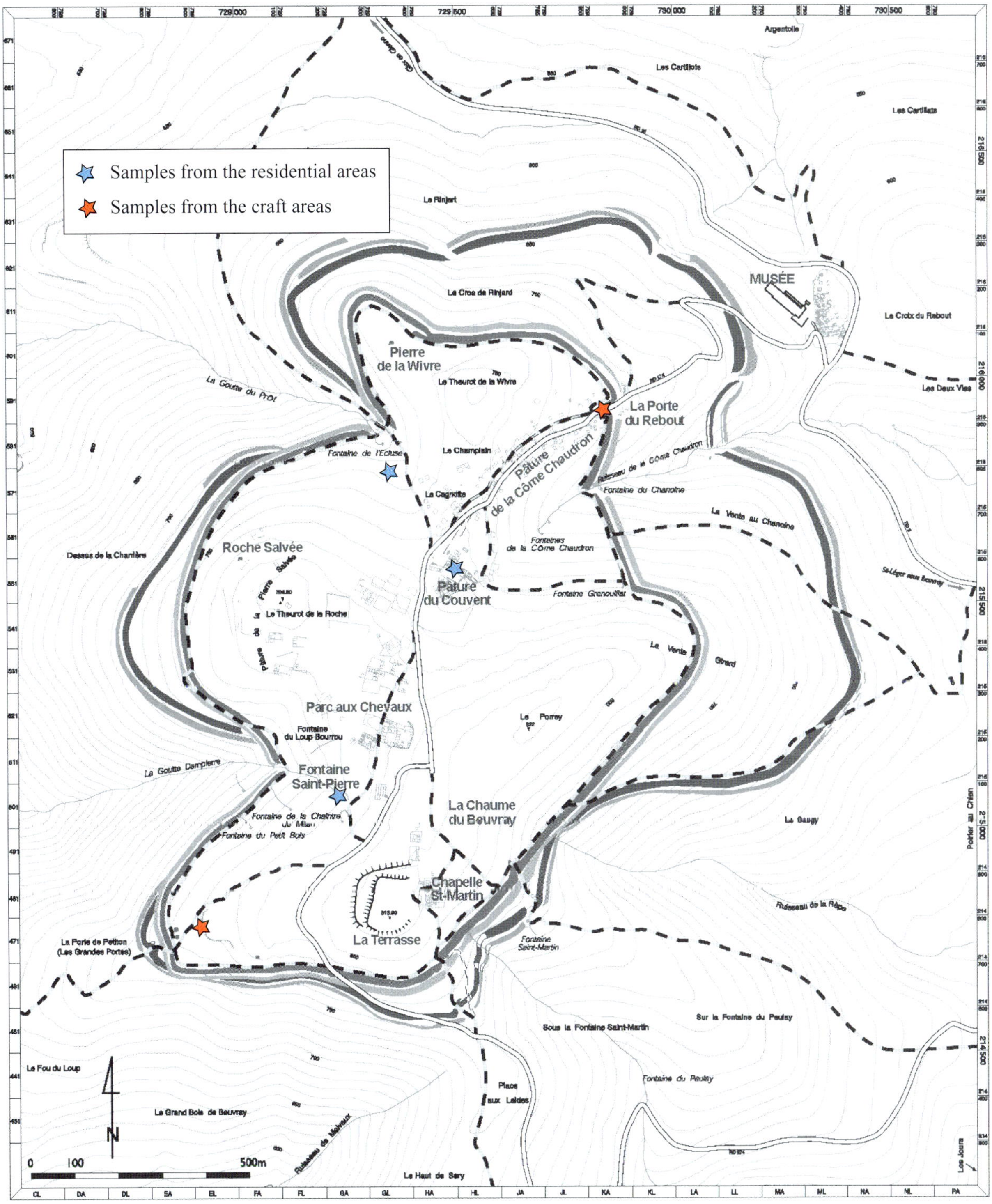

Fig. 8.16. Oppidum of Bibracte, location of the excavation and the archaeobotanical samples.

Of course, the comparison of samples with different conditions of preservation may be biased by taphonomical (deposition and preservation) aspects, but concerning diversity, the biases are limited. In fact, in the four archaeobotanical assemblages from upper class contexts, we can recognise two groups, one group with a great diversity of species (more than 15) and one group with fewer species recorded (less than 15). Waterlogged and dry contexts were present in both groups. At Bibracte, preservation does not seem to have had a great influence on the recorded number of gathered fruits. Indeed,

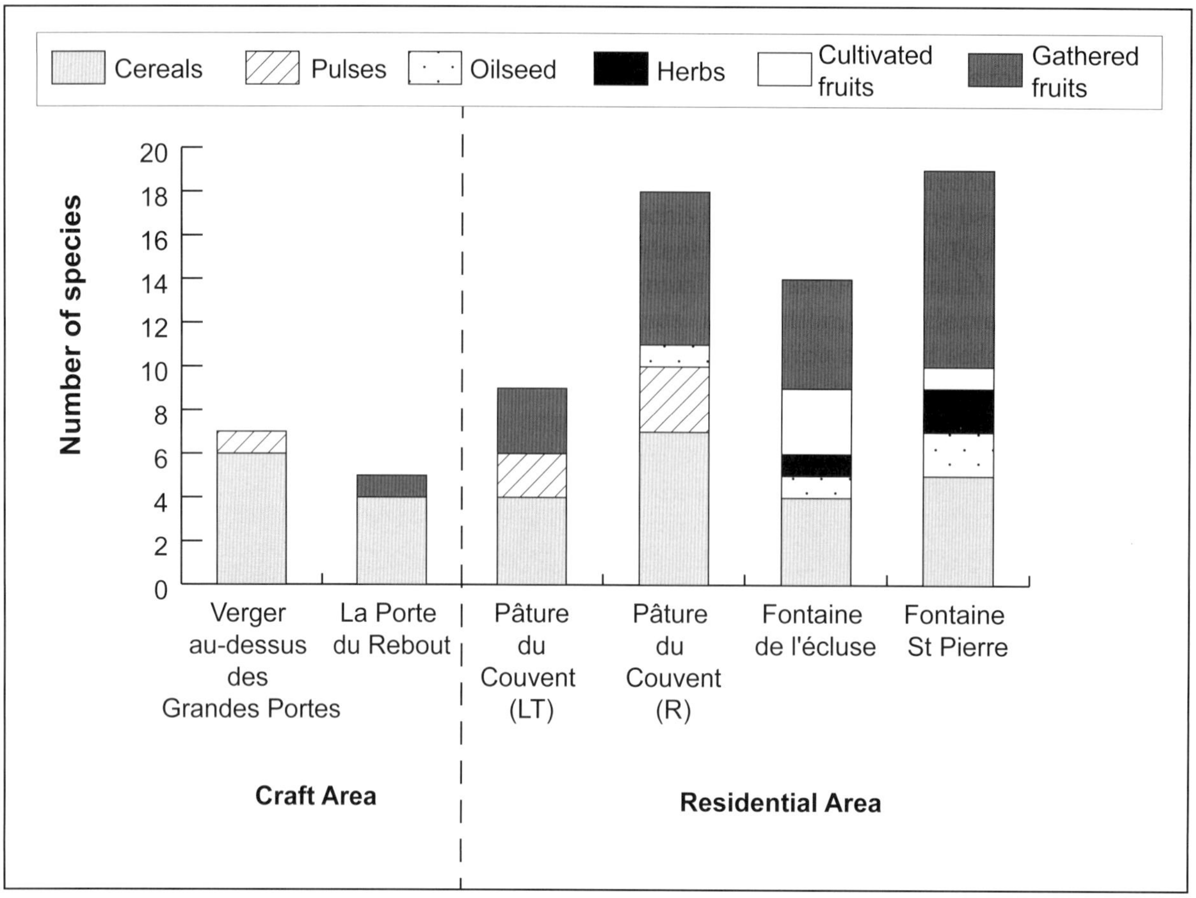

Fig. 8.17. Oppidum of Bibracte, number of species and type of food in the assemblages from the late La Tène and the early Roman periods.

gathered fruits are more numerous in the sample of La Fontaine Saint Pierre, but they are also well represented in the dry contexts of La Pâture du Couvent attributed to the early Roman period. On the contrary, aromatic plants are – due to their fragility and the problems of correct determination – only recorded under wet conditions. Oilseed and cultivated fruits are better preserved under those conditions, too. Consequently, it is impossible here to use pulses, cultivated fruits, aromatic plants and oilseeds as social markers, the taphonomic biases being too strong. Thus, no food plant can be related to a social class. Nevertheless, the predominance of gathered fruits in elite areas can be noticed. This observation is in contrast to the modern occidental image of gathered fruits as substitution foods more related to poorer classes, as seen in Cruz-García discussion of this notion (Chapter 8.5).

At Bibracte, quantitative criteria seem to show that the diet of the local elite was more diversified. According to archaeobotanical researches on contemporary late Iron Age settlements in Burgundy (Wiethold 1999c; 2004; 2008), food plants are not as diversified (from two to eight edible species, and one (cereals) to three (cereals, pulses and gathered fruits) groups of plant consumed). The settlement of Le Petit Chauvort at Verdun-sur-le-Doubs (Saône-et-Loire, Burgundy) with the greater diversity of food plant adds some subtleties but the character of a settlement strongly influenced by trade activity (Barral 1999) may explain that point. A study by Bouchette (1999) in northern Burgundy of both waterlogged and charred contexts shows a low diversity of food plants (four cereals) according to analysis of charred remains, but a greater diversity in the waterlogged remains (15 species for three groups). Thus, taphonomical bias appears as a discriminating factor in using quantitative criteria. Unfortunately, contemporary settlements are not documented as well as Bibracte and the function of sites and their hierarchical status in the society often remains unknown.

	Craft area		Residential area			
	Verger au dessus des Grandes Portes	La Porte du Rebout	Pâture du Couvent	Fontaine St Pierre	Pâture du Couvent	Fontaine de l'écluse
Number of sample	5	4	67	3	13	1
Datation	Late La Tène	Late La Tène	Late La Tène	Early Roman	Early Roman	Early Roman
Cereals						
Triticum dicoccum Schübl..	2	2	1	1	2	1
Triticum spelta L.	1			2	1	1
Triticum monococcum L.				1	1	
Triticum aestivum	1	1	2		2	
Hordeum vulgare L.	1	2	1	1	1	
Panicum miliaceum L.	1	1	1	2	1	2
Setaria italica (L.) P. Beauv.	1				1	1
Pulses						
Vicia faba L. var. minor			1			
Vicia sativa agg.			1		1	
Lens culinaris Medik.					1	
Pisum sativum L.	1				1	
Oil seed plants						
Papaver somniferum L.				1		
Brassica rapa L.				1		1
Olea europaea L.					1	
Aromatic herbs						
Apium graveolens L.				1		
Anethum graveolens L.				1		
Coriandrum sativum L.						1
Cultivated fruits						
Prunus domestica cf. subsp. *insititia* (L.) C.K. Schneider				1		
Pyrus cf. *communis* L.						1
Ficus carica L.						1
Vitis vinifera subsp. *vinifera* L.						1
Wild fruits						
Corylus avellana L.		1		1	1	1
Prunus spinosa L.			1	1	1	
Fragaria vesca L.				1	1	1
Rubus idaeus L.				1	1	
Rubus fruticosus agg.			1	1	1	1
Rubus caesius L.				1		
Sambucus nigra L.			1	1		
Sambucus ebulus L.				1		1
Malus sp.				1	1	1
Pyrus sp.					1	
Number of species	7	5	9	19	18	14

Fig. 8.18. Species from the six archaeological contexts in Bibracte from Wiethold 1993; 1994a; 1994b; 1996a; 1996b; 1997; 1998; 1999a; 1999b. 1 = occurrence 2 = prevailing species. Taxonomic terms from Zohary *et al.* 2012.

Do we see any difference between the main residential area and the other sites investigated according to whether the recorded species can be characterised as traditional or exotic food?

Whatever the social context may have been at Bibracte, the vegetal diet was mainly based on cereal consumption. Using the data available thus far, no difference between the species from residential places and the craft areas can be discerned. The cereals recovered in Late Iron Age and early Gallo-Roman Bibracte are the same species recorded from other excavations in Burgundy (Wiethold 1998; 1999c; 2003; 2004). Again, we cannot distinguish any difference in pulse consumption. The cereals and pulses known from the *oppidum* of Bibracte are traditional species and varieties that had already been cultivated and consumed in Burgundy since the Bronze and the early Iron Age (Wiethold 2007; Labeaune and Wiethold 2007; Wiethold and Labeaune 2005). Gathering is principally dependent on the environment, so wild fruits are not considered. Archaeobotanical remains of cultivated olive (*Olea europaea* L.) and fig (*Ficus carica* L.) in the elite district reveal the use of imported 'exotic' foods. Coriander (*Coriandrum sativum* L.) at Bibracte, recorded only from the waterlogged sediments of the Fontaine de l'Ecluse (Wiethold 2009), is considered a spice plant of Mediterranean origin, but recent archaeobotanical findings have proven an introduction or importation as early as the second Iron Age (Bouchette 1999; Wiethold 2002). Those botanical remains, dated to the transition La Tène D2–early Gallo-Roman and the early Roman period, were recovered from the residential area at the Pâture du Couvent and from a fountain, which probably also had a special function as sanctuary and important place frequently visited by the inhabitants. On the contrary, the assemblages from craft contexts are dated to the first half of the first century BCE. The integration of samples dating to different chronological phases does not seem to represent a real bias, because the principal diet obviously does not change much during the fifty years after the Roman conquest (Wiethold 1998; 2003). The botanical indicators of a 'Romanisation' are extremely scarce at Bibracte and the daily diet seems to maintain the plant food traditions of the preceding Late La Tène period (Wiethold 2011). Indeed, the spectra of cereals, pulses and gathered fruits remain the same. The occurrence of oilseeds and aromatic condiments, restricted to the residential area, seems to be more influenced by the chronological bias of the samples and the different preservation conditions. The scarcity of waterlogged and mineralised assemblages in other settlements in Burgundy does not help us to elucidate the presence of oilseed and herbs in Bibracte. Thus, their integration in the data set seems to be more problematic. Olive (*Olea europaea* L.) and coriander (*Coriandrum sativum* L.) are common in the Roman diet (Zohary *et al.* 2012). First evidence of coriander in central and eastern Gaul is reported from middle and late La Tène contexts (Bouchette 1999; Wiethold 2010). Nevertheless, coriander is an introduction from the Mediterranean and mainly related to Roman culture. During the first century BCE, in Gaul and the northwestern provinces, it can be considered an important spice and condiment indicating Roman or Romanised food habits. The data provided by more frequent archaeobotanical records seems to be strongly related to the development of the Roman Empire (Livarda and van der Veen 2008). At Bibracte, the presence of characteristic Roman finds in layers dated to the transition La Tène D2–early Gallo-Roman can be assumed as a botanical signal for the 'Romanisation' process. In this early archaeological context, coriander and olive still keep their status as 'exotic foods', probably connected to a higher social status (van der Veen 2003).

Qualitative criteria such as the question of the consumption of traditional food plants, whether imported or not, cannot be used efficiently on our data because of chronological and taphonomic bias. Plant food diversity portrays a general tendency indicating a greater diversity in the residential contexts. Up to now, in Bibracte, human alimentation seems to be very traditional and changes in the economy occur more slowly than in other aspects of daily life. Only the use of huge quantities of wine, imported in amphorae and, probably, also the introduction of the use of olive oil seem to reflect rapid changes in food consumption to demonstrate the newly adopted Romanised lifestyle.

Discussion

The present study seeks to examine the variation in elite diet at the transition from the Late Iron Age to the Gallo-Roman period. We used the occurrence of plant species to understand the influence and effects of social status and wealth on diet. Unfortunately, this approach is limited because it blurs the diversity of cooking and food preparation and it does not enable us to consider food quality, different cooking habits and taste. A second limitation results from the nature of archaeobotanical remains themselves. Only a non-quantifiable part of the crops is recovered because archaeobotanical material derives from accidental events (Cappers 1995) and is often found as a secondary deposit not in its original context. Consequently, archaeobotany cannot reach any conclusions about luxury food based on food quantity.

Vegetal food is the base of the diet in rural societies (Grognier and Magne 1841; Carrere and Forrest 2009), but it constitutes only a part of the diet. The role of meat was perhaps of minor importance, but its study is considered of value because, as pointed out before, social status may be a factor of variation in vegetal food/meat proportions. Meat consumption might have been a social marker. Of course, archaeozoology is also not free from taphonomical biases, but bones resist decay better than seeds. It cannot be applied to all sites, because acid soil causes the disintegration of bones. The small size of bones from some species represents a recurrent problem; hence, without sieving, evidence of some animals does not occur in archaeozoological assemblages (Méniel 2002), and this is particularly true for remains of birds and fish. Unfortunately, archaeozoological analysis did not provide remarkable results at Bibracte because the preservation of bones is poor due to the site's acid soils. Consequently, it is impossible to evaluate diet in all its aspects in the *oppidum*, nor is it possible to test any hypothesis about vegetal food/meat proportions by using archaeological data. Méniel's research on the relationship between hunting and the presence of the Gaulish elite highlights the special place dedicated to hunting in late Celtic societies. His research was based on forty archaeozoological assemblages from Late La Tène sites across Europe. The main species recorded were red deer, roe deer and wild boar and hare. Wild animals are always recorded in small proportions; they are more frequent in rural settlements, which contained aristocratic artefacts. There were always many reasons to hunt. We may first think of providing meat for consumption, but obtaining important raw materials like skins, fur, antlers or bones may have been another important aspect. A third reason may have been the protection of the arable fields and the harvest and, finally, hunting may have been an activity of a special group in society. For example, during medieval times, hunting is strongly association with men in the aristocracy, hunting also being a training for war (Montanari 2010). The meat supply is not a convincing hypothesis to explain hunting because animal husbandry generally provides enough meat (Méniel 2002), with the exception of times of crisis (Collin Bouffier and Sauner 2006). Archaeological contexts where wild animal bones appear in higher quantities are related to the presence of elites. On the other hand, we can refer to the presence of hunting trophies like bears' bones or wild boar canine teeth in upper-class burials. According to Méniel (2002) the practice of hunting can be characterised as an indicator of an aristocratic elite, although the consumption of game cannot.

Diet is not only composed of food but also of beverages. Some researchers are interested in ancient beer (see, for instance, Laubenheimer *et al.* 2003; Samuel 1996; 2000; Stika 1998; 2011; Bouby *et al.* 2011), but beer leaves tenuous remains only, either because of perishable containers or because of the inability to recognise beer containers. Trying to observe differences between beverages consumed by the upper class and by the lower class, we focus on wine consumption indicated by amphorae because of their various advantages. Amphorae are nearly indestructible and at the end of the first century BCE, they are clearly identifiable importations. Moreover, they suppy additional information on diet and dietary habits, because they contained various products such as wine, oil, *garum* (a fermented fish sauce), or food preserved in brine. In Bibracte, huge amounts of amphora sherds were excavated. They are mostly wine amphorae, which came from several places in Italy, from Greece or from Spain. During the first century BCE, the presence of Brindisi amphorae illustrates the early importation of olive oil from Italy, while the rise of olive oil importation from Spain took place in the early Roman period (Olmer 2002). The amphorae of type D7/11 are dated to the end of the first century BCE; they were used to transport and to store fish sauces like *liquamen*, *muria* and *garum*, and also for fish, either dried or pickled in brine. The studies on amphorae published up to 2008 do not deal with the complete ensemble of amphorae of the *oppidum*, although some data are available from the residential areas (La Pâture du Couvent; La Fontaine Saint Pierre; Olmer 2002) and from the craft area (La Porte du Rebout; Olmer 1999). The calculation of the minimum number of amphorae shows a greater abundance of amphorae in the residential area of La Pâture du Couvent than in areas with craft activities (La Porte du Rebout), which probably indicates a higher presence in the upper class area and a more abundant consumption of imported products by the elite, with their ability to access goods, than by common people (Chapter 8.1). Dietler and Herbich's (2006) observations among the Luo people in Kenya show how drinks are closely related to social status, notably the consumption of imported alcool as restricted to the upper class living in town.

Conclusions

This attempt to analyse food plants reflecting social and hierarchical differences enables us to conclude that the number of species and the number of groups of plant foods are relevant criteria indicating food diversity. In Bibracte, the upper-class diet appears to have been more diversified. Even if oilseed plants and aromatic herbs have been discovered only in the residential area, we do not consider the presence of those vegetal food groups to be relevant criteria for social and hierarchical differenciation because of taphonomical and methodological biases (preservation, chronology). Consumption of traditional or imported food plants do not provide valuable data for the same reason.

It would be more advantageous to use this interpretation grid in future on a more homogeneous corpus of archaeobotanical data. Moreover, the investigation of diet cannot be limited to species determination, which constitutes only one aspect of human diet and which masks variations in cooking and food preparation. An integrated approach to social aspects of diet should also include the study of animal bones, ceramic vessels, amphorae and ancient texts to obtain a complete picture of changes and persistent elements in human alimentation and society.

8.5. SYMBOL OF POVERTY? CHILDREN'S EVALUATION OF WILD FOOD PLANTS IN WAYANAD, INDIA

Gisella S. Cruz-García

Introduction

Social status, clearly expressed by food plant choices, is an agent for differential access to food plants. The social position and cultural identity of a person would influence the choice he/she makes regarding the consumption of specific groups of food plants, as well as the values he/she attributes to them. The same happens the other way around: the consumption of certain wild food plants will associate the person with a specific socio-cultural group. The consumption of wild food plants, as a practice, is immersed in a more holistic socio-cultural context and, therefore, affected by social interactions.

Wild food plants are essential under conditions of scarcity or food shortage; hence they are usually referred to as 'famine foods' (Grivetti and Ogle 2000). Certainly, stress conditions will mostly affect the poorest who would be the most dependent on such resources during this difficult period. In several social contexts, the consumption of wild food plants is stigmatised as related to a low social status, as opposed to the consumption of modern and exotic foods that are a symbol of a higher social status.

The change of cultural values and dietary traditions around wild food plants due to social stigmatisation has been reported on in several parts of the world. For example, in Swaziland (Malaza 2003), the dependency on food imports and exotic vegetables has eroded wild food plant consumption. Wilken (1970, 294), referring to his work in Tlaxcala, Mexico, said that: 'social status restricts the use of a group of foods', whereas Price (2005) – who studied wild food plants in northeast Thailand – explained that social status restrictions on the consumption of wild food plants will determine to what extent they become part of the diet.

This is also the case of Wayanad District, a biodiversity hotspot situated in the Western Ghats of India, where tribal people – as opposed to non-tribal communities – consume the highest quantity of wild food plants. Indeed, it has been reported in this area (Narayanan *et al.* 2004) that there is a decrease in both knowledge and consumption of wild food plants, mainly due to the eroding availability of these plant resources and changing values attached to their consumption. In addition, there is an increasing social stigma associated with the collection and consumption of wild food plants, which are often seen as symbols of poverty and 'tribalness' (Cruz-García 2005). Further, it was reported that poor and tribal people attempt to hide their collection and consumption from outsiders (Narayanan *et al.* 2004). The assumption guiding the research was that this behaviour expresses feelings of inferiority and shame about poverty as well as about 'tribalness', as opposed to the consumption of more socially-accepted 'modern' (often exotic) foods.

While the previous study in Wayanad documented the range and diversity of wild food plants consumed by the different socio-cultural groups in the area and noted with alarm the erosion of biological diversity and indigenous values, it did not investigate the aspect of value attribution for wild food plants in any depth. The objective of this study was to understand children's cultural evaluation of wild

Fig. 8.19. Boys collecting wild food plants in Wayanad after school. Photo: Gisella S. Cruz-García.

Fig. 8.20. Paniya girls collecting wild food plants in the agricultural fields of Puthoorvayal. Photo: Paul Peters.

food plants. Children were selected because they are very sensitive to the accelerated process of change that is occurring in the area. Clearly 'children's views, values and visions reflect broader social contexts' (Grodzins 2002, 276), and these values are usually transmitted by their parents through the enculturation process (Cruz-García 2006). Children not only collect wild food plants with their parents or family for preparing a meal, but also on their own or with other children; for instance 'on the spot' while going to school or visiting friends (Fig. 8.19 and Fig. 8.20).

'Value', defined by Kluckhohn as 'conceptions of the desirable' or undesirable (Graeber 2001), is a very abstract term that cannot be approached as such. In order to elicit children's value categories related to wild food plants considering their own ways of understanding the world, it was necessary to use more simple related terms. In this way, children were asked if they 'like', 'dislike' and consider wild food plants 'important', as well as the reasons for these preferences. In addition, observing practices such as collection and consumption of wild food plants is a starting point for assessing knowledge and cultural evaluation (Cruz-García 2006), and the social context where a practice occurs shapes children's choice regarding collection and consumption of plant species. Therefore 'values' were also explored through children's perceptions of others' attitudes towards these practices in concrete public events (around which values and attitudes are expressed). Children were asked whether they like or dislike that other people see them collecting wild food plants, and whether they invite their friends to eat with them when their mother prepares wild food plants at home (Fig. 8.21).

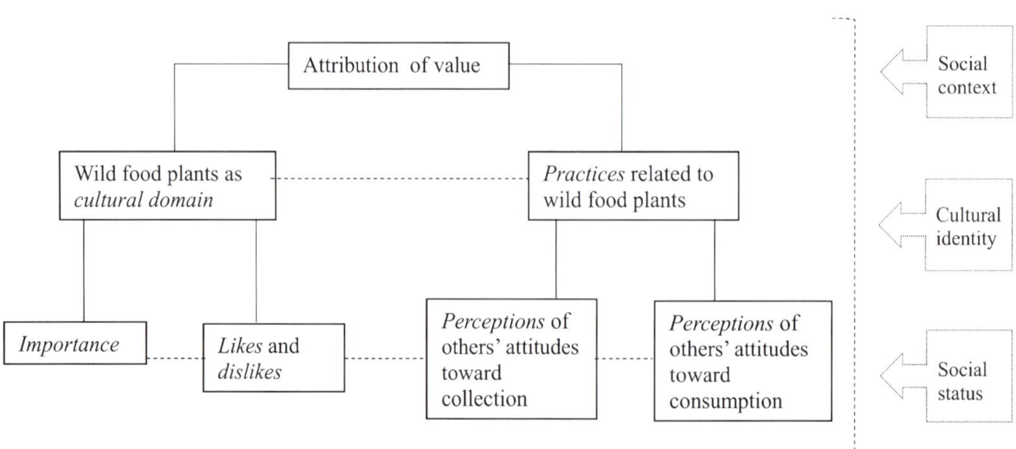

Fig. 8.21. Attribution of value chart.

Wayanad District and Socio-Cultural Groups

Wayanad District (Fig. 8.22), located in the Western Ghats in the northeastern part of the State of Kerala, India, is home to a great wealth of flora and fauna (Kumar *et al.* 2003). Located between 700–2100 m above sea level, it has an area of 2136 km^2, where 37% of the land area is under forest cover and 55% is cultivated (coffee, tea, rubber, rice, banana plantations, among other crops) (Josephat 1997).

During the post-independence period, the Indian government officially declared more than 50 million people as 'scheduled tribes' in the country (Pfeffer 1997). In Wayanad, the tribal population represents 17% of the total population of the district, and is the largest tribal population in the state of Kerala (Josephat 1997). There are five dominant tribal groups - Kurichiya, Kuruma, Paniya, Adiya and Kattunaikka - and seven minor communities (Kumar *et al.* 2003). The present study includes Paniya and Kuruma tribes and non-tribal rural communities pertaining to the lower economic class.

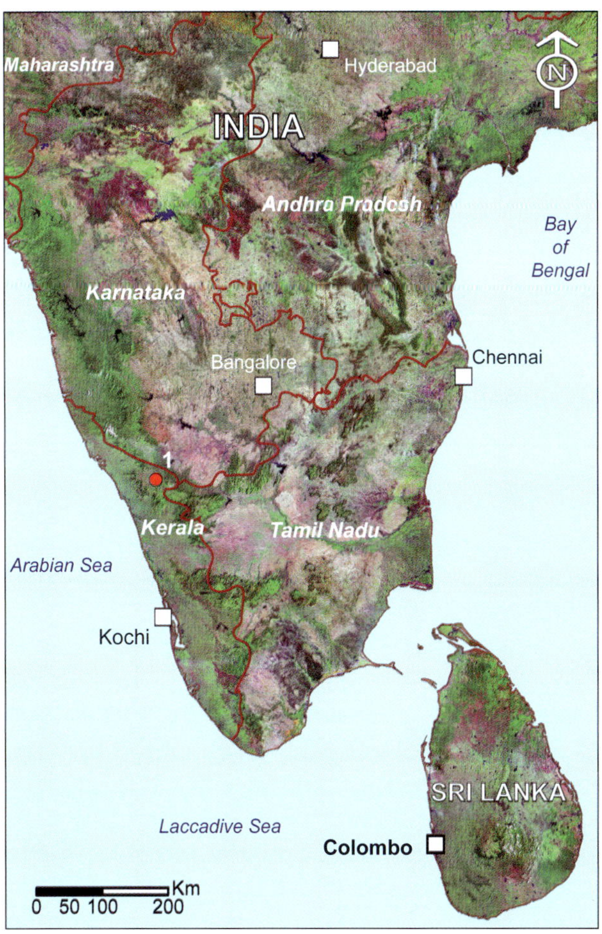

Fig. 8.22. Location of Wayanad (1), Western Ghats, India. Map: R. Lugon, J.-C. Loubier and A. Chevalier.

The Paniya, a landless tribe, constitutes 46% of the total tribal people in Wayanad and is the largest scheduled tribe in Kerala. Most of them depend upon wage labor in the paddy fields and on farms of the landowning classes. They consume the highest quantity of wild food plants (152 species) in relation to the other socio-cultural groups that have been studied in the area and are considered 'famed wild leaf eaters' (Narayanan and Kumar 2007). The Kuruma are a landowning tribe that constitutes 14.6% of the tribal population in Wayanad. In comparison to tribals, non-tribals are better off economically and have more capacity to purchase food in markets. Their children attend the same schools as tribal children do, having the opportunity to interact with them.

Methods

Fieldwork was conducted in 2004 with a research population of 66 children from seven to 14 years old. From a local point of view, 'children' above 15 years of age ('teens') are not considered 'children' any more. Morover, according to Ruddle's (1993) findings, after this age children have already finished their 'training' in knowledge acquisition about natural vegetation. The sample was stratified according to socio-cultural group (Paniya, Kuruma and non-tribal rural communities), sex and age.

Both qualitative and quantitative data gathering and analysis tools were used. Research instruments included interviews with closed-ended and open-ended questions, carried out separately to keep each informant from influencing the others' answers. The most sensitive questions came at the end of the interviews and were asked carefully; sometimes more than once from a different angle in order to identify contradictory answers and attitudes. Nevertheless, some children did not answer some questions, especially those related to their reasons for finding 'important', liking and disliking plants, for instance: *Why do you think wild food plants are important, why do you like or dislike them, and why do you dislike that other people see you collecting or consuming wild food plants?*

A list of specific topics was designed *a priori* for each interview section, but the process was flexible and probing was used to permit exploration of the topics depending upon the experiences and perceptions of the informants. Interviews were pretested with informants that were not later represented in the sample, in order to determine whether the questions generated the desired information (Bernard 2002). The experience showed that the best way to approach people from a specific socio-cultural group is through those who belong to the same group. For instance, Prasad (Kuruma, 17 years old) was asked to interview only tribal children who appeared to be very comfortable with him. This presumably encouraged them to provide more valid answers. The interviews in general were conducted either in Malayalam, the official language of the State, or in tribal dialect (Paniya and Kuruma).

Wild Food Plant Importance

Most children (94%) consider wild plants to be very important as food, one child considered them to be moderately important, and three children considered them not to be important. Of the 55 valid answers, 83% of children explained why they think wild food plants are important. Most children emphasised that these plants are 'healthy' (71%), which was the most important reason across all socio-cultural groups. Consensus drops off sharply after that, where the second most frequent reason given (expressed by 27% of children) was that wild food plants are 'tasty'. In individual interviews children always referred to taste when explaining why they like a specific plant species, kind of plant or why they prefer a type of food. Twenty percent of children consider wild food plants to be 'fresh', 13% consider wild food plants as 'medicine', 9% said that they 'contain no chemical residues' (usually referring to pesticides and fertilisers), 9% regard them as 'nutritious', 6% as 'less expensive', 2% said that they help them to 'resist diseases', and, finally, 7% gave other reasons (Fig. 8.23). The fact that children talk about chemical residues in plants (usually referring to pesticides) is, in itself, probably attributable to the fact that the conversion from paddy fields to banana plantations is leading to over-application of agrochemicals (Narayanan *et al.* 2004).

There were no significant interpretable differences between the reasons children gave for finding wild food plants to be important according to social-cultural group or sex. However, older children had a larger number of value categories which shows that they use different terminology than younger children. Categories such as 'taste' and 'fresh' are more important for younger children (9–11 years old), whereas the belief that wild food plants are not contaminated with agro-chemical residues and are less expensive are only important for children 12 years old and older. Also, younger children consider wild food plants as 'medicine', while older children also regard the consumption of wild food plants as a way to 'resist disease'.

Wild Food Plant Preferences': 'Likes' and 'Dislikes'

All children said that they like [some] wild food plants and most of them gave the reasons why they like them (only two children did not provide any reason). It is important to note that some of the statements that children gave referred to a specific type of wild food plants, or to specific species, while others referred to wild food plants in general. The reasons why children 'like' wild food plants were grouped into value categories. Some of these value categories were also found when analysing wild food plant 'importance'. The category 'taste' is the most frequent reason, mentioned by 88% of the children. The second value category is 'good for health', mentioned by 34% of the children. This is followed by 'appearance' (22%), which encompasses the sub-categories shape, size, beauty and color. Other categories are their role as 'medicine', 'availability', 'appear in clusters' (referring to the fact that some species grow aggregated in groups of plant individuals), 'ease to collect or prepare', 'texture', 'smell' and 'nutritional value', among others (Fig. 8.24).

Almost half of the children (44%) reported that there are wild food plants that they dislike. Most explained why they dislike them, either referring to these plants in general or to some species (24 valid answers), and the reasons they provided were also grouped into value categories (Fig. 8.25). The most relevant reasons for disliking certain wild food plants are their 'bitterness' (42%) and 'to have no taste' (25%), which were grouped together under the category 'taste'. Some 25% of the children said

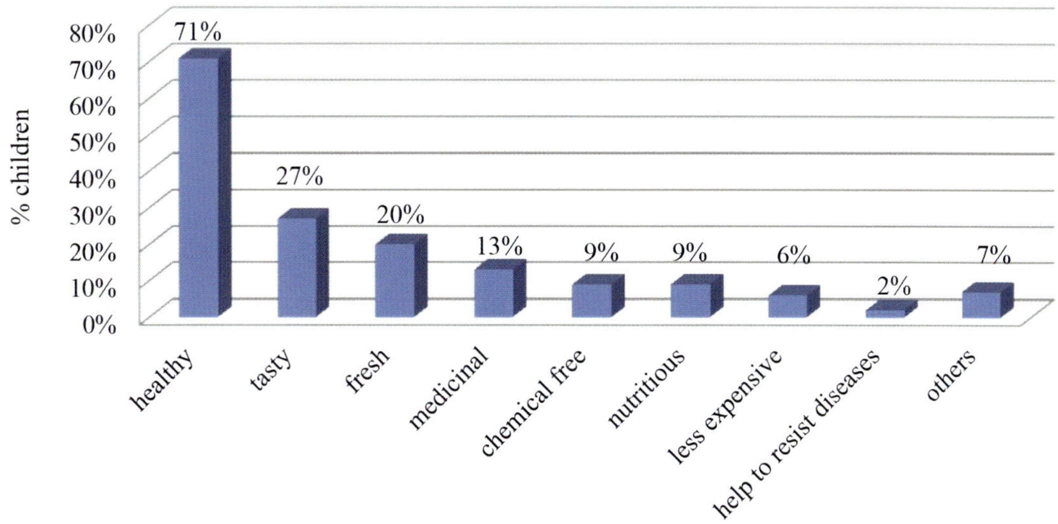

Fig. 8.23. Children's reasons for considering wild food plant important (n=55).

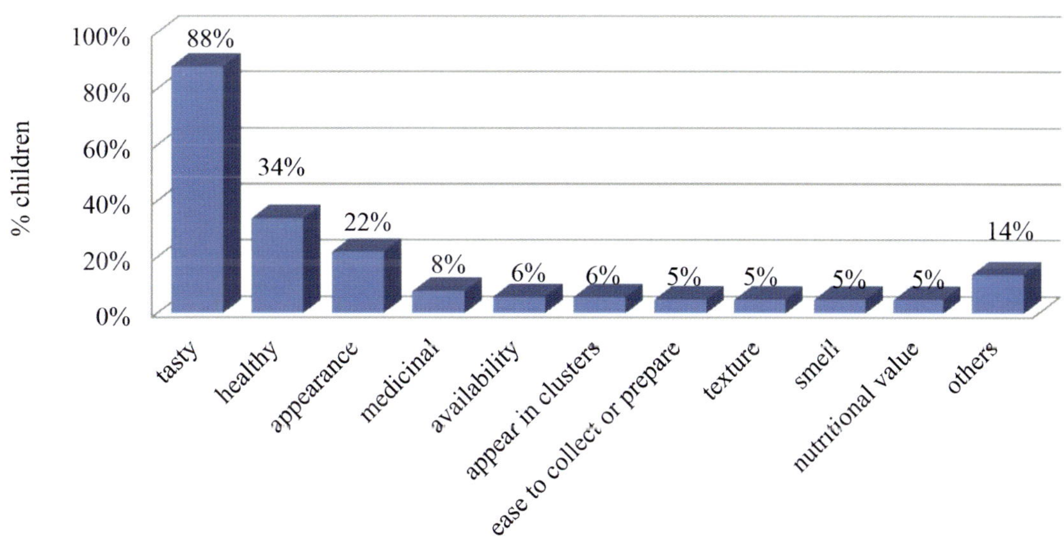

Fig. 8.24. Reasons children like wild food plants (n=64).

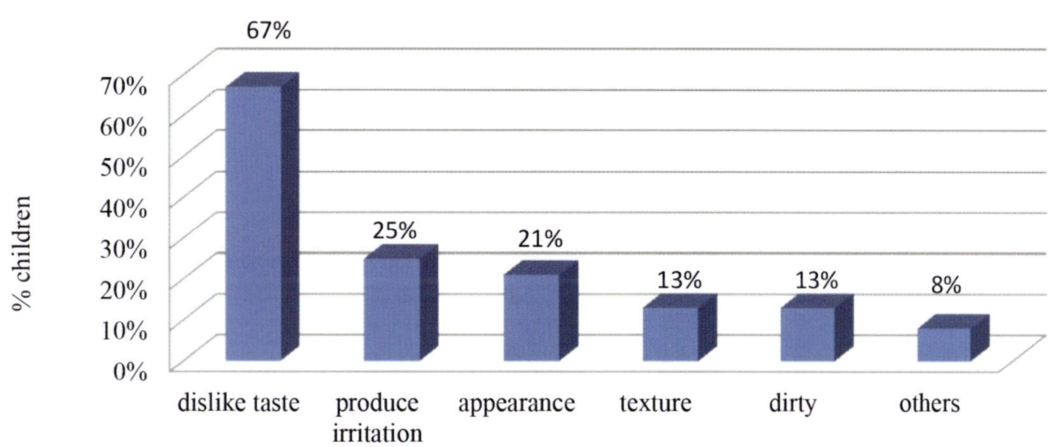

Fig. 8.25. Reasons children dislike wild food plants (n=24).

that certain plants produce 'irritation', including 'itchiness' and the presence of 'spines'. Some 21% said that they 'look bad'; for example, two children explained that they 'dislike the plants' colour' and that 'their leaves are full of holes', which was classified as 'appearance'. The next category is 'texture' (13%); for instance some children said certain plants are 'not fully soluble', 'too slippery' and 'sticky'. Some children also stated that wild food plants are 'dirty' (13%).

Perceptions of Others' Attitudes Toward Wild Food Plant Collection and Consumption

Children were asked if they dislike having other people see them collecting wild food plants. Not everybody replied to this question (58 valid answers). Fifty five percent said that they do not care if others see them, while 45% dislike it (mainly tribal children). More than three fourths of the children explained why they dislike having other people see them collecting wild food plants (20 valid answers). Social stigma was clearly reflected on the reasons provided (Fig. 8.26). The most common reason given was 'I feel ashamed' (35%), followed by 'it is low caste food' and 'others eat good quality food' (25%). Children also mentioned 'others make fun of me' (20%) and 'it is not good for girls to collect wild food plants' (15%). Certainly tribal children gave reasons related to caste, ethnic differences and discrimination. Older children expressed more clearly the problem of social stigma in their statements compared to younger children. In general, girls dislike more than boys to be seen collecting wild food plants and 'shame' was mostly expressed by tribal girls.

'I feel ashamed' was only mentioned by tribal girls using the following expressions: 'I feel ashamed because *they* [other people] scold and insult us'; 'I hide what I have collected and wait for others to go away before I continue collecting'; 'I hide myself'; 'I feel ashamed because they will tell everybody that I am collecting wild food plants'; 'I hide *it* [plants], I feel ashamed [because] they are rich people and will insult us'; 'the problem is not to collect flowers, the problem is to collect wild food plants that make us feel inferior'; 'they insult us in front of others, I will not go to collect again'. The statements 'it is low caste food' and 'others eat good quality food', provided by mainly tribal children, were explained with expressions such as 'ours is low caste food, for people who have no money'. Girls said 'others eat good quality food', and explained that 'good quality food' refers to fish, meat and food obtained in the market by 'rich people'. 'Others make fun of me' was also mentioned by tribal children. The statement 'it is not good for girls to collect wild food plants' mainly related to gathering in the forest, and may refer to gender norms and social status of non-tribal children, but it could also be understood that it is

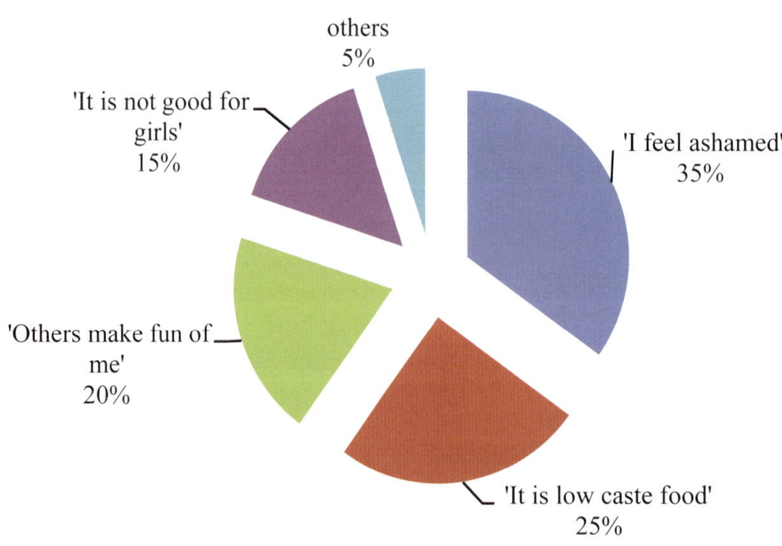

Fig. 8.26. Reasons children dislike being seen by others when collecting wild food plants (n=20).

considered to be dangerous for girls to go alone to the forest.

Children were asked if they would invite their friends to eat with them when their mother prepares wild food plants at home. Most children replied to this question (62 valid answers). Seventy three percent would invite their friends, while 27% would not wish to invite them. Mostly non-tribal children (39%) and Paniya children (36%) would dislike inviting their friends to consume wild food plants. Only two Kuruma children would not wish to invite anyone. Four children did not explain why they would not invite their friends to eat wild food plants (13 valid answers). Some – but not all – children gave reasons related to social stigma such as 'my friends do not take food from tribals', 'my friends' parents would not allow them to come home', or 'my friends would make fun of me'. There are no differences in relation to age and sex.

Discussion and Conclusions

Children express simultaneously positive and negative values related to wild food plants and their related practices. On the one hand, most children think these plants are important, emphasising that they are 'healthy', and all children said that they like [some] wild food plants mainly because of their 'taste'. On the other hand, almost half of the children reported that they dislike [some] wild food plants, and, in addition, they dislike having other people see them collecting wild food plants. Whereas disliking plant species *per se* was mainly related to 'taste', producing 'irritation' and 'appearance', disliking being seen collecting wild food plants was, to a major extent, related to social stigma.

Accelerated processes of modernisation and acculturation are leading to the change of cultural values associated with wild food plant use. Despite the fact that wild food resources play a vital role in tribal and non-tribal people's diet in Wayanad, their collection and consumption is increasingly associated with poverty and social marginality, as opposed to the consumption of socially accepted 'modern' (often exotic) foods. For instance, children feel ashamed if somebody sees them collecting these plants, immediately hiding themselves or what they have collected, because wild food plants are 'low caste food' and 'others make fun' of them. The Paniya children, specially the girls, are those that express more feelings of inferiority and shame, which is remarkable given the fact that the Paniya are the most dependent on wild food plants for satisfying their basic food needs. They are the poorest group and this poverty is apparently what makes them feel shame in comparison to other children.

Social status and cultural identity are reflected in the relations between people and plants, considering that rural societies are undergoing a continual process of change. The social context where collection and consumption occurs also influences a person's choice and attribution of value to food species. Studying wild food plant evaluation provides a means to understand the social processes embedded in a very heterogeneous society that is characterised by a strong hierarchical system. The 'untouchability' of scheduled tribal groups, although not as strongly expressed through explicit social rules as decades ago, is now implicitly observed in behavior (*e.g.* 'tribal food is dirty', 'my friends do not take food from tribals'). In this, the importance of what people of the same or higher socio-cultural group (hierarchically) are going to 'think about them' – emphasised in this study as the perceptions of others' social attitudes toward wild food plants with respect to collection and consumption – is essential for their social acceptance, relations and status, which is also expressed through what they eat.

It has been recognised that wild food plants, which constitute a major source of nutrients such as vitamins, minerals and secondary metabolites, are crucial for ensuring food security and dietary diversity of rural societies across the globe (Heywood 1999; Ogle 2001). However, the stigmatisation of wild food plant species restricts their access to local families. This might have negative consequences for the food security and nutritional diversity of these families, in particular the poorest and most vulnerable households such as the Paniya. In this regard, as explained by Cruz-García (2005), the implementation of extra-curricular educational programs that motivate children to re-valorise their local culture and resources while stimulating the interaction of tribal and non-tribal children, have been shown to be successful in counter-acting social stigmatisation of wild food plant species.

8.6. MORE THAN SIMPLY FALLBACK FOOD? SOCIAL CONTEXT OF PLANT USE IN THE NORTHERN GERMAN NEOLITHIC

Wiebke Kirleis and Stefanie Klooß

Introduction

People use plants in different social contexts. Past plant assemblages may act as 'social markers' and give insight into a society's hierarchical structure. For the area of Bibracte (modern Mont Beuvrey in France), it was the diversity of available and used plants that indicates a high social status (Durand and Wiethold, Chapter 8.4). In this paper, people's plant choices are discussed for different social contexts in the northern German Neolithic based on the first results of new archaeobotanical analyses on charred plant remains. Of particular interest is the role of gathered and domestic plants. Does plant gathering simply compensate for harvest failures? Do gathered plants just serve as fallback foods? If symbolic meanings of plants are considered, there might be more: For the northern German Neolithic, a *rite de passage* is discussed, in the case when gathered plants dominate the plant assemblage of a megalithic burial.

A separation of different spheres of plant use (economic versus ritual) seems reasonable to reconstruct social activities and social space in the Neolithic. But, as we are dealing with charred plant remains, our hypothesis has to be examined in light of the formation process (van der Veen and Jones 2006). Thus, we set up a research orientation and test whether the plant selection in question can be explained as a result of people's plant choices in different social contexts. We apply a structural approach to site types of varying function: domestic sites and ritual sites, such as burials and enclosures, to gain a deeper understanding of the shaping of Neolithic social space.

Two late Early Neolithic to Middle/Late Neolithic archaeological sites situated in modern Schleswig-Holstein are compared: a settlement site and a megalithic tomb. Differing assemblages of charred plant remains are observed when comparing the two site types. What exactly do we observe? In the settlement, we mainly see crop plants. In contrast, the tomb reveals mainly gathered plants – crop plants are underrepresented, if present at all. First of all, we state that plant remains from settlements give insight into daily food production and into consumption; second, in tomb sites, we see either burial gifts or remains from the burial ritual, for example, a meal for the dead or the waste from a community celebration related to the burial. Our focus is on the absence (or near absence) of crop remains in the burial context and the symbolic meaning of gathered plants in the burial ritual with all its social implications.

First Farmers in Northern Germany

In northern Germany, the first settlers practicing agrarian food production belong to the Funnel Beaker Culture (FBC) starting *c.* 4100 cal. BCE (Behre 2008a; Dörfler 2001; Hartz *et al.* 2000; Hoika 1993; Kirleis *et al.* 2011; 2012; Müller 2009a). In contrast to central Europe, we observe a delay of about 1400 years in the uptake of husbandry practices in the north: the people of the Central European Linearband ceramics (also know as LBK cultures) had been living on the basis of agriculture since as early as 5500 cal. BCE. The distribution of the FBC ranges from the eastern Netherlands via northern Germany to southern Scandinavia and western Poland. Chronologically, the northern German Neolithic is subdivided into the Early and Middle Neolithic with the FBC, the Younger Neolithic with the Single Grave Culture (SGC), and the Late Neolithic. Findings of the FBC cover the time span from 4100–2800 BCE (Fig. 8.27, Müller *et al.* 2010). In the Late Neolithic, Bell Beaker influences (BBC) and dagger assemblages were common in the region (Rassmann 1993; Vandkilde 2007).

As we can learn from the northern German evidence, the Neolithisation (in an economic sense) is a process of adaptation that lasted for several generations. The first phase is characterised by small-scale clearings around settlements mainly along the coast and a few inland sites. Agricultural activity that is added to Mesolithic practices is shown by evidence for domestic animals like goat, sheep and cattle and for cereals (*e.g.* Fischer 2002; Hoika 1993). The effect on the landscape remains on a low level. It is not before the late Early Neolithic that husbandry practices change. A rapid increase of ribwort plantain (*Plantago lanceolata* L.) is observed in pollen records; this is a plant associated with soil perturbation due to agriculture occurring on fallow land (Behre 1981). Pollen grains of the cereal type and further human indicators are traceable as well. This 'Neolithic landnam' gives notice of the beginnings of the formation of a cultural landscape during the late Early Neolithic, around 3500 cal. BCE (Dörfler *et al.* 2012; Iversen 1941; Kalis and Meurers-Balke 1998; Kirleis *et al.* 2011; 2012; Lütjens and Wiethold

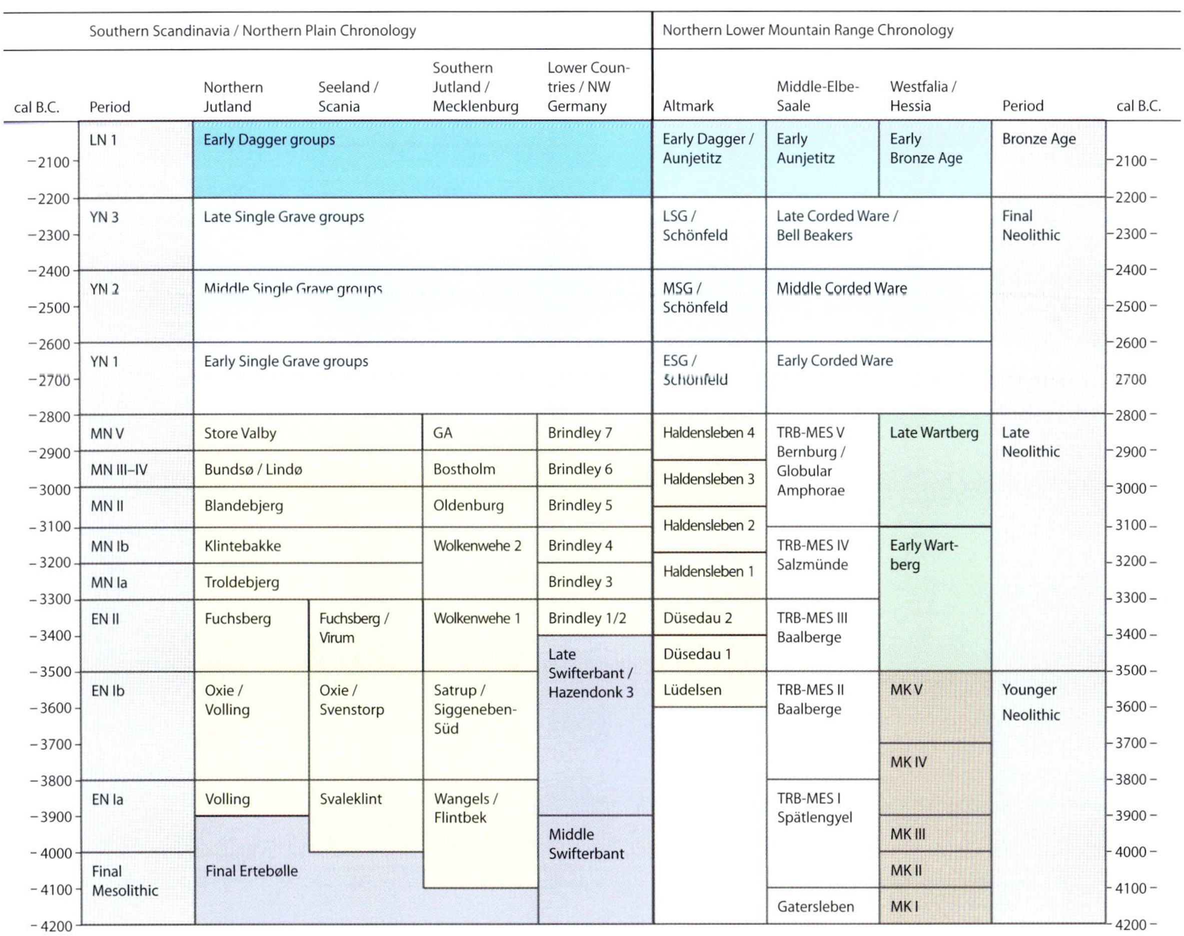

Fig. 8.27. Periodisation of the Northern European Neolithic (Müller *et al.* 2010).

Abbreviations: LN Late Neolithic – YN Younger Neoltihic – MN Middle Neolithic – EN Early Neolithic – LSG Late Single Grave groups – MSG Middle Single Grave groups – ESG Early Single Grave groups – E Early – FB Funnel Beaker – MK Michelsberg – TRB–MES Funnel Beaker Middle Elbe Saale - GA Globular Amphorae.

	Rastorf LA 6 (EN II)		Flögeln-Eekhöltjen (MN II-IV)		Oldenburg-Dannau LA 191, charred remains (MN II-IV)		Wangels LA 505, charred material, without handpicked material (MN V)	
Number of analysed soil samples	12		111		10		12	
	n	%	n	%	n	%	n	%
Cultivated plants								
Hordeum vulgare L., hulled			501	19.87	1	0.01		
Hordeum vulgare L., naked	17	22.08	51	2.02	5594	67.58	49	7.79
Hordeum vulgare L., rachis segments					4	0.05	84	13.35
Triticum monococcum L.			11	0.44	7	0.08		
Triticum monococcum L., glume bases					4	0.05		
Triticum dicoccum Schübl.	3	3.90	462	18.33	1887	22.80	17	2.70
Triticum dicoccum Schübl., glume bases	5	6.49			699	8.45	410	65.18
Triticum aestivum L./*T. durum* Desf.	7	9.09	7	0.28			2	0.32
Triticum aestivum L., rachis							6	0.95
Triticum sp.			47	1.86				
Cerealia indeterminata	12	15.58	1378	54.66			24	3.82
Papaver somniferum L.							2	0.32
Sum	44	57.14	2457	97.46	8196	99.02	594	94.44
Gathered plants								
Corylus avellana L.	13	16.88	45	1.79	6	0.07	1	0.16
Chenopodium album L.			1	0.04			4	0.64
Daucus carota L.	1	1.30						
Iris pseudacorus L.							1	0.16
Rubus fruticosus L.	2	2.60	4	0.16			1	0.16
Rubus idaeus L.					2	0.02		
Schoenoplectus lacustris (L.) Palla					1	0.01	1	0.16
Sum	16	20.78	50	1.98	9	0.11	8	1.27
Sum cultivated and gathered plants	60	77.92	2507	99.44	8205	99.13	602	95.71
Weeds								
Apiaceae p. p.					1	0.01		
Bromus secalinus L.					3	0.04	2	0.32
Descurainia sophia (L.) Webb ex Prantl							1	0.16
Echinochloa crus-galli (L.) Beauv.			1	0.04				
Galium aparine L.	5	6.49						
Lamiaceae	1	1.30	1	0.04				
Panicum miliaceum L.			1	0.04				
Phleum sp.							15	2.38
Poaceae p. p.					57	0.69	5	0.79
Polygonum convolvulus L.			5	0.20	1	0.01		
Polygonum persicaria L./*P. lapathifolium* L.			1	0.04	4	0.05	1	0.16
Rumex sanguineus L.-type					1	0.01		
Solanum dulcamara L.					5	0.06		
Spergula arvensis L.			1	0.04				
Stellaria media (L.) Vill.			1	0.04				
Vicia sp.	1	1.30	3	0.12			2	0.32
Sum weeds	7	9.09	14	0.56	72	0.87	26	4.13
Others								
Alnus glutinosa (L.) Gaertner							1	0.16
Tilia sp.	10	12.99						
Sum others	10	12.99					1	0.16
Sum	77	100	2521	100	8277	100	629	100

Fig. 8.28. Charred plant remains from four northern German Late Early to Middle Neolithic (EN II–MN V) settlement sites (Behre and Kučan 1994, 26–30; Kroll 1981; 2001; 2007; Steffens 2009, 28; Zimmermann 2008).

1999; Nelle and Dörfler 2008). The archaeological record is supplemented with a new kind of feature: megalithic tombs as an impressive feature in the landscape, as well as plough marks and cart tracks that indicate technical innovations in agriculture (Mischka 2010; Steffens 2009; Hoika 1981). The beginning of the use of the ard most probably was accompanied by a gradual establishment of a new division of labour. Although social structures of the first farming communities are difficult to discern, social changes are expressed through the erection of thousands of megalithic tombs in the short period of some 300 years (*c.* 3600–3300 cal. BCE, see Furholt and Müller 2011; Klassen 2004; Mischka 2011; Mischka and Demnick 2011; Müller 2009a and b; 2011). The new burial habits show a different perception of identity and of landscape. In addition, the collective burial rituals sustain social stability in times of increasing differentiation (Müller 2010).

Archaeobotanical analyses on megalithic tombs are rare. There is only one archaeobotanical report available for northern central Germany from a non-megalithic tomb dating to 3440–2840 cal. BCE (Hellmund 2008). Nonetheless, it is worthwhile applying a systematic sampling strategy to tease out the information on plant use in a ritual site, even if the find densities hardly exceed one find per litre - the more so as low find concentrations are generally observed for the northern European Neolithic (Regnell and Sjögren 2006b).

With regard to domestic sites, only four modern archaeobotanical investigations on soil samples covering the time span from the Late Early Neolithic to the Middle Neolithic are available (Fig. 8.28; Behre and Kučan 1994; Kroll 1981; 2001; 2007; Steffens 2009, 28; Zimmermann 2008). They show that the main crop plants in the FBC are naked barley (*Hordeum vulgare* L., naked) and emmer (*Triticum dicoccum* Schübl.). In Flögeln, hulled barley (*Hordeum vulgare* L., hulled) is more important than naked barley. Evidence for einkorn (*Triticum monococcum* L.) and naked wheat (*Triticum aestivum* L./*T. durum* Desf.) is sparse. Only in Rastorf LA 6 does naked wheat occur to a somewhat greater extent. Seeds of opium poppy (*Papaver somniferum* L.) were found in relatively high numbers in the wet layers of Wangels LA 505 (Kroll 2007).

In general, there are two ways of getting access to food plants. Either food plants are grown as crops or they are gathered in the vicinity of the settlement. If we separate the plant assemblages into crop plants and gathered plants, the data from the four different settlements shows similar tendencies: In Rastorf LA 6 (EN II) cultivated plants represent 57% of the plants. In Oldenburg-Dannau LA 191 (MNII/III–IV) and Flögeln (MN II–IV) almost all food plant remains (98 to 99%) stem from cereals. The situation at Wangels LA 505 (here: MN V) is slightly different because the plant remains are partly charred and partly waterlogged. If the waterlogged material is excluded, gathered plants hardly occur (about 1%). This relation is turned upside down if the waterlogged seeds and fruits for Wangels LA 505 are included (see Kroll 2001 for a complete taxa list). Thus, domestic seeds and fruits show an average proportion of 7% and gathered plants one of 93%. However, if (water-logged) chaff is considered, we again end up with a balanced ratio that indicates cereal processing as part of economic and social activities. To sum up, the archaeobotanical records for the settlement sites mainly reveal crop plants. This is not astonishing because cereal processing, food production and food consumption take place within the domestic sites.

New Sites and Archaeobotanical Results

Here we present the two late Early to Late Neolithic sites investigated and the results of the analyses of the charred plant material (Fig. 8.29).

The late Early to Middle Neolithic Settlement Oldenburg LA 77 in Schleswig-Holstein

The late Early to Middle Neolithic (EN II-MN I) settlement Oldenburg LA 77 was located on a former island in a fjord, in the south of the Baltic Sea. The settlement was used between *ca.* 3500 and 3100 cal. BCE. The main archaeological feature on the sandy ridge is an amorphous cultural layer of about 30 cm in depth. In addition, there are several pits, a sunken floor and a well that was used as a rubbish pit later on. Around 3000 cal. BCE, the fjord and the island were cut of from the Baltic Sea as a result of the development of a beach wall. Meanwhile, a change from marine to freshwater conditions occurred in the surrounding environment. The area was revisited in the Younger Neolithic (YN) around 2800 to 2200 BCE (Brozio 2010; 2011).

Fig. 8.29. Map of the sites in northern Germany as mentioned in the text. 1) Oldenburg LA 77; 2) Albersdorf–Brutkamp LA 5; 3) Rastorf LA 6; 4) Flögeln; 5) Oldenburg–Dannau LA191; 6) Wangels LA 505. Map: R. Lugon, J.-C. Loubier and A. Chevalier.

The sampling strategy for archaeobotanical analyses is based on quadrant- and cut-wise extraction of 10-litre soil samples. As the site has been sampled intensively and analyses are still ongoing, here we shall show the results from 16 soil samples from pits and one from a sunken floor as part of a former house (a total of 170 litres of soil) as a first insight into results from an important site. The samples were floated using a sieve with 300 µm mesh width to recover the charred plant remains. The samples contained 1009 charred plant remains altogether. About three-quarters of the material originates from the sunken floor sample. The mean find concentration of about 11 finds per litre of soil is high, if compared to other Neolithic sites in northern Europe (Bogaard and Jones 2007; Greig 1991, 300; Regnell and Sjögren 2006a and b).

Cultivated plants dominate the plant assemblage (Fig. 8.30 and Fig. 8.31). Cereals contribute 98% to the charred remains, 1.4% belong to weed species and 1.2% from gathered plants. The main component of cereals is naked barley. It is followed by emmer, far behind. Single grains of einkorn present a third domestic plant. It is mainly the fruits of the cereals that are documented. Threshing remains such as rachis fragments and glume bases occur to a lesser extent. Weeds are scarce. Poaceae, in particular annual bluegrass (*Poa annua* L.), represent the most finds. In addition, false cleavers (*Galium spurium* L.), vetches (*Vicia* sp.) and sedges (*Cyperaceae*) are present. The spectrum of gathered plants consists of three species: common hazel (*Corylus avellana* L.), brambles (*Rubus fruticosus* L.) and lambsquarters (*Chenopodium album* L.). The latter, although often classified as a weed species, is integrated into the ensemble of gathered plants because its seeds and leaves are suitable as food. The sunken floor material consists of pure cereal remains (738 fruits

Cereals	EN II-MN I [n]	EN II–MN I [%]
Hordeum vulgare L., naked	498	49.4
Hordeum vulgare L.	64	6.3
Hordeum sp., rachis fragments	4	0.4
Triticum monococcum L.	2	0.2
Triticum dicoccum Schübl.	35	3.5
Triticum dicoccum Schübl., glume bases	48	4.8
Triticum sp.	28	2.8
Cerealia indeterminata	304	30.1
Sum cereals	983	97.4
Gathered plants		
Corylus avellana L., pericarp fragments	8	0.8
Rubus fruticosus L.	2	0.2
Chenopodium album L.	2	0.2
Sum gathered plants	12	1.2
Weeds		
Vicia sp.	1	0.1
Galium spurium L.	1	0.1
Poa annua L.	8	0.8
Poaceae p. p.	3	0.3
Cyperaceae	1	0.1
Sum weeds	14	1.4
Total sum	**1009**	**100**

Fig. 8.30. Charred plant remains from the late Early to Middle Neolithic settlement site Oldenburg LA 77 (EN II–MN I).

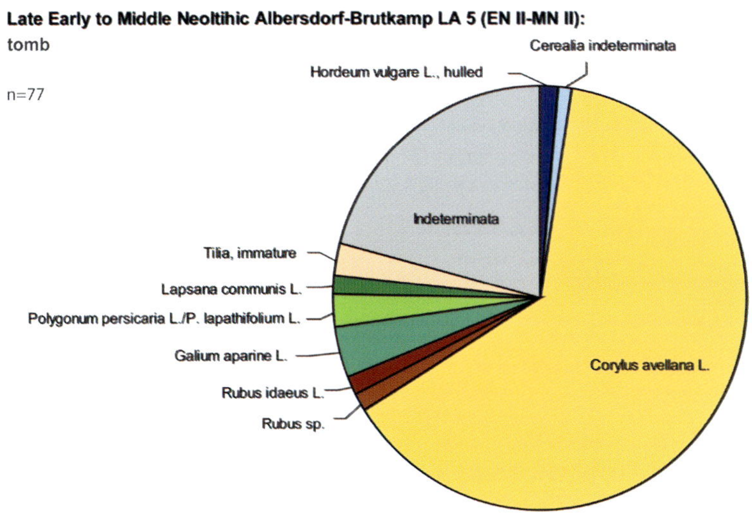

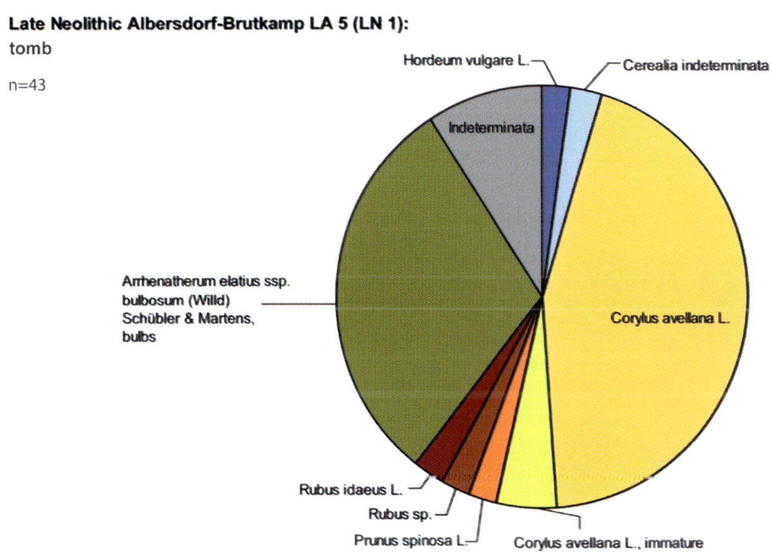

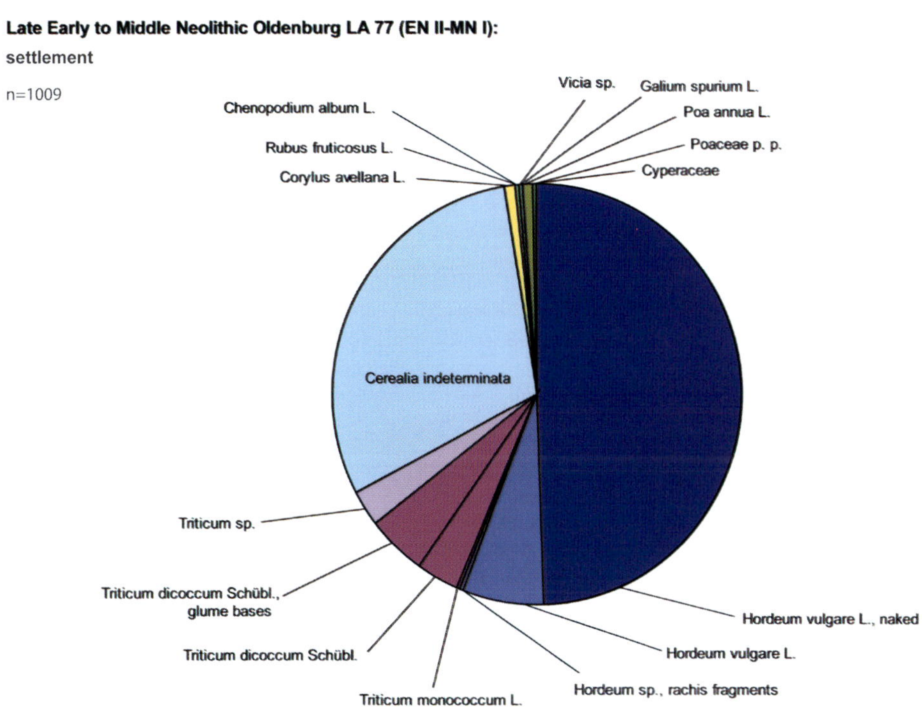

Fig. 8.31. Comparison of charred plant assemblages of the settlement Oldenburg LA 77 and the megalithic tomb Albersdorf-Brutkamp LA 5.

dominated by naked barley (492), four barley rachis fragments and 24 emmer glume bases). The remaining material from pits has to be interpreted as scattered left-overs from daily food preparation and consumption.

The Late Early to Middle Neolithic Megalithic Tomb Albersdorf-Brutkamp LA 5 in Schleswig-Holstein

The megalithic tomb Albersdorf-Brutkamp LA 5 is a huge polygonal dolmen with an entrance passage (Fig. 8.32). The periphery of the stone covering the grave chamber measures about ten metres and it weighs more than 20 tons. Although the grave chamber has not yet been excavated, it is most possible that its floor was paved with pebble. The pebble layer was covered by burnt flint stone fragments (Dibbern and Hage 2010; Kelm 2006, 78). The tomb expresses collective burial rituals of the late Early to Middle Neolithic (EN II-MN II). In the Late Neolithic (LN 1), flint daggers were deposited in the grave mound. In 2009, the southeastern entrance area of the tomb was excavated. Twelve quadrants were cut in layers of 20 cm (Fig. 8.33). The stratigraphy is shown in Fig. 8.34. The topsoil layer is followed by the covering layer of the tomb. Red-brown sand with pebble and flint is situated underneath the covering layer; below the red-brown sand, a layer of light yellow in-situ sand was found. A stone kerb surrounds the grave chamber and the cairn that was most probably cleared in Late Neolithic times. Then, its content was deposited on the original cairn within the outer stone kerb (Dibbern and Hage 2010).

For archaeobotanical analyses, ten litres of soil were sampled per quadrant and cutting layer. In addition, 20-litre soil samples were taken from special

Fig. 8.32. Entrance area of the megalithic tomb Albersdorf–Brutkamp LA 5 (photograph: H. Röhrs).

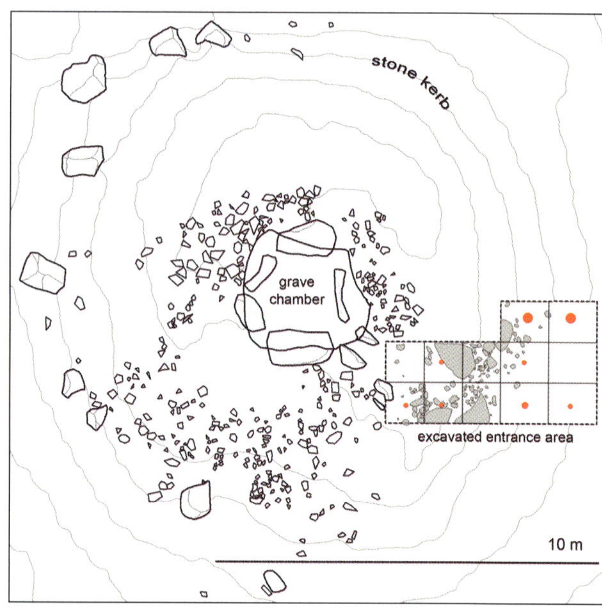

Fig. 8.33. Excavation area of the megalithic tomb Albersdorf–Brutkamp LA 5 and distribution of charred bulbs of false oat grass (Dibbern and Hage 2010, supplemented).

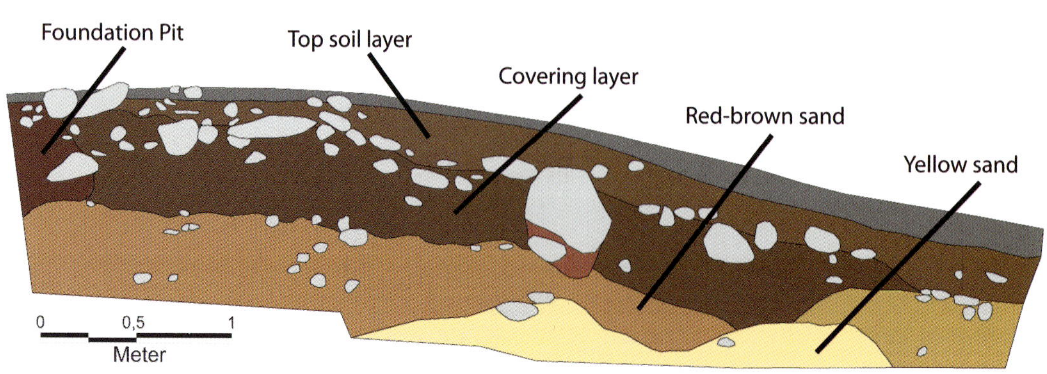

Fig. 8.34. Profile of the megalithic tomb Albersdorf–Brutkamp LA 5 (H. Dibbern and F. Hage).

features. Altogether, 69 soil samples (690 litres of soil) were floated, again using a sieve with 300 µm mesh width to recover the charred plant remains. A total of 140 charred plant remains were discovered in 51 samples. The mean find concentration is about 0.2 finds per litre of soil. Such poor find concentrations are common for the Northern European Neolithic (Bogaard and Jones 2007; Greig 1991, 300; Regnell and Sjögren 2006a and b).

Charred plant remains from the tomb can be separated into a late Early to Middle Neolithic (EN II-MN II) and a Late Neolithic (LN 1) assemblage (Fig. 8.27, Fig. 8.31 and Fig. 8.35). Taking into account the main plant groups, the composition for both phases is similar. Crop plants are hardly present (about 2%) and wild, collected plants dominate both plant assemblages with about 60%. However, weed species occur in the Middle Neolithic only.

For the late Early to Middle Neolithic, two single cereal grains were detected. One could be identified as naked barley. Three different weed species were present amounting to 7% of the plant assemblage: pale smartweed (*Polygonum lapathifolium* L.), cleavers (*Galium aparine* L.) and nipplewort (*Lapsana communis* L.) Gathered plants are dominated by common hazel (*Corylus avellana* L.) with 58%. Two charred seeds of the genus *Rubus* (including raspberries, blackberries, dewberries, etc.) complement the gathered plants spectrum. In addition, two lime seeds, an inflorescence axis of alder and some vegetative plant remains were found.

Late Neolithic evidence for cereals is based on one barley caryopsis. Due to a bad state of preservation a second caryopsis is indeterminable. None of the common Neolithic weed species was found. The amount of gathered plants amounts to 67% of the

Cereals	EN II–MN II [n]	EN II–MN II [%]	LN 1 [n]	LN 1 [%]
Hordeum vulgare L., hulled	1	1.2		0.0
Hordeum vulgare L.		0.0	1	2.3
Cerealia indet.	1	1.2	1	2.3
Gathered plants		0.0		0.0
Corylus avellana L., pericarp fragments	49	58.3	19	44.2
Corylus avellana L., immature fruit		0.0	2	4.7
Prunus spinosa L.		0.0	1	2.3
Rubus sp.	1	1.2	1	2.3
Rubus idaeus L.	1	1.2	1	2.3
Weeds		0.0		0.0
Galium aparine L.	3	3.6		0.0
Polygonum persicaria L./*P. lapathifolium* L.	2	2.4		0.0
Lapsana communis L.	1	1.2		0.0
Others		0.0		0.0
Tilia, immature seed	2	2.4		0.0
Alnus, inflorescence axis	1	1.2		0.0
Arrhenatherum elatius ssp. *bulbosum* (Willd) Schübler & Martens, bulbs		0.0	13	30.2
Bud	2	2.4		0.0
Spur shoot	3	3.6		0.0
Straw	1	1.2	1	2.3
Fragment of gymnosperm needle	1	1.2		0.0
Indeterminata	16	19.0	4	9.3
Sum	84	100	43	100

Fig. 8.35. Late Early to Middle Neolithic and the Late Neolithic charred plant remains from the megalithic tomb Albersdorf–Brutkamp LA 5 (EN II–MN II and LN 1).

general plant assemblage. Again, the hazel (*Corylus avellana* L.) is dominant (49%). The genus *Rubus* is present. Furthermore, a charred kernel of sloe was detected. There are also important finds of bulbs from tuber oat-grass (*Arrhenatherum elatius* ssp. *bulbosum* (Willd) Schübler & Martens), listed under the heading of 'others'. The thirteen bulbs represent 30 % of the charred plant remains. The fact that the bulbs are charred hints at past fire-use activity of some kind. There are three possible interpretations for the presence of the bulbs: 1) the cover layer of the tomb consists of sods that originate from an area in the surroundings that underwent fire clearance beforehand or, more likely, 2) the tomb itself was covered by *Arrhenatherum* that was burnt when the tomb was destroyed or ritually prepared for the deposition of the flint daggers and the bulbs were worked into the soil after this; and a more exciting possibility, 3) the starch-rich, lowest swollen stem internodes were gathered for dietary purposes, combined here with an intentional deposition during Late Neolithic ritual activities. It may be preferable to interpret the presence of *Arrhenatherum* bulbs in grave deposits as being related to intentional deposition in ritual contexts for their high symbolic value (Engelmark 1984; Jensen *et al.* 2010; Preiss *et al.* 2005; Viklund 2002).

Neolithic Plant Assemblages from the Settlement and the Tomb Compared

From the examples of the two investigated sites we observe that, within the vast array of edible plants available in nature, only certain taxa were selected for food. When interpreting the archaeobotanical record, we have to consider that depositional factors and factors related to the preservation of plant remains influence the archaeobotanical sample (see Chapter 2). We have to distinguish several single deposition events at the tomb and deposition of finds covering a larger time span in the settlement as background noise. But it is findings from the sunken floor that increase the find numbers (Fig. 8.30 and Fig. 8.35). In general, in contrast to cereals, a lot of gathered fruits are not carbonised because they do not acquire any treatment with fire, as they are eaten raw. Thus, gathered plants are underrepresented in charred material. However, hazel nutshells are notably overrepresented in charred plant assemblages due to their density and weight. There is also a difference in the way they are handled: nutshells would be discarded while cereal grains would be prepared and eaten. Another reason for nutshell overrepresentation may be functional, if they were intended to be used as fuel for fire (Jones 2000). Other 'waste' from food production has its own function; for example, threshing remains serve as animal fodder (van der Veen and Jones 2006; Jones 2000) and, therefore, hardly occur in the archaeobotanical record of the two sites.

Gathered plants are often regarded as minor substitution foods (Chapter 8.5). However, throughout the Neolithic, collected fruits were a welcome addition to people's daily diet in northern Germany and supplemented it with extra nutrients such as starch, minerals and vitamins. In ritual contexts, they even gained a relevant symbolic meaning. As a prerequisite to assessing the importance of cultivated versus gathered plants, we have to explain our classification of the gathered plant group. We broaden the common definition by adding a classical weed species. Our allocation of common goosefoot (*Chenopodium album* L.) is based on the assumption that it contributes to the daily diet. The compilation of Behre (2008b) shows its relevance in the food supply from the Linearband-ceramic period to modern times. Note, for example, the findings of 54,518 seeds in a pot from the Neolithic lakeshore settlement of Niederwil, Switzerland, which point to the use of goosefoot as a gathered plant and the evidence from the intestines of seven European Iron Age bog bodies (van Zeist and Boekschoten-van Helsdingen 1991; Behre 2008b). Today, in India, the leaves and young shoots of this plant and even the seeds are used in dishes (Board 2004, 146). The gathered plants are collected in the 'wild'. However, it is particularly the immediate surroundings of the settlements that offer suitable habitats for gathered plants, as most of them are light-demanding species like hazel (*Corylus avellana* L.) or raspberry (*Rubus idaeus* L.) that naturally grow at the forest border. Plant gathering takes place in an intermediate space: it is open land where the crops are grown and where animals are grazed on fallow land. Most possibly, this area has been structured by hedges since Neolithic times (Kreuz 1992). Thus, it is a transition zone between the settlement as inner economic sphere and the dense primeval forest. Both archaeological sites, Oldenburg LA 77 and Albersdorf-Brutkamp LA 5, show evidence for plant gathering. However, their importance depends on the sites' context. The hazel nutshells may be overrepresented in the tomb site as mentioned above, but it is the apparent absence of crops which makes the difference.

Social Spheres of Plant Use

In the two Neolithic sites presented, people's social status is difficult to grasp from an archaeobotanical point of view. However, the low number of archaeobotanical finds that are common for the Neolithic in general (one to eleven finds per litre of soil) may indicate a very careful manner of food production. The invention of crop production in the Neolithic probably is accompanied by an appreciation of each single cereal grain. Thus, we might imagine a sphere of cautious cereal processing and food production where wastage of valuable food is minimised.

Although there is an extreme difference in absolute numbers of the charred finds for both examples given here, we can suggest two different social spheres of plant use. The ratio of cultivated to gathered plants for the settlement site Oldenburg LA 77 is cult/gathMN= 81.9. For the megalithic tomb Albersdorf-Brutkamp LA 5 in the late Early to Middle Neolithic, it is cult/gathMN=0.04, and for the Late Neolithic it is cult/gathLN=0.08. Thus, in the settlement, we do get some insight into crop plant production and consumption. It is the cereals that are overrepresented, as threshing remains or waste they are present but rare. The cereals from the sunken floor either got charred accidentally during food preparation or the whole house that contained the cereals burnt down. The corresponding threshing remains of the assemblage show that emmer was stored in spikelets, whereas for barley the whole ears were stored. Other finds such as gathered plant remains occur as additional background information from unspecific pits (Fig. 8.31). On a regional level, for the northern German Early and Middle Neolithic, an overrepresentation of crop plants in settlements is shown for a total of six domestic sites, whereas four tomb sites display the reverse picture with a dominance of gathered plants; if only relevant investigations with >50 seeds/fruits for settlements and >10 seeds and fruits for tombs are considered and chaff remains are disregarded completely (Kirleis *et al.* 2012, Fig. 9).

On a supra-regional level, there has been a long debate on the relevance of crops versus gathered plants in the British and Irish Neolithic, where material from settlement sites and causewayed enclosures was investigated. Studies by Moffett *et al.* (1989) and Robinson (see Hey *et al.* 2003) have revealed a clear dominance of hazel nutshells in the plant spectra at several sites. At the Yarnton Floodplain site in Oxfordshire, gathered plants were interpreted as representing a special collection that may not represent everyday diet. Based on this, a debate about the sedentary or mobile character of Neolithic society was initiated. However, since plant remains from several house structures have now revealed cereal storage (*e.g.* Fairweather and Ralston 1993; Monk 2000), the importance of cereal cultivation is unquestioned. Nevertheless, collected fruits like crab apple, blackberry, sloe and hawthorn have been regularly found in charred plant assemblages. In contrast to the northern German assemblages, the gathered plant spectra are dominated by several hundred to thousands of hazel nutshell fragments, and taphonomic factors must not be overlooked. For the causewayed enclosures and other 'ritual' sites in Britain and Ireland, it remains an open question whether the plant remains – be it crops or gathered plants – represent ceremonial burning and deposition or arise from disposal of everyday burnt waste (Jones and Rowley-Conwy 2007; Bishop *et al.* 2009). For the settlements, both the cultivation of cereals and the gathering of wild food plants can be confirmed as a usual aspect of Neolithic economy (Robinson 2000).

What does this tell us about the social organisation of the Funnel Beaker societies? As both cereal grains and threshing remains are present in the domestic sites, we may assume that, during the Funnel Beaker period, subsistence economy on the household level was the prevalent husbandry strategy. Gathered plants served as additional food that occurred regularly. Their regular presence supports the model of household-based economies, as ethnographic studies show that plant gathering often is carried out by children or on the family level and thus including the children (Chapter 8.5).

On the basis of the material from the Albersdorf-Brutkamp LA 5 tomb, it is extremely difficult to separate ritual activities from unintentional factors, as we are dealing with poor samples in this case. However, one option for reconstructing past plant use might be that the deposition of the plant remains was primarily related to the consumption of fruits from collected plants as part of a social gathering connected to a burial ritual. Still, we cannot exclude the possibility that the gathered plants, which are all light-demanding species, come from the ruderal vegetation within or at the edge of the clearing that surrounds the megalithic tomb. The social status of the people buried inside in the tomb is most probably expressed through the impressive grave monument itself that shows the importance of the site (Fig. 8.32). As mentioned above, it additionally functions as a place of social identification that furthers the affiliation of individuals to a particular group. The burial ritual may be linked to a food rule related to a particular social group (see Chapter 8.1). The charred nutshells of hazel and seeds of raspberry show that the people involved continued to favour plant gathering, in particular if the world of the ancestors is of concern. Proposing that the burial ritual might be connected with former Mesolithic traditions is difficult to prove because, unfortunately, evidence for the Early Neolithic I time-span (EN I; 4100–3600 cal. BCE) is weak in general. In the Late Neolithic, two flint daggers were deposited in the grave mound at Albersdorf-Brutkamp LA 5. Late Neolithic bulb finds of false oat grass (*Arrhenatherum elatius* ssp. *bulbosum* (Willd) Schübler & Martens) hint at a ritual fire on the grave mound. The gathering of the starch-rich, lowest swollen stem internodes of false oat grass for dietary purposes might be considered, as well as its occurrence in the natural vegetation (Engelmark 1984; Jensen *et al.* 2010; Preiss *et al.* 2005). In Europe, evidence for Neolithic bulbs of false oat grass is rare. Finds are usually related to cremation graves and thus date to the Bronze Age and more recent periods (Roehrs *et al.* 2013). In this case, *Arrhenatherum* bulbs are interpreted as having been awarded a high symbolic value and thus were intentionally deposited in ritual contexts (Viklund 2002). This is in analogy to the deposition of onions from asphodel (*Asphodelus* spp.) which occur in Hellenistic graves. Asphodel is mentioned as the lily of the Elysian Fields in Greek mythology (after Engelmark 1984). To conclude from this, the occurrence of *Arrhenatherum* bulbs in the Late Neolithic Albersdorf grave mound may hint at an exceptional social context.

Only one other archaeobotanical report on a central German Neolithic tomb is available. A mass find of cotton thistle (*Onopordum acanthium* L.) seeds was discovered at the non-megalithic stone chamber grave of Kreienkopp II close to Ditfurt in Sachsen-Anhalt, dating to 3440–2840 cal. BCE (Hellmund 2008). Ruderals dominate the plant assemblage, while einkorn (*Triticum monococcum*), emmer (*T. dicoccum*) and barley (*Hordeum vulgare*) were also present, mainly as chaff. In addition, gathered plants such as blackthorn (*Prunus spinosa*) and hackberry (*P.* cf. *padus*) occurred. All the finds are interpreted as grave goods. As most of the *Prunus* stones had been opened by rodents before charring, it is assumed that the fruits had been deposited before the burning of the grave chamber and, therefore, due to this delay, grains are rarely preserved in the burial context. Finally, we must note that too few Neolithic tombs have been investigated so far to come up with a general picture of the relevance of plants in Neolithic burial contexts. However, these very first results show that plant assemblages in a burial ritual differ from assemblages of economic plant use in domestic sites.

Conclusions

In the Funnel Beaker period when agriculture was established for the first time in northern Germany, a subsistence economy on household level was prevalent. Gathered plants, often regarded as substitution foods, contributed to daily diet on a regular basis. Surplus production had not yet been developed, and we may be able to trace the very first steps of division in labour that follows after the introduction of the ard. From the two examples, we deduce that, apart from being connected to the economy, the meaning of husbandry and gathering can be placed within the ratio of domestic/social activities at the sites. The most important driving-factor for plant choice within a specific cultural context seems to be whether we are dealing with a domestic or a ritual context. If the settlement is compared with the tomb, collected fruits are over-represented in the grave deposits and indicate a *rite de passage*. Meanwhile, the settlement shows mainly evidence for crop plant use and can be separated from the tomb, the former being the primary economic sphere. Ongoing research on the important settlement site and on further domestic and ritual sites will show whether or not we can generalise about the separation of different social spheres of plant use in the Neolithic.

8.7. LEGAL CONSTRAINTS INFLUENCING CROP CHOICE IN CASTILLE AND ENVIRONS FROM THE MIDDLE AGES TO THE 19TH CENTURY: SOME EXAMPLES

José Luis Mingote Calderón

In the past, laws unquestionably influenced choices in many agricultural issues. These legal regulations show the dialectical relationship between top-down imposition and what we can glimpse through them of local practices. Such documents do indeed tell us something about peasant customs, whereas these rules are under broad general powers and, at the end, they must be sanctioned by royal power. That is to say, these laws mark a theoretical horizon for action that is not always precisely respected in day-to-day life. However, with the passage of time over the centuries (from the twelfth to the nineteenth century), these legal texts display clear patterns concerning some of the ways of life documented by ethnographers who worked on Spanish pre-industrial society. Options to grow particular crops were, for instance, strongly influenced by the enforcement of regulations. Throughout the Middle Ages and up to the nineteenth century there were, however, many examples (particularly in times of crisis) of the difficulties of streamlining the regulatory process. So, in spite of the fact that life was governed by laws and regulations, in Spain there were many circumstances in which these regulations were not systematically followed. This is, at least, what in the thirteenth century a canon from Segovia recalled when referring to the way vineyards should have been worked:

> 'since I started to attend the church I never saw any more this way of working'
>
> ('*esta costumbre non la ví guardar en ninguno desde que fuy en la eglesia*' García Sanz *et al.* 1981, 133).

This contribution deals with examples of the limitations imposed by the numerous regulations (municipal ordinances) farmers have been faced with regarding agricultural choices. Data comes from private and public legislation from Castile and León. Both tenant contracts for renting fields and municipal ordinances (*Ordenanzas*) from the twelfth century to the end of the 19th century are the main types of documents analysed. In particular, the texts selected consist of renting contracts from the clergy of the Cathedral of Segovia (thirteenth century), ordinances from the fourteenth to the sixteenth centuries and private contracts (nineteenth century).

The Clergy of Segovia's Cathedral: Examples from Renting Contracts

Contracts for land renting included agreements concerning the way fields should remain at the end of the contract period. There were also indications on the manner in which the field had to be worked. Not only did the new tenant have to follow the practices carried out by the previous renter, but the crops cultivated also had to be the same, namely, wheat, rye and barley:

> 'and at the end of the contract I shall leave 20 *obradas* sown (a traditional unit of measure for land area equivalent to ca 4000 m^2): eight with wheat, ten with rye and two with barley'
>
> ('*e devo dexar [at the end of the contrac] sembradas veynt obradas: las ocho de trigo e las X de centeno e las dos de çevada*' García Sanz *et al.* 1981, 129).

There were also many rules concerning the range of crops tenants had to sow and pay back to the landlord, mainly rye, wheat and barley:

> 'I took for sowing wheat 20 *fanegas* (unit of capacity); rye 20 *fanegas*, barley 16 *fanegas*'
>
> (*'Item tomé pora simiente trigo, XX fanegas; centeno XX fanegas; cevada XVI fanegas'* García Sanz *et al.* 1981, 110).

A distinction was made between crops for human food and species mainly used for fodder. There were species like rye that could be used for both humans and animals, while others such as lentils were generally fed to animals:

> 'Ten *moyos* (unit of capacity) for fodder and sowing' and 'one and a half *fanegas* of lentils for fodder'
>
> (*'X moyos de centeno pora ceva e sembrar'* and *'e fanega e media de lenteias pora çeva'* García Sanz *et al.* 1981,142, 136).

Texts also bring together examples of the obligation to grow a particular crop of a particular species well adapted to certain soils of a region. This is the case of rye in the area of Segovia:

> 'after having accepted what the division established, I found lands only valid for rye which can have an extension between nine or ten *obradas*'
>
> (*'Otrossí depués que lo tomé fallé tierras que son pora centeno en que puede aver de IX obradas fata X'* García Sanz *et al.* 1981, 119).

Whenever changes or modifications to the agreement were made, these were explicitly recorded. For example, there are documents which record decisions made to use specific species for grafting or to grow particular crops such as almond trees:

> '...and in these vineyards there are five cherry trees which I grafted with 'borrinos' (local name for wild cherry tree) and two new almond trees that I planted...
>
> (*'ý en estas viñas ay çinco çerezos que yo enxerí de borrinos'* [...] *'e dos almendros nuevos que yo pus'* García Sanz *et al.* 1981, 127).

Bylaws from the Fourteenth and Sixteenth Centuries: Some Cases

As shown in texts, in many regions the choices to grow crops were clearly limited by the existing bylaws. In addition, there were also laws that prohibited cultivation. However, these regulations could also not be respected and therefore the wrongdoer could be penalised with the loss of the harvest. In 1583, in Buitrago (Madrid province; Fig. 8.36) triennial rotation was enforced and subsequently the failure to comply with the law could lead to the use of that particular harvest by other people.

> 'It was ordered that the *quiñones* (portions of land for sowing) could not be sown every year but every three years, leaving one in fallow under penalty of 600 mrs (*maravedí*, a former Spanish monetary unit) for those not following this rule and of free access to use the harvest of these lands by other people and the same applies to all *tercios* (thirds in which the space is divided up) in all the localities of the region...'
>
> (*'Otrosi hordenaron e mandaron que los quiñones no se puedan senbrar un año tras otro más que de que se sienbren a treçero año que pase un año de claro sin que se sienbre so pena que el que de otra manera lo senbrare tenga de pena seisçientos mrs. rrepartidos según dho es, e qualquiera pueda comer libremente el pan que estuviere en los tales quiñones e lo mysmo sea e se entienda en los terçios de los lugares de esta tierra'* Fernández García 1966, 28).

Examples are also available on the obligation to grow specific products in gardens. A clear case comes from *Ordenanzas* relating to Villatoro (Ávila province, in 1503; Fig. 8.36) where people were forced to grow onions, leeks and cabbages:

> 'whoever had land in the *soto* (site on the banks and meadows with trees and shrubs), either private or rented, is compelled to sow vegetables, at least two *eras* (physical space whose dimensions are unknown) of onions and two furrows of leeks and five of cabbages'
>
> (*'que qualquiera que toviere heredad en el soto, suyo o aRendada, que sea obligado a poner ortaliza, a lo menos dos eras de çebollas e dos surcos de pueros e çinco surcos de verças'* Blasco 1933, 399).

Another example comes from Pedraza de la Sierra (Segovia province, in 1400 and 1474; Fig. 8.36) where people were obligated to sow garlic:

> 'everyone having a vegetable garden should grow two *braços* (surface unit) of garlic under penalty of 60 *maravedís* (former Spanish monetary unit) for the *concejo* (formal meeting of all residents, married men only)'
>
> (*'e cada vno que huerto tuviere que ponga dos braços de ajos so la dicha pena de los dichos sesenta maravedís para el dicho conçejo'* Franco Silva 1991, 131, 139–140).

Finally in La Alberca (Salamanca province, in 1515, 1568, 1616 and 1668; Fig. 8.36) farmers were obliged to grow leeks and cabbages:

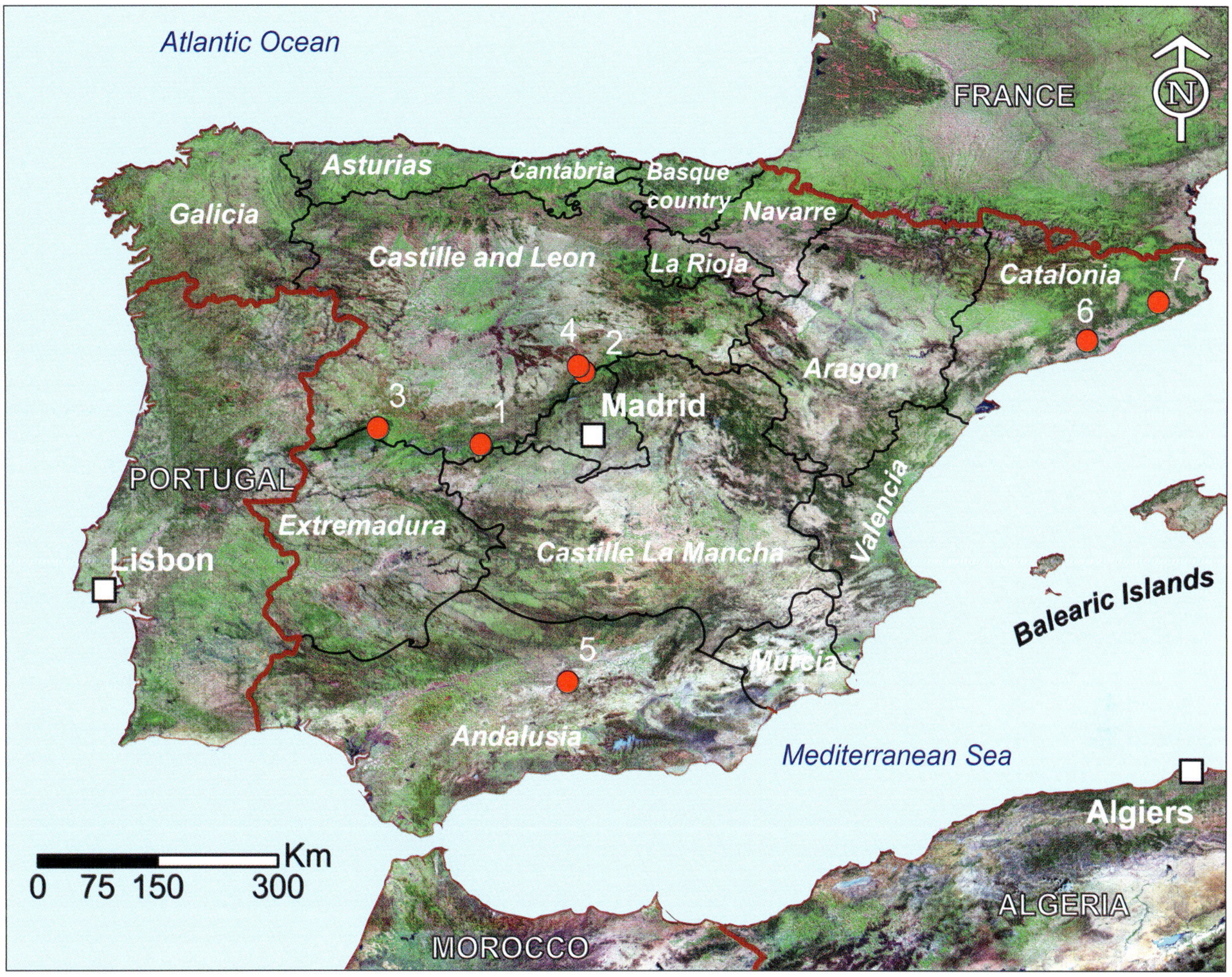

Fig. 8.36. Map of Spain with the locations mentioned in the text: 1) Villatoro; 2) Pedraza de la Sierra; 3) La Alberca; 4) Buitrago; 5) Martos; 6) Gélida; 7) Santa Coloma de Farnés. Map: R. Lugon, J.-C. Loubier and A. Chevalier.

> 'each should plant five *arroyos* (unit of surface) with leeks and two *rregaderas* (unit of surface) with cabagges'
>
> ('*que... cadauno ponga cinco arroyos de puerros e dos rregaderas de coles e dende arriva el que mas quisiere poner*' Berrogain 1930, 420).

In the same area when the commons were privatised, owners had to plant chestnuts:

> 'Each lucky neighbour possessing a piece of land has to plant 10 or more chestnuts if he wishes or if the trees fit in the space'
>
> ('*Cada vecino al que le tocó en suerte un pedazo de tierra debe poner diez castaños o mas, si quiere y caben en ese espacio*' Berrogain 1930, 433).

In order to obtain certain rights within the community, there were detailed regulations which established the quantity and type of crops to be grown. In Segovia, in 1514, these regulations involved both wheat and vine (Riaza 1935, 486–487). On many occasions, the agrarian products were stolen, and this situation led to either the abandonment of the land or the decision to stop growing the species generally stolen (commonly vines and garden crops). This is clearly stated in the bylaws of Pedraza de la Sierra in Segovia province (Franco Silva 1991, 128).

Examples from Nineteenth Century Contracts

The limitations imposed by laws also continued during the nineteenth century. Again, there were rules for establishing the way fields (vineyards, cereal fields) had to be left at the end of the renting period. This is well attested in many areas of the Iberian Peninsula. For instance, in Santa Coloma de

Farnés (Barcelona; Fig. 8.36), the tenant needed to follow the customs of the region maintaining the existing crops to avoid losses in productivity:

> 'The tenant farmer has to cultivate the rented land following the rules and customs of a good farmer, conserving the equal distribution of trees, crop systems, etc., and performing the corresponding farm work so the fields maintain their condition, state and current value'
>
> (*'El colono ó arrendatario ha de cultivar la tierras que se le arrienden á uso y costumbre de buen labrador, conservando igual distribución, árboles, sistemas de cultivo, etc., y practicando las labores correspondientes, para que no desmerezcan de sus condición, estado y valor actual'* Espejo 1900, 179).

The concept 'good farmer' reflects 'the way it should be' in agricultural work and this is well known to everyone.

The landowner could also provide the seed to be sown to the tenant, leading to the cultivation of certain crops such as wheat, barley, chickpeas, broad beans, einkorn or bitter vetches. A good case is represented by the practices carried out in Albacete where the landowner provided 75% of the sowing seed in the sharecropping contracts (Espejo 1900, 102–103). Another example comes from Gélida (Barcelona; Fig. 8.36) where, after the grape phylloxera epidemic, farmers were forced to plant American stocks provided by the landowners:

> 'the farmer will plant at his own expense the resistant American stocks given by the landowner'
>
> (*'plantará á sus costas las cepas americanas resistentes que el propietario le entregará', lo que deberá estar hecho en un plazo de seis años'* Espejo 1900, 149).

In some cases, as for example in Sevilla, the contract established the division of the field into thirds with clear indications on the kind of crop that had to be sown in each. Altering the system was forbidden (Espejo 1900, 40–41). Another example from Sevilla illustrated how tenants had to follow the rules imposed by owners and, at the end of the contract, give back the allotment cultivated in the same state as it was delivered at the beginning of the agreement. In this particular case, the ban on growing cereals in the home garden is clearly stated:

> 'the garden existing in the property has to be given back to the owner in the same state as it was delivered at the beginning of the renting period. The sowing of cereals is not allowed'
>
> (*'la huerta que existe en la finca se entregarán (sic encargarán) los colonos de ella sin aprecio, entregándola á su salida en igual forma, no pudiendo sembrar en la misma semilla de raspa'* Espejo 1900, 48).

A further example comes from Martos (Jaén; Fig. 8.36) where there is a prohibition on sowing some cereal and legume varieties under the olive trees:

> 'The tenant undertakes not to sow wheat, barley, einkorn or bitter vetch under the olive tress'
>
> (*'el otorgante se obligará á no sembrar los suelos de los olivares de trigo, cebada, escaña ni yeros'* Espejo 1900, 78).

In some cases, as for example in Aragón, the improvements carried out on fruit trees and garden products were quantified. This implied that there was a certain amount of flexibility and that some changes were starting to be introduced (Espejo 1900, 202).

Conclusions

In Medieval society, farmers were heavily influenced by sometimes strict laws and norms that impeded them from choosing the type of crops to be sown. This problem was already noted by Jovellanos in his *Informe sobre la Ley Agraria* of 1784–1787. The jurist recalled the spread of norms and rules and pointed to the need to eliminate them:

> 'What could be judged of the many laws and bylaws that have limited the freedom of owners and tenants to use their lands, laws that forbid turning arable fields into pasture land or the other way round, that limit plantations or that prohibit removing vine stocks or trees?'
>
> (*'¿qué se podrá juzgar de tantas leyes y ordenanzas municipales como han oprimido la libertad de los propietarios y colonos en el uso y destino de sus tierras, de las que prohíben convertir el cultivo en pasto, ó el pasto en cultivo, de las que ponen límite a las plantaciones, ó prohíben descepar las viñas y montes'* Jovellanos 1968, 37).

These regulations were generally assumed by the population who saw in them a body of knowledge which was easily understood and shared by everybody. The local community was considered as a proper unit in which the same rules involving agricultural decisions were equally applied to everybody. This was a way to avoid problems and conflicts. It was common to find that farmers sharing a particular cultivation area grew the same crops and more importantly, most of the farming work related to those crops had to be carried out by all the tenants and this was carefully controlled. The interests of the landlords and owners overrode those of the tenants. The latter grew whatever the owner decided and in the way they powerfully imposed. The same applied to the distribution of agrarian production and its marketing.

Peasants under a feudal regime (under royal power, church or manor domain) with communitary guidelines (expressed in the *concejo* and their regulations) may implement rules that later are still present in the beginnings of a capitalist agriculture, in the nineteenth century. The change in the organisation of society is deep-reaching and one aspect has changed drastically: the control exercised on daily life in the old regime, by the community itself, for the *vecinos* (among neighbours), is diluted in behavior patterns that no longer have a legal status.

8.8. LATE CLASSIC MAYA PROVISIONING AND DISTINCTION IN NORTHWESTERN BELIZE

David J. Goldstein and Jon B. Hageman

Introduction

Our investigation into food preparation and use at two small Late Classic Maya settlements in northwestern Belize indicates that the Ancient Maya developed an active set of codes or foodways for the use of certain foods (Douglas 1971, 61; Keller Brown and Mussell 1997). Feasting, when contrasted with daily food use in less ceremonial settings, exemplifies the kind of contrast in food use that Douglas (1971) argues is emblematic of social codes used to actively establish and maintain social hierarchy. Our work focuses on this coding as expressed by two residential units within a Late Classic (CE 600–900) lineage. At our case study site of Guijarral, a rural site in northwestern Belize, we recovered archaeobotanical datasets associated with periodic feasting near ancestor shrines as distinct from daily domestic activities near housemounds. The consumption of specific foods in specific places created and reinforced within-group social inequality during the Late Classic. Our data include plants generally held to represent comestibles outside of the agricultural complex considered as 'traditional' by Mayanists (Coe 1994; Fedick 1996; Reina 1967; Sharer 2005). In keeping with the concept of how the historical and ecological realities impacted the resource base in our study area, the species encountered indicate a heavy reliance on successional forest species. That these species were used for performing both daily food production and feasting is in line with results from other lowland tropical areas, where successional or disturbance response species provide the backbone to subsistence (Rappaport 1984; Moran 1990; Baleé and Erickson 2007).

The Pre-Hispanic Context

The ecological history constructed for northwestern Belize indicates a trend of increasing deforestation from the Middle Preclassic through Late Classic 1 (1000 BCE–CE 700). Pollen cores indicate a marked increase in maize and other pollens associated with human agricultural disturbance (Marchant *et al.* 2002), such as grasses and asters. With the exception of some plant families with a high abundance of economically important plants such as the *Sapotaecae*, primary forest taxa are absent (Dunning *et al.* 1999, 654). Similar data from across the Yucatan Peninsula during Late Classic 2 (CE 700–850) are interpreted as demonstrating further agricultural intensification. Sediment core data correlate increasingly heavy phosphate loads with the presence of an increased land clearance, and this is attributed to a surge in shifting agriculture and settlement (Curtis *et al.* 1996; Hodell *et al.* 1995).

In northwestern Belize (Fig. 8.37), this intensification of land clearance is evinced through geo-archaeological investigations at the mouths of drainages intersecting with *bajo* (seasonal swamp) margins. These excavations yielded buried peats containing aquatic plant pollen. Dunning *et al.* (1999, 655) use this evidence to argue that, by the beginning of the Early Classic (CE 250), the *bajos* in northwestern Belize had been transformed from perennial to annual wetlands through infilling processes, and that much of the region's sloping upland terrain was 'largely devoid of soil cover' (Dunning *et al.* 1999, 656). Though land clearance was widespread and transformed much of the

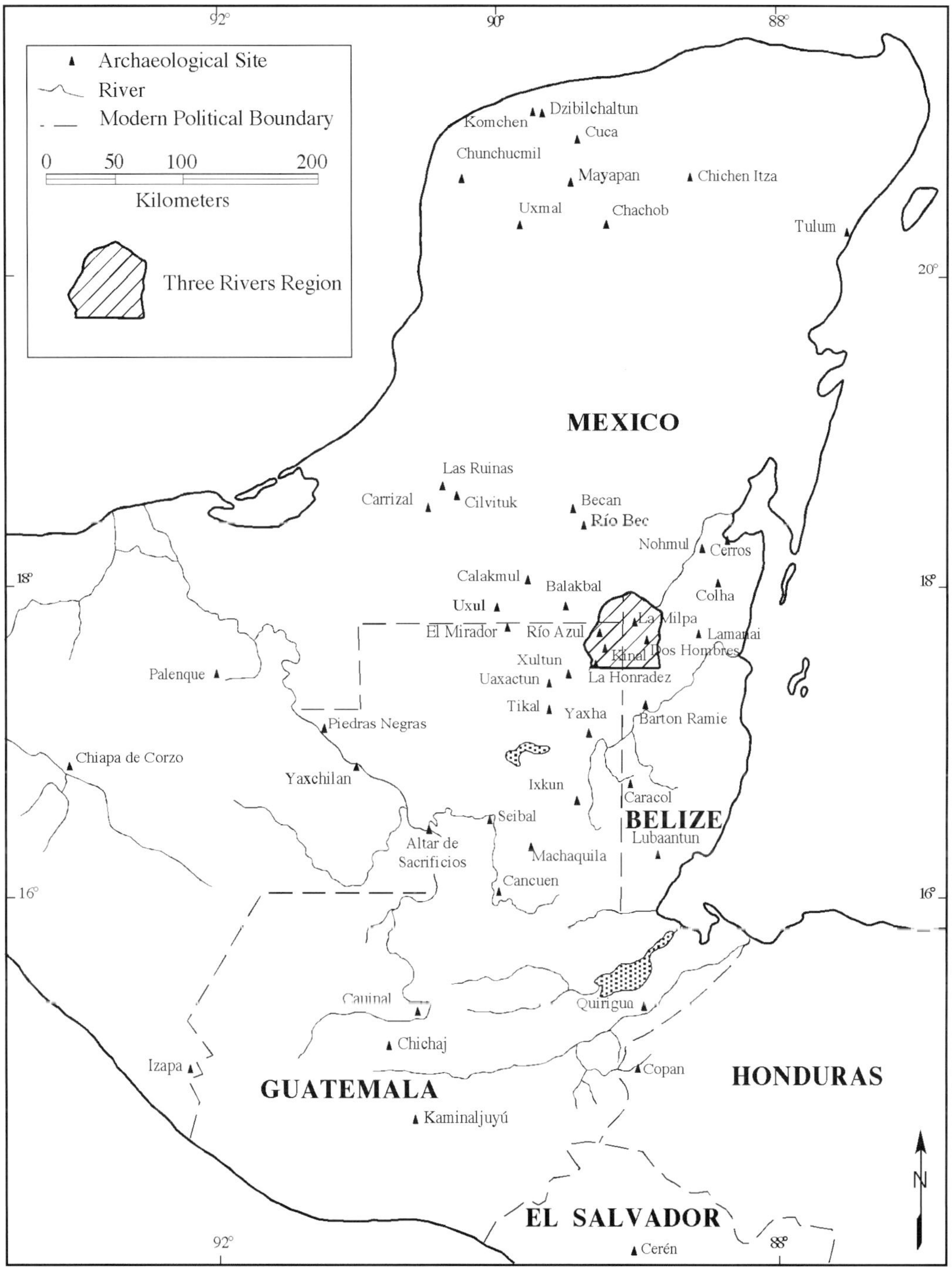

Fig. 8.37. Three Rivers Region of northwestern Belize relative to Central American countries and major Ancient Maya centres.

landscape during this era populations would not peak for several centuries.

Settlement pattern studies indicate a significant and profound population increase shortly after CE 700. Around CE 650 the population density, based on settlement extension and functional architecture, is estimated at about 110 persons per km^2. 150 or so years later, this density is estimated at 510 persons per km^2 and includes increased construction in the regional major centres (Adams *et al.* 2004). As populations increased, houses and farms were constructed in ever more marginal landscapes, where large-scale modifications to the land were implemented and added to earlier infrastructure to increase agricultural productivity. Agricultural land became scarce, and corporate groups such as lineages arose to secure resources and manage their use (Hageman and Lohse 2003).

In sum, by CE 700 the environment of northwestern Belize had been severely degraded as a result of widespread and long-term human impact. Populations in the area quintupled between CE 700 and 850, while facing diminishing amounts of arable land. Agricultural terraces appear in greater numbers during this period and are ostensibly interpreted as ways to conserve soils and permit marginal areas to sustain more intensive agricultural use. Shifting agriculture is thought to have been practiced on these terraces. Lineages emerged in parts of northwestern Belize as a potential means of securing productive resources and minimising risk, particularly in areas where resources were in high demand (Hageman 2004; Hageman and Lohse 2003).

Guijarral: A Late Classic Maya Rural Settlement

The small ritual centre and settlement of Guijarral is situated eight km east of the Late Classic centre of La Milpa, in the Three Rivers Region of Guatemala, Belize, and Mexico (Fig. 8.37). The site is near the edge of the western Rio Bravo Escarpment, straddling an intermittent drainage that flows from the escarpment at the east to a western *bajo* (upland swamp: Fig. 8.38). The settlement includes several housemounds, a few small plaza groups, and a single two-courtyard ceremonial or elite architectural enclosure. This last group contains two shrines in the form of small pyramids at its centre. Research at the site included the record of 140 hillslope and cross-channel terraces in the 0.5 km^2 area. Regionally, this concentration represents a major investment in landscape modification. This phenomenon parallels ethnographic examples cross-culturally as suggesting that Guijarraleños were invested in the maintenance and extraction of all productivity potential from their surroundings, and concerned with sustaining dependable and consistent production levels (Dunning *et al.* 2003; Hageman 2004).

Guijarral's developmental history is long-lived and socially complex. From the relative chronology of ceramic finds, we know that the site and some of its

Fig. 8.38. Contour map of Guijarral site core and study area.

surrounding households were occupied during the Late Preclassic and Early Classic. This continuity at the surrounding and less impressive *plazuela* groups (plazas bounded by two or more buildings; oriented toward cardinal directions) reveal an 800-year occupation period. The smaller of the two pyramids, in the Guijarral site core, was constructed at the end of the Early Classic, at about CE 550, but the remaining construction dates to Late Classic 2 (CE 700–850). As with later and contemporary Maya, the house of the lineage head, and the seat of group power, is denoted by the shrine.

The research area is circumscribed on the west by a *bajo*, to the south by a large intermittent drainage, and on the east and north by the escarpment. The terraces and shrines, emblematic of organised labor investments in the landscape, reflect the boundaries of the area's resources and a corporate claim to the landscape, respectively. Comparable to other Late Classic Maya sites, feasting is indicated through

	Feasting Middens			Domestic Middens	
	Xunantunich Group D	Paco 15	Guijarral Site Centre	Xunantunich Plazas	Grupo Chispas
Bowls, Plates, Food Prep and Serving Pots	72.45%	64.26%	64.55%	50.46%	50%
Jars and Food Storage Pots	26.50%	27.70%	35.44%	48.82%	50%

Fig. 8.39. Distribution of ceramic forms at Xunantunich (LeCount 2001), Paco 15 (Fox 1996), and the study area (Hageman 2004). Figures represent percentages of the ceramic assemblage at each site. Not all ceramics were used for food. Note that domestic middens have 1:1 ratio of food storage to food preparation and serving vessels.

a predominance of food preparation and serving vessels in middens associated with ancestral shrines (Fig. 8.39). This holds for Guijarral, where almost 65% of the ceramics recovered from the site centre were from food preparation and serving vessels, compared to 50% from regular domestic contexts (Fox 1996; Hageman 2004; LeCount 2001).

Plant Use at Guijarral in the Late Classic

Given that the ceramics associated with specific architectural features indicated the presence of ancient feasting, we were interested in what was being consumed at these feasts and the degree to which these items differed from those foods consumed as part of the daily diet. In 2005 and 2006 we excavated middens in two residential groups at Guijarral. The first was the site centre (referred to here as GSC to distinguish the primary residential group from the rest of the site), identified with ancestor shrines and feasting. The second was Chispas, a small residential plaza group atop a hill some 100 metres away from GSC. Excavations covered eight metres^2 at each midden. We excavated the middens in 10 cm levels, collecting four metres of soil from each 1 m × 1 m × 0.1 m level. We then used flotation and dry standard series fine-screens to recover archaeobotanical remains. While all of the soil samples were examined, our report represents the materials recovered from 21 cm below the ground surface as bioturbation (disturbance of the sediment or soil by the activity of plants or animals) was less evident and this material is likely better preserved.

We recovered 3,738 charred or mineralised items. Of these, 1,710 were recovered from GSC's middens associated with the feasting ceramic assemblage and 2,028 were from the Chispas' middens associated with primarily domestic, non-feasting ceramics. Between the two contexts, 190, or 4.1% of the total individual plant remains, were unidentifiable. This leaves some 20% of our materials as unknowns needing further study to arrive at determinations.

Guijarral Site Centre (GSC) Feasting Plant Assemblage

We have assigned seven different taxonomic determinations (n=205; 10% of GSC assemblage) to plant remains recovered exclusively in the feasting contexts at GSC (Fig. 8.40). An additional 12 items require determination (n=109; 5% of GSC assemblage). Together the remains from these taxa comprise 8% of all the taxa recovered from both sites. Many of these taxa are known to have been utilised by the historic and ancient Maya. Yet, we recovered several plant remains that are known to be used ethnographically and ethnohistorically by the Maya in feasting, including *Amaranthus* sp. (amaranth, n=3), *Guazuma* sp. (guácima, n=5), *Orbigyna* sp. (babassu, n=1) *Psidium* sp. (guava, n=2), and *Zuelania* sp. (arbol caspa, n=2). Ecologically speaking, for the lowland rainforest of northwestern Belize these are not uncommon taxa. What is of interest, however, is that these are disturbance taxa associated with human clearance and/or treefalls. In no case are they strictly associated with open air agriculture, or specifically *milpa* (traditional Mesoamerican shifting agriculture) agroecosystems. The recovered taxa are common in successional and modern agroforestry throughout the region today, including the use of *Zuelania* sp., *Psidium* sp., and *Orbigyna* sp. (Atran 1993). Aside

Family	Determination	Guijarral Raw Data	Guijarral Ubiquity (n=11 Lots)
Asclepidaceae	*Asclepias* sp.	191	81.8%
Unknown	UKN #35-FS19	45	45.5%
Unknown	UKN #28-FS7	28	27.3%
Unknown	UKN #34-FS19	6	27.3%
Amaranthaceae	*Amaranthus* sp.	3	18.2%
Flacourtiaceae	*Zeulania* sp.	2	18.2%
Myrtaceae	*Psidium* sp.	2	18.2%
Sterculiaceae	*Guazuma* sp.	5	18.2%
Unknown	UKN #39-FS19	4	18.2%
Unknown	UKN #44-FS20	3	18.2%
Arecaceae	*Orbigyna* sp.	1	9.1%
Burseraceae	UKN #195-FS4	1	9.1%
Malvaceae	*Malva* sp.	1	9.1%
Unknown	UKN #24-FS7	3	9.1%
Unknown	UKN #38-FS19	2	9.1%
Unknown	UKN #45-FS20	7	9.1%
Unknown	UKN #49-FS20	3	9.1%
Unknown	UKN #50-FS20	4	9.1%
Unknown	UKN #56-FS29	3	9.1%

Fig. 8.40. Taxa exclusive to feasting at Guijarral.

from their incidence, South American lowland rainforest agriculture models accept the importance of these species in agricultural and cultural contexts (Denevan 2007). These species offer shade cover, regenerate quickly and survive well as renewable resources in terrace-based agricultural systems, such as those recovered at Guijarral (Clement 2007). The trees have predictable and consistent fruiting seasons while providing timber, renewable fuel resources, and protection from erosion even when the field systems directly associated with them are not undergoing active cultivation (Atran 1993).

These taxa have documented medicinal and ceremonial uses among modern Maya, and may have been used similarly in the past. *Zuelania* sp. (n=2) may be *Zuelania guidonia,* (Sw.) Britt. and Millsp., common in contemporary forests of northwestern Belize. For the modern Yucatec Maya, the bole of the tree serves as a game where people are invited to climb a greased tree trunk during certain festivals, *e.g.* carnival (Roys 1931; Atran *et al.* 2004). The leaves, when ground into a fine paste, present an alkaloid that is used as a diuretic and pain reliever. *Psidium* sp. is known in the form of *Psidium guajava,* L. (guava), for its fruits. *Guazuma* sp., potentially *Guazuma ulmifolia,* Lam., has been used in the Maya lowlands for producing fermented beverages, and by ethnographic extension may represent potential parts of ritual feasts (Roys 1931; Atran *et al.* 2004). *Guazuma* sp., a relative of *Theobroma cacao,* L. (cacao), is sometimes used to make a ritual beer. We also have seeds from what might be *Bursera* sp. in our feasting context, UKN #195–F.S.4. Different incenses are made from the sap, fruit, bark, and leaves of *Bursera copal,* L., (Stross 1997; Roys 1931; Atran *et al.* 2004). While several archaeological examples of *Bursera* sp. seeds are known, most are from construction fill or agricultural canals at sites such as Cuello, Pulltrouser Swamp and Tikal (Lentz 1999). The connection of these remains with the feasting ceramic assemblage suggests that these items were associated with festal events.

Other food consumption species include *Amaranthus* sp. This is a commonly used source of greens that can be boiled and eaten as an important dietary source of fibre and iron in the rainforest (Bye 1981; Roys 1931). *Amaranthaceae* species native to the region are associated broadly with agricultural activity and specifically human disturbance, as weeds growing in open fields. The presence of amaranth seed here indicates the potential collection and processing of this resource as a foodstuff in the performance of daily and festival events. *Orbigyna* sp., likely *O. cohune,* (Mart.) Dahlg. ex Standl., or cohune palm fruits are an additional reminder that non-domesticated and ordinary plants were associated with human activity in the Mesoamerican rainforest (Atran 1993). These provide palm oil and also nut meats not dissimilar in flavor and texture to coconut, *Cocos nucifera,* L. Each of these plants could have played substantial roles as ordinary commodities that were transformed into extraordinary items in feasting contexts (Kaličk 1997).

Domestic Non-Feasting Plants from Chispas

We determined the presence of nine taxa (n=73; 4% of Chispas assemblage) that occur only in the middens together with domestic non-feasting ceramics (Fig. 8.41). They comprise 2% of the all the plant remains recovered from both sites. Of these taxa, only *Potamogeton* sp. (pondweed) has been determined (Fig. 8.42). *Potamogeton* sp. is found

Family	Determination	Chispas Raw Count	Chispas Ubiquity (n=9 Lots)
Fabaceae	UKN #11-FS1	5	33.3%
Fabaceae	UKN #6-FS1	27	22.2%
Unknown	UKN #14-FS1	3	22.2%
Alismataceae	*Potamogeton* sp.	3	11.1%
Verbenaceae	UKN #5-FS1	4	11.1%
Pachychilidae	*Pachychilus* sp.	2	11.1%
Unknown	UKN #26-FS7	3	11.1%
Unknown	UKN #31-FS7	3	11.1%
Unknown	UKN #32-FS7	11	11.1%
Unknown	UKN #51-FS13	13	11.1%

Fig. 8.41. Taxa exclusive to domestic contexts at Chispas.

in standing water or ponds, easily found growing in the nearby *bajo* during the rainy season (Roys 1931; Atran *et al.* 2004). Recovery of seeds in house middens is not completely unexpected, as these were historically collected, stewed and consumed as leafy greens (Roys 1931).

Eight other taxa are present only in the domestic middens at Chispas. These taxa, including two legumes/pulses (*Fabaceae*) and one from the vervain family (*Verbenaceae*), are potential indicators for foodstuffs or daily-use materials, heretofore unrecognised by other investigators in non-elite household settings. In the case of UKN#11 and #6 they were present in several of our excavation units. Their increased ubiquity potentially signifies that their presence is not a chance occurrence and instead is related to human preferences and activities at this residence. UKN#11, based on its morphology, is likely a tree legume; these are trees common around the site today, whose sap, bark, wood, and fruits are/were used by the modern and historic Maya (Roys 1931; Atran 1993).

Plants Common to Both Feasting and Non-Feasting contexts

14 determined taxa appear in both midden types (n=660: 20% of the total assemblage (Chispas + GSC) Fig. 8.41). These include species of locustberry (*Byrsonima* sp.), yarumo (*Cecropia* sp.), South American mountain bamboo (*Chusquea* sp.), *maize* (*Zea mays* L.) and unknown seeds from the Families *Fabaceae, Asteraceae*, and *Solanaceae*. Additionally, we recovered evening-primrose (*Oenothera* sp.), Family *Onagraceae*, seeds (n=67) from both sites. This plant is not found in the region today, though it is the most common determined taxon from both

Family	Determination	Chispas Raw Count	Chispas Ubiquity (n=9 Lots)	Guijarral Raw Data	Guijarral Ubiquity (n=11 Lots)	Overall Raw Count	Overall Ubiquity (n=20 Lots)
Onagraceae	*Oenothera* sp.	43	77.80%	24	81.80%	67	80.00%
Unknown	UKN #20-FS7	51	77.80%	10	27.30%	61	50.00%
Poaceae	UKN #37-FS19	6	22.20%	9	54.50%	15	40.00%
Unknown	UKN #13-FS1	22	44.40%	15	27.30%	37	35.00%
Malphigiaceae	*Byrsonima* sp.	12	33.30%	11	18.20%	23	25.00%
Fabaceae	cf. Fabaceae	2	22.20%	2	18.20%	4	20.00%
Poaceae	cf. Poaceae	1	11.10%	5	27.30%	6	20.00%
Asteraceae	cf. Asteraceae	1	11.10%	3	18.20%	4	15.00%
Poaceae	*Chusquea* sp.	1	11.10%	4	18.20%	5	15.00%
Solanaceae	UKN #4-FS1	2	22.20%	1	9.10%	3	15.00%
Unknown	UKN #7-FS1	3	22.20%	27	9.10%	30	15.00%
Cecropiaceae	*Cecropia* sp.	1	11.10%	1	9.10%	2	10.00%
Poaceae	*Zea mays*	2	11.10%	9	9.10%	11	10.00%
Unknown	UKN #33-FS23	1	11.10%	1	9.10%	2	10.00%
Unknown	Lithics	79	66.70%	232	100.00%	311	100.00%
Unknown	Unid	69	100.00%	125	100.00%	194	100.00%

Fig. 8.42. Taxa common to both feasting (Guijarral) and domestic (Chispas) contexts.

contexts. Some authors have suggested that this plant is part of the pre-Hispanic ceremonial snuff traditions of the Caribbean and South America (Newsom and Wing 2004). Whatever its use may have been, its ubiquity in the Guijarral area during the Late Classic indicates that the plant and an understanding of its ecology were integrated in domestic, private settings, as well as potentially more ritual affairs.

Byrsonima sp. (locustberry) is one of several trees that produce edible and oil-bearing nuts in the Neotropics, and *Byrsonima crassifolia*, (L.) DC., (craboo) is known to be used as a famine food as well as for producing fruit preserves and a fermented beverage (Puelston 1971; Roys 1931). *Cecropia* sp. (n=2), may represent *Cecropia peltata*, L., (trumpet tree), which is common to house gardens, abandoned fields, fallow areas and any disturbed area of the forest. As a local medicinal plant; it is used to cure common ailments, *e.g.*, flu, fungal activity and fever (Roys 1931; Atran *et al.* 2004). In all, given the modern and – potentially – the ancient ecological setting, none of these three species appearing in either context is particularly surprising. Instead their presence reinforces the idea that disturbance forest taxa may have predominated what was available and useful to the Late Classic Maya in the area.

These plants, associated with both feasting and domestic contexts, likely indicate similar resource consumption patterns for the households around GSC. Furthermore, they demonstrate an overlap between daily and festival foods. The presence of *Z. mays* is no exception. We have very limited evidence for *Z. mays* at either site. We recovered only seed fragments from Guijarral and cupules, cob fragments, only from contexts at Chispas. Hence, we interpret *Z. mays* to have been the common denominator for food use at both sites. Rather than being restricted to the activities of one or the other sites, as described by Spanish Chroniclers (Coe 1994; Roys 1931), maize use may have been a basic staple to foodways activities at both *loci*.

Social Implications of Plant Remains Recovered from Guijarral and Chispas

There is no question that recovering plant remains from archaeological contexts in the rainforest presents challenges to traditional archaeobotanical recovery. This situation is especially so when compared with the archaeology of deserts or the temperate climate zone. The data presented here are limited in their quantity, but their presence is a testimony to Ancient Maya foodways, and this deserves interpretation and consideration.

Plants from both our feasting and domestic middens are related to smallholding agriculture typical of contemporary regional farming systems as well as broader Neotropical rainforest subsistence patterns (Baleé and Erickson 2007; Clement 2007; Denevan 2007; Fedick 1994). Maya archaeologists often invoke an idealised vision of Ancient Maya food production presented by authors like Rice (1990) and Reina (1967). In this model, infield and outfield systems played major roles in traditional agriculture, similar to how they are used in the region today. The data from GSC and Chispas, however, demonstrate the importance of other kinds of agroecological systems in the immediate area, if not regionally, in the Late Classic. Apart from the traditional use of gardens and fields in production, the Late Classic Guijarral relied, to a large extent, on a variety of successional species growing between houselots and field systems. These spaces generated catchment areas, including the fallow terraces. Apart from gardens being the reslient catchment area for use in times of scarcity (Killion 1992), local foodways included a variety of successional species and disturbance taxa across social boundaries.

In this way, successional or resilient plant resources were extensively used in potentially different kinds of households, serving to increase the breadth of the local diet (Winterhalder and Goland 1997). Food resources likely included plants from fallow terrace agriculture, *e.g.* tolerated invasives (weeds) near habitations and sucessional fruiting trees. These resources were, in turn, integrated into marked food preferences. As these items were seasonal fruits they were used to mark scheduled rituals that included foodway expressions, *e.g.* feasts.

Returning to Douglas's (1971) ideas, our data may demonstrate remnants of distinction processes of foodstuff selection in feasting versus domestic consumption at both GSC and Chispas. Given their archaeological and historic context, it is clear that these two sites were linked through a social bond, if not by sightlines alone. Looking more closely, the production and consumption residues, *e.g.* garbage, signal that processes of choice and selection of foodstuffs were encoded at both sites. Kalčik (1997, 46) revisits Douglas' codes and applies them directly to cultural material by arguing that food preferences are often the best indicators of cultural identity. The fact that foodstuff commonalities existed between the two residential groups represents potential indicators of taste culturally common to both the GSC residents and those of Chispas. More than that, the fact that successional trees and what a modern observer could consider 'wildfoods' were potentially a part of the subsistence base taken from the local ecosystem strengthens the bond between culture and landscape even further.

Alternatively, the distinctions between foodstuffs, again taken from the same breadth or catchment base, deployed in each locale under different circumstances may demonstrate how distinction was asserted or socially constructed during the Late Classic. Kalčik (1997, 48) argues that patterns of acceptance and avoidance of comestibles, especially in contexts of kinship ties, can be used as cultural indicators of social stratification and modalities of kinship. In our case, the distinctions are between feasting activities and daily meals, as evidenced through the material ceramic record when compared with the archaeobotancial remains. Using these datasets in tandem we can begin to distinguish commonalities from more auspicious occasions. We believe that we are beginning to locate the potential threads of Late Classic Maya foodways that coded for the ties that bound the residents at the sites in our research areas. These practices marked not only significant events, like ancestor veneration, but also the places in which those events were staged, and how resources from these places were managed. The association of plants used by contemporary and colonial Maya for feastal purposes, the ancestor shrines and the large houses, speak to the use of these plants as part of ritual activity that not only proclaimed the unity of the lineage, but also distinguished the practices of the denizens of this residential group from those of other residences.

Conclusion

The residents of Guijarral transformed their social organisation and their environment to safeguard their resource base and to produce more food during the Late Classic. In so doing, successional and invasive plant resources became of greater importance to local residents. This suite of plants is concomitant to human-induced landscape change and also incorporated into diet and ritual cycles in the area. As Guijarraleños organised themselves into a landholding corporation grounded and legitimised in the ancestral past, their foodways, ritual or mundane, culturally incorporated the products of this landscape into their consumption practice. Some of these practices included domesticated plants used by Maya in colonial-era and modern feasts, as well as tree crops, plants associated with fallow fields, and wild plants. Some of these fallow and wild plants were used exclusively for feasting, others for domestic consumption, and still others were utilised in both contexts. The plants used, many of which are wild and successional species from the transformed landscape, were implicated in negotiating the status of the lineage head through feasting, thereby associating them with long-term processes of land tenure and cultural preference in Late Classic Maya society.

8.9. CONCLUSIONS

Alexandre Chevalier

Apart from the common biological attitude toward food, such as the innate preference for sweetness and rejection of bitterness (Cohen 1987; Rozin *et al.* 1997), and the individual pre- and post-natal induced tastes influenced by a mother's diet and breast-feeding, all the ethnographic, social, historical and anthropological research shows that humans' food choices are culturally shaped (Douglas 1971; Fischler 1993; 1994; Lévi-Strauss 1962; 1964; 1997; Mead 1997; Rozin 1987). Food plants have to be transformed into an acceptable cultural (Lévi-Strauss 1997) and social (Lahlou 1998) element before being used or ingested. Plants have to be humanised first, then socialised: whether wild or domesticated, plants are culturally meaningful and embedded in social relationships (Barthes 1997; Douglas 1984, 1997; Farb and Armelagos 1983). The choice of a plant for food is also highly dependant on what the consumer's individual and social identity and status are (Tajfel and Turner 1986; Turner 1989), on the social context of the plant's use, and on the cultural, social and symbolic values attributed to a specific plant (Goody 1982; Harris 1986; Meigs 1997). Plant choices will not only reflect the economic and social status, as well as the cultural identity of the eater, but also the means of maintaining and possibly reinforcing his/her social status and state and his/her cultural identity through social representation and wealth redistribution (Bourdieu 1979; Mennell 1985; Mennell *et al.* 1992; Mintz 1986; 1996; Montanari 2000a; 2000b; Tajfel and Turner 1986; Turner 1982).

The different contributions in this chapter enrich our understanding of cultural and social choices concerning food plants. They all highlight the fact that there is no ecological determinism (Meggers 1954; 1979) or practical necessity (Harris 1974) in the choice of plants. Nor is the choice of plants an adaptative process due to the environmental constraints, but a deliberate cultural and social choice based on a given ecological setting and plant availability. The Late Classic Maya residents of Guijarral in Belize (Goldstein and Hageman, Chapter 8.8), for instance, integrated successional and invasive plant resources in their diet when their exploitation of nature led to its degradation and to landscape changes, but they specifically chose the ones according to their social status and context, and the others which would be involved in negotiating the status of the ruling lineage. Another example is given by medieval feudal Spain, where cultural codes are translated into laws, forcing peasants to cultivate, grow and use specific plants (Mingote, Chapter 8.7). Mingote shows that these regulations were still present up to the nineteenth century, no longer as legal rules, but completely embedded into the peasant's cultural world as social codes. Both Kirleiß and Kloß (Chapter 8.6) and González Reyero (Chapter 8.3) show that wild plants have a strong symbolic meaning for high-status groups within their own cultures, respectively in Late Bronze Age northern Germany, and in Iberian Spain. The same wild plants are shown as of low social status in present-day Nothern India according to Cruz-García (Chapter 8.5). In turn, luxury items and food diversity are key to understanding high social status groups in the Gaulish world (Durand and Wiethold, Chapter 8.3). Feasts utilising specific food plants and the presence of fermented drinks can be seen as a means to state, justify and reproduce the power of the ruling class over these societies, as proposed by González Reyero, Durand and Wiethold, Goldstein and Hageman, and Chevalier (Chapter 8.2).

Consciously or not, humans elaborate mentally and socially what is good to eat and what is taboo independently of what Descola (2011) calls conjectural ecology, in other words, the constraints of nature on humans.

Several authors adapted Levi-Strauss' statement about objects as 'good to think with', in order to reflect the fact that humans chose what they will take from their natural surroundings according to how they perceive this nature, and according to their own social norms and laws. Descola (2011, 26), for instance, states that nature that is good to think with and Hastorf (2003, 545) writes that food is good to think with. In turn, we would say that food plants are not only good to think with, but also that humans think what is good to eat: based on their perception of plants and nature, they decide what to include in their world and what to eat, how to do this, and what to leave aside, according to their social norms.

Chapter Acknowledgements

The archaeobotanical research carried out by A. Chevalier, presented in Chapter 8.2 by A. Chevalier and J. Dulanto was funded by the Roche Reseach Foundation, the Schmidheiny Foundation and the University of Geneva Borzinscki Trust. Most critically, the authors thank Prof. Deborah Pearsall at the University of Missouri-Columbia for her scientific support, and the University of Missouri-Columbia Anthropology Museum for the use of their facilities, without which this research would not have been possible. We also thank the PATL team and its director, Prof. K. Makwoski at the Católica University in Lima, for giving access to the archaeobotanical material from Pampa Chica.

The study by G. S. Cruz-García (Chapter 8.5) was carried out with the support of the Educational Programme of the Community Agrobiodiversity Centre in Kalpetta (CAbC), Wayanad, of the M. S. Swaminathan Research Foundation, Kerala, India. Wayanad's children are acknowledged for their invaluable cooperation and collaboration.

Research by W. Kirleis and S. Klooß (Chapter 8.6) was carried out at Kiel University, Germany, within the framework of the priority program '*Early monumentality and social differentiation*' (SPP 1400) in the subproject '*Agriculture and Environment as Basis for Early Monumentality*' (KI-1494/1). We gratefully acknowledge the funding by the DFG (German Research Foundation). Special thanks go to Ines Reese, Kiel, who upgraded the quality of the figures presented here.

The research by D. J. Goldstein and J. B. Hageman (Chapter 8.8) was made possible with financial support from Northeastern Illinois University College of Arts and Sciences, as well as infrastructural support from Dr. Fred Valdez, the R. E. W. Adams Research Facility, and the University of Texas at Austin. Most critically, the authors thank a series of donors to the R. E. W. Adams Research Facility for the acquisition and upkeep of their Flote-Tech A flotation machine, without which this research would not have been possible. We thank our students Gardner Brandt, Meghan Walker, Bethany Arthion, Sylvia Deskaj, Abigail Middleton, Kim Knudsen, and Josh Halperin Givens who participated in both the fieldwork and the post-season laboratory work.

Chapter Notes

1 A line drawn to join the lowest points along the entire length of a stream bed or valley in its downward slope. The term roughly corresponds to the valley bottom.

2 Pampa Chica was first described by Patterson in 1966, but excavated only in 1994 by Jalh Dulanto (see Dulanto 1994; 2008) within the 'Proyecto Arqueológico Tablada de Lurín' directed by Prof. Krysztof Makowski at the PUCP in Lima.

3 Based on seven radiocarbon dates calibrated with Oxcal v3.5. Gd-11197 : 2640±70BP = 960-540BCE; Gd-11202 : 2540±60BP = 810-420BCE; Gd-11200 : 2460±60BP = 780-410BCE; Gd-7653 : 2440±40BP = 760-410BCE; Gd-11192 : 2410±70BP = 770-370BCE; Gd-7651 : 2330±40BP = 710-230BCE; Gd-7648 : 2210±40BP = 380-160BCE.

4 Polysemy means that a sign (a word, a phrase, a symbol, etc.) can bear multiple meanings (semes).

5 Organic remains of grapevines can easily be identified by their pips (sedes) to species level (*Vitis vinifera*). However, it is very difficult to discriminate the wild sub-species (*V. vinifera* subsp. *sylvestris*) from the cultivated one (*V. vinifera* subsp. *vinifera*): in fact, this is the factor that encouraged the authors to opt for the cultivated sub-species (Alonso 2000; Pérez Jordà 2000; Pérez Jordà *et al.* 2007).

Chapter Bibliography

Abad, L. and Sala, F. (1993) *El poblado ibérico de El Oral (San Fulgencio, Alicante)*. Trabajos Varios del SIP 90. SIP, Diputación de Valencia, Valencia.

Adams, R. E. W, Robichaux, H. R., Valdez Jr., F., Houk, B. A. and Mathews, R. (2004) Transformations, Periodicity, and Urban Development in the Three Rivers Region. In A. A. Demarest, P. M. Rice and D. S. Rice (eds.) *The Terminal Classic in the Maya Lowlands: Collapse, Transition, and Transformation*. University Press of Colorado, Boulder, pp. 324–341.

Almagro-Gorbea, M. (1983) Pozo Moro. El monumento orientalizante, su contexto socio-cultural y sus paralelos en la arquitectura funeraria ibérica. *Madrider Mitteilungen* 24, 177–293; Tafel 12–34.

Alonso i Martínez, N. (2000) Registro arqueobotánico de Cataluña occidental durante el II y I milenio A.N.E. *Complutum* 11, 221–238.

Alva, W. and Donnan, C. (1993) *Royal tombs of Sipán*. Fowler Museum of Cultural History, Los Angeles.

Appadurai, A. (1986) Introduction: commodities and the politics of value. In A. Appadurai (ed.) *The Social Life of Things: Commodities in Cultural Perspective*. Cambridge University Press, Cambridge, pp. 3–63.

Atran, S. (1993) Itza Maya tropical agro-forestry. *Current Anthropology* 34, 633–700.

Atran, S., Ximena, L. and Ucan Ek', E. (2004) *Plants of the Petén Itza' Maya*. University of Michigan Press, Ann Arbor.

Balakrishnan, R. (2005) *Rural women and food security in Asia and the Pacific: prospects and paradoxes*. FAO, Regional Office for Asia and the Pacific, Bangkok.

Baleé, W. and Erickson, C. (2007) *Time and Complexity in Historical Ecology: Studies in the Neotropical Lowlands*. Columbia University Press, New York.

Barral, P. (1999) Carnet de fouilles : Verdun-sur-le Doubs, monnaie et commerce. *L'Archéologue* 41, 76–77.

Barral, P. and Richard, H. (2009) *Fouilles de la fontaines Saint-Pierre au Mont Beuvray (1988-1992, 1996). Aménagements d'une source sur l'oppidum de Bibracte*. Collection Bibracte 17. Centre archéologique européen du Mont Beuvray, Glux-en-Glenne.

Barthes, R. (1997) Toward a psychosociology of contemporary food consumption. In C. Counihan and P. Van Esterik (eds.) *Food and culture: a reader*. Routledge, New York/London, pp. 20–27.

Bastien, J. W. (1978a) *Mountain of the Condor: metaphor and ritual in an Andean ayllu*. Monographs of the American Ethnological Society 64. West Pub. Co, St. Paul.

Bastien, J. W. (1978b) Mountain/Body metaphor in the Andes. *Bulletin de l'Institut français d'études andines (Lima)* 7 (1–2), 87–103.

Behre, K.-E. (1981) The interpretation of anthropogenic indicators in pollen diagrams. *Pollen et Spores* 23, 225–245.

Behre, K.-E. (2008a) *Landschaftsgeschichte Norddeutschlands. Umwelt und Siedlung von der Steinzeit bis zur Gegenwart*. Wachholtz, Neumünster.

Behre, K.-E. (2008b) Collected seeds and fruits from herbs as prehistoric food. *Vegetation History and Archaeobotany* 17, 65–73.

Behre, K.-E. and Kučan, D. (1994) *Die Geschichte der Kulturlandschaft und des Ackerbaus in der Siedlungskammer Flögeln, Niedersachsen, seit der Jungsteinzeit*. Probleme der Küstenforschung im südlichen Nordseegebiet 21. Isensee, Oldenburg.

Bernard, H. R. (2002) *Research methods in anthropology. Qualitative and quantitative approaches*. Altamira Press, Walnut Creek (CA).

Berrogain, G. (1930) Ordenanzas de La Alberca y sus términos: Las Hurdes y Las Batuecas. *Anuario de Historia del Derecha Español* 7, 381–441.

Bessière, F. and Guichard, V. (2010) Chronique des recherches sur le Mont Beuvray : 2006–2008. *Revue archéologique de l'Est* 59 (1), 211–239.

Bishop, R. R., Church, M. J. and Rowley-Conwy, P. A. (2009) Cereals, fruits and nuts in the Scottish Neolithic. *Proceedings of the Society of Antiquaries of Scotland* 139, 47–103.

Blánquez J. (1996) Lugares de culto en el mundo ibérico (Nuevas propuestas interpretativas de espacios singulares en el sureste meseteño). *Revista de Estudios Ibéricos* 2, 147–172.

Blasco, R. (1933) Ordenanzas municipales de Villatoro (Avila). *AHDE* 10, 391–434.

Board, N. (2004) *Handbook on herbs cultivation and processing*. Asia Pacific Business Press Inc., Delhi.

Bogaard, A. and Jones, G. (2007) Neolithic farming in Britain and Central Europe: contrast or continuity? In A. Whittle and V. Cummings (eds.) *Going over: the Mesolithic-Neolithic transition in north-west Europe*. Oxford University Press, Oxford, pp. 357–375.

Bonet, H. (1995) *El Tossal de Sant Miquel de Llíria. La antigua Edeta y su territorio*. SIP, Diputación de Valencia, Valencia.

Bouby, L., Marinval, P. and Boissinot, P. (2011) Never mind the bottle. Archaeobotanical Evidence of Beer-brewing in Mediterranean France and the Consumption of Alcoholic Beverages during the 5th Century BC. *Human Ecology* 39, 351–360.

Bouchette, A. (1999) Les plantes cultivées et sauvages. In A. Poyeton (ed.) *Etablissements ruraux du deuxième Age du fer à St. Julien-du-Sault, 'Les Boulins'*. DFS, Programme d'interventions dans les carrières de granulat de l'Yonne. AFAN-SRA Bourgogne: Passy-Dijon, Nancy, pp. 55–67.

Bourdieu, P. (1979) *La distinction: critique sociale du jugement. Le sens commun.* Les editions de Minuit, Paris.

Bourdieu, P. (1980) *Le sens pratique.* Les éditions de Minuit, Paris.

Braudel, F. (1967) *Civilisation matérielle, économie et capitalisme, XVe-XVIIIe siècle.* Armand Colin, Paris.

Brillat-Savarin, A. (2009) *Physiologie du goût.* Flammarion, Paris.

Broncano, S. (1989) *El depósito votivo ibérico de El Amarejo, Bonete, Albacete*, Ministerio de Cultura, Madrid.

Brozio, J.-P. (2010) Neue Untersuchungen zu Trichterbecherzeitlichen Organisationsformen in Ostholstein. *Archäologische Nachrichten aus Schleswig-Holstein* 16: 30–33.

Brozio, J.-P. (2011) Von Siedlungen und Grabenwerken der Trichterbechergemeinschaften. Tielthema: Frühe Bauern im Norden. *Archäologie in Deutschland* 2: 22–25.

Burger 1993

Burger, R. and Salazar-Burger, L. (1991) The second season of investigations at the Initial Period center of Cardal, Peru. *Journal of Field Archaeology* 18, 275–296.

Burger, R. and Salazar-Burger, L. (1993) The place of dual organization in Early Andean ceremonialism: a comptarative review. In L. Millones and Y. Onuki (eds.) *El mundo ceremonial andino.* Senri Ethnological Studies 37. National Museum of Ethnology, Osaka, pp. 97–116.

Buxó, R. (1997) *Arqueología de las Plantas.* Grijalbo Mondadori, Barcelona.

Buxó, R. (2008) The agricultural consequences of colonial contacts on the Iberian Peninsula in the first millennium BC. *Vegetation History and Archaeobotany* 17: 145–154.

Bye, Jr., R. A. (1981) Quelites – Ethnoecology of Edible Greens – Past, Present, and Future. *Journal of Ethnobiology* 1 (1): 109–123.

Caesar, G. J. (1996) *La guerre des Gaules.* Traduction by L.-A. Constans. 14th edition reviewed and corrected by A. Balland, 1996. Editions Belles Lettres, Paris.

Cañete, C. and Vives-Ferrándiz, J. (2011) Almost the same: dynamic domination and hybrid contexts in Iron Age Lixus, Larache, Marocco. *World Archaeology* 43 (1), 124–143.

Cappers, R. T. J. (1995) A palaeoecological model of the interpretation of wild plant species. *Vegetation History and Archaeobotany* 4, 249–257.

Carrere, I. and Forrest, V. (2009) Et si le Néolithique s'arrêtait dans les années 1950? In D. Fabre (ed.) *De Méditerranée et d'ailleurs.... Mélanges offerts à Jean Guilaine.* Archives d'Ecologie Préhistorique, Toulouse, pp. 173–190

Chapa, T. and Mayoral, V. (1998) Explotación económica y fronteras políticas: diferencias entre el modelo ibérico y el romano en el límite entre la Alta Andalucía y el Sureste. *Archivo español de arqueología* 71 (177–178), 63–78.

Chapa, T., Pereira, J., Madrigal, A., and Mayoral, V. (1998) *La necrópolis ibérica de Los Castellones de Ceal (Hinojares, Jaén).* Universidad de Jaén, Jaén.

Chevalier, A. (1999) Archaeology + Ethnobotany = Palaeoethnobotany ? *Bulletin de la Société des Américanistes* 63, 149–162.

Chevalier, A. (2002) *L'exploitation des plantes sur la côte péruvienne en contexte formatif.* Ph.D. thesis, Université de Genève, Geneva.

Clement, C. (2007) Pre-European Forest Cultivation in Amazonia. In W. Baleé and C. L. Erickson (eds.) *Time and Complexity in Historical Ecology: Studies in the Neotropical Lowlands.* Columbia University Press, New York, pp. 165–185.

Coe, S. D. (1994) *America's First Cuisines.* University of Texas Press, Austin.

Cohen, M. N. (1987) The significance of long-term changes in human diet and food economy. In M. Harris and E. Ross (ed.) *Food and evolution: toward a theory of human food habits.* Temple University Press, Philadelphia, pp. 261–283.

Collin Bouffier, S. and Sauner, M.-H. (2006) *Substitution de nourriture, nourriture de substitution en Méditerranée.* Actes du colloque (Aix-en-Provence, 14–15 Mars 2003). Presses Universitaires de Provence, Aix-en-Provence.

Counihan, C. and Kaplan, S. (1998) *Food and gender: identity and power.* Harwood Academic Publishers, Amsterdam.

Cruz-García, G. S. (2005) Children's knowledge and valuation of wild food plants: The influence of an educational program with tribal and non-tribal children in Western Ghats, India. In F. Ertuğ (ed.) *Ethnobotany: at the junction of the continents and the disciplines.* Proceedings of the Fourth International Congress of Ethnobotany (ICEB 2005). Yeditepe University, Istanbul, pp. 65–67.

Cruz-García, G. S. (2006) The mother – child nexus. Knowledge and valuation of wild food plants in Wayanad, Western Ghats, India. *Journal of Ethnobiology and Ethnomedicine* 2 (39).

Curtis, J. H., Hodell, D. A. and Brenner, M. (1996) Climate Variability on the Yucatan Peninsula (Mexico) during the Past 3500 Years, and Implications for Maya Cultural Evolution. *Quaternary Research* 46, 37–47.

Delgado, A. (2010) De las cocinas coloniales y otras historias silenciadas: domesticidad, subalternidad e hibridación en las colonias fenicias occidentales. In C. Mata Parreño, G. Pérez Jordà and J. Vives-Ferrándiz Sánchez (eds.) *'De la cuina a la taula'. IV Reunió d'economia en el primer mil·lenni AC.* Saguntum extra 9. Laboratorio de Arqueología, Valencia, pp. 8–43.

Delgado, A. and Ferrer, M. (2007) Cultural contacts in colonial settings: the construction of new identities in Phoenician settlements of the western Mediterranean. *Stanford Journal of Archaeology* 5, 18–42.

Denevan, W. L. (2007) Pre-European Forest Cultivation in Amazonia. In W. Baleé and C. L. Erickson (eds.) *Time and Complexity in Historical Ecology: Studies in the Neotropical Lowlands*. Columbia University Press, New York, pp. 153–163.

Descola, P. (2005) *Par-delà nature et culture*. Gallimard, Paris.

Descola, P. (2011) *L'écologie des autres. L'anthropologie et la question de nature*. Quae, Versailles.

De Wet, J. M. J. and Harlan, J. R. (1975) Weeds and Domesticates: Evolution in the Man-Made Habitat. *Economic Botany* 29: 99–107.

Dibbern, H. and Hage, F. (2010) Erdwerk und Megalithgräber in der Region Albersdorf. Vorbericht zu den Grabungskampagnen am Diecksknöll und Brutkamp. *Archäologische Nachrichten aus Schleswig-Holstein* 16: 34–37.

Dietler, M. (1996) Feasts and commensal politics in the political economy: food, power, and status in prehistoric Europe. In P. Wiessner and Schiefenhövel (eds.) *Food and the Status Quest: An Interdisciplinary Perspective*. Berghahn Publishers, Oxford, pp. 87–125.

Dietler, M. and Hayden, B. (2001) *Feasts. Archaeological and ethnographic perpectives on food, politics, and power.* Smithsonian Series in Archaeological Inquiry. Smithsonian Institution Press, Washington/London.

Dietler, M. and Herbich, I. (2006) Liquid Material Culture. Following the Flow of Beer among the Luo of Kenya. In H.-P. Wotzka (ed.) *Grundlegungen. Beiträge zur europäischen und afrikanischen Archäologie für Manfred K. H. Eggert*. Francke, Tübingen, pp. 395–407.

Dillehay, T. D. (1995) Introduction. In T. D. Dillehay (ed.) *Tombs for the Living: Andean Mortuary Practices*. Dumbarton Oaks Research Library and Collection, Washington, D.C., pp. 1–26.

Dörfler, W. (2001) Von der Parklandschaft zum Landschaftspark. Rekonstruktion der neolithischen Landschaft anhand von Pollenanalysen aus Schleswig Holstein. In R. Kelm (ed.) *Zurück zur Steinzeitlandschaft. Archäologische und ökologische Forschung zur jungsteinzeitlichen Kulturlandschaft und ihrer Nutzung in Nordwestdeutschland*. Albersdorfer Forschungen zur Archäologie und Umweltgeschichte 2. Boyens, Heide, pp. 39–56.

Dörfler, W., Feeser, I., van den Bogaard, C., Dreibrodt, S., Erlenkeuser, H., Kleinmann, A., Merkt, J. and Wiethold, J. (2012) A high quality annually laminated sequence from Lake Belau, northern Germany: Revised chronology and its implications for palynological and tephrochronological studies. *The Holocene* 22 (12), 1413–1426.

Douglas, M. (1971) Deciphering a Meal. In C. Geertz (ed.) *Myth, Symbol, and Culture*. Norton, New York, pp. 61–78.

Douglas, M. (1975) *Implicit meanings*. Routledge, London/ Boston.

Douglas, M. (1984) *Food in the social order: studies of food and festivities in three American communities*. Russell Sage Foundation, New York.

Douglas, M. (1997) Deciphering a meal. In C. Counihan and P. Van Esterik (eds.) *Food and culture: a reader.* Routledge, New York/London, pp. 36–54.

Dulanto, J. (1994) Prospección arqueológica en el Valle de Lurín. *Willay* 41.

Dulanto, J. (2002a) The Archaeological Study of Ancestor Cult Practices: The Case of Pampa Chica, a Late Initial Period and Early Horizon Site on the Central Coast of Peru. In H. Silvermann and D. Small (eds.) *The space and place of death*. AP3A 11. American Anthropological Association, Arlington (VA), pp. 97–117.

Dulanto, J. (2002b) Pampa Chica: prácticas de culto a los ancestros en la costa central del Perú. *Gaceta Arqueológica Andina (Lima)* 26, 37–67.

Dulanto, J. (2008) *Festivales de muerte: Pampa Chica y los orígenes de la civilización en los Andes centrales*. Ph.D. thesis, University of Illinois at Urbana-Champaign.

Dunning, N. P., Scarborough, V., Valdez Jr., F., Luzzader-Beach, S., Beach, T., and Jones, J. G. (1999) Temple Mountains, Sacred Lakes, and Fertile Fields: Ancient Maya Landscpaes in Northwestern Belize. *Antiquity* 73:650–660.

Dunning, N. P., Jones, J. G., Beach, T. and Luzzadder-Beach, S. (2003) Physiography, Habitats, and Landscapes of the Three Rivers Region. In V. L. Scarborough, F. Valdez, Jr. and N. Dunning (eds.) *Heterarchy, Political Economy, and the Ancient Maya: The Three Rivers Region of the East-Central Yucatán Peninsula*. University of Arizona Press, Tucson, pp. 14–24.

Eckman, K. (1994) *Rural households and sustainability: Integrating environmental and gender concerns into home economics curricula*. Women in Agricultural Production and Rural Development Service, Human Resources, Institutions and Agrarian Reform Division. FAO, Rome.

Engelmark, R. (1984) Two useful plants from Iron Age graves in central Sweden. In E. Baudou (ed.) *Papers in Northern Archaeology*. Archaeology and Environment 2, Umeå, pp. 87–92.

Espejo, Z. (1900) *Costumbres de derecho y economía rural consignadas en los contratos agrícolas usuales en las provincias de la Península española, agrupados según los antiguos reinos*. Imprenta del Asilo de Huérfanos del Sagrado Corazón de Jesús, Madrid.

Fairweather A. D. and Ralston, I. B. M. (1993) The Neolithic timber hall at Balbridie, Grampian Region, Scotland: the building, the date, the plant macrofossils. *Antiquity* 67, 313–323

Farb, P. and Armelagos, G. (1983) *Consuming passions: the anthropology of eating*. Pocket Books, New York.

Fedick, S. L. (1994) Ancient Maya Agricultural Terracing in the Upper Belize River Area. *Ancient Mesoamerica* 5, 107–127.

Fedick, S. L. (1996) *The Managed Mosaic*. University of Utah Press, Salt Lake City.

Fernández García, M. (1966) *Fuentes para la historia de Buitrago y su Tierra. Volumen I. Ordenanzas.* Ed. del autor, Madrid.

Fischer, A. (2002) Food for feasting? An evaluation of explanations of the Neolithisation of Denmark and southern Sweden. In F. Anders and K. Kristiansen (eds.) *The Neolithisation of Denmark.* J. R. Collis Publications, Sheffield, pp. 341–364.

Fischler, C. (1993) *L'homnivore.* Odile Jacob, Paris.

Fischler, C. (1994) *Manger magique* (Mutations/Mangeur; vol. 149). Autrement, Paris.

Flandrin, J.-L. and Montanari, M. (eds.) (2000) *Food: a culinary history from Antiquity to the présent.* Penguin, New York.

Fox, J. G. (1996) Playing With Power: Ballcourts and Political Ritual in Southern Mesoamerica. *Current Anthropology* 37, 483–496.

Franco Silva, A. (1991) Pedraza de la Sierra. El proceso de formación de unas ordenanzas de villa y tierra. *Historia. Instituciones. Documentos* 18, 97–142.

Fritsch, B., Furholt, M., Hinz, M., Lorenz, L., Nelson, H., Schafferer, G., Schiesberg, S., and Sjögren, K.-G. (2010) *Dichtezentren und lokale Gruppierungen - Eine Karte zu den Großsteingräbern Mittel- und Nordeuropas.* Website: www.jungsteinsite.uni-kiel.de (accessed 29/11/2010).

Furholt, M. and Müller, J. (2011) The earliest monuments in Europe - Architecture and social structures (5000–3000 BC). In M. Furholt, F. Lüth and J. Müller (eds.) *Megaliths and identities. Early monuments in Neolithic societies from the Atlantic to the Baltic. 3rd European Megalithic studies group meeting 13th–15th of May 2010 at Kiel University.* Verlag Dr. Rudolf Habelt, Bonn, pp. 15–32.

Gallant, T. (1991) *Risk and survival in Ancient Greece: reconstructing the rural domestic economy.* Stanford University Press, Stanford (CA).

García Cardiel, J. (2009) Renacer entre leones: una nueva perspectiva de los leones de Pozo Moro (Chinchilla, Albacete). *Lucentum* 28, 51–68.

García Sanz, A., Martín, J.-L., Pascual, J.A. and Pérez Moreda, V. (1981) *Propiedades del cabildo segoviano, sistemas de cultivo y modos de explotación de la tierra a fines del siglo XIII.* Documentos y estudios para la historia del occidente peninsular durante la Edad Media 2. Universidad de Salamanca, Salamanca.

Goldstein, P. (2003) From stew-eaters to maize-drinkers. The *chicha* economy and the Tiwanaku expansion. In T. Bray (ed.) *The archaeology and politics of food and feasting in early states and empires.* Kluwer Academic, New York, pp. 143–172.

Goldstein, D., Coleman Goldstein, R. and Williams, P. (2008) You Are What You Drink. A Sociocultural Reconstruction of Pre-Hispanic Fermented Beverage Use at Cerro Baúl, Moquegua, Peru. In J. Jennings and B. Bowser (eds.) *Drink, Power, and Society in the Andes.* University Press of Florida, Gainesville, pp. 133–166.

Gómez-Pompa, A., Flores Guido, J. S., and Aliphat Fernández, M. (1990) The Sacred Cacao Groves of the Maya. *Latin American Antiquity* 1:247–257.

González Reyero, S. (2008) Approaches de l'image ibérique: Nature, identité et changement social. *Perspective, Revue de l'Institut National d´Histoire de l´Art* 1: 49–66.

González Reyero, S. (2010) Nature and the symbolization of landscape in the Iberian iconography (III–II BC). Images and social dynamics. In L. Levêque and M. Ruiz del Árbol (eds.) *Heritage, Images Memory of European Landscapes.* L´Harmattan, Paris, pp. 41–57.

González Villaescusa, R. (2000) Aportación al estudio de los paisajes agrarios de la Edetania. In M. Olcina and J. A. Soler (coords.) *Scripta in honorem Enrique A. Llobregat Conesa.* Vol. 1. Instituto Alicantino Juan Gil-Albert, Alicante, pp. 325–340.

Goody, J. (1982) *Cooking, cuisine and class: a study in comparative sociology.* Cambridge University Press, Cambridge.

Gosden, C. (2008) *Arqueología y colonialismo: el contacto cultural desde 5000 a.C. hasta el presente.* Bellaterra, Barcelona.

Goudineau, C. and Peyre, C. (1993) *Bibracte et les Eduens, à la découverte d'un peuple gaulois.* Editions Errance, Paris.

Graeber, D. (2001) *Toward an anthropological theory of value: The false coin of our own dreams.* Palgrave, New York.

Grau, I. (2002) *La organización del territorio en el área central de la 'Contestania' Ibérica.* Universidad de Alicante, Alicante.

Grau, I. (2007) Dinámica social, paisaje y teoría de la práctica. Propuestas sobre la evolución de la sociedad ibérica en el área central del Oriente peninsular. *Trabajos de prehistoria* 64 (2), 119–142.

Grau, I. (2010) Vajillas mediterráneas y prácticas de comensalidad en el área central de la contestania ibérica. In C. Mata Parreño, G. Pérez Jordà and J. Vives-Ferrándiz Sánchez (eds.) *'De la cuina a la taula'. IV Reunió d'economia en el primer mil·lenni AC.* Saguntum extra 9. Laboratorio de Arqueología, Valencia, pp. 264–271.

Grau, I. and Olmos, R. (2005) El Vas dels Guerrers de la Serreta. *Recerques del Museu d'Alcoi* 14, 79–98.

Grau Almero, E. and Duque Espino, D. (2007) Los paisajes rurales protohistóricos: una síntesis arqueobotánica. In A. Rodríguez Díaz and I. Pavón Soldevila (coords.) *Arqueología de la tierra: paisajes rurales de la protohistoria peninsular.* Universidad de Extremadura, Cáceres, pp. 297–326.

Grau, I., Olmos, R. and Perea, A. (2008) La habitación sagrada de la ciudad ibérica de La Serreta. *Archivo Español de Arqueología* 81: 5–29.

Greenfield, H. and Southgate, D. (2003) *Food composition data: production, management and use.* FAO, Rome.

Greig, J. (1991) The British Isles. In K. E. Behre, K. Wasylikova and W. van Zeist (eds.) *Progress in old world palaeoethnobotany. A retrospective view on the occasion of 20 years of the International Work Group for Palaeoetnobotany*. Balkema, Rotterdam/Brookfield, pp. 299–334.

Grivetti, L. E. and Ogle, B. M. (2000) Value of traditional foods in meeting macro- and micronutrient needs: The wild plant connection. *Nutrition Research Reviews* 13 (1), 31–46.

Grodzins, A. (2002) Children and trees in north India. *Worldviews* 6 (3), 276–299.

Grognier, L.-F. and Magne, H. (1841) *Cours de multiplication et de perfectionnement des principaux animaux domestiques*. Bouchard-Huzard, Paris/Lyon.

Guillaumet, J.-P., Szabó, M. and Barral, P. (2002) L'urbanisme à Bibracte. In D. Maranski and V. Guichard (eds.) *Les âges du Fer en Nivernais, Bourbonnais et Berry oriental: regards européens sur les âges du Fer en France. Actes du XVIIe colloque AFEAF (Nevers, 20-23 mai 1993)*. Collection Bibracte 6. Centre archéologique européen du Mont Beuvray, Glux-en-Glenne, pp. 277–292.

Gumerman [IV], G. (1997) Food and complex societies. *Journal of Archaeological Method and Theory* 4 (2): 105–139.

Gurung, J. D. (2002) Getting at the Heart of the Issue: Challenging Male Bias in Nepal's Department of Forests. *Mountain Research and Development* 22, 212–215.

Hageman, J. B. (2004) The Lineage Model and Archaeological Data in Late Classic Northwestern Belize. *Ancient Mesoamerica* 15, 63–74.

Hageman, J. B. and Lohse, J. C. (2003) Heterarchy, Corporate Groups, and Late Classic Resource Management in Northwestern Belize. In V. L. Scarborough, F. Valdez, Jr., and N. Dunning (eds.) *Heterarchy, Political Economy, and the Ancient Maya: The Three Rivers Region of the East-Central Yucatán Peninsula*. University of Arizona Press, Tucson, pp. 109–121.

Harris, M. (1974) *Cows, Pigs, Wars and Witches: The Riddles of Culture*. Random House, New York.

Harris, M. (1986) *Good to eat: riddles of food and culture*. Allen & Unwin, London.

Hartz, S., Heinrich, D. and Lübke, H. (2000) Frühe Bauern an der Küste. Neue ^{14}C-Daten und aktuelle Aspekte zum Neolithisierungsprozeß im norddeutschen Ostseeküstengebiet. *Prähistorische Zeitschrift* 75: 129–152.

Hastorf, C. (2003) Andean luxury foods: special food for the ancestors, deities and the élite. *Antiquity* 77 (297), 545–554.

Hayden, B. (1998) Feasting in prehistoric and traditional societies. In P. Wiessner and W. Schiefenhövel (eds.) *Food and the status quest: an interdisciplinary perspective* Bergahn Books, Providence/Oxford, pp. 127–147.

Hellmund, M. (2008) The Neolithic records of *Onopordum acanthium, Agrostemma githago, Adonis* cf. *aestivalis* and *Claviceps purpurea* in Sachsen-Anhalt, Germany. *Vegetation History and Archaeobotany* 17, 123–130.

Hey, G., Mulville, J. and Robinson, M. (2003) Diet and culture in southern Britain: the evidence from Yarnton. In M. P. Pearson (ed.) Food, culture and identity in the Neolithic and early Bronze Age. *BAR International Series* 1117. Archaeopress, Oxford, pp. 79–88.

Heywood, V. (1999) *Use and potential of wild plants in farm households. Vol. 15*. FAO Farm Systems Management Series FAO, Rome.

Hodell, D. A., Curtis, J. A. and Brenner M. (1995) Possible Role of Climate in the Collapse of Classic Maya Civlization. *Nature* 375, 391–4.

Hoika, J. (1981) Jungsteinzeitliche Siedlungsreste aus Oldenburg-Dannau, Kreis Ostholstein. *Offa* 1981: 53–83.

Hoika, J. (1993) Grenzfragen oder: James Watt und die Neolithisierung. *Archäologische Informationen* 16, 6–19.

Iborra, M. P., Mata, C., Moreno, A., Quixal, D. and Pérez Jordà, G. (2010) Prácticas culinarias y alimentación en asentamientos ibéricos valencianos. In C. Mata Parreño, G. Pérez Jordà and J. Vives-Ferrándiz Sánchez (eds.) *'De la cuina a la taula'. IV Reunió d'economia en el primer mil·lenni AC*. Saguntum extra 9. Laboratorio de Arqueología, Valencia, pp. 99–114.

Isbell, W. and McEwan, G. (eds.) (1991) *Huari administrative structure: prehistoric monumental architecture and state government*. Dumbarton Oaks Round Table, May 17–19 1985. Dumbarton Oaks, Washington DC.

Iversen, J. (1941) Landnam i Danmarks Stenalder. En pollenanalytisk Undersøgelse over det første Landbrugs Indvirkning paa Vegetationsudviklingen [Land occupation in Denmark's Stone Age. A pollen-analytical study of the influence of farmer culture in the vegetational development]. *Danmarks Geologiske Undersøgelse II. Raekke* 66. 1–68.

Jacomet, S. and Kreuz, A. (1999) Archäobotanik. Aufgaben, Methoden und Ergebnisse vegetations- und agrargeschichtlicher Forschung. Ulmer Verlag, Stuttgart.

Jäger (2006) *Thesaurus Cultus et Rituum Antiquorum. Vol 3*. Paul Getty Foundation, Los Angeles.

Jennings, J. and Bowser, B. (2008) *Drink, Power, and Society in the Andes*. University Press of Florida, Gainesville.

Jensen, P. M., Høyem Andreasen, M. and Mikkelsen, P. H. (2010) Bulbous oat grass – a magic plant in prehistoric Jutland and Funen. In C. Bakels, K. Fennema, W. A. Out and C. Vermeeren (eds.) *Van planten en slakken/Of Plants and Snails*. A collection of papers presented to Wim Kuijper in gratitude for forty years of teaching and identifying. Sidestone Press, Leiden, pp. 103–114.

Joannes, F. (2000) The social function of banquets in earliest civilizations. In J.-L. Flandrin and M. Montanari (ed.) *Food: a culinary history from Antiquity to the présent*. Penguin, New York, pp. 32–37.

Jones, G. (2000) Evaluating the importance of cultivation and collecting in Neolithic Britain. In A. Fairbairn (ed.) *Plants in Neolithic Britain and beyond*. Neolithic Studies Group Seminar Papers 5. Oxbow Books, Oxford, pp. 79–84.

Jones, G. and Rowley-Conwy, P. (2007) On the importance of cereal cultivation in the British Neolithic. In S. Colledge and J. Conolly (eds.) *The origin and spread of domestic plants in Southwest Asia and Europe*. Left Coast Press, Walnut Creek, pp. 391–419.

Jones, M. (1985) Archaeobotany beyond subsistence reconstruction. In G. Barker and C. Gamble (eds.) *Beyond domestication in prehistoric Europe*. Academic Press, London, pp. 107–128.

Jones, M. (1988) The Arable Field: A Botanical Battleground. In M. Jones (ed.) *Archaeology and the Flora of the British Isles*. Oxford Committee for Archaeology, Oxford, pp. 86–92.

Josephat, M. (1997) *Wayanad. District handbooks of Kerala*. Natarajan I.A.S, Kerala.

Jovellanos de, G. M. (1968) *Informe sobre la Ley Agraria*. Edición de Materiales, Barcelona.

Kalčik, S. (1997) Ethnic Foodways in America: Symbol and the Performance of Identity. In L. Keller Brown and K. Mussell (eds.) *Ethnic and Regional Foodways in the United States: The Performance of Group Identity*. Tennessee University Press, Knoxville, pp. 37–65.

Kalis, A. J. and Meurers-Balke, J. (1998) Die 'Landnam'-Modelle von Iversen und Troels-Smith zur Neolithisierung des westlichen Ostseegebietes, ein Versuch ihrer Aktualisierung. *Prähistorische Zeitschrift* 73 (1), 1–24.

Keller Brown, L. and Mussell, K. (1997) Introduction. In L. Keller Brown and K. Mussell (eds.) *Ethnic and Regional Foodways in the United States: The Performance of Group Identity*. Tennessee University Press, Knoxville, pp. 3–13.

Kelm, R. (2006) Die frühe Kulturlandschaft der Region Albersdorf. Grundlagen, Erfassung und Vermittlung der urgeschichtlichen Mensch-Umwelt-Beziehungen in einer Geestlandschaft. EcoSys Suppl. 45a. Verein zur Förderung der Ökosystemforschung zu Kiel e.V, Kiel.

Killion, T., ed. (1992) *Gardens of Prehistory: The Archeology of Settlement Agriculture in Greater MesoAmerica*. University of Alabama Press, Tuscaloosa.

Kirleis, W., Feeser, I. and Klooß, S. (2011) Umwelt und Ökonomie. Titelthema: Frühe Bauern im Norden. *Archäologie in Deutschland* 2, 34–37.

Kirleis, W., Klooß, S., Kroll, H. and Müller, J. (2012) Crop growing and gathering in the northern German Neolithic. *Vegetation History and Archaeobotany* 21, 221–242.

Klassen, L. (2004) *Jade und Kupfer: Untersuchungen zum Neolithisierungsprozess im westlichen Ostseeraum unter besonderer Berücksichtigung der Kulturentwicklung Europas 5500-3500 BC*. Aarhus University Press, Århus.

Klooß, R. (2008) Ein Fundplatz des Übergangs vom Mittelneolithikum A zum Mittelneolithikum B aus Wangels LA 505, Kreis Ostholstein. Ein Vorbericht. In W. Dörfler and J. Müller (eds.) *Umwelt - Wirtschaft - Siedlungen im dritten vorchristlichen Jahrtausend Mitteleuropas und Südskandinaviens*. Internationale Tagung Kiel 4.–6. November 2005. Wachholtz, Neumünster, pp. 127–134.

Koopman, J. (1997) *Gender and participation in agricultural development planning: Key issues from ten case studies*. Working document for the 'Workshop on Gender and Participation in Agricultural Development Planning: Harvesting Best Practices', Rome, 8–12 December 1997. Women in Development Service. FAO, Rome.

Kreuz, A. (1992) Charcoal from ten early Neolithic settlements in Central Europe and its interpretation in terms of woodland management and wildwood resources. In J.-L. Vernet (ed.) *Les charbons de bois, les anciens écosystèmes et le rôle de l'Homme. Colloque organisé à Montpellier du 10 au 13 septembre 1991*. Bulletin de la Société Botanique de France 139 (2/3/4). Société Botanique de France, Paris, pp. 383–394.

Kroll, H. (1981) Mittelneolithisches Getreide aus Dannau. *Offa* 38, 85–90.

Kroll, H. (2001) Der Mohn, die Trichterbecherkultur und das südwestliche Ostseegebiet. Zu den Pflanzenfunden aus der mittelneolithischen Fundschicht von Wangels LA 505, Kr. Ostholstein. In R. Kelm (ed.) *Zurück zur Steinzeitlandschaft, Archäologische und ökologische Forschung zur jungsteinzeitlichen Kulturlandschaft und ihrer Nutzung in Nordwestdeutschland*. Albersdorfer Forschungen zur Archäologie und Umweltgeschichte 2. Boyens, Heide, pp. 70–76.

Kroll, H. (2007) The plant remains from the Neolithic Funnel Beaker site of Wangels in Holsatia, northern Germany. In S. Colledge and J. Conolly (eds.) *The Origins and Spread of Domestic Plants in Southwest Asia and Europe*. Left Coast Press, Walnut Creek, pp. 349–357.

Kulirani, F. (2003) The shrinking livelihood strategies of the Paniyar. In C. Norström (ed.) *Livelihood Strategies among forest - related tribal groups of South India: contextual analysis of local livelihood strategies at the Centre for Indian Studies*. Interdisciplinary Conference (Mysore, India). Swedish South Asian Studies Network, Lund University, Lund, pp. 1–10.

Kumar, A., Girigian, G. and Venugopal, R. (eds.) (2003) *A report on agri-export zone: prospects of Wayanad District*. Wayanad Agriculture Development Committee and Community Agrobiodiversity Centre – M.S. Swaminathan Research Foundation, Kerala.

Kuznar, L. A. (2001) An introduction to Andean religious ethnoarchaeology: Preliminary results and future directions. In L. A. Kuznar (ed.) *Ethnoarchaeology of Andean South America*. International Monographs in Prehistory, Ann Arbor, MI, 38–66.

Labeaune, R. and Wiethold, J. (2007) L'habitat du Ier âge du Fer dans le Dijonnais, d'après les fouilles récentes: résultats archéologiques et carpologiques. In P. Barral, A. Daubigney, C. Dunning, G. Kaenel, and M.-J. Roulière-Lambert (eds.) *L'âge du Fer dans l'arc jurassien et ses marges. Dépôts, lieux sacrés et territorialité à l'âge du Fer. Actes du XXIXe Colloque AFEAF (Bienne, 5-8 mai 2005)*. Annales Littéraires de l'université de Franche-Comté, Environnement, société et archéologie 11. Presses Universitaires de Franche-Comté, Besançon, pp. 73–100.

Lahlou, S. (1998) *Penser manger*. PUF, Paris.

Laubenheimer, F., Ouzoulias, P. and van Ossel, P. (2003) La bière en Gaule. Sa fabrication, les mots pour le dire, les vestiges archéologiques: première approche. *Revue archéologique de Picardie* 1–2, 47–63.

LeCount, L. (2001) Like Water for Chocolate: Feasting and Political Ritual among the Late Classic Maya at Xunantunich, Belize. *American Anthropologist* 103, 935–953.

Lentz, D. L. (2000) Anthropocentric Food Webs in the Precolombian Americas. In D. Lentz (ed.) *Imperfect Balance*. Columbia University Press, New York, pp. 89–120.

Lévi-Strauss, C. (1962) *La pensée sauvage*. Plon, Paris.

Lévi-Strauss, C. (1964) *Le cru et le cuit* (Mythologiques; 1). Plon, Paris.

Lévi-Strauss, C. (1997) The culinary triangle. In C. Counihan and P. Van Esterik (eds.) *Food and culture: a reader*. Routledge, New York/London, pp. 28–35.

Livarda, A. and van der Veen, M. (2008) Social access and dispersal of condiment in North-West Europe from the Roman to the medieval period. *Vegetation History and Archaeobotany* 17 (S1), 201–210.

López Pardo, F. (2006) *La torre de las almas. Un recorrido por los mitos y creencias del mundo fenicio y orientalizante a través del monumento de Pozo Moro*. Universidad Complutense, Madrid.

Lütjens, I. and Wiethold, J. (1999) Vegetationsgeschichtliche und archäologische Untersuchungen zur Besiedlung des Bornhöveder Seengebietes im Neolithikum. *Archäologische Nachrichten aus Schleswig-Holstein* 9, 30–67.

Malaza, M. (2003). Modernization and gender dynamics in the loss of agrobiodiversity in Swaziland's food system. In P. Howard (ed.) *Women and plants. Gender relations in biodiversity management and conservation*. Zed Press and Palgrave Macmillan: London and New York, pp. 243–257.

Marchant, R., Almeida, L., Behling, H., Berrio, J. C., Bush, M., Cleef, A., Duivenvoorden, J., Kappelle, M., De Oliveira, P., Teixeira de Oliveira-Filho, A., Lozano-García, S., Hooghiemstra, H., Ledru, M.-P., Ludlow-Wiechers, B., Markgraf, V., Mancini, V., Paez, M., Prieto, A., Rangel, O. and Salgado-Labouriau, M. L. (2002) Distribution and ecology of parent taxa of pollen lodged within the Latin American Pollen Database. *Review of Palaeobotany and Palynology* 121 (1), 1.

Mata, C., Badal, E., Collado, E. and Ripollés, P. P. (2010) *Flora ibérica. De lo real a lo imaginario*. Trabajos Varios del SIP 111). SIP, Valencia.

McAnany, P. (1995) *Living with the Ancestors: Kinship and Kingship in Ancient Maya Society*. University of Texas Press, Austin.

Mead, M. (1997) The changing significance of food. In C. Counihan and P. Van Esterik (eds.) *Food and culture: a reader*. Routledge, New York/London, pp. 11–19.

Meggers, B. (1954). Environmental limitations on the development of culture. *American Anthropologist*, 56 (5), 801–824.

Meggers, B. (1979). Climatic oscillation as a factor in the prehistory of Amazonia. *American Antiquity*, 44 (2), 252–266.

Meigs, A. (1997) Food as cultural construction. In C. Counihan and P. Van Esterik (eds.) *Food and culture: a reader*. Routledge, New York/London, pp. 95–106.

Méniel, P. (2002) La chasse en Gaule une activité aristocratique ? In V. Guichard, F. Perrin and J.-C. Decourt (eds.) *L'aristocratie celte à la fin de l'âge du Fer (du IIème avant J-C. au Ier après J-C.). Actes de la table ronde (10-11 juin 1999)*. Collection Bibracte 5. Centre archéologique européen du Mont Beuvray, Glux-en-Glenne, pp. 223–230.

Mennell, S. (1985) *All manners of food: eating and taste in England and France from the Middle Ages to the present*. Blackwell, Oxford.

Mennell, S. (1997) On the civilizing of appetite. In C. Counihan and P. Van Esterik (eds.) *Food and culture: a reader*. Routledge, New York/London, pp. 315–337.

Mennell, S., Murcott, A. and van Otterloo, A. (1992) *The sociology of food: eating, diet, and culture*. Sage, London/Newbury Park (CA).

Mintz, S. W. (1986) *Sweetness and Power: The Place of Sugar in Modern History*. Penguin, New York.

Mintz, S. W. (1996) *Tasting food, tasting freedom*. Beacon Press, Boston.

Mischka, D. (2010) *Flintbek LA 3, Biography of a monument*. Website: www.jungsteinsite.uni-kiel.de (accessed 05/01/2010).

Mischka, D. (2011) The Neolithic burial sequence at Flintbek LA 3, north Germany, and its cart tracks: a precise chronology. *Antiquity* 85, 742–758.

Mischka, D. and Demnick, D. (2011) Totenlandschaften für ein Leben in Ewigkeit. Titelthema: Frühe Bauern im Norden. *Archäologie in Deutschland* 2, 26–29.

Moffett, L., Robinson, M. A. and Straker, V. (1989) Cereals, fruits and nuts: charred plant remains from Neolithic sites in England and Wales and the Neolithic economy. In A. Milles, D. Williams and N. Gardner (eds.) *The beginnings of agriculture*. BAR International Series 496. Archaeopress, Oxford, pp. 243–261.

Monk, M. A. (2000) Seeds and soils of discontent: an environmental archaeological contribution to the nature of the early Neolithic. In A. Desmond, G. Johnson, M. McCarthy, J. Sheehan and E. Shee Twohig (eds.) *New agendas in Irish prehistory: papers in commemoration of Liz Anderson*. Bray, Wordwell, pp. 67–87.

Molinos, M., Chapa, T., Ruiz, A., Pereira, J., Risquez, C., Madrigal, A., Estebán, A., Mayoryl, V. and Llorente, M. (1997) *El Santuario Heroico del Pajarillo (Huelma, Jaén).* Diputación Provincial de Jaén, Universidad de Jaén, Consejería de Cultura de la Junta de Andalucía y Centro Andaluz de Arqueología Ibérica, Jaén.Montanari, M. (2000a) Food systems and models of civilization. In J.-L. Flandrin and M. Montanari (ed.) *Food: a culinary history from Antiquity to the présent.* Penguin, New York, pp. 69–78.

Montanari, M. (2000b) Peasants, warriors, priests: image of society and styles of diet. In J.-L. Flandrin and M. Montanari (ed.) *Food: a culinary history from Antiquity to the présent.* Penguin, New York, pp. 178–185.

Montanari, M . (2010) *Le manger comme culture.* Editions de l'université de Bruxelles, Bruxelles.

Moore, J. (1989) Pre-Hispanic beer in coastal Peru: technology and social context of Prehistoric production. *American Anthropologist* 85 (3), 545–562.

Moran, E. F. (1990) Levels of Analysis and Analytical Level Shifting: Examples from Amazonian Ecosystem Research. In E. F. Moran (ed.) *The Ecosystem Approach in Anthropology.* University of Michigan, Ann Arbor, pp. 279–308.

Müller, J. (2009a) Neolithische Monumente und neolithische Gesellschaften. In H.-J. Beier, E. Classen, T. Doppler and B. Ramminger (eds.) *Neolithische Monumente und neolithische Gesellschaften.* Varia Neolithica 6. Beier & Beran, Langenweissbach, pp. 7–16.

Müller, J. (2009b) Monumente und Gesellschaft. *Archäologische Nachrichten aus Schleswig-Holstein* 15: 30–33.

Müller, J. (2010) Ritual cooperation and ritual collectivity: The social structure of the middle and younger Funnel Beaker North Group (3500–2800 BC). In M. Furholt, F. Lüth and J. Müller (eds.) *Megaliths and identities. Early monuments in Neolithic societies from the Atlantic to the Baltic.* 3rd European Megalithic studies group meeting 13th–15th of May 2010 at Kiel University. Verlag Dr. Rudolf Habelt, Bonn, pp. 273–284.

Müller, J. (2011) *Megaliths and Funnel beakers: societies in change 4100-2700 BC.* Stichting Nederlands Museum voor Anthropologie en Praehistorie, Amsterdam.

Müller, J., Brozio, J.-P., Demnick, D., Dibbern, H., Fritsch, B., Furholt, M., Hage, F., Hinz, M., Lorenz, L., Mischka, D. and Rinne, C. (2010) *Periodisierung der Trichterbecher-Gesellschaften. Ein Arbeitsentwurf.* Website: www.jungsteinsite.uni-kiel.de (accessed 29/11/2010).

Nabhan, G. P., and Carr, J. L. (1994) *Ironwood: An Ecological and Cultural Keystone of the Sonoran Desert.* Conservation International, Washington D.C.

Narayanan, M. K. R. and Kumar, N. A. (2007) Gendered knowledge and changing trends in utilization of wild edible greens in Western Ghats, India. *Indian Journal of Traditional Knowledge* 6 (1): 204–216.

Narayanan, M. K. R., Swapna, M. P. and Kumar, N. A. (2004) *Gender dimensions of wild food management in Wayanad, Kerala.* Community Agrobiodiversity Centre – M.S. Swaminathan Research Foundation (MSSRF-CAbC) and Uttara Devi Resource Centre for Gender and Development, Kerala.

Nelle, O. and Dörfler, W. (2008) A summary of the late- and post-glacial vegetation history of Schleswig-Holstein. In J. Dengler, C. Dolnik and M. Trepel (eds.) *Flora, Vegetation, and Nature Conservation from Schleswig-Holstein to South America. Festschrift for Klaus Dierßen on Occasion of his 60th Birthday.* Mitteilungen der Arbeitsgemeinschaft Geobotanik in Schleswig-Holstein und Hamburg 65, 45–68.

Newsom, L. A. and Wing, E. (2004) *On Land and Sea.* University of Alabama Press, Tuscaloosa.

Ogle, B. M. (2001) *Wild vegetables and micronutrient nutrition. Studies on the significance of wild vegetables in women's diets in Vietnam.* Uppsala University.

Olmer, F. (1999) Les amphores. In O. Buchsenschutz, J.-P. Guillaumet and I. Ralston (1999) *Les remparts de Bibracte, Recherches récentes sur la porte du Rebout et le tracé des fortifications.* Collection Bibracte 3. Centre européen d'archéologie, Glux-en-Glenne, pp. 163–174.

Olmer, F. (2002) Les amphores de Bibracte. Premiers apports des fouilles récentes. In D. Maranski and V. Guichard (eds.) *Les âges du Fer en Nivernais, Bourbonnais et Berry. Regards européens sur les âges du Fer en France. Actes du XVIIème colloque de l'AFEAF (Nevers, 20-23 mai 1993).* Collection Bibracte 6. Centre archéologique européen du Mont Beuvray, Glux-en-Glenne, pp. 327–336.

Olmer, F. (2003) *Les amphores de Bibracte, 2. Le commerce du vin chez les Eduens d'après les timbres d'amphores.* Collection Bibracte 7. Centre archéologique européen du Mont Beuvray, Glux-en-Glenne.

Olmos, R. (1992) *La sociedad ibérica a través de la imagen.* Ministerio de Cultura, Dirección General de Bellas Artes y Archivos, Madrid.

Olmos, R. (1996a) Pozo Moro: ensayos de lectura de un programa escultórico en el temprano mundo ibérico. In F. J. Martínez Quirce (coord.) *Al otro lado del espejo. Aproximación a la imagen ibérica*, Pórticos, Zaragoza, pp. 99–114.

Olmos, R. (1996b) La sociedad ibérica en el espejo de su imagen. Una mirada multiforme a la imagen ibérica'. In F. J. Martínez Quirce (coord.) *Al otro lado del espejo. Aproximación a la imagen ibérica*, Pórticos, Zaragoza, pp. 9–17.

Olmos, R. (1998) Indigenismo y romanización en la imagen ibérica de época republicana. In J. Mangas Manjarrés (coord.) *Italia e Hispania en la crisis de la república romana: actas del III Congreso Hispano-Italiano.* Universidad Complutense, Madrid, pp. 433–440.

Olmos, R. (2002) Los grupos escultóricos del Cerrillo

Blanco de Porcuna (Jaén): Un ensayo de lectura iconográfica convergente. *Archivo Español de Arqueología* 75 (185–186), 107–122.

Olmos, R. (2004) Imaginarios y prácticas religiosas entre los iberos. Perspectivas de un proceso historico. *Archiv für Religionsgeschichte* 6, 111–134.

Olmos, R. (2008) La simbolización del espacio sagrado en la antigua Iberia. In X. Dupré Raventós, S. Ribichini and S. Verger (eds.) *Saturnia Tellus: definizioni dello spazio consacrato in ambiente etrusco, italico, fenicio-punico, iberico e celtico*. Proceedings of the International Congress (Roma, 10–12 November 2004). Consiglio Nazionale delle Ricerche, Roma, pp. 251–265.

Olmos, R. (2010) Qu'est-ce que l'art protohistorique? Débat avec les points de vue de Henri-Paul Frankfort, Christoph Huth, Ricardo Olmos et Miklós Szabó, avec Stéphane Verger, *Perspective* 2010/2011 (2), 11–30.

Olmos, R. and Rouillard, P. (eds.) (1997) *Formes archaïques et arts ibériques. Formas arcaicas y arte ibérico*. Colección de la Casa de Velázquez, Madrid.

Olmos, R., Cabrera, P. and Montero, S. (coords.) (2005). *Paraíso cerrado, jardín abierto: el reino vegetal en el imaginario religioso del Mediterráneo*. Polifemo, Madrid.

Olmos, R., coord. (1999) *Los iberos y sus imágenes*. Instituto de Historia del CSIC – Micronet, Madrid.

Olmos, R. and Tortosa, T. (2010) Aves, diosas y mujeres. In T. Chapa and M. I. Izquierdo (eds.) *La Dama de Baza: Un viaje femenino al más alla*. Actas del Encuentro Internacional Museo Arqueológico Nacional, 27 y 28 noviembre 2007. Museo Arqueológico Nacional, Madrid, pp. 243–258.

Orlove, B. (1974) Reciprocidad, desigualdad y dominación. In G. Alberti and E. Mayer (eds.) *Reciprocidad e intercambio en los Andes peruanos*. Perú Problema 12. Instituto de Estudios Peruanos, Lima, pp. 290–321.

Ossio, J. (1992) *Parentesco, reprocidad y jerarquía en los Andes*. Fond Editorial de la Pontificia Universidad Católica del Perú, Lima.

Page, V. (2003) *El Cigarralejo*. Museo monográfico de Arte Ibérico, Murcia.

Perea, A. (2011) *La fíbula de Braganza / The Braganza Brooch*. CSIC, Madrid.

Perea, A. and Armbruster, B. (2011) Tomb 100 at Cabezo Lucero: new light on goldworking in fourth-century BC Iberia. *Antiquity* 85 (327): 158–171.

Pérez Jordà, G. (2000) La conservación y la transformación de los productos agrícolas en el Mundo Ibérico. *Saguntum* 3, 47–68.

Pérez Jordà G., Alonso A. and Iborra, P. (2007) Agricultura y ganadería protohistóricas en la Península Ibérica: modelos de gestion. In A. Rodríguez Díaz and I. Pavón Soldevila (eds.) *Arqueología de la tierra. Paisajes rurales de la protohistoria peninsular*. Universidad de Extremadura, Cáceres, pp. 327–372.

Perrin, F. and Decourt, J.-C. (2002) *L'aristocratie celte à la fin de l'âge du Fer (IIe s. avant J.-C. - Ier s. après J.-C.)*. Collection Bibracte 5. Centre archéologique européen du Mont Beuvray, Glux-en-Glenne.

Pfeffer, G. (1997) The scheduled tribes of middle India as a unit: problems of internal and external comparison. In G. Pfeffer and D. Kumar Behera (eds.) *Contemporary society tribal studies. Volume 1: Structure and process*. Ashok Kumar Mittal, New Delhi, pp. 3–27.

Platt, T. (1978) Symétries en miroir. Le concept de yanantin chez les Macha de Bolivie. *Annales. Économies, Sociétés, Civilisations* 33 (5–6), 1081–1107.

Preiss, S., Matterne, V. and Latron, F. (2005) An approach to funerary rituals in the Roman provinces: plant remains from a Gallo-Roman cemetery at Faulquemont (Moselle, France). *Vegetation History and Archaeobotany* 14, 362–372.

Price, L. (2005) Wild food plants in farming environments with special reference to northeast Thailand, food as functional and medicinal, and the social roles of women. In A. Pieroni and L. Price (eds.) *Eating and healing, traditional food as medicine*. The Haworth Press, New York, pp. 65–99.

Prieto, I. M. (2000) El recorrido en torno a la sepultura turriforme de Pozo Moro y secuencia narrativa de sus relieves: algunas propuestas. *Espacio, Tiempo y Forma, Serie II, Historia Antigua* 13, 325–356.

Rappaport, R. A. (1984) *Pigs for the Ancestors*. Yale University Press, New Haven.

Rassmann, K. (1993) *Spätneolithikum und frühe Bronzezeit im Flachland zwischen Elbe und Oder*. Beiträge zur Ur- und Frühgeschichte Mecklenburg-Vorpommerns 28. Archäologisches Landesmuseum Mecklenburg-Vorpommern, Lübstorf.

Regnell, M. and Sjögren, K.-G. (2006a) Use of wild plants and wood. In K.-G. Sjögren (ed.) *Ecology and Economy in Stone Age and Bronze Age Scania*. Oxbow Books, Oxford, pp. 80–105.

Regnell, M. and Sjögren, K.-G. (2006b) Introduction and development of agriculture. In K.-G. Sjögren (ed.) *Ecology and Economy in Stone Age and Bronze Age Scania*. Oxbow Books, Oxford, pp. 106–169.

Reina, R. E. (1967) Milpas and Milperos: Implications for Prehistoric Times. *American Anthropologist* 69, 1–20.

Riaza, R. (1935) Ordenanzas de Çiudad y de Tierra. *AHDE* 12, 468–495.

Rice, D. S. (1990) Roots: Resourceful Maya Farmers Enabled a Mounting Population to Survive in a Fragile Tropical Habitat. *Natural History* 2, 10–14.

Rivière, C. (1983) Pour une approche des rituels séculiers. *Cahiers Internationaux de Sociologie, Nouvelle Série* 74, 97–117.

Robinson, M. A. (2000) Further consideration of Neolithic charred cereals, fruits and nuts. In A. S. Fairbairn (ed.) Plants in Neolithic Britain and beyond. Neolithic Studies Group Seminar Papers 5. Oxbow Books, Oxford, pp. 85–90.

Rodríguez-Ariza, M. O. and Esquivel, J. A. (2007) Una

valoración de la paleovegetación del Sureste de la Península Ibérica durante la Prehistoria reciente a partir de las aplicaciones estadísticas en Antracología. In J. Molera, J. Farjas, P. Roura and T. Pradell (eds.) *Avances en Arqueometría 2005. Actas del VI Congreso Ibérico de Arqueometría*. Universitat de Girona, Girona, pp. 263–272.

Roehrs, H, Klooß, S. and Kirleis, W. (2013) Evaluating prehistoric finds of *Arrhenatherum elatius* var. *bulbosum* in north-western and central Europe with an emphasis on the first Neolithic finds in northern Germany. *Archaeological and Anthropological Sciences* 5 (1), 1–15.

Rostworowski de Diez Canseco, M. (1986) *Estructuras andinas del poder: ideología religiosa y política*. 2nd edition. Instituto de Estudios Peruanos, Lima.

Roys, R. L. (1931) *Ethno-Botany of the Maya*. The Department of Middle American Research, The Tulane University, New Orleans.

Rozin, P. (1987) Psychobiological perspectives on food preferences and avoidances. In M. Harris and E. Ross (ed.) *Food and evolution: toward a theory of human food habits*. Temple University Press, Philadelphia, pp. 181–205.

Rozin, P., Haidt, J., McCauley, C. and Imada, S. (1997) Disgust: preadaptation and the cultural evolution of a food-based emotion. In H. Macbeth (ed.) *Food preferences and taste: continuity and change*. Berghahn Books, Providence, pp. 65–82.

Ruddle, K. (1993) The transmission of traditional ecological knowledge. In J. T. Inglis (ed.) *Traditional ecological knowledge: concepts and cases*. International Program on Traditional Ecological Knowledge and International Development. Research Centre, Ottawa, pp. 17–31.

Rueda, C. (2011) *Territorio, culto e iconografía en los santuarios iberos del Alto Guadalquivir (ss. IV a.n.e.-I d.n.e.)*. Universidad de Jaén, Jaén.

Ruiz, A. (1998) Los príncipes iberos: procesos económicos y sociales. *Saguntum* 1, 285–300.

Ruiz, A. (2008) Iberos. In F. Gracia (coord.) *De Iberia a Hispania*. Ariel Prehistoria, Barcelona, pp. 733–844.

Ruiz, A. and Molinos, M. (1999) *The Archaeology of the Iberians: Culture Contact and Culture Change in Iron-Age Europe*. Cambridge University Press, Cambridge.

Ruiz, A. and Molinos, M. (2007) *Iberos en Jaén*. Universidad de Jaén, Jaén.

Ruiz, A. and Sanmartí, J. (2003) Models comparats de poblament entre els ibers del nor i del sud. In J. Guitart, J. M. Palet and M. Prevost (eds.) *Territoris antics a la Mediterrània i a la Cossetània oriental*. Actes del Simposi Internacional d'Arqueologia del Baix Penedès (El Vendrell, 8–10 de novembre de 2001). Departament de cultura de la Generalitat de Catalunya, Barcelona, pp. 39–57.

Ruiz, A., Molinos, M., Gutiérrez, L. M. and Bellón, J. P. (2001) El modelo político del pago en el Alto Guadalquivir (s. IV–III a.n.e.). In A. Martin i Ortega and R. Plana Mallart (eds.) *Territori polític i territori rural durant l'edat del Ferro a la Mediterránea Occidental*. Monografies d'Ullastret 2. Museu d'Arqueologia de Catalunya, Girona, pp. 11–22.

Ruiz-Arranz, M., Davis, B., Stampini, M., Winters, P. and Handa, S. (2002) *More calories or more diversity? An econometric evaluation of the impact of the PROGRESA and PROCAMPO transfer programs on food security in rural Mexico*. Vol. 02–09. ESA Working Paper. FAO, Rome.

Sahlins, M. (1978) *Culture and Practical Reason*. University Of Chicago Press, Chicago.

Sahlins, M. (2008) *The Western Illusion of Human Nature: With Reflections on the Long History of Hierarchy, Equality and the Sublimation of Anarchy in the West, and ... on Other Conceptions of the Human Condition*. Prickly Paradigm Press, Chicago.

Samuel, D. (1996) Archaeology of Ancient Egyptian Beer. *Journal of the American Society of Brewing Chemists* 54 (1), 3–12.

Samuel, D. (2000) Brewing and baking. In P. T. Nicholson and I. Shaw (eds.) *Ancient Egyptian Materials and Technology*. Cambridge University Press, Cambridge, pp. 537–576.

Sanmartí, J. (2004) From local groups to early states: the development of complexity in protohistoric Catalonia. *Pyrenae* 35 (1), 7–42.

Sanmartí, J. and Belarte, C. (2001) Urbanización y desarrollo de estructuras estatales en la costa de Cataluña (siglos VII–III a. C.). In L. Berrocal and Ph. Gardes (coords.) *Entre celtas e íberos: las poblaciones protohistóricas de las Galias e Hispania*. Casa de Velázquez, Real Academia de la Historia, Madrid, pp. 161–174.

Scholliers, P. (2001) *Food, drink and identity*. Berg, New York/Oxford.

Sharer, R. (2005) *The Ancient Maya*. Stanford University Press, Stanford.

Smith, J. J. (1993) Using ANTHROPAC 3.5 and a spreadsheet to compute a free-listing salience index. *Cultural Anthropology Methods* 5 (3): 1–3.

Solomon, F. (1995) 'The beautiful grandparents': Andean ancestor shrines and mortuary ritual as seen through Colonial records. In T. Dillehay (ed.) *Tombs for the Living: Andean mortuary practice* Dumbarton Oaks, Washington D.C., pp. 315–353.

Steffens, J. (2009) *Die neolithischen Fundplätze von Rastorf, Kreis Plön*. Universitätsforschungen zur prähistorischen Archäologie 170. Habelt, Bonn.

Stika, H.-P. (1998) Bodenfunde und Experimente zu keltischem Bier. In M. Fansa (ed.) *Experimentelle Archäologie in Deutschland. Bilanz 1997*. Archäologische Mitteilungen aus Nordwestdeutschland. Beiheft 19 Isensee, Oldenburg, pp. 45–54.

Stika, H.-P. (2011) Early Iron Age and Late Mediaeval malt finds from Germany – attempts at reconstruction of early Celtic brewing and the taste of Celtic beer. *Archaeological and Anthropological Sciences* 3, 41–48.

Stross, B. (1997) Mesoamerican Copal Resins. In L. Schele (ed) *U Mut Maya VI: Reports and Readings Inspired by the Advanced Seminars led by Linda Schele at the University of Texas at Austin.* University of Texas, Austin.

Sutton, D. (2001) *Remembrance of Repasts: An Anthropology of Food and Memory.* Berg Publishers, London.

Szabo, M., Timar, L. and Szabo, D. (2007) La basilique de Bibracte: un témoignage de l'architecture romaine en Gaule centrale. *Archäologisches Korrespondenzblatt* 37 (3), 389–408.

Tajfel, H. and Turner, J. (1979) An integrative theory of intergroup conflict. In S. Worchel and W. Austin (ed.) *The social psychology of intergroup relations.* Brooks/Cole, Monterey (CA), pp. 33–48.

Tajfel, H. and Turner, J. (1986) The social identity theory of intergroup behavior. In S. Worchel and W. Austin (ed.) *Psychology of intergroup relations.* Nelson-Hall, Chicago, pp. 7–24.

Tortosa, T. (2006) *Los estilos y grupos pictóricos de la cerámica ibérica figurada de la Contestania.* Anejos de Archivo Español de Arqueología. CSIC, Mérida.

Tortosa, T. and Santos, J. A. (2003) *Arqueología e Iconografía. Indagar en las imágenes.* L'Erma di Bretschneider, Roma.

Turner, J. (1982) Toward a cognitive redefinition of the social group. In H. Tajfel (ed.) *Social identity and intergroup relations.* Cambridge University Press, Cambridge, pp. 15–40.

Turner, J. C. (1989) *Rediscovering the social group: A self-categorization theory.* Blackwell, Oxford.

van der Veen, M. (2003) When is food a luxury? *World Archaeology* 34 (3), 405–427.

van der Veen, M. and Jones, G. (2006) A re-analysis of agricultural production and consumption: implications for understanding the British Iron Age. *Vegetation History and Archaeobotany* 15, 217–228.

Van Zeist, W. and Boekschoten-van Helsdingen, A. M. (1991) Samen und Früchte aus Niederwil. In H. A. T. Waterbolk and W. van Zeist (eds.) *Niederwil, eine Siedlung der Pfyner Kultur.* Haupt, Bern, pp. 49–113.

Vandkilde, H. (2007) A review of the early late Neolithic period in Denmark. *Offa* 2004/05, 75–110.

Vázquez-García, V. (2008) Gender, ethnicity, and economic status in plant management: Uncultivated edible plants among the Nahuas and Popolucas of Veracruz, Mexico. *Agriculture and Human Values* 25, 65–77.

Viklund, K. (2002) Issues in Swedish archaeobotany – a guide through twenty years of archaeobotanical research at the University of Umeå. In K. Vikund (ed.) *Nordic Archaeobotany - NAG 2000 in Umeå.* Archaeology and Environment 15. University of Umeå, Umeå, pp. 193–202.

Vitali, D. and Wiethold, J. (1996) Restes végétaux carbonisés à la Pâture du Couvent (amidonnier germé stocké dans un bâtiment d'époque Augustéenne). *Revue Archéologique de l'Est* 46 (2), 271–287.

Wachtel, N. (1966) Structuralisme et histoire: à propos de l'organisation social du Cuzco. *Annales. Économies, Sociétés, Civilisations* 21 (1), 71–94.

Wiethold, J. (1993) Restes végétaux de l'époque de La Tène finale sous forme de sédiment organique du bassin de la Fontaine Saint-Pierre. *Revue Archéologique de l'Est et du Centre-Est* 44 (2), 351–360.

Wiethold, J. (1994a) Analyse de macrorestes végétaux du Mont Beuvray. In J.-L. Flouest (ed.) *Bibracte, Centre archéologique du Mont Beuvray, rapport scientifique intermédiaire, activité 1994.* Centre archéologique européen, Glux-en-Glenne, pp. 253–265.

Wiethold, J. (1994b) Analyse des macrorestes végétaux du Mont Beuvray. In J.-L. Flouest (ed.) *Activités 1993, prévisions 1994, rapport de campagne de recherche 1993.* Centre européen d'archéologie du Mont Beuvray: Glux-en-Glenne, pp. 247–254.

Wiethold, J. (1996a) Late Celtic and early Roman plant remains from the oppidum of Bibracte, Mont Beuvray. *Vegetation History Archaeobotany* 5, 105–116.

Wiethold, J. (1996b) Fonctionnement socio-économique de l'oppidum du IIe s .av. J-C. au Ier ap. J-C., Analyse de macrorestes végétaux du Mont Beuvray. In J.-L. Flouest (ed.) *Document final de synthèse 1993-1995, rapport triennial.* Centre archéologique européen du Mont Beuvray, Glux-en-Glenne, pp. 1–13.

Wiethold, J. (1997) Etude archéobotanique. In V. Guichard (ed.) *Rapport annuel d'activité scientifique du Centre archéologique européen du Mont Beuvray*, Centre archéologique européen du Mont Beuvray, Glux-en-Glenne, pp. 249–263.

Wiethold, J. (1998) Recherche archéobotanique en France de centre-est. In V. Guichard (ed.) *Rapport annuel d'activité scientifique 1998 du Centre archéologique européen du Mont Beuvray.* Centre archéologique européen du Mont Beuvray, Glux-en-Glenne, pp. 217–240.

Wiethold, J. (1999a) Les matières organiques. In O. Buchsenschutz, J.-P. Guillaumet and I. B. M. Ralston (eds.) *Les remparts de Bibracte, recherches récentes sur la porte du Rebout et le tracé des fortifications.* Collection Bibracte 3. Centre archéologique européen du Mont Beuvray, Glux-en-Glenne, pp. 217–219.

Wiethold, J. (1999b) Les plantes cultivées et cueillies. *Gallia* 55, 63–65.

Wiethold, J. (1999c) Restes végétaux carbonisés provenant du site du Petit-Chauvort. In Ph. Barral (ed) *Verdun-sur-le-Doubs, ' Le Petit Chauvort ' (Saône-et-Loire). Rapport de fouille archéologique, opération de fouille programmée 20/06/1998-25/07/1998.* Dijon, pp. 15–22.

Wiethold, J. (2002) Pflanzenreste aus einem spätlatènezeitlichen Brunnen vom oppidum Fossé de Pandours, Col de Saverne (Bas-Rhin) – Vorbericht zu den archäobotanischen Analysen – In S. Fichtl and A.-M. Adam (ed.) *L'oppidum médiomatrique du Fossé des Pandours au Col de Saverne (Bas-Rhin). Rapport triennal 2000-2002.* Université Marc Bloch, Strasbourg, pp. 177–186.

Wiethold, J. (2003) How to trace the 'Romanisation' of central Gaule by archaeobotanical analysis? In F. Favory and A. Vignot (eds.) *Actualités de la Recherche en Histoire et Archéologie agraires*. Annales Littéraires 764/Environnement, sociétés et archéologie 5. Presses Universitaires de Franche-Comté, Besançon, pp. 269–282.

Wiethold, J. (2004) Les analyses archéobotaniques. In M. Joly and Ph. Barral (eds.) *Mirebeau-sur-Bèze. La Fenotte ' 21 416 003)(Côte d'Or). Fouille programmée triennale 2002–2004 sous la responsabilité de Martine Joly. Rapport triennal de la campagne 2002–2004. Vol. 2 Documentation.* Besançon/Dijon, pp. 335–338.

Wiethold, J. (2007) Restes végétaux carbonisés provenant des sites de l'Âge du Bronze (St. Apollinaire, 'Le pré de Crot 1'; Couternon, 'Larrey'; Varois-et-Chaignot, 'Le pré du Plancher'). In R. Labeaune, F. Ducreux, Ph. Barral, J. Wiethold and C. Bémilli (eds.) *Liaison Routière Dijon / Arc-sur-Tille. Les occupations protohistoriques. Vol. 5 Synthèses.* AFAN-INRAP-SRA de Bourgogne-Conseil général de la Côte d'Or, Dijon, pp. 58–63.

Wiethold, J. (2008) Les macro-restes végétaux: Imphy 'La Garenne'. Rapport d'étude carpologique et anthracologique 2008/8. In P. Stephenson (ed.) *Imphy ' La Garenne' Vol. 2, Etudes spécialisées (hors étude architecturale)*. Inrap Grand Est-sud, Dijon, pp. 237–253.

Wiethold, J. (2009) Restes végétaux gorgés d'eau provenant d'un niveau organique de l'état 2. In Barral, Ph. and Richard H. (eds.) *Fouilles de la fontaines Saint-Pierre au Mont Beuvray (1988–1992, 1996). Aménagements d'une source sur l'oppidum de Bibracte* (Bibracte 17). Centre archéologique européen du Mont Beuvray, Glux-en-Glenne, pp. 135–150.

Wiethold, J. (2010) L'histoire et l'utilisation de la coriandre (*Coriandrum sativum* L.), à partir du deuxième âge du Fer jusqu'au début de l'époque moderne. Culture, utilisation, sources écrites et données carpologiques. In I. Théry-Parisot, C. Delhon and S. Thiébault (eds.) *Des hommes et des plantes. Exploitation du milieu et gestion des ressources végétales de la préhistoire à nos jours.* XXXe rencontres internationales d'archéologie et d'histoire d'Antibes. Éditions APDCA, Antibes, pp. 141–159.

Wiethold, J. (2011) Les recherches carpologiques sur l'oppidum de Bibracte: contribution à l'étude de l'agriculture et de l'alimentation végétale au deuxième âge du Fer en Bourgogne. In J. Wiethold (ed.) *Carpologia. Articles réunis à la mémoire de KarenLundström-Baudais* (9–12 juin 2005, Glux-en-Glenne). Bibracte 20. Centre archéologique européen Bibracte, Glux-en-Glenne, pp. 221–252.

Wiethold, J. and Labeaune, R. (2005) Pluvet 'Larrivoux'. Un habitat de plaine du premier âge du Fer: premiers résultats sur les macro-restes végétaux. In C. Petit (ed.) *Occupation des plaines alluviales dans le Nord de la France de l'âge du Fer à l'époque gallo-romaine. Actes de la table ronde (Molesme, 17–18 septembre 1999).* Annales Littéraires 786/Série Environnement, sociétés et archéologie 8. Presses Universitaires de Franche-Comté, Besançon, pp. 197–211.

Wilken, G. (1970) The ecology of gathering in a Mexican farming region. *Economic Botany* 2, 286–295.

Winterhalder, B. and Goland, C. (1997) An Evolutionary Ecology Perspective on Diet Choice, Risk, and Plant Domestication. In K. J. Gremillion (ed.) *People, Plants and Landscapes: Studies in Paleoethnobotany*. University of Alabama Press, Tuscaloosa, pp. 123–159.

Zimmermann, W. H. (2008) Phosphate mapping of a Funnel Beaker Culture house from Flögeln-Eekhöltjen, district of Cuxhaven, Lower Saxony. In H. Fokkens, B. J. Coles, A. L. van Gijn, J. P. Kleijne, H. H. Ponjee and C. G. Slapp Endel (eds.) *Between foraging and farming. An extended broad spectrum of papers presented to Leendert Louwe Kooijmans.* Analecta Praehistorica Leidensia 40. Faculty of Archaeology, Leiden University, Leiden, pp. 123–129.

Zohary, D., Hopf, M. and Weiss, E. (2012) *Domestication of Plants in the Old World.* Oxford University Press, Oxford.

Zuidema, T. (1964) *The ceque system of Cuzco: the social organization of the capital of the Inca.* E.J. Brill, Leiden.

9 Conclusions – Plants for Thoughts

Alexandre Chevalier, Leonor Peña-Chocarro and Elena Marinova

The human need for plants, both physiologically crucial for human survival and socially essential to holding power, has been and remains a key driver for economic development and social interaction on the positive side, but also for social domination, massacres and slavery on the negative side (Mintz 1986).

Nature has been transformed by humans in order to allow plants to be grown in a wide range of natural circumstances including, at times, rather unsuitable areas, just as elaborate techniques of agriculture, crop selection and transformation, food preservation and storage, as well as techniques for eliminating toxins, have been developed over the centuries by human beings to secure their everyday plant intake.

Plants have travelled along with humans across mountains, deserts, seas and continents to be grown and utilised outside their original ecology and geography, not only for their essential nutritive aspects but also, and often for, social reasons. For millennia, the human quest for particular species has led to wars, either to gain access to habitats where they grew or to control their trade, and eventually to colonial occupations of vast territories by many different cultures all around the world.

This happened not only to plants central to the human diet and to human survival, but also to mere 'ornamental' plants; in other words, plants which had such powerful social or cultural symbolism that that humans were willing to die for them. We have but to consider, for instance, the 'tulip wars' or the many sailors whose lives were sacrificed in the quest for spices from the East Indies; as well as the many peoples all over the world on whom untold suffering was (and still is being) inflicted.

Human history is inseparable from that of plants. There is not just a single history of plant use and choices, but as many histories as cultural groups and identities across time and regions that have existed. All the subjects presented in this book highlight the multiple ways nature has been perceived, as well as the manifold answers and choices humans made from – and within – these perceptions, since the beginning of agriculture to present-day farmers around the world. Although it does not cover all the topics related to human plant uses, this volume achieves the EARTH programme goals in investigating the knowledge, skills, perceptions and experiences of farmers – both of the past and of the present – from a human-centred and deeply inter-disciplinary approach.

From the European Early Neolithic to present-day farmers, across most European countries and from India to the Americas, the volume shows the astonishing diversity of choices, each embedded in particular cultural backgrounds. Moreover, these choices continue to evolve and it would be pointless to stop our observations at some precise chronological moment.

Corresponding to this diversity of plants and choices, a diversity of theoretical frameworks exists, expressed by the authors according to their own educational and scientific background, and experiences: ranging from behavioural ecology with the defenders of an objective nature, such as Harris (1974), Douglas (1975), Winterhalder *et al.* (1999; 2006), to historical ecology with advocates of a subjective nature such as Lévi-Strauss (1976; 1983), Sahlins (1976), Descola (1996; 2005; 2011), Lentz (2000), Balée (2002) or Balée and Erickson (2006), among many others.

The human-nature relationship is still the object of lively – often heated – debate and, probably, will continue to be, due to our changing relationship with the natural world. This volume should be considered as a contribution to this debate, not a solution to it.

Clearly, plant choice is not only an adaptive process depending on environmental constraints, but is motivated by cultural and social norms as well as by the particular availability of plants.

Humans have overcome environmental constraints by adapting plants to new environments through the selection of varieties that eventually became full species or by modifying the landscape by creating specific ecosystems, such as terraces, sunken fields, floating gardens, or greenhouses, as is demonstrated in Volume 3 of the EARTH Series. They have defeated many of the pests and fungal diseases by technical or chemical means, by influencing the DNA of the plants through selection processes, if not by direct manipulation. They have crafted specific tools to address the cultivation, harvesting and processing of these plants, as referred to in Volume 2 of the EARTH Series.

Nevertheless, these technical solutions are particular to each cultural group, for each one has a specific perception of nature according to their cultural roots. What is more, each individual or group has particular cultural reasons for choosing one technical solution rather than another and for cultivating, using or exploiting one particular plant rather than another. Examples are legion in all contributions and in the two other EARTH Series volumes, as well as in on-going work outside this series. For example, work on Bronze Age barley bread production in the Aegean world explores the links between culinary practices and myth, culture and social structure on one hand and, on the other hand, shows the reinterpretation of the same diet in other geographical or chronological contexts (see Procopiou, Chapter 7.6 in van Gijn *et al.* 2014); another example is the implementation of large and systematic irrigation systems in the Balearic Islands by Arabic and Berbers rulers as soon as they conquered the islands and replaced the Visigoths (see Kirchner and Retamero, Chapter 1.5 in Retamero *et al.* 2014).

To our industrialised, Western eyes, traditions – or to put it more precisely – cultural continuities, seem to define non-industrialised farming societies. This statement appears to be true when non-industrialised farming societies are compared to societies and cultures which rapidly produce novelties. In absolute terms this is, however, completely wrong: cultural identities are dynamic; they evolve through time and change according to their context. It is believed that food is one of the most permanent elements of identity, or at least the element most symbolically related to the culture and identity of the user. From one generation to the next, for instance, the use of plants can change drastically and integrate new elements from other cultural settings or re-appropriate old cultural elements. What was rejected under 'normal' conditions, such as acorn or chestnut flour preparations among the Western bourgeois social class, was accepted as famine food during wars, rejected once the food supply became stable again and, more recently, accepted by the upper classes as highly praised 'alternative' food. What is at work here is the cultural and social identity of the eater, the context of the use and the constant balance between this identity and the individual necessity of using a plant, either physiologically or socially. When J. L. Mingote shows that what was a conflict-creating, mandatory cultivation of some plant species in Medieval Spain became a social norm a few centuries later (Chapter 9.6 in van Gijn *et al.* 2014), we are in the situation of a process of the creation of social identity through the integration and adaptation of norms, involving positive differentiation against outgroup pressure, as shown by Tajfel and Turner (1986).

Non-industrialised societies are far from being static and this is also reflected in their choices of plants. This is, in fact, one of the major achievements which this volume has to offer readers, most especially, perhaps, those in an urban context: not to believe that people in the past ate more or less the same food and utilised the same plants year-round. This would lead us to think such societies were limited in both their food and plant choices. This may well be true, if we compare them with the opportunities and ease with which people in the northern hemisphere today can have exotic food plants brought from the tropics in less than twenty-four hours. However, it is wrong, not only in terms of absolute choices in the past but also in current non-industrialised regions of the world, as the contributions of this volume clearly show.

Many urbanised societies have oriented their choices towards a few cultivated plants – tomatoes, potatoes, rice, zucchinis or string beans – when enormous bounty is just out there, in and outside fields. We might almost have used the word 'restricted' to characterise our current Western use of plants, either for cultural reasons, economical necessity or loss of biodiversity, but this is absolutely not true. Humans are among the most effective agents for creating plant diversity, either directly by favouring or, since at least the seventeenth century in Europe, intentionally creating new plant varieties, or by being a sort of troublemaker who creates successional vegetation (and, therefore, plant diversity), which have been used intentionally or opportunistically (see Goldstein *et al.* Chapter 8.8 in this volume, or Descola 1996). Even in our industrialised societies, the diversity of plant products may be higher than before industrialisation, as Brush (2004) has proposed. We are simply not aware of it most of the time.

Indeed, at some point during the industrialisation process, urbanised societies lost their links with the realities and necessities of farming and plant production. We do not realise today what it is like to have one's life depend on climate events that might well prevent us from producing enough food plants to avoid suffering from hunger and disease. We cannot imagine the absolute necessity of exploiting several ecological niches and plant varieties to secure one's everyday food, or to produce surpluses to trade and thus have access to plant products that are necessary for health and survival. Unless we have had very specialised training, most of us cannot just go outside and gather edible wild plants in case of necessity or pleasure. What was a dietary necessity sixty years ago in most of the European mountainous regions is now a family Sunday pastime in most European countryside places. Of course, such activities certainly provide an identity reminder to the group, perhaps declaring a strongly-felt link with former settlers or ancestors or membership in a group: mushroom, wild berries or dandelion gatherers currently act for cultural or social reasons, but no longer because of physiological needs. Even present-day farmers in industrially developed countries do not understand what life was like for their early nineteenth century counterparts: production risks are currently close to zero (if not for their bankers), but state subsidies, production stability mechanisms and trade have completely shaped their – and our – minds and our relationship towards nature and plants.

This is, in fact, the achievement and the strength of the contributions in this volume: to identify and to index past and current non-industrialised farming diversity and choice practices, to remind us that mankind's history is tightly bound to the history of plants and that the human-nature relationship is a relationship of co-evolution and mutual influences – both in the past and for the future.

Chapter Bibliography

Balée, W. (ed.) (2002) *Advances in historical ecology*. Columbia University Press, New York.

Balée, W. and Erickson, D. (2006) Time and complexity in historical ecology. In W. Balée and D. Erickson (eds.) *Time and complexity in historical ecology*. Columbia University Press, New York, pp. 1–17.

Brush, S. (2004) *Farmers' Bounty. Locating Crop Diversity in the Contemporary World.* Yale University Press, New Haven.

Descola, P. (1996) *In the society of nature: a native ecology in Amazonia.* Cambridge University Press, Cambridge.

Descola, P. (2005) *Par-delà nature et culture.* Gallimard, Paris.

Descola, P. (2011) *L'écologie des autres. L'anthropologie et la question de nature.* Quae, Versailles.

Douglas, M. (1975) *Implicit meanings.* Routledge, London/Boston.

Harris, M. (1974) *Cows, Pigs, Wars, and Witches: The riddles of culture.* Random House, New York.

Lentz, D. (2000) Definitions and conceptual underpinnings. In D. Lentz (ed.) *Imperfect balance: landscape transformations in the Precolumbian Americas.* Columbia University Press, New York, pp. 1–11.

Lévi-Strauss, C. (1976). Structuralisme et empirisme. *L'Homme* 16 (2–3), 23–38.

Lévi-Strauss, C. (1983). *Le regard éloigné.* Plon, Paris.

Mintz, S. (1986) *Sweetness and Power: The Place of Sugar in Modern History.* Penguin, New York.

Retamero, F., Schjellerup, I. and Davies, A. (eds.) (2014) *Agricultural and pastoral landscapes in pre-industrial society: choices, stability and change.* EARTH Series vol. 3. Oxbow, Oxford.

Sahlins, M. (1976) *Culture and Practical Reason.* University Of Chicago Press, Chicago.

Tajfel, H. and Turner, J. (1986) The social identity theory of intergroup behavior. In S. Worchel and W. Austin (eds.) *Psychology of intergroup relations.* Nelson-Hall, Chicago, pp. 7–24.

van Gijn, A., Whittaker, J. and Anderson, P. (eds.) (2014) *Exploring and explaining diversity in agricultural technology.* EARTH Series vol. 2. Oxbow, Oxford.

Winterhalder, B., Lu, F. and Tucker, B. (1999) Risk-sensitive adaptive tactics: Models and evidence from subsistence studies in biology and anthropology. *Journal of Archaeological Research* 7 (4), 301–348.

Winterhalder, B. and Kennet, D. (2006) Behavioral ecology and the transition from hunting and gathering to culture. In D. Kennet and B. Winterhalder (eds.) *Behavioral ecology and the transition from hunting and gathering to culture.* University of California Press, Berkeley / Los Angeles, pp. 1–21.

Annexes

CONTRIBUTORS

Dr Patricia C. Anderson
Université Nice Sophia Antipolis
CEPAM (Cultures and Environments, Prehistory, Antiquity, Middle Ages), UMR 7264 and GDR 2517, CNRS (Centre for National Scientific Research)
St. Jean d'Angély 3
06357 Nice Cedex 4 • France
Email: *patricia.anderson@cepam.cnrs.fr*

Dr Mohammed Ater
Université Abdelmalek Essaâdi
Faculté des Sciences, LDICOSYB
BP 2062
93002 Tétouan • Morocco
Email: *mater20@gmail.com*

Dr Yildiz Aumeeruddy-Thomas
Centre d'Ecologie Fonctionnelle et Evolutive (CEFE)
UMR 5175 CNRS (Centre for National Scientific Research)
1919 route de Mende
34293 Montpellier Cédex 5 • France
Email: *yildiz.thomas@cefe.cnrs.fr*

Professor Corrie Bakels
Leiden University
Instituut voor Prehistorie
PO Box 9515
2300 RA Leiden • The Netherlands
Email: *c.bakels@arch.leidenuniv.nl*

Dr Emmanuelle Bonnaire
Direction de l'Archéologie Préventive
227 rue Jean Perrin
59500 Douai • France
Email: *ebonnaire@douaisis-agglo.com*

Dr Laurent Bouby
CNRS, Université Montpellier II
Centre de Bio-Archéologie d'Ecologie (CBAE)
UMR 5059 CNRS (Centre for National Scientific Research)
163, rue Auguste Broussonet
34090 Montpellier • France
Email: laurent.*bouby@univ-montp2.fr*

Dr Maria Bruno
Washington University in St. Louis
Department of Anthropology
Campus Box 1114 / One Brookings Drive
St. Louis MO 63130-4899 • USA
Email: *mcbruno@artsci.wustl.edu*

Dr Bui Thi Mai
Université Nice Sophia Antipolis
CEPAM (Cultures and Environments, Prehistory, Antiquity, Middle Ages), UMR 7264 and GDR 2517, CNRS (Centre for National Scientific Research)
St. Jean d'Angély 3
06357 Nice Cedex 4 • France
Email: *buithi2009@gmail.com*

Dr Alexandre Chevalier
Institut Royal des Sciences Naturelles de Belgique/ Koninklijk Belgisch Instituut voor Natuurwetenschappen
29 rue Vautier
1000 Bruxelles • Belgium
Email: *alexandre.chevalier@naturalsciences.be*

Dr Gisella S. Cruz-García
Wageningen Universiteit
Centre for Crop Systems Analysis
P.O. Box 430
6700 AK Wageningen • The Netherlands
Email: *ragracancha@gmail.com*

Dr Linda Scott Cummings
Paleoresearch Institute
2675 Youngfield Street
Golden CO 80401 • USA
Email: *linda@paleoresearch.com*

Dr Rob Cuthrell
Berkeley University of California
Department of Anthropology
232 Kroeber Hall
Berkeley CA 94720-3710 • USA
Tel: +1 510 643 - 11 12
Email: *rcuthrell@berkeley.edu*

Dr Jalh Dulanto
DePauw University,
Department of Sociology and Anthropology
205E Asbury Hall
Greencastle IN 46135 • USA
Email: *jalhdulanto@depauw.edu*

Dr Frédérique Durand
Centre de Recherche sur la Préhistoire et la Protohistoire de la Méditerranée (CRPPM)
UMR 5608 CNRS (Centre for National Scientific Research)
39, Allées Jules Guesde
31000 Toulouse • France
Email: *frederique_durand@aliceadsl.fr*

Dr Füsun Ertuğ
Rıdvan Paşa Sok., Refik Bey Apt. 13/14
81086 Göztepe/İstanbul • Turkey
Email: *etnofertug@gmail.com*

Dr Jaime Gil
Universidad de Las Palmas de Gran Canaria
Departamento de Ciencias Históricas
Calle de Pérez del Toro 1
35003 Las Palmas de Gran Canaria • Spain
Email: *jgil@gmx.fr*

Dr David John Goldstein
National Park Service – CHRI/BUIS/SARI
2100 Church Street #100
Christinasted VI 00820 • USA
Email: *david_goldstein@nps.gov*

Dr Susana González Reyero
Consejo Superior de Investigaciones Científicas (CSIC)
Albasanz 26-28
28037 Madrid • Spain
Email: *susana.gonzalezreyero@cchs.csic.es*

Dr Cozette Griffin-Kremer
Université de Bretagne Occidentale
Centre de Recherche Bretonne et Celtique
20 Rue Duquesne
29238 Brest Cedex 3 • France
Email: *griffin.kremer@wanadoo.fr*

Dr Jon B Hageman
Northeastern Illinois University
Department of Anthropology
5500 N. St. Louis Avenue
Chicago IL 60625 • USA
Email: *j-hageman@neiu.edu*

Dr Ann-Marie Hansson
Stockholms Universitet
Arkeologiska forskningslaboratoriet (AFL), Wallenberg-laboratoriet
106 91 Stockholm • Sweden
Email: *amhan@tele2.se*

Dr Andreas G. Heiss
Universität für Bodenkultur Wien (BOKU)
Institut für Botanik, Arbeitsgruppe Archäobotanik
Gregor Mendel-Strasse 33
1180 Wien • Austria

Universität Wien
Vienna Institute for Archaeological Science (VIAS)
Althanstrasse 14 – Geozentrum
1090 Wien • Austria
Email: *andreas.heiss@erbsenzaehler.at*

Dr Younes Hmimsa
Université Abdelmalek Essaâdi
Faculté des Sciences, LDICOSYB
BP 2062
93002 Tétouan• Morocco
Email: *hmimsa.younes@caramail.com*

Dr Bouchaïb Khadari
Conservatoire Botanique National Méditerranéen de Porquerolles
UMR 1098 CNRS (Centre for National Scientific Research)
76a, Av. Gambetta
83400 Hyères • France
Email: *khadari@supagro.inra.fr*

Professor Wiebke Kirleis
Universität Kiel
Institut für Ur- und Frühgeschichte
Johanna Mestorf-Straße 2-6
24098 Kiel • Germany
Email: *wiebke.kirleis@ufg.uni kiel.de*

Stefanie Klooß
Universität Kiel
Institut für Ur- und Frühgeschichte
Johanna Mestorf-Straße 2-6
24098 Kiel • Germany
Email: *s.klooss@ufg.uni-kiel.de*

Dr Elena Marinova
Katholieke Universiteit Leuven
Centre for Archaeological Sciences (CAS)
Celestijnenlaan 200e
Heverlee 3001 • Belgium
Email: *elena.marinova@bio.kuleuven.be*

Dr Philippe Marinval
Centre de Recherche sur la Préhistoire et la Protohistoire de la Méditerranée (CRPPM)
UMR 5608 CNRS (Centre for National Scientific Research)
39, Allées Jules Guesde
31000 Toulouse • France
Email: *marinval@cict.fr*

Dr José Luis Mingote Calderón
Museo Nacional de Antropología
Centro de Investigación del Patrimonio Etnológico
C/ Alfonso XII, 68
28014 Madrid • Spain
Email: *jluis.mingote@mcu.es*

Dr Jacob Morales
Universidad de Las Palmas de Gran Ganaria
Departamento de Ciencias Históricas
Calle de Pérez del Toro 1
35003 Las Palmas de Gran Canaria • Spain
Email: *jacobmor@terra.es*

Dr Manuel Pardo de Santayana
Universidad Autónoma de Madrid
Departamento de Biología (Botánica)
Calle Darwin 2
Campus de Cantoblanco
28049 Madrid • Spain
Email: *manuel.pardo@uam.es*

Dr Leonor Peña-Chocarro
Consejo Superior de Investigaciones Científicas (CSIC)
Escuela Española de Historia y Arqueología en Roma
Via di Torre Argentina 18
00186 Roma • Italy
Email: *leonor.chocarro@csic.it*

Consejo Superior de Investigaciones Científicas (CSIC)
G.I Arqueobiología. Instituto de Historia
Albasanz 26-28
28037 Madrid • Spain
Email: *leonor.chocarro@cchs.csic.es*

Dr Mauro Rottoli
Musei Civici di Como
Laboratorio di Archeobiologia
Piazza Medaglie d'oro, 1
22100 Como • Italy
Email: *archeobotanica@alice.it*

Dr Marie-Pierre Ruas
Muséum National d'Histoire Naturelle
Archéozoologie, Archéobotanique – Sociétés, Pratiques et Environnements (AASPE)
UMR 7209 CNRS (Centre for National Scientific Research)
55, rue Buffon
Paris 75005 • France
Email: *ruas@mnhn.fr*

Dr Marie Russel
Institut National de la Recherche Agronomique (INRA)
Unité Collège de direction (CODIR)
147, rue de l'Université
75338 Paris Cedex 07
Email: *marie.russel@paris.inra.fr*

Dr Matthew Sayre
University of South Dakota
Department of Anthropology
326 East Hall
Vermillion SD 57069 • USA
Email: *matthew.sayre@usd.edu*

Camille Selleger MA
27, rue Moillebeau
1209 Genève • Switzerland
Email: *camilleselleger@yahoo.fr*

Dr François Sigaut († 2012)
Ecole des Hautes Etudes en Sciences Sociales (EHESS)
Centre de Recherches Historiques
54, boulevard Raspail
Paris Cedex 6 75270 • France

Dr Javier Tardío
Instituto Madrileño de Investigación y Desarrollo Rural, Agrario y Alimentario (IMIDRA)
Finca El Encín, Apdo. 127
28800 Alcalá de Henares, Madrid • Spain
Email: *javier.tardio@madrid.org*

Dr Jean-Frédéric Terral
Université Montpellier II
Centre de Bio-Archéologie et d'Ecologie (CBAE)
UMR 5059 CNRS (Centre for National Scientific Research)
163, rue Auguste Broussonet
34090 Montpellier • France
Email: *terral@univ-montp2.fr*

Dr Soultana-Maria Valamoti
Aristotle University of Thessaloniki
Department of Archaeology
Thessaloniki 54006 • Greece
Tel: +30 23 10 99 73 10
Email: *sval@hist.auth.gr*

Dr Julian Wiethold
Institut National de Recherche Archéologique Préventive (INRAP)
Direction interrégionale Grand-Est Nord
12, rue de Méric / CS 80005
57063 Metz cedex 2 • France
Email: *julian.wiethold@inrap.fr*

Dr Lydia Zapata Peña
Euskal Herriko Unibertsitatea
Departament de Geografia, Prehistoria y Arqueología
Tomás y Valiente s/n
01006 Vitoria-Gasteiz • Spain
Tel: +34 94 416 88 18
Email: *l.zapata@euskalnet.net*

THE EARTH PROGRAMME

Steering Committee, 2004–2009

Dr Patricia C. Anderson (Co-Chair)
CNRS, CEPAM Nice • France

Professor Corrie Bakels
Universiteit Leiden • The Netherlands

Professor Miquel Barceló (2004-2006)
Universitat Autonoma de Barcelona • Spain

Professor Sofus Christiansen († 2007)
Københavns Universitet • Denmark

Professor Gabriel Cooney (from 2008)
University College Dublin, College of Arts & Celtic Studies • Ireland

Professor Mareile Flitsch
Völkerkundemuseum der Universität Zürich • Switzerland

Dr Karl-Erik Frandsen
Københavns Universitet • Denmark

Dr Michael Given (Co-Chair 2004-2006)
University of Glasgow • United Kingdom

Dr Eric Huysecom
Université de Genève • Switzerland

Professor Urve Miller
Stockholms universitet • Sweden

Dr Leonor Peña Chocarro (Co-Chair from 2006)
Consejo Superior de Investigaciones Científicas (CSIC), Rome • Italy / Madrid • Spain

Professor Georges Raepsaet (2004-2005)
Université libre de Bruxelles • Belgium

Professor Bjarne Rogan
Universitetet i Oslo • Norway

Professor Erik Thoen
Universiteit Gent • Belgium

Professor Marijke van der Veen (from 2007)
University of Leicester • United Kingdom

Dr Irmeli Vuorela († 2012)
Helsingin yliopisto • Finland

Coordinators

Dr Marie Russel (Scientific Coordinator 2004–2009) INRA (Institut National de la Recherche Agronomique) • France

Dr Andreas G. Heiss (Publication Coordinator from 2010) Universität für Bodenkultur Wien (BOKU)/Vienna Institute for Archaeological Science (VIAS) • Austria

Members of Team 1, 'Crop Choices and Diversity'

Dr Laurent Bouby
CNRS, Université Montpellier II • France

Dr Bui Thi Mai
CNRS, Université Nice Sophia Antipolis • France

Dr Alexandre Chevalier (Team Leader from 2008)
Institut royal des Sciences Naturelles de Belgique / Koninklijk Belgisch Instituut voor Natuurwetenschappen • Belgium

Dr Linda Scott Cummings
Paleoresearch Institute, Golden, CO • USA

Dr Füsun Ertuğ (Team Leader 2004–2008)
İznik • Turkey

Dr Cozette Griffin-Kremer
Université de Bretagne Occidentale, Brest • France

Dr. Mária Hajnalová
Univerzita Konštantína Filozofa, Nitra • Slovakia

Dr Ann-Marie Hansson
Stockholms Universitet • Slovakia

Dr Andreas G. Heiss
Universität für Bodenkultur Wien (BOKU)/ Vienna Institute for Archaeological Science (VIAS) • Austria

Dr Stefanie Jacomet
Universität Basel • Switzerland

Dr Elena Marinova
Katholieke Universiteit Leuven • Belgium

Dr Jacob Morales
Universidad de Las Palmas de Gran Ganaria • Spain

Dr Carol Palmer (Team Leader 2006–2008)
University of Leicester • United Kingdom

Dr Leonor Peña-Chocarro (Team Leader 2004–2006)
Consejo Superior de Investigaciones Científicas (CSIC), Rome • Italy / Madrid • Spain

Dr Mauro Rottoli
Musei Civici di Como • Italy

Dr François Sigaut († 2012)
Ecole des Hautes Etudes en Sciences Sociales (EHESS), Paris • France

Dr Julian Wiethold
Institut National de Recherche Archéologique Préventive (INRAP), Metz • France

Dr Lydia Zapata
Euskal Herriko Unibertsitatea, Vitoria-Gasteiz • Spain

Members of Team 2, 'Skills, Processes and Tools'

Dr Álvaro R. Arizaga Castro
Universidade de Santiago de Compostela • Spain

Dr Sabine Beckmann
University of Crete • Greece

Yorgos Brokalakis
University of Crete • Greece

Dr Carolina Carpinschi (Team Leader 2004–2008)
CDHT-CNAM-EHESS, Paris • France

Professor Annelou van Gijn (Team Leader from 2005)
Universiteit Leiden • The Netherlands

Professor Detlef Gronenborn (Team Leader 2004-2005)
Universität Mainz • Germany

Dr Maria Gurova
Bulgarian Academy of Sciences, Sofia • Bulgaria

Dr Ouafaa Kadim
Lebanon

Dr Helena Knutsson
Uppsala universitet • Sweden

Dr Grith Lerche
Den Kongelige Veterinær og Landbohøjskole, København • Denmark

Dr José Luis Mingote Calderón
Museo Nacional de Antropología, Madrid • Spain

Dr Hara Procopiou
Université Paris I • France

Dr Thomas K. Schippers
Université de Provence, Aix-Marseille I • France

Bernard Simonel (Team Leader 2008)
CNRS, Université Nice Sophia Antipolis • France

Dr Inja Smerdel
Slovenski etnografski muzej • Slovenia

Pascal Verdin
Institut National de Recherche Archéologique Préventive (INRAP), Nice • France

Professor John C. Whittaker
Grinnell College • USA, Iowa

Members of Team 3, 'Agricultural Landscapes'

Dr Uzi Avner
Ben-Gurion University at Eilat • Israel

Dr Andrea L. Balbo
Consejo Superior de Investigaciones Científicas (CSIC), Barcelona • Spain

Dr Karoline Daugstad
Norges teknisk-naturvitenskapelige universitet, Trondheim • Norway

Dr Althea Davies
University of Stirling • United Kingdom

Dr Dagmar Dreslerová
Archeologický ústav AV ČR, Praha • Czech Republic

Dr Aline Durand
Université de Provence, Aix-Marseille I • France

Dr Christian Keller
Universitetet I Oslo • Norway

Dr Valter Lang
Tartu Ülikool • Estonia

Dr Esa Mikkola
Museovirasto, Helsinki • Finland

Dr Per Grau Møller
Syddansk Universitet, Odense • Denmark

Dr Fèlix Retamero (Team Leader from 2004)
Universitat Autònoma de Barcelona • Spain

Dr Inge R. Schjellerup (Team Leader from 2004)
Nationalmuseet København • Denmark

Dr Luc Vrydaghs
Universiteit Gent • Belgium

SCIENTIFIC NETWORKING WORKSHOPS CONTRIBUTING TO THE CONTENTS OF THIS BOOK

Plenary Meetings

February 2007: 'Stability and balance'
Organiser: Leonor Peña-Chocarro, Centro de Ciencias Humanas y Sociales (CSIC), Madrid • Spain
Venue: Hotel Alixares, Granada • Spain
Additional funding: Unicaja Granada

February 2006: 'Agricultural decision-making'
Organiser: Annelou van Gijn, Universiteit Leiden • The Netherlands
Venue: Universiteit Leiden • The Netherlands

March 2005: 'The dynamics of non-industrial agriculture: a human-centred approach'
Organisers: Michael Given, University of Glasgow, G. Sprott, D. Dornan and E. Edwards, Museum of Scottish Country Life, East Kilbride, Scotland • United Kingdom
Venue: Museum of Scottish Country Life, East Kilbride, Scotland • United Kingdom

Scientific Workshops

Team 1

July 2007: 'Understanding local diversity: from routine practice through to times of crisis and transfer'
Organiser: Stephanie Jacomet, IPNA, Basel University, Basel • Switzerland
Venue: Oberwalliser Mittelschule (OMS) St. Ursula, Brig, Valais • Switzerland
Additional funding: Fonds zur Förderung von Lehre und Forschung

May 2006: 'The role of wild plants within an agricultural system: stability and balance'
Organiser: Füsun Ertuğ, İstanbul Üniversitesi, Prehistorya ve Arkeoloji Bölümü, Istanbul • Turkey
Venue: İlçe Halk Kütüphanesi (Library of Buldan), Buldan • Turkey
Additional support: Buldan Nature Preservation Association

June 2005: 'Assessing and interpreting crop diversity'
Organiser: Leonor Peña-Chocarro, Centro de Ciencias Humanas y Sociales (CSIC), Madrid • Spain
Venue: Casa de Cultura, Proaza • Spain
Additional funding: CSIC

Team 2

June 2008: 'Skills, Processes, and Tools'
Organiser: I. Smerdel, Slovenski etnografski muzej, Ljubljana • Slovenia
Venues: Slovenski etnografski muzej, Ljubljana, and Pliskovica konferenčna dvorana, Lipica • Slovenia
Additional funding: Slovenski etnografski muzej, Republika Slovenija, ministrstvo za kulturo

September 2007: 'Innovation, resilience and revival'
Organiser: Hara Procopiou, Université de Paris I • France
Venue: Municipality of Milos, Plaka, and Milos Conference Centre George Eliopoulos, Adamas • Greece
Additional funding: Costopoulos Foundation and the municipality of Milos

May 2006: 'Local knowledge and technological stability'
Organiser: B. Simonel, Université Nice Sophia Antipolis • France
Venue: Maison d'Hôtes Kasbah Tizourgane, Chouka Ait Baha, and Hôtel Les Amandiers, Tafraoute • Morocco

September 2005: 'The social and cultural dimensions of tool use: perspectives on choice'
Organiser: Carolina Carpinschi, CDHT-CNAM-EHESS, Paris
Venues: Ecomusée Rural du Pays Nantais, Vigneux-de-Bretagne, Musée du Pays de Retz, Bourgneuf-en-Retz and Musée du Vignoble Nantais, Le Pallet • France
Additional support: René Bourrigaud

Team 3

April 2008: 'Agricultural landscapes: synthesis'
Organiser: Aline Durand, Maison méditerranéenne des Sciences de l'Homme, Aix-en-Provence • France
Venue: Maison méditerranéenne des Sciences de l'Homme, Aix-en-Provence • France

May 2007: 'Agricultural landscapes: crisis and change'
Organiser: Valter Lang, Tartu Ülikool, Tartu • Estonia
Venue: Estonian Open Air Museum and Hotel in Sagadi • Estonia

June 2006: 'Agriculture, environment and society in equilibrium: managing slopes; livestock and pastoralism'
Organiser: Inje Schjellerup, Nationalmuseet, København • Denmark
Venue: Nationalmuseet, København • Denmark

October 2005: 'The changing patterns of the agricultural landscapes'
Organiser: Felix Retamero, Universitat Autònoma de Barcelona • Spain
Venue: Hotel Agamenon, Es Castell, Menorca • Spain

Editorial meetings

September 2006 and September 2011: Organiser: Patricia C. Anderson, CNRS, CEPAM, Nice • France
Venue: St. Vallier de Thiey • France

March 2009 and May 2010 : Team 1
Organiser: Alexandre Chevalier, Institut Royal des Sciences Naturelles de Belgique, Brussels • Belgium
Venue: Institut Royal des Sciences Naturelles de Belgique, Brussels • Belgium

February 2009: Team 3
Organiser: Dragomir Dreslerová, Archeologický ústav AV ČR, Praha • Czech Republic
Venue: Archeologický ústav AV ČR, Praha • Czech Republic

February/March 2009: Team 2
Organiser: Pascal Verdin, CEPAM, INRAP, CNRS, Sophia Antipolis • France
Venue: Palais des Congrès, Juan-les-Pins • France

August 2008 : Team 1
Organiser: Leonor Peña-Chocarro, Centro de Ciencias Humanas y Sociales (CSIC), Madrid • Spain
Venue: Centro de Ciencias Humanas y Sociales (CSIC), Madrid • Spain

Summer schools taught by members of the Earth Programme

September 2006 and August 2008: Summer school 'Researching non-industrial agriculture' (Proaza, Spain)[1]
Organiser: L. Peña-Chocarro, Centro de Ciencias Humanas y Sociales (CSIC), Madrid • Spain
Venue: Casa de Cultura, Proaza • Spain
Additional funding: CSIC, Caja Astur

Note

1 We want to thank Valentina Palacios, Pepe Álvarez, Margarita Fernández Mier, Laura Panera and Juaco López Álvarez for giving the EARTH Team and the invited students so much insight into traditional agrarian practices in the Asturias region of Spain.

Index

Page numbers in italics are figures; with 'n' are notes.